PETER FRANKOPAN

THE EARTH TRANSFORMED

Peter Frankopan is professor of global history at Oxford University. He is the author of *The First Crusade: The Call from the East, The Silk Roads: A New History of the World* and *The New Silk Roads: The Present and Future of the World*. He lives in Oxford.

THE EARTH TRANSFORMED

ALSO BY PETER FRANKOPAN

The New Silk Roads
The Silk Roads
A Chronology of the Crusades
The First Crusade

THE EARTH TRANSFORMED

An Untold History

PETER FRANKOPAN

VINTAGE BOOKS

A DIVISION OF PENGUIN RANDOM HOUSE LLC

NEW YORK

FIRST VINTAGE BOOKS EDITION 2024

Copyright © 2023 by Peter Frankopan

All rights reserved. Published in the United States by Vintage Books,
a division of Penguin Random House LLC, New York, and distributed in Canada
by Penguin Random House Canada Limited Toronto. Originally published in
hardcover in Great Britain by Bloomsbury Publishing PLC, London and in
the United States by Alfred A. Knopf, a division of Penguin Random
House LLC, New York, in 2023.

Vintage Books and colophon are registered
trademarks of Penguin Random House LLC.

The Library of Congress has cataloged the Knopf edition as follows:
Name: Frankopan, Peter.
Title: The Earth transformed: an untold history / Peter Frankopan
Description: First edition. | New York: Alfred A. Knopf, 2023.
Identifiers: LCCN 2022057303 (print)
Classification: QC903 .F736 2023 (print) | DDC 304.2/5
LC record available at https://lccn.loc.gov/2022057303

Vintage Books Trade Paperback ISBN: 978-0-593-08213-3
eBook ISBN: 978-0-525-65917-4

vintagebooks.com

Printed in the United States of America
1st Printing

To Jessica

When God created the first human, He took him and led him round all the trees of the Garden of Eden and said to him … 'Pay attention that you do not corrupt and destroy My world; if you corrupt it, there is no one to repair it after you.'

Midrash Ecclesiastes Rabbah, 7:13

The drought is so excessive,
And we are tormented by the heat.
I have not stopped offering sacrifices …
To the powers above and below I have made sacrifices and buried offerings.
There are no spirits I have not honoured.

King Xuan of Zhou (r. 827–782 BC), 'Yunhan' (雲漢)
from Shijing (詩經, Classic of Poetry)

[God] has raised the heavens and set up everything in balance,
Therefore do not transgress the balance that has been established.

Qur'ān, 55:7–8

A change in our climate … is taking place … Both heats and colds are become much more moderate.

Thomas Jefferson, Notes on the State of Virginia (1785)

The poorest nations, already beset by man-made disasters, have been threatened by a natural one: the possibility of climatic changes.

Henry Kissinger, Address to the Sixth Special Session of
the United Nations General Assembly (April 1974)

I've seen it, I've read some of it … I don't believe it.

Donald Trump, 45th President of the
United States of America, on the
US National Climate Assessment 2018

Contents

Note on Transliteration

Historians can tie themselves in knots trying to work out the best way to transliterate names of peoples, places and individuals. I have tried to use my judgement as best I can to make the text readable and, in doing so, to recognise that some readers may on occasion prefer a more faithful rendition, most notably when transliterating non-European languages. Nevertheless, I ask for forbearance from the reader who demands consistency. In return, I hope to inform, enlighten and help provide new perspectives on how we might look at the world we live in.

Maps

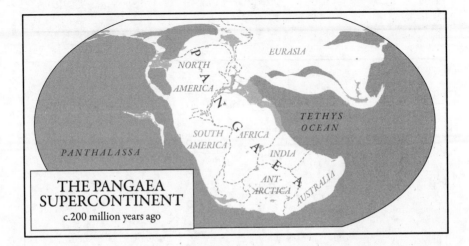

THE PANGAEA SUPERCONTINENT
c.200 million years ago

EURASIA

NORTH AMERICA

P A N G A E A

SOUTH AMERICA

AFRICA

INDIA

ANT-ARCTICA

AUSTRALIA

TETHYS OCEAN

PANTHALASSA

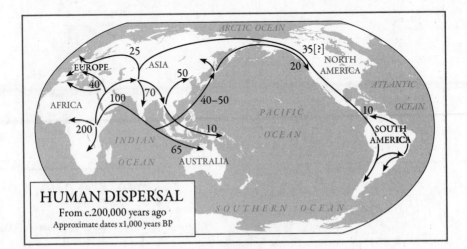

HUMAN DISPERSAL
From c.200,000 years ago
Approximate dates x1,000 years BP

ARCTIC OCEAN

EUROPE

ASIA

AFRICA

25

40

100

70

50

40–50

35[?]

20

NORTH AMERICA

ATLANTIC OCEAN

PACIFIC OCEAN

10

SOUTH AMERICA

200

10

65

INDIAN OCEAN

AUSTRALIA

SOUTHERN OCEAN

EASTERN ASIA
Civilisations and
settlements

0   500   1,000 km

LAOHUSHAN DAKOU
GOBI DESERT

QIJIA

Shimao

Xinzhai Kaifeng

SHAN-
TUNG

*TIBETAN
PLATEAU*

QINLING
MTS

ERLITOU

Taxila

H I M A L A Y A S

Harappa

Judeirjo-daro Banawali

Mohenjo-daro
Chanhu-daro
Dholavira
Surkotada

Ganges

Brahmaputra

Narmada

Yufucun

BAODUN

Mekong

Yellow

Yangtze

LIANGZHU

Hattusa

Çayönü

Çatalhöyük

Urfa Göbekli Tepe

Tell Halula

Tell Leilan

Dur-Šarrukin

Emar

Qermez
Dere

Nineveh
Assur

Ebla

Euphrates

M E S O P O T A M I A

Tigris

Ecbatana

Palmyra

Mari

*Caspian
Sea*

Z A G R O S   M T S

*Mediterranean
Sea*

Tyre

*SYRIAN
DESERT*

Sippar
Babylon

Akkad(?)

Susa

Jericho 'Ain Ghazal
Jerusalem Tall el-Hammam

Nippur

Uruk
Ur

Lagash

Giza
Memphis

'Ain Jammam

AN - NAFUD

The Gulf

Nile

Naqada

Thebes
Nekhen

E G Y P T

Red Sea

NUBIA

KUSH

WESTERN ASIA
Civilisations and
settlements

0   250   500 km

S A H A R A   D E S E R T

Ghat

Iwelen

E G H A Z Z E R   B A S I N

Agadez

Katsina
Kano
Zaria

Sokoto

Benin

Ilé-Ifè

Bénue

Niger

Lagos
Porto-
Novo Jakin
Ouidah

Tadmakka (Essouk)

SONGHAY

Niamey

GAO

Grand-Popo Accra (Christiansborg)
Agoué Cape Coast

Gao
Kobé

SONGHAY

White Volta

Volta

Taghaza

MALI

Timbuktu

Hamdu Bahi
Nando

Black Volta

Elmina
Axim
Assinie

Tichitt

Walata

GHANA

MÉMA
Mopti
Dienné
Ségou San

Kumbi
Saleh

Banako
Niani

MALI

Aoudaghost

Dhar Néma

MALI

GHANA

Senegal

FUTA Dinguiraye
JALON Timbo

Sherbro

Akjoujt

TAKRUR

JOLOF

Bissau

Cacheu

Arguin

Nouakchott

Saint-Louis

Dakar

A T L A N T I C

O C E A N

WEST AFRICA
Civilisations and settlements

0   250   500 km

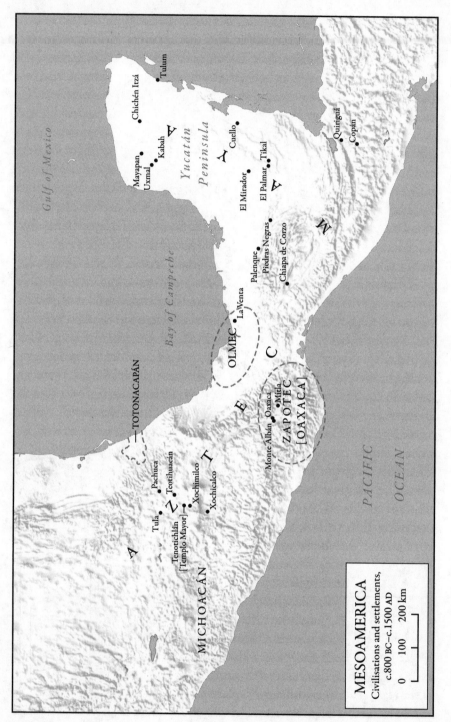

MESOAMERICA
Civilisations and settlements,
c.800 BC–c.1500 AD

0   100   200 km

Gulf of Mexico

Bay of Campeche

PACIFIC OCEAN

Yucatán Peninsula

Tulum
Chichén Itzá
Mayapan
Uxmal
Kabah
Cuello
El Mirador
El Palmar
Tikal
Palenque
Piedras Negras
Chiapa de Corzo
Quiriguá
Copán

La Venta
OLMEC

TOTONACAPÁN

Pachuca
Teotihuacán
Tula
Tenotichlán
[Templo Mayor]
Xochimilco
Xochicalco

Monte Albán
Oaxaca
Mitla
ZAPOTEC
[OAXACA]

MICHOACÁN

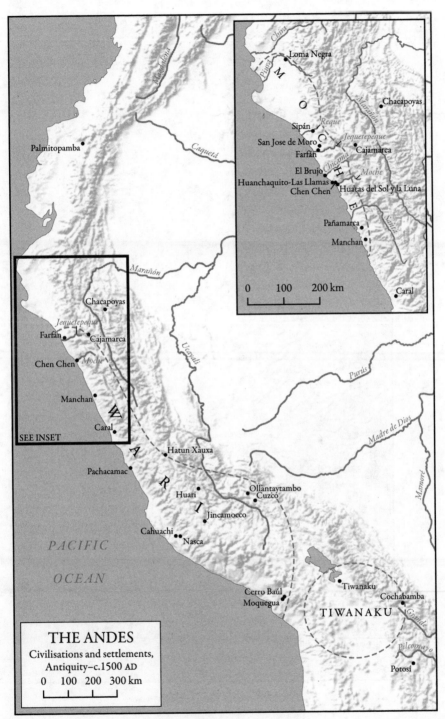

Loma Negra

Chacapoyas

Sipán
San Jose de Moro
Farfán   Cajamarca

El Brujo
Huanchaquito-Las Llamas
Chen Chen   Huacas del Sol y la Luna

Pañamarca
Manchan

0   100   200 km

Caral

Chacapoyas

Jequetepeque
Farfán   Cajamarca
Moche
Chen Chen

Manchan

Caral

Hatun Xauxa

Pachacamac

Huari   Ollantaytambo
               Cuzco

Jincamocco

Cahuachi
        Nasca

PACIFIC

OCEAN

SEE INSET

Cerro Baúl   Tiwanaku
Moquegua   Cochabamba

TIWANAKU

Potosí

THE ANDES
Civilisations and settlements,
Antiquity–c.1500 AD

0   100   200   300 km

GOBI DESERT

YOUZHOU

Liadong

Sogyong
[Pyongyang]

Songdu

XIHAI

SONG DYNASTY (c.1100)

Taiyuan

Lanzhou

Yanan

Yellow

Jinan

*Yellow
Sea*

Kaifeng

Chang'an

C H I N A

Yangzhou

Xiangfan

Chang Jiang

Suzhou

Chengdu

Yangtze

Hangzhou

Ya'an

*East
China
Sea*

Tanzhou

Nanchang

Dali

Ganzhou

Fuzhou

Quanzhou

Guangzhou

Salween

Đại La

*Philippine
Sea*

DAI VIET

Viên Chan [Vientiane]

A N N A M

*South
China
Sea*

SUKHOTHAI

Sukhothai

Indrapura

Kamphaeng Phet

K H M E R

Mekong

AYUT-

Thong

Ayutthaya

Champassak

C H A M P A

THAYA

Angkor

Vijaya

Vat Nokor

Phnom Penh

PANDURANGAL

Maha
Rosei

Chaiya

*Gulf
of
Thailand*

S R I V I J A Y A

KADARAM

Panai

THE RISE OF SOUTH
AND SOUTH-EAST
ASIAN EMPIRES, c.1000 AD

0   250   500 km

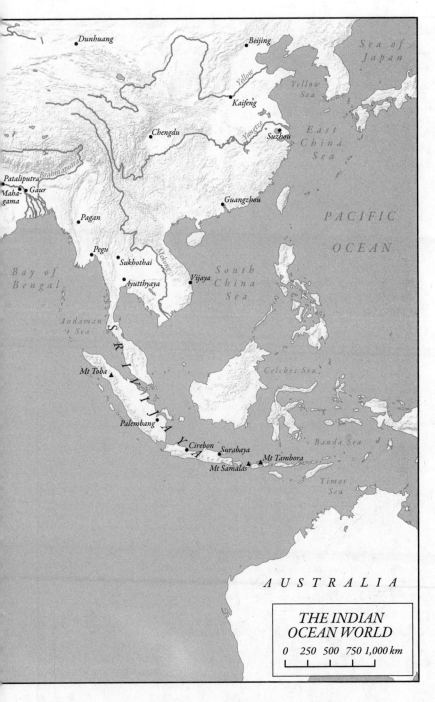

Dunhuang

Beijing

Sea of Japan

Yellow

Kaifeng

Yellow Sea

Chengdu

Yangtze

Suzhou

East China Sea

Pataliputra *Brahmaputra*

Maha-
gama Gaur

Pagan

Guangzhou

PACIFIC OCEAN

Pegu

Sukhothai

Mekong

South China Sea

Bay of Bengal

Ayutthyaya

Vijaya

Andaman Sea

S R I V I J A Y A

Celebes Sea

Mt Toba

Palembang

Cirebon

Surabaya

Mt Tambora

Banda Sea

Mt Samalas

Timor Sea

AUSTRALIA

**THE INDIAN
OCEAN WORLD**

0   250   500   750 1,000 km

THE SPREAD OF RELIGIONS
AND THE BLACK DEATH

→ Black Death   → Mongol campaign

0   1,000   2,000 km

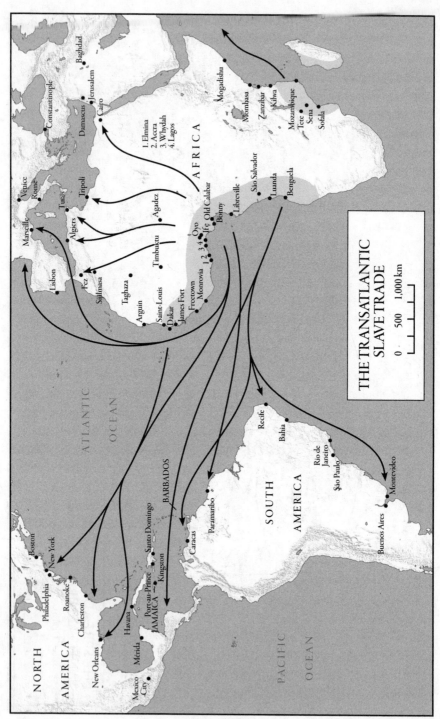

THE TRANSATLANTIC
SLAVE TRADE

0   500   1,000 km

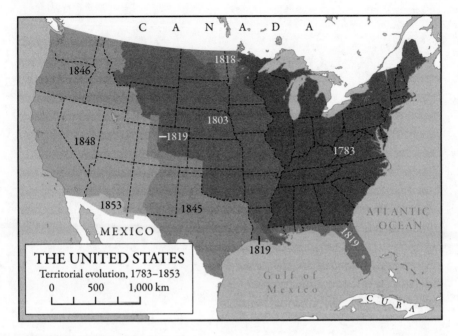

THE UNITED STATES
Territorial evolution, 1783–1853
0   500   1,000 km

THE EARTH
TRANSFORMED

Introduction

Three things exercise a constant influence over the minds of men: climate, government and religion.

Voltaire, *Essai sur les mœurs et l'esprit des nations* (1756)

'Man's first disobedience', wrote John Milton at the start of *Paradise Lost*, was to eat the fruit 'of that forbidden tree' in the Garden of Eden. The decision 'brought death into the World, and all our woe'. The loss of paradise turned the earth from a place of beauty and plenty into one of sorrow and sadness, where 'peace and rest can never dwell, hope never come' and where life was turned into 'torture without end'.[1]

Milton's epic poem, first published in the second half of the seventeenth century, was a retelling of the story that appears at the start of the Book of Genesis explaining how humans came to be the architects of their own demise. By allowing themselves to be tempted by the 'infernal Serpent', Adam and Eve condemned all future generations to lives of ecological challenge, ones where the environment was no longer always benign, where food was not always easy to come by and where humans had to work, rather than receive benefits from God. Paradise had been lost.

In today's world, the ways that our species works the land, exploits natural resources and treats sustainability are topics of vehement discussion – not least since many believe human activities to be so extensive and so damaging that they are changing the climate. This book sets out to look at how our planet, our enclosed garden (the literal meaning of the word 'paradise'), has changed since the beginning of time, sometimes as a result of human endeavours, calculation and miscalculation, but also thanks to a host of other actors, factors, influences and impulses that have shaped the world we live in – often in ways we do not think about or understand. This book will explain how our world has always been one of transformation, transition and change because, outside the Garden of Eden, time does not stand still.

My first encounter with the human impact on the environment and climate change came with a children's current affairs programme called *John Craven's Newsround* which was shown every day in the UK when I was a young boy. *Newsround* was a flagship BBC project that was a lifeline, connecting younger viewers to the world beyond the British Isles. One of the few programmes my parents allowed my siblings and me to watch when we were growing up, it introduced me to the suffering of people at the hands of the Khmer Rouge, to the complexities of the Middle East and to the realities of the Cold War.

One of the themes that came up regularly in the late 1970s and early 1980s was the subject of acid rain. I remember being transfixed by the horror of trees without leaves and by the thought that human activity was responsible for the degradation of nature. The idea that the factories belched emissions that devastated forests, killed animals and contaminated the ground came as a shock to me. Even as a young boy, it seemed obvious that the choices we made to produce goods and products had impacts that had long-term effects on us all.

These misgivings were compounded by a fear of devastation that was a hallmark of my childhood. I am part of a generation that was brought up to believe that the world might see global nuclear war between the United States and the Soviet Union that would result in large-scale death not only from the detonation of countless intercontinental ballistic missiles (ICBMs) but from the nuclear winter that would result from mushroom clouds released by warheads on impact. One film, *When the Wind Blows*, which came out in the mid-1980s, painted a poignant and awful picture of what lay ahead: sadness, suffering, hunger and death – all because of humanity's ability to invent weapons of mass destruction that would not only kill millions through firestorms and explosions, but would change the earth's climate so drastically that survival alone would be a miracle.

The detonation of scores of nuclear weapons promised to throw so much debris into the atmosphere that we would have to learn to live in sub-zero temperatures. Sunlight would be blocked by blankets of dust and particles with the result that plants would die. Animals would succumb as a result too – leaving those who survived the blasts not only freezing cold but hungry. Fallout from radiation would contaminate flora and fauna, poisoning all forms of life. The aim was to get through the apocalypse and to hope to be one of the survivors. In due course, we hoped, the climate would reset. Then it would be a case of seeing how many people were left alive and where, and starting again.

The fears of my generation were swelled by disaster, the most dramatic of which was the explosion in 1986 of the reactor at Chernobyl in what is now Ukraine. The reports of the catastrophic failure – strenuously denied for days by the Soviet authorities – were a reminder that miscalculations, misjudgements and incompetence could affect the world we lived in. In the months that followed, I studied maps of the fallout, was careful about what I ate and became acutely aware of the dangers posed by the potential for climatic change.

We used to spend our summers by a lake in the middle of Sweden. We said that we would flee there if there was ever a chance of a nuclear war breaking out. As most people know, Sweden is not the warmest country in the winter as it is; but I was reassured by the idea that being out of the way of soldiers, tanks and missiles would be a benefit. I was also comforted by the knowledge that blueberries (still my favourite fruit) were resilient to the cold. So I had a little bag packed by the side of my bed that I would update each year with my necessities for when (not if) changes to the world's climate would demand adaptation: a bar of chocolate; a Swiss Army penknife so I could make bows and arrows; some woollen gloves; a deck of cards and three balls; two pens (in case one ran out of ink); and some paper.

As it happened, my preparations were never needed – although it turns out that this was often because of luck rather than skill. As we now know, missile launches almost took place because of bears breaking down wire fences in search of food; because of misunderstandings about military exercises that made one side believe an attack was imminent; and because of weather balloons being misidentified as ballistic weapons systems. I grew up in a world of close shaves, near disasters and human error.

To be sure, there were many other things that scared me growing up: the 1970s and 1980s were a time of injustice, hatred, instability, terrorism, famine and genocide. But ecological devastation, climate and climate change were constantly in the background as current problems that would get worse in the future. Few things were certain for my generation. One thing was clear: we were all but guaranteed to live on a planet that was more hostile, more unstable and more dangerous than the one we had grown up in. I assumed that that would be because of the catastrophe of global war or large-scale accidents.

It did not cross my mind that the end of the Cold War would lead to an age of ecologies being placed under ever greater stress, or that increased global economic co-operation would result in massive rises in levels of carbon emissions and a warming world. I was brought up to believe that disaster stemmed from the horrors of war; after all, that was what I was taught in the classroom. Peace and harmony, on the other hand, were supposed to be the solution – not part of the problem. And so, a journey that began many years ago watching *Newsround* has led me to think about human interventions in the landscape, about how the climate might have changed in the past, and above all about the role that climate has played in shaping the history of the world.

We live in a world teetering on the brink of disaster because of climate change. 'Every week brings new climate-related devastation,' said António Guterres, Secretary General of the United Nations in 2019. 'Floods. Drought. Heat waves. Wildfires. Superstorms.' This was no apocalyptic prediction, he said, for 'climate disruption is happening now, and it is happening to all of us'. As to what the future has in store, he went on, there is little hope. Lying in wait is nothing less than 'catastrophe for life as we know it'.[2]

There are many problems facing humanity, said Barack Obama in his penultimate State of the Union address as President of the United States of America; 'and no challenge – no challenge – poses a greater threat to future generations than climate change'.[3] 'Today's ecological crisis, especially climate change,' said Pope Francis in 2019, 'threatens the very future of the human family.' The situation looks bleak. 'Future generations stand to inherit a greatly spoiled world,' he added. 'Our children and grandchildren should not have to pay the cost of our generation's irresponsibility.'[4]

Agreements reached by governments to deal with carbon emissions and global warming represent 'the minimum steps to be taken to protect the Earth, our shared homeland', noted President Xi of the People's Republic of China in 2020. 'Humankind can no longer afford to ignore the repeated warnings of nature.' It is vital, therefore, to 'launch a green revolution and move faster to create a green way of development and life, preserve the environment and make Mother Earth a better place for all'.[5]

Others have put the threat personally as well as forcefully. 'You have stolen my dreams and my childhood with your empty words. And yet I'm one of the lucky ones,' said Greta Thunberg at the UN Climate Action Summit in September 2019. 'People are suffering. People are dying. Entire ecosystems are collapsing. We are in the beginning of a mass extinction, and all you can talk about is money and fairy tales of eternal economic growth. How dare you!'[6]

If climate change is going to be – or already is – the theme that will dominate the twenty-first century, sparking water shortages, famines, large-scale migrations, military conflict and mass extinction, then understanding what the future holds should be essential not just for politicians, scientists and activists, but for everyone. As a historian, I know that the best way to address complex problems is to look back in time, as this helps provide context and perspective for current and future challenges. And history can also teach valuable lessons that help formulate questions and sometimes even answers relating to some of the big issues that lie ahead of us.

This is particularly true when it comes to the relationship between human activities, the environment and the natural world in the regions and places that I have spent decades researching. In many if not all, availability and use of water, the expansion of food production and the geographic challenges and opportunities of local as well as long-distance trade are not just important factors but fundamental elements that underpin the broad sweep of history. As Fernand Braudel put it, the study of the past does not just involve the competition between humans and nature; it *is* the competition between humans and nature.[7]

When I first studied the Sasanian and ʿAbbāsid empires, I quickly learned that the success and stability of the state were closely linked to the irrigation of fields that enabled a rise in agricultural yields and supported larger populations.[8] Looking at the histories of China led me to research which argues that the rises, falls and replacement of imperial dynasties stretching back more than a millennium are closely correlated with changes in temperature, with colder phases serving as periods of demographic decline, conflict and the replacement of imperial rulers with new regimes.[9]

Likewise, reading poetry such as *Meghadūtam* ('The Cloud Messenger'), likely written in the fifth century AD by the famous fifth-century Sanskrit poet Kālidāsa, made clear how the monsoons and the rains, as well as the seasons, play a fundamental role in the literature, culture and histories of South Asia.[10] I learned long ago too that in the more recent past, Soviet policy in Central Asia in the 1950s was not only environmentally disastrous but had a significant impact on the Cold War and plays a role in the use of coerced labour in the region today.[11] I also know from experience how pollution in places I visit regularly is acrid, damaging and dangerous – with cities like New Delhi, Bishkek and Lahore ranked among the worst in the world for air quality. In Tashkent, the capital of Uzbekistan, the air was classified as dangerous 80 per cent of the time in the course of 2020.[12]

So I have set out to examine environmental history and to understand more clearly what the past tells us about human behaviour, about anthropogenic change in the natural world and about how extreme weather events, long-term weather patterns and climatic change have influenced and impacted history. I have wanted to assess why we seem to have arrived at the edge of a precipice where the future of our species – as well as those of a significant part of the animal and plant worlds – is at risk. Rather as a doctor should have full knowledge of an illness before trying to devise a cure, so too is investigating the causes of the current problems essential if we are to suggest a way to deal with the crises now confronting us all.

Historians are living in something of a golden age thanks to a rash of new evidence and new types of materials that help improve understanding of the past. Machine learning, computer models and data analysis are not just providing new lenses to look at other periods in history but revealing a plethora of information that was unknown and unseen. For example, networks of villages in the Amazon rainforest dating back many centuries and which were set out to mirror the cosmos have been identified thanks to Light Detection and Ranging (LIDAR) technology.[13] Advances in cost-effective laboratory visible-near infrared/ shortwave infrared spectroscopy data have enabled groundbreaking work to reach conclusions about the social change in the Mapungubwe landscape at the confluence of the Shashi and Limpopo rivers during the twelfth century.[14] Isotope data from human burials and from pig teeth in what is now Papua New Guinea help shed light not only on settlement patterns but on proportions of marine foods that people were eating more than 2,000 years ago.[15] And new technology has helped identify the mineralisation of plant materials and seeds preserved in refuse pits and cesspits in 'Abbāsid-era Jerusalem, providing support for hypotheses about the westward diffusion of crops in the early Islamic period.[16]

Some of the most exciting advances have come in the way we understand the climate. These include inventive ways of using written sources that have been ignored or poorly utilised in the past. For example, clam shells from the coast of Peru enable climate reconstructions through changes in the chemistry of shells that allow researchers to identify yearly, monthly and even weekly ocean temperatures.[17] Records of cherry blossom festivals in Japan that go back to the start of the ninth century and note the date of the flowering of cherry trees help establish when spring arrived each year over the course of many centuries.[18] Registers kept by the harbour authorities in Tallinn in Estonia covering the last 500 years show the arrivals of the first ships each year, and consequently not only reveal when the sea became ice-free, but indicate patterns in longer and warmer springs.[19] Driftwood from the Svalbard archipelago in the Arctic shows considerable variability of sea ice between 1600 and 1850 which in turn points to unusual climatic patterns in this period.[20]

Above all, new and exciting 'climate archives' are being added to all the time. Many will feature in this book. We will consider information from growth rings in trees from the Altai mountains in Central Asia and from the build-up of mineral deposits from caves in Spain that show changes in temperature and rainfall. We will look at air bubbles trapped in ice cores in Greenland and in glaciers in the European Alps that provide evidence of volcanic eruptions as well as of human activities such as metallurgy and the burning of crops, forests or fossil fuels; we will encounter fossilised pollen from Oman and pollen deposits in lake varves in Anatolia that provide insights into changes of vegetation, both through natural causes and because of human intervention; we will come across carbonised and desiccated seeds in South-East Asia, dried nutshells from northern Australia and digested and partly digested foods from Palestine that provide evidence of diets as well as disease. We will look at climate conditions conducive to the spread of parasitic pathogens in the Americas and at evidence for crop cycles in West Africa – as well as at phylogenetic trees of plague in Ethiopia, Kyrgyzstan and Cambridgeshire.

Many new sources of climate data are becoming available that allow us better to understand the natural world deep into the past. For example, a team of researchers are working on an eighty-metre-deep sedimentary layer in south-east Kazakhstan that provides a record of soil moisture – as well as offering insights into the role that Central Asia plays in global climate evolution in general and into the land–atmosphere–ocean water cycle in the northern hemisphere in particular. This is of considerable significance not only for studies of the past but also for future long-term global climate analysis.[21] So too is new research on the Tibetan plateau, where modelling based on findings from high, treeless areas suggests a major decrease in plant diversity in alpine habitats in the coming centuries.[22]

Such new sources of evidence have led to revolutionary new ideas being developed about the past. New climate data provides insights into a tumultuous period in the middle of the third century in the Roman empire, with some scholars seeking to link reduced levels of solar activity, increases of sea ice and several major volcanic eruptions with rapid cooling, disrupted food production and a series of political, military and monetary crises precisely in this period.[23] Data about the persecution of Jews in Europe drawn from almost a thousand cities between 1100 and 1800 shows that a decrease in the average growing-season temperature of about one-third of 1 degree Celsius is correlated with a rise in the probability of Jews being attacked in the subsequent five-year period – with those living in and near locations with poor soil quality and weaker institutions more likely still to be the victims of violence during times of food shortages and higher prices.[24]

And a comparison of cold temperatures and wheat prices in Europe has led to new models being proposed about which cities were more resilient than others to price shocks; this has in turn spurred hypotheses that cooler weather in England in the early modern period led to agricultural revolution, which in turn prompted and rewarded the development of new technologies that led to an energy transition and ultimately gave rise to an age of European global empires.[25]

Not surprisingly, eye-catching arguments like these are a matter of lively discussion and sometimes heated debate among historians, with particular concerns voiced about historical and environmental determinism and about the problems of distinguishing between correlation and causation.[26] There are other challenges of interpretation. A case in point comes from the Indian subcontinent, a region that is ecologically and culturally diverse and home to a wide array of 'settled villages, hunter-gatherers, swidden cultivators, nomadic pastoralists and fisherfolk', as well as possessing astonishing species diversity and great climatic and ecological variety; as such, some scholars argue not only that are there dangers in making generalisations about the subcontinent as a whole, but that comparisons with other parts of the world are simply not appropriate.[27]

Another related issue is that those who write about climate and its impact often focus heavily on societal collapse, usually with a narrow band of signature examples – most notably the Maya, Easter Island and the 'fall' of the Roman empire – which have all been attributed to climate change in recent bestselling books.[28] Apart from the problems of oversimplifying complex narratives into narrow explanations (which authors are sometimes at pains to note), some believe that the urge to impart lessons – about the exhaustion of natural resources, about the failures to adapt to changing environmental conditions and to the consequences of not living sustainably – is a case of the tail wagging the dog, which is to say viewing the past through the prism of contemporary concerns.[29]

Much depends therefore on lightness of touch when dealing with new kinds of materials – just as good history requires sound judgement when dealing with written sources and with material culture. The problem then is not that climate science, data or new approaches are themselves flawed or misleading; rather they need to be handled carefully and put in contexts that are balanced, persuasive and appropriate.[30]

By and large, the weather, climate and environmental factors have rarely been seen as a backdrop to human history, let alone as an important lens through which to view the past. There are a handful of cases where climate features prominently, though usually not always plausibly. The famous story of King Xerxes ordering the waters of the Hellespont to be given 300 lashes after a storm brought down bridges that slowed his invasion of Greece in 480 BC would seem to be an apocryphal tale told to highlight the illogical rage of a barbaric, tyrannical ruler rather than a reliable statement of fact.[31]

That two attacks ordered by Kublai Khan, the grandson of the great Činggis (or Genghis) Khan, on Japan in the late thirteenth century were thwarted by 'divine winds' or 'kamikaze', sent by the gods to frustrate the invaders, says more about how these events came to be seen in Japanese history than about the reason for the failure of Yuan dynasty which controlled most of what is now China to conquer Japan.[32] Most celebrated of all, though, is the onset of a hard Russian winter, which in popular imagination played a decisive role both in derailing Napoleon's ill-fated attack on Moscow in 1812 and in grinding German forces to a standstill and then to disaster after Hitler's attack on the Soviet Union in 1941. Both popular tropes obscure the fact that overambitious objectives, inefficient supply lines, poor strategic decisions and worse execution of plans on the ground were what doomed both invasions as much as, if not more than, the snow.[33]

In the main, though, we ignore climate and long-run climate patterns or changes altogether when we look at history. Most people can name the great leaders and major battles in the past, but few can name the biggest storms, the most significant floods, the worst winters, the most severe droughts, or the ways that these influenced harvest failures, provoked political pressures or were catalysts in the spread of disease. Reintegrating human and natural history is not just a worthwhile exercise; it is fundamentally important if we are to understand the world around us properly.[34]

Assessing the role of weather, extreme events, long-run climate patterns and changes in climate requires a detailed understanding of how the global climate system and subsystems are connected. The earth's climate is shaped by several closely related factors. First is the global weather system, which is constantly modulating because of changing atmospheric conditions, ocean currents and ice-sheet behaviour, as well as because of geological and plate tectonics and oscillations in the flow of liquid iron in earth's outer core. The tilt of the planet's axis, the mild eccentricity of the earth's orbit around the sun and the uneven distribution of energy between the equator and the poles also affect weather and climate patterns – as do the interactions between all these factors.[35]

The main source of seasonal climate anomalies is the El Niño–Southern Oscillation (ENSO), which describes the relationship between the atmospheric and oceanic conditions in the equatorial Pacific, including the direction and strength of trade winds, surface water temperature and air pressure. The ENSO cycle of alternating warm El Niño and cold La Niña events is the dominant year-to-year climate signal on earth.[36] It affects volumes of rainfall in South America, but also affects conditions in South Asia, East Africa and Australia – although the Indian monsoon can also be influenced by episodic climatic shifts in the North Atlantic.[37]

Other subsystems also play a significant role in temperature and climate conditions and variations spanning years or even decades. For example, the North Atlantic Oscillation (NAO), which describes the balance of sea-level pressure between the Azores and Iceland, creates periods of cyclonic and anti-cyclonic patterns which affect western Europe. It also plays a role in dictating winter precipitation in the Mediterranean and Black Sea, as well as driving cold air from Siberia and polar regions into central and western Europe.[38] Antarctica and Greenland produce meltwater that causes ocean surface cooling – although recent research indicates that the impact on the Southern Ocean is much more significant for global temperature and sea level than the impact on the Arctic.[39]

Solar activity has an important function in global climatic conditions, in particular because of the sun's magnetic activities. The most prominent of these are sunspots and aurora which follow cycles typically lasting eleven years.[40] Solar activity is also modulated by long-term variations which result in more active and more settled patterns, known as grand maxima and grand minima.[41] The most recent example of the latter is known as the Maunder Minimum, which took place between around 1645 and 1715 when sunspot activity was extremely rare.[42]

Volcanic activity is also an important factor in forcing alterations in climate. In 1991, for example, a major eruption of Mount Pinatubo in the Philippines injected between fifteen and twenty megatons of sulphur dioxide into the atmosphere which was then oxidised to form stratospheric particles of sulphate aerosols; these then propagated, increasing the opacity of the stratosphere. Among the startling results was a reduction of direct sunlight by 21 per cent and a reduction in insolation that led to an averaged global temperature cooling in excess of 0.5 °C.[43]

These figures mask important regional patterns. While the North Atlantic dipped to 5 °C cooler than average, the following winter in Siberia, Scandinavia and central North America was significantly warmer than normal; the year after the eruption saw extensive flooding in the south of the United States and significant water shortages and droughts in sub-Saharan Africa, South and South-East Asia as well as many parts of central and southern Europe. Nevertheless, taken as a whole, the impact was dramatic. The reduction of shortwave solar radiation resulted in a globally averaged sea surface temperature reduction of 0.4 °C – or around one hundred times the total annual worldwide energy consumption.[44]

Volcanic eruptions bring other wide-ranging consequences for the natural world. These include producing phytoplankton blooms through the input of lava into the ocean and localised warming up of deep waters that rise to supply the sunlit layer with nutrients.[45] As we shall see, eruptions can lead to sharp reductions in agricultural production which in turn can result in economic, social and political disruption. We shall also look at the impact that eruptions have by changing habitats for disease-carrying species, or by being catalysts for different enzootic cycles for pathogens or by opening up what one scholar has called 'epidemic highways'.[46]

One crucial element in volcanic eruptions is that their timing can matter as much as their magnitude and scale. New research using supercomputers and thousands of simulations has shown that eruptions that take place during the summer produce a much higher impact on global climate than those that occur during the winter and spring.[47] The locations of large-scale eruptions are also important, with models now showing that volcanoes outside tropical regions have produced stronger hemispheric cooling than tropical volcanoes over the last thirteen centuries.[48] Studies of volcanoes and volcanic areas have also shown a substantial increase in recent years in the measurement of CO_2 flux, with emissions of degassing volcanoes far greater than those produced by relatively short-lived eruptions.[49]

There are other phenomena where the climate has a significant impact on the natural world. Heavy rains on the Indo-Gangetic Plain to the north of the Indian subcontinent can increase stress loads on the earth's crust, leading to a decrease in micro-seismicity (or minor tremors) in the adjacent Himalaya region.[50] Evidence linking strong typhoons in east Taiwan with seismic activity beneath the island suggests not only that weather conditions may trigger geological responses but that they might do so in small, modest and regular ways that prevent single, large and devastating earthquakes.[51]

Climate and temperature shape biodiversity too: the number of species decreases drastically from the equator to the poles – with some estimating that tropical forests contain more than half the species of flora and land-based fauna on earth. However, it now emerges that while tropical forests are astonishing in the range of animals and plant that they host, this is the result of gradual changes over extended periods of time; in fact, new species form faster in cold, dry, unstable and extreme environments.[52]

Historians have long noted that solar activity, long-term weather cycles and the impact of volcanic activity seem to form patterns that span decades and even centuries. Some of these periods have been given names to encapsulate a sense of uniformity, mainly based on the behaviour of the sun and the resultant impact on the complex global climate subsystems. The Roman Optimum (c.100 BC–AD c.200) and the Medieval Climate Anomaly (c.900– c.1250) are two such examples, purportedly times of favourable, warmer-than-average but above all stable conditions, while the Little Ice Age (c.1550–c.1800) was a time of distinctly cooler temperatures, lower solar irradiance and global crisis.[53]

That, at least, is the theory. One of the challenges posed by climate science is that new evidence from other areas and increasing levels of accuracy can show that what holds in one part of the world does not hold in another. While the central and eastern Pacific Ocean appears to have been unusually cold in the fifteenth century, for example, the same does not seem to have been the case elsewhere, much as north-western Europe and the south-eastern parts of North America appear to have endured harsher conditions in the cooler seventeenth century than other regions. In fact, there is no evidence for globally coherent warm or cold periods over the two millennia before the industrial revolution.[54]

Just how much care needs to be taken can be shown by looking at the period c.1220–50. During this relatively narrow window, the hydroclimatic situation was rather favourable for cereal production in the eastern Mediterranean and the southern Levant (roughly modern

Israel, Palestine, Jordan) but was much less so just a few hundred kilometres away in the central Mediterranean, Sicily and southern Italy.[55] It is important, in other words, not to extrapolate too much from site-specific information and apply it to other locations where evidence has not been gathered either because they have not been so intensively studied or because they do not offer suitable corroborative material.

Synoptic frequency of extreme marine heat across ocean basins from 1900–2019

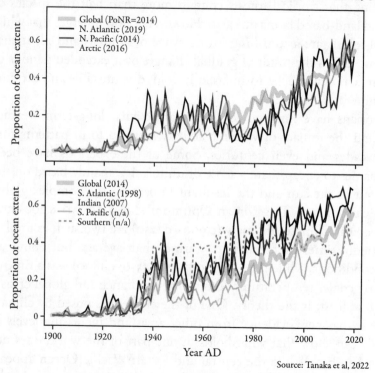

Source: Tanaka et al, 2022

The question of regional climate coherence is an issue too in today's world where global warming is affecting 98 per cent of its surface area, apart from the Antarctic, where warming has not yet been observed over the entire continent.[56] Warming patterns do not affect all parts of the earth the same way nor do they do so at the same rate. Indeed, as a recent report noted, while the majority of countries around the world will experience the 'pernicious effects' of climate change, a small number may actually benefit.[57]

Nevertheless, even before considering the accuracy and margins of error of future climate models, analyses of the here and now makes for a sobering read. The Atlantic Meridional Overturning Circulation (AMOC) – a system of interconnected surface and deep currents in the Atlantic Ocean that is largely responsible for the relative warmth of the northern hemisphere – is at its weakest in more than 1,500 years.[58] Early-warning indicators from multiple sea surface temperature and salinity data from across the Atlantic Ocean basin suggest that currents may be nearing a shutdown, something that would cause severe disruption to the global climate system and increase the chances of a cascade of further transitions – including the distribution of rain in tropical monsoons and the melting of the Antarctic ice sheet.[59] Some scientists argue that these risks represent nothing less than 'an existential threat to civilization'.[60]

These current and major changes to global climate are almost entirely due to human impact on the environment. Anthropogenic impacts began to have a radical effect from the second half of the eighteenth century, following the invention of the steam engine and the energy and industrial revolutions that transformed production and societies and marked the start of a fundamentally different human relationship with the natural world. The era that began with the Industrial Revolution has been named the 'Anthropocene', following the suggestion of the Nobel-prize-winning chemist Paul J. Crutzen in 2002 to describe a period that has seen levels of carbon dioxide and methane emissions rise sharply and consistently.[61] More recently, a panel of distinguished international scientists agreed to treat the Anthropocene as a formal gatepost in history – and voted to consider this as starting in the middle of the 20th century, a point when carbon emissions from human activities began to rise extremely quickly.[62]

Burning fossil fuels like coal and oil releases water vapour, carbon dioxide (CO_2), methane (CH_4) and nitrous oxide (N_2O), which trap heat and are therefore known as greenhouse gases. Rising populations, increased energy demand, falling production prices and massive investment in infrastructure have seen dramatic surges in fossil-fuel use, leading to substantial rises in emissions and to sharp temperature rises. For 800,000 years until the start of the industrial revolution, there were around 280 parts of CO_2 for every million air molecules. In 2018, this had risen to over 408 parts per million – levels that have not been so high since the Pliocene era, over 3 million years ago, when sea levels were almost twenty-five metres higher than today and when average temperatures were 2–3 °C warmer than today.[63] By the summer of 2022, levels were higher still, measured by the Mauna Loa Atmospheric Baseline Observatory in Hawai'i at a monthly average of 421 parts per million.[64]

There are multiple knock-on effects. Global warming leads to melting of the ice caps, which in turn raises sea levels. A single iceberg, known as A68, that broke free from the Larsen-C ice shelf in Antarctica in 2017 was dumping 1.5 billion tonnes of fresh water into the ocean every single day before it disappeared in 2021.[65] The collapse of ice-shelves mean that glaciers – which hold a large fraction of the world's freshwater – flow into the oceans; this has obvious implications for the world's largest cities, many of which are located on the coast. Modelling using artificial intelligence and highly accurate readings of elevations suggests that land that is currently home to 300 million people will flood at least once a year by 2050, with Asian populations most heavily affected. As it is, around 1 billion people already live on land less than ten metres above high-tide levels, and 230 million live in coastal communities at less than one metre's elevation above water.[66]

The UK's energy infrastructure is highly exposed to even modest rises in sea level, with all nineteen of the country's nuclear reactors situated in coastal locations, as are all the major fossil-fuel power stations in Scotland, Wales and Northern Ireland.[67] Some estimates suggest that $3–11 trillion worth of assets globally are at risk in the coming decades.[68]

There could be 'catastrophic outcomes', notes the International Monetary Fund (IMF), if steps are not taken to reduce emissions, including lower agricultural yields, frequent disruption of economic activity, destruction of infrastructure, deterioration of health and increased prevalence of infectious diseases.[69] According to UNICEF, 1 billion children – almost half the world's children – are already at 'extremely high risk' of the impacts of the climate crisis.[70]

The scale of the challenge in minimising the consequences of the rapid warming in the coming decades is hard to exaggerate. Recent modelling has suggested oil and gas production must decline globally by 3 per cent each year until 2050 – and that 60 per cent of oil and fossil methane gas and 90 per cent of coal must remain unextracted to keep within a 1.5 °C carbon budget.[71] The fact that not a single one of the world's major economies, including the entire G20, had climate plans in 2021 that met their own commitments under the Paris Agreements of 2015 indicates that we should be preparing for the worst rather than the best case – even though some commentators stress that doomsday scenarios provide little or no scope for adaptation, technological innovation or successful mitigation of some or all of the worst potential problems.[72]

There is of course much to be said about the dangers of over-relying on the twin temptations of crystal-ball gazing and catastrophising. However, recent models suggest an even bleaker future than many have predicted with temperature rises of around 4 °C by 2100. In fact, a report by the US National Highway Traffic Safety Administration in 2018 said that there was little point in enforcing fuel-efficiency standards in cars because this would have little practical impact in the long run, since moving away from fossil fuels would require 'the economy and the vehicle fleet' to operate in ways that were 'not currently technologically feasible or economically practicable'.[73] This was taken by many as a statement that the planet's fate is considered sealed – at least in some parts of the US government.[74]

Less easy to argue with are the problems that are already very much part of the present and not of the near or distant future. The energy revolution has had a disastrous impact on human health, with air-pollution levels in some cities at levels ten times higher than World Health Organisation minimum standards: indeed, 92 per cent of the world's population live in places which exceed these limits.[75] Dirty air is not simply the result of fossil fuels being burned for energy; it also comes from open-air burning of rubbish. An estimated 40 per cent of global waste is burned in the open air, resulting in substantial injections of particulate matter and polycyclic hydrocarbons into the atmosphere.[76]

Air pollution is lethal. In 2015, it resulted in approximately 9 million premature deaths worldwide.[77] The latest figures put the number of annual deaths attributable to air pollution in India at more than 1.6 million, with the highest number of fatalities in states with low per capita incomes.[78] Indeed, in 2017, the death toll from air pollution in war-torn Afghanistan was almost eight times higher than the number of civilian casualties.[79]

While chronic levels of pollution affect large parts of the developing world, those living in wealthy countries also pay the price for the failure of governments to understand fully or deal with the dangers posed. In Europe as a whole, 8 per cent of mortalities can be attributed to exposure to particulate matter with a diameter of less than or equal to 2.5 μm ($PM_{2.5}$) and to exposure to nitrogen dioxide (NO_2) – almost 500,000 deaths per year, in other words.[80] Recent research goes further still, suggesting that 18 per cent of total global deaths in 2018 were caused by the effects of fossil-fuel pollution.[81]

As with so many other issues, air pollution is also closely connected to socio-economic status and to income levels – including in wealthy, developed countries. Businesses that produce pollutants are more likely to be located in communities that are home to populations with high numbers of minorities and lower incomes.[82] The damage goes much further too: particulate matter has a strong and highly detrimental impact on cognitive function, reducing memory, orientation, verbal fluency and visual-spatial ability.[83] Breathing in nitrogen oxides and fine particulate matter as a child or an adolescent is a risk factor for mental illness in adulthood and for dementia, as well as for an elevated risk of self-harm.[84] Exposure to pollution for even a single day in childhood can profoundly affect the cardiovascular and immune systems in later life, resulting in a detrimental impact on long-term health.[85]

According to a recent study by the World Bank, the cost of health damages associated with exposure to air pollution is $8.1 trillion – or more than 6 per cent of global GDP.[86] Human behaviour, lifestyles and impact on the environment do not just end up killing people; these factors also affect how they behave, how they think and how they communicate with each other.

Human impact on the natural environment has been devastating almost everywhere, in almost every way, from water contamination to erosion, from plastics entering the food chain to pressure on animal and plant life that has reached such a high level that the most recent United Nations report talks of declines in biodiversity at rates that are unprecedented in human history, and that threaten an erosion of 'the very foundations of our economies, livelihoods, food security, health and quality of life worldwide'.[87]

Human activities compromise the world's rivers, seas and oceans with plastic debris present in every major ocean basin over a decade ago – impacting wildlife because of contamination, intestinal blockage, internal injury and entanglement.[88] The scale of pollutants is mind-boggling, with an estimated nine trillion microplastic fibres shed from synthetic clothing being discharged from washing machines every week in the UK alone.[89] These can be now found in astonishing volumes all around the planet, with one survey of the Arctic finding forty microplastic particles on average for every cubic metre of seawater.[90] Studies in the US suggest that people consume and inhale 74,000–121,000 pieces of microplastics per year, while separate research shows the presence of microplastics in the placentas of pregnant mothers, and in high concentrations in the stools of infant children, as well as in all human blood.[91]

Pressure on the environment is now so great that 40 per cent of the world's plants are considered endangered.[92] This is partly the result of the collapse of insect populations, which in turn stems from deforestation, heavy use of pesticides, urbanisation and climate change, developments that now not only threaten animal and plant food chains but have potentially catastrophic implications for agriculture and food production.[93] Some estimate that almost $600 billion worth of global crops are already at risk annually because of pollinator loss.[94]

With millions of hectares of tropical forests being cleared every year and chronic overfishing in the world's oceans, animals are responding to climate change by changing their habitats – and their shapes and sizes. Some animals are responding to warmer climates by altering the ways that they thermoregulate, or cool themselves down, with limbs, ears, beaks and other appendages changing shape and size as a result of increased temperatures.[95] Maternal heat stress reduces growth in calves, especially of organs linked to the immune system – which has obvious implications for both dairy and meat production.[96]

Land-based species on mountainsides are moving upslope to avoid warming lowlands, while fish have been driven deeper by warming sea surfaces. Terrestrial taxa are moving towards the poles by seventeen kilometres per decade and marine species are moving more than four times as far.[97] Many species of butterflies and moths in the Himalayas have moved a thousand metres higher – or more – in search of better habitats.[98] Marine animals such as fish, crustaceans and cephalopods (including octopus, squid and cuttlefish) are moving an average of fifty-five metres deeper in the Mediterranean in search of cooler waters.[99]

As it is, the average population sizes of vertebrate species that have been monitored closely have fallen by almost 70 per cent in the last fifty years.[100] Bird numbers in North America have declined by nearly 3 billion since 1970, while more than 40 per cent of amphibian species are at risk.[101] Models that estimate potential extinction rates not only show dramatic collapses, but likely underestimate falls in species abundance and distribution.[102]

Declines are not uniform – and in fact, although some species and ecosystems are collapsing, others are not doing so badly and, in some cases, are thriving, as the case of tree species in boreal forests in eastern Canada shows.[103] Moreover, declines in some species open up opportunities for others.[104] Some scientists also point out the importance of evaluating at the local rather than the global level, and suggest that catastrophic falls in some species populations should be assessed as clusters of extreme decline (or increase) rather than taken to illustrate general, widespread and potentially misleading patterns.[105]

Nevertheless, there is a wide consensus among scientists about an ongoing 'biological annihilation' taking place in front of our eyes that is now regularly referred to as a 'mass extinction event'.[106] Research on polar oceans suggests that changes to food webs are already under way with profound implications not only for marine but for global ecosystems.[107] Many warn of a 'cascading erosion of biodiversity' and 'co-extinctions' affecting all levels of flora and fauna.[108] The 'sixth mass extinction' is different to those that have gone before because this time an animal species is responsible – humans.[109] One recent report spelt things out bluntly: 'The scale of the threats to the biosphere and all its lifeforms – including humanity – is in fact so great that it is difficult to grasp, even for well-informed experts.'[110]

This book is not about what will happen in the future. Nor is its aim to challenge the overwhelmingly uniform voices of the scientific community, either on the basis of the current global conditions or through examining what steps might be taken to mitigate some or even many of the worst problems posed by climate change, whether through adaptation or the introduction of new technologies. Its aim, rather, is to look at the past and to understand and explain how our species has transformed the earth to the point that we now face such a perilous future.

Originally, I had thought I would write only about the history of how the climate shaped the world around us and about the ways in which modulations in global temperatures, rainfall and sea level – along with extreme events, such as major storms, volcanic eruptions and meteor strikes – have influenced the past, setting out moments, periods and themes that explain what an important role climate has played in global history.

However, it became clear very early on when I started thinking about this book that opening the door to climate, to changes in weather patterns and to human intervention in the natural world led on to a much bigger set of questions and challenges about the relationship between agricultural surplus and the origins of the bureaucratic state; about the connection between pastoralists and nomads on the one hand and sedentary societies in villages, towns and cities on the other; about the role and development of religions and belief systems as a function of climate, environment and geography; about race and slavery and their part in the process of extraction of resources; about the spread of foods, pathogens and disease; about demographics, poverty and consumption patterns in the centuries since the industrial revolution; about globalisation, the standardisation of industry, agriculture, food and fashion in the last century; about why the twenty-first century is at a moment of crisis.

So this book has three goals. The first is to reinsert climate back into the story of the past as an underlying, crucial and much overlooked theme in global history and to show where, when and how weather, long-run climate patterns and changes in climate – anthropogenic and otherwise – have had an important impact on the world. The second is to set out the story of human interaction with the natural world over millennia and to look at how our species exploited, moulded and bent the environment to its will, both for good and for ill.

And the third is to expand the horizons of how we look at history. Study of the past has been dominated by the attention paid to the 'global north', that is the wealthy societies of Europe and North America, with the history of other continents and other regions often relegated to secondary significance or ignored entirely. That same pattern applies to climate science and research into climate history, where there are vast regions, periods and peoples that receive little attention, investment and investigation – something that speaks volumes about long-accepted perspectives of the past as well as about how academic funding (and intellectual silos) develop and deepen in practice.

If that points to one fundamental reason to reassess history, so too does the heavy overemphasis by historians on towns and cities and on states that resemble each other in terms of leadership, bureaucracy and behaviour. Indeed, 'civilisation' itself literally refers to the life of cities, of those who lived in them and who projected power and ruled from them. This is reflected in most written historical materials – narrative accounts, records of land sales, tax receipts and so on – that served to reinforce the administration of hierarchies. Much of history has been written by people living in cities, for people living in cities, and has focused on the lives of those who lived in cities. This skews the way we look at the past and at the world around us.[111]

And yet 'civilisation' is by far the single greatest factor in environmental degradation and the most important cause of anthropogenic climate change – because of the demands that the populations of cities place on energy and on the consumption of natural resources, including food and water. Although cities occupy only 3 per cent of the earth's land surface, urban areas accommodate more than half the world's population. Cities are not only responsible for a substantial proportion of global warming, but they will also be greatly affected by it in coming decades.[112]

It is no coincidence, then, that the last century, which has seen a rapid expansion in the number, size and populations of cities, has also seen the most serious depletions of the environment and the most rapid growth in the rates of consumption. As cities rise, so too do pressures on nature, biodiversity and sustainability through changes in land use and cover, because of the modification of hydrologic systems and as a result of the impact of altered and compromised biogeochemical cycles.[113] In the years 2001–18 alone, built-up areas in China expanded by 47.5 per cent, while in the US they increased by 9 per cent. Indeed, the world's population living in cities is forecast to rise by some 3 billion to around 7 billion by 2050 based on current demographic trends.[114] To put that in historical perspective, just over 15 per cent of the global population lived in towns and cities in 1900; by 2050, more than 70 per cent will do so.[115]

New technologies speeding up and driving down costs of production have spurred radical shifts in manufacturing, transportation and consumption patterns. More than 75 per cent of all the virgin plastics ever created are estimated to have become waste, of which about 9 per cent has been recycled, 12 per cent incinerated and the remainder – about 5 billion metric tons, or around 60 per cent of all plastics ever produced – has accumulated in landfill or in the natural environment.[116]

By some measures, while human-made mass – such as concrete, construction materials and metals – was equal to about 3 per cent of global biomass a century or so ago, today it exceeds it. On average today, further human-made mass that is more than the equivalent of the body weight of every person on earth is produced every single week, a phenomenon closely linked to the rise of cities and megacities, and to high levels of consumption of food, water, energy and non-perishable goods.[117] This in turn is connected to globalisation and to supply chains and networks that form both a virtuous circle of hyperconnectivity, standardisation, high velocities of exchange and low prices, and a vicious circle of extraction, resource depletion and environmental damage.

Throughout history, on the other hand, peasants, pastoralists and nomads, indigenous peoples and hunter-gatherers, who have had to understand the limitations of the land and adapt to even modest changes, have been either written out of the story of the past or typified as barbaric, erratic and primitive. One who has no need for the city, wrote Aristotle, 'is either an animal or a god'.[118] Nomads in Central Asia had been 'abandoned by heaven', stated a Chinese author a few centuries later; Ibn Faḍlān, writing in the tenth century, agreed after meeting mobile pastoralists: they 'live in poverty, like wandering asses', he wrote. 'They do not worship God, nor do they have any recourse to reason.'[119]

These attitudes persist today in many parts of the world, often manifested in the creation and funding of wildlife reserves that evict local populations to create what look to city inhabitants like natural paradises since they are devoid of humans. One good example of this comes from the Grand Canyon, 'a natural wonder which ... is absolutely unparalleled throughout the rest of the world', according to President Theodore Roosevelt after his visit in 1903. 'Man can only mar it,' he added in a telling statement of how the 'natural' can be seen as pure and unblemished only if kept free of human intervention. Just over a decade later, the Grand Canyon became a national park, imposing restrictions and controls over lands that had been lived on by the Havasupai and other indigenous peoples for more than 700 years.[120]

In today's world, we see regular, aggressive, overtly racist campaigns against indigenous peoples, hunter-gatherers and foragers such as Bushmen in Botswana, the Baka in West Africa, Adivasi peoples in India or traditional nomads in large parts of Central Asia, invariably laced with insults about supposedly 'primitive' ways of life. This is ironic given that indigenous populations maintain forest levels well, storing more carbon as a result, as well as developing strategies that

support biodiversity conservation and good long-term stewardship of
the environment.[121]

One of the difficulties of writing history is that there are inevitably major
gaps in coverage. It is true that scholars are using new and increasingly
sophisticated ways to interpret the oral histories of societies which did
not produce literary forms, such as in the American south-west or in
the Mount Saint Elias region in what is now northern Canada and
Alaska.[122] However, the lack of written materials from many parts of
the world, such as Australia or southern Africa, means that inevitably
a book like this cannot be fully balanced in terms of its geographic
focus. The fact that most climate research by scientists is focused on and
based in countries that are well explored and well resourced compounds
the problem of imbalance. This is particularly ironic given that the
greatest impact of climate change will be felt in the poorest regions and
countries, precisely those whose voices have been silent or ignored by
history for decades, centuries and millennia.[123]

Such problems cannot be solved by one book. But what a book can
do is provide a broader perspective, and introduce themes, regions and
questions that can help push the boundaries of history and of historical
research in the future. Perhaps, too, it can offer some grounds for
optimism, as well as constructive suggestions for how to best navigate a
time not only of profound climatic but also of technological, political
and economic change.

Writing this book has taught me a great many lessons about how
we conceptualise the world around us. But it has also made me realise
that the reason we are at such a dangerous intersection is the result of
trends that have deep roots in the past. As far back as written records
go, people worried about human interaction with nature and warned of
the dangers of overexploitation of resources and of long-term damage
to the environment. It may well be that we are now on the verge of
finally becoming victims of our own success as a species, and that
the stresses and strains that our behaviour has put on ecosystems has
pushed us close to or even beyond a tipping point that has catastrophic
consequences. We cannot say, however, that we were not warned.

I

The World from the Dawn of Time
(c.4.5bn–c.7m BC)

In the beginning when God created the heavens and the earth, the
earth was a formless void ...

Book of Genesis, 1:1

We should all be grateful for dramatic changes to global climate. Were
it not for billions of years of intense celestial and solar activity, repeated
asteroid strikes, epic volcanic eruptions, extraordinary atmospheric change,
spectacular tectonic shifts and constant biotic adaptation, we would not be
alive today. Astrophysicists talk of habitable regions around stars that are
not too hot and not too cold as being in the 'goldilocks zone'. The earth
is one of many such examples. But conditions have changed constantly
and sometimes catastrophically since the creation of our planet around
4.6 billion years ago.[1] For almost all the time that the earth has existed, our
species would not and could not have survived. In today's world, we think
of humans as architects of dangerous environmental and climate change;
but we are prime beneficiaries of such transformations in the past.

Our role on this planet has been an exceptionally modest one.
The first hominins appeared only a few million years ago, and the
first anatomically modern humans (including Neanderthals) around
500,000 years ago.[2] What we know of the period since then is patchy,
difficult to interpret and often highly speculative. As we get closer to the
modern day, archaeology helps us understand more reliably how people
lived; but to know what they did, thought and believed we have to wait

till the development of full-writing systems around 5,000 years ago. To put that into context, accounts, documents and texts that allow us to reconstruct the past with nuance and detail cover around 0.0001 per cent of the world's past. We are not just fortunate to exist as a species, but in the grand scheme of history we are new and very late arrivals.

Like rude guests who arrive at the last minute, cause havoc and set about destroying the house to which they have been invited, human impact on the natural environment has been substantial and is accelerating to the point that many scientists question the long-term viability of human life. That in itself is not unusual, however. For one thing, our species is not alone in transforming the world around us, for other species of biota – that is to say, flora, fauna and microorganisms – are not passive participants in or simple bystanders to a relationship that exists solely or even primarily between humans and nature. Each is actively involved in processes of change, adaptation and evolution – sometimes with devastating consequences.

This is one reason why some scholars have criticised the idea and the name of the 'Anthropocene', which prioritises humans into 'a distinguished species' that has claimed the right to identify what is and is not wild, to classify 'resources' as ones that can be used – sustainably or otherwise. Such, argue some, is the 'arrogance that greatly overestimates human contributions while downplaying those of other life forms almost to the point of nonexistence'.[3]

For around half the earth's existence, there was little or no oxygen in the atmosphere. Our planet was formed through a long period of accretion, or gradual accumulation of layers, followed by a major collision with a Mars-sized impactor – which released enough energy to melt the earth's mantle and create the earliest atmosphere from the resultant exchange between a magma ocean and vapour that was anoxic, that is to say, lacking in oxygen.[4]

New models suggest that 1 to 5 billion lightning flashes that occurred per year on early earth may have been the source of large volumes of prebiotic reactive phosphorus that played an important role in the emergence of terrestrial life. The earth's biogeochemical cycles eventually resulted in a radical transformation. Although there is considerable debate about how, when and why oxygenic photosynthesis occurred, evidence from organic biomarkers, fossils and genome-scale data suggests that cyanobacteria evolved to absorb and take energy from sunlight, using it to make sugars out of water and carbon dioxide, releasing oxygen as a by-product.[5]

Around 3 billion years ago – if not earlier – enough oxygen was being produced to create 'oases' in protected nutrient-rich shallow marine habitats.[6] Whether because of chemical reaction, evolutionary development, sudden superabundance of cyanobacteria, volcanic eruptions or a slowdown in the earth's rotation (or a combination of all five), atmospheric oxygen levels accumulated rapidly around 2.5–2.3 billion years ago, resulting in an episode known as the Great Oxidation Event. This was a key moment that paved the way for the emergence of complex life as we know it.[7]

It also led to dramatic changes in climate, as rapidly increasing oxygen reacted with methane, producing water vapour and carbon dioxide. Alongside the effects of a supercontinent being formed from the collisions of landmasses, the earth's greenhouse was weakened, leading to the planet being covered completely in ice and snow.[8] Changes in the earth's orbit around the sun, known as the Milankovitch cycle, may also have played a role in this process.[9] So too, giant meteorite impacts may also have played an important role in the formation of the continents.[10] The glacial episodes may have been weaker or stronger over the course of several hundred million years, but in general the effect of 'Snowball Earth' was so dramatic that some scientists refer to this period as a whole as a 'climate disaster'.[11]

This process was precarious and complex, and is the subject of considerable advances in current research.[12] As with later glaciations, however, it resulted in profound changes for the planet's plant and animal life.[13] One outcome appears to have been the evolution of small organisms into larger sizes, capable of moving at faster speeds to compensate for the high viscosity of cold seawater.[14] It has recently been suggested that the formation of 8,000-kilometre long belts of 'supermountains' may have played a role in the rise of atmospheric oxygen and in stimulating biological evolution as a result of phosphorus, iron and nutrients being deposited into oceans as mountains eroded over the course of hundreds of millions of years.[15]

The fossil record of complex, macroscopic organisms begins with the Ediacara Biota period which started 570 million years ago and which saw at least forty recognised species developing into multicellular animals that were symmetrical – presumably helpful for functions such as mobility.[16] It marked a period of extraordinary diversification in the variety of animals living in the oceans and in their evolution, development and adaptation, with some creatures like trilobites developing respiratory organs on their upper limbs.[17]

Near the end of the Ordovician period, around 444 million years ago, a sudden cooling, perhaps triggered by tectonic shifts that produced the Appalachian mountains, led to sharp falls in temperature and initiated shifts in deep ocean currents, as well as declines in sea level that shrank habitats for marine planktonic and nektonic species. That cooling produced one pulse of extinction; another came when temperatures moderated, sea levels rose and ocean current patterns stagnated, with a resultant sharp fall in oxygen levels.[18] Traces of mercury and indications of significant acidification suggest that volcanic activity was a key factor in the second stage of a process that ultimately brought about the extinction of 85 per cent of all species.[19]

This was just one of several spectacular episodes that wiped out life for all but a small proportion of living organisms. The moon may have played a part in the changes in the millions of years that followed. Probably formed from debris as a result of a Mars-sized object impacting with earth, the moon has a gravitational pull that plays a major role in ocean tides; as such, some scholars have argued that it helps transport heat away from the equator and towards the poles, fundamentally shaping earth's climate.[20]

As the moon used to be much closer to the earth following its formation, these forces were considerably stronger and therefore had a greater impact on the earth's climate and also perhaps on its wildlife: recent modelling suggests that big tidal ranges may have been responsible for forcing bony fish into shallow pools on land, thereby prompting the evolution of weight-bearing limbs and air-breathing organs.[21] The moon played a role not only in the transformation of the earth, in other words, but also in the development of life on this planet.

It still exerts an important influence. The reproductive cycles of many marine creatures are closely synchronised with lunar phases, with migration and spawning in fish, crab and plankton species triggered by the moon's glow.[22] Coral genes change their activity level according to the waxing and waning phases of the moon.[23] Lunar phases seem to influence the timing of the mating season of wildebeest in the Serengeti, and have been linked to spontaneous delivery in cows.[24] Many primates become more active at night during a full moon – perhaps because higher light levels offer greater chances of predator evasion.[25] It has been noted too that albatrosses are more active on moonlit nights.[26] Though little studied, lunar phases and moonlight seem to be closely connected to the annual migrations of billions of seasonal animal species, especially birds, whose fuelling opportunities are heavily light-dependent.[27]

Indeed, there seem to be important links between human behaviour, activity and even fertility and lunar rhythms. Studies of indigenous communities in Argentina that do not have access to electricity (and therefore offer helpful control conditions) show that sleep starts later and is shorter on nights before the full moon, when moonlight is available in the hours following dusk. This suggests too that pre-industrial communities that did not have access to artificial light may likewise have had sleep patterns strongly influenced by lunar activity.[28] Long-term data from women's menstrual cycles shows a correlation with lunar light and lunar gravity, with some scholars arguing that human reproductive behaviour was originally synchronous with the moon, but has been modified more recently by modern lifestyles.[29]

While the role of the moon in influencing and disturbing human behaviour is often reflected in popular culture and even in language – with the word 'lunatic' suggestive of a relationship between mental illness and the moon – any causal links are usually downplayed by scientists.[30] However, some researchers have emphasised that manic episodes in patients with bipolar disorders have a remarkable synchrony with three distinct lunar phases.[31] The moon, in other words, plays an important role in ocean currents, global temperatures and climate, as well as in reproductive cycles, and life on earth in general.

More work needs to be done to assess the role that lunar tides play in the ionosphere–thermosphere weather system, as well as in past evolutionary processes or during extinction events.[32] The latter were not uncommon. The most deadly was the so-called Great Dying, which took place 252 million years ago. The primary cause was an epic volcanic episode in what is now Siberia that produced enormous volumes of magma.[33] It is possible that a key moment came when lava stopped erupting above ground and began to form as sheets of magma that trapped gases underground until pressure was released in a series of gigantic, violent eruptions.[34] Whatever the precise circumstances, the ultimate result was the injection of enormous amounts of greenhouse gases into the atmosphere that destabilised the biosphere. Soil and seawater temperatures may have risen initially by 8–10 °C and then later by an additional 6–8 °C, with temperatures at the equator perhaps as high as 40 °C. The result was the extinction of 96 per cent of marine life, three-quarters of land animals and all the earth's forests.[35]

Other massive volcanic events brought about significant transformation, such as at the end of the Triassic era, around 200 million years ago, when a period of changing marine conditions led to sharp drops in sea levels and water-column freshening, which in turn resulted in less saline, shallow-water complex communities of microorganisms.[36] This was accompanied by huge wildfires and abrupt injections of volcanogenic gases into the atmosphere that quadrupled carbon dioxide levels, acidified the oceans and prompted another wave of mass extinction of plant and animal life.[37]

Such events produced major ecosystem remodelling, as flora and fauna responded to change and diversified rapidly.[38] New kinds of plant assemblages and food demanded adaptation, one of which, in the case of the Triassic, was the evolution of more powerful jaws which generated the strength for shearing bites that enabled efficient feeding. This was especially important in the context of harder, tougher plant materials that became more common – and a key factor in dictating which herbivores thrived and which died out.[39]

The single most famous moment of large-scale transformation in the past, however, was caused by an asteroid or comet strike that impacted the earth 66 million years ago on the Yucatan peninsula, near what is now the town of Chicxulub in Mexico, and led to the demise of the dinosaurs.[40] This was just one of many major extraterrestrial impacts since the formation of the earth, one of the earliest identified examples of which dates to around 3 billion years ago and has an impact crater near Maniitsoq in west Greenland.[41]

While the local devastation of the strike near Chicxulub must have been dramatic – including high levels of thermal radiation from the impact plume, hurricane-force winds and likely giant tsunamis and landslides that scoured the ocean floor – the consequences of this particular strike were global: around 325 gigatons of sulphur and 425 gigatons of CO_2 were forced into the atmosphere at speeds of more than 1 kilometre per second. The result would have included firestorms caused by ejecta heating on re-entry to the earth's atmosphere, short-term cooling from the dust blocking the sun's light, long-term warming from the release of enormous volumes of CO_2 and high levels of ocean acidification.[42]

What made this strike so deadly was the size of the impactor, which recent research suggests may have been part of a comet from the Oort cloud at the edge of the solar system that was around twelve kilometres in diameter – but also how and where it hit the earth.[43] Astrophysicists were able to understand how catastrophic the effect of the size of an object could be in 1994 when the Shoemaker–Levy 9 comet hit Jupiter. On that occasion, parts of the comet disintegrated into smaller pieces before impact, with the largest final fragment only a kilometre or so in size. It was enough to produce impact scars of some 100,000 kilometres – almost eight times the diameter of the earth – and to leave watching scientists visibly shocked by the scale of the strike and its after-effects.[44]

This has obvious implications for the Chicxulub strike, for other similar impacts in the past and for those that will take place in the future – especially given that new research has suggested that estimates of the chances of similar long-period comets hitting earth should be raised by a factor of ten.[45] The specific angle of entry likewise mattered, with new simulations showing that a steeply inclined trajectory amounted to a worst-case scenario that produced maximum lethality for life on earth thanks to the catastrophic volumes of debris pushed up into the atmosphere.[46] The timing was significant too: because the Chicxulub impact occurred during the boreal spring/summer and soon after the spawning season for fish and most continental taxa, follow-on implications for flora and fauna were particularly acute.[47]

That the impact occurred at about the same time as enormous volcanic eruptions may have made matters even worse; and indeed some scientists have argued that volcanic activity played a role in ecosystem collapse and contributed to mass extinctions.[48] Either way, the outcomes included 10–16 °C cooling in mean surface air temperature on land and sharp drops in seawater temperatures, especially at shallower depths – and mass extinction of plant and animal life.[49]

Such events were spectacular and devastating. They also each played a part in the extraordinary series of flukes, coincidences, long shots and serendipities that ultimately brought about the rise of humankind, as well as the many species of flora, fauna and biota that exist today. All life on earth today descends from animals, plants and organisms that survived not one but multiple mass extinction events, as well as a near inexhaustible series of smaller-scale episodes of major changes in climatic and atmospheric conditions that created the world that we think of as familiar.

Deep ocean temperature from the Palaeocene to the present

Data: NOAA   Source: Hunter Allen and Michon Scott

Even the most cataclysmic event that brought such large-scale change had consequences that we think of as fundamental hallmarks of contemporary global ecosystems, although their roots lie tens of millions of years in the past. For example, pollen-grain analysis from South America shows that the Chicxulub strike helped create tropical rainforests as we now know them. Before the impact, tropical forest trees were spaced far apart, allowing light to reach the forest floor. After it, they grew much more densely, perhaps as a result of the extinction of large herbivores, with greater shading and enabling the flourishing of legumes and pods that take nitrogen from the air thanks to interactions with bacteria. Ashfall from the impact added weatherable phosphorus minerals to terrestrial ecosystems, which in turn were crucial in stimulating soil fertility and forest productivity. This would also have enhanced the relative advantage for flowering plants over conifers and ferns, thereby creating a springboard for a surge in biodiversity and the conditions for the enormous rainforests that are such an important part of the carbon cycle today.[50]

There were other, more modest events of climatic change that produced substantial results without causing mass extinction. A good example is the Paleocene–Eocene Thermal Maximum, a period of significant warming around 56 million years ago that followed the release of massive amounts of carbon into the ocean-atmosphere system, leading to a rise in global temperatures of at least 4–5 °C lasting around 200,000 years.[51] Some have suggested that tropical temperatures rose to as high as 40 °C.[52] So large were the carbon dioxide volumes that studies have speculated that concentrations were sixteen times the levels of CO_2 during the pre-industrial period.[53]

Although the source of the carbon is a matter of debate, volcanic eruptions would again seem to be the most likely cause of the destabilisation that brought about large shifts in geographic ranges of marine and terrestrial organisms, spurred rapid evolutionary processes and affected food chains.[54] It also brought about a boom in diversity of flora – at least in tropical regions – as well as increasing levels of precipitation around the world, including in North America, South Asia, North Africa and Antarctica.[55] The latter became home to lush forests until thick continental ice sheets began to form, a process that has been linked to a substantial decline in the concentration of atmospheric carbon dioxide levels that affected much of the terrestrial southern hemisphere.[56]

Other shifts in regional and global climates occurred following forty-two enormous eruptions since the demise of the dinosaurs, each of which was more than 150 times more powerful than the Mount Pinatubo eruption of 1991. The most notable was at Fish Canyon Tuff in what is now Colorado around 28 million years ago, the single largest volcanic eruption in the last 500 million years.[57] Asteroid and meteorite strikes also transformed the natural environment, such as one caused by an object 2 kilometres in diameter whose impact 800,000 years ago threw debris across the eastern hemisphere, including much of Asia, Australia and Antarctica and whose crater has only recently been identified in what is now Laos, partly because it was hidden under a volcanic lava field created by later eruptions.[58] And change was also brought about by periods of long-term warming such as during the Piacenzian phase of the Pliocene (around 3 million years ago), which saw temperatures more than 3 °C warmer and sea levels twenty metres higher than today, when more carbon dioxide was present in the atmosphere than at any time until the twentieth century thanks to a large-scale reorganisation of global weather patterns.[59]

Geology and movement of tectonic plates also played a significant role in shaping and reshaping the earth, and in creating the geographic distribution of water, land and life as we know it. Over the course of millions of years, a single giant supercontinent broke up, perhaps because of mantle plumes at the core–mantle boundary causing motion, perhaps because of the negative buoyancy of oceanic plates exerting pressure from above, or perhaps a combination of the two.[60] In some cases, surges of hot material from a supervolcano caused plates to split and rotate – as happened with the Indian plate that broke away from Africa just over a hundred million years ago.[61]

Ultimately, of course, these movements led to the distribution of the world's continents in the positions they occupy today. However, the process of their formation and relocation was one that had important implications. For one thing, not all landmasses remained above sea level. Indeed, an elevated area surrounding what is now New Zealand and New Caledonia was part of a single continuous body of land of which almost 95 per cent became submerged and was so vast that some have called it the earth's 'eighth continent'.[62]

In this instance, the disappearance of a large landmass beneath the waves was the result of stretching and thinning. It was a different story when a piece of continental plate around the size of Greenland broke off from what later became North Africa, crashed into southern Europe and was eventually forced underneath it.[63] Impacts like this resulted in enormous forces being created and in land becoming crumpled, creating the great mountain ranges of the world. These include the Andes in South America and the Himalayas, formed when the Indian subcontinent smashed into Eurasia around 50 million years ago, pushing land that had been at sea level upwards – with the result that marine fossils can be found at or near the summits of some of the highest peaks on earth.[64]

The formation of these extensive mountain ranges has in turn played a part in changing and shaping local, regional and even global climate patterns. It is widely assumed, for example, that the location and size of the Rockies affects precipitation patterns and storm-track development on the east coast of North America, in the North Atlantic and perhaps as far away as Norway.[65] It has long been argued too that the uplift of the Himalaya and Tibetan plateau determines rainfall distribution over Africa, although recent sensitivity modelling suggests that the influence is a weak and modest one.[66] Rather, changes in land cover and dust emission both appear to play a much more important role in the strength of monsoon precipitation in Asia, at least over the course of the last few thousand years.[67]

The reconfiguration of global landmasses has had important implications for flora and fauna, and also specific consequences for the development of human societies. For example, evolutionary change over the course of millions of years led to very sharp differences in the numbers and distribution of large mammal species in Eurasia compared with the Americas. Of particular significance was the lack of animals suitable for domestication in the latter by the time of early human settlements around 25,000 years ago: this had a profound impact not only on how societies understood and interacted with the natural world, but also on farming techniques, the capacity to generate food surplus, the emergence of social hierarchies and even the immunological response to disease – one of the key by-products of close interaction with domesticated animals.[68]

The break-up of the supercontinent and the creation of the continents that began around 250 million years ago did more, however, than create maps that are familiar to us today. For example, one outcome was the closing off of a huge body of water called the Tethys Sea just over 20 million years ago causing it to shrink and eventually diminish into the Mediterranean. This led to a reorganisation of global climate patterns that included the aridification of large parts of Africa and the onset of the long-term glaciation of Antarctica.[69] Changing conditions brought about the 'Messinian salinity crisis' around 5.6 million years ago, which resulted in the desiccation by evaporation of the Mediterranean and created plant and animal pathways between Europe, Africa and the Middle East that lasted until Atlantic waters forced their way through the Straits of Gibraltar about 300,000 years later and rapidly refilled the Mediterranean basin in an event known as the Zanclean flood.[70]

From the perspective of the twenty-first century, however, what was even more significant was the fact that the continental rifts, collisions and changes to major ocean basins resulted in the creation of giant hydrocarbon deposits worldwide: almost all the world's 877 giant oil and gas fields (namely those containing 500 million barrels or more)

cluster in just twenty-seven key regions around the world.[71] The location
of these fields underpins a fossil-fuel economy worth trillions of dollars
a year – but it is also the main driver of climate change in the modern
era: the energy revolution that began with the burning of fossil fuels
accelerated rapidly with the development of motors, engines and power
stations that are fired by oil and gas. Contemporary anthropogenic
climate change, global warming and pollution are all fuelled, in other
words, by shifts that took place over the course of hundreds of millions
of years.

Indeed, these long-term developments are not only linked to current
environmental problems; they are also central to the story of global
economic, social and political powershifts in the modern age. For
example, the bulk of the coal that powered the industrial revolution was
formed from plant debris in the Carboniferous and early Permian period
around 300 million years ago caused by a massive fall in atmospheric
CO_2 levels.[72]

The location of these deposits was therefore of crucial significance
once coal-powered mechanisation offered new and extraordinary
opportunities to boost production and productivity. Indeed, some
scholars have argued that one of the reasons for the Great Divergence –
the point where Europe leapfrogged past states such as Qing dynasty
China and elsewhere in Asia – was because coal fields in the former
were closer to potential manufacturing centres, had access to more
labour pools to extract them and could therefore be exploited both
more quickly and more cheaply.[73] As we shall see, there were many
other factors involved in the rise of European powers, but luck of the
geological draw was profoundly important once the energy revolution
opened up new possibilities at a time of intensifying globalisation.

It helped open up new ecological frontiers too. For example, the
rise of cities and railways across the midwest in the United States and
beyond was facilitated by vast fossil-fuel deposits of coal, oil and gas
in states like Illinois, Iowa and Nebraska, and then in a further giant
belt stretching south from the Dakotas and Wyoming and through
Colorado down into New Mexico.[74] In the second half of the nineteenth
century, 'instant cities' began to pop up in the American heartlands,
with industrialisation and urbanisation rising hand in hand. This
helped create a manufacturing powerhouse, but it also resulted in an
important redistribution of population from the coast to the interior.[75]

Conversely, more recently pressure on jobs in the coal-mining industry, driven by government incentives for cleaner energy production and by sharply reducing costs of renewable energy, has influenced responses at the ballot box during presidential elections, with a strong surge in support for pro-coal Republican party candidates. The location of coal reserves and of the populations who have been involved in its extraction, historically and today, influences who does – and who does not – get into the White House every four years.[76]

That geological chance plays an important role in the modern world can be underlined by many other parallels. For example, during the Cretaceous period, 145–65 million years ago, the world was much warmer and sea levels were much higher than they are today. Billions of marine microorganisms that died formed sedimentary levels that ultimately created oil reservoirs. But their demise produced other outcomes. In the southern part of the United States, massive chalk formations grew up from plankton and sea life that died out as the world cooled and sea levels dropped. This led to swathes of extremely fertile land, especially after rainfall dissolved nutrient-poor carbonate minerals.

The arc of lands that marks a swoosh across the south-eastern states, known as the Black Belt thanks to its rich, dark soils, proved to be ideal for intensive crop production, especially of cotton. Following the arrival of Europeans in the Americas and the establishment of the transatlantic slave trade, these lands became heavily populated by Africans who were shipped in enormous volumes and grotesque conditions to do work that was labour-intensive. Despite the abolition of slavery in 1865, large numbers of black Americans were excluded from voting until the Voting Rights Act that outlawed discriminatory voting practices was passed a century later. Americans with African ancestry make up a majority of the population in many counties in the Black Belt region today, especially those with high unemployment levels and poor education and healthcare provision. Votes cast not just in this part of the United States but in specific counties have an important role in determining the outcomes of presidential elections.[77] Climatic change is not just a topic for the present and the future; it is also a fundamental part of the past.

It is a similar story with the distribution of resources in other parts of the world. The story of oil and gas is one that has played a central role in global geopolitics for the last century. Massive reserves in Saudi Arabia,

Iran, the Gulf and elsewhere in the Middle East and North Africa are closely linked to the story of military intervention, to the imposition of autocratic and theocratic regimes and to a host of wider issues. American engagement with this region may not define US presidencies over the last fifty years; but it is no coincidence that hostage crises, weapons sales, invasions, terrorism and nuclear deals have been essential elements to US foreign policy since the 1970s, if not before. Had there been no oil and gas in the Middle East, things would have been very different.[78]

The same is true for Britain, Germany and Japan in the nineteenth and first half of the twentieth centuries. One of the curious serendipities of the creation of the British empire was that while it famously covered almost a quarter of the world's surface area by the time of the First World War, the vast imperial landmass contained few significant oil deposits. So it was necessary to find reliable sources and to establish control over them in order to nourish the sinews of empire. The resultant decisions about military and political interventions reshaped the Middle East after the First World War and produced consequences that resonate to this day.[79] Likewise, the lack of oil resources available to both Germany and Japan influenced strategic decisions during the Second World War, not least the major military pushes towards the Caucasus and into South-East Asia respectively that ultimately overstretched supply lines and capabilities.[80]

Similarly the distribution of other resources, natural and otherwise, has played a fundamental role in human history and will continue to do so. The world's accessible reserves of precious metals, including gold, are the result of the earth being bombarded by a flux of meteorites after its formation.[81] This has shaped the fortunes of those living in locations – for good and for bad – where gold has been abundant and where costs of extraction have been low, leading to forced and free population movements and in some cases to military confrontation.

Heavy metals, including rare earths – minerals that are in fact not scarce but are rarely found in concentrations that make their extraction practicable – likely originate as by-products of explosions of supernovas typically thirty times the weight of the sun.[82] Many of these can be subsequently associated with alkaline igneous activity and magmatic systems on earth.[83] Here too geology and chance have dictated how easily these can be mined, and helped determine political developments, military rivalries and the evolution of societies and states: some predict that the twenty-first century will be shaped by a scramble for a new set of elements like beryllium, dysprosium and yttrium which had little value or use a few decades ago but are now essential components in many hi-tech devices. New technologies will enhance competition in the future – one reason for the renewed interest in lunar and planetary missions, and in particular for mineral prospecting and extraction on extraterrestrial objects.[84]

Nor is the distribution of the earth's resources a matter only of energy or precious metals – for the environmental lottery applies to a host of other materials and substances, including flora and fauna. The importance of spices, most notably from South and South-East Asia, helped stimulate trading networks that brought these regions into intensive contact with the Middle East, Africa and the Mediterranean, as well as with China, Japan and beyond. The habitat of the silk worm, likewise, was instrumental in the manufacture of textiles that were light, tough and expensive and prompted high demand from thousands of kilometres away. As we shall see, the spread of animals and plants as a result of short-, medium- and long-distance trade – both deliberately and inadvertently – is a central part of the world's ecological history and one in which humans have played a disproportionately significant role.

One challenge, then, is determining how our species understands the natural world and perhaps, most importantly, conceptualises our own place within it. Conservationists sometimes imply that there is a way in which time can be made to stand still, that the rainforests should remain untouched, that grasslands should be left intact, that 'nature' should be spared human intervention. Yet plants and animals have their own methods of causing change and even degradation and destruction. Nature is not a harmonious, benign and complementary concept that preserves balance, for ecosystems have always been transformed and reshaped by many non-human forces.

What humans do, on the other hand, is to bring about change through conscious modification of landscapes, deliberate interventions into ecosystems and intentional, ill-conceived decisions that lead to overexploitation. These can bring about inadvertent outcomes too, including chain reactions after the introduction of species into new environments, or the spread of pathogens and disease that have dramatic effects not only on human life but on flora and fauna too.

In that sense, the evolution of our species has been the single most important development in the history of this planet. Extinction events in the past have been caused by volcanoes and by comets, but humans have managed to develop technologies to bring about mass extinction independently. Some would argue that the unsustainable way that we live in the twenty-first century and the impact of global warming is putting the existence of humanity at risk, and of countless animals and plants. Our interactions with each other are partly responsible for this, in the form of travel and transportation, as well as the globalisation of goods, produce and ideas.

However, we have also developed the capabilities to ensure our own destruction by other means. Accidents like the failure of the reactor at the Chernobyl nuclear power plant, the *Exxon Valdez* oil spill and the leak at the Union Carbide plant in Bhopal, India, provide warnings that our new technologies can lead to large-scale environmental disasters. The development of nuclear weapons has shown likewise, both at Hiroshima and Nagasaki at the end of the Second World War and at test sites in the former Soviet Union, in North America and in the Pacific.[85]

The power of nuclear arsenals means that we could achieve the same outcome as an extra-terrestrial strike ourselves – including by accident.[86] False alarms happen worryingly often, as in 2018 when a ballistic-missile alert was sent to television, radio and mobile phones in Hawai'i.[87] The laws of probability suggest that it is a matter of when, and not if, a catastrophic accident resulting from human error, escalation of political rivalries or geopolitical miscalculation takes place.

It is perhaps ironic that the greatest threat following a major nuclear confrontation, triggered intentionally or otherwise, would not be from the massive missile-delivered payloads but from the rapid global cooling that would follow in a so-called nuclear winter. Soviet and US modelling of precisely this scenario played an important role in encouraging arms-control agreements in the 1980s and efforts to control proliferation of nuclear weapons and technology.[88]

These are areas that have become important once again in the contemporary world – meaning that risk levels of human-induced disaster are greater than at any point in history since the emergence of our species. So how is it, then, that humans have become so central to the present and future of a planet where we have only been present for what amounts to the batting of an eyelid in the billions of years of history since the earth was formed?

2

On the Origins of Our Species
(c.7m–c.12,000 BC)

I will here attempt to give a brief, but I fear imperfect, sketch of the progress of opinion on the Origin of Species.

Charles Darwin, *On the Origin of Species by Means of Natural Selection* (1870)

The origin of *Homo sapiens* is something of a mystery, as well as a matter of considerable debate. The human lineage diverged from that of apes around 7 million years ago and likely much earlier, if recent estimates on mutation rates in chimpanzees are correct.[1] The fossil record strongly suggests that all human ancestors lived in Africa, although identifying the location is neither easy nor a matter of agreement. Early hominins included several species of *Australopithecus*, the most famous of whom was christened Lucy after her discovery in Ethiopia in 1974, and an even more complete specimen nicknamed Little Foot that was excavated at the Sterkfontein Caves near Johannesburg and dates back to about 3.67 million years.[2] Recent analysis of Little Foot's shoulder blades, joints and collarbones shows a spine configuration well suited for tree climbing and hanging from branches, in turn providing insights into likely habitat.[3]

The genus of *Homo*, from which our own species descends, emerged perhaps around 3 million years ago, with the first possible example found in 2013 of a jawbone with teeth still in place dating back about 2.8 million years.[4] While the appearance of the first members of the

Homo genus is sometimes seen as a momentous transformation, the transition and emergence were neither sudden nor profound, with many features shared with australopithecines.[5] Indeed, some scholars have argued that some early *Homo* species should be reclassified as *Australopithecus*.[6]

Recent research suggests that there was considerable diversity within the *Homo* genus – and as a result there is little consensus about how to describe or explain variations and differences and even about what constitutes a species.[7] How *Homo rudolfensis*, *Homo habilis* and *Homo erectus* developed and how they were different is a source of considerable conjecture, as is the question of how they relate to later *Homo heidelbergensis*, from which both *Homo neanderthalensis* and *Homo sapiens* may have evolved. Nevertheless, all seem to have shared characteristics that distinguished them from other primates, including bipedal locomotion, erect posture, larger brains and perhaps the use of specialised tools.[8]

It has long been argued that the evolution of hominins was accelerated by shifting climate patterns that brought about a change from forests to drier conditions, resulting in evolutionary pressures that favoured an upright posture, larger brains and changes in tooth size and either necessitated or rewarded the ability to cover longer distances – or both.[9] These changes were not uniform; nevertheless, key factors in the resilience, success and expansion of the genus as a whole were dietary flexibility during times of environmental unpredictability and an ability to reduce mortality risks.[10]

Certainly, stone and bone tools from a unique cave site in southern Africa point to adaptation in the face of substantial changes in ecosystems for multiple species of hominins.[11] These echo findings from East Africa where volcanism has left stratigraphic sequences of fossil and archaeological remains which are possible to date with some accuracy, and which demonstrate cases of problem solving through social co-operation.[12] Examples include the use of core–flake–hammerstone technology from the Oldupai Gorge site in modern Tanzania, and evidence of the use of tools for hunting and butchering, defleshing and demarrowing animals such as the *Parmularius*, a now extinct member of the bovid family that includes the wildebeest, as well as aquatic species such as fish, crocodiles and turtles that were also incorporated in the hominin diet.[13]

This was important because such food sources provide specific nutrients needed for human brain growth, in particular poly-unsaturated fatty acids, and as such were vital in the evolution of greater and new capabilities.[14] Indeed, some scholars have argued that changes in neurochemistry of the basal ganglia in the brain were the critical factor in the emergence of hominins from apes, as well as in subsequent developments of the cerebral cortex that help explain social behaviour, including communication, empathy and altruism.[15]

Some have also argued that the increase in brain size is correlated with rising population levels and reflects the need to adapt to growing group size (the 'social brain hypothesis'). Others however stress that the pressures of new or changing ecological environments were crucial, requiring intelligence for adaptation in order to ensure survival. A pivotal genetic mutation in *Homo sapiens* that dramatically increased the number of brain cells may also have brought cognitive advantages that had far-reaching consequences. While this is a lively area of scholarly discussion, there is wide agreement that ecological and social competition were 'motors' of evolution, driving biological, neurological and behavioural changes.[16]

Key to this too was the human ability to use and control fire, whose benefits include cooking, warmth and protection from predators and the encroachment of wildlife. Natural fire occurs primarily as a result of the hundreds of millions of lightning flashes that occur annually over the earth.[17] Changes to the human body, to teeth and brain size, and in the development of language and of more complex cognitive skills have been linked to the introduction of the cooking of foods – although each was part of steady progression spread over long periods of time.[18] Evidence of burned wood, bone, grasses and other material from cave sites in southern Africa, the Levant and eastern China suggest that at least some *Homo* populations had learned to control fire between 1 million and 500,000 years ago, with traces becoming common at many sites in Europe, Africa and Asia by around 400,000 years ago.[19]

One of the most challenging problems facing early *Homo* was fluctuating climate patterns with stratigraphic, lake, faunal and other data suggesting regular dynamic shifts between aridity and moisture across hundreds of thousands of years. These were driven by tectonic and volcanic impacts, by the eccentricity of the earth's orbital cycle, by solar activity and by changes in moisture availability linked to the strength of monsoon intensity, the size of large lakes and the El Niño–Southern Oscillation (ENSO). Taken together, these required versatility, and played important roles in shaping dietary, cognitive and social adaptability in early *Homo*.[20]

Africa was the melting pot for different species – so much so that the genetic diversity in African groups and individuals for the most recent human species is greater than in any other part of the world.[21] By around 2 million years ago, some *Homo* species and populations had migrated to other continents, likely in two dispersal events.[22] The earliest attested example – of *Homo erectus* – can be found in the Caucasus.[23] There is evidence of *Homo erectus* in China and Java around 1.8 million years ago and then of the colonisation of western Europe over a million years ago.[24]

The context, nature and timing of these processes are unclear, as are the differences within as well as between different *Homo* populations. While a range of explanations seek to explain these divergences, our own species may have started to emerge as distinct from *Homo neanderthalensis* as much as 800,000 years ago – considerably earlier than the first fossil records from sites in North-West Africa and Ethiopia that are usually dated to 315,000–195,000 years old, though this is a matter of fierce debate.[25]

What is not contested is that recent research has underlined that the story of hominin population is more complicated and more varied than has long been thought. New hominin populations have recently been identified, some of which must have mixed with modern humans.[26] Scholars have argued that specific haplotypes (sets of genetic variants that can be inherited) in Native American populations reflect the fact that the ancestors of some of these were among the first settlers of the Americas, while it has also been suggested that some passed on genes that helped modern humans survive at high altitude.[27]

Neanderthals inhabited western Eurasia over many millennia, and later evidently reproduced with our own species – though not everywhere: Neanderthal ancestry is found in human genomes today, and in ancient populations as far apart as northern Europe, Siberia and China, with little present in sub-Saharan Africa.[28] Admixture between Neanderthals and *Homo sapiens* introduced groups of genes into modern humans which, for example, affect the immune systems – some have been linked with resistance to COVID-19 infection, and others to increased susceptibility.[29]

Neanderthal population distribution appears to have been heavily impacted by climatic stress, more so than our own species was (at least by around 45,000 years ago), because larger bodies, bigger brains and consequently high energy and food requirements perhaps resulted in lower levels of resilience, especially during periods of rapid cooling.[30] This might explain long-term patterns where large parts of Europe were colonised, then abandoned and then recolonised by Neanderthals, depending on weather conditions being more or less favourable.[31] This might also explain separate radiations of Neanderthals attested by mitochondrial DNA (mtDNA) from cave sediments in western Europe and southern Siberia.[32] Indeed, it has been argued that cold episodes provided crucial triggers for major divergences within Neanderthal populations, as well as prompting variations in mtDNA of other hominin groups living in Europe and Asia.[33]

Coping mechanisms for extremely challenging climate spells, such as around 450,000 years ago when ice-cap and permafrost advances in Europe made much of the continent uninhabitable, may have included drastic adaptations for *Homo* groups including Neanderthals: it has even been suggested that fossilised bones discovered in Atapuerca, northern Spain, provide indications of bodies entering a hypometabolic state. Some of our ancestors, in other words, appear to have hibernated during cold winters. This hypothesis is highly speculative; but if the findings are accurate, it may be that adults were able to do so successfully on occasion, while adolescents appear to have found the process more difficult.[34]

Neanderthals and *Homo sapiens* shared many characteristics and abilities including similar auditory and speech capacities.[35] Nevertheless, *Homo sapiens* is a relatively homogeneous species, with mtDNA and Y-chromosome genealogy pointing to the fact that all modern humans descend from a small or even a very small population size, followed by a highly effective and successful expansion.[36] Changes in climate over the last 200,000 years seem to have been crucial in demographic dispersals as well as in repeated redistributions of genes – as geographic sinks formed that were peopled by groups like the Hadza in Tanzania, the Baka/Biaka and Mbuti/Efe Pygmies in western and eastern Central

Africa respectively, that are comparatively unmixed descendants of an early, modern human population.[37] More favourable climatic conditions in specific regions and pockets brought about genetic, cultural and behavioural change, while those living in more challenging locations in Africa found that survival was difficult or even impossible.[38]

The ability to adapt, learn and pass on information was important – and was acquired successfully: as one recent study based on a site in southern France shows, early humans 170,000 years ago understood how to position a hearth in caves in the optimal location for smoke dispersal (and therefore to lower exposure to smoke levels) but ideal for their social, heating and cooking needs.[39]

One key point in the history of our species came around 130,000 years ago with a sudden and dramatic reorganisation of global temperatures, sea level and weather which were modulated by changes in ocean circulation and deep-water ocean CO_2 storage that were part of long-term glacial cycles linked to variations in the earth's orbit and relative strength of the sun.[40] These were likely impacted by a so-called Heinrich event, in which large numbers of icebergs are discharged from the Laurentide Ice Sheet through the Hudson Strait into the North Atlantic Ocean, leaving layers of debris in the ocean sediment.[41] This would have significantly reduced North Atlantic Deep Water formation and interfered with ocean circulation, leading to heat accumulation in the southern hemisphere. In any event, the result was a rapid rise in polar temperatures, which led to the 'greening' of the Arctic as plants expanded their habitats northwards.[42] In due course, the melting of ice sheets led to global sea levels going up by several metres, and perhaps to as much nine metres above present-day levels.[43]

Evidence from chironomids (non-biting midges) trapped in sediment layers in Denmark has helped with temperature reconstructions which suggest that this period was one of sustained warming for a period lasting several thousand years.[44] It also corresponds to changes to vegetation and drainage networks across the Sahara that opened up a network of corridors linking East Africa to the Mediterranean and led to the first *Homo sapiens* settlements in the Levant.[45] The identification of six phases of lake formation, each with distinct sedimentary markers which can be detected using luminescence dating, and which mirror stone-tool usage by early humans, suggests that this process took place not in one go but in a series of pulses.[46]

There is still considerable uncertainty about the timings, routes and nature of dispersal of modern humans from Africa – although pioneering work in whole-genome sequencing technology is providing new insights into genetic and regional variation, into mixing with Neanderthals and Denisovans and into the existence of other human populations that disappeared but are known through skeletal records.[47]

Human expansion and settlement were shaped above all by finding areas that were ecologically benign. These included a wide variety of habitats, ranging from warm wooded environments to savannah grasslands, as well as coastal areas rich in marine life – although it seems that very open environments were deliberately avoided.[48] One particularly attractive location was the narrow strip of woodland between the Mediterranean coast and the flanks of Jordan's Rift Valley, a location with stable water supplies and wildlife that could be exploited with relative ease. There is evidence that this region was inhabited concurrently with a Neanderthal population that had moved southwards from Eurasia, as evidenced by burials and skeletal and dental remains, as well as by signs of interbreeding with the modern humans who had also established themselves there.[49]

Existence could be precarious, as this case proves: around 73,000 years ago, a period of extreme aridity and glaciation brought about such difficult times that the populations in the Levant appear to have failed and died out. The cause may have been a massive eruption at Mount Toba in what is now Indonesia, the largest on earth in the last 2 million years. The injection into the stratosphere of volcanic fine-particle sulphate aerosols was so great that a volcanic winter resulted that lasted for several years, leading to a sharp drop in temperatures in many locations around the world. Deposits were left across millions of square kilometres, with more than 1 per cent of the earth's surface being covered with at least ten centimetres of ash.[50] Heavy layers of ash fell on the Indian subcontinent, with levels in several alluvial basins in 1 to 3 metres thick; there is evidence too of large-scale deforestation and the transformation of woodland into grasslands.[51] It has been argued that pressures on food supplies and survival were so great as to drive a major reduction in the human gene pool globally.[52]

Although devastating, the impact of the Mount Toba eruption was not uniform. Global climate model simulations suggest that the effects in Europe, North America and Central Asia were pronounced, but were much more modest in the southern hemisphere.[53] This would explain why sediment from Lake Malawi does not indicate any significant temperature change through this period.[54] Archaeological sites in South Africa meanwhile show that humans thrived through the Toba eruption and the glacial conditions that followed. There are signs too from a valley in central India – which was covered with 1 million cubic metres of tephra (pyroclastic material ejected by volcanoes during eruptions) – that the human population was able either to overcome the challenging conditions or, if not, to reoccupy sites not long afterwards that had been previously lived in.[55] Recent research has proposed that the eruption was so powerful that sulphate particles were formed in the atmosphere of such size that the cooling effect was both shorter and less dramatic than previous models have suggested, thanks to heavy particles falling to earth relatively quickly.[56] Ironically, the human survivors of this episode were fortunate, in other words, that the eruption was as powerful as it was.

Even before the Toba eruption, a more successful series of dispersals of *Homo sapiens* from Africa had already begun, again either provoked or influenced by environmental pressures. One speculative hypothesis suggests migration across the Red Sea from what is now Eritrea to the coast of Yemen was linked to all-time low marine-life and sea levels around 85,000 years ago, leading to shortages in seafood and protein sources which prompted a search for more favourable habitats. In due course, humans spread out into South-East Asia, China and beyond, reaching Australia by around 65,000 years ago.[57]

By this time, humans had already long been working on how to manage ecosystems, for example developing forest-management techniques that included deliberate use of fire to clear areas, influencing vegetation composition and in some cases causing soil erosion.[58] These became increasingly important not only because of the effects of the eruption but also because of a new period of cooling lasting several millennia that was under way in the northern hemisphere in response to a warming of the South Atlantic Ocean.

This also brought about more humid conditions and higher levels of rainfall in southern Africa and consequently a platform for increases in population size, density and increased socialisation. Linking cultural patterns to rapid changes in climate is a problematic exercise. It is striking, nonetheless, that what must have been challenging conditions coincided with the emergence of symbolic expression through engraved ochres, tools and jewellery that bear testimony to important innovations not only in behaviour but also in the development of new technologies.[59] Beads found at Blombos Cave in Western Cape in South Africa were made of estuarine molluscs around 70,000 years ago, with use-wear patterns consistent with friction from rubbing against thread, skin or other beads.[60] These were broadly contemporaneous with the earliest drawings, chunks of ochre engraved with complex geometric patterns that were found in the same location.[61]

It was a similar story with other shifting behaviours and innovations, for example with practical tools, including the emergence of projectiles and weapons such as the bow and arrow that create greater acceleration than hand-cast spears, enhancing wounding capacity while also improving personal security by enabling the ability to wound or kill at greater distances.[62] Such devices had the further benefit of increasing the size of potential targets, which in turn had further effects on social organisation for human populations.[63]

All such transformations were gradual, not uniform, and are not always easy to interpret or understand. For example, it has been suggested that the expansion and then collapse of social networks across eastern and southern Africa around 50,000–33,000 years ago, which can be attested by ostrich eggshell beads, was influenced by global and regional climatic change, and moreover that a later improvement in conditions led to a renewal of contact between communities, including the spread of standardised cultural behaviour.[64] Such hypotheses are compelling, but they depend on relying on wisps of evidence that can be interpreted in many different ways.

Clearly, however, the social world of hunter-gatherers became more complex, with rituals starting to develop that helped manage and regulate relationships, alongside an emerging context for symbolic expression. By around 50,000 BC, Neanderthals had been experimenting with artistic forms, including creating ornamental works – perhaps as jewellery – as well as undertaking cave paintings such as at Cueva de Ardales in Spain.[65]

By around 40,000 years ago, geometric and iconographic representations were becoming common and are well documented in the Near East and North Africa.[66] Different kinds of artistic forms also began to be used. The earliest cave art attributed to *Homo sapiens* comes in the form of two cave paintings depicting two warty pigs from the limestone karsts of South Sulawesi, in modern Indonesia, which were created around 45,000 years ago according to uranium-series dating.[67]

In Europe, where the first *Homo sapiens* are first attested by that time, early cave art depicts humans and animals interacting with each other.[68] Some rare examples can also be found of therianthropes – that is, mythical figures – that are half human and half animal; these hint at the development of narrative fiction, religion, folklore and ideas about the supernatural.[69]

It has recently been argued that the blossoming of cave art that took place 42,000 years ago was linked to an abrupt period of dramatic climate change. An episode known as the Laschamps Excursion saw simultaneous shifts of precipitation and wind patterns in the Pacific and Southern Ocean, a sharp decline in the strength of earth's magnetic field and a spell of unstable solar activity that included multiple massive flares that also affected global weather patterns. The results included glacial expansions in the Andes and aridification in Australia that were so profound that they led to the wholesale extinction of large animals. The expansion of the Laurentide Ice Sheet and the cooling of North America and Europe, combined with electrical storms and auroras that produced spectacular light shows in the skies, forced modern humans to seek protection in caves for longer periods and perhaps provided inspiration for more creative forms of social engagement and artistic expression. That, at least, is one intriguing hypothesis – which has not convinced many specialists.[70] Another is that early artists looking for inspiration might have been inspired, intentionally or otherwise, by low concentrations of oxygen levels at the back of deep, dark caves that caused altered states of consciousness, resulting in hallucinations and out-of-body experiences.[71]

The period around 40,000 years ago also marked the disappearance of the Neanderthal population in Europe, with the sudden harsh subarctic conditions appearing to be a contributing factor, and perhaps even a decisive one.[72] The Neanderthals had already begun to experience a rapid demise across Eurasia, with a range of explanations advanced to explain the demographic collapse – including changes to vegetation, rising levels of disease and interbreeding. While each may have played a role, it may be that their fate was sealed by the fact that *Homo sapiens* proved better at exploiting food resources that became scarce as temperatures become colder – although recent research suggests that Neanderthals and *Homo sapiens* coexisted for far longer than had previously been assumed in at least some locations.[73] The tenfold expansion of the modern human population in Europe around this time also meant that there were not only better competitors but many more of them. It seems that the Neanderthals had died out entirely around 35,000 years ago.[74]

Neanderthals were not the only ones who suffered as a result of shifting climatic patterns and, perhaps just as importantly, of the resultant changes in ecosystems. It is no coincidence, for example, that the expansion of savannahs from Indo-China across South-East Asia to what is now Indonesia had corresponded with maximum hominin diversity.[75] As conditions changed, almost all hominin species failed to survive – even though the transitions were gradual and spread across thousands of years. Nevertheless, what is most striking about *Homo Sapiens* was the ability to cope with and even flourish in different conditions across rainforest and marine habitats.[76] The fact that modern humans set off for new pastures around this time [about 35–40,000 years ago] settling in New Guinea, the Bismarck archipelago and the Solomon Islands, shows both resourcefulness and remarkable initiative.[77]

Of course, it is all very well to suggest that modern humans' adaptability, versatility and durability were key ingredients for success, but it is clear that these alone were not enough. So while it is tempting to focus on the fact that a community at the Madjedbebe rock shelter in the Kakadu region of northern Australia was able to thrive and expand during a long period of cold weather that brought about major water shortages to most of the continent around 30,000 years ago, it is more revealing that most of Australia was abandoned at the same time.[78]

It was a similar story in other parts of the world during the Last Glacial Maximum which started at roughly the same time and was the result of the expansion of ice sheets in the northern hemisphere, a decrease in Pacific sea temperatures and a decline in atmospheric CO_2 levels.[79] Population levels in central Europe collapsed as human populations either died out or relocated in search of more hospitable conditions.[80] Refuges like the Grotta del Romito in southern Italy became extremely important settlements and provide invaluable case studies in helping show how people lived, coped and managed through difficult times.[81]

The western Mediterranean was particularly affected through this cold period, which lasted for several thousand years, with the formation of clay dune and pollen records indicating significant falls in precipitation levels. Archaeological evidence, meanwhile, shows repeated migration from unstable 'high-risk' parts of what is now southern Spain to the north, as well as presenting a picture of substantial demographic decline.[82] While much of the period of the Last Glacial Maximum (c.26,500–19,000 years ago) was dry, it was punctuated by a 3,000-year phase of massive autumnal and early-winter snowfall in the European Alps that resulted in the expansion of glaciers – and to conditions that were unforgiving if not unsurvivable.[83]

Evidence from marine cores from North Africa throughout the Last Glacial Maximum show collapsed rainfall levels but also very sharp cooling, with temperatures 10–15 °C lower in that region than today, with evidence from Greenland suggesting an even greater fall, with temperatures perhaps as much as 21 °C lower than 2023.[84] Land temperatures fell by around 6 °C based on studies of noble gases dissolved in groundwater measured across six continents.[85] Sea levels in the South China Sea dropped by 100–120 metres and by even more in the East China Sea.[86] The winter monsoons became considerably stronger and also changed distribution of rainfall patterns – resulting in marked shifts in vegetation, including increased desertification in the north of what is now China and an expansion of deciduous forests in the south.[87]

It was not just humans who struggled in these cold conditions. Sites in Morocco reveal that Aleppo pine and evergreen oaks benefited at the cost of cedar trees, which found it all but impossible to grow because of the cold and the lack of rainfall.[88] Further south, in what is now the Democratic Republic of Congo, forest cover was reduced – although assessing how and when is more complex than adopting a story of simple contraction and expansion.[89] Changing ecosystems and vegetation in South-East Asia, meanwhile, brought about losses of grazing lands for elephants, rhinos and tapirs and the extinction of hyenas in this region.[90]

These were just the latest in a series of massive losses of animal life. Between 50,000 and 10,000 years ago, there were spectacular declines of megafauna – that is, large mammals such as woolly rhinoceroses and sabre-toothed cats. Over this period, at least 97 of the 150 known genera weighing more than 44 kilos became extinct. The causes of these losses have been much debated, with the main explanations being that they were precipitated by human expansion principally through hunting, by anthropogenic change of habitats, by environmental and climatic change, by the lack of access to freshwater sources or by a combination of all four.[91] The result was a set of animal extinctions that were of a speed and scale that had been unprecedented for millions of years.[92]

The fact that disappearance rates of large animals in South America were high despite evidence of stable environmental conditions and that they appear mainly to have taken place after the arrival of the first humans around 15,000 years ago provides a telling pointer to the conclusion that our ancestors played an important role in large-scale losses in this continent and perhaps in others too.[93] Some scholars have suggested that the decline in large animals forced humans to hunt smaller prey and to develop new technologies better suited to doing so; indeed the lack of availability of large game in the southern Levant has been linked to foragers broadening their dietary range and even to the beginnings of agriculture.[94] It is worth stressing, however, that this is a complicated picture and that the regions that suffered the most significant climatic disruption across this period also experienced the greatest losses.[95]

The extinctions of such large numbers of species and in such high volumes naturally had an impact on terrestrial ecosystems. There were striking changes in flora distribution, including declines in large-seeded tree species as well as fruit dispersal, as a result of the disappearance of large herbivores. These changes had an effect on carbon storage in Amazonian forests, leading to a significant reduction in capacity.[96] Evidence from Australia also suggests that megafauna disappearance, whether because of hunting and expansion or environmental change, brought about relaxed herbivore pressure on vegetation that helped transform ecosystems – as did increased and deliberate use of fire in the landscape by the human population.[97]

Deglaciation around 19,000 years ago began a new set of modifications to landscapes. Ice sheets across North America began to melt, resulting in megafloods that were among the largest flood events in the world's history and whose courses were shaped by the deformation and tilting of the earth's crust – changing the elevation of topography by hundreds of metres in the process.[98] The retreat of ice sheets in the northern hemisphere and of glaciers over several thousand years resulted in the addition of enormous volumes of freshwater into the oceans. The results included major sea-level rises around the world that averaged eighty metres, the disturbance of terrestrial and marine eco systems and a net release of carbon dioxide and methane into the atmosphere.[99]

To give just one example of the scale of change, around 2 million square kilometres of Australia became submerged between around 15,000 and 8,000 years ago, eventually reducing the land mass by a third. The coastline was 160 kilometres further offshore than it is today, with LIDAR revealing human settlements that had to be abandoned as sea levels rose. Assessing what socio-economic and cultural consequences these had is difficult, but they must have been substantial.[100] Even though population concentrations appear to have been relatively small, the fact that the coastline receded by an average of more than twenty metres per year shows that changes must have seemed both never ending and ominous as the sea kept advancing and land kept disappearing.[101]

The impacts elsewhere in the world were also substantial. Most scholars date the arrival of the first modern humans in the Americas to around 22,000 years ago, with genomic data showing that ancestral Native Americans diverged from Siberians and East Asians before crossing the Bering Straits into what is now Alaska, helped by lower sea levels which revealed an archipelago of islands that turned the straits into a series of stepping stones.[102] Subsequent genetic divergences split those who reached the Americas into two further distinct groups.[103]

The nature, causes and migration routes of the dispersals are currently the subject of major re-evaluation, partly thanks to new discoveries, but also thanks to better and more accurate techniques used for interpreting materials.[104] For example, radiocarbon and luminescence dates for a high-altitude cave in the Astillero mountains in the state of Zacatecas in Mexico, as well as cut marks on rabbit and deer bones from the Tehuacán Valley, have recently pushed the date for the earliest settlers in the Americas back by some 10,000 years, to around 30,000 BC, while recent research on the Colorado plateau suggests an earlier date still, to around 37,000 years ago.[105] Forest sanctuaries along the Pacific coast of North America seem to have played vital roles in ensuring the survival of the first peoples to reach the Americas by providing refuges that offered shelter, food and other resources.[106] What is most striking, however, is not just the dates of expansion of populations, but the fact that sparse examples above all show how precarious life was for the first settlers. Waves of groups may have arrived and simply have died out time and time again, until conditions improved to enable more successful, long-term settlement.

If this was the most likely scenario, then it is perhaps no coincidence that a marked improvement in environmental and climatic conditions corresponds to much greater evidence of successful migration and permanent settlement across North America and the Caribbean, as well as in Central and South America.[107] One attraction for the expansion of groups southwards may well have been that post-glacial warming began first in parts of the southern hemisphere, making such regions more hospitable and attractive.[108]

Between 16,000 and 10,000 years ago, there were considerable rises in global temperature – that came on top of the shift in the glacial cycle from an ice age to an interglacial period – with estimates suggesting a likely range of 4–7 °C for the increase in surface air heat and lesser but still significant rises in the warmth of oceans.[109] An abrupt warming in the northern hemisphere around 14,700 years ago seems to have been caused by the release of heat from warm waters in the deep North Atlantic Ocean which in turn was linked to increased salinity at depths that had previously preserved the static stability of the water column.[110]

Records from speleothems and from marine and lake pollen cores in the Levant reveal that there was a rise in precipitation around 10,000–7,000 years ago – something that has been linked with prompting an expansion of foragers, who fanned out into new locations across this region, bringing with them lithic tools.[111] Indeed, some have suggested that improving ecological conditions played a role not only in encouraging wider dispersal patterns of the existing population, but in attracting new settlers, including from North Africa.[112]

Macrofossil and pollen records taken from hundreds of sites around the world show that changes to vegetation as the Last Glacial Maximum subsided were substantial – especially at mid- to high latitudes in the northern hemisphere, in South America and in tropical and temperate southern Africa, as well as across the Indo-Pacific region and Oceania. This warming phase saw significant uplifts in atmospheric CO_2 concentrations, rising from 190 to 280 parts per million during the period of deglaciation. However, temperature increases did not lead to uniform ecological change and in some cases even produced counter-intuitive outcomes.[113] For example, studies on plants in the contemporary world which show that some species respond to warming through adaptation remind us how important it is to think not just about human reactions when faced with changing conditions – and also to bear in mind the nuances and complexities of understanding climate change as a whole.[114] This is significant given that the rate of global warming in the coming decades is projected to be in the order of sixty-five times as high as it was during the last major deglaciation.[115] Understanding how, where, when and why plant – as well as human and animal – life reacts to change is fundamental to making sense of what the future is likely to hold in store locally, regionally and globally.

A new climate shock that occurred around 12,900 years ago brought about an abrupt reversal of the long-term warming process. The causes of this event, known as the Younger Dryas, remain controversial. The prevailing view is that the cooling was the result of large ice-sheet discharge of freshwater into the North Atlantic.[116] However, some scholars have proposed that a supernova explosion in the Vela constellation depleted the ozone layer, prompting atmospheric and surface changes that led to cooling.[117] A remarkable platinum spike at the aptly named Wonderkrater site in Limpopo province, South Africa, has persuaded some that a meteor or asteroid strike was responsible.[118] The presence of volcanic gas aerosols in well-dated sediment at a site in Texas, however, suggests a large-scale eruption as the most likely cause of change – with the Laacher See volcano in Germany a particular suspect.[119] There certainly were comet strikes elsewhere around this time, such as in the Atacama desert in northern Chile, where one impact generated so much heat that sandy soil was turned into glass.[120] It is not impossible, of course, that multiple events may have been responsible for the rapidly changing climate – and that they combined to produce effects that, although not as dramatic as major extinction events, were nevertheless substantial.

Although the causes are not clear, there were significant anomalies in global temperatures. In Greenland, which was 15 °C (+/- 3 °C) colder than today, the Younger Dryas saw a period of rapid warming, with abrupt changes evident in as little as one to three years.[121] On the other hand, carbonate geochemistry from lake sediments in northern China provides evidence of abrupt cooling during a thousand-year-long period that seems to have been the result of the atmospheric coupling of the North Atlantic and East Asian weather patterns.[122] So does the pollen record from New Zealand, although this also suggests that parts of the southern hemisphere were getting colder before the Younger Dryas event.[123]

One obvious result was an increase in the variability of sea-ice cover and unstable global climatic conditions generally.[124] Another was a change in monsoon rainfall, although there were significant spatial differences, with the calcium carbonate record from Lake Qinghai on the Tibetan plateau suggestive of a decline in precipitation, while palaeohydrological records from the middle reaches of the Yangtze indicate higher rainfall levels following the Younger Dryas event.[125] High-resolution studies from Lakes Malawi, Tanganyika and Bosumtwi indicate abrupt shifts in the wind-driven circulation of these lakes, while a range of other evidence suggests an abrupt northward translation of the African monsoon system that resulted in marked increases in precipitation in the northern tropics and drought conditions in the south.[126]

Naturally, the impacts on flora and fauna were profound. There was a new wave of large-scale animal extinctions – once again as a result of hunting, climate stress or a combination of the two.[127] There were substantial changes to vegetation distribution in Europe, Canada, Africa and elsewhere.[128] Not surprisingly, there are also clear indications of contractions in the human populations, for example in North America.[129] DNA evidence suggests genetic substitutions in Europe and elsewhere that are indicative of major demographic reductions around this period.[130] A marked decrease in the number of sites occupied by humans in Japan also provide a telling signature of population collapse.[131]

In the Levant, a response to more hostile conditions was the establishment of small settlements inhabited by permanent or semi-permanent populations. Such developments may have enabled a pooling of resources and skills, but also served as a collaborative solution for security and defence needs against other groups at a time of rising food shortages and pressures. Staying put may also have been an effective way of protecting land that yielded wild cereals and of ensuring that the best sites were not taken over opportunistically by others.[132]

After around a thousand years of these cold conditions came a reset some 11,900 years ago. Climate archives show that the first signs of a warming process began in the western tropical Pacific and the southern hemisphere, which then spread into the North Atlantic over the course of around two centuries.[133] Ice-core records from Greenland suggest that, once temperatures started to rise, they did so extremely quickly, going up by more than 10 °C in sixty years.[134] The scale of warming, as well as the acceleration over such a short period, again has implications for understanding the potential consequences of climate change in the contemporary world, not least since many current projections assume global warming will be gradual and constant, rather than sudden and dramatic.

The end of the Younger Dryas marked the start of a new epoch, first named the Holocene – literally 'an entirely new era' – in the 1860s by the French palaeontologist Paul Gervais, who was struck by the changes to the sediment layer following the deglaciation that marked the end of the Ice Age. Scholars today now typically pinpoint the start of the Holocene by stable isotope data from Greenland ice cores that measure the ratio of stable oxygen isotopes to 11,700 years ago.[135]

Since that time, there have been fluctuations and changes, including ones that have seemed and have even been substantial. Seen in the long run, the end of the Younger Dryas is little more than a staging post in history, just another moment in the fluctuations between cooling and warming phases marked by the expansion or retreat of ice sheets, by alterations of monsoon and precipitation patterns and by changes to air surface and ocean temperatures that, while not uniform, brought challenges and in many cases resulted in adaptive changes for vegetation, animal and marine life. One good example comes from the sudden appearance of Pacific nitrogen in the Arctic basin as the land bridge linking what is now Russia and Alaska flooded with the rise in sea levels, changing ecosystems for plankton and the food chains that build upwards from them.[136]

From a human perspective, however, it marked a watershed. The onset of a long period of warmer, stable conditions corresponded with some profound changes in demographic expansion, settlement patterns and innovations – the most significant of which was the emergence of agriculture. Even before the onset of the Holocene, there is evidence that rising levels of complexity were producing innovations in technology and in cultural behaviour. In North Africa, for example, new lithic and bladelet technologies start to appear around the time of the deglaciation.[137] This was also the case across many parts of Asia, such as in northern Mongolia, where rock art in the form of petroglyphs, often depicting ibex, began to appear in the early Holocene.[138]

The earliest examples of human cemeteries are suggestive of changing ideas about the body, about life and about identity – as is the fact that burials provide evidence of the deliberate removal of healthy teeth, likely at an early age and above all but not only among women; this may have been part of an attempt at enabling group identification within relatively restricted geographic zones, and as such reflects the adoption of ideas about the distinction between kinship and social groups.[139] The transition from mobile hunter-gatherer societies into more sedentary economies required new ways not only of acting, but also of thinking.[140]

There is abundant evidence too from the Levant from before the Younger Dryas of some communities adopting either semi- or even fully sedentary lifestyles, and building structures with stone foundations, including storage buildings. The presence of grinding stones at several locations has been interpreted by some scholars as indicating not only the collection of wild cereals but also their cultivation.[141]

Many scholars have sought to associate behavioural change with environmental factors during the Younger Dryas itself, arguing for example that the earliest experimentations with plant cultivation were responses to the need to increase food production during periods of declining resource abundance.[142] These were translated into more systematic and successful efforts to grow food once climate conditions became more favourable. These in turn prompted innovations in tool manufacture as well as the generation of surpluses which enabled human groups to select and establish themselves in environmentally favourable locations that did not require seasonal mobility and allowed ever larger communities to live together.

Over time, this laid the basis for villages, towns and cities and as such spurred the development of writing systems, religions, complex economies, new social and political structures – indeed for 'civilisation' in its literal meaning of urbanised settlements. Such transitions, as we shall see, were not straightforward or without their costs. And, as we shall also see, other forms of engagement with flora and fauna, including pastoral nomadism – that is to say, the domestication of livestock – developed and were often both complementary to and in competition with sedentary societies.

Nevertheless, looking at the period up to the start of the Holocene in the round, it is hard not to be both surprised and impressed by how modern humans defied the odds time and again, managing to survive volcanic eruptions, intense or reduced solar activity and meteorite strikes – as well as dramatic changes in climate induced by tectonic, geological or axial movements that brought about sharp swings in habitats. Many species were unable to navigate these challenges – including all other hominins, who became extinct one by one.

Not all our ancestors made it, of course, with doomed expansions and settlements a part of the human story of trying and failing, as well as succeeding and surviving. And yet enough pockets flourished by adapting, by innovating and perhaps most importantly by being lucky enough to avoid being in the wrong place at the wrong time. By around 11,000 years ago, modern humans had managed to spread out across all the world's continents, except for Antarctica: some estimate three-quarters of terrestrial nature was inhabited, used and changed by this time, meaning that lands that were untouched were as rare then as they are today.[143] Striking as that is, what was more significant was that this meant that species risk was diversified, significantly raising the long-term chances of survival.

Even more fortuitously, then, while the last eleven millennia have not been uniformly clement or favourable, they have been less unstable and variable than at many times in the past.[144] While contending with potentially catastrophic global warming in the course of the twenty-first century should not be downplayed, current projected rises of 1.5–2 °C are modest in the grand scheme of climatic change, not only in the history of the earth but in that of humans too, and look paltry indeed compared to the very many and regular double-digit rises and falls that have occurred in the past.

Fortunately, our own limited sense of time is helpful in avoiding too many sleepless nights. When we think of history, we conceptualise time with reference to human achievements and to events we choose to relate to. By necessity this makes our frames of reference both highly self- and species-centred and laughably narrow. Casting our minds back to the age of the Victorians is to imagine another world – inviting us to investigate how people thought and behaved, to wonder what they wore, wrote and listened to, and to marvel at their lives. Thinking back four millennia to the time of kingdoms in Mesopotamia and Egypt, to the cities like Harappa or Mohenjo-daro, to communities in the Niger Delta or Middle Yangtze, or to sites like Kotosh in the Andes or Tlapacoya in central Mexico seems almost impossibly exotic and difficult, a challenge for historians, archaeologists, anthropologists and others to imagine how societies in the past were structured and functioned.

However, periods spanning decades, centuries and even millennia are the briefest of blinks of an eye in terms of the history of this planet; and in that sense they are perhaps not even important ones either: the earth has been spinning around the sun for billions of years, and will continue to do so for billions of years more. In one way, then, the Holocene is the perfect tag for the modern age – a time that is geologically and climatically distinct, but which also corresponds to the era of modern humans, an era we invariably fail to place within the bigger picture.

Since its beginnings – whenever or wherever those may be – our species has expanded, colonised, reproduced, created and dominated, but it has also destroyed, devastated and exterminated. It has done so better than almost any other organism that has lived in the last 4.5 billion years. It is no coincidence that a more benign climate provided the springboard for all that to happen.

3

Human Interactions with Ecologies
(c.12,000–c.3500 BC)

Those who are too lazy to plough in the right season will have no
food at the harvest.

Book of Proverbs, 20:4

The Holocene marked the start of much more favourable climate conditions
in many locations around the world. For one thing, the shift from around
10,000 years ago began a long period of stable weather patterns, which
in itself was significant, reducing the number and frequency of shocks –
although there were of course significant spatial differences between and
within continents. Generally, however, temperatures rose, as did levels of
rainfall. Crucially, CO_2 content in the atmosphere also increased sharply
from levels during the Last Glacial Maximum that were so low that some
scholars have argued that photosynthesis would have been limited and
would therefore have affected the capacity, durability and reliability of
plants as food resources. As one influential research paper has put it,
agriculture may not have been impossible before the Holocene, but it
suited conditions perfectly after its onset.[1]

Sedimentary sequences and pollen spectra from Guxu Lake in eastern
China show clear evidence of warmer and more humid conditions, as
does the expansion of evergreen–deciduous broadleaf mixed forests,
starting around 8000 BC.[2] Radiocarbon-dated sediment cores from
eastern Guatemala indicate progressive increases in precipitation and
erosion levels spanning a period of 5,000 years after c.7500 BC.[3] North

Africa experienced profound changes that were closely correlated to and perhaps explained by increased summer insolation and Arctic sea-ice loss – which seems to have driven strong solar heating in the northern hemisphere generally.[4] Much of the Sahara was transformed into grass savannah, opening up new areas for human occupation and transforming plant and animal ecosystems in the process.[5] This has often been attributed to the rotation of tilt of the earth's axis that occurs roughly every 25,000 years, which enhanced the summer monsoon, although some now propose that the so-called Green Sahara was the result of a southward shift of the Mediterranean winter precipitation system.[6]

The effects of these changes were particularly important in the Levant, above all in the Fertile Crescent where population levels rose during the Younger Dryas – probably a reflection of migrations from less hospitable locations as conditions elsewhere became harsher.[7] Higher levels of competition presumably put pressure on flora and fauna food resources, something that would have been made worse by the apparent declining availability of wild cereals and pulses before environmental conditions began to improve at the start of the Holocene.[8] One response was for settlers to pay more attention to harvesting wild plants; another was the systematic exploitation of wild herds of sheep, goat and gazelle.[9]

Many scholars today stress the importance of gender in the development of agriculture, in terms both of crop cultivation and of food storage and preparation.[10] Women played crucial roles in rituals of gathering, preparation, display and use of plants, while burials of women inside houses can also be understood not as reflecting domestic roles, but as evidence of their role as primary intercessors with ancestors and the spiritual world.[11] The sheer number of figurines depicting women as deities across multiple cultures in Mesopotamia and elsewhere underlines the fact that gender roles were enormously significant in the development of ideas about fertility and sustenance as well as about life and death.[12]

The centrality of women to early agriculture is also evident from the osteological analysis that reveals upper-body strength at levels that would have surpassed today's elite female athletes.[13] Musculoskeletal stress levels in both men and women suggest that there was a sharing of labour when it came to early sowing and reaping of crops leading some to suggest that separate gender roles did not emerge until around 3000 BC.[14]

Across the Fertile Crescent and the Zagros mountains in Iran, evidence of rising domestication of crops and animals from around 8000 BC suggests that knowledge was being spread as coping skills became the bedrock of new ways of living.[15] Domestic cereals like single-grained einkorn wheat, emmer wheat and hulled barley, as well as lentils, peas and bitter vetch, became established parts of human diets and spread first across the Fertile Crescent itself and then into Anatolia, Egypt and beyond.[16] Around the same time, the first signs of herding appear in South-West Asia, suggesting that sheep and goats were domesticated first, with cattle and pigs being brought under human control soon afterwards.[17]

These processes were slow and likely haphazard. It took time for new skills to be learned and refined, and when they were, they took time to be taught and to spread. And while it can be tempting to think of a 'transition' from hunter-gathering and foraging to agriculture and domestication, the reality was much more opaque – so much so that the value of the labels that describe apparent lifestyles is itself questionable. Development and adoption of new skills did not replace existing knowledge or make it redundant: in other words, the question is not so much where one draws lines between concepts like collection of food, cultivation, domestication and agriculture but why it is necessary to do so in the first place.[18]

What is clearer is that the start of the Holocene saw the emergence and expansion of new ways of dealing with food and with animals, and of seeking to optimise their benefits. That climatic and environmental change played a catalytic role in stimulating these developments would help explain why it is possible to identify similar patterns emerging in several parts of the world more or less concurrently. For example, despite reaching the Americas long after arriving in other continents, the human population also began to exploit the land more intensively around this time, with scholars arguing that reductions in numbers of large animals through hunting and because of climate impact were closely linked to efforts to secure more sedentary sources of food. Indeed, some have even argued that the extinction of large herbivores (weighing 200 kilos or more) may have been helpful to early crop domestication because their removal meant lower levels of disruption to plant habitats – and therefore lower risk and better reward for cultivators.[19]

In China, the evidence for the dating of more intensive engagement with crops is much debated and remains controversial. It is possible, for example, that early rice cultivation took place at Shangshan in the lower Yangtze River region around 8000 BC, though there is more certainty from other sites in central China, such as at Jiahu on the upper Huai River a thousand years later.[20] Nevertheless, the emergence of new settlement patterns, especially in the form of villages in both north and south China, point to fundamental socio-economic and ideological changes as populations shifted from the exploitation of wild plants to their management and cultivation and then to their domestication.[21]

The availability of reliable wild and domesticated food supplies not only facilitated the rise of sedentary life and of villages, but also enabled the concentrations of people that they could support. During earlier periods, groups living together in the Levant occupied sites that were small and were inhabited seasonally. Now, however, the size and number of settlements started to grow steadily over time – with communities numbering in the hundreds and sometimes even in the thousands by around 7000 BC. Naturally, this spurred a series of revolutions in social behaviour and in the adoption of new ideas, for example about the treatment of the dead – as attested by more elaborate burial practices, which included painting, retrieval and circulation of bones and skeletal parts, and of skulls in particular.[22] In at least some locations, it seems that vibrancy was injected by migration patterns, with new arrivals not only bringing new ideas and technologies with them, but also boosting gene pools. Diverse populations helped spur cultural dynamism, even if the ways in which they did so varied: in Upper Mesopotamia and some parts Asia Minor, for example, early sedentary communities were arranged around biological family lines, while others were marked by different kinship and social structures.[23]

Objects, totems and idols from multiple locations indicate rising interest in asking and answering cosmological questions about the human relationship with nature, the divine and unseen forces. These included stone stelae at Göbekli Tepe in south-eastern Anatolia, as well as sculptures of birds, snakes and anthropomorphic figures found at other sites in the same region.[24] Some were in human form, such as a stone carving of a man wearing a necklace, clasping his hands, found at Urfa in what is now Turkey, or a remarkable, very striking human-shaped idol, carved from wood and standing more than five metres in height, that was discovered by gold miners in the Shigir peat bog near Sverdlovsk more than a century ago.[25] Monumental buildings in north-western Arabia in which animal horns and cranial elements were arranged deliberately meanwhile provide evidence for an early cattle cult that should be understood alongside the rock art of this region.[26] A cemetery by Lake Onega in the largest-known burial ground in north-east Europe that has yielded large numbers of ornaments and pendants made of elk teeth – likely used to create rattling sounds for drumming, ceremonial dances and shamanic and other rituals – has been linked with changes in cultural and socio-economic practices that were driven by climatic shifts around 8,000 years ago.[27]

It has been argued that the advent of farming demanded the development of concepts of possession and of private property that had previously not been necessary – which is both logical and plausible, but for which it is difficult to provide hard evidence.[28] Rather more easy to demonstrate is the introduction of new skills and technologies, not least in architecture, with the construction of different types, sizes and uses of buildings.[29] Again, none of this was quick: the changes here were spread not just over decades and centuries, but over millennia.

In North Africa, the Sahara and the Nile Valley, settled communities also became more common as a result of more plentiful resources – in this case, fish life in particular, which was abundant in the vast chains of lakes and river networks that once criss-crossed the heart of the continent. Large volumes and variety of fish and molluscs are attested in the archaeological record, such as in the Takakrori rock shelter in south-west Libya in what is now the heart of the Sahara desert, as are remains of amphibian, reptile and bird species normally associated with aquatic environments.[30]

Changing ecosystems and shifts to a more sedentary way of life brought about new habits and lifestyles, with goods that were not easily transportable like grindstones becoming more common as people stayed in or close to fixed locations.[31] It is perhaps no coincidence, then, that the changing settlement patterns saw growing experimentation with new technologies. In north-west China, for example, new types of microlithic tools, blades and arrowheads were introduced and became increasingly common from around 8000 BC.[32] It was a similar story in the Americas, where assemblages of distinct and new tools that emerged at the start of the Holocene are linked by most scholars to changing ecologies and ways of living.[33]

The rapid diffusion of ceramics in many different regions around the same time suggests too that this was a time of innovation and rising complexity. What had previously been very limited production of pottery in Japan now expanded dramatically, perhaps as a result of new strategies of storage, perhaps because of rising sedentism, perhaps because of population growth – or a combination of all three. The spread of pottery technology across northern Eurasia and to Alaska suggests that a new era demanded new solutions.[34]

Some of the oldest examples of ceramics come from Niger and the Dogon plateau in Mali in West Africa and date to c.8000 BC.[35] In their earliest phase, ceramics were very rare in African settings. One explanation for this is that while the introduction of ceramics in general is sometimes seen as a moment of pivotal technological importance, the reality is that many low-tech containers had done a perfectly good job for the transportation and storage of food and water in the past. Reed baskets may not have been as durable as ceramic vessels, for example, but they required lower levels of investment of time, energy and competence. Likewise, ostrich eggshells had not only been widely used for decorative purposes for tens of thousands of years in Africa but had served effectively as water containers. The introduction of ceramics did not therefore represent a breakthrough in capabilities nor open new opportunities. Indeed, this is precisely why some scholars suggest that the earliest ceramics were primarily devised for social and ritualistic use.[36]

Ceramics were being produced in other parts of the continent by 7000 BC, most notably across a broad arc stretching from central Sahara to the Upper Nile.[37] By this time, ceramics were also widely attested across settled communities in much of the Levant, south-east Anatolia and in what is now the Nefud desert in Saudi Arabia but which at that time was studded with lakes and rivers and was an attractive habitat for human activity.[38] Curiously, the use of pottery for aquatic products in particular seems to be a recurrent theme across many global settings, as indicated by chemical and isotopic residue analysis from Korea, for example.[39]

Shifts to more sedentary ways of living were not instantaneous, and nor were they uniform geographically. And of course, while some communities formed in small settled clusters and others were more substantial, the distinctions between those who lived in permanent settlements and foragers or hunter-gatherers were often blurred. Certainly, there were wide ranges of different lifestyles, resource-gathering techniques and energy expenditure decisions, as well as competition and co-operation that was dependent on location, season and population size. In other words, there was no 'one size fits all', but rather considerable nuance and difference depending on time, place and habitat.

Nor were settlements always successful. While attention is usually paid to groups and locations that settled and flourished, the reality is that in many cases attempts to live from the fruits of the land ended in failure.[40] Even in the Levant, where it has long been thought that the rise of village settlements was both inexorable and rarely punctuated by disruptions, recent research has revealed that abandonment of sites happened often, reflecting the fragilities and realities of efforts to build sustainable communities.[41] It is likely too that competition for resources and for the best locations brought about intergroup rivalry, conflict and violence – which in turn led to innovations in defensive fortification measures, as more contemporary examples from New Guinea in the south-western Pacific show.[42]

Nevertheless, perhaps the most important result of the improving climate conditions that marked the onset of the Holocene was a sharp rise in human population.[43] Certainly, the reliability and diversity of cereals reduced the risk of hunger, as did the construction and management of storage pits in which surplus could be kept for winter months and mitigate seasonal shocks. New tools and more durable storage vessels evidently helped with productivity jumps too, enabling larger communities to support themselves, especially in ecologically favourable locations.

The increase in calorie availability and the reduction in energy expenditure that resulted from being able to live on the produce of dependable and abundant resources in close proximity – from cereals or marine stocks – would seem in itself to go a long way to explaining rising population sizes.[44] So too does rising sedentism: modern ethnographies indicate that women in hunter-gatherer and forager societies such as the Agta in the Philippines typically have fewer children and longer gaps between childbirths than those in sedentary communities.[45] The fact that shorter interbirth intervals and higher levels of fertility are associated with increased carbohydrate consumption which in turn is linked to higher body mass index is also significant.[46] In other words, changes to living patterns as well as new engagement with plant-food sources may have influenced gender relationships and in particular maternal roles and expectations. This fits neatly with skeletal data from burial sites in Iberia from the early Holocene onwards that supports the hypothesis that changes to settlement and nutrition patterns were instrumental in driving higher numbers of births – and therefore in boosting population sizes.[47]

There were trade-offs to the rise in reliance on cereals and to rising densities of habitation – not least the impact on health. For one thing, work involving heavier tools for grinding and processing grain placed greater demands on the human body, resulting in cases of osteoarthritis rising as crop cultivation became more widespread; for another, the sugars in cereal carbohydrates brought about degradations in tooth enamel, with resulting high incidence of dental cavities.[48] The decline in dental health seems to have had a greater impact on women, perhaps because of higher fertility levels and resultant hormonal fluctuations, immunological competence during pregnancy and concurrent and subsequent variations in saliva composition.[49]

The harnessing of oxen around 6000 BC, initially for threshing, was significant therefore, helping relieve pressure on human labour forces thanks to time and energy boosts that helped provide the basis for a major uplift in food consumption.[50] The use of large animals in agriculture in turn spurred innovations such as carts and ploughs that helped further increase production, allowed more land to be cultivated more quickly and therefore supported larger populations. These developments had societal consequences too. For one thing, they brought about another change in gender roles as women shifted into domestic household tasks while men increasingly took over horticultural and agricultural activities.[51] For another, the construction and ownership of ploughs and carts produced rewards and surpluses for those who owned them. As one leading scholar has put it, in these intensifications 'lay the seed of social inequality'.[52] So did those of gender inequality.[53]

An even bigger biological price to pay came from the costs of living close together. These included faecal pollution and poor sanitation that could cause bacterial disease, and conditions that favoured the spread of viruses and parasites from person to person.[54] Stored food resources attracted rodents, which were important vectors for zoonotic disease – that is to say, illnesses that jumped from animals to humans – as were cattle, goats and sheep, as they became domesticated and used as sources of food, milk, clothing and textiles.[55] These included diseases such as mumps, chicken pox, rubella and whooping cough, which all jumped from animal species and spread easily from person to person thanks to higher population densities.[56]

Set in this context, population increases were all the more remarkable, with reproductive increases more than setting off the negative drag of morbidity and mortality caused by disease. There were silver linings too for communities where infectious pathogens thrived: while they were often ravaged in the short run, repeated outbreaks eventually resulted in populations becoming 'disease-experienced', with partial immunity being built up thanks to frequent exposure.[57]

Ironically, this provided long-term advantages when climatic conditions opened up new regions, or when population sizes necessitated outward migrations. Pathogen exposure and lack thereof could have significant impacts in terms of competition for land, power and resources. One obvious example is the arrival of the Spanish in the Americas after Columbus' Atlantic crossing in 1492 when the immunological naivety of the indigenous peoples to smallpox is often cited as the cause of catastrophic demographic and political collapse.[58] Another comes from the way that Bantu peoples spread out into much of west and central West Africa around 3,000 years ago, when built-up resistance to malaria was a key factor in the expansion of culture, language, identity and genetics.[59] As we shall see, this immunity was to play a crucial role in the re-shaping of the world in later periods.[60]

Global mean surface temperature (GMST) change over the past 24 kyr.

Source: Osman et al, 2021

As the Holocene progressed, further waves of migration kept opening up new ecological zones. Those who moved brought not just abstract concepts about farming with them or even new tools, but crops too. New corridors opened up into Central Asia, the Mediterranean basin, the Nile Valley and Europe that disseminated wheat and barley, as well as other crops from south-west Asia. By around 7000 BC, agriculture had appeared in Crete and Greece, and not long afterwards as far afield as Andorra and Spain, Bosnia and Sicily. Calcified tissue on the teeth roots of sheep and goats, as well as cut marks to bone remains, shows that they were being farmed in the Fergana Valley in Central Asia by 6000 BC.[61] By 5500 BC, farming culture was established across much of central Europe, as well as in northern France, with villages studding the landscape from southern Spain to the Nile Valley to the Caucasus.[62] Such movements of ideas and of people are best understood not as single, large waves but as many cases of small-scale mobility that together add up to a large-scale signal – a hypothesis that is convincingly supported by DNA evidence.[63]

Those who brought crops and new concepts to Europe also made their own genetic impact. Advances in genomics allow us to identify separate and distinct groups of early farmers and others who moved west, with farmers from Anatolia and the Levant beginning to arrive around 7000 BC, while populations from the Caucasus reached eastern Europe around the same time. They introduced gene variants responsible for the evolution of light skin pigmentation in modern Europeans.[64]

In fact, genetic variants associated with light skin arose in Africa, which is home to a vast range of skin colours. Darker skin provides UV protection in equatorial zones, while at higher latitudes lighter skin maximises vitamin D production. As early human populations had moved out of Africa, responses including those to different intensities of sunlight triggered mutations that were then passed along to subsequent generations. As the incidence of the SLC24A5 gene was particularly common with west Eurasian populations, when these groups moved into Europe there was a slow but substantial replacement of the relatively dark skin pigmentation of earlier arrivals who had been poorly adapted to high-latitude conditions.[65] The genetic signal of lighter skin also helps track population movements into Central Asia, north India and East Africa – where haplogroup data demonstrates important migrations back from Asia to what is now Ethiopia and Tanzania, in the period from around 7000 BC onwards.[66]

The ecological lottery again played a role in differences in other parts of the world. Although California has many climatic, topographic and ecological parallels with the Fertile Crescent, Its plant domesticates had much lower rates of self-fertilisation, as well as no large-seeded grasses. Taken alongside the difficulties of controlling pollination in maize to achieve optimal outcomes and maintain genetic purity, it is perhaps not surprising that plant-intensive subsistence only developed much later in the western flank of North America.[67]

Agricultural revolutions took place at different times in different places and with different effects. In most parts of Africa, for example, unlike other continents, herding of goats, sheep and cattle preceded crop farming, with the domestication of indigenous plants occurring only in the last 5,000 years – starting with wild pearl millet, groundnut, yams, sorghum and other crops, with distribution varied and dependent on suitable climatic conditions.[68] South Asia and eastern North America present pictures that are dissimilar to the Near East, with archaeobotanical evidence pointing to food production being undertaken by mobile and small population groups, rather than by larger farming communities.[69] In China, although there is evidence of sedentary villages around 7000 BC, plant and animal domesticates do not become common until around two millennia later.[70] By that point, intensive agricultural practice known as Banpo had begun to spread throughout the Yellow and Wei River systems.[71]

Around 6200 BC, however, a large-scale shock event took place that prompted a new series of challenges and changes. In what is sometimes referred to as a 'Goldilocks abrupt climate change event' because of its importance in helping model future disaster planning, an ice dam over Hudson Bay collapsed, resulting in the catastrophic drainage of Lakes Agassiz and Ojibway.[72] The results included an acceleration of melt of the Laurentide Ice Sheet that contributed to the weakening of Atlantic circulation patterns, cooling across large parts of the northern hemisphere of 1–3 °C that lasted 160 years and releases of freshwater into the oceans in pulses that caused global sea-level rises of up to one metre.[73]

In the grand scheme, such changes hardly seem dramatic enough to require major adaptation. However, the consequences did prove momentous. Sediment data from the central Sahara shows significant falls in lake levels and increasing aridity around this time, presumably because of changes to the monsoon system that resulted from interruption to Atlantic circulation patterns.[74] These changes appear to have affected South Asia too, where severe and sudden desiccation at precisely this time underlines the relationship of the Indian summer monsoon with the North Atlantic in general but also with the climate event of 6200 BC specifically.[75] Pollen and stalagmite records from Bigeum Island, South Korea, at this time that reveal substantial changes to vegetation bear witness to the global impact and also demonstrate the fragility of ecosystems when subject to what seem to be relatively modest changes.[76]

Survival was dependent on adaptation – and good fortune. Adaptation was more achievable in some locations than others, but much must also have depended on the ability to devise coping mechanisms in the face of sudden change. It might be, for example, that climate shock provided the context for the move of livestock herders from the Levant into Africa – with mobility to be understood as a response to uncertainties of resource supply and to perceptions of increased competition and risk, perhaps from other humans, from predators or even from the environment itself.[77] This was also broadly when new genetic markers appear in East Africa, including lighter skin pigmentations as populations with historic ancestry in Africa returned and brought subsequent mutations back with them.[78]

Sharp decreases in human activity in places like western Scotland, north-eastern Spain and some parts of the Danube are substantial enough to have led scholars to ask whether they were abandoned in the aftermath of the event or whether the populations in each simply could not and did not survive.[79] Some studies have sought to link the event with an impetus towards further expansion by early farmers from Anatolia and parts of Greece to new pastures in Macedonia, Thessaly and the Balkans, although the dating, circumstances and nature of these dispersals is not clear.[80]

The effects were also felt in South-West Asia, where this was a time of stress with disruptions at multiple sites in southern Turkey, northern Mesopotamia, Syria and even Cyprus – although how this was influenced by or connected to new climate conditions is not clear. Those living at Çatalhöyük, one of the best-known and most intensively studied sites in Turkey, shifted settlement from its east to its west mound, and also seem to have moved habitations further apart.[81] Changes in hydrogen isotopic composition of animal fats preserved in pottery vessels found at the site show that there were also major shifts in food production at this time.[82] It was a period of cultural change too, not least in terms of evolving ideas about cosmology, religion and the divine.[83]

At Tell Sabi Abyad in northern Syria this was also a time of transition, with the introduction of new architectural and pottery styles, and evidence of a more diversified population that included mobile pastoralists as well as sedentary farmers. The discovery of stamp seals and abstract tokens used to control access to goods or services points to changing ideas about personal property – and presumably to the need for higher levels of identification and protection during a challenging period.[84] While many speculative theories about societal collapse have been associated with the climatic shift around 6200 BC, more sober assessment shows that the most useful conclusions to draw are ones about the human propensity for resilience and adaptation.[85]

The same cannot be said for a major natural disaster that took place at about the same time as the sudden cooling of the northern hemisphere that reshaped the geography of Europe. In around 6150 BC, a 190-kilometre shelf of sediment off the coast of Norway was dislodged, likely by an earthquake, creating a giant tsunami that swept south through the North Sea.[86] The scale of the wave and the devastation it inflicted can be shown by models which estimate that it would have reached twenty-one kilometres inland – twice as far as the Fukushima tsunami in 2011.[87] It swamped and submerged an area known as Doggerland that connected Britain with the European continent, although the most recent research suggests that the tsunami may have created an archipelago of islands that only later disappeared as sea levels rose around 5000 BC.[88] Nevertheless, whether or not the geographical detachment was completed in stages, the consequences not only for European but for global politics in the early and modern eras were substantial: Britain's separation from continental Europe was crucial in everything ranging from the development and dominance of British sea power to the military outcome of the Second World War to the exceptionalism that helped drive the Brexit vote to leave the European Union in 2016.[89]

At the time, the wave would have killed everything and everyone in its path. Moss-stem analysis shows that the slippage of the sediment shelf and the tsunami took place late enough in the year to sweep away peoples on the North Sea coast of Norway who typically spent the warmer months on high ground hunting reindeer but wintered close to the coast. Those who managed to survive would have struggled to make it through the winter because of the loss of their dwellings, boats, equipment and supplies.[90]

In due course, things slowly improved into what is sometimes called a climatic optimum, with warm and humid conditions favouring the development of agricultural economies. Archaeobotanical evidence shows the spread of domesticated breeds of plants such as maize varieties across South America from around 6000 BC into many areas of the continent including Colombia and north-western Ecuador, coastal and then highland areas of Peru and eventually the central Andes.[91] As in other parts of the world, shifts to farming brought about changes in dental health there and over time even to tooth shape.[92] And as elsewhere, the return of more benign circumstances, combined with stable food production – themselves supplemented by terracing and irrigation canals – brought about rising population sizes, although these were not uniform geographically or chronologically across the continent.[93]

The period up to c.3000 BC was one marked by considerable ecological and demographic change. One reason for this was an unusually high level of volcanic activity, attested by the volumes and concentrations of sulphur traces in the ice-core record.[94] The effects of solar minima and in particular the ≈2,400-year Hallstatt cycle – a time when the sun's reduced activity was connected with cooling, with the expansion of glaciers and with the release of iceberg rafts from the Arctic into the Atlantic Ocean – also played a role in impacting global climates.[95]

The widespread adjustment of complex and mutually dependent global weather patterns was influenced by other factors too. It seems that a warming phase in the southern hemisphere had resulted in glacial melt, decreasing Pacific Ocean surface temperatures; this in turn had an effect on the El Niño–Southern Oscillation (ENSO), which reduced in both intensity and variability up until around 3000 BC.[96] This may have been more significant than the cyclical changes in earth's orbital geometry, which is usually assumed to have played a key role in the strength of the ENSO, and therefore in global climate patterns.[97]

Evidence from Central Asia shows the retreat of the monsoon from c.4000 BC, from which point Indian proxies suggest greatly reduced rainfall as well.[98] The concurrent weakening of the West African monsoon system, followed by a dramatic decline around three centuries later, turned the period up to c.3000 BC into one of rising and consistent aridity.[99] This brought about a major population shift as lakes and rivers in the Sahara started to dry up – rapidly in East and North-East Africa, and more gradually in places like northern Chad, to judge from pollen and sedimentological records from Lake Yoa.[100] Oases that had reliable water and pasture sources evidently offered attractive solutions, as did the Nile Valley which in time also drew in significant numbers of migrants presumably searching for more hospitable locations.[101] Other adaptations seem to have included increased reliance on sheep and goats, and seasonal use of shelter.[102]

Some parts of North-West Africa, including coastal areas of what is now Mauritania, continued to experience considerable rainfall.[103] However, the effect was such that the Sahara began to desertify, expanding as it did so to create new environmental conditions – and becoming the increasingly impregnable barrier for human movements that we know today. This led to genetic divergences that resulted in strong differentiation in Y-chromosome haplogroup compositions in the respective populations in northern and sub-Saharan Africa that do not much persist in the modern world but define and reinforce the realities of the peoples of Africa as representing the highest levels of diversity anywhere on the planet.[104] In other words, climate change divided Africa not only geographically and environmentally but genetically.

These sharp changes brought about what have been called 'cascades of adjustments' in vegetation, fauna, humans and pathogens. Among these were new cultural practices such as cattle burials and livestock bone deposits set within built structures at sites in eastern Egypt and south-western Libya that evidently had a ritualistic context – though whether these are related to religious beliefs or to emerging ideas about protection from disease at a time when pathogen environments were changing rapidly is not clear. The fact that aridity negatively affects the availability of fodder and water and is closely related to increased susceptibility to disease, for both humans and livestock, would logically appear to be important.[105]

When the ENSO started to kick back in around 5,000 years ago, it brought a new set of effects as it did so that were not globally uniform. North America experienced droughts and severe water shortage, with lakes either falling in water levels or drying up completely.[106] Pollen and tree records provide evidence of aridity and changes to vegetation as atmospheric circulation patterns shifted.[107] Precipitation levels in the Mediterranean changed too, leading to the 'Mediterraneanisation' of landscape ecologies – that is to say, transforming them into how we think of them today – dominated by evergreen shrubs and trees that are adapted to summer drought and cold, wet winters.[108]

It was a different story in South and South-East Asia, where northern India and northern Indo-China seem to have been particular beneficiaries of higher precipitation, although mapping out likely patterns elsewhere is subject to considerable uncertainty because of the limited study of potential proxies that might help – such as lake sediment, pollen, stalagmite data and more.[109]

Abundant evidence from elsewhere however helps show that parts of East Asia saw wetter conditions as Africa became drier. The Inner Mongolian plateau became warmer and experienced higher levels of rainfall from around 4000 BC in a new climate pattern that lasted for over two millennia. Temperature and precipitation levels in the Tibetan plateau remained stable, however, as they did in Xinjiang.[110] Analysis of fossilised single-cell organisms and coral off the coasts of Indonesia, Papua New Guinea and the Great Barrier Reef suggests cooling sea surface temperatures – which together with evidence for lower sea levels around 3500–3200 BC provide further indications that this was a period of considerable climate shift in some parts of the world, if not in all.[111]

How human populations reacted and responded to the challenges is not immediately clear, and nor should it be. As we shall see, extreme events could have catastrophic consequences and sear the experiences into collective memories. However, on a day-to-day, month-to-month and year-to-year basis, even very substantial alterations to weather patterns can hardly have been discernible. Decisions to move or migrate, to adopt new habits and to engage differently with each other and with the environment would not have been instantaneous or immediate, but reached or developed in the face of continual incremental change.

Likewise, ecological shifts brought about by changing rainfall, by temperature differences, by transformations of habitats for flora and fauna were not rapid either, but spread over decades and centuries. As such, then, it is more productive to think about how societies started to conceive of the world around them, of how people thought about the human relationship with nature, with animal and plant life, and how they conceptualised the roles of our species relative not only to changing climatic circumstances but to the wider question of resource exploitation.

Our ability to make sense of such questions becomes a lot easier from c.3500 BC because of the growing complexity of human societies that in turn demanded new ideas, new solutions and new tools. Changing climates did not create the need for political systems, pave the way for the rise of towns, cities or states or lead to the development of writing systems. All were the product of rising population numbers, greater demands on water and food resources in particular and the need for social organisation. And yet the role of the environment was central to all. If societies were going to survive and flourish, humans would need not only to dominate nature, but to bend it to their will.

4

The First Cities and Trade Networks
(c.3500–c.2500 BC)

He is the man who built Uruk.

Sumerian King List (c. 2100 BC)

By around 3500 BC, the human impact on the environment was not just changing but was becoming so significant that it had itself become a factor in changing ecological habitats of flora and fauna. Spatial modelling suggests that more and more land that was suitable for agriculture was being exploited around the world, with pastoralism also spreading further into more arid regions. For some scholars, this level of activity represents nothing less than the start of the anthropogenic transformation of the earth.[1]

Indeed, some have gone further, arguing that human activities had an impact on the climate itself. For one thing, recent studies have suggested that population levels in China and Europe were much higher around 5,000 years ago than is usually assumed, raising questions about how much land was needed to support such substantial numbers of people, as well as about the ways in which land was used.[2] Climate records show that this was a time of significant additions of heat-trapping CO_2 to the atmosphere, which some scholars have also linked to deforestation.[3] The signature of high levels of methane found in ice cores in Greenland dating back to c.3000 BC meanwhile has likewise been associated with intensifications in wet-rice farming that lead to a complex soil process involving plant–microbe interactions and an increase in emissions.[4]

This has led to the formulation of what has become known as the early anthropogenic hypothesis, which posits that by around 5,000 years ago human activities and behaviour had so great an impact on global climate as fundamentally to change it.[5] Recent modelling has been used to support this argument, with the authors suggesting that a new glacial period was averted because of the warming generated by higher than expected greenhouse-gas concentrations in the atmosphere, with farming considered the most likely cause of these emissions.[6]

The problem with such hypotheses is that differentiating between correlation and causation is not straightforward – that is to say, identifying broadly simultaneous warming patterns on the one hand and demographic and behavioural changes on the other is not the same as proving that the latter was responsible for the former. Linking the two can be tempting, but it is fiendishly difficult to prove that the association was anything more than coincidental.

For example, changing climate has been suggested as a major factor (if not the prime cause) in the downturn of settlements of Trypillia culture that were spread out over the region between the Carpathians and the Dniepr in eastern Europe and which had become both more numerous and larger from around 5000 BC. Some of these became so substantial in size that they have been described as megasites, and deserving of the title the 'earliest known cities'. Often regarded as the largest sites in Eurasia and the world, settlements were typically centred on large buildings that were used for ritual activities and for the storage and consumption of surplus. Each site was part of a major trading network that involved exchanges of flint, manganese, copper and salt, sometimes over long distances.[7] Day-to-day needs were satisfied by a combination of cereals, domesticated and wild animals and dairy products.[8]

By around 3500 BC, these sites showed signs of stress, with buildings not being maintained, lower levels of occupation and the emergence of social hierarchies that have been interpreted as signs of social inequality that stemmed from diminishing availability of resources.[9] As it happens, these pressures are attested not only by lower levels of pollen – which indicate lower levels of agricultural productivity – but also by colder temperatures, which some have identified as a key factor in forcing population dispersal.[10]

It may be that demographic decline was precipitated not by climatic change, however, but by other factors, including the exhaustion of the soil through overexploitation.[11] Or perhaps the explanation lies in disease, with high population densities, combined with the large-scale presence of animals, providing ideal conditions for the emergence and spread of infectious disease.[12] It is perhaps no coincidence, therefore, that new research has found unambiguous evidence of *Yersinia pestis*, the aetiological agent of plague, in the Eurasian steppe around this time – as well as multiple lineages branching out between c.4000 and c.3000 BC.[13]

The possibility of multiple causes of demographic decline, depopulation and societal change would go some way to explaining why most large settlements in central Europe disappeared around 3300 BC, with evidence pointing to a compressed timeframe for catastrophic change.[14] The fact that this was a time of growing uncertainty would also help explain an intensification of ritual practices at many locations in Europe, particularly since the changes in social behaviour are widely thought to be reflections of increasingly challenging circumstances.[15]

These were hardly the only places in the world that were going through rapid and major societal transitions. The first signals of growing links between populations, material cultures and languages in southern China, Taiwan, what is now the Philippines, Indonesia, Melanesia and the Pacific begin around 3000 BC. This is often seen as a process of expansion from north to south, albeit one that should be measured across many centuries, often involving indigenous peoples adopting and spreading customs, ideas and technologies, rather than a sudden and rapid surge.[16]

It was a similar story on the Atlantic and Gulf coasts of North America, where around 3000 BC communities began to create circular middens known as shell rings from faunal and floral remains, centred on a plaza.[17] While many interpretations have been offered for these structures, they were evidently central features in settlements, and, as recent evidence suggests, focal points for local trade and within networks that stretched for hundreds of kilometres.[18] Surveys using remote sensing and machine learning have now revealed more such rings in coastal forests and marshes that suggest the existence of a larger number of communities, as well as even more intensive patterns of exchange starting to build up around this time than had previously been thought.[19]

Although questions about the early dispersal of Transeurasian languages, including Japanese, Korean, Tungusic, Mongolic and Turkic, are contested, the triangulation of genetics, linguistics and archaeology points convincingly to the conclusion that common agropastoral and basic vocabularies were spread by farmers and the associated expansion of farming across north-east and eastern Eurasia in this period.[20]

Europe also saw waves of mass movement of the peoples of the Yamnaya culture from the forest-steppe region above the Black Sea to the west. This had a profound demographic impact but also brought about sharp cultural change, again including the spread of languages.[21] Those coming from the Pontic–Caspian steppe in relatively small numbers settled on islands in the Aegean and in what is now Greece, where they built the first monumental palaces and urban centres in Europe.[22]

The expansion of new pottery styles identified as belonging to peoples of the Corded Ware and Bell Beaker cultures spread rapidly from c.2750 BC, covering much of Europe and North-West Africa in a matter of two centuries. What was even more striking, however, was the population turnover that also took place: in what amounts to one of the most comprehensive movements of people ever recorded, mitochondrial genome data shows the almost complete replacement of the gene pool of Europe.[23] In the case of Britain alone, it is estimated that the population turnover was a minimum 90 per cent.[24] In parts of central Europe, it was 100 per cent.[25] Emulation of culture was important, but migration much more so.[26]

In Mesopotamia, the Nile Valley, the Yellow and Yangtze River regions in China, the Indus Valley in South Asia and the valleys of Andes, however, the story was one not of population dispersal but of accretion. Although it is true that the period c.3500–3000 BC saw demographic growth generally, more significant was the expansion of the numbers of villages as well as the rising numbers of those who lived in them. While people were being pushed out of permanent and semi-permanent settlements north of the Black Sea, in other places they were being pushed together. The fact that this happened in many locations at around the same time suggests that similar impulses drove similar trends. So too does the speed of a process that gave rise not only to large numbers of villages, but to the rapid expansion in size of some of these into substantial urban centres.[27]

As one scholar has observed, while it is important to note where cities emerged, it is also worth paying attention to where they did *not* do so – above all in western Europe, the Amazon basin and eastern North America. That the locations where large conurbations did start had different soil types, drainage, rainfall, temperature and even altitude means that it was not coincidental that they all had one thing in common: these early cities rose in 'circumscribed land', that is to say locations that were locally bountiful but hemmed in by hostile geographic features such as deserts, mountains and the sea.[28] As such, the motors that drove the rise of cities – and therefore of 'civilisation' – were powered by pressures that forced populations into narrow bands of environmentally hospitable and productive land, where the ability to expand ecological footprints were limited. Early cities rose, in other words, as products of necessity where co-operation was essential to prospects of survival.

Increasing levels of local and regional trade within and between these hotspots paved the way for the consolidation of social interaction and for the centralisation of political authority. From around 4000 BC, the homogenisation of styles across the Fertile Crescent, such as in pottery and settlement plans, provides a strong indication that commercial and cultural exchanges were rising steadily – a process that finds a parallel in the regions of the Black Sea, Mediterranean, Aegean and Anatolian plateau at the same time.[29] This was accompanied and perhaps even spurred on by the development of sophisticated metallurgical industries that had already spread from Anatolia to Mesopotamia, Iran, Pakistan and south-east Europe and to the Black Sea – and later to Sicily and the Iberian peninsula. Along with the introduction of new crops that diversified diets, the emergence of new commodities and materials that could themselves be traded, either as staple foods or as luxury, exotic items, opened up possibilities for new ways of life and also for the expression of status.[30]

These trends can also be found in East Asia, South Asia, North Africa and South America – in the Hongshou cultures along the River Liao, the Dawenkou cultures of the Lower Yellow River valley and groups in the Indus Valley, along the Nile and in the Norte Chico ('Little North') of coastal Peru. All also saw similar patterns of demographic growth, agricultural improvements, greater crop diffusion and the creation of long-distance trade networks and the spread of ideas and technologies such as those related to metallurgy – following similar trajectories to Mesopotamia, albeit at a later date.[31]

The establishment of permanent settlements required that ideas take shape about personal possessions, including movable and non-movable goods, and about access to and control over land and the resources it could yield. The development and existence of social hierarchies in the ancient as well as the contemporary world are often associated with urban settings and above all with property ownership – whether of fields, crops, animals or goods – whose relevance is enhanced by higher population densities: wealth accumulation and transmission enable the creation of social elites and therefore shape political structures and decision-making. Wealth disparities became a signature of the populations that urbanised earliest and most intensively.[32]

That is not to say that equality is a hallmark of non-urban societies. Indeed, recent research on four present-day pastoral groups in East Africa, West Africa and South-West Asia found substantial levels of intergenerational transmission of assets as well as status.[33] Even though the effect is more modest, life chances are measurably influenced by intergenerational wealth transmission in hunter-gatherer populations too.[34] Nevertheless, and taking into account substantial differences between the pre-modern and modern worlds, the rise of cities spurred a series of dramatic changes, involving a range of issues from kingship to religion, from the rise of bureaucracies to ideas about slavery. The rise of cities also went hand in hand with ideas about civic identity and with the emergence of 'states', a term that needs to be understood in very broad terms because of the obvious problems of over-simplification.[35]

In northern parts of Mesopotamia, curiously, some of the largest settlements had grown up around 3000 BC not in better-watered areas but mostly away from the major rivers in a series of alluvial plains and basins. These communities appear to have been egalitarian and communally oriented, with no centralised leadership and little evidence of the marking of status.[36] Some of these settlements were large, such as at Brak in what is now Syria, which grew to cover as much as 130 hectares, although how people lived, how many the town and land supported and how these changed over time is unclear.[37]

It was a different story in southern Mesopotamia where larger, more complex and more hierarchical settlements began to build up. The earliest and most significant of these was Uruk, although stories written later in Sumerian, as well as in the Old Testament, offered competing, retrospective claims for locations ranging from Nippur to Eridu to Babylon.[38] Much later accounts identify Uruk's founder as Enmerkar, who was said to be the son of the sun god Utu and of the good cow Ninsumuna; he was 'the man who built Uruk' according to the *Sumerian King List*.[39] Above all, however, Enmerkar's achievements were not architectural but political – revolving around his success in taking up a leadership role and being able to direct others and also to take credit for the outcomes. Enmerkar and his wife Enmerkar-zi, reads a text written around 3,000 years ago, 'know how to build towns, make bricks and brick pavements'. They knew too how to make wooden ploughs and yokes, cables and threshing sledges. And they knew how to make 'irrigation canals and all kinds of irrigation ditches'.[40]

Whether Enmerkar's skill set was quite as wide as this source claims is open to question. But that they are attributed to him – and also, strikingly, to his wife – is revealing for what it says about the emergence of elites in places like Uruk and other major urban centres. Some scholars see the role played by those able to acquire high status and wealth for themselves as crucial in the process of cities becoming larger and more efficient. The ruling class who dominated land ownership, owned herds and controlled production were able to provide incentives as well as to use coercion to accumulate more for themselves, producing ambitions and efficiencies that shaped the physical setting of cities and determined socio-political structures.[41] The results can be seen in the first instance in the construction of temples and in lavishly decorated buildings that are sometimes referred to as palaces.[42] It is significant that these temples were controlled by priestly hierarchies who concentrated power in their own hands while also regulating, monitoring and even controlling trade.[43]

Again, this finds obvious echoes elsewhere, not least in Egypt, where the same dual principles of secular and priestly authority evolved and likewise built up monopolies of power by a ruling class that dominated land ownership and affirmed their status through the erection of temples, stone buildings and elaborate burial monuments – most famously, the pyramids. In the case of Egypt and the Nile Valley, by 3500 BC ideas about kingship emerged in tandem with urban centres at Naqada, Hierakonpolis, Abydos and elsewhere, which in due course became consolidated with one ruler overseeing multiple regions and cities.[44] The process of centralisation may have been helped by demographic growth and perhaps also by environmental factors: either way, pressure on available land supply led to intensification in its exploitation, which helps explain the evolution of a more unified state structure.[45]

A parallel process took place across multiple chronologically overlapping cultures in what is now China, where dispersed, politically decentralised cultures such as the Yangshao faded and were replaced by the Liangzhu, Longshan and Hongshan in what is now respectively south-east, central and north-east China, that were dominated by theocratic elites operating from increasingly large urban centres. A prime example was Liangzhu city itself, centred on a palace complex and protected by large walls, where burial tombs show social differentiation and rising levels of stratification, with the wealthy being buried alongside rare, exotic and valuable objects that were connected with ritual and hence imply that their owners had a status that was not based simply on wealth.[46] In Liangzhu, as well as in the many other cultures in East Asia at this time, jade played a uniquely important role as a rare and valuable material that served as a marker of wealth and status.[47]

The East Asian examples were unusual too, as the belief systems that developed were very different to those that dominated in Egypt, Mesopotamia and (later) South Asia. At the apex of the social pyramid in all these cases were rulers who enforced their position by claiming a connection with a supernatural world that often gave them unique genealogical pedigrees that cemented their authority as well as being a source of divine powers. Mesopotamian kings derived power from being outsiders and found advantages in being 'strangers in their own lands,' for example, while the rulers of the Chu state in what is now northern Jiangsu and Anhui province in China traced their lineage to a deity, the god of fire, Zhurong (祝融), who was said to control the water spirit and ensure fertile soils and good harvests.[48] This was part of a cycle that is familiar in other regions and periods and one that some anthropologists argue is linked to the growing complexity of societies on the one hand and the origins of religion on the other.[49]

In this model, as societies become larger in size and more specialised in their work, rulers and priests become the interpreters of everything from natural disaster to environmental challenges, from resource surpluses to shortfalls, from military defeats to premature deaths, helping explain punishments or bounties that were being administered by unseen gods. Environmental and natural calamities in particular were closely linked to 'moralising gods' who, out of anger or simply from boredom, handed out punishments for transgressions and apparent lack of respect. It is striking, though perhaps not surprising, that regions that were vulnerable to changes in weather conditions – above all droughts, but also floods and storms – developed cosmological systems based on 'moralising gods' who used such events to punish, show their displeasure and teach lessons.[50]

As we shall see, deities who were prone to infighting or unleashing disasters on humans played an important role in Babylonian and Egyptian literature and theology, as they did later in post-Harappan India and in the Mediterranean: it was claimed that Greek gods had provoked wars, had petty vendettas with each other (and with individual humans) and rained down misfortune on those who dishonoured them while showering others with rewards. In East Asia, however, things were rather different, as is clear from both literature and religious practice, where the primary themes are not of destruction and punishment, but altogether more abstract and pacific. Perhaps one reason why the early Chinese did not quarrel with their gods as others did elsewhere, one scholar has argued, was 'because there was less to quarrel about, ecologically and environmentally', thanks to climate conditions being more benevolent and predictable. And perhaps even distinctive ideas about ancestral status were shaped by the cold climates of north China, where the long period required for the decay of the body enabled extended pre-burial rituals – and more opportunities for priests and those connected to curating ideas about belief systems to emphasise their own importance.[51]

The use of temples, sacred spaces and special treatment in death for those of high status to affirm links with the divine and to assert a relationship between elites and supernatural beings evidently served as a tool of control and as a marker of social stratification. Such barriers were reinforced, not only in Mesopotamia, Egypt, the Indus, East Asia and Mesoamerica but all over the world (albeit in different time periods that corresponded to locally arising complexity) by the use of human sacrifice. The deliberate and ritual killing of an individual to placate supernatural beings is known from early Arab, Turkic, Inuit, American, Austronesian, Chinese and Japanese cultures, as well as from Mesopotamia. While many suggestions have been offered to explain the purpose and aims of human sacrifice, perhaps the most compelling is that such acts legitimise the authority of rulers by demonstrating their power over the population and also their exclusive ability to please the gods.[52]

This was emphasised not only in life but also in death. Dead rulers were often presented with offerings, such as Uranamma, who received daily offerings in Nippur.[53] Elite funerals also played an important role, with lavish ceremonies accompanied by music and feasting accentuating the status of those of high social class and wealth.[54] The Royal Tombs at Ur in southern Mesopotamia that were first excavated in the 1920s and 1930s yielded thousands of skeletons, including some of young retainers who had been buried with their prestigious kings and queens around the middle of the third millennium BC. They had been struck with an instrument with a small acute end that must have been hit with enough force to penetrate the skull. CT scans show that the attendants' bodies had been preserved so that they could 'continue' their service in the next world – as does the fact that they had been dressed and arranged around the corpses of their masters and mistresses.[55]

While there are ambiguities about how far we can generalise about social hierarchies from one place to the next and also across different periods, what is clearer is that the scale and positioning of public buildings testify not only to central planning but to the effective mobilisation of mass labour.[56] The most celebrated example is undoubtedly the construction of the Great Pyramid at Giza, which involved at least 10,000 men working for thirty years – something that presented considerable logistical challenges, not least that of feeding the workforce every day.[57]

Elite control of the infrastructure of new cities was such that workers neither owned nor had rights over agricultural lands that supported the populations. Instead, as later Mesopotamian texts attest, they depended on institutions, temples and the wealthy for compensation in the form of rations, or by way of a share of the produce they were responsible for.[58] Cuneiform texts from cities like Lagash, Nippur, Ur and elsewhere record distributions of barley and other cereals to workers of all kinds – including shepherds and weavers, agricultural workers and brewers.[59]

The scale of landholdings of powerful institutions and individuals grew to astonishing proportions. By around 2300 BC, some temples are attested as controlling thousands and even tens of thousands of hectares of land, with the crown overseeing many times that size in multiple regions across Mesopotamia.[60] In this respect, what mattered was control of workforces and the ability to direct and monopolise their labour. That perhaps explains why food production, distribution and even cooking were centralised, with ovens and hearths located not in houses but in temples.[61] It might also help explain why there are few signs of settled rural populations living near the city: it would seem that if the workforce lived on the land, they did so on a temporary basis.[62]

One way to maintain control therefore may have been to build walls around cities. It has been suggested that at least in some cases early city walls were built to protect against environmental factors and above all water and flooding, though they later assumed the role of serving as defensive fortifications.[63] As has been pointed out recently, the scale of city walls at locations like Uruk was far greater than any conceivable military needs might have required, raising the question of whether they were built not as fortifications against attacks but as symbolic expressions of power in their own right.[64] This has a natural parallel in Egypt, where investment in and expenditure on military forces appear to be a reflection of royal prestige rather than related to defence at a time when external threats were limited.[65]

Another suggestion, which might not be mutually exclusive, is that walls served to curtail freedom of movement to ensure the labour supply was constant: this, after all, was one way to guarantee that there were enough people to maintain the grain supply and therefore to secure the long-term future of the city itself. Cities depended on food supply; as such, therefore, the availability of a large enough workforce to exploit the land was not just crucial to their short-, mid- and long-term success but essential to their basic viability.[66] Ancient cities in Mesopotamia, and presumably elsewhere, were likely characterised by more deaths than births because of disease prevalence and lack of sanitation. As such, constant streams of new arrivals were needed for urban populations to stay stable, let alone grow.[67]

In practice, this meant attracting inflows of migrants, as these were important to the maintenance of urban populations, and essential for their expansion.[68] Extensive evidence of slavery in southern Mesopotamia suggests that not everything was left to chance, with forced labour being used to support irrigation systems and, in the case of women, to work on textile production.[69] All this meant that by around 3000 BC cities like Nippur, Adab, Kish, Ur and Umma were all home to substantial populations – although none was as large as Uruk, which covered 250 hectares and was home to somewhere between 20,000 and 40,000 people.[70]

Higher levels of regional exchange, coupled with demographic increases and rising levels of production that were facilitated by new tools, new techniques and the development of irrigation canals, demanded higher levels of administrative oversight and competence by emerging cities and states. This created a model of the city or the state as a parasite, where growth was powered by the labour force and the benefits harvested by an elite who set up and maintained barriers to cement their own positions and restrict access at the same time.

The purpose of an emergent bureaucracy was to standardise, homogenise and harmonise – and, in doing so, drive ever greater rewards to the centre. This required institutions and tools that would make it possible to measure and account for productivity and to monitor goods and labour in ways that prevented leakage from the system. One obvious example was the standardisation of weights and measures; another was the expansion of administrative norms into neighbouring towns and settlements. But the most obvious and important innovation was in the development of writing systems that emerged around this time in the Fertile Crescent.

Clay tokens representing units of goods had begun to be used at the start of the Holocene in Mesopotamia for basic accounting purposes. By around 3500 BC, cylinder seals were being used to mark bales of goods as well as other produce.[71] Two centuries or so later, these were being replaced by two-dimensional pictographic signs, inscribed by a stylus on clay tablets, that recorded numerals and goods; the signs developed over a few centuries into a writing system of phonetic signs that started with the transcription of the names of individuals before expanding further.[72] These were stages in a process that led not only to the creation of alphabets but to the opening of new horizons in which uses of language and communication widened dramatically.[73]

Events, stories, thoughts and myths could now be written down, allowing them to be disseminated widely and passed on in a standardised form. But these breakthroughs all traced back to needs that were much more mundane and unromantic: the desire to record goods and trade; the drive to charge and collect tax; and the efforts by elites to monopolise exchange networks and their hold over power. We have the needs of tax collectors 5,000 years ago to thank for the great works of literature that have been produced since: without accountants there would perhaps have been no Torah, Bible or Qur'ān, no Shakespeare, no *Bābur-Nāma* and no Tolstoy.

Nor would there have been any Babylonian epics – combinations of myths, stories and histories that relate to this early period of history in Mesopotamia – or indeed literatures from other parts of the world where the same processes of urbanisation along with the emergence of hierarchies and priesthoods, of political leadership and of bureaucracies took place. Societies in Mesoamerica and in the Nile, Indus Valley and Yellow and Yangtze River regions in China – and in other regions too – did not pass the same staging posts at the same time as southern Mesopotamia or as each other when it came to developing writing systems; but the pathways were similar in each case, even if the Indus scripts have yet to be interpreted or fully understood.[74]

Elites elsewhere also developed in much the same way as in Mesopotamia, with the ruling class in Egypt, for example, monopolising the labour of the rural peasantry and retaining power within a self-selecting priestly class.[75] Bureaucracies emerged that became obsessive compilers of information, with endless lists recording the names of towns, peoples, foods and cereals – just as they did in Mesopotamia.[76] The desire to tie the labour force to the land to ensure high levels of agricultural production was a core focus of officials in Qin dynasty China, with registers being maintained to prevent peasant mobility and to tie workers to the land.[77]

By c.2500 BC, cities in southern Mesopotamia like Uruk, Eridu, Umma, Uqair and Ur had blossomed and grown, as had urban centres like Naqada and Nekhen in Egypt, Mohenjo-daro and Harappa in the Indus Valley, and Liangzhu as well as Baodun and Yufucun in what is now Sichuan province, China. Urbanisation seems to have spurred change in neighbouring regions, leading to higher population densities and social stratification among nearby peoples – presumably as a reaction to opportunities presented by bigger markets with growing demands.

A good example is Nubia, where pastoralists began to settle around what became the city of Kerma in what is now Sudan and within a few centuries formed the centre of the powerful kingdom of Kush, which was a rival to the Middle Kingdom of Egypt.[78]

This was important, for higher population densities are also closely connected with the emergence of social stratification. This explains why things looked rather different in the Indus Valley, where there is less evidence of the consolidation of political power into the hands of a single ruler, dynasty or elite. In this region, there is also less evidence of the elaborate tombs and monumental architecture celebrating individuals and specific families, and generally a considerably lower footprint of temples and religious buildings – suggesting a different spiritual and social arc for the people of this region.[79] It seems, therefore, that the communities living in the alluvial plains of the Indus and its tributaries were more egalitarian than those living further west between the Tigris and Euphrates or along the Nile.[80]

This also appears to have been the case in the Norte Chico region in South America in what is now Peru. A rapid transformation around 3000 BC led to the first large, permanent settlements that were characterised by monumental public architecture and large ceremonial buildings.[81] Although it has been argued that coastal locations, supplied by bountiful marine resources, formed the catalyst for change and also the basis of expansion, it is now clear that the largest and most important towns were situated in valleys in the interior and in the highlands, with access to arable farmland.[82] The most significant of these was Caral, which may have served as an administrative centre controlling a wider region.[83] Similarities of layouts, storage areas and ritual structures across locations which in some cases were situated at considerable distances from each other suggest local exchange networks that began to create a common culture between villages and regions from c.3000 BC onwards.[84] New developments in irrigation and soil management and an expansion in production, processing and consumption of crops – especially maize – helped support larger settlements and spectacular demographic growth.[85] Some believe the population rose by a factor of thirty in the course of three centuries.[86]

Many urban settlements were centred on mounds that were important for celebrating ceremonials and sacred rites. Although only a small number of people appear to have been allowed on to these mounds (which might point to elite power and status), other evidence suggests that social stratification was weak.[87] What is most striking, however, is that there are no signs of warfare or violence in the regions that made up the Norte Chico: all the sites that have been identified in a mosaic of settlements are not only situated in 'distinctly non-defensive locations', but also lack fortifications, walls and other pointers to suggest that conflict and confrontation were significant or regular occurrences.[88] One obvious explanation for this is that links between communities were weak, especially between different geographic and climatic settings, with limited exchange between those living on the coast, those in the lower valleys of the interior and those in the highlands.[89]

A key differential, however, also seems to lie in the population size as a whole and, more importantly, in settlement densities: as in the Indus Valley, Andean civilisations were spread out over much greater distances than Mesopotamia and Egypt. Even within villages and cities, space was at less of a premium than in other more ecologically pressed niches. As with the Trypillia megasites, then, the fact that a ruling class does not seem to have emerged in the same way in the Indus and the Andes as it did elsewhere may reflect a culture of plenty of production: greater quantities and qualities of land, and the absence of anxieties about food shortage, in turn meant that the need to protect assets and demonstrate status was simply not so acute.[90]

The correlation of status and hierarchy on the one hand and competition for resources and space on the other goes a long way to explaining competition between cities. This was especially the case in Mesopotamia, where cities were clustered closely together. Rivalries were sometimes fierce and protracted – with Lagash and Umma engaging in a border conflict that started around 2500 BC and lasted for 150 years.[91] The problem was not so much land as water. Unlike the Nile, whose floods were regular (even if not always constant), or northern Mesopotamia, where annual rainfall supported farming, agricultural production in southern Mesopotamia relied on the flows of the Tigris and Euphrates and on irrigation and canal construction. Access to prime land and water was what mattered.[92]

Also in short supply, however, were materials for tools, weapons and jewellery and luxury materials for temples, palaces and homes. Metals like copper and tin were rare, while even basic resources such as stone and timber were in short supply and hard to come by. Perhaps not surprisingly, sources of new minerals and other materials were usually 'revealed' by the gods to kings, who used such messages from the divine to send out expeditions to look for, exploit and bring back much needed supplies.[93] This sharpened animosities towards and rivalry with other cities, who had their own shortages to confront, with the result that struggles for control of locations which were rich in minerals and resources could be acute. That spurred socio-economic development and innovation.

It also led to the creation of networks of satellite settlements and colonies that controlled access to regions where metal and stone were abundant, where wool and raw materials for textiles could be acquired (or taken); these evidently also served as strategic outposts to protect the commercial interests of the mother cities and as gateways to long-distance trade networks that by 2500 BC extended not only to include the Nile, Anatolia and the Indus region but into the Caucasus and Mediterranean as well. Needless to say, these networks were not one-way, but instead enabled the spread of goods, ideas and technologies to join North Africa, the Middle East and at least some parts of South Asia into an interlinked system of exchange.[94]

They also helped shape ideas about kingship, partly because rulers sponsored trading expeditions (and therefore benefited most from their outcomes) and partly because the acquisition of exotic, rare and unusual materials, goods and objects allowed them to boost their own status – visually and otherwise – as well as providing rewards for retainers. Additionally, because gift exchange between rulers was an important part of an expression of power and of diplomatic alliance-building, what one scholar has called a 'brotherhood of kings' arose, helping expand rituals, develop new ceremonials and even elaborate the nature of kingship itself.[95] These contacts played an important role in driving higher levels of social stratification both at an abstract level and practically, with oversight and the benefits of trade in luxury items enabling the rise of elites in South-West Asia, in many parts of China and elsewhere.[96]

These processes helped fuel the rise of elites and therefore of states: locations such as southern Mesopotamia, which were well-suited for crops and pasturelands, offered opportunities for the production of surplus food and goods. This set a premium on controlling the labour force and on extracting profits, as this embedded and deepened hierarchies, rewarding those at the top of social pyramids. Additionally, these were protected and enhanced by writing systems and by the creation of import and export markets for luxury goods.[97]

Men were prime beneficiaries, for patriarchy was a core feature of states and empires in all periods of history. The realities and the glamourisation of violence, of warfare, all helped fuel what one scholar has called 'the triumph of aggressive cultures'. Group violence increased as societies became more complex, while rewards and prestige flowed to those who were proficient at combat or at leading others in battle. As states rose, violence, patriarchy and centralisation went hand-in-hand: and all had impacts on the environment - and on the role of women.[98]

The development of trading networks, acquisition of raw materials, tools and technologies and the increasing competence of central bureaucracies all produced remarkable efficiency gains which allowed cities to attract and support larger and larger populations. So too did improvements in agriculture and irrigation which opened up new lands for production in what must have seemed like a never-ending cycle of expansion that funded spectacular monuments ranging from ziggurats in Mesopotamia to pyramids in Egypt to bathhouses in the Indus Valley.

The trouble was that scale creates pressure. More people required more food, meaning either that land had to be worked harder, that marginal areas needed to be exploited or that the footprint of land being sown became larger and therefore further away from where its produce was consumed. Naturally, this created competition for the best lands not only between cities but also from within them. As social differentiation widened, so must consumption have changed, with the wealthy demanding greater qualities, quantities and ranges of goods and food – putting further pressure on ecological resources.

These were ideal conditions for political opportunism. But they were also the ideal conditions for environmental and ecological stress. It did not take much to expose the vulnerability of densely populated communities to episodes of fragility. These could be anthropogenic and result from expensive confrontations that depleted resources, took workers from the fields and reduced food supplies as fewer crops were sown or harvested. But they could also come as a result of climate events or environmental degradation. Agrarian economies were particularly dependent on rainfall and water availability – which are of course closely linked to climate patterns. When abundant, the great rivers such as the Tigris and Euphrates, the Yellow River, the Yangtze and the Indus and their tributaries meant that land could be cultivated easily. Scarcity on the other hand meant shortages, crop failure and famine, alongside associated risk of disease.[99]

It is perhaps no surprise, then, that some of the earliest ideas that were set down in writing were stories about the origins of the earth – and about the perfect conditions that had been created for humans. In the Sumerian creation story 'Enki and the World Order', the god Enki 'stood up full of lust like a rampant bull, lifted his penis, ejaculated and filled the Tigris with flowing water'.[100] The Babylonians had a similar if less vivid version. *Enūma Eliš*, often known as *The Epic of Creation*, discovered in the ruined library of Ashurbanipal at Nineveh in the nineteenth century, explains that at the start of time all that existed was water swirling in chaos. This divided into fresh, sweet water and bitter, salty water which ultimately turned into the source of the rivers Tigris and Euphrates.[101] Eventually, the gods decided to create the first man, Lullu, who was directed to serve them by maintaining order on earth. Nevertheless, it was the gods who were responsible for 'pasturage and watering places', ensuring there was 'pure ploughland in the countryside' and enough grain for all. It was the god Enibulu, in the end, 'who heaps up a surplus for people, who brings rain of abundance over the broad earth, and makes vegetation grow profusely'.[102]

It was a similar story in South Asia, where four great collections of hymns, incantations and ritual formulas known as the Vedas record the early histories and beliefs of the Aryan people in the sacred language of Sanskrit. The earliest and most famous are the *Rig Veda* – or 'Verses of Knowledge' – that were written down around 2,500 years ago, but record songs and poems dating perhaps a millennium further back in time. Among the collection is the story of how Indra, one of three main deities, released rains that had been held captive in clouds by Vritra, the demon of drought, and forced them to pour down on to the scorched earth. No longer bound by evil spells, 'the clouds discharge their liquid', leading rivers to swell and those working the land to rise with 'deep delight and thankful heart' that their leafless fields would soon be home to crops waving in the wind and that 'Mother Earth' would transform from a state of being 'brown and bare' to being clothed with 'a robe of brilliant green'.[103]

The story that is best known, however, is that of the Abrahamic tradition that is set out in the Book of Genesis. 'In the beginning when

God created heaven and the earth, the earth was a formless void and darkness covered the face of the deep.' After creating land, sea, vegetation and 'swarms of living creatures', God created humans, ordering that they should 'have dominion over the fish of the sea, and over the birds of the air, and over the cattle and over all the wild animals of the earth, and over every creeping thing that creeps upon the earth'.[104]

Man and woman were told to 'be fruitful and multiply, and fill the earth and subdue it'. Behold, said God, I have provided plants, fruit and food for you, and made sure that every beast on earth, every bird in the air and 'everything that has the breath of life' give you all you need in terms of food and nourishment. Things did not work out quite as intended, and rather than having a life where all was provided for, Adam and Eve and their descendants were cast out of the Garden of Eden, forced to sweat and to work the land, struggling against thorns and thistles.[105] The world, in its original form, was benign and generous. After the fall, it was difficult and dangerous. In Jewish, Christian and Muslim teaching, the earth was created by God for humans to enjoy; but their disobedience led to punishment – and ejection from a world of perfection in paradise to one of labour and of climatic and environmental risk.

Shifting weather patterns were one of a number of potential triggers that could have serious consequences for societies operating at the fringes of their ecological and agricultural sustainability. Moreover, the bigger the state, city or society, the greater the vulnerability at a time of unexpected risk, as they became dangerously exposed to even relatively modest pressure or confluence of pressures. The idea of corporations or states being 'too big to fail' is one that entered public consciousness during the financial crisis in the first decade of the twenty-first century. But, as history shows only too well, the past has been shaped by precisely the opposite happening: empires forming and then collapsing – usually at speed. Rather than being too big to fail, increasing size and complexity create their own momentum that can lead to a house of cards tumbling down, chaotically, quickly and decisively.

5

On the Risks of Living Beyond One's Means
(c.2500–c.2200 BC)

May your holy walls, to their highest point, resound with mourning!

'The Curse of Akkad' (c.2100 BC)

Around 2300 BC, the ruler of the city of Akkad (in Mesopotamia in modern Iraq), Sargon, embarked on a campaign of conquest, bringing a host of cities across Mesopotamia under his control and establishing what is usually referred to as an empire.[1] He was not modest in celebrating his achievements, leaving triumphant inscriptions recording the names of those he had defeated and proclaiming himself as 'king of the world'.[2] 'I cut through mighty mountains with copper pickaxes,' he boasted, climbed high mountains and sailed around the seas three times.[3] He had 'neither rival nor equal', according to the *Chronicle of the Early Kings*, a history written much later, and succeeded in dominating all his enemies and destroying places that defied him 'so that there was not even a perch for a bird left' standing.[4]

Sargon's conquests were not universally welcomed, as the accounts of the many revolts that he faced during his reign make clear. As usually happens with the expansion of states, the centralisation of power as well as of resources inevitably involved consolidation of trade routes that stimulated and facilitated exchange between locations but also led to dislocations of supply, with Sargon's relentless pulling down of city walls attesting to a deliberate policy of ensuring that potential rivals remained weak and also that goods, materials and labour were diverted to the centre of the empire. These funded new temple constructions that further validated the authority of the ruler and the elite – who also benefited from control over the distribution of agricultural production that saw millions of litres of crops like barley and emmer being dispatched by barge to regional centres favoured by the Akkadian leadership.[5]

By the time Sargon's grandson, Naram-Sin, took power two and a half decades after Sargon's death in 2253 BC, some cracks were already starting to show. One account talks of a major rebellion taking place when 'all four corners of the world together revolted against me', while another talks of the ruler losing divine favour, defying the gods and insulting them by attacking sacred sites.[6] He was also the subject of one of the famous texts of the ancient world which associated his reign with catastrophe.

Known as 'The Curse of Akkad', the account explains the punishment meted out by the deities because of Naram-Sin's insulting behaviour: for the first time since cities were built and founded, it says, 'the large arable tracts yielded no grain, the inundated tracts yielded no fish, the irrigated orchards yielded no syrup or wine, the thick clouds did not rain'. Crop failure led to price inflation in the markets of all the cities of the empire. Death was everywhere, leading to people not being buried. 'People were flailing at themselves from hunger.'[7]

Climate data seems to offer firm support for a hypothesis of an 'evaporation event' that resulted in drought that would have had a dramatic effect on areas that were ecologically sensitive. Sediments from the northern Red Sea, for example, indicate environmental changes around 2200 BC that have been connected to shifts in the North Atlantic Oscillation, solar variability or perhaps both.[8] Fossilised corals off the coast of Oman show prolonged winter dust-storm seasons which have been linked to likely agricultural failure in Mesopotamia.[9] A spike in magnesium concentrations in the stalagmite record of the Gol-e-Zard Cave in Iran, dateable very precisely using uranium-thorium chronology, testifies to the start of a long and difficult period of aridity that lasted for several centuries.[10]

Places like Tell Leilan in what is now north-eastern Syria that had previously been home to a prosperous and lively population saw a rapid decline in activity – which reflected dramatic upheaval elsewhere.[11] Increasingly difficult conditions resulted in land degradation, plummeting cereal yields and desertification that was closely linked to a process of rapid deurbanisation and abandonment of towns and large settlements as precipitation fell by 30–50 per cent. Building projects were stopped mid-construction, while there are signs that some locations were abandoned suddenly.[12]

The results were catastrophic, prompting the start of what one scholar has called a 'Dark Age'. Famine and economic breakdown were everywhere. Refugees fled from their homes and flooded into southern lowlands before the empire was overrun by fierce nomadic peoples known as the Gutians, described by one near contemporary source as 'an unbridled people with human intelligence but canine instincts and monkeys' features: like small birds they swooped on the ground in great flocks. Nothing escaped their clutches.'[13] Political disintegration and chaos followed swiftly. Changes in climate had brought about nothing less than the collapse of the Akkadian empire.[14]

For some, the collapse of the Akkadian empire has become an important and salutary example for the modern age. Living in the shadow of impending environmental catastrophe, it serves as a stark warning of how mighty civilisations can fold in on themselves quickly and comprehensively. Indeed, it has become so embedded in public and scholarly consciousness that geologists have chosen 2200 BC to serve as a boundary moment between distinct geological periods: pollen, diatom, testate amoebae assemblages and other indicators from all seven continents point to aridity and drought conditions setting in around this time.[15]

This moment has even been named as the start of the Meghalayan Age, named after a cave in north-eastern India where the change in oxygen atom isotopes is particularly revealing of a decrease in monsoon rainfall. According to the International Commission on Stratigraphy, climatic changes around 2200 BC provoked a megadrought that caused the collapse of civilisations not only in Mesopotamia but in many other places too, such as in Egypt, Greece, Syria, Palestine, the Indus Valley and the Yangtze River Valley. As such, the Commission stated, 2200 BC was a defining moment not only in geological history but in history itself.[16]

In fact, however, there is considerable uncertainty about global climate uniformity, its potential causes and its effects. While multiple studies of North and South America, as well as of Europe, Egypt,

East Africa, Mesopotamia, the Indus Valley and several regions in what is now China, suggest that conditions became unusually dry in many parts of the world around the same time, the process of aridity was not uniform either geographically or chronologically.[17] Its causes too are disputed and by no means clear. While many studies have linked declining precipitation to polar ice-raft release and subsequent falling sea surface temperatures in the North Atlantic, recent research suggests that the relationship was minimal, if it existed at all.[18] Another explanation might be provided by a model of amplified feedbacks caused by vegetation changes and atmospheric dust loads, caused above all by the transition from a green to a drying Sahara: this appears to have driven a redistribution of moisture away from northern Africa, the Middle East and northern China and even had an impact on the El Niño–Southern Oscillation.[19]

Establishing a single date as a significant boundary and talking of it as an 'event' can be a useful short cut inviting consideration of wider conclusions. Doing so hides the complexity of changes, and it also obscures the fact that shifts in climate patterns took place over decades and centuries, rather than in a sudden, dramatic moment – as did the consequences. As a result, some have warned that the desire to identify patterns of 'collapse' around 2200 BC may influence both the choice and interpretation of data, including cases where evidence is limited, absent or contradictory.[20] That the archaeological sources are not conclusive nor emphatic in many locations presents obvious similar problems – with some cautioning that, if there really was a new era, it is attested by a whimper rather than a bang.[21]

Ideas of societal or civilisational 'collapse' can easily oversimplify the timing, nature and reality of what actually took place. In the southern half of the Iberian peninsula, for example, rather than being a time of decline, it was one of social and economic transformation, driven by high levels of genomic turnover that are indicative of migration, as well as being a period of sharp rises in mobility and regional connectivity that saw increases in exchange with Sicily and with Africa.[22]

In the Indus Valley, meanwhile, the archaeological evidence suggests that cities continued to function and even grow with little or no problem for some 300 years after their predesignated catastrophic collapse. The longevity of cities like Harappa and Mohenjo-daro from 2200 to 1900 BC have puzzled some scholars who expected to see higher levels of stress, economic and social pressure and civic breakdown long before the first clear signs of these began to become apparent.[23]

Despite the widespread assumption that Indus urban settlements had developed in tandem with a glacial-fed Himalayan river (often identified as the mythical Sarasvati), in fact they had blossomed in an abandoned river valley and were sustained by monsoonal rains.[24] The weakening of the monsoon substantially changed the dynamics of agriculture and of population distributions, as rivers gradually dried or became seasonal.[25] Changing weather patterns were nuanced, producing more erratic and less extensive precipitation during the summer, but higher levels of winter rain. This led to a series of adaptations, including settlement downsizing as well as crop diversification and a shift towards traditional winter crops like wheat and barley.[26] Far from collapsing, then, challenging climatic conditions resulted in the development of coping strategies and adaptation.

Another response that took place over time was a significant relocation from flood-deficient valleys to the more reliable plains of the Himalayan foothills – where the result was population growth and higher urban densities. Demographic expansion was both rapid and unplanned, resulting in increasingly disorganised settlement plans. One outcome was a marked deterioration in sanitation conditions, which in turn provided a perfect context for infectious diseases to spread. These included mycobacterial tuberculosis, which had been circulating in South Asia populations for millennia, as well as leprosy, which is attested to by skeletal records from this period. Strikingly, though perhaps not surprisingly, disease prevalence was highest in communities that were socially and economically marginalised.[27]

The pressures of changing demographics, of closer and dirtier living conditions and of the spread of infectious disease and declining health proved a heady cocktail – perhaps made worse by a second wave of rainfall reduction.[28] By around 2000 BC, evidence of mangrove shells and other archaeological indicators at the well-excavated site of Dholavira in India suggest much thinner habitation levels, including signs pointing to a lack of new buildings and of maintenance, as well as a noticeable deterioration in the quality of craftsmanship.[29] Within a century, rising levels of interpersonal violence can be detected, alongside population dispersal to the south and the collapse of Indus Valley civilisations.[30] Climate certainly played a role in this process, and perhaps was even a catalyst to change; but it was one of many factors in a complex picture that requires a carefully balanced explanation.

In the case of the different Chinese cultures, again much is in the eye of the beholder as we try to judge what the effects might have been of climate variations. There are significant regional imbalances in the work that has been done in recent years to assess what these variations might have been – to say nothing of gauging their potential impacts. It seems, for example, that there were very different circumstances to the north and to the south of a boundary broadly stretching from the Qinling mountains to the lower reaches of the Yangtze River.[31]

If assessing regional variations is problematic, so too is linking real or imagined change across and within different regions in East Asia. One recent hypothesis has used speleothem and other palaeoclimatic and archaeological data to show that a two-decade period around 2300 BC brought about such high levels of rain that the Liangzhu culture of the Yangtze River Delta was engulfed by floods, and that extensive hydraulic complexes such as large earthen dams that had offered effective defences for flood mitigation and/or irrigation were overwhelmed. The result was the inundation of low-lying land, thus impeding rice farming especially on the Taihu Plain region of the Lower Yangtze Delta, and causing Liangzhu city to fail and its population to disperse.[32]

According to traditional Chinese chronologies, the rise of the Xia dynasty is often dated to c.2070 BC and linked to a devastating flood which later historians state resulted in the Yellow River being dredged by the dynasty's founder, Yu 'the Great', thereby protecting the people from being affected in the same way in the future. Recent research suggests at least one major flood around 1920 BC, likely caused by the collapse of a dam following an earthquake.[33] Several studies have sought to associate the rise of the Xia with management of river flooding and with a transition towards drier conditions that are apparent from stable isotope records from caves across precisely the regions where a new power centre emerged in this period.[34]

Whether this really provided the context for consolidation of lands and peoples into a single entity, and if it did, why it did so, is not clear. Likewise, while the Xia dynasty – likely associated with the Erlitou culture and centred on a newly founded city at Xinzhai (新砦) – seems to have brought an end to regional competition and constant warfare, correlating this process with changing climate conditions is controversial, not least since most scholars reckon the earliest date for Erlitou to be c.1750 BC.[35]

While there is evidence of contractions within advanced agricultural societies such as the Qijia culture of the Gansu–Qinghai region and the Laohushan–Dakou culture of central Inner Mongolia around 2000 BC, it is by no means clear that there were major changes across many parts of what is now China as a result of changing climate. Indeed, there is little to suggest that c.2200 BC marked a watershed moment or period in East Asia.[36]

In this sense, it might then be asked why other previous and lengthy periods of drought had not brought about similarly damaging effects – and whether it is right to ascribe so fundamental a role to drought on this occasion. It is important to note that the same climate records attest to a similarly arid spell lasting many decades around three centuries earlier, and that the cities and polities of Mesopotamia and elsewhere were able somehow to navigate the challenges more successfully at that time: indeed, one of the studies that identified a spike in magnesium concentrations around 2200 BC found a similar surge 250 years earlier which did not spark socio-economic or political chaos.[37] Rather than marking a climatic or geological boundary, in other words, the significance of a date of c.2200 BC seems to have become magnified by apparent and supposed societal and social consequences, rather than by a more sober analysis of realities on the ground.

In any event, it is not clear what 'collapse' at that date means: while it can be tempting to focus on large-scale architecture, symbols of royal power and the powers of the bureaucracy and to correlate their disappearance and reduction with societal failure, it is perhaps more useful to question how life changed in practice for agricultural workers, women, children and families if this really was a time of profound change. Poorly maintained buildings or dispersal from cities may have been bad news for those at the political centre, whose power, authority and wealth lay in the ability to control trade, religion, status and labour, but it is less obvious that there were negative outcomes for everyone. Indeed, for many, being freed from obligations may even have represented a liberation of sorts. Lower availability of different textiles, or of rare and tastier foods, for example, that meant a narrowing of dietary habits may have been disappointing, but that did not spell the end of the world.

That the implosion of a central administration, the disappearance of elites and of the symbols of their authority, had different consequences for different parts of society of course says much about how commentators and historians think about the world around them, both today and 4,000 years ago – and about the way that structures are so often looked at from the top down rather than the bottom up. In fact, as some have argued, even declines and contractions (to say nothing of purported collapses) are more interesting as examples of resilience and resistance than as signs of cataclysmic disaster for which there is often little evidence.[38] So while deurbanisation and population resettlement might at first glance appear to be negative indicators of fortunes, they can be more productively seen as logical and effective solutions to problems of under-supply and over-demand – and to the failure of centralised governments to overcome bumps in the road.

This is a moot point when it comes to the Akkadian empire specifically. Although Naram-Sin's reign is linked with catastrophe because of 'The Curse of Akkad', in fact it was also a time of territorial expansion resulting in the conquests of Armanum and Ebla, an achievement commemorated in an inscription on a monument set up in the city of Ur which proclaimed that Naram-Sin had done something that no other king had been able to do 'since the creation of mankind'.[39] Naram-Sin took to calling himself the King of the Four Corners of the World, hardly an immodest title to choose for someone who was supposed to have presided over disaster and societal failure.[40] In fact, modern scholars of Mesopotamian history now argue that military and administrative reforms under Naram-Sin were what transformed Akkad from a successful kingdom into an empire.[41] Far from overseeing collapse, in other words, the subject of the Curse of Akkad oversaw a strengthening of the imperial centre.

The attribution to Naram-Sin of disaster and the motivation behind the conception of the Curse reflect later Sumerian desires to draw lessons that were less about environmental stress and sudden climate change than about the nature of kingship and above all about the relationship between rulers and the gods. What was important to those who read or heard this story in later centuries was that dishonouring the gods – as Naram-Sin was alleged to have done when he desecrated the Ekur temple in Nippur and had himself declared a living god – had consequences: gods could choose to be benevolent, but they needed to be kept happy. If not, there was trouble.[42]

Moreover, while it is tempting to see the causes of later political instability as linked to climate factors, there had been frequent rebellions against Sargon's successors long before rainfall patterns changed, including a major uprising in 2276 BC and the Great Revolt in 2236 BC in which simultaneous coalitions of northern and southern cities rose up against Naram-Sin.[43] Food shortages had also been common, to judge from the *Chronicle of the Early Kings*, which not only mentions famine but connects it to immoral behaviour – for example when the god Marduk punished Sargon for disturbing the soil of Babylon by causing food shortages to take place and misery to follow.[44]

These dangers were familiar in the world of antiquity, which was why particularly prominent episodes were passed on from generation to generation. A notable example is a giant flood that is discussed at length in Babylonian texts such as *The Sumerian Flood Story*, *The Epic of Atrahasis* and *The Epic of Gilgamesh* as well as in the Egyptian *Book of the Heavenly Cow* and, perhaps most famously, in the Book of Genesis. All these accounts relate similar versions of a great flood that was unleashed on humans as divine punishment for their wickedness and impiety, with the aim of wiping them out altogether.

Many versions single out sympathy for an individual who was warned in advance of the impending deluge: in the case of the *Atrahasis*, the eponymous hero was advised to build an ark for his safety, and told to load a pair of every kind of animal before 'the flood roared like a bull, like a wild ass screaming', as the winds howled and darkness fell, blocking out the sun.[45] The Book of Genesis tells a strikingly similar story. God had turned to Noah and declared, 'I am going to put an end to all people, for the earth is filled with violence because of them. I am surely going to destroy both them and the earth. So make yourself an ark of cypress wood.' Noah had done so, enabling him, his family and the animals to remain safe as rains kept coming, causing floodwaters that rose to cover 'all the high mountains under the entire heavens'.[46] There was a long tradition, then, that the gods – or God – rewarded those who lived well and obediently with environmental stability; those who did not, had to contend with environmental disaster.

The fact that a single, devastating event played such a central role in collective memory across a wide region and for such a long period of time says much about the shock of sudden catastrophic events. The same was true of the destruction of Sodom and Gomorrah, towns that God decided to punish because they were permeated with sin. 'Burning sulphur' was rained down from heaven, leaving the cities with 'dense smoke rising from the land, like smoke from a furnace' – and Lot's wife turned into a pillar of salt.[47]

This too may be based on a real event, in this case a cosmic airburst or multiple airbursts caused by a comet or meteorite dating to around 1650 BC, which flattened the city of Tall el-Hammam in the southern Jordan valley: the city's massive four-metre-thick walls were destroyed, as was much of the palace complex. Heating reconstructions suggest that temperatures exceeded 2000 °C, with all settlements within a twenty-five-kilometre radius being abandoned for centuries, partly because of the effect of the hypersalinity of the soil following an airburst-related influx of salt. The transformation of Lot's wife in the Book of Genesis into a pillar of salt may not be coincidental; skeletal remains at the site indicate extreme fragmentation suggests that the idea of people being turned into ash may have a basis in fact, rather than being conjured up by an active imagination or by poetic licence.[48] Whether Tall el-Hammam and Sodom were one and the same is a matter of conjecture; however, the close similarities – and the fact that the story was not only part of the Book of Genesis but features prominently in the Qur'ān – shows that in the milieu of the Levant and the Middle East memories of catastrophic events lived long in the mind, and were closely associated with divine punishment for sinfulness and, in the case of the Qur'ān, for homosexuality in particular.[49]

Famine and disease, however, were both more common and more devastating than spectacular natural disaster and – despite contemporary belief to the contrary – were usually the result of human miscalculation. In the case of the Akkadian empire, the problems owed much to the realities of an expanding and extractive imperial realm and to the way relentless centralisation led to political fragmentation and fracture as well as to acute supply problems.

It was not for nothing, therefore, that 'The Curse of Akkad' specifically mentions that vast amounts of agricultural produce, as well as cattle, goats and exotic goods, were shipped to the imperial capital – so much, indeed, that the city's most prominent deity, Innana, 'did not know how to receive these deliveries of food'.[10] It is not hard to see that those who were obliged to transmit resources to the political centre would seek ways to challenge and mount effective resistance for local political opportunistic ends or because the diversion of food put regional populations under pressure – or both.

It did not take much, then, for a precarious balance to be upset. A single bad harvest could cause problems that included not only famine but political turbulence and social upheaval. And in that sense the crucial question should not be what were the effects of a substantial change in climate around 2200 BC, but rather what steps were taken to mitigate the challenges that this posed. What mattered, in other words, was whether rulers, elites, priests, bureaucrats and workers could adapt, not least to rising environmental pressures, and whether those choices and steps were appropriate and effective. It was not so much that climate brought down the Akkadian empire as that the Akkadian empire collapsed under its own weight, splintering into a new constellation of city states that in many ways were a reversion to the time before Sargon's unification. That might have been bad news for Sargon's descendants and their retainers, but it was good news for others, as local powers were restored to local hands.

In Egypt, too, this was a time of tumult, as a rock-cut tomb belonging to a southern provincial governor named Ankhtifi makes clear. No one could have accused Ankhtifi of being shy. 'I am a brave man who does not have an equal,' declare several inscriptions in his tomb. 'I have done more than my ancestors did, and my successors will never equal me, not for a million years, in all that I have done,' says another. However, the inscriptions also paint a devastating picture of life in the period that followed the reign of Pepi II (d. c.2184 BC). We learn that Egypt had descended into chaos, with rival factions fighting bitter battles. Famine was everywhere, affecting the whole of Upper Egypt so badly that many starved to death. Others were so hungry that they turned to cannibalism and ate their own children. People everywhere roamed about naked and without footwear. Anarchy was all around.[51]

This seems to be confirmed by another source, *The Admonitions of Ipuwer*, which talks of crop failure, and refers to all luxuries – from spices to clothes – as being in short supply. 'The storehouse is empty,' while scribes were being 'killed and their writings taken away'. Stray cattle wandered about with no one to look after them, and foreign soldiers were rampaging everywhere.[52]

It is tempting to attribute the breakdown in Egypt to drier conditions and to the diminution or even failure of the Nile floods: geological and archaeological evidence shows a sharp decrease in summer precipitation in the Ethiopian Highlands that resulted in catastrophically low discharges into the Blue Nile drainage basin as well as occasional intensive rainfall that produced dangerous flood conditions.[53]

While water shortage (and occasional excess) may well have produced problems, pressure also came from the breakdown in long-distance trade that accompanied Akkadian collapse, cutting off important sources of revenue. It is perhaps no coincidence, then, that one of Ankhtifi's inscriptions also mentions that elites had been targeted by the wider population and had goods and assets taken from them by force: the wealthy and well-connected had been extremely successful in winning privileges, tax exemptions and land grants from the crown in the centuries and decades before.[54] Those who had been rich now gasped for water, says *The Admonitions of Ipuwer*, while those who had begged for dregs now had overflowing bowls. Those who had had no food now owned barns filled with goods; enslaved people had now become enslavers. The old order, in other words, had been overthrown.[55]

That is what these sources seem to suggest anyway. Whether they are reliable, however, is another matter. While the archaeological evidence points to a time of economic contraction, it does not suggest a time of upheaval, tumult and dislocation.[56] The purpose of the inscriptions in Ankhtifi's tomb is to underline that he was a saviour, bringing hope, comfort and relief to the people, as well as restoring stability to his own province.

And in fact the key question is precisely how what seemed to be robust and durable states could be so unstable and precarious. One cause of weakness was failure of adaptation; but another was the domino effect that made bad situations worse. That failure to respond was a crucial element in the centralised political systems that were built on the demands of entrenched elites which not only lacked the necessary skills to respond to changed circumstances, but actively resisted doing so – for fear of losing power, wealth and status. The cascades of dislocation spilled far beyond the local: one of the key factors in the rise of cultures and civilisations in the Nile, Mesopotamia and the Indus in particular had been their interconnectivity. Exchanges across and between these well-populated complex societies had stimulated trade as well as sparking cultural rivalry and assimilation. Something similar had happened in East Asia, where models of rulership, settlement and trade had drawn geographically separate cultures into frequent and even intensive contact.

Indeed, it might even be argued that the opportunities and ambitions within these clusters and competition between and within them were decisive features in their development and evolution: for example, wheat and barley had spread eastwards across Eurasia in the fourth millennium BC, with rice and millet being introduced in the opposite direction in the centuries that followed.[57] Crops from Africa such as sorghum and finger millet were being planted in north-west India by around 2000 BC, while melons and citron, fruits of South and South-East Asian origin, were being planted in the Levant, Egypt and eastern Mediterranean perhaps earlier still.[58] There was also significant contact between India and Australia around this time, to judge from substantial genetic evidence, as well as from changes in tool technology and the way food was collected and cooked; the appearance in the fossil record of the dingo – one of the animals most closely associated with Australia in the modern era but which came from Asia – was another product of regular if not intensive contact.[59]

Key to the expansion and interlocking of these networks was the uneven distribution of valuable raw materials, such as copper, tin, carnelian and lapis lazuli, but also of technologies such as metallurgy. The need for copper for body ornaments, weaponry and tools made metals highly desirable for social differentiation and rituals as well as for more practical purposes. Some scholars even go so far as to talk of an age of 'Bronzisation' as a shorthand to describe the ways in which closer interactions brought peoples, cultures and regions together in a process that mirrors many understandings of globalisation.[60]

These mechanisms that had enabled the spread of goods, foods and ideas now broke down as environmental, social and economic pressures took their toll. By around 2000 BC, trade between Mesopotamia and the Indus Valley which had been based on ivory, copper, textiles and perishables declined sharply, to judge from the written and archaeological evidence.[61] The drying up of markets must have had an impact on the social and economic landscape too, just as rising aridity did on the physical environment.

Such compressions and collapses are a recurrent part of human history. Often, climatic and environmental change were factors that played a role in unfolding crises. As the economist and Nobel laureate Amartya Sen has argued so persuasively, famines are caused by poor decision-making, by weak leadership or by intentional suppression of measures that might have mitigated or even prevented food shortages. Famines, he suggests, are the result not of lack of availability but of pricing problems, in particular affecting those who cannot afford food at times of sharply rising prices. Such prices, furthermore, are made worse not only by inflationary pressure in the face of poor harvests or supply shocks, but by hoarding practices, either for profit or from fear.[62]

While Sen's work is concerned with the twentieth century, the same principles can be applied when we consider dislocations in the more distant past. Certainly, weather conditions and extreme events could create problems, not only in the ancient world but in history more broadly. Most of the time, however, these were navigated or solved successfully – by adjustments, by coping strategies that included migration, resettlement and the development of new ways of living. In the case of the events of 2200 BC, then, we should be looking at the wider question of why societies struggled, rather than being tempted by the convenience of a simple and even a seemingly logical answer that starts and finishes with climate. And one of the most interesting things to look at is how problems could be contagious, and how troubles in one location could produce vulnerabilities in others.

6

The First Age of Connectivity
(c.2200–c.800 BC)

If you do not quickly arrive here, we ourselves will die of hunger.

Ugaritic tablet (c.1200 BC)

The breakdowns that took place around 2200 BC onwards are illustrative of the fact that interlocking regional networks could telegraph shocks as well as facilitate the transport of goods and ideas. The contraction of these trading systems was not permanent, with old connections eventually being re-formed or new ones being made.

In the period that followed the crises of c.2200 BC, a new set of nodes and links began to emerge and expand, resulting in the recreation of an interconnected geographic region that in due course spread from the eastern Mediterranean to the Indian Ocean and beyond. To start with at least, fragmentation had resulted in a series of city states blossoming, the most notable of which was the kingdom of Ur in Mesopotamia, whose rulers sought to connect the city and themselves to the great kings of Uruk, namely Enmerkar, Lugalbanda and Gilgamesh, and to position themselves as their heirs.[1]

One of the most significant trends of the centuries which followed the crises, however, was the movement towards political centralisation and of imperial dynasties establishing themselves across substantial areas. These were characterised by bureaucratic administrations that produced efficiencies through streamlining behaviour and practices and devising rules and laws that set out the responsibility of rulers, as well as of their subjects. A good example comes from the code of Ur-Nammu, ruler of Ur around 2050 BC, which stipulates the punishment for anyone bearing false witness when testifying during disputes.[2]

The best-known set of laws, however, was laid down by Hammurabi, ruler of the first Babylonian empire, which grew from around c.1900 BC to become economically and militarily dominant across much of Mesopotamia. These laws are known from several sources, the most striking of which is a diorite stele originally set up in the temple of Marduk in the capital city of Babylon and which is now in the Louvre in Paris. 'I have inscribed my precious pronouncements', reads Hammurabi's inscription, 'in order that the strong [do] not wrong the weak [and] to provide justice for the hungry and the widowed.' The code consists of 282 rules concerned with subjects ranging from trade and tariffs to marriage and divorce, from theft to non-payment of debt. They also detail the punishment for those who deliberately cause environmental damage, such as cutting down another man's orchard without the owner's permission.[3]

New states sprang up elsewhere too in ecological and environmental niches that supported rising populations, were located along trade routes, or had access to raw materials that were in wide demand elsewhere – or all three. The Assyrians established a series of merchant colonies in eastern Anatolia, for example, trading tin and textiles in return for silver and gold with local kingdoms, the most powerful of which, Hatti, grew to absorb others and form what today is often confusingly called the Hittite state, a term derived from Old Testament references to an entirely different people from Palestine. The Assyrians were themselves architects of a state in their own right that was centred on the cities of Nineveh and Assur and on what is now northern Iraq, north-eastern Syria and south-east Turkey, which was simultaneously a peer, ally, rival and competitor with the other great kingdoms of the Near Eastern world such as those of the Mitanni, the Ebla and the Kassites – the latter controlled Babylonia following the fall of the Old Babylonian empire around 1600 BC.[4]

New technologies opened up opportunities for exchange networks to blossom and expand further. Crucial to this was the development of Crete in the first instance and then of myriad towns scattered across the Aegean, mainland Greece and Asia Minor which spurred the evolution of sailing ships and (of course) the nautical skills required to navigate as well as to handle them in difficult conditions.[5] While there is considerable discussion among scholars about the nature of social and political structures on Crete in particular, the fact that substantial building complexes were being constructed by c.1900 BC suggests at the very least a widening of horizons of all kinds at a time when contacts between the island and other locations were increasing.[6] As elsewhere, the creation of writing scripts in Crete (Linear A) and then Mycenae (Linear B) and the evolution of increasingly elaborate burials again attest to institutions and elites forming as societies became more populous, more stratified and more closely connected to other worlds through the trade in rare, exotic and expensive commodities – above all metals, but also foods as shown by new research on traces of starch grain found on ceramic vessels.[7]

Much the same happened in Lower Nubia where the weakening of Egypt produced a vacuum that others were able to fill. By far the most successful were the kings of Kush who became so powerful that, once the pharaonic rulers in Egypt had stabilised their authority to the north c.1850 BC, they had to resort to building massive chains of fortresses to protect themselves from attack – as well as erecting abusive inscriptions about their new rivals. So successful was the Kush state that it grew to cover large parts of the Nile Valley and had an influence as far as the Horn of Africa. The 'empire of Kush', notes one historian, grew to become the 'largest political entity in Africa at the time'.[8]

States create hierarchies, but they also create stability and therefore usually help galvanise favourable conditions for trade by providing the security required to protect merchants able and willing to move produce to more distant markets. One Assyrian ruler in a typical inscription dating to around 1900 BC boasted that he had established freedom for his people and protected them from attacks and disruption.[9] This was an achievement in itself and clearly one that affirmed the role of the king as a saviour and protector. But it was also an essential prerequisite for the development of exchange – and was closely connected to the creation of social structures.

Indeed, it was precisely the expansion of closer trading networks that served as a catalyst for complex, hierarchical societies in West Africa and also for what subsequently became the heartlands of the Kush state.[10]

Together with the development of early metallurgical traditions in the Akjoujt region of what is now Mauritania and the Eghazzer basin in Niger, these changes to social organisation provided a similar backdrop that was essential for peoples, communities, goods and ideas to become linked over increasingly wide geographic areas.[11] Taken together, this paved the way for the consolidation of the power and resources of mobile pastoral elites in the central Libyan Sahara.[12]

In the Indian subcontinent, on the other hand, one of the key drivers of change was large-scale population movement. The centuries following the decline of the Indus Valley civilisations c.1900 BC were a period of profound demographic upheaval as a result of the arrival of new peoples migrating from the steppes. While the causes, nature and precise timing of the waves of migration are questions of considerable sensitivity and frequently a source of controversy, it is now clear that these arrivals largely postdate the dispersal of populations from Harappa and elsewhere, rather than being either coterminous or a cause of downfall.[13]

The genomic data leaves little scope for doubt about the extent of the mixing of the existing population with new waves of arrivals, and leads to the emphatic conclusion that the overwhelming majority of the migrants were men. So substantial were the migrations that almost 18 per cent of men in India today belong to haplogroup R1a which is widely represented across populations in Scandinavia, central Europe and Siberia.[14] Languages spread too, with Proto-Indo-European tongues providing another marker of the realities of an interconnected world more than 3,500 years ago.[15] Linguistic similarities between Balto-Slavic and Indo-Iran subfamilies of Indo-European languages are reflections of the movement of people and also of common pasts that lead to the steppe lands of Eurasia.[16] As recent pioneering research has shown, however, expansions of people and languages were complex and there are often surprises when genetic and linguistic data is assessed independently of long-held assumptions, with entirely different patterns identifiable in Europe and South Asia, not to mention Anatolia and the Near East.[17]

The arrival of these new populations in the Indian subcontinent in a series of waves is often associated with the introduction or rapid evolution of new ideas.[18] Chief among these is the caste system, which has long been associated with the rise of Vedic religion, its beliefs system and social practices.[19] However, the fact that genetic lineages point to the gradual merging of immigrants and indigenous populations suggests that new concepts were not imposed or adopted decisively, but adapted and developed over time.[20]

These ideas were set out in a language known as Sanskrit (from *saṃskṛta*, meaning 'perfectly formed' or 'pure and perfected'), with the oldest documents – the Vedas – written down around 1500 BC. Establishing the precise origin and nature of these ideas and charting how they had evolved is not straightforward, not least because the Vedic texts – consisting of the *Rig Veda*, *Sama Veda*, *Yajur Veda* and *Atharva Veda* – are the religious foundational texts of Hinduism. They are held to set out *sanatana dharma* – an eternal order that adherents believe is subject to neither influence nor change. Sanskrit accounts of the three Vedic worlds of earth (*bhūrloka*), heaven (*svarloka*) and 'mid-space' (*bhuvarloka* or *antarikṣa*) which contained air and light suggest calm and serenity, as do hymns in the *Atharva Veda* saying 'the earth is our mother and we are its children'; the constant warnings about respecting forests, animals and water sources speak of harm and damage being regularly done to these – they are not paeans to careful attention being paid to each and to all.[21] The revelation of the Vedas (literally 'knowledge') was made possible only when Indian sages who were suitably sensitive spiritually were finally able to hear and record vibrations that had resounded through the universe since the beginning of time.[22]

As we shall see, alongside the ideals, duties and responsibilities that these texts set out as virtues are detailed instructions on how to engage with the natural environment and how to understand the relationship between the human world and that of animals and plants – as well as that of the divine. This spread of ideas, customs and behaviour is echoed in other parts of the world around the same time. In Egypt, for example, a dynasty known as the Hyksos rose to power c.1640 BC and dominated the country, ruling from a capital city at Avaris over territory that stretched from the Nile Delta south as far as Cusae, and controlling western trade routes linked by oases to Nubia.[23] Although the main written account of this period by the Ptolemaic priest Manetho describes the Hyksos rulers as outsiders who attacked from the Levant and took control by force, strontium isotope analysis of skeletal remains and tooth enamel reveals that, while the Hyksos did indeed originate from outside Egypt, their seizure of power was a local uprising by members of a community who had first moved to the region around 2000 BC, rather than the result of a contemporary foreign invasion.[24]

As with migrants to the Indian subcontinent, they were a source of innovations, ranging from clothing to ceramics, from weaponry to burial practices, which did not involve traditional amulets or scarabs that originated in western Asia.[25] Here again there seems to have been a gender imbalance, in this case with a sex bias towards women, with considerably more evidence for men with Egyptian names marrying women with non-Egyptian names than the other way around.[26] Although the Hyksos were eventually overthrown by a dynasty from Thebes in Upper Egypt, innovations they introduced had a lasting impact, locally as well as beyond. These may have included the alphabet, perhaps spread to Egypt by miners who had initially developed or at least adopted alphabetic script by c.1800 BC.[27] Other technologies that were likely introduced from the Levant and South-West Asia included the composite bow and the horse-drawn chariot, all of which proved to have very significant implications, paving the way for expansions of political, social and economic authority.[28]

Something similar may have happened in the Mitanni kingdom in what is now northern Syria. Scholars have noted that, unusually, despite the language of the kingdom being Hurrian – a non-Indo-European language – rulers took Old Indic throne names and, when drawing up treaties with their Hittite neighbours and rivals, invoked some of the most prominent and important deities known from Vedic texts, including Indra, Varuna and the Nāsatya twins.[29] One hypothesis is that the Mitanni kingdom was founded by mercenaries, perhaps charioteers or horsemen, who were part of the same movement of peoples that led to Indo-European settlement in the Indian peninsula in broadly the same period.[30] It is more likely, however, that linguistic and theological ideas were spread by contact and by absorption, rather than by large-scale migrations – for which evidence is very limited indeed in the case of Anatolia and the surrounding region.[31] This certainly seems the most logical explanation for the transmission of ideas about astronomy and cosmology, and for similarities of Mesopotamian ideas about the three paths of heaven (to the north, middle and south) to those that appear in Vedic literature.[32]

This was the case too in the Andean civilisations from around 1800 BC onwards, with rising levels of cultural and economic exchange between communities on the coast, in the valleys of the interior and in the highlands. Over time, ceremonials, temple design, ceramic styles and foodstuffs grew increasingly standardised as interactions between regions became more frequent and more intensive.[33] Again, these contacts resulted in the expansion of languages – in this case Aymara and Quechua – with some scholars linking these processes specifically to the adoption and spread of agricultural practices.[34]

As had been the case in the Near East, increasing contact, higher levels of exchange and the adoption of new ideas, methods and technologies all played important roles in the creation of large states that incorporated peoples of different origins and developed cultures that drew them together, with a single focal point, their elites having simultaneous roles as spiritual interpreters and beneficiaries of status, power and wealth. In the Andean world, this had happened by around 1000 BC, with Chavín de Huántar starting to emerge as the ceremonial centre for most of what is now north and central Peru, as well as extending its influence deep into the south, bringing peoples, communities and topographies together within a unified cultural, religious and political envelope.[35]

This paralleled what happened in East Asia. By 2000 BC, an enormous walled settlement had become established at Shimao in the Central Plains area of China which was one of the largest in the world at the time. At its heart was a pyramid palace complex that was at least seventy metres tall, dominating the city and surrounding area. Thousands of jade artefacts reveal that Shimao was the centre point of a vast local and long-distance trade network – while human sacrifices on a massive scale provide evidence of elite control and show that the city was a place of power.[36]

Recent excavations have shown the importance of Shimao, in the process turning long-accepted ideas about early Chinese civilisations on their head. The picture in East Asia starts to become a little clearer first with the vast city of Erligang and then, from c.1200 BC, when a state emerged under the Shang dynasty that extended its authority across what is now north and central China – although the precise dating remains uncertain because of contradictions in the literary and archaeological records.[37] Once more, the result of an overarching political structure was a homogenisation and standardisation of cultural styles, as well as of religious practices and social structures.[38] The expansion of agriculture was fuelled by labour gangs supporting a royal household that was strongly incentivised to maximise crop yields as its source of wealth, largesse and authority. As elsewhere, an emerging bureaucracy and administrative elite developed writing scripts – in this case inscriptions carved into bones – enabling instructions and decisions to be recorded and driving a familiar process of cementing and centralising power.[39]

Curiously, then, while historians are often quick to discuss the role of climate in the collapse or decline of empires, kingdoms and states, they are rather more reluctant to do so when it comes to patterns of consolidation, expansion or efflorescence, such as in the long period that followed the crises of c.2200 BC. In some ways, of course, this is not surprising: the principal regions that saw large-scale permanent settlements were environmentally and ecologically well suited to supporting populations and, crucially, to facilitating their growth – either through increasing the size of cultivated land or through more intensive interventions such as irrigation and canalisation.

The benign disease environment was also an important factor in enabling demographic expansion and conversely in explaining why other regions did not blossom in the way that Mesopotamia, the Nile, parts of China and the north-western flank of South America did in the millennium after 2200 BC. In South-East Asia and in many areas of West Africa, malaria acted as a brake on population size, as did the lack of availability of seeds, grains and foods that were so important elsewhere.[40] In the case of equatorial Africa, a vicious cycle developed with the Bantu migrations that started c.1500 BC: although the adoption of banana and plantain allowed larger permanent settlements and both geographic and demographic expansion, these became centres for malarial infection, producing the same effect as the so-called 'white man's grave' 2,000 years later when Europeans involved in the slave trade died in droves. Populations which had no resistance to malaria were replaced by those whose ancestors had developed Duffy Negativity, the genetic variation that provided immunity from *vivax* malaria; but this took many centuries to occur, and meanwhile prevented the rise of towns and cities in the same way as in some other parts of the world.[41]

In other cases, climatic changes, extreme weather events or natural disasters could have devastating implications in the emergence of diseases. The enormous eruption of the Thera/Santorini volcano c.1600 BC, one of the largest in the last 5,000 years (it had the power of 2 million atomic bombs of the type dropped on Hiroshima), is an obvious example of the latter. In fact, the most important impact of the event was not the catastrophic and famous tsunami that engulfed the island of Crete, nor even the reorientation of the Mediterranean civilisations that essentially occurred as a direct result.[42] Rather it was that it may have played a role in pathogen emergence, and specifically of the variola virus, which appeared in the Nile region, perhaps as a consequence of the evolutionary forces resulting from the eruption, its fallout and ejection of gases and acids.[43]

If this was indeed the case, it is hard to overstate the importance of looking at the bigger picture of how events in the world's natural history have had major long-term impacts that affected the world not just of the past but of the present too – or, in this case, of the near present: variola virus, better known as smallpox, claimed the lives of 300 million people in the twentieth century alone, and perhaps as many as half a billion between 1850 and the eradication of the disease in 1977.[44] Among its first victims were members of the ruling dynasty of Egypt, including Pharaoh Ramses V, whose mummified body bears clear traces of the virus's unmistakable pustules, especially on his cheeks.[45]

Such one-off events as major volcanic eruptions could produce dramatic and unexpected results. More broadly, however, the real problem was not flooding, or long periods of drought that lasted decades, although these could act as triggers. What presented the greatest risk was population load. Harvests that failed repeatedly became an issue if there were too many mouths to feed; crops that were washed away by unexpected and unusual amounts of rain could mean disaster for communities that had to make adjustments in consequence; hoarding as a result of panic or in order to take advantage of inflated prices could make a disaster out of a precarious situation.

The demand for authority and wealth, driven by the centre, encouraged the exploitation of resources – and often brought about their exhaustion. What little timber had existed in Mesopotamia, for example, had been quickly used up, leading by 2000 BC to the need to import wood from as far away as Oman and the western coast of India.[46] Demand for metals such as copper, tin and lead in Shang dynasty China for the production of ritual vessels as well as weapons and tools caused limited local deposits to become depleted quickly and gave rise to a dependence on long-distance sourcing networks, although claims that these reached as far as Africa appear to be misguided.[47]

The weakness of course lay primarily in the size of individual cities, as this was where risks were greatest in times of shortage. Urban settlements were also centres for potential crisis because they were where uprisings could start as citizens who were disaffected, hungry or worse sought to take matters into their own hands. However, the size of the territory under the control of a single ruler and administration was an obvious source of weakness too, for it did not require much for newly or recently incorporated peoples to rise up, especially if they had been conquered militarily. Locations that were peripheral had most to lose from domination by faraway centres and provided the greatest opportunities for old or new elites to offer alternative visions and solutions that could generate their own steam.

In today's world, we refer to systems that link peoples and places together as globalisation. As has become clear in recent years, as well as having significant environmental costs, the processes that enable goods and produce to be made or grown, fuels to be extracted and resources tapped at the lowest possible prices and shipped at the optimum speed relative to outlay create positives and negatives. There are winners and losers in equations that involve threats and competition as well as efficiencies and benefits.

As we have also learned, however, these globalised forces create another more specific and difficult challenge, namely that close connections can suddenly tip from being part of a solution to the cause of problems. Interdependence means fragilities can easily be magnified – and can spread quickly and seemingly out of control.

A case in point are the widespread calamities that took hold c.1200 BC and brought about concurrent crises in Egypt, the eastern Mediterranean, the Near East and Mesopotamia. These included chronic food shortages which we learn were relieved, at least to start with, by shipments of grain from Egypt following the order given by the pharaoh, Merneptah, to keep the people of these regions alive.[48] Send food urgently, begged one Hittite King; 'it is a matter of life or death'.[49]

Records from Emar in what is now Syria are short and simple, begging for help: 'If you do not quickly arrive here,' they urged, 'we ourselves will die of hunger. You will not see another living soul from your land.'[50] Other clay tablets from the city state that families were so desperate that they were forced to sell their children in order to survive.[51] Sources from Babylon and Assyria also describe challenging times, with references to crop failure, food shortage and outbreak of disease. In time, Egypt too was affected by poor harvests – and by repeated strikes by workers at the royal tomb in the Valley of the Kings primarily as a result of inadequate or overdue rations.[52] By around 1170 BC, shortages had resulted in rampant inflation as grain prices rose eightfold, and on occasion even more steeply.[53]

This was a time of considerable disruption thanks to a group or perhaps groups who were identified at the time as coming from 'islands in the midst of the sea' and are usually referred to today as the Sea Peoples, a name coined by the French scholar Emmanuel de Rougé in the late nineteenth century. These included those who are referred to in the Hebrew Bible as Philistines, who migrated from southern Europe to the Middle East.[54] They are reported to have brought chaos to a wide area in the eastern Mediterranean. 'Now the ships of the enemy have come,' reads one cuneiform tablet purporting to relate the words of the King of Ugarit. 'They have been setting fire to my cities and have done harm to the land.'[55]

Egypt suffered the same way, to judge from texts engraved in the mortuary temple of the pharaoh Ramses III at Medinet Habu and in papyrus accounts from this period. 'Those who came together on the sea, the full flame was in front of them at the river-mouths, while a stockade of lances surrounded them on the shore,' says the inscription on the wall of Ramses' tomb. 'They were dragged in, enclosed, and prostrated on the beach, killed and made into heaps from tail to head. Their ships and their goods were as if fallen into the water.'[56] The Sea Peoples devastated one community after another as they laid their hands upon all the lands of the earth, mourned one contemporary account.[57]

These events took place during a challenging period of lower rainfall, attested by multiple proxies across these regions, including pollen data, charred plant remains, low lake levels in the Dead Sea and the Sea of Galilee, a drop in discharges in the Nile and drops in sea surface temperatures in the Mediterranean which are often associated with reduced precipitation levels.[58] However, once again, the time horizon for these changes was an extensive one. All were part of long-term trends that dated back centuries – perhaps even some 400 years earlier to the volcano on Thera/Santorini and the modifications that such a large eruption can produce in local and global weather patterns.[59]

Lower rainfall may have made life more and more difficult, and have pushed the margin of error to become ever narrower and more precarious, but other factors may have been much more important in bringing about the end of the Bronze Age.[60] For example, divisions and disputes among the social elite have been suggested as a leading cause of upheaval in the Hittite empire around this time.[61] As another scholar has noted, citing Claudius in Shakespeare's *Hamlet*, 'When sorrows come, they come not single spies / But in battalions,' the fifty-year period between 1225 and 1175 BC saw an 'earthquake storm' along the geological fault lines in the Aegean and eastern Mediterranean, with multiple destructive quakes measuring 6.5 or higher on the Richter scale.[62]

More significant than identifying a single, primary cause of multiple collapses, therefore, is the principle of contagion: the failure of one part of a network, whether because of poor harvest, earthquake damage or infighting between blood relatives, could cascade into the interruption, dislocation and even the system breakdown of the network as a whole. Elaborate long-distance exchange networks that facilitated trade and underpinned the status of the wealthy and powerful were also mutually dependent: the upside of hypercoherence was increased availability of luxury goods; the downside was infectious hyperfragility.[63]

It did not take much for networks to fall apart, in other words; and when they did, social structures, states and empires could wobble and even collapse. This was recognisable across many regions and periods of history. One example is the fall of the Roman empire in western Europe many centuries later, when modest pressure led to a downward spiral into what historians used to refer to as the Dark Ages; another is the collapse of the Soviet bloc at the end of the twentieth century, when what had looked like an impregnable empire dissolved almost overnight. Economists too are interested in models showing how minor shocks can create ripples that lead to failure and breakdown in supply chains – a principle known as the bullwhip effect.[64]

Seemingly all-powerful, well-structured empires and networks could unravel and fall apart rapidly under relatively little pressure – because of the availability of new commodities, for example, or, in the opposite case, because of resource shortage. Climate and weather were often aggravating factors rather than the direct cause of cataclysmic change. Moreover, while single shocks could occasionally have profound effects, what was more usual was that shifts that were spread out over the long term slowly eroded the ability and willingness of peoples to co-operate, leading to dispersals of population in waves rather than in one go.

A good example of this is the breakdown of the Poverty Point culture in the lower Mississippi Valley around 1300 BC. The central point of the communities in this region, which were socially, politically and ethnically diverse and likely spoke a range of different languages, was a massive earthen mound marked by concentric half-circle ridges that had been built with impressive skill, starting around three centuries earlier.[65] Stratigraphic analysis from Jaketown indicates that exceptional rainfall caused flooding that breached a levee in a particularly dramatic single incident c.1310 BC.[66]

However, what was more decisive in triggering large-scale population movements was the fact that flooding occurred regularly and unpredictably for several hundred years.[67] Ironically, communities could survive single traumas better than persistent threats that undermined cohesion and encouraged dispersal to new locations. Once environmental challenges had stabilised or disappeared, sites were often reoccupied and repurposed – as was the case in the lower Mississippi Valley as Early Woodland peoples recolonised settlement spots that had become benign, as they had been in the past.[68]

It was certainly true that difficult times which arose from poor harvests were not easy for ruling regimes to negotiate, for under the right circumstances these offered opportunities that were ripe for exploitation by opportunistic rivals. The trick, then, for rulers was to make sure no such rivals existed, or if they did, that they were either kept under control or managed well. When that did not happen, the outcome could be dramatic, as was the case in central and northern China in 1046 BC, when the Shang dynasty was overthrown by the Western Zhou. Later accounts attributed the fall of the last Shang Emperor Di Xin (商帝辛)

to his personal shortcomings, blaming his indolence, cruelty and love of luxury, principally in order to serve as a mirror for princes and to offer warnings that had a contemporary resonance.[69] But the point was clear: a single bad apple could rot the entire crop.

Although the fall of the Shang marked a new chapter in Chinese history, it effectively amounted to a management change that adapted, added to and introduced new strategies and cultural practices. Further west, however, the picture was rather more muted. It took time for exchange networks to recover. Some places, like Egypt, did better than others, because they were less dependent on outside imports, were better organised or simply proved more resilient under pressure – or perhaps a combination of all three. There were other beneficiaries too, with a handful of locations that were propitiously located emerging as pockets that survived intact and eventually wove new tapestries of connections (and rivalries) – the rise of kingdoms and city states in the Aegean such as Mycenae, Athens, Sparta and others that became the backbone of the ancient Greek world offering a case in point. Groups of cities and states rose and fell together, stimulating and spurring each other on through exchange and competition – creating a mutual dependence that could also result in failure, as well as success.

The period around 1200 BC was also one of population movements, colonisations and change in Near and Remote Oceania in the Pacific, where archaeological sites with large numbers of shellfish as well as bones of giant tortoises, land crocodiles and bird species attest to waves of migration of the Lapita culture first to Vanuatu, New Caledonia and Fiji and later to Tonga and then Samoa.[70] It has been suggested that these movements may have been responses to changing and reversing wind patterns that made sailing west to east both easier and quicker – meaning that less time was needed at sea, which in turn meant reduced food and water requirements.[71]

It is striking, then, that 3,000 years ago – if not earlier still – questions about the threats of overpopulation, of fragility and of the risks posed by interdependence had already started to concern thinkers, priests, administrators and rulers. There was already not only awareness but acute concern about unforeseen problems that could suddenly appear and threaten disaster. Learning how to identify, anticipate and deal with these challenges became a fundamental concern for those living millennia ago. At the heart of the question of how to conceptualise the world was the problem of how to understand the human relationship with the divine – and with nature – in order to anticipate challenges, be ready to deal with them and, better still, prevent them from appearing in the first place.

7

Regarding Nature and the Divine
(c.1700–c.300 BC)

The Lord is my shepherd, I shall not want; he makes me lie down
in green pastures.

Psalm 23

The dangers of environmental degradation, overconsumption of
resources and unsustainable population load were not lost on people
living thousands of years ago. For example, *The Epic of Atrahasis*, an
Old Babylonian text of which the earliest clay tablets date to around
1700 BC, reveals an acute awareness of the vulnerabilities that came
from pushing ecological boundaries beyond their limits. The gods had
created humans, says the narrator of the text, because they realised that
they had too much work to do themselves: 'mortal man' was created
therefore, with the specific instruction that he should 'bear the load
of the gods'.[1] The trouble was that the gods had not accounted for the
length of a natural lifespan, which soon resulted in overpopulation. It
was not long before the number of people became 'too numerous' and
'the country was as noisy as a bellowing bull'.[2]

This annoyed the god Ellil, who 'had to listen to their noise' and soon
complained that he was 'losing sleep over the racket'. This happened
on several occasions, leading to the gods taking matters into their
own hands and deciding to kill off most of the population in order to
achieve some peace and quiet. As a result, they sent terrible droughts
that brought famine as well as other 'solutions' to the problem of the

excessive noise and the excessive number of people, including disease and pestilence.[3]

The most dramatic attempt was a great flood, an episode attested by archaeological evidence and presumably the basis for later Egyptian accounts and for those that feature in biblical versions.[4] As well as speaking of the inundation as divine punishment, implicit in each version of the flood is the theme of population loss and population control. In the case of *The Epic of Atrahasis*, the issue is specific and explicit: there had been too many people living on the earth, so steps needed to be taken to reduce numbers and to stop the same thing from happening again.

The mother goddess Nintu, says that account, had been particularly upset by the flood and the deaths of so many people because it deprived her of the regular offerings of beer that she had much enjoyed receiving from humans keen to maintain her goodwill. Once it became clear that Atrahasis and his ark had been saved, however, the gods came up with a new way of dealing with overpopulation: they tweaked the design of humans, to create a new, updated race who were not as fertile as the last. Their scheme included women who were unable to have children, higher levels of miscarriage and the consecration of virgins to serve the deities. All the tweaks were intended to reduce reproductive levels and to allow the population to be smaller, more sustainable – and quieter.[5]

While the flood story was a dramatic one, the connection between overpopulation and fragility is a powerful one that fits within the wider picture of concern about resource depletion, environmental degradation and the detrimental impact of human intervention within and on the physical landscape. Such concerns were well founded, for prolonged periods of reduced agricultural production, whether because of lower rainfall or as a result of soil depletion that came from land being overworked, presented acute problems, particularly in ecologically sensitive regions like southern Mesopotamia.

Both the nature and the scale of the problem are clear from a study of crop yields, which shows that these declined precipitously and consistently as greater demands were made of the land. For much of the third millennium BC, good agricultural land had produced around 2,000 litres per hectare per year; by around the time of Naram-Sin's reign in c.2200 BC, they were producing only just over half of that amount. Things got worse in the centuries that followed as the Akkadian empire was replaced by a smattering of new states that rose in its place, the most brilliant of which was the Third Dynasty of Ur. By 1700 BC, yields had fallen to around 700 litres per hectare per year, although a significant proportion was producing less than 400 litres per hectare.[6] Overexploitation meant that land was producing just 20 per cent of what it had been able to achieve a few centuries earlier.

For some scholars, collapse in production was the direct result not only of rising population levels but of a rapacious central bureaucracy, whose own need and aspirations placed demands that forced farmers to push their lands to their ecological limits. In particular, the problem was caused by excessive irrigation that ruined soils by causing salinity levels to rise. The search for short-term gain resulted in outcomes that were not only unsustainable but damaging. As we have seen in the cases of implosions of states and the breakdown of trade networks, rising complexity produces diminishing returns as well as magnifying vulnerability. Far from success breeding success, in other words, what seemed like never-ending success as states expanded and land under cultivation increased in extent was often an exercise in the seeds of ruin being sown inadvertently.[7]

There is considerable evidence in the sources too of the awareness of the dangers posed by overexploitation of nature. The writer of *The Epic of Gilgamesh*, for example, counsels that drought followed deforestation.[8] Perhaps most striking is the moment when Gilgamesh kills Humbaba, the forest god, whose murder stands as a metaphor for the deforestation of Mesopotamia and humans' simultaneous encroachment on and lack of respect for the natural world.[9] Sensitivity to the natural environment and to the delicacy of ecological equilibrium is also implied in the personalisation of flora and fauna, which have the power of speech as well as a capacity for feeling, with trees, rivers and wild animals weeping with sorrow when Gilgamesh's friend and companion Enkidu is killed by vengeful gods.[10]

Gilgamesh's exploits stand as a metaphor for struggles against natural disasters that were caused by the gods. The eponymous hero was revered by the people of Uruk, we are told, because he restored order by rebuilding holy places that had been destroyed by flooding; his was the task of fighting against a devastating drought, sent down from the heavens, that 'dried up the groves, reedbeds and marshes' and lowered the river by seven cubits – around three and a half metres.[11] Man was in a struggle not only against nature, in other words, but against hostile forces that sought to challenge, reduce and even destroy humanity. Other Mesopotamian texts, like the *Enūma Anu Enlil*, contain multiple references to celestial and atmospheric signs, which show a sharp interest in trying to make sense of meteorological conditions and even in trying to influence them.[12]

The rise of the Neo-Assyrian empire in the centuries that followed the collapse of the Bronze Age brought about profound changes to the environmental, socio-economic and political landscapes of northern Mesopotamia in particular. Urban centres began to grow to astonishing sizes, with some new cities founded – like Dur-Šarrukin – and others expanded dramatically, such as Nineveh, which grew to some 750 hectares during the reign of Sennacherib (705–681 BC).[13]

This required different thinking about and engagement with the land. One change was the introduction of massive investments in irrigation systems that completely transformed the physical and natural environments as webs of canals were built that transported snowmelt from the mountains and rivers to where it was most needed – namely, the major cities that were few in number but vast in size (and therefore vulnerable to both food and water shortages) and the fields that surrounded them.[14] It was not only the succession of imperial capitals that became home to the Assyrian rulers that benefited; so too did more rural areas that were home to lower population densities.[15]

These interventions were an important part of an ideological transformation of the landscape that emphasised the power of the Assyrian kings, who were able to tame rivers, redirect their supply, enable cultivation of new lands and ensure enough food for all. Not surprisingly, then, these achievements were the subject of boastful inscriptions that praised the king and the works he had overseen. '[I] Sennacherib, great king, strong king, king of the world', reads one, transformed fields that were wasteland into ones that bore abundant fruit. People had used to turn their eyes to the skies in hope of rain; he dug wells, created irrigation channels, built canals and constructed sluices and dams to control the water supply. These works were so glorious and effective, it adds, that 'I had a royal image of myself expressing humility' set up at the mouth of the canal. The inscription ends with an entreaty from its royal author: if any of his descendants dismantles or does not maintain these works in the future, 'may the great gods overthrow his dynasty'.[16]

Domination over nature was supplemented by the creation of networks of roads, tracks and way stations that can be identified today from satellite photos. All these were maintained by the crown, as well as by those taken prisoner during Assyrian wars of expansion.[17] Perhaps as many as 1.5 million people were forcibly resettled by the authorities in the two centuries after 800 BC, to reduce the risk of local rebellions, to exploit the empire's resources systematically and to live within ecological limits.[18]

Those who were dispersed travelled in comfort. Letters such as that addressed to Tiglath-Pileser III (r. 745–727 BC) recorded that the King had instructed deportees to be given 'provisions, sackcloth, leather bags, sandals and oil'. If the King's donkeys had been available, it notes, these would have been provided, as would carts for them to pull. Assyrian royal art shows men, women and children often travelling on animals or on wagons – and never tied up.[19] In their dealings with those who were resettled and deported, kings were frequently compared to gardeners, uprooting precious trees and replanting them in the best possible environment for them to flourish.[20]

As we have seen before and shall see again, in pre-industrial societies exploitation of the land was dependent on human labour. As such, a key driver in territorial expansion was not only the earning of prestige or the acquisition of good land, but the boosting of manpower. This was not always done by force: when the Assyrian army besieged Jerusalem in 701 BC, envoys were sent on behalf of King Sennacherib to appeal to the city's inhabitants: 'Make your peace and come out to me!' If they did, they were told, they would be taken away to a land just like their own, 'a land of grain and wine, a land of bread and vineyards, a land of olive trees and honey', where they would be looked after and cherished.[21]

The elision of royal power and control over nature extended into other areas too, notably in the form of the king hunting wild and dangerous beasts which represented his mastery over the animal kingdom and the metaphorical protection he gave in defending his people from threats. Reliefs that show the great Ashurbanipal (r. 669–631 BC) wrestling with a lion with his bare hands were intended to show the ruler's superhuman bravery and skill, setting him apart from normal men and thereby affirming his authority.[22]

Presiding over giant feasts was another way that the king could show his wealth both literally and through the range and quantity of food at his disposal. A stele found in 1951 at Kalḫu (known today as Nimrud, near Mosul) records a spectacular event that demonstrates this perfectly. We learn from the inscription that King Ashurnasirpal II held a ten-day celebration for precisely 69,574 guests, laying on '1000 barley-fed oxen, 1000 young cattle … 14,000 common sheep … 1000 lambs, 500 deer, 500 gazelles, 1000 large birds, 500 geese, 500 fowls, 1000 suki-birds, 10,000 fish and 10,000 locusts', as well as 10,000 earthenware jars of beer and the same number of goatskins filled with wine.[23]

The creation of non-native ecosystems within capital cities that included elements such as simulated marshes and imported flora were likewise intended to demonstrate dominance over the natural world.[24] Perhaps the best example was the astonishing garden made by Sennacherib at Nineveh, near modern Mosul, one of the wonders of the ancient world; associated thanks to a series of later manipulations, misunderstandings and confusions with an entirely different city, it seems to have become known as the Hanging Gardens of Babylon.[25]

Despite many changes across hundreds and indeed thousands of years in the Middle East, there were continuities too, for example in agriculture, architecture, political structure and the economy. Empires rose and fell; regions were sometimes more and sometimes less intensively linked with each other over short and long distances. Even languages came and went – with Assyrian cuneiform and language largely disappearing following the sack of Nineveh in 612 BC at the hands of a powerful Babylonian state that rose in its place, until it too was supplanted, this time by the Achaemenids of Persia.[26]

One of these continuities lay however, in cosmology. Societies differed in their allegiances to individual gods, in the specifics of how to win their favour and of how to interpret advice and admonitions. But the approaches were very similar when it came to trying to make sense of the world and to interpret and intervene – particularly in relation to climate and its vagaries. Astronomical diaries recorded astrological phenomena, as a way both of tabulating knowledge and of creating a framework to identify and then try to understand anomalies.[27] The job of interpreting was taken up by seers and priests in the Mesopotamian world, who claimed responsibility for explaining signs and omens to help humans understand the whim and will of the gods.

Animals were offered in sacrifice in order to win the favour of the deities and to ensure smooth channels of communication. Sometimes questions were whispered to sheep as they were being slaughtered so that messages could be delivered quickly and directly to the gods.[28] If animals had birth defects, such as no bladder, it was inferred that 'the flood in the river will be cut off, the rain in the sky will be scarce'.[29]

Not all of these omens were linked to weather and climate. Will an enemy army gather, get organised, march and fight with us? asked one powerful ruler of the Neo-Assyrian empire that rose from the ashes of the Bronze Age collapse. No, came the reply. Would a young man be likely to recover from epilepsy? came another question. Yes, was the response. Often, however, there were warnings that the price of goods would rise or that there would be shortages. If the moon seemed to have a halo, for example, grain supplies would fall. If Mars was erratic, barley would become more expensive. Many omens were concerned with the harvest, with rainfall, with starvation or with locust swarms that could devastate crops. Predictions therefore could give warnings to the ruler, to tell him he should be cautious and prepare accordingly.[30] Naturally, this was textbook forward planning by seers – who could always say that their admonitions had not been heeded if disaster struck, but that they had been if it did not.

It was a similar equation for those interpreting the natural world in Shang dynasty China, where the power of elites was closely connected to the ability to control and explain reality and to commune with a complex pantheon of gods that included Di (帝), the high god, and nature powers like the river, mountain and sun, as well as mythical and semi-mythical beings and ancestors. Oracle bones made mainly from ox scapulae and turtle shells were used as divination tools with the king or his diviners applying a hot rod to produce a series of heat cracks and then interpreting the 'answers' accordingly. 'Auspicious', reads one example. 'We will receive harvest.'[31]

The relationship with the supernatural world was administered by the Shang ruler, who had a shamanistic role in undertaking rituals and ceremonies and acting as an intermediary with the deities as well as with royal ancestors.[32] This required suitable gifts to be given up in offerings to ensure rain, good harvests and other benign conditions, and to appease hostile or malevolent spirits. The offerings could be substantial, as is clear from a set of oracle bones that record '100 cups of wine, 100 Qiang prisoners, 300 cattle, 300 sheep and 300 pigs' being given up by one ruler on a single occasion.[33] Human sacrifice of prisoners, usually in groups of ten but sometimes numbering hundreds, followed by interment of their decapitated, dismembered bodies in royal burials alongside retainers and relatives, was an important part of creating a dynamic that underlined the king's unique political, social, cultural and spiritual position.[34]

As in Mesopotamian cultures, a wide range of topics were investigated, including whether those who were ill would be cured, what the outcomes might be of conflict with neighbouring peoples, what the gender of offspring would be and what the status was of the king's health. Many questions were concerned with the weather and in particular with rain, for which Di was primarily responsible. Di was feared for his ability to 'send down' disasters on humans and for his power to control the winds, to order rain or to send thunder.[35] 'From today', reads one bone oracle, 'Di will order rain.' If the king goes hunting on a given day, reads another, 'he will have no disasters and it will not rain'.[36]

The Shang rulers' concern with avoiding disaster, overseeing sufficient rainfall and ensuring good harvests was shared by the Zhou dynasty, which took power in 1046 BC, and which celebrated new or perhaps existing rituals that demonstrated royal patronage over the land and the fruits it yielded. These included the ploughing of the sacred field at the start of the agricultural year.[37] 'Be attentive in your tasks,' warned one classic poem, and 'prepare your spades and hoes'. The king 'has himself come to scrutinize you' as you work; as a result of his involvement and thanks to heaven, there will be a plentiful harvest.[38]

Assuming responsibility for good outcomes in the natural world served as a powerful claim to be the conduit between heaven and earth, something that the Zhou rulers emphasised with gusto as they shuffled the cosmological deck of cards to introduce a new concept of Tian (天) – broadly equating to heaven or nature, a universalised concept which blended a supreme deity with a cosmic moral force that ruled the world and took an interest in the affairs of human beings. One reason for doing so may have been the need for political propaganda to assert the legitimacy of their rule, while pacifying those whose loyalties lay with the Shang – a substantial challenge made clear by a major rebellion that broke out soon after the Zhou took power.[39]

Crucial in this revised model was an even greater emphasis on the role of the ruler maintaining a bilateral bond, under which authority was vested by heaven and maintained by the king. According to the *Classic of Poetry*, the 'Zhou is an old people / but its Mandate is new'. The Shang had fallen, says the author of the poem, because they had neither understood nor fulfilled their commitment to protect the will of heaven. Luckily, the Zhou had arrived to protect everyone, in the first instance under its founder, King Wen. Although it would not be easy, Wen and his descendants would protect the 'Mandate of Heaven' and, in doing so, bring peace, prosperity and harmony to all.[40]

Careful attention was needed, therefore, in times of adverse conditions. Lack of rain was a particular concern. The poem 'Yunhan' (雲漢) from the celebrated *Shijing* (詩經, *Classic of Poetry*) records the anguish of King Xuan at a time of drought around 800 BC.

> The drought is so excessive,
> And we are tormented by the heat.
> I have not stopped offering sacrifices.
> From the suburban altar I have gone to the temple hall.
> To the powers above and below I have made sacrifices and buried offerings.
> There are no spirits I have not honoured …
> The drought is so excessive,
> And it cannot be alleviated.
> It is terrifying and perilous,
> Like lightning and thunder.
> Of the dark masses that still remain in Zhou
> There will not be a single one left.[41]

The ruler was at pains to ensure that he had made the appropriate offerings and had followed the right rituals in order to demonstrate not only that the drought was not his fault but also that he had done, and was continuing to do, all he could to alleviate the challenging circumstances.

Excessive rain could also cause problems, as another poem written almost 3,000 years ago during the period of the Western Zhou dynasty notes: heaven was 'sending down rain' after being 'arrayed in anger', leading to widespread famine and to people having to leave their homes and become 'wandering fugitives'.[42] Such events could easily undermine the authority of rulers, both because of the practical problems that pressure on food production posed for social stability and because of questions about the competence, efficacy and suitability of those entrusted with ensuring smooth relations with gods, spirits and the divine.

This was one reason why efforts were made to understand and record natural phenomena. The legendary Emperor Yao, one of the first figures recorded by Chinese sources, was said to have instructed 'astronomical officers to observe sunrise and sunset, and the stars and the planets, and to make a solar and lunar calendar with 366 days and also calculate the leap month'.[43] Weather observations and celestial occurrences, such as eclipses, were avidly recorded in Mesopotamia in the form of astronomical diaries dating at least to the mid-seventh century BC if not earlier, to help anticipate and explain unusual celestial and environmental occurrences.[44]

Efforts were also made to interpret signs for guidance to what the future might hold, with predictions based on the combined positions of various stars relative to the sun being taken as a warning that 'cities will start vying with each other, city walls will be destroyed [and] the people will be dispersed'.[45] Some hedged their bets, such as Akkullanu, royal astrologer to King Ashurbanipal, who wrote to the ruler in the middle of the seventh century BC 'about this year's rains that were diminished and no harvest was reaped'. In fact, he claimed, 'this is a good omen for the life and wellbeing of the king my lord'. Moreover, it was an omen of opportunity: if the King 'takes the road against the enemy, he will conquer whatever [land] he will go to and his days will become long'.[46]

If that was a case of fancy footwork and trying to avoid blame for agricultural failure, others were more careful not to leave things to chance. 'I brought the Nile to the upland in your fields,' boasted one pharaoh in Egypt, 'so that plots were watered that had never known water before.'[47] The rewards of investing in infrastructure like irrigation channels were for the general population, but the credit was taken by the ruler.

The fusion of benign climate, divine favour and reward for good leadership was one with deep roots in many parts of the world. Writing in the eighth century BC, for example, the Greek author Hesiod argued that when people 'deal in justice straight and fair' and observe the law, 'their city flourishes, their people bloom'. In particular, 'famine and folly pass the righteous by' and 'the just feast on what well-worked fields supply. The earth abounds for them.' Hesiod did not just recommend moral qualities but extolled the benefits of hard work: 'It's work that prospers men, and makes them rich ... it's working which endears you to the immortals.' In contrast, those who commit evil deeds are not just punished personally; 'often a whole city pays the price'.[48]

The elision of natural order and power would become a fundamental part of Chinese political and religious philosophy and of imperial political ideology in particular over the course of three millennia. It effectively fused the role of emperor with good outcomes – environmental and otherwise. To justify the mandate, rulers had to rule well. This equation worked both ways: credit could be taken for periods of plenty, prosperity and peace; but, equally, disequilibrium could be interpreted as a sign of celestial order being disturbed and therefore a justification for punishment and hardship. Shrewd leaders could of course use this opportunistically, taking advantage of challenging circumstances to blame others and remove rivals, real or imaginary. Nevertheless, ideas about balance – and the need to maintain order – were hugely important and influential, and could be applied to the state, the ruler and each individual.

Many offered suggestions for how best to do this – so many, in fact, that the period from the sixth century BC onwards, a period when the number of cities and their populations rose sharply in Eastern Zhou dynasty China, became known as the time of the Hundred Schools of Thought.[49] Among figures prominent in this period was Kong Qiu (better known in the west as Confucius), who counselled that people should aspire to live moral lives and show filial devotion, respect for ritual and benevolence towards others. Just as a society as a whole should reflect the celestial order, so too should rulers protect it, while all should live in a way that defended stability and harmony.[50]

Others, like Laozi ('Old master') and Zhuangzi, reacted to this, attacking the vanity of those who claimed to know and practise virtue. In fact, they argued, what governed the universe was an abstract theory based on the concept of Dao ('the Way'), a force that unified existence in a state of harmony. 'Man takes his law from the Earth; the Earth takes its law from Heaven; Heaven takes its law from the Dao. The law of the Dao is its being what it is.'[51] Being virtuous required more than being humane. It demanded the abandonment of luxurious living, renunciation of war and a life of simplicity and innocence. The best three 'treasures' were compassion, frugality and 'daring not to be first in the world'.[52]

This all sounded very nice, argued Mozi, but examples needed to be set by those in positions of authority. 'When the state is well ordered, the laws and punishments will be justly administered.' Those of poor moral character should be weeded out, while 'no special consideration' should be shown to those 'of their own kin, no partiality for the eminent and rich, no favouritism for the good-looking and attractive'. It was the duty of the ruler to provide and maintain a structure. Balance did not just happen on its own, in other words.[53]

Such discussions and debates were not entirely dissimilar in India, where the earliest hymns of the Vedic Aryans articulated notions of heaven and earth and of *ṛta* (ऋत – a fixed or settled order) and explained how the universal laws that governed *ṛta* were sustained by the guardian deity Varuna. Early Indian literature took the form of *sruti* (श्रुति – sayings that were later written down) that could be understood only by a *rishi* (ऋषि – a sage or a wise man) who had acquired the requisite skills through a process of contemplation and revelation. Such texts included the Vedas and the *Bhagavad Gita*, a 700-verse scripture that is preserved as part of the epic *Mahābhārata*.[54]

The collections of hymns that make up the four Vedas (*Rig, Yajur, Sama* and *Atharva*), conventionally thought to have been composed c.1500–1000 BC, explore the relationship between the terrestrial and the celestial, extolling the way the combination of heaven and earth can 'shower down honey for us' as well as 'great glory, reward and heroic strength'. Verses in the *Rig Veda* bid heaven and earth to 'bring unto us gain, reward and riches'.[55]

Advice in the Vedas was practical as well as existential and spiritual, and included prayers to keep serpents away alongside guidance on how to use plants such as the *apāmārga* (*Achyranthes aspera*) to overcome disease, and how to ward off curses, avoid bad dreams, remove the threat of hunger and thirst and stop losing at dice.[56] Some herbs could help keep the air free of foul smells and pollutants, an important quality given that fresh air was important for human health.[57]

These texts make frequent references to wider conceptualisations about the human relationship with the natural environment. 'The oceans are the treasure of wealth, so you should protect them,' says the *Yajur Veda*, which also warns that water should be kept clear, clean and fresh and not be polluted. Do not pollute the water, it adjures, and do not harm or cut the trees.[58] Animal welfare was important, not least because some creatures were useful: 'No person should kill animals that are helpful to all.' This advice was expanded for the sake of rulers, along with an explanation: 'O King, you should never kill animals like bullocks that are useful in agriculture, or like cows which give us milk, and all other helpful animals – and you must punish those who kill or do harm to such animals.'[59]

Many of the verses are concerned with the idea of deities as beneficent beings choosing to hand out riches and rewards, a theme that recurs repeatedly in the Vedas. One hymn sings of Dawn, personified as a beautiful woman, appearing each morning in her chariot to usher in a new day as she rides across the skies. 'Without fail she continually creates wealth. Making her path toward prosperity, the much praised goddess, bringing all desirable things, becomes radiant.'[60]

Such ideas were elaborated on in a series of later works that built on the teachings of the Vedas and the early Upaniṣads; these were composed around 600 BC and set out norms for ideal ways of living.[61] In one, we learn of the story of King Bṛhadratha, who gave up the throne to his son and went to live in the forest to ponder the purpose of his existence and the 'essence of the self'. He came to understand that life was transient, and that 'flies, gnats and other insects, as well as herbs and trees', would first grow and then die. But what truly concerned him was 'the drying up of great oceans', the 'submergence of the earth' and constant change that meant things were not in their right place. Seeing the world in this way, he said, 'what is the use of the enjoyment and pleasures in a world such as this?' Life was barely worth living, he told a famous holy man he met living in the forest: 'I am like a frog in a dry well.'[62]

Taken as a whole, scriptural texts, historical narratives, law codes, myths and legends and philosophical tracts that are collectively known as *smriti* (स्मृति – or recollections) set out frameworks for making sense of the universe, as well as advising on the most appropriate ritual practices and pushing at the boundaries of philosophy, ethics and epistemology. When it came to the natural world, much of the advice was again practical and prescriptive: the planting of trees was to be praised, while the polluting of water was to be condemned. Being surrounded by animals and by fresh, green trees and away from other humans was an effective way of being at peace – enabling, as the great *Rāmāyana* epic suggests, all living beings to exist in a state of blissful harmony.[63]

These texts set out complex theological explanations and vested authority in those who were trusted to understand, relate and interpret them. In South Asia, this privileging of information concentrated spiritual authority in the hands of a priestly class who were able to engage with the wishes of Brahmanaspati (the Lord of Prayer) and Vāch (the goddess of speech). *Brāhmans* (holy men or priests) controlled rituals, techniques and knowledge that enabled intercessions with and even control over the supernatural – including control over favourable climatic conditions. Some would ask which god should be honoured and thanked for creating all humans and all animals, for the 'snow-covered mountains' (the Himalayas), the great river (the Rasā that marks the boundary of the earth) and the shining deep waters of the seas. Only those who knew the hymn would be able to provide the answer: Prajāpati.[64]

That such views were written down, survived and were disseminated and added to naturally attests to their significance as the basis of what later became known as Hinduism.[65] What is rather harder to determine, however, is how they were perceived and understood beyond the elites who clearly invested time, energy and resources preserving teachings that contained universal truths about the earth, the gods and how to win the favour of the latter.

Challengers emerged who set out rival cosmologies, and many of them placed a premium on how people lived and behaved. In some cases, there were reactions against the idea that knowledge and wisdom lay in the hands of others rather than in the self. Among the most prominent and successful of these were the teachings of the Buddha, the son of a nobleman who spurned a life of luxury to search for enlightenment, which he discovered under a sacred peepal tree at Gayā in what is now Bihar, northern India.

Many of the sermons and views attributed to the Buddha are best understood as being formulated by his followers and successors, *bhikkhus* (भिक्षु), who were instructed to wear yellow robes and to continue his preaching after his death, which modern scholars now usually date to around 400 BC.[66] One central element of Buddhist philosophy was that the path to enlightenment was a highly personal one, centred on the doctrine of the 'Four Noble Truths'. These truths held that life was a journey of suffering; that this suffering was the result of desire and craving; that suppressing such desires removed sorrow; and that suppression could be achieved only through strict discipline resulting from intense concentration and meditation.[67]

While some scholars caution that treating Buddhism as a reaction to Vedic Brahmanism can be misleading, the fact that the journey towards enlightenment was a highly personal one which involved each individual discovering truths for themselves – rather than relying on others to intervene on their behalf – was significant.[68] The theoretical stance of Buddhism was one that did not require *brāhmans*, in other words; as the *Sutta Nipāta* pointedly notes, claims by others of moral superiority were ill placed:

No *brāhman* is such by his birth
No outcast is such by his birth
An outcast is such by his deeds.
A *brāhman* is such by his deeds.

What mattered, therefore, were personal actions rather than deep understanding of the workings of the universe that could be revealed by ancient Sanskrit texts.[69]

It was not for nothing, then, that many of the teachings set out by the Buddha and his followers were practical rather than theological. A husband should treat his wife by honouring her; treating her with respect; remaining faithful to her; giving her charge of the home; and regularly giving her gifts and adornments. A master should not overwork his servants; should ensure they are well fed and well paid; should look after them through illness; should share 'especially tasty luxuries' with them; and should let them take holidays at appropriate intervals.[70] Being charitable, kind, restrained and reserved was the source of personal treasure – treasure, furthermore, that could not be stolen by thieves or given to others.[71]

While Buddhism is often co-opted today and celebrated for its supposed ideas about ecological awareness and environmental sustainability, these are modern constructs rather than reflections of reality. In fact, for early Buddhists, the natural world was a place that was transitory; far from being a place of wonder, the physical world, like human life, was primarily about 'suffering, decay, death and impermanence'.[72]

That a canon of Buddhist texts was curated, preserved and disseminated provides a clear indication, of course, that not everything was left to the individual. Guidance could be sought to help steer one towards the goal of enlightenment, meaning that even within a notionally egalitarian world special status was reserved for monks and holy men. These texts contain warnings too of the dangerous and devious teachings of the 'Six Tirthakas' (false teachers), contemporaries of the Buddha whose views ranged from dismissing efforts to identify reason and order in the universe to denunciations of ideas about moral goodness. There was no benefit or reward in giving alms, performing sacrifices or giving offerings, said Ajita Kesakambala, for example: those who believed that doing 'good deeds' would result in future rewards were fools. 'There is no passing from this world to the next,' and anyone who taught that there were connections between the material and supernatural world were talking nonsense, indeed lying.[73]

Some sages advocated more stark re-thinking about the world, about life in general and about nature in particular. According to Vardhamāna Mahāvīra ('Great Hero'), the inspiration behind the Jain religion, the lover of possessions is deluded and falls into great pain – an idea that finds an echo in Buddhism. Jainism went further, however, by arguing that all organisms including plants were alive, as were the elements themselves. 'Injury to the earth is like striking, cutting, maiming or killing a blind man,' states one of the canonical texts of Jain scripture, the Ācārāṅga Sūtra. 'Knowing this, a man should not sin against earth, or cause or permit others to do so.'[74]

Jain ecological cosmology was extensive and demanded an asceticism that presented human beings as participants in a rich panoply of living entities that included 'earth and water, fire and wind, grass, trees and plants and all creatures that move'. In hurting any of these, 'men hurt themselves'.[75] As such, it was not enough to be mindful and to seek enlightenment and detachment. What was required was an unlimited awareness and heightened consciousness which revealed that trees felt pain as they were cut, stripped and sawn, that iron anvils trembled in pain and suffering when struck by the hammer and tongs of a blacksmith.[76]

It was essential, according to Jain teaching, to understand that humans are part of a living, connected environment and ecosystem and that they could be affected, hurt and damaged by human actions. 'A man who lights a fire kills living things,' states one text, for insects could hop into a fire inadvertently, worms living in wood that was used as fuel could be burned, and cutting down plants not only killed them but disrupted habitats for other plants and animals. So anyone who possessed wisdom and understood the meaning of life 'should never light a fire'.[77]

Some modern scholars have often associated the efflorescence of new ideas in India, which also extended into grammar, law, literature and drama, with wider social and political change driven by agricultural surplus in the first instance and by a new wave of urbanisation starting in the sixth century BC.[78] This was accompanied by a period of centralisation and consolidation as a succession of small kingdoms sprang up and then coalesced into sixteen large kingdoms (*mahājanapadas*), before these too expanded and merged.[79]

This presented new challenges and resulted in substantial shifts in thinking, and it was striking that some of the new ideas involved challenging traditional brahmanic teaching – in particular, those to do with animal sacrifice. Buddhist texts, for example, disdained the slaughter of animals, with the Buddha warning that animal offerings achieve no great results. Indeed, animal sacrifices are said to have produced negative outcomes, including longing, hunger and decrepitude, that themselves quickly multiplied to create ninety-eight further ailments.[80]

Conversely, notes the *Dīgha Nikāya*, offerings could be given where no animals and plants were hurt. Behold, it says, 'no cows were killed, no goats were killed, no cocks and pigs were killed, nor were the diverse living creatures subjected to slaughter, trees were not cut down for sacrificial posts nor were grasses mown …The sacrifice was pursued with clarified butter, fresh butter, curds, honey and molasses.'[81]

This was a provocative and revolutionary approach that effectively called into question the rights of the Vedic priesthood to control ritual and access to the divine, while asserting, if only implicitly, that their methods of doing so were unnecessary and suspect. In turn, efforts were made to articulate a model that explained the roles of the different members of society. The most prominent were centred on *varṇadharma*, a sacred set of laws that held that four hierarchical classes (*varṇas*) had been created, each of which was subject to their own *dharma* and order. Dividing society into *brāhmans* (priests and teachers), *kṣatriyas* (nobles and warriors), *vaiśyas* (merchants and agriculturalists) and *śūdras* (servants and menial labourers) amounted to a detailed political philosophy that not only dealt with abstract concepts about engagement with the natural world, but also offered formal explanations and guidance about the workings of secular society that went hand in hand with laws setting out the responsibilities of rulers (*rājadharma*).[82]

It may be that ideas about reincarnation developed in response to increasing spiritual competition, although both the timing and context are hard to pinpoint. Most scholars agree that there is no concept of rebirth in the Vedas and that the earliest expression comes in one of the early Upaniṣads. 'When people depart from this world,' it says, 'it is to the moon that they all go,' since the moon is 'the door to the heavenly world'. Those who are unable to pass through to the other side are turned into rain, which then falls on the earth and 'they are born again in these various conditions – as a worm, an insect, a fish, a bird, a lion, a boar, a rhinoceros, a tiger, a man, or some other creature – each in accordance with his actions and his knowledge'.[83]

This can be interpreted as humans trying to make sense of their fortunes and misfortunes in life, and seeking explanations for the events outside their control that took place in an earlier manifestation.[84] It is notable that similar hypotheses were being posited in the eastern Mediterranean around the same time. Herodotus, writing in the fifth century BC, observed that Egyptians and then Greeks had adopted the belief 'that the soul of a human being is immortal, and that each time the body dies the soul enters another creature just as it is being born'.[85] Some, like Pythagoras, were said not only to have lived previous lives, but even to have known who they had been – with the great mathematician apparently able to offer 'indisputable proofs that he had been Euphorbos son of Panthoos, the opponent of Patroklos', who had featured in *The Iliad*.[86] Another, Empedokles, had a wider set of experiences, noting that 'in the past, I have been a boy and a girl, a bush, a bird and a mute fish swimming in the sea'.[87]

Ideas about the immortality of the soul, about reincarnation and about goodness were bound up with wider questions of how one's actions should be judged both in this life and after death; this was a time when philosophers like Plato held that one would be rewarded or punished accordingly.[88] Such questions were part of broad and vibrant philosophical reconfigurations, many of which reflected on interaction between humans, their interaction with the divine, with the natural world or all three. Among the cosmological views adopted in ancient Greece were the beliefs that the universe was animate, sentient and intelligent, held together by harmonious forces.[89]

This inevitably informed ideas about all living things, which according to Plato, were all 'endowed with soul and reason'.[90] Those with ascetic beliefs took this further, arguing that eating animals and plants was wrong as this usually involved killing them. Food 'that has had life' was scorned, therefore, in favour of milk, cheese, honey, wine, oil and leafy vegetables which could be consumed without harming a living animal and therefore guilt-free.[91]

Such lifestyle choices seem remarkably up to date, even fashionable in the world of today where decisions about food in particular are often influenced by opinions of animal and plant welfare – which are themselves sometimes set within a wider context of resource availability and sustainability. As today, however, it would be wrong to think that such attitudes were mainstream. Not only did many other philosophers not equate animals and plants with human beings, some suggested an entirely different hierarchical model. For Aristotle, plants existed for the benefit of animals; animals existed for the sake of people; and inferior people served as slaves to those who were superior.[92] Socrates was even blunter, at least according to Plato: there was nothing to be learned from trees, nature and the countryside; the only place one could gain knowledge was in the city, from other men.[93]

Nevertheless, anxieties about environmental degradation and pollution had their champions both on an individual basis and, in some places, institutionally. As early as c.700 BC, Hesiod was warning about the dangers of urinating or defecating in springs or rivers, stressing that doing so could lead to disease and suffering. Xenophon was one of many who advocated the benefits of good treatment of ecological habitats: the earth rewarded those who treated it well, he wrote, 'for the better she is served, the greater the benefits she gives in return'.[94] Others, like Theophrastus, took to observing which habitat suited which plants best, and devised and recorded typologies in order to systematise knowledge so that it could be useful for future generations.[95]

Further east, what started as a local revolt by the Persian leader Cyrus in the mid-sixth century BC eventually ended with him overthrowing Neo-Babylonian rulers and expanding the frontiers into an enormous empire that stretched from the Mediterranean to the Indus Valley. The new masters championed a set of beliefs centred on a supreme deity, Ahura Mazda (literally 'Lord Wisdom'), and on a reverence of nature so profound that some scholars have even referred to the faith – known to its followers as Mazdaism or the Good Religion, but to most as Zoroastrianism – as the world's 'first environmentalist religion'.[96]

Zoroastrians sought to focus on goodness and purity, especially articulated in the form of fire, water and soil. The insistence on avoiding pollution was commented on by others. Because they greatly revere rivers, wrote Herodotus; 'they do not urinate, or spit into them, nor do they wash their hands there, or allow anyone else to either,' while the later Roman writer Strabo noted that Persians did not bathe in water, nor 'throw anything unclean into it'.[97]

The inspirational figure behind the religion was one Zarathustra (or Zoroaster to Europeans), a figure about whom little is known, with speculation about the dates when he was alive ranging from c.1800 to c.600 BC.[98] Many of the teachings attributed to him were recorded in a set of texts, the earliest of which is a collection of seventeen hymns known as Gathas. The oldest of these in turn contains an appeal to Ahura Mazda to look after cattle and pastures:

> What help shall my soul expect from anyone,
> In whom am I to put my trust as a protector of my cattle ...
> How is he to obtain the cattle which brings prosperity, O
> Wise One,
> He who desires it, together with the pastures?

Only those who behold the sun, live their lives well and with righteousness, would deserve such benefits, says the hymn.[99]

Another hymn offers thanks to Ahura Mazda, 'who created water and good plants', as well as order, light, earth 'and all good things'.[100] In yet another hymn, the soul of an ox offers a lament to the Wise Lord, pondering the 'cruelty and tyranny' imposed by humans who seek to kill it. 'Humans do not understand how to treat the lowly,' laments the ox, who goes on to note that even Zarathustra's appointment as his protector left him at the mercy of 'the powerless word of a powerless man'.[101] Nature should be a source of joy, says another hymn.[102] However, as these texts make clear both implicitly and explicitly, the relationship between people and the natural world was an imperfect one, not least because of the destructive habits and tendencies that shaped human behaviour.[103]

The championing of the teachings of Zarathustra was a feature of Persian kingship and of a new ruling class that reserved authority and status in the hands of the few and, as in other periods and regions, provided the elites with a powerful tool for the concentration of cosmological interpretation and spiritual intercession. The emergence of new hierarchies, as well as debates and arguments, discussions and competitions for prominence, took place in many different places in Asia, North Africa and the Mediterranean concurrently in a way that makes it tempting to see not only patterns of development but also ideas, influences and reactions within and between peoples, cultures and civilisations.

This included Judaism, which had been heavily influenced by contemporary accounts of the Jewish experience at the hands of the Assyrian and Neo-Babylonian empires in the seventh and sixth centuries BC. The extent of these influences can be shown on the one hand by the fact that the Assyrians are mentioned 150 times in the Bible, always negatively, and on the other by the sack of Jerusalem and the destruction of the Temple by King Nebuchadnezzar in the 580s BC, and by the resulting Babylonian captivity which lasted for decades.[104] The release from bondage following the rise of the Persians and the fall of the Neo-Babylonian empire, and the edict permitting the rebuilding of the Temple in Jerusalem, resulted in the Persian leader, Cyrus, being deemed by the Prophet Isaiah to have been 'anointed by God'.[105]

These experiences had a profound impact on how Jewish scholars thought about the natural world around them and about the past. As many scholars have pointed out, the narrative portions of the Torah were written following the destruction of the Temple and the exile of the Jews. These include the discussion of the relationship between God, nature and mankind in the Book of Genesis, the causes and meaning of the enslavement and departure of the people of Israel from Egypt recounted in the Book of Exodus, and the consequences of breaking the Covenant that God had agreed with Moses outlined in the Book of Deuteronomy. Jewish experiences had resulted in texts which were first set down around this time and which emphasised the importance of devotion, of obedience and of following the priestly teaching.[106]

Jewish writers drew on a rich set of stories, stretching deep into the past. These include the accounts of Noah and the flood that derived from events that had happened many centuries, millennia even, earlier, as we have already seen, and of the Tower of Babel, which has many similarities with a tale that appears in the Sumerian *Enmerkar and the Lord of Aratta*.[107] Other events from the distant and not-so-distant past were woven into texts that were written down and went on to form the basis of scriptural texts as we know them today. One was the story of Joseph, who rose to become the right-hand man of the Pharaoh and one of the most powerful men in Egypt; this likewise would appear to draw on the life of a high-ranking official, likely a chancellor named Baya who lived around 1200 BC.[108] The way in which the expulsion of the Hyksos rulers from Egypt influenced Jewish thinking has been a subject of lively scholarly discussion in recent years.[109]

Perhaps the most interesting element in Jewish thought, however, was the antipathy that was displayed towards urban space – one element in a wider rejection of the high culture of the Near East.[110] The desert, not the city, was the place where the faithful could get to know God directly and also be tested by Him – both in Judaism but in the other Abrahamic faiths too, namely Christianity and Islam. Another important part of the evolution of Judaism was the rejection of polytheism in favour of a monotheistic faith, while the condemnation of animal sacrifice and the worship of idols likewise reflect a strident formulation of a set of beliefs that stood in sharp contrast to the attitudes of those who had oppressed the Israelites from imperial capitals.[111]

The equation of Jewish engagement with the natural world thus diverged in important ways from traditional views of the environmental framework of the Levant. Although the Garden of Eden had served as an idyll of plenty, it was a paradise lost with the fall of Adam and Eve. Humans had been created to work or serve in this garden, in the same way that priests serve God in the Temple. This relationship continued, nevertheless, with God being served through the cultivation of the earth and the careful stewardship of its resources.[112]

This was typified by evocations of idealised rural rather than urban settings, and with God himself as a pastoralist, looking after his flock – most obviously in Psalm 23: 'The Lord is my shepherd, I shall not want; he makes me lie down in green pastures. He leads me beside still water; he restores my soul.'[113] Key figures in scriptural narratives are regularly portrayed as farmers and shepherds, including Abraham, David and all the patriarchs.[114]

Indeed, the Covenant made between God and his people following the redemption from slavery in Egypt and settlement in Canaan is specifically related to the divine bestowal of an ecological agreement: in return for their following the commandments, including disavowing the worship of idols and other deities, God agreed to make the land fertile and productive, and to ensure peace for Israel. The terms of the Covenant also involved environmental punishment: as Moses reminded his people, the Promised Land was not like Egypt, where 'you planted your seed and watered it with your foot like a vegetable garden'; rather it was a place 'watered by the rains of heaven'. As such, any breach of the agreement with God would result in rains being withheld, with obvious consequences for food, health and happiness.[115] The elision between God and nature was such that the Almighty appeared in the guise of a cloud so that Moses and others could hear what he had to say, having previously marked the 'sign of the covenant' in the form of a rainbow.[116]

Conversely, those who see themselves – rather than God – as masters of the earth and its resources were to be punished. Annoyed by Pharaoh's bragging about how he controlled the River Nile and had improved it, God threatened that 'Egypt will become a desolate wasteland' and so be taught a lesson. 'I will make the land of Egypt a ruin,' came the warning in the Book of Ezekiel, and ensure that 'The foot of neither man nor beast will pass through it; no one will live there for forty years.' Cities would be ruined and become depopulated, and Egypt would turn into a 'lowly kingdom' that was 'so weak that it will never again rule over the nations'. Not only could God bend nature to his will, in other words; but in doing so he would instil the virtue of humility among men who thought too much of their own powers and too little of those of the Lord.[117]

The extent and vibrancy of the formulations and reformulations of philosophy, religion and behaviours from the eighth to the third centuries BC in the Yellow and Yangtze River valleys, the eastern Mediterranean, the Levant and the Ganges Valley have been much remarked on, most notably by the German philosopher Karl Jaspers who in an influential book published soon after the Second World War characterised this period as an 'Axial age'. This was a time, suggested Jaspers, when humanity was taking 'a deep breath' and pondering profound questions that stimulated new levels of consciousness, which Jaspers contended had happened independently across these five 'islands of light', namely in India, China, Persia, Palestine and the eastern Mediterranean.[118]

Other scholars have followed suit, arguing that this was a period of 'cultural crystallisation', an 'Age of Transcendence', marked by a new ability described as 'standing back and looking beyond'.[119] It was a revolutionary era 'in the realm of ideas and their institutional bases which had irreversible effects' not only in many regions but 'on human history in general', leading to an 'extensive re-ordering of the internal contours of societies as well as their internal relations' and changing 'the dynamics of history' as a result.[120]

Not all are convinced by the argument of multiple simultaneous watersheds during which archaic values were abandoned or modernised, dismissing the hypothesis as too simplistic and noting that some of the apparent transformations took place hundreds, or even thousands of years before the supposed 'age' of change.[121] What does appear to have been significant, however, was not a notional cognitive shift in ways of thinking as much as a change in the mechanics of how knowledge was codified and disseminated.

It was the proliferation of texts and, no less importantly, their preservation, transmission and copying that was crucial. Lists, narratives and scriptures as well as other written materials established corpuses of information that could be learned, discussed, added to and interpreted. While religion was clearly important, it was by no means the only branch of knowledge that benefited from the boost in more people writing, as is clear from seminal works of philosophy, mathematics and sciences being produced notably but not only in the eastern Mediterranean.[122]

An additional factor seems to have been rising levels of urbanisation and of affluence which fuelled a counter-intuitive shift away from material rewards towards self-discipline and selflessness – a common feature across many of the belief systems that blossomed around the sixth century BC, including Buddhism and Jainism in South Asia, Stoic philosophy in ancient Greece and the teachings of Confucius, Laozi and others in East Asia. The emphasis is on moderation and on the suppression of desires through fasting, abstinence, suffering and compassion at the expense of greed, indulgence and consumption. These ideas may in turn have stemmed from the emergence of a new class of scholars and thinkers who had the resources and influence to spend time contemplating questions existential – and thus represent the interests, as well as lifestyles, of elites expanding their spheres of authority beyond economic power and into more ethereal domains.[123]

Perhaps not surprisingly, this then spilled over not only into giving advice to rulers but into castigating them for adverse natural phenomena. In one chapter of the *Shangshu* (尚書), the Book of Documents compiled around the fifth century BC (though sometimes drawing on older materials), an explicit link is drawn between good harvests and the stability provided by a conscientious ruler, by an 'enlightened administration' and by the reliance on 'talented men'.[124] The connection is forcefully made between the shortcomings of a king's character and weather conditions: excessive rains were a reflection of a lack of discipline, excessive wind a sign of foolishness, excessive cold an indicator of poor judgement, excessive heat the result of laziness and drought the result of arrogance. This was one reason why for many centuries kings led ritual ceremonies to mark the start of the agricultural years.[125] Weather was therefore connected to morality, giving precious little climatic room to manoeuvre; anything meteorologically unexpected or harmful was the fault of a single person – the ruler. This equation proved remarkably durable in Chinese cultures over the centuries.[126]

Conversely, this opened up opportunities for intermediaries, such as official rainmakers who had devised elaborate rituals to encourage rain, through dances, offerings and prayers. Stories of extraordinary personal suffering became more common in accounts written from around the third century BC about the sacrifices made by Chinese kings as they tried to protect others. Presumably intended to create idealised images of previous rulers for contemporary audiences, these tales often spoke of deprivation so acute that it led to organ failure, to bodies shrivelling and to physical deformities that took away the ability to walk. Many involved exposure to the sun, fervent prayer and threats of self-immolation, including by leading officials, whose willingness to suffer was intended both as a cipher for their own selflessness and as oblique criticism of their masters.[127]

This was part of a familiar pattern of elite acquisition and guarding of political powers through control of ritual in general, but also through control of water and climate in particular across regions and indeed periods. The royal bath ceremony of late nineteenth-century Madagascar provides one compelling example; another is provided by pre-colonial Balinese society, where water and ritual were used to dramatise status and authority, as a central element of what one influential scholar memorably described as a 'theatre state'.[128] Pre-Buddhist ascetics undertook retreats in preparation for monsoon rains in a practice that was then adopted as a requirement of Buddhist monastic discipline and developed into a series of rituals known as the varṣāvāsa.[129]

Ideas about rains evolved over time, with beliefs in serpent deities in northern India, particularly in the Mahāyāna tradition, finding echoes in East Asia, where it was held that dragons controlled the rains and that these powers could be appropriated by suitably qualified intermediaries. Such intermediaries not only used rituals to showcase their skills, but produced literatures attesting to these abilities that shrewdly also emphasised aristocratic and royal patronage and eminence as well as drawing putative connections with Buddhist 'dragon kings'.[130]

In Central America, population density and colonisation levels had remained low until the start of the 'Classic period', which scholars usually date to AD c.250. As they rose, control over water became increasingly important, both because the thin karstic soils of the Maya lowlands were dependent on seasonal rainfall that could be erratic and because of an absence of natural water sources.[131] Rain petitioning became an important element in religious thinking, social hierarchy and climatic engagement. Archaeological evidence suggests that rain-related ceremonies took place in caves, with finds of whistles, flutes, bone rasps and turtle shells used as drums.[132] In some cases, children were offered as human sacrifices to the rain deity, part of a tradition that stretched back to c.1500 BC.[133]

Fertility rituals connected to the land and to clement weather and climate conditions were widespread, if not as ubiquitous, as ceremonials based on elaborate cosmological explanations were devised to demonstrate human agency in influencing the weather via engagement with the divine. Those who controlled the knowledge protected their privileged positions jealously: in the case of the Eleusinian Mysteries in ancient Greece, divulgence of secrets was punished by sentence of death.[134]

While there were wide variations in interpretations and in attempts to ensure benign climate conditions, the principles were the same: take the credit for plentiful harvests and assign blame for extreme weather events that led to food shortages. It was hardly a surprise, therefore, that this equation was accompanied in many locations by a growing awareness that resources were finite and that overexploitation could lead to depletion and exhaustion that not even the gods could resolve – let alone their intercessors on earth.

In 524 BC, Duke Mu of Shan, a senior official at the court of the Zhou King, warned of the dangers of unsustainable cutting down of 'the forests of the mountains'. It would not just be bad for the environment if trees disappeared and 'thickets and marshes have come to the last days of their existence'; if these things happened, 'the forces of the people will be weakened, the farmland where cereals and hemp grow will be uncultivated and they will lack resources'. This was not simply a question of caring about those who would suffer as a result. Any right-minded individual 'should be concerned about this problem' and furthermore should do so 'without relaxation'. After all, it was not hard to understand that this would create problems for all in society, not just for those who worked the land.[135]

In some cases, laws were passed to prevent access to and exploitation of natural resources. In 243 BC, for example, the Emperor Ashoka, grandson of Chandragupta Maurya who had founded an empire that covered much of northern India, issued an edict that forbade setting fire to forests.[136] This was partly about establishing imperial control over resources and over locations that had gained (and retained) reputations of teeming with lawbreakers, robbers and ascetics: indeed, the Buddha himself had been involved in a celebrated altercation with Aṅgulimāla, a violent brigand whom he met as he passed through the forests, persuading the latter to give up his ways and follow the Buddhist path.[137]

However, Ashoka's motivations seem to have been intertwined with a wider philosophical attitude towards animals and plants. Rock edicts set up across his territories announced that 'no living beings are to be slaughtered or offered in sacrifice'. For centuries, these inscriptions add, 'killing or harming of living beings' (including humans) had increased. It was vital that all should practise 'self-control and purity of heart' rather than being ruled by 'various desires and passions'. To support compassionate behaviour, the Emperor declared that he had 'made provision for two types of medical treatment: medical treatment for humans and medical treatment for animals. 'Wherever medical herbs suitable for humans or animals are not available, I have had them imported and grown. Wherever medical roots or fruits are not available I have had them imported and grown. Along roads I have had wells dug and trees planted for the benefit of humans and animals.'¹³⁸

This remarkable investment in welfare, assuming it is a reflection of reality, was furthered by the announcement that the royal household would become vegetarian. 'Formerly, in the kitchen,' stated Ashoka, 'hundreds of thousands of animals were killed every day to make curry.' From now on, however, 'only three creatures, two peacocks and a deer, are [to be] killed, and the deer not always. And in time, not even these three creatures will be killed.'¹³⁹

Ashoka took other steps to protect wildlife, issuing proclamations protecting 'parrots, mynas, aruna, ruddy geese, wild ducks, bats, queen ants, terrapins, boneless fish, tortoises, porcupines', along with many other animals – including 'all four-footed creatures that are neither useful nor edible'. The sale of fish was forbidden on designated days, as was the castration of 'billy goats, rams, boars and other animals' and the branding of horses and bulls.¹⁴⁰

How these orders were enforced, by whom, where and when is not clear. Nor is the line between Ashoka's religious and spiritual beliefs and his royal powers entirely distinct, for an important element of such commands was plainly that of asserting rights not only over animals and plants, but also over behaviour, customs and standards. Nevertheless, the fact that they were carved into rocks across Ashoka's lands is itself significant in so far as it demonstrates rulers claiming a mandate not only over their human subjects, but also over the environment, its animals and plants as well as its natural resources.

Another example of this comes from the *Arthaśāstra* of Kauṭilya, a text written in the first century AD which drew on sources at least a hundred years older.¹⁴¹ A lengthy section sets out the role and responsibilities of judicious leaders. The advice proffered includes recommendations to grant land exempt from taxes and fines to scholars and holy men, as well as gifts that could not be sold or mortgaged to officials with important functions – among them revenue officers, supervisors, elephant and horse trainers and courtiers. Land that was not being cultivated should be confiscated and redistributed, and all help should be given to those working it to ensure that revenues were generated, 'for a king with a depleted treasury devours' and preys on his own subjects.[142]

Among the recommendations for a functioning bureaucracy was the appointment of a 'Superintendent of Agriculture', 'who must be either proficient in agricultural science, geometry and plant science, or assisted by experts in these'. The fact that special prayers were to be offered using a set formula – praising Prajāpati, Kāśyapa, Deva and Goddess Sītā, 'at the first sowing of any kind of seed', with the first fistful prepared by being immersed beforehand 'in water containing gold' – suggests that while Ashoka's own devotion to Buddhism had been extensive, the invocations either existed alongside Vedic beliefs or were displaced by them in the decades after his death and the fall of his empire.[143]

In any event, the point here is that Kauṭilya – one of the most important thinkers and authors writing in classical India two millennia ago – declared that rulers should protect and manage all the key resources of the state. To this end, careful attention needed to be paid to domesticated animals, human crafts, agriculture, mining and commerce, as well as to wildlife, forests, forest produce and water. Those who damaged or overexploited these resources should be punished with death.[144] The subtext was clear: good governance led to optimal outcomes. And underpinning this was the virtue of social, economic and environmental sustainability.

It was a message that echoed what some had been writing thousands of kilometres away around the same time. Expressions of idealised nature, in which names of plants serve as implicit and explicit metaphors for moral qualities, do not so much occur regularly in the poetry of the Han dynasty in China that took power in 206 BC as define collections such as *The Songs of the South* (楚辭).[145] To put it another way, wrote Xenophon in the *Oikonomikos* a few centuries earlier, 'the earth willingly teaches righteousness to those who can learn; for the better she is treated, the more good things she gives in return.'[146] He did not need to say what this also meant: that those who extracted too much and overexploited resources did not just damage the earth, they demonstrated their failings as humans too.

8

The Steppe Frontier and Formation of Empires (c.1700–c.300 BC)

History has become an organic whole.

Polybius (second century BC)

As we have seen, human settlement and social patterns had changed sharply with the cultivation of crops. The domestication of goats, sheep and then cattle had likewise brought a series of mini-revolutions as a result of the availability of stable sources of protein, sources of materials such as wool and leather hides that could be used for clothing, storage and new technologies. New sources of energy were also released as animals were put to work, enabling greater agricultural returns for lower human input – which in turn opened up ways to invest this additional time.

No less important than these developments was the domestication of the horse. As one scholar has put it, horses – which can travel at around ten times the speed of humans – were 'essential to the creation of large empires in the classical age'.[1] Indeed, their central role in global history extends far beyond antiquity, not least in the Americas, where the impact of the introduction of horses on indigenous peoples is difficult to overestimate.[2] Above all, the domestication of horses transformed human engagement with, use of and treatment of the environment.

Although both Europe and Asia were home to large horse populations at the start of the Holocene, ecosystem change brought about a sharp decline in the former regions, where the demise of steppe-tundra landscape and the return of forests as a result of warming climate patterns offered less suitable habitats for horses and perhaps even caused an increase in the frequency of genetic variants coding for darker coat colour as a result of lower exposure to sunlight in dense forest environments.[3]

While horse populations in most parts of Europe seem to have become fragmented and genetically isolated, it was a different story on the steppes of Central Asia where conditions were much more climatically favourable.[4] Although dating and location are matters of considerable scholarly debate, among the first to tame and breed horses were the Botai people who lived in what is now Kazakhstan around 3500–3000 BC. Hundreds of thousands of horse bone fragments attest to widespread dependence on the creatures, with tests on pottery showing fatty deposits that suggest horses were prized for their milk.[5] At least some peoples mastered horseback riding in this period, as well as training horses to pull chariots and wagons, although heavy, rugged cheekpieces, as well as scrape marks on horse teeth, show that harnessing and controlling horses was not easy at all.[6]

It did not take long for horses to be introduced to other regions, such as Mesopotamia, where at least initially they were regarded as more of a curiosity than a widely utilised animal, and where accounts record them as often having been fed to lions that were kept in captivity as entertainment for local rulers.[7] In general, across western Asia from the time they first start to appear in written and archaeological sources, horses were considered unreliable and dangerous.[8]

Horse-drawn chariots nevertheless became crucial in enabling an expansion of control of territory under Egyptian pharaohs in the New Kingdom (c.1550–c.1050 BC), while accounts like the *Annals of Thutmose III* include booty lists that show how highly prized chariots were as markers of elite status and as tools of political control.[9] This can be shown too by the fact that six complete 'high-performance machines' were discovered as prized possessions buried alongside Tutankhamun.[10]

Poems written under the Zhou dynasty celebrated heroes driving their obedient dappled grey horses in large red painted chariots charging against enemies, rumbling and crashing 'like a roll of thunder'.[11] Horses and chariots were also high-status gifts, conferred to show favours and honour, with one-third of bronze inscriptions that record appointments to office or grants from the crown after c.900 BC recording chariot paraphernalia.[12]

'I confer on you a chariot with bronze fittings, trappings of soft leather painted red for the horses,' one ruler announced, with 'a canopy of tiger skin with red lining, bronze bells for the yoke bar, gilded bow holder and fish-scale quiver, a team of four horses with their bits and bridles and bronze ornaments, gilt girth straps and a scarlet banner with two bells'.[13]

Across Eurasia, horses were given an important ritual status. The Hittite *Code of the Nesilim*, for example, a legal text written around 1600 BC, held that while anyone found having sex with pigs, dogs or cows should be tried and sentenced to death, bestiality with horses should not be considered a punishable act.[14] Horses were also revered further east, appearing in the *Rig Veda* in India as offerings in ritual sacrifice to the gods, as an animal mastered and ridden by deities, or transporting gods by pulling them in chariots.[15]

What is most striking about the expansion of horse culture, however, is both how far and how fast it spread. By around 1200 BC, horse sacrifice was being practised across large swathes of Eurasia, from Kazakhstan to north-western China and Mongolia.[16] In South Asia, horse immolation was also a key means of reinforcing the authority and power of the priesthood in general and of the ruler in particular. The *Śatapatha Brāhmaṇa*, a text compiled c.900–700 BC, sets out in detail the practice of *aśvamedha*, the sacrificial killing of a horse, that was one of the most important rites in Vedic ritual. As the text makes clear, horses were of particular significance and value – so that while Prajāpati, the lord of creation, treated by the author as the supreme deity, assigned other sacrifices to other gods, he claimed horse sacrifice for himself.[17]

By c.1200 BC, it seems that horse-handling, training and herding techniques had become increasingly sophisticated, resulting in both higher levels of dietary reliance and economic exploitation. Perhaps most significantly, however, zooarchaeological evidence points to large increases in the numbers of horses being reared, alongside the emergence of widespread horseback riding in Mongolia and across Central Asia.[18] Human skeletal remains from this period also show substantial increases to the muscle, tendon and ligaments connected to foot, hip and elbow joints that are most logically connected with horseback riding – providing an indication of the extensive and rapidly growing practice of riding.[19]

Although the causes of this transition are not clear, that it took place around the same time as a climatic reorganisation that affected large parts of the Levant and beyond may not be coincidental. As stable nitrogen isotope values taken from bone collagen show, before the emergence of horse riding, meat and milk products consumed on the steppes were primarily from sheep, goats and cattle.[20] The fact that the domestication of horses took place and expanded dramatically at a time when the open steppe regions stretching from the Black Sea across Central Asia into Mongolia were particularly dry has been noted by scholars, who have suggested that pastoral subsistence and a turn to horses as a source of food, protein and milk and as a labour supplement were responses to changing climate conditions. High aridity levels in central and western Mongolia in particular would have produced dryland grasses and steppe plants on which horses thrive, while the long grazing ranges and resilience of horses faced with water shortages explains why more time and effort went into horse herding.[21]

Moreover, while linking the adoption of new equine strategies and techniques with a period of rising aridity seems both plausible and logical, identifying cause and effect is an altogether more complex and difficult exercise. In any event, what mattered more were the ecological impacts that horse riding brought with it: a transition to a more diverse pastoral economy from one which had previously been primarily based on sheep; larger and larger herds of horses, which in turn had implications for the amount of pasture required; and major shifts in settlement patterns from semi-sedentary to more dispersed levels of living, and higher levels of mobility.[22] Isotopic data suggests that while ruminants were intensively grazed close to settlements, horses were moved over extensive areas across the steppe.[23]

These changes were instrumental in creating networks of long-distance contacts that enabled and encouraged the spread of ideas, beliefs and rituals, as well as exchanges of goods and technologies.[24] Mobile pastoralists had already played an important part in the transmission of crops across Eurasia over the course of many centuries – millennia even. That the earliest evidence comes from carbonised seeds of barley, wheat and broomcorn millet found in mortuary contexts (often human cremations) suggests that trade may initially have been linked to rituals rather than driven by the importance of foodstuffs to diet or calorie consumption.[25]

By c.1500 BC, domesticated grains such as millet, wheat and barley had been spread from South-West Asia to the east of the continent, including into the mountain regions of Inner Asia.[26] Millet seems to have been particularly important, likely thanks to its short growing season and resilience, especially when faced with hot conditions.[27] Foodstuffs were not the only thing moved along the networks of herders and nomads, for so were commodities such as copper and tin, and technologies such as ceramics and metallurgy which saw the appearance of bronze knives, axes and other artefacts produced in similar ways and in similar styles across the steppe as part of a 'Transeurasian exchange' mechanism.[28]

The introduction of horse riding must have played a part in the intensification of contacts between diverse ecological zones, partly as a result of herders fanning out in search of greater spaces for grazing, which led to competition for the best pasture lands and water sources, and partly because it brought peoples into closer and more regular contact with each other. One effect can be seen in substantial genetic turnovers and genetic admixtures across Central Asia around this time.[29] Another is apparent from a phase of rapid socio-economic transformation, including widening of social hierarchics in mobile populations shown by increasingly elaborate burial rituals in the steppe lands, as well as by evidence that points towards rising levels of craft specialisation.[30]

As some scholars have pointed out, there were significant variations that were closely linked to local ecological conditions and constraints. Different settlement patterns developed in locations where summer rainfall was abundant or where soils were richer; in such instances, agriculture could be intensified and in some cases expanded through the development of irrigation systems.[31] There were significant regional differences too, with plentiful evidence for farming in the eastern steppe contrasting sharply with limited and even minimal evidence further west.[32]

Nevertheless, as analysis of tooth enamel and skeletal remains shows, nomadic peoples developed a highly diversified range of economic strategies, rather than relying on uniform models of pastoralism.[33] As well as showing that agriculture was an important part of pastoralist lifeways, the production of the food sources that could and presumably did generate surpluses played a vital role in rethinking uses of the land and engagement with the environment by peoples living on the steppes; it was no less important in helping scholars understand the rise of nomadic empires – which provides one of the great themes of Eurasian and global history.[34]

In antiquity, nomads came on to the radar of writers based in cities intrigued by their way of living but also by their resilience. Herodotus explains how 'Scythian' horse riders were able not only to hold off the mighty Persian armies, but also to harass them effectively and to melt away to fight another day. When asked why he kept retreating and refusing to fight in open battle, the Scythian ruler purportedly replied, 'What I am doing now is not far removed from my usual way of life during peacetime. I'm not going to fight you and I'll tell you why. If we had towns we might worry about the possibility of them being captured, and if we had farmland we might worry about it being laid to waste; but we don't have either.' While this exchange clearly reflects artistic licence, the articulation of the idea that nomads could benefit from an indefinite margin of retreat when in conflict with sedentary rivals captures the reality of the relationship between the peoples of the steppes and those within striking distance.[35]

It was not quite true that nomadic peoples did not have permanent settlements. Groupings like the Xiongnu, Uighur and Liao that rose to prominence over later centuries and established control over vast territories all created or expanded large urban areas that included monumental structures, gardens, orchards and irrigation systems.[36] Drone surveys at Mohuchahangoukou in what is now Xinjiang in China have revealed hundreds of fields, canals, dams, cisterns and houses that show how mobile pastoralist communities combined complex and varied lifestyles that were sensitive to small environmental and ecological change and adopted strategies both to try to take advantage of these and to reduce risks as appropriate.[37]

Those who wrote about mobile pastoralists in whatever period and whatever region or continent invariably described them as barbaric and chaotic. Nomads look like monkeys, do not revere the gods, dig up truffle in the foothills and eat raw flesh, wrote one Sumerian author. They had 'the heart of wild birds and beasts', were 'unbridled people without human intelligence', 'abandoned by heaven' and 'the seedbed of evil', interested only in stealing from other people, just like wolves – to cite just a handful of examples across almost three millennia.[38] Such accounts were written by people living in towns and cities for the benefit of those living in towns and cities and represent the views of metropolitan intellectuals revelling as much in their own self-importance as in their derision of the ways that others lived.

There is some evidence that nomads were unimpressed by haughty city-dwellers who used sweet words and offered fine treasures – but always because they had ulterior motives. Do not be tempted to go to live with these people who inhabit fortresses which 'rub shoulders with the sky' and where souls get snatched, one son was told by his father. Much better, the father said, to stay away from them.[39]

Crass characterisations of nomads by urban elites do not just obscure but ignore the fact that hybrid, flexible and adaptable lifestyles required sophisticated processes of administration which combined decision-making with effective policies that ensured common purpose as well as social cohesion for groups living together. Far from being ignorant, impulsive and lacking reason, steppe societies were efficient, structured and highly successful. Indeed, even Herodotus says as much: the Scythians – a catch-all for all nomadic peoples on the steppes – could hardly be considered admirable; however, they 'have managed one thing, and that is the most important in human affairs, better than anyone else on the face of the earth: I mean their own preservation'.[40] These societies did not resemble those of city-based states, controlled by bureaucratic elites, but instead ones shaped at a local level rather than by a centralised authority, where the interests of powerful aristocrats were instrumental in processes of decision-making.[41]

What is of course most striking about the relationship between sedentary and mobile societies was their interdependence. Those who herded animals had to be located close to consumers who needed or wanted products – ranging from dairy to meat, from wool to leather, from pack animals to horses. These needs were not uniform and themselves reflected local demands, tastes and climates. For example, milk and vegetables are more important in diets across the Middle East relative to Central Asia, where meat consumption is notably higher.[42]

Nevertheless, the patterns of exchange worked in similar ways. Pastoralists provided sources of food, materials and goods, and in return could tap into sources for luxury goods and objects that were key elements helping to underpin social stratification and the authority of tribal leaders both in the exhibition of their own status and in the way they could reward their own family, kinship groups and wider networks.[43] While pasture was owned by groups collectively, animals were owned by individuals; as a result, prestige was also derived from the composition of herds, as well as from their size.[44]

***

One area of intensive interaction was in the northern and western regions of what is now China, where regular and vibrant exchanges saw animal products and animals being sold in one direction, and silks, gold and bronze ornaments, ceramics and other materials specifically designed in 'nomadic style' for nomadic markets going in the other.[45] Interaction with western Asia is also attested by trade in carnelian beads that were almost certainly finished in Mesopotamia and that feature prominently in graves in the Wei and Yellow River basins around this period.[46] These exchanges not only brought materials across long distances but also introduced ideas and designs – such as griffin motifs which spread from the Aegean, eastern Mediterranean and Persia across to East and South Asia.[47]

The discovery of large numbers of cowrie shells, perhaps from the south coast of China and possibly from the Indian Ocean, shows that this was a time of growing movement of goods and people in general and not only along land corridors. That such shells are found in graves belonging to people of modest means points to extensive circulation networks and also to widening socio-economic expansion.[48] While it has been often argued that cowrie shells were used as a form of currency, their value would seem to be more naturally associated with ornamental, ritual and funerary functions.[49]

Geography, environment and climate nevertheless played a crucial role in creating ecological as well as cultural boundaries, especially in the relationship between mobile and sedentary societies in East Asia. While Chinese sources regularly describe nomadic peoples and cultures in negative and even derogatory terms, the expansion of animal husbandry into zones that were within reasonable distance of cities and towns clearly played a role in the ability of urban populations to feed themselves and have access to materials for textiles and leather – which were of more practical and immediate importance than the movement of luxury goods and the transmission of artistic styles.

Although scholars have long noted the role played by nomads in the creation and flourishing of long-distance trade networks that spanned hundreds of kilometres, some have recently sought to emphasise that forms of mobile pastoralism were varied and that ideas about behavioural continuity over long periods (and still more between or within regions) are oversimplistic.[50] Indeed, some scholars have gone further still, seeking to emphasise that nomads did not so much offer an alternative way of life to those in settled communities as form part of an integrated world that was mutually beneficial. Rather than being the antithesis of city living, nomads in many cases galvanised their rise through the twin motors of supplying much needed goods and materials on the one hand and of presenting a threat on the other that encouraged social, political and military consolidation.[51]

The fundamental role that nomads played in the context of the Warring States period in China, and particularly during the fourth and third centuries BC, has been persuasively and eloquently put by one leading scholar who has underlined the accelerating importance in this period of steppe populations in general and of horses in particular. This was a time when rivalries and hostilities between rulers across a range of Chinese states produced a near-endless appetite for resources, starting with horses but including other animals too, such as pack animals for moving military equipment, as well as grooms, trainers, saddles, clothing and leather which had many uses, not least as armour.[52]

These demands, which can only have been increased further still by cavalry reforms and almost constant competition between states, are reflected in sources whose mounting interest in the peoples of the steppe provides a telling clue to their military and strategic importance, as well as to the increasing scale of transactions around 2,500 years ago.[53] Another pointer lies in changes in tactics and in military planning, of which the construction of massive walls was perhaps the most important: it may be that the intention was not to keep barbarian invaders out, as is usually thought, but to extend territories under the control of the state, in order to protect sources of supply.[54]

This was crucial in other settings too. Writing in the fifth century BC about the origins of Greece and how some cities had become powerful, Thucydides had observed that being able to protect fertile land was key, as was the ability and need to build up agricultural surpluses. The construction of walls was vitally important, he notes, and enabled the powerful to dominate and subjugate others, growing wealthier and stronger still in the process.[55]

As was true of Mesopotamia, investment in fortifications was centred not only on defensive motivations but also on control over the best land and over labour forces. In practice, cities, states and polities could increase capacity in one of only two ways: improvements in productivity or the annexation of additional agricultural land.[56] The dynamics of competition drove technological advances as well as a process of political centralisation. This was because, first, threats from neighbours and near neighbours demanded suitable defensive and offensive strategies that strengthened the position of those whose status derived from their ability to solve problems. Challenges from rivals inevitably prompted those at the centre to demand greater control over resources in order to support policies of expansion or resistance. Secondly, as cities, states and settlements became larger and more complex, they developed institutions that united, harmonised and homogenised to create distinct identities and encouraged integration, sociability and incentives for co-operation. These ranged from bureaucracies to writing systems, from education to religion. Perhaps counter-intuitively, though, military threats and war played important roles: both could be traumatic, expensive and further privilege elites; but they were extremely helpful in bonding people together and propelling the development and evolution of norms and what some scholars call 'ultrasociability'.[57] External threats concentrated the mind; but they also provided reasons to co-operate, and, as such, were vital in the formation and intensification of common identities.

These rhythms were common in many parts of the world. Strikingly, however, they were at their most vibrant and noisy in locations that were close to the steppes. The domestication of horses, benign ecological conditions and the intensification of exchange between peoples of the steppes and those living in cities suggests that the catalyst for rapid transformation across many parts of Asia was the interplay between agrarian states and 'mirror' nomadic confederations. This was shaped by the introduction of sophisticated cavalry tactics as well as by innovations such as saddles, stirrups and compound bows that enabled pastoral peoples to raid and exploit nearby states – in turn resulting in the latter adopting and adapting horse tactics themselves in a competition that almost resembled an arms race. This increased the effectiveness of the state but also provided greater incentives for nomads to improve their capabilities, as well as increasing the size of their potential rewards.[58]

The competition within and between the two resulted in centripetal forces that drove dramatic expansion in each: in the course of the first millennium BC a succession of weaker states and pastoral groups were eliminated in a process that saw the largest agrarian states quadruple in size in this period. The same impulses led to the agglomeration of nomadic groups into broad groupings whose identification in the sources often belie complex and diverse identities that are simplified by referring to the name of the dominant groupings – such as the Xiongnu on the Eurasian steppe, or, in later periods, the Puyŏ in Manchuria or Chosŏn in the Korean peninsula. These confederations all became significantly more powerful, overseeing more and more territory, just as agrarian states did so.[59]

One reason for this was that large confederations were all the more important as suppliers of horses, meat, dairy, textiles, leather and other materials as towns, cities and sedentary populations that were naturally constrained by the limits of the productivity of the land grew; another was that rising demand for and control of prime grazing land and substantial flocks and herds gave steppe leaders greater leverage with their own followers just as it did when it came to bargaining with those who wanted or needed produce. And of course these enhanced resources and manpower also boosted their ability to raid and to threaten to raid other tribes and neighbouring states – all of which could prove both lucrative and a source of prestige. The benefits of war as a tool for cohesion applied to all.[60]

The result was a balance that was both ambiguous and precarious – and not only in the context of antiquity, but right into the modern era. It is no coincidence, for example, that the first empires in history grew up in regions close to the steppe belt. One study has observed that even if quantification and classification can be fraught with problems, the approximation that some 85 per cent of large empires over more than three thousands years developed in or close to the Eurasian steppe tells its own story of how important were geography and the interactions between peoples who used different kinds of land in different ways.[61] Far from being a desolate wasteland that was home to wandering nomads, the steppes consistently proved to be the catalyst for imperial expansion.

New approaches to understanding the past, such as predictive models and statistical analysis, underline the importance of ecological and geographical factors in the formation of large, complex human societies. In particular, close correlations become evident: empires arise more commonly in regions where agriculture has been practised for longer and where warfare was more intense, creating pressures for societies to respond. The steppe belt played an important role in creating the opportunities and processes that generated momentum for centralisation and growth.[62]

Indeed, it is not hard to make the case that the engine of the steppes was not just a significant factor but the single most important factor in Chinese history. As one eminent historian has noted, the source of empire in Chinese history almost exclusively came from the northern frontier, which provided the source of all but one of the major 'unification events' over the course of more than 3,500 years. The only exception was that of the Ming dynasty, which had its genesis in central China – although in that case, too, context was important, as we shall see.[63]

The key factor was that of terrain. The outsize role played by the northern lands in Chinese history is explained by their flatness. The lack of natural barriers facilitated military conquest and political consolidation, both of states and of nomad confederations: indeed, before AD 1800, all major invasions came from the north. In other words, the geographical characteristics of these flat lands meant not only that there was a persistent vulnerability to nomad incursions (or worse), but also that geography played a crucial role in the emergence of a single, centralised state. Certainly, there were moments of disunity in Chinese history. But the overwhelming momentum was towards unification and a political centre that sought to defend, control and expand its periphery.[64]

This drive towards centralisation in turn produced a 'hub-and-spoke system', based on a ruler and their court that devolved limited power beyond the core, whose strengths lay in the efficient distribution of information, standardised bureaucracy and (rather unusually) a single language and writing system. Its weaknesses, however, were high levels of volatility in population levels and economic output, as well as limited resilience in the face of internal pressures. This meant frequent domestic upheavals, the overthrowing of dynasties and regular depositions of individual rulers.[65] Nevertheless, this marked a sharp difference with Europe, for example, where large states were rare and short-lived. As recent simulations have demonstrated, the location of mountains in East Asia in general and in what is now China in particular was also a factor in a relentless process of unification that intrigued contemporaries too: Lü Buwei, the chancellor of the Qin kingdom in the third century BC, noted with interest that while tens of thousands of states had once existed, now only a handful survived. His explanation – that those that had been swallowed up had failed because of 'the ineffective use of reward and punishment' – was not entirely convincing. More revealing was the fact that after his death these were reduced further still as the Qin themselves overwhelmed neighbours and rivals to establish what is often described as the first imperial state in Chinese history.[66]

It was not just that the relationship between the steppe lands that stretch from the northern lip of the Black Sea across Central Asia as far as the Pacific Ocean and regions that bordered or were close to them provides insights into the evolution of empires; it also helps explain where large states did *not* develop. Herding of livestock – including horses – was perfectly suited to arid grasslands, but it was difficult and even unviable in tropical climates, because of poor pasture and food sources and also because of the prevalence of livestock diseases.[67]

This helps explain why empires struggled to take hold in India, a location once described by the French historian Georges Duby as 'le lieu de non-Empire'. There were some obvious, fleeting exceptions in certain periods, such as under the great Ashoka in the third century BC, or under the Guptas 500 years later. For the most part, however, the motors of consolidation, expansion and centralisation that were so important in East and South-West Asia, and that were accelerated by the domestication of horses and by political competition, were either weak or non-existent in the Indian subcontinent.[68]

Ecological factors played a key role in this: in South and South-East Asia, river instability in the face of variable climate patterns could be challenging, with rivers frequently silting up, deep-water channels becoming blocked and unpredictable formations of river deltas: the Mekong and Irrawaddy expand across more than fifty metres of land a year at their river mouths, while the Solo River in Java carries around six times as much sediment as the Rhine – despite being 60 per cent shorter.[69] This made city living precarious, with major urban settlements across South-East Asia regularly failing and being abandoned even up to the early modern period – including major locations such as Melaka, Baruas and Palembang.[70] In South Asia, even major commercial centres and political capitals failed because of their inability to adapt to environmental changes such as insufficient water supplies, riverine shifts and other hydrological challenges. Cities like Cambay, Kanauj, Debal, Kayal and Gaur were just some of the large conurbations that were effectively abandoned as a result.[71] 'Inevitably,' as one leading commentator has put it, 'India became a graveyard of cities' – a cipher for much of South and South-East Asia.[72]

The uncertainty of climate conditions made the task of choosing where and how to create irrigation channels difficult, which helps explain why agricultural surpluses seem to have been rare – and further emphasises how difficult it was for cities to flourish.[73] Experiences such as those of Ibn Baṭṭūṭa, the great fourteenth-century explorer, were instructive. Visiting Delhi in the 1330s, he described it as the 'capital of the country of India, a very illustrious city, large, combining beauty and power'. It was 'greatest of the cities of India [*mudun al-hind*] and even of all the cities of the Islamic East'. Passing through Delhi six years later, however, he found it 'entirely abandoned ... without fire, smoke, or torch'. Once an 'immense city', now it was desolate 'and its population completely scattered'.[74]

While this may be exaggerated – he is describing Delhi after an attack by the Mongols – the idea of cities suddenly collapsing or being abandoned was one that finds echoes in many other accounts from different periods – as Bābur, the founder of the Mughal dynasty in the sixteenth century, explains in the *Bābur-Nāma*. 'In Hindustan the destruction and building of villages and hamlets, even of cities, can be accomplished in an instant. Such large cities in which people have lived for years, if they are going to be abandoned, can be abandoned in a day, even half a day, such that no sign or trace remains.' Naturally this had an impact on the location of cities and on the infrastructure that was built (or not built) accordingly. 'If they have a mind to build a city,' wrote Bābur, 'there is no necessity for digging irrigation canals or building dams ... There is no making of houses or raising of walls. They simply make huts from the plentiful straw and innumerable trees, and instantly there is a village or city.'[75]

This was a similar picture to that painted by one traveller who visited India in the 1560s and whose work was widely read in England. 'Every yeere at Buttor they make and unmake a Village,' he wrote, 'with houses and shoppes made of strawe, and with all things necessarie to their uses, and this village standeth as long as the ships ride there, and till they depart for the Indies, and when they are departed, every man goeth to his plot of houses, and there setteth fire on them, which thing made me to marvaile.'[76]

The wet monsoon climate had other consequences in South and South-East Asia, not least the existence of large tropical forests that were labour-intensive and time-consuming to clear. The lack of grasslands meant that horse breeding, rearing and herding were difficult, especially on a large scale. Horse procurement was a matter of enormous interest to the authorities, therefore, above all to support the military, where horses played a major role both as cavalry and in the chariot corps. Special attention was paid to where to acquire horses, with the finest imported from the north-west, from what is now Iran, Afghanistan and Pakistan. 'Of those fit for battle,' notes Kauṭilya in the *Arthaśāstra*, 'the best are those bred' in Gandhāra, Sindh, Punjab and Vanāyu (likely Persia); those from other regions were either of middling quality or worse.[77] Put another way, the potential of rulers in India was linked to their ability to source horses, and in turn to their geographic proximity to regions where they could do so.

Not surprisingly, therefore, one of the central stories in the development of trade links between India and its neighbours at least until the industrial age was that of the import of horses, principally from Central Asia.[78] Not only that: just as Chinese history was shaped by the activities of steppe nomads in the north, India's fate was greatly affected by similar influences. When peoples of the steppe expanded south, the impact was emphatic. One obvious example comes with the creation of the Kushan empire around 2,000 years ago, which peaked under Kanishka (r. AD c.127–50). Another was that of Bābur and the Mughals themselves – a dynasty whose roots and gaze were firmly linked to Central Asia.[79]

This did not mean that there was no centralisation or territorial expansion, as was the case in the complex relationship between those living in the steppe lands and those nearby; it simply meant that these processes were less intense and less wide-ranging. If this marked one difference, the disease environment marked another, with tropical climates proving a crucible in which infectious diseases could flourish. Demographic trends took on distinctive forms depending on region, on pathogen habitat and on the zoonotic jumps that play such an important role in the transmission of disease from animals to humans. Our understanding of the history of pathogens in general in Asia is hampered by the very limited academic research into this subject – in sharp contrast to the studies in Europe and the Americas – and also because most works on palaeogenetics have to rely on European samples.[80]

This leaves many questions about how and where different strains of malaria, tuberculosis, leprosy, smallpox and plague developed, and how these affected different regions and subregions, why and when.[81] Likewise assessing actual, likely or possible mortality levels is difficult to do with accuracy or certainty, meaning that hypotheses about what kind of brake disease applies to population growth must by necessity be both vague and cautious. Nevertheless, differing geography, ecology and climate provide a backdrop and a context to help explain wide regional variations in how societies developed in antiquity and later.

This was true in the Americas too, where there was considerable variation between structures that emerged more than two millennia ago. In South America some early entities that emerged in Andean cultures, such as the Moche, Wari and Tiwanaku, seem to have been driven by competition and rivalry for control of resources, which helped elite groups assert their authority and use warfare, religion and other social norms to create overarching identities in much the same way as had been done in Mesopotamia and elsewhere in the past.[82]

As trade built up across the Gulf of Mexico around 2,500 years ago, regional chiefdoms sprang up. Not all of these expanded into states, or entities resembling states, through competition and conquest absorption; but all the states that did emerge rose from chiefdoms that became better at delegating authority, more effective at developing administrative capabilities – or both.[83]

One such case was a culture that blossomed c.800–500 BC, today known as Olmec – a name that refers to an artistic style rather than to the social, ethnic or linguistic identities of the inhabitants of parts of what is now Central America and Mexico. The most important city was La Venta, centred on a main plaza adorned with a thirty-metre-tall pyramid, a 300-metre-long platform used for rituals that demonstrated elite power, and burial chambers known as Massive Offerings that were designed and adorned to represent the model of the cosmos and the centrality of the Maize God – the begetter of food.[84] As at other sites, such as Cuello and Chiapa de Corzo, particular importance was attached to greenstones, such as jade, serpentine and gneiss.[85]

Maize was the magic ingredient in many parts of Central America over a long period of time. According to the *Popul Vuh*, the creation story of the Maya peoples, the gods had transformed the world from a place of silent, calm dark water into a place that could provide for their own needs. The efforts of the gods to create animals to provide for them proved frustrating; first they made animals such as deer, birds and pumas who were unable to speak properly; then they embarked on a botched attempt to use mud and earth to make a person, which also did not end well. 'It was still not good. It merely came undone and crumbled. It merely became sodden and mushy.' This early human could say a few words 'but without knowledge. Straightaway it would merely dissolve in water, for it was not strong.'[86]

A third effort was also in vain, with beings made from wood; but although they multiplied and had daughters and sons, they were ignorant and incapable of understanding, and crawled around 'on their hands and knees'. Fortunately, the final effort proved successful, when the gods used the marvellous ingredient of maize to make the first humans. These beings 'were able to walk and converse. They were able to listen. They were able to walk and hold things in their hands. They were excellent and chosen people.' They had consciousness and were able to become knowledgeable. Finally, here were creatures who could cultivate the earth not only for their own benefit but to satisfy the aims and needs of the gods.[87] Maize was not just a valuable food, in other words, but the very secret of intelligent life.

In West Africa too, the rhythms of the evolution of urbanism, co-operation and ideas about social systems beat in their own fashion rather than mirroring others. Like most of the rest of the continent, West Africa has been all but excluded from the mainstream of academic research on early cities.[88] Indeed, current scholarship on the Sahel – the belt of transition between the Sahara to the north and savannah lands to the south that spans the continent – emphasises that societies in this region were distinctive polities, which functioned in very different ways to those in other parts of the world at the same time.[89]

Evidence of political centralisation, social hierarchies and states absorbing each other in sub-Saharan Africa is rare. Capital cities, elite dwellings and administrative machinery designed to consolidate centralised power and co-opt people into solving problems posed by external threats are conspicuous by their absence.[90] The Tichitt Tradition, which is probably the best example of early complex society in West Africa, is attested by dozens of major settlements with drystone architecture along the Tichitt and Oualata escarpments of southern Mauritania, extending into the inland Niger Delta by c.1000 BC. Although it has long been thought that increasing population and decreasing amounts of arable land drove centralisation, excavations at Dhar Nema, one of the easternmost settlements, suggest that this was not the case, with little sign of pressure on land availability. Likewise, the range of different lifestyles attested to both here and in the Mema to the south provides a reminder that categorising populations as 'sedentary', 'mobile' or 'semi-mobile' brings with it the danger of oversimplification.[91] The relationship between large settlements and the regions that surrounded them was complex, not only in antiquity but in later periods too.[92]

The same is true of social change, which naturally was fluid over time and regularly if not constantly adjusting – often as a result of climatic change. One of the key elements in the crystallisation of Tichitt culture seems to have been the drying Sahara, which prompted population accretion in humid locations and resultant adaptations that included the domestication of bulrush millet. An acute phase of aridity c.500 BC precipitated the gradual dispersal of families, groups and individuals, who left in search of less precarious environmental conditions.[93] It has been persuasively argued that this diaspora later provided the core of a Ghanaian empire that rose and flourished from AD c.300 to c.1100, a period that was climatically benign in West Africa.[94]

The primary differences between these regional centres lay in the way in which they were – or were not – linked together. In the Americas, vibrant networks were too far removed from each other to interact in a meaningful way or even at all, until the arrival of Europeans in the late fifteenth and early sixteenth centuries, meaning that cultures in Central and South America developed independently of each other. In West Africa, in contrast, there were intensive interactions locally and regionally – with luxury goods brought over long distances playing a role in shaping tastes and styles.[95] But the intensity of these connections, the tapestry of interconnected regions, the challenges and opportunities they created and the effects that they produced were of an entirely different magnitude elsewhere.

The Greek historian Herodotus put it well when he noted that, leaving aside the obscure reasons why women's names had been given to Europe, Africa and Asia, it was folly that each had been given separate designations in the first place – since in reality, this is all one single landmass.[96] His interest was not in geography, but in the mosaic of peoples who lived in these regions and who fought, traded and jostled with each other, and in the connections, borrowings and habits of each. His aptitude was praised by the Roman orator and author Cicero, who called him 'the father of history', although others, such as Plutarch, took a rather different view, saying that his inaccuracies and prejudices meant he should be known as 'the father of lies'.[97]

What is perhaps more important here is the range of the vision of scholars and writers in the ancient world – even if the information they provide often proves to be suspect, as in the case of the Chinese chronicler who stated that in the Roman empire, 'when wind and rain are not harmonious', the emperor would abdicate 'and a worthy man rules in his stead'.[98] Such comments betray creative and even wishful thinking, and are particularly interesting in light of concern about the problems that adverse climate conditions could produce. But they attest to overlapping and interlocking networks that ultimately linked the Mediterranean through the Gulf and Red Sea to Central, South and East Asia.

This brought about the spread of ideas and beliefs as well as the exchange of goods and the movements of people. We know that ritual horse sacrifices took place in ancient Rome, for example, that were very similar to those that took place in India: this included the selection of a race-winning stallion as the offering, the symbolic division of the horse's body and the connection of the event to spiritual and material wellbeing.[99] Such resemblances revealed much about consumption patterns and demand for similar products in different parts of the world. The point struck the historian Polybius during the second century BC when he considered the age that he lived in. The past, he wrote, had once 'consisted, so to speak, of a series of unrelated episodes'. This had changed in a new chapter of increasing globalisation. Now 'history has become an organic whole: the affairs of Italy and Africa are connected with those of Asia and Greece, and all events bear a relationship and contribute to a single end'.[100] In the distant past, as well as today, globalisation had consequences for production, for ecological exploitation and for treatment of the environment.

These links are not exclusively concerned with wildlife, ecology and beliefs, but those all provide an important thread that runs for more than two millennia into the past. For example, while many centuries of scholarship have led to a traditional view that early Jewish writings were essentially framed by Mediterranean contexts, it is striking to find that elephants, peacocks, monkeys, pomegranates, textiles, ivory and other flora, fauna and produce from Asia feature prominently in them. Thus the wealth of Asia and its rulers is repeatedly referenced in the Hebrew scriptures – and so too, it seems, are Indian wisdom and folktales.[101]

Ideas about heaven and hell, the apocalypse, the last judgement, angels, a universal god and resurrection of the body owe natural and obvious debts to Zoroastrianism which itself drew on earlier traditions.[102] Seminal stories, such as the tale of King Solomon trying to resolve a dispute between two women about the parentage of a child, bear an uncanny similarity to a story that appears in the Tibetan translation of the Buddhist *Tripitaka*. Studies of Tamil *aham* poetry meanwhile point to striking parallels with the Song of Songs – including loan words from South Asia that seem to have found their way into the Hebrew versions of this story.[103] Influences from Asia are of fundamental importance in Judaism, in Christianity and later in Islam; these are usually overlooked by scholars, whose reference point is almost invariably focused on the Mediterranean – and no further east.[104]

Ironically, this narrowness of vision was not how those writing millennia ago saw things. 'The Chaldaians and the Indian wizards are the first people to my knowledge who ever said the soul of man is immortal, and one of the most important Greeks they convinced was Plato,' wrote the author Pausanias in the second century AD.[105] They might also have provided food for thought for scholars like Empedokles, who argued that living things should not be killed as their souls were later reincarnated, sometimes as lions, sometimes as 'laurel leaves with goodly foliage'.[106]

Such exchanges ran in both directions, with concepts about planetary models established by Greek astrologers being adopted in India around the time Pausanias was writing, while astronomical texts based on Hipparchus and others not only proved influential but were clearly in circulation in the centuries that followed.[107]

Common ideas evolved about beliefs and practices too. For example, tattooing has long been practised in human history, with cases in the Tyrolean Alps and the Chinchorro culture in South America and references in Chinese texts such as the *Shang Shu* revealing that tattooing on bodies was not only known across continents but dates back thousands of years.[108] Over time, however, tattoos became indicators of something different. In Chinese societies, tattooing was thought of as something done by barbarians and not by 'civilised' people.[109] Further west, however, it became synonymous with slavery specifically. Prisoners of war, captives and enslaved peoples in Egypt and Mesopotamia were branded with the name of the religious sect of their owner. This association may be why the Torah banned the practice as 'the symbol of servitude', and although a reference in the Book of Isaiah sanctions body markings, it was only as a sign of submission to God.[110] Greeks who were captured during the wars against the Persians in the fifth century BC were tattooed by their captors, something that Athenians and others did to those they vanquished in the Aegean and as far away as Sicily.[111]

Regions that were closely connected by trade networks experienced dislocation, shortage and price shocks when supply and demand in one location rose or fell. A vast archive of cuneiform tablets from Babylon provides price data for commodities including barley, wheat and wool prices, as well as for the prices that enslaved people paid for their freedom. Study of this material reveals that the campaigns of Alexander the Great across swathes of the Middle East and Central Asia led to dramatic inflation and prices that stayed high for more than two decades – perhaps because of Alexander's fiscal policies and his liberal distribution of silver to his followers, or more likely perhaps because of the large-scale handouts that were given to those competing to take power after his death in 323 BC.[112]

An interwoven maritime world also took shape, with Buddhist *Jātaka* tales describing large ships up of to 75 tons, with merchants and their goods criss-crossing the Bay of Bengal as early as 300 BC. It was the Austronesians of maritime South-East Asia and Oceania, however, who achieved 'mastery of the monsoons', learning from their experiences of sailing to and colonising islands across the Pacific as far as Easter Island, and eventually building 400- to 500-ton vessels that were fifty metres in length and capable of handling difficult sailing conditions. As one Chinese source in the third century AD put it, the ships' design and the seamanship of their crews meant that 'they do not avoid strong winds and violent waves, and therefore can travel very swiftly'.[113] What the author did not say was why this was so significant: not being at the mercy of the winds, the currents and the weather did not just open up seasons; it opened up new parts of the world.

This had happened in regions of the Pacific where large population expansion in places like Fiji resulted in major transformations of landscape and of the natural environment from around 1000 BC. New research has shown that this was a time of deliberate land clearance and of the introduction of new plant species intentionally or inadvertently brought by new arrivals from overseas. There were local beneficiaries too, such as a common native bee species that flourished as a result of new landscapes and stimuli.[114]

Perhaps the best example of the radical impact of advances in maritime technology was the settlement of populations from Borneo and what is now Indonesia in Madagascar – some 4,000 miles away – around 2,000 years ago. The imprint of paternally inherited Y-chromosomal lineages and maternally inherited mtDNA lineages was substantial, producing an admixture that is uniquely helpful in understanding wider questions about movements of people, genetics and anthropology alike.[115] It was not just genetics that flowed along these connections, for so too did language: while Malagasy – the principal language of Madagascar – borrows from Bantu languages of Africa, its predominant ancestry is from South-East Asia: 90 per cent of its basic vocabulary is shared with Maanyan, a language spoken in southern Borneo and around the Barito River specifically.[116]

The link between human migration, languages and changes to ecosystems was an important and profound one. Recent work on comparisons of linguistics and archaeobotany has resulted in new models being put forward that explain expansions of vegeculture from the Sahel to southern Africa by populations that spread the Bantu languages along savannah corridors.[117] Likewise, research on walnuts shows how the spine of Asia had become criss-crossed with routes for anthropogenic plant dispersal and changes to ecosystems as landscapes were modified, forests planted and settlements established or expanded either by trade, migration or conquest. Tracking of the dispersal of a species native to Iran and Trans-Caucasus to Central Asia, to western and eastern China and to the Mediterranean also reveals the correlation between the population structure of the common walnut and human linguistic diversity: the overlap between the spread of walnut genes and the evolution of languages helps show how plants and human agency changed biodiversity and ecosystems over thousands of years – and had linguistic consequences too.[118]

Of course, humans were not the only agents for the dispersal of seeds and plants, for this role was played by animals of all kinds whose movements, lifestyles and response to climatic conditions likewise had a major impact on global natural history.[119] The difference was that such redistributions were haphazard; this was often the case when it came to human activities and the spread of pathogens, microbes, insects and rodents that was typically both inadvertent and invisible. But the search for more resilient crops, richer sources of carbohydrates and protein, tastier foods, materials with medicinal properties and stronger or more versatile pack animals to provide energy boosts led to constant changes to ecosystems and to proliferation: hundreds of millions of people today rely on the cultivation of tubers, corn and fruit crops that originated in South-East Asia, Near Oceania and South America. These not only became parts of staple diets but came to dominate tropical agriculture around the world because of their resilience in varied and challenging climatic and ecological conditions, and because they deliver reliable food outputs with low labour inputs.[120]

It was not just political systems and human society that centralised, simplified and consolidated, in other words. Plants, crops, vegetables, fruits, spices and flowers – as well as animals like camels, horses, buffalo and donkeys – were not just part of trade and exchange networks that linked many regions of Asia, Europe and Africa; they were also part of widespread and constant reconfigurations of the environment, flora and fauna.[121] These were primarily driven by one motor: the desires and needs of human beings.

9

The Roman Warm Period
(c.300 BC–AD c.500)

You can imagine how much stronger I felt after reaching my vineyards.

Seneca (first century AD)

One of the greatest challenges that humans face is the problem of water shortage. Hydrological stress is of course closely linked to consumption demands, and is in turn magnified by population concentration. Communities whose sources of water fail as a result of lower than normal rainfall, the deviation of river courses or exhaustion or overexploitation of resources can be quickly exposed to hunger, disease and death. One way of coping with this was to anticipate problems through social reforms around resource ownership and allocation; another was to create reserves, of both water and food, to mitigate risks of catastrophe.

Those at the top of social hierarchies had the most to lose from periods of upheaval that could often follow times of unexpected climatic shifts – and the most to gain from planning ahead. The Ptolemaic rulers of Egypt were a case in point. Descendants of Alexander the Great's generals who had taken power after his death, the Ptolemies were careful to reach accommodations with the priestly classes that enabled them to retain and reaffirm control during crises.[1]

What is striking about the decrees set up around the kingdom, such as on the famous Rosetta stone, is not just that they provide evidence of how this relationship functioned, but also that the erection of inscriptions shows a statistically significant correlation to major volcanic eruptions: large eruptions in the tropics could have consequences for the annual Nile flood that is evidenced not only by accounts such as a papyrus from the early third century BC which lamented that 'most of the farmers were killed and the land has gone dry' but also by a raft of land sales that are suggestive of panicked attempts to raise money to meet tax and other obligations and of socio-economic stress more broadly.[2] On this occasion, the rulers of Egypt were not able to convince elites to side with them as they had managed to do in the past, having to contend instead with a major revolt, centred on the city of Thebes, in which the priesthood played a leading role. It took several decades before order was finally restored.[3]

While volcanic eruptions could have dramatic consequences for the Nile and for those who relied on its annual floods, it is also important to stress there were many occasions when water levels fell well below average and yet disasters were averted, food stocks proved plentiful enough to avert large-scale famine or emergency measures proved effective – as had happened around thirty years earlier when Ptolemy III had been able to secure 'the salvation of the population' by importing food 'at great expense'. What mattered, in other words, was how regimes coped with unforeseen difficulties, and how decisions were made – financial, social and political – that enabled challenges to be risen to, instead of allowing them to spark a descent into chaos.

Of course, not all climate events were equal. When it came to volcanoes, for example, there were specific and even substantial differences depending on where the eruption took place, its magnitude and also, as we have seen, what time of year it occurred, with events that take place during the summer months of the northern hemisphere having greater consequences. Nor were impacts uniform even within relatively narrow geographic regions. As scholars of the Mediterranean have shown so eloquently and forcefully, ecological differences varied not so much from one part of the sea to another as from one bay to the next.[4] As such, generalising about the impact of shifts, changes and modifications of climate variations – still less assuming that conditions exist that can be termed 'normal' – is fraught with problems.

Nevertheless, very major eruptions could produce major effects. In 43 BC, two years after a powerful but short-lived eruption, the

Okmok volcano in Alaska exploded with massive force over a period lasting many months and perhaps as long as two years. The impact was recorded in the Mediterranean, with many authors remarking that the sun's light became weak and feeble, that there seemed to be three suns in the sky – presumably because of the amount of particles in the sky – and that fruit shrivelled rather than ripened because of the cold.[5]

Data from Arctic ice cores, from carbonate deposits in north-eastern China and from tree rings in Scandinavia, Austria and California shows a sudden, sharp and severe change of climatic conditions. Simulations suggest that some areas of southern Europe and northern Africa saw falls in temperature of perhaps as much as 7 °C, which would help explain reports of sharp declines in crop yields in Egypt, the traditional breadbasket of the Mediterranean, as a result of the failure of the Nile flood in both 43 and 42 BC. Water shortage, famine, plague, inflation, migration, land abandonment and depopulation weakened the state significantly.[6] As such, social and political pressure built up steadily on Queen Cleopatra, Egypt's ruler at the time; these factors are crucial to understanding the fall of the Ptolemies, the annexation of Egypt by the Romans and the moment that marked the transition of Rome into an empire.[7]

Some contemporary scholars note that descriptions about unusual climatic conditions are generic and are often borrowed by one writer from another, sometimes long after the events in question. Moreover, and perhaps more importantly, they have warned of the obvious dangers of historical determinism that suggests a neat line between a seismic activity in one part of the world and political change in another.[8] Certainly, the fact that there were no similar upheavals in other parts of the world means that recent excitable news headlines that a volcano brought about the demise of Cleopatra need to be tempered and given context.

While the fallout of Okmok must certainly have added to the challenges facing Cleopatra, other factors explain the sequence of events that ended with the Queen's death in 30 BC. Ruling Egypt was difficult at the best of times – which explains incestuous and consanguineous marriages: Egyptian rulers would often marry their sisters, mothers, brothers, fathers or close relatives in order to concentrate power within the family and prevent other elites gaining access to power, thereby becoming a threat.[9] The Ptolemies faced the additional hurdle of being perceived as outsiders, something that could offer advantages at times but could also act as an Achilles heel at others – which meant that Egypt was particularly vulnerable to dislocation and to crises.[10]

Most important, however, was the timing of the Okmok eruption. Just a few months earlier, Julius Caesar had been assassinated on the steps of the Senate in Rome. His death had plunged the city – and the Roman Republic – into a bitter civil war, as factions wrestled for control. As they did so, they looked to all quarters for allies and for support. In such circumstances, backing the right horse could bring great benefits; backing the wrong one could spell disaster. Weakened by the situation at home, Cleopatra threw her lot in with Mark Antony, judging that his military track record and popularity would make him the likeliest of those involved to emerge on top in the struggle for power – a decision that looked eminently sensible, at least to start with. The problem came when the hunting down of Julius Caesar's assassins did not bring matters to a close, but instead triggered a further period of intense struggle between the most powerful figures in Rome, which this time saw Mark Antony being isolated and picked off by the ambitious young Octavian, bringing Cleopatra and Egypt down with him.[11]

In many ways, Octavian's advance was a classic piece of Roman political opportunism: the domination of the Mediterranean basin by a small city on the shin of Italy was an anomaly. Tying together so many disparate regions, geographies, cultures and peoples was an extraordinary achievement and – as the last 1,500 years have shown – unique not least in its durability. Many have sought to explain the Roman achievement in social, economic, military or cultural terms, noting for example the looseness of Roman identity that allowed so many different languages to be spoken, such a wide range of religious beliefs to be practised, such diverse customs to be followed. But the most convincing is that Rome and its citizens excelled in one thing above all else: they were better organised and quicker out of the blocks than their rivals, more able to take advantage of favourable circumstances than others, more capable of translating openings into tangible gains. Put simply, Rome built an empire because it was able to outmanoeuvre all its competitors and potential competitors.[12]

The Romans were also lucky. For one thing, the fact that the Mediterranean is calmer and less treacherous to cross than other major seas and bodies of water meant that taking control of the entire littoral was cheaper and less risky than in other regions – and moreover offered opportunities for networks to be expanded or added to that boosted trade, widened intellectual horizons and enabled the spread of common cultural values.[13] For another, climatic conditions were exceptionally favourable precisely as Rome was taking on its neighbours, near neighbours and those further afield. This included a long humid period starting from c.200 BC that coincided with the expansion of Greek and Phoenician colonies as well as with the rise of Rome and its principal competitor in the Mediterranean, Carthage.[14]

This period has become known as the Roman Warm Period (or Roman Climatic Optimum), which lasted for over three and a half centuries – precisely the time that Rome emerged supreme in the Mediterranean, Europe, North Africa and the Near East. It was not only easily the most humid time of the last 4,000 years but, as organic evidence from pollen, marine and lacustrine sources shows, also by far the most intensively productive in the history of the Mediterranean during the last four millennia.[15] Among other things, this helped boost agricultural outputs in southern Europe and North Africa which in turn improved demographic growth, manpower for conquest and the stability that helped political authorities legitimise and cement their own powers in the process.[16]

Set in this context, the Okmok eruption was not only extraordinarily well timed from the point of view of the Roman conquest of Egypt that followed just over a decade later, but it was also a rare exception during a period of more than 300 years of unusually low levels of volcanic activity, few extreme weather events and predictable climate patterns. As well as consistent rainfall in regions such as north-eastern France, the fact that the Nile flood occurred with almost metronomic regularity, producing particularly good outcomes every five years on average, was of considerable importance given Egypt's role as a breadbasket for the Mediterranean and the empire that emerged under Octavian – now renamed Augustus by the grateful Romans.[17]

δ¹⁸O isotopic ratio profiles at three sites in Italy from 1000 BC–1000 AD

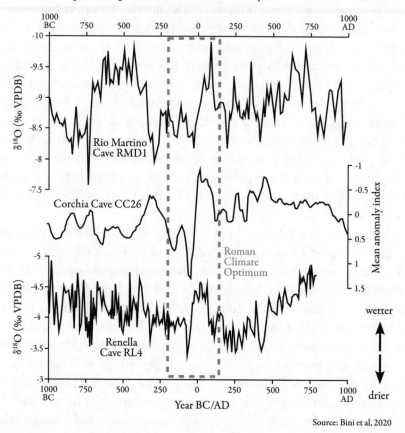

Source: Bini et al, 2020

Although proxy data from other parts of the world have been subject to considerably less assessment and analysis, it is notable that Rome's apogee was paralleled by similar developments elsewhere. The Han dynasty in China followed a comparable trajectory of rising demographics, increasing centralisation and agricultural expansion. Officials in the Red River Delta and in Jiuzhou in what is now Tranh Hóa province in northern Vietnam set about trying to teach 'correct betrothal and marriage procedures' (as well as how to wear hats and sandals) and to encourage those previously outside Han domains to learn about 'feelings of respect and morality'. But they also introduced land-reclamation techniques and new tools that helped make crop production quicker, easier and more abundant.[18]

In the Mississippi Valley, farmers began to develop more intensive techniques to support extended agriculture around 2,000 years ago, including water management, crop selection and food-storage facilities. The uplift in production, combined with greater food security, has been linked with the rise of Mississippian chiefdoms and a series of social revolutions that started a long transformation of the American south-west.[19]

In Central America, meanwhile, interventions in river courses led to the consolidation of farming villages in the Teotihuacan Valley and to the astonishing growth of the city of Teotihuacan from the first century AD into one of the biggest cities on earth – similar in the size of urban footprint to Rome itself.[20] One of the causes of the city's spectacular growth seems to have been an influx of people from the Basin of Mexico and the Puebla–Tlaxcala Valley who were migrating in search of hospitable and ecologically sustainable locations.[21] Strontium isotope analysis of bones reveals new settlement by people who had not been born in the Teotihuacan Valley, and who brought with them ideas, artefacts and building styles from Oaxaca, west Mexico and the Maya heartlands of Central America.[22] Ironically, the trigger for these movements may have been the eruption of the Xitle volcano, which archaeomagnetic dating suggests took place around the time of Rome's annexation of Egypt.[23] The boom that followed saw lavish new ceremonial structures being built or enlarged, including the enormous Sun Pyramid, with some scholars suggesting that a master plan was devised for the city's development that was then followed over the course of decades and even centuries.[24]

This was a time of empire, in other words, in multiple, disconnected parts of the world. That alone is significant, for empires often rise in tandem as neighbouring blocs form in competition, emulation and under threat from each other. It is instructive that similar processes of expansion took place during a period of propitious climate conditions generally and perhaps more importantly during a long period of stability.

Tempting though it is to attribute the rise of disparate empires to climate, more significant was the fact that each entity was able to develop the required administrative and logistical skills for maintaining social, economic and political equilibrium. Each had its own responses in doing so. In Han dynasty China, a single writing system and limited linguistic diversity served as a propeller to drive harmonisation and the strengthening of the imperial core. In Roman territories, things were rather different, with several alphabets in use, sometimes alongside each other, as well as a plethora of languages being employed not only by the population at large but also in literature – including Aramaic, Celtic, Punic and a veritable galaxy of languages in Asia Minor.[25]

In Rome's case, the achievement of uniting different peoples into a single entity was all the more impressive given the difficulties posed by geographies beyond the littoral of the Mediterranean itself. Dominating the Italian peninsula was one thing, but succeeding in taking and maintaining control of the Iberian and Balkan peninsulas, Asia Minor and North Africa, as well as large parts of the northern European plain that today comprises much of Germany, France and the Low Countries, points not only to strategic acumen but also to an extraordinary capability for keeping regions together that were so climatically, ecologically, socially and culturally different. This feat of consolidation was one that remains unrivalled in the many centuries that have followed the fall of the Roman empire.

Rome's success fostered ideas of civic service, as well as personal ambition and conspicuous consumption, particularly by members of the elites. Those with the right qualities for public affairs, urged Cicero, 'should cast aside all hesitation, strive for election to magistracies and take part in government'.[26] You should not seek glory just for yourself, the same author told his brother, even though you would be able to 'immortalise the memory of your name'; it helped that the glory would also be shared 'with me' and the next generation of the family.[27]

Cicero was full of such advice that put a gloss on idealised views of the world, not least of nature. Nothing could beat agriculture, he wrote; it was not possible to think of anything 'more fruitful, more enjoyable, more worthy of a free man', referring those who were interested to his writing where he discussed this in detail. He did not mention, however, that there was a difference between tending nature as a hobby and doing so for a livelihood, or that deriving enjoyment from backbreaking work was dependent on being able to choose to do so rather than being obliged to do so.[28]

Booty captured in war, tribute from vassals and the benefits of prosperity that came with intensifications of local exchange on the one hand and an acceleration of long-distance trade in luxury goods – including foods – on the other led to a surge in spending power.[29] According to Pliny the Elder, the finest house in Rome in the first century BC was not even one of the best hundred just over a century later.[30] Lavish banquets became a common way for the wealthy to show off and enhance their status, just one of many competitive displays that supported social and political positioning – as it has been in other cultures, regions and periods, with the Babylonian Talmud providing one example among many.[31] Indeed, as Rome had expanded, legislation had been introduced that targeted extravagant expenditure on entertaining.[32] Intellectuals regularly complained about heavy spending on domestic art and architecture, usually arguing that it showed moral failings as well as poor judgement.[33] But benefits cascaded downwards too, albeit unevenly. Juvenal put it pithily when he commented that where Romans had once chosen their own destiny and their own leaders, 'they now hoped for just two things: bread and circuses'. Rome's rulers just needed to keep food flowing in and entertainment on tap and all would be well.[34]

Rome serves as a useful case study, for, as with all empires, states and cities, political, military and economic success – along with urbanisation – came at a high ecological cost. As goods, people and ideas were drawn to the centre, strains were put on natural resources which had to be brought from a distance to supply consumers and industries that needed or demanded them. The Stoic philosopher Seneca might have marvelled at human ingenuity in being able to detect and extract metals and minerals that were 'hidden' in the earth; but, as he made clear elsewhere, there was a price to pay for the relentless industrial progress. Leaving Rome with its 'oppressive atmosphere', along with the stench of 'smoking cookers', clouds of ash and 'poisonous fumes', he wrote to a friend, led to an immediate boost not just to his health but to his mood. 'You can imagine how much stronger I felt', he noted, 'after reaching my vineyards.'[35]

As ice-core data from Greenland and northern Russia shows, the first two centuries of the Roman empire saw a quadrupling of lead pollution, perhaps linked to the exploitation of mines in northern Spain and in Germany along the Rhine following successful military campaigns. While some historians have termed this a 'peace dividend' which provides a strong indication of rising prosperity and production (especially of coins), what is more striking is that both created pressures as far as environmental sustainability was concerned.[36]

This was not lost on many commentators, who noted that deforestation had seriously depleted wood supplies in many regions. The forests in what is now Tuscany had been cut down and exhausted, wrote Strabo around 2,000 years ago, to provide wood both for ships and for houses in and around Rome, including over-the-top villas that were of 'Persian magnificence' – a nod to opulence, excess and bad taste.[37] Pliny the Elder, writing not long afterwards, noted sadly that too many people undermine nature with the sole purpose of self-enrichment; it should hardly come as a surprise, therefore, that the earth should occasionally show its displeasure, through disasters such as earthquakes. Rather than content themselves with the bounteous food and natural wealth that the world provides, humans were too busy being overwhelmed by avarice to stop overexploiting its resources.[38]

Such concerns built on a long tradition in the Mediterranean. Centuries earlier, Plato had noted that 'many great convulsions' had taken place over the course of thousands of years, changing landscapes in the process. Overwork of the land combined with these changes meant that 'what now remains compared to what then existed is like the skeleton of a sick man, all the fat and the soft soil having wasted away, and only the bare framework of the land being left'.[39] Others went further, drawing explicit links between changes to the environment caused by human activity and alteration to the climate itself. Writing in the fourth century BC, Theophrastus observed that by the city of Larissa in Thessaly 'Formerly, when there was much standing water and the plain was a lake, the air was thick and the country warmer; but now that the water has been drained away and prevented from collecting, the land has become colder and freezing is more common.' Once there had been fine tall olive trees not only in the surrounding countryside but in Larissa itself, 'whereas now they are found nowhere'. Moreover, whereas vines had never frozen before, now they did so regularly.[40]

The question of sustainability was one that concerned Roman authors too, with Lucretius Carus comparing the earth to a mother who begat the human race, but now was 'like an ageing woman' whose 'child-bearing years' have come to an end; in other words, the earth had finite resources that had to be carefully looked after.[41] 'Our poor earth', says Lucretius, 'is worn out, exhausted [and] gives birth to no more great ages.' Fields no longer produce plentiful crops, but only 'mean and stingy' returns. The farmer goes about his business with a sigh, often finding his labour yields nothing. 'When he compares the present to the past, the past was better, infinitely so.'

This had nothing to do with a changing climate, and everything to do with human incompetence – at least according to the Roman author Columella. 'Again and again, I hear leading men of our state blaming the problems of harvests first on the soil, and then on the weather in recent years.' It was preposterous, he went on, that anyone should blame this on the natural environment which was 'endowed with perennial fertility by the creator of the universe'. Crop failures happened not because of the 'fury of the elements, but rather because of our own fault'. The land had been badly looked after. Romans had only themselves to blame. Rather than working it with the greatest care, as had been done in the past, it had been handed over to those unable and unqualified to care for it properly and uninterested in doing so: coerced labourers – and hopeless ones at that.[42]

It was not only Greek and Roman authors who were anxious about overexploitation. 'Live in complete harmony with nature,' advises the *Yajur Veda*, a Vedic Sanskrit text dating back 3,000 years.[43] 'The world is beautiful and verdant and God has appointed you as His stewards over it,' says a ḥadīth, one of the collection of sayings attributed to Muḥammad, the founder of the Islamic faith, who died in the first half of the seventh century. God had created the Garden of Eden and shown it to the first human with the words 'Look at My works, how beautiful and praiseworthy they are! And all that I have created, it was for you that I created it!' – so states a passage in the *Kohelet Rabbah*, a collection of commentaries on the books of the Torah that was written around the ninth century. 'Pay attention that you do not corrupt and destroy My world,' God warned, because 'if you corrupt it, there is no one to repair it after you.'[44] 'No living creatures like birds, beasts, insects and fish should be killed wantonly,' wrote Kaibara Ekken, the leading Japanese neo-Confucian scholar of the early modern period, and 'not even grass or trees should be cut down out of season. All are objects of nature's love, having been created and nurtured by it.' Cherishing and protecting flora and fauna was 'the way to serve nature' and to protect the relationship between heaven and earth. It was wrong, he added, to love birds or beasts 'to the neglect of human beings'. Doing so was 'inhumane'.[45] Conservation, ecological stewardship and concern for the natural environment has deep roots, in other words.

It is not easy to judge how seriously to take complaints about ecological degradation in the Roman world, not least because of the plentiful evidence about the concern of landowners with arboriculture in general and tree planting in particular, which speaks to consideration for the environment, widespread knowledge of techniques such as coppicing and high levels of awareness of the value of biodiversity as well as of balancing self-sufficiency with keeping the markets supplied with firewood. Piecing together pollen research and literary evidence from multiple locations around the empire suggests serious depletion of woodland, at least in some places in some periods of the Roman empire; but more and better evidence would be needed to make the case strongly and convincingly that deforestation took place in uncontrolled and damaging ways.[46]

Although making precise calculations is not easy, there were obvious practical and logistical challenges in supplying all that was required to heat, feed and supply populations in megacities like Rome that ran into the hundreds of thousands, even if commonly cited estimates as high as 800,000–1,000,000 are implausible.[47] While the stories of emperors and their intrigues preoccupied historians at the time and have continued to do so since, the ins and outs of how mouths were fed and homes were heated were important, as was the issue of how reliable – and sustainable – such supplies were over time.[48]

As some scholars have pointed out, 'every brick, coin, tile, glass object and iron tool dating to the Roman period' can be interpreted as the product of wood fuel. Such was the scale of energy production at Rome's peak that more lead particulates were trapped in the ice of Greenland than in any other era before the start of the industrial revolution.[49] All empires leave an ecological footprint; Rome's was substantial.[50]

In addition to regional variations in the needs of people living in different climatic regions, there would have been variations in diets, personal preferences and availability of particular crops and foodstuffs. Meat consumption not only shows a wide diversity between different parts of the Mediterranean, but also varies in quantity, the result presumably of rising and falling availability, as well as of changes in appetites and broader social trends.[51]

Nevertheless, taking as a base case the assumption that adults need to consume 2,000–3,000 calories per day to be healthy, it is possible to reach some rough approximation about consumption requirements. In addition, it is reasonable to assume per capita quantities of wood that would be needed each day – not for house, temple or ship building (as contemporary commentators complained), but for basic heating and cooking. Energy was also required for animal fodder, as well as for industries such as pottery, glass, tile production and metallurgy. Taken together, these suggest a plausible estimate of 1.5 hectares required per capita, split between food production, forest and pasture. Clearly, with an empire whose population size was somewhere in the order of seventy million in the middle of the second century AD, these requirements not only presented logistical challenges when it came to delivering supplies to where they were needed, but also placed significant pressures on short-term solutions to meet demand.[52]

Long before this time, soil erosion had become a problem in some parts of the Roman Mediterranean.[53] Deficiencies in food supply have also been linked with the high incidence of malaria in the Tiber around Rome, although some have suggested this was a result of overfishing and the depletion of mosquito-eating fish in the same area.[54] Studies of dental caries and tooth wear from skeletal remains in different parts of the empire reveal differences in the eating habits – or rather differences in available foods – for those living in cities and towns compared with those in the countryside, with considerable variation suggesting that some populations were better fed and healthier than others.[55]

Demand for exotic animals for food, to be killed in the arena for entertainment or as curiosities was enormous. Cassius Dio claims that 11,000 animals were killed during a set of games lasting ten days that were laid on by the Emperor Trajan.[56] Animals were routinely hunted to the brink of extinction and sometimes beyond for the amusement of the citizens of Rome and other cities in the empire.[57] Likewise whales and other marine animals were fished to the point where stocks appear to have fallen precipitously low.[58]

There were ways to increase food and energy supplies. Improvements to tools could make some difference, although the gains delivered by such uplifts tended to be minimal. Shifts to other fuel sources – such as coal, which was widely used in Roman Britain – could have a greater impact, though much depended on the location of coal deposits and the ease of extraction.[59] A few foods could be preserved through salting, smoking and drying, one reason why fish-salting factories sprang up in the Mediterranean.[60] Most other food had to be consumed soon after slaughter, harvest or collection to prevent it spoiling, which presented challenges of supply, delivery and price. At times of crisis, such as during the reign of Diocletian (284–305), the authorities sometimes even stepped in to try to cap prices; as was usual in other periods and regions, interventions by the state made little difference and are more useful as a sign of panic among the political elite, which often had limited understanding of how markets work, rather than as actions that had meaningful (let alone positive) outcomes.[61]

Yields are not just a function of factors such as soil types, agricultural competence and weather, for much is also dependent on societal factors, such as land ownership, the size of holdings and the location relative to market and supply networks. These could be improved – and could certainly be affected – by political decisions.[62] There were clearly incentives to open up new lands, through either conquest or cultivation. This too offered obvious problems. Territorial expansion effectively delivered resources from one set of mouths to another, sometimes displacing populations in the process and sometimes not. Either way, this did not increase production, but rather amounted to re-distribution to a new set of beneficiaries.

Developing new regions for cultivation sounded simple in theory; in practice, it was usually more complicated. The major determinant of production in all contexts was the availability of labour, rather than of land itself. Control of the former was far more significant. In the case of Sudanic kingdoms in the Middle Nile, for example, the absence of animal traction, the limited use of the plough, abundant land (albeit of varying quality) and low population levels meant that there were a series of caps on how far intensification of agriculture could reasonably be taken.[63]

Where such problems could be solved, it could often be a case of diminishing returns, not only since the best land tended to have been developed first, but also because clearing and settling land took time and persuasion. Moreover, such lands were by definition likely to be peripheral to existing territories – so were either difficult to exploit or likely to be frontier zones that might require protection. Additionally, since new regions were logically and by necessity set at a distance from large urban settlements, the practicalities of getting goods to where they were wanted and needed were not simple – and not cheap.

Officials of the Northern Wei dynasty that ruled northern China for a century and a half after AD 386 took deliberate steps to 'extend the cultivated acreage, and to accumulate a stockpile of cereals' – while careful to deploy military forces to protect newly transformed fields. Around the same time, plans were made by rulers to the south to reclaim pastures that currently serviced pigs, sheep and horses, as well as to drain marshlands that had not yet been 'opened up for agriculture' but which promised to be productive. Wealthy families presented an obvious obstacle to development, being unlikely to want to give up the abundant fish stocks that they controlled. Nevertheless, it was essential that they be made to do so, for food inequality was a source of considerable problems, not least the 'many who wander hither and thither in search of food'.[64]

States could and did intervene by passing legislation and punishing transgressors severely. One law mentioned by sources from the Liu Song dynasty stated that developing land without authorisation was 'tantamount to robbery with violence', and that 'Those who steal more than ten feet of land are to be beheaded in public.' This had clearly not worked, for people had simply ignored edicts, causing untold environmental damage by cutting down vegetation on mountains and building dams across rivers. It had to stop, wrote a senior official: 'These incursions are serious abuses that damage good government.'[65]

One obvious way to mitigate risks of overconsumption was to invest in infrastructure that expanded capacity, provided a safety cushion or both. Good examples come from elaborate water-management systems that were particularly important for supplying cities, towns and even pastoralist populations. Hundreds of water-storage buildings, wells and water tanks in the kingdom of Kush (centred in what is now Sudan) on the one hand point to the extension and enhancement of royal power and to attention being paid to guaranteeing and improving taxable revenues; on the other, however, the construction, which would have required substantial co-ordination and mobilisation of labour, also demonstrates that such interventions were required to protect not only the livelihood of the population but also that of the state, its administrative apparatus and its leadership by providing insurance against shortfalls or changes in water availability.[66]

The same equation could be found in big cities, such as Constantinople, which was founded by the Emperor Constantine on a well-established site on the European banks of the Bosporus in AD 330. The major investment in a massive building campaign coupled with the arrival of a sizeable new urban population required a substantial upgrade to the existing forty-seven-kilometre aqueduct system that already fed water into the city. This was extended dramatically in the fourth and fifth centuries, covering more than ninety bridges and tunnels that fed water into giant cisterns and reservoirs. The network was around 500 kilometres in length, and possibly more, in what was the longest water-supply system of any city in the ancient world.[67] This network did not just have to be built; it had to be maintained – regularly – to prevent the build-up of calcium carbonate and contamination by clay deposits.[68]

If this provides one example of how much advance planning and investment went into supporting large populations in order to create solutions to overcome the ecological challenges of resource shortage, then the case of Tikal in northern Guatemala provides another. Sited in a location that was essentially dependent on seasonal rainfall, Tikal would not have been sustainable without human interventions to protect its citizens from considerable environmental shortcomings. While the monumental architecture was designed to emphasise the wealth and power of those who built and maintained the structures (and impresses visitors to this day), the viability of the city was based on a sophisticated series of hydrological networks that were designed to trap water. Plazas and courtyards were linked by a convex microshed water system that directed rainfall to large storage tanks lined with stones and imported clays that could hold almost a million cubic metres of water and were controlled by sluice gates that released water into causeways as needed.[69]

This proved important not only for the ecological stability of the city but also as a tool of social, political and economic control. Perhaps most important of all, however, was that it cemented Tikal's primacy relative to other urban and non-urban centres: household water-catchment tanks could hold enough water to get through one dry season; but they proved no use if rains failed for more than one year – because of regional variation or because of more fundamental changes to climate patterns.

The thought and investment that had gone into Tikal's sophisticated hydrological plan meant that the city, its inhabitants and elites could navigate short-term problems and could do so better than others. This naturally offered additional and obvious advantages, at least for those who stood to benefit from larger markets, higher tax takes and expansions of the workforce.[70]

In many ways, the primary story of the intensification of land usage, the greater extent of exploitation and the increasing degrees of social complexity lay in the fact that levels of environmental stress remained surprisingly low. There were times of shortage, certainly; but by and large the world of around 2,000 years ago was remarkably resilient, at least in terms of avoiding large-scale famine and high levels of mortality as a result of food shortages. Of course, that did not mean that there was not a price to pay for ecological expansion. Some scholars have sought to use statistical data to argue that the biological standards of living were at their lowest when population densities were at their highest. Rome's political and military ascent came not only at the price of life expectancy but even at the cost of falls in average height, if data that is geographically and chronologically uneven is to be believed.[71]

Indeed, cities were far more lethal than changes to climate patterns. While the rich grumbled about the stench and the hustle and bustle of metropolitan life, it was far more worrying that people who lived close together, often in unsanitary conditions, provided the ideal conditions for diseases to take hold and spread quickly.[72] Two examples might be provided by the Antonine plague that took hold in AD 165 and the so-called plague of Cyprian a century later: according to some modern historians, these wreaked devastation, leading to demographic collapses that triggered price and wage shocks and severely depressed agricultural yields, land values and industrial production – and perhaps even helped the spread of Christianity.[73] Others, however, are more sceptical about the impacts of these pandemics, about which even basic epidemiological details are vague.[74]

What is clearer is that the outbreaks of disease took place immediately after changes in oxygen isotope levels and spikes in sulphur contents in ice cores – which point to falls in temperature as a result of substantial volcanic eruptions, multiple ones in the case of the plague of Cyprian in the second half of the third century.[75] This was seemingly exacerbated by highly erratic precipitation (at least in north-eastern France and central

Europe), expansion both of Arctic sea ice and of European glaciers and reduced levels of solar activity.[76] The period that began shortly before the Antonine plague was one of 'climate disorganisation covering three centuries', notes a leading scholar, and one marked by climate instability which strained the Roman empire's 'reserves of strength and intervened dramatically in the course of events'. Taken as a whole, the spell from AD c.150 to c.650 'witnessed one of the most dramatic sequences of climate change in the entire Holocene'.[77]

The timing of climatic and pathogenic challenges was broadly contemporaneous with dramatic societal change – and not only in the Roman empire, which went into a tailspin in the 230s: over the next five decades twenty-six emperors came to the throne, and around double that number of would-be usurpers sought to seize power. In East Asia, the Han dynasty of China came to an end with the abdication of the Emperor Xian in AD 220 and was followed by a turbulent period known as the Three Kingdoms, which was in turn followed by that of the Sixteen Kingdoms as the state that the Han had united kept on disintegrating into smaller units. In Persia, in AD 224, Ardashir I pushed all rivals to one side and established himself as the founder of a new dynasty before expanding his territories, above all into oasis towns located along the Silk Roads. Not long afterwards, in north-eastern India, a similar process of political, economic and territorial consolidation took place that would eventually create the Gupta empire, the pre-eminent state in South Asia and an intellectual powerhouse that produced great scholars, scientists and mathematicians such as Kālidāsa, Āryabhaṭa and Varāhamihira.

The question, however, is not so much about whether climate change was a factor in these dramatic upheavals but rather about identifying the precise role that it played in some or all of these cases. While the overarching picture of shocks to weather patterns might seem to be both consistent and compelling, it is worth re-emphasising that the Mediterranean alone is composed of a wide variety of geographies, environments and climate systems; although it can be tempting to extrapolate universal models from one set of data, doing so can be both simplistic and misleading.[78] To give one example, assemblages of ceramics from across Egypt show increases in both the production and the consumption of wine – which suggests temperatures that were not cold and did not place agricultural production under extreme threat.[79]

'The tumultuous social and political changes also need to be set into wider chronological and causal contexts. In Persia, the usurpation of Ardashir was a crucial moment and understandably a watershed for conceptualisations about change. As in Rome and elsewhere, the game of imperial musical chairs was primarily of interest to those who had vested interests in who won, or lost, power. Just what and how much this meant at the time – and to whom – was another matter. The family connections of whoever sat on the throne of Persia were of course significant, but they were perhaps not as important to much of the population as commentators made them out to be at the time – and often have done since.

The same was true of the success of the Gupta empire, which according to an inscription from Kaushambi covered the ancient Aryan homelands (Áryāvarta) across northern India. So extensive were Gupta-controlled territories that by the fourth-century ruler Samudra, Gupta was being acknowledged as master of regions as far away as Sri Lanka, Nepal and Gandhara, and being referred to as 'a god dwelling on earth'.[80] Rather than being persuaded by climatic factors and the patterns we are tempted to see, it is more useful to see how political opportunism, astute leadership and logistical reforms combined to stimulate exchange and to fuel the rise of a new ruling class – and a new ruling family.'[81]

Similarly, while the spectacular splintering of lands that had been consolidated under the Han dynasty is one of the features of the two centuries that followed the abdication of Emperor Xian, the reality was that the forces that destabilised the empire had been building for many decades. Underpinning these were the familiar problems of rising complexity and diminishing returns as stagnation set in, coupled with the failure of the state to adapt and adjust; the incompetence, inefficiency and corruption that were hallmarks of a bureaucratic class which became part of the problem rather than the solution; and an imbalance between revenues and expenditure that not only led to fiscal pressure and disillusion but also incentivised challengers to chance their arm in taking advantage personally, stopping the rot or both.

It was much the same in the Roman empire, where almost constant civil wars from the late third century AD onwards exacerbated existing strains. These included a collapse in public building and changes in the attitude of local elites towards civic-mindedness, perhaps because of a move away from wanting or needing to legitimise their social and political status, or perhaps because of depleting resources that enabled them to do so. Either way, a substantial slowdown in economic and cultural activity cannot be explained only through the prism of climate: even though agricultural production was closely linked to reliable weather patterns, much also depended on availability of labour, on direct and heritable control over land and on demand for non-agricultural crafts, skills and goods, as well as on the resilience of urban markets and of the spending power of consumers.[82]

Large states could prove to be surprisingly brittle therefore, and could break down under relatively modest pressures and imbalances, with cascading consequences. For example, in the Roman empire, the provisioning of supplies from the Mediterranean to the Rhine by imperial agents largely reflected military deployments as well as requirements for maintaining control of the frontier. This meant that not only was the location of markets closely correlated to army centres but consumption patterns were linked to the spending of soldiers individually and collectively. Most important of all was the fact that, as the state began to implode, these supply and trade networks broke down entirely: when grain stopped reaching the army, regional economies went into freefall, urban centres contracted and the empire's defences weakened.[83]

Pressures that built up from the periphery flowed back towards the centre, aggravating and accelerating the process of collapse. Such problems could not be remedied by simple fixes, largely because their creation in the first place was the result of direction and oversight by the political centre. Structures that had been built by state intervention were dependent on the smooth functioning of bureaucratic and logistical functions, and when these were strained or failed, it was difficult if not impossible to stem the tide. It was no coincidence, therefore, that in due course it was the western half of the Roman empire that collapsed, as this was where the arm of the state was at its longest and most active. In the eastern half, on the other hand, where cities and markets built up organically, there was much greater capacity to deal with difficulties.[84]

In other words, while it may well be that worsening climate affected agricultural yields at least in some locations, and perhaps even disrupted food supplies, there were other, more significant forces that better explain some of the concurrent turmoil and reconfigurations that began to unfurl in different parts of the world around the end of the second and start of the third centuries. Climatic changes were more decisive and important, however, in the case of events on the steppes of Central Asia from c.350 onwards.

Records of juniper tree rings from what is now Qinghai province in north-western China attest to three periods of intense drought from around 350–550 that together represent the most challenging conditions of the last 2,000 years.[85] These are usually associated with large-scale migration of constellations of nomads, with the Huns considered the prime motor that drove tribes westwards across the steppes north of the Caspian and Black seas towards the Roman frontier and also southwards through the Caucasus towards Persia. These waves carved their way through Rome's western provinces, overrunning what is now Germany, France and Spain and reaching into North Africa – and even sacking Rome in AD 410.[86]

The results included urban decline, wholesale collapse of long-distance trade and supply networks, large-scale dispersals of populations and impoverishment not only in locations that were attacked: in fact, the effects were felt from Egypt to Britain.[87] Indeed, they may even have been felt beyond the Roman world itself, with the disappearance of the Meroitic kingdom in what is now Sudan perhaps best explained in the context of the world that it abutted experiencing a long phase of social, economic and political cardiac arrest.[88] Certainly, decline did not have to be spectacular or dramatic: as multi-isotope studies on skeletal remains from cemeteries in the Carpathian basin show, many changes following the invasions by Huns and others in the fifth century were characterised not by violence but by adaptation of diet, lifestyle and farming techniques.[89] So many converted to Hunnic lifestyles, and so many Huns adopted local practices, that some wags have labelled these the effects of 'Tiller the Hun'.[90]

It did not take much for towns to suffer from dislocation, falling into disrepair and becoming depopulated not because of violent attacks but because of the steady pressure on resources. Urban centres were suffocated rather than battered to death. The descent into what scholars used to refer to as the Dark Ages resulted not from chaos and indiscriminate bloodshed but from the splintering of the state and the disappearance of centralised authority that previously glued the western provinces together.

While barbarian invasions have long enjoyed a particular, gruesome role in public imagination (as well as in modern scholarship), some historians are now seeking to argue that in addition to serving as a catalyst to migrations from the steppes, challenging climate conditions may have played a role in the deterioration in the Roman empire too – not only in the Levant and Asia Minor, where lower rainfall would clearly have affected agricultural output, but also in the western provinces.[91] The cocktail of adverse climate conditions, reduced levels of local, regional and long-distance trade and rising demand from the centre for money and manpower proved to be lethal.[92]

Of course, evaluating precisely how aridity was linked to movements of peoples, to invasions and to military conquests is not straightforward. Nor is it clear that a detailed set of climate data from one part of East Asia can be applied to the entirety of Central Asia and beyond to support a putative hypothesis that lower-than-average rainfalls lay behind decisions to move to different locations. After all, nomads were just as likely to try to adapt to changing conditions as they were to migrate or move over long distances; and in any event, if the same climatic stress was uniform across large areas, then migration did not necessarily solve the problem anyway. The explanation may lie in the fact that the southern winter pasture in the western part of the steppe was significantly more benign than that in the east and therefore able to support far greater populations of animals and their owners.[93] The shunting of tribes towards the Roman and Persian frontiers was not about the efforts of peoples of the steppes to reach let alone conquer these empires; rather, it was part of a process of competition for the best land and locations to ensure survival at a time of harsh climatic conditions.

Evidence from New Mexico and from New Zealand suggests that this period as a whole was one of considerable variation in the El Niño–Southern Oscillation (ENSO) in the equatorial Pacific and that it is plausible that this had an effect on moisture availability on the steppes. Ironically, what might have been most important, however, was not the periods of aridity and drought, but rather a long 'pluvial' period of above-average rains that interrupted the waves of dryness: this would have produced better conditions for nomads, greater levels of pasture for their livestock and a rapid build-up in numbers of horses at the disposal of a leadership structure that had already been moulded by crisis. In other words, it was not the drought or rains that was important but the relationship between the two.[94]

Where climate could also have a significant impact was in decisions that rural populations made when faced with short-term pressures. *The*

Book of Wei, a history of the Northern Wei dynasty that ruled northern China for a century and a half from AD 386, alongside other histories of this period, records when harvests were poor as well as extreme weather events, such as severe droughts and floods. These caused authorities considerable concern – primarily because they provoked rural populations to seek refuge in cities, above all Pingcheng (modern Datong in Shanxi province), whose population may have been as high as 1 million by the middle of the fifth century. Located on the Hai River at a transitional point between semi-arid pastoral grasslands and the humid zone that favoured intensive cultivation, Pingcheng was well suited as a point of exchange with the steppes, as well as being able to benefit from a substantial agricultural hinterland.

It was not impossible and not even necessarily difficult to support a large and growing population as long as needs were predictable. For example, in AD 469 and then again six years later, imperial orders were issued instructing that grain be sent to the city from elsewhere in Wei territories. While some take this as an indication of worrying shortages at a time of an 'extraordinary cooling phase in northern China', the bigger problem was that uncertainty made forward planning difficult. It was not the drop of temperature by just over 1 °C that is attested to in the late fifth century that presented a challenge, but being sure of having food supplies where they were needed at the right time and at affordable prices. The pressures of trying to calculate what this meant are clear from the shortages of 486, when, according to some estimates, more than half of the inhabitants of Pingcheng left the city in a panic.[95]

As such, while the decision by the Xiaowen Emperor to move the capital to Luoyang in 493 was partly driven by political motivations – taking advantage of the collapse of the Liu Song state to the south – the fact that this relocation enabled the imperial court to make a fresh start in a location that was more ecologically sustainable was important: relocating in lands to the south whose economic and calorific value had been magnified by pressure on food availability not only made sense but was an obvious way to try to cope with changing circumstances. The move was no mean feat given that Pingcheng was home to more than 6,000 temples and almost 80,000 monks and nuns. Heavy investment went into Luoyang and the Longmen complex of caves, grottoes and shrines dedicated to Buddha, Buddhism and Buddhist teaching as a statement of intent that this was a site that was the subject of imperial benefaction and attention. The Longmen caves were later added to and expanded by craftsmen during the Tang dynasty (618–907). Still, the scale of the transformation of this site – and of Luoyang as a major centre – is made clear from the fact that later commentators, such as Peng Gang, an official of the Ming dynasty (1368–1644), claimed that the subsequent 'breakup of the Northern Wei state was the result of its excessively large number of stone buddhas'.[96]

That was meant, of course, as a warning to contemporary rulers who spent money on vanity projects that drained finances and also served as distractions for the population. Balancing the need to protect the power of elites with the need to administer the state, to gather taxation and to administer justice involved striking compromises. The difficulties of doing so could be aggravated by any number of challenges, including uprisings, war with neighbours, weather shocks that led to crop failure and also deeper shifts in climate patterns that brought about deeper change.

In that sense, therefore, much depended on how risks were evaluated in advance, and how threats were responded to in real time: in all disasters, the burden of calamity falls disproportionately on the poor – with modern research underlining that the decisive factor in vulnerability to socio-economic stress and famine is specifically the proportion of the population who have no assets. Put simply, the greater the number of poor people, the greater the risk of food shortage, famine and state breakdown.[97] Not surprisingly, then, it did not take much to upset the equilibrium. The consequences could not just be dramatic; in some cases, they could fundamentally transform the earth.

10

The Crisis of Late Antiquity
(AD c.500–c.600)

What avails a thousand boxes of pearls to him who is starving of cold?

King Senka (AD c.530)

Changes to climate conditions that affected humans of course drew the most attention from commentators. However, the impact on flora and fauna could be both significant and dramatic in the short and the long term. Locust migrations are closely connected with weather patterns. Studies of outbreaks of oriental migratory locusts (*Locusta migratoria manilensis*) over almost two millennia in China have revealed that locust abundance peaks in cold and dry periods, with separate research suggesting links between above-average frequencies of floods and droughts – at least in the lower Yangtze River.[1] Work on swarms in the twentieth century in the Huang River and Huai River regions meanwhile points to a strong correlation with El Niño episodes, and outbreaks in the first or second year that follows.[2]

Hormonal changes take place in locusts in response to rises in their population density, prompting morphological changes, for example to colour, body and wing size. As these occur, the locusts gather in swarms numbering in their millions, tens of millions, reaching as much as two billion, devastating crops, green plants and tree bark – they can double their weight in a single day.[3] Not surprisingly, locusts and the damage they could do feature prominently in ancient texts, ranging from the

Upaniṣads to the records of Mesopotamian, ancient Egyptian and Chinese civilisations – and beyond.[4] Biblical texts made much of the threats posed by swarms, often correlating the damage they did to food supplies with divine punishment – perhaps not surprisingly since the insects even appeared to come from the heavens. The Romans worried about locusts that were periodically swept across from Africa to Italy; their arrival forced the population to 'look for remedies in the *Sibylline Books*', the collection of oracular sayings held in the capital, to see how best to survive the famine and the troubles that they could cause.[5]

Changes in climate patterns could trigger more frequent outbreaks of vector-borne disease following abnormally high levels of rainfall. Indeed, three of the most lethal pandemics of the last two millennia – the Justinianic plague of the sixth century, the Black Death of the 1340s and the plague of the seventeenth century – were heavily influenced by climate variation. In each case, warmer springs and wetter summers produced the sharp spikes in the prevalence of the bacterium that caused bubonic plague, although this was not enough on its own to lead to mass human fatalities: for that to happen, routes transmitting the disease from plague foci needed to exist – alongside hosts who lived close enough to infect each other.[6]

This was true for a range of other diseases, including malaria – although, as with other pathogens, this needs to be understood through microanalysis at a local level, given the significant variables of geography, hydrology, climate and human activities, as well as competition for breeding sites between mosquito species, many of which are incapable of transmitting malaria to humans.[7] Awareness of the risks of disease and the threats to health haunted people in the past, with reports and enquiries about health a standard part of the formula of letter writing across cultures – although writers were much more likely to present themselves as poorly and suffering from ailments than admit to being in robust spirits. Still, as one letter written in what is now China in AD c.400 notes, alongside an unusually upbeat report that 'Mother is blessed with good health, and everybody else here is in good health,' no one could take this for granted: 'we are constantly worried about another outbreak of malaria'.[8]

It was not always easy for humans to stay ahead of change. For example, a shift in the favourable conditions in the Andes that had facilitated intensive farming in the Patacancha Valley in southern

Peru led to significant falls in pollen levels from pasture grasses and quinoa, an important food crop, as well as from ambrosia – a plant that flourishes in disturbed soil and therefore provides a useful indication of human cultivation and intervention in the landscape. The contraction of agriculture, likely the result of sustained falls in temperature that can be detected from glacial evidence as well as from the Huascarán ice core, prompted moves towards pastoralist lifeways in order to cope with more challenging environmental circumstances.[9]

Adapting to circumstance was not easy, especially in locations where the ecological profile was precariously balanced. In the south-west United States, for example, water availability was one of the key factors in determining settlement and subsistence strategies and demographic trends (including migration), as well as shaping agricultural capabilities.[10] As such, a period of severe aridity known as the Second-Century Drought, attested to by tree-ring records, placed severe stress on animal and plant life: changes in climate that influence plant growth impact the animals that rely on them – and on humans that rely on both.[11] The pressures were so great that in order to survive, the human population had to use firewood to melt cave ice for water.[12]

The cause seems to have been connected to unusually high levels of solar irradiance that were also responsible for the Roman Climate Optimum, and for warm temperature anomalies across the northern hemisphere.[13] Other regions saw higher-than-average rainfall – such as the Basin of Mexico, where an abrupt rise of some 60 per cent was followed by a period lasting for some three centuries that featured the largest and most sustained levels of precipitation of the past 2,000 years.[14] In this case, the increase in water availability helped fuel a rise in irrigation agriculture as well as in population size – which in turn underpinned a boom in construction of monumental architecture both in Teotihuacan and at Cholula in the Puebla Valley. This was not a process of unfettered growth and prosperity, as is clear from the cessation in building around AD 250 and the near contemporaneous desecration of the Feathered Serpent Pyramid in Teotihuacan: while the cause of this turmoil is not entirely clear, it is naturally best understood as a case of an important recalibration in the city's political system, as a sign of uneasiness in the relationship between the elites and others and as a crisis of expectations, perhaps in the face of pressure on resources.[15]

In the case of the Basin of Mexico, additional water supplies resulting from higher rainfall patterns brought considerable benefits and opportunities. Elsewhere, however, this presented major problems. While it is true that vegetational growth is closely linked to climate patterns, with higher agricultural productivity accompanying warm, wet summers, the problem arose when such growth was too rapid.[16] This was especially the case in tropical regions, which required clearing and maintenance in order to support human settlement.

The present-day Congo rainforest area offers a good example of the difficulties that can be caused by a change in climate in general and in precipitation in particular. Initial migration into these regions by Bantu-speaking populations was facilitated by a reduction of forest cover in the centuries starting around 400 BC that enabled settlers to expand rapidly over a vast area, attested by pottery finds at numerous sites and in several distinctive styles.[17] Four centuries later, starting around AD 1, however, a marked trend of wetter climatic patterns led to extremely fast growth of prolific trees which transformed the forest into an increasingly damp and humid space that must have impeded food productivity, resulted in deteriorating living conditions and created substantial resource difficulties for settlers. This was at least part of the cause of a chronic decline in population density, although recent scholarship has also pointed to epidemic disease as playing a role in demographic collapse in this region around AD 400.[18] Increases in rainfall and temperature made life easier in some locations, in other words, but harder – if not impossible – in others.

As we have seen, the rise of empires could reshape landscapes owing to the demands made by rising demographics, urbanisation and consumption patterns. The fall of empires could do the same: as Rome's western provinces went into a prolonged period of decline, their botanical footprint was transformed too. Fields that had once been given over to alfalfa to supply the imperial armies' horses and livestock gave way to new crops, or even to none. From the point of view of vegetation, Rome's loss was rye's gain, as cultivation expanded rapidly in Europe between the fourth and eighth centuries. A series of 'outlaw plants' spread naturally on a south-to-north axis and curiously did not seem to accompany the great tribal shifts from east to west. So while most commentators understandably focus on the human aspects that accompanied the decline and fall of the Roman empire, such as the collapse of trade networks, lower mobility and reduced literacy levels, it is also important to note that ecological change was another important outcome.[19]

It was not surprising, then, that authors, poets, musicians and others created works that presented nature in an idealised form – invariably one that was stable, benign and under human control and agency, even if that was not always made entirely clear. This involved some sleight of hand: take 'Rhapsody on Dwelling in the Mountains' by Xie Lingyun (385–433), which describes the joys of living in seclusion alongside animals and plants in various idyllic settings; the fact that he was in the process of undertaking a massive project of expansion and renovation of a family estate – and was able to afford to do so – meant that the scenes he described were the product of human intervention in the landscape, not the result of man finding peace in an existing, natural setting.[20] Creating gardens that were seen as mirrors of the heavenly paradise, as in Constantinople, the capital of the Byzantine empire, in Syria, Jordan and Iraq following the Arab conquests, in Agra in northern India in the Mughal period or in contemporary Gloucestershire, was all very well. As one leading scholar has put it, formal, cultivated gardens were greatly admired and much in demand, but they needed people to cultivate them, and thus reflected the wealth and status of those able to afford such luxuries.[21]

The ambiguity about idealised presentations of the natural world is apparent too from Buddhist texts. Many Buddhist texts are hostile to nature, such as the *Vessantara Jātaka*, whose message was not one of respecting nature but rather of taming it. Moreover, later romanticisations of nature were often reflections of its loss as a result of human activities and overexploitation; in the case of Japan, focus on plants was closely connected with monastic control over natural resources that could be monetised, namely fields and forests.[22]

Attitudes were similarly double-sided in relation to deserts and barren spaces, especially in theological thought. Remote, desolate locations were idealised as places to commune with the divine across multiple religions – places where God's voice could be heard in the Abrahamic faiths (Judaism, Christianity and Islam). Caves, secluded and isolated sites set far away from cities and towns, were also important in Hinduism, Buddhism and Jainism. Across many faiths, communing with the divine or with higher spiritual planes was connected with getting away from other humans – often by seeking out places that were not verdant, green and offering fruits of nature but the opposite; arid, harsh and punishing. Ironically, the fame of individuals who moved to caves and deserts attracted followers and imitators, with the result that monasteries and complexes of cells turned these places of self-isolation into bustling communities in their own right, such as in Sinai, at Dunhuang or at Elephanta off the coast of Mumbai. This did not defeat the point of escapism or the belief in finding spiritual fulfilment, but it certainly made it harder.

Indic literatures written fifteen hundred years ago contained a wide variety of poetic expression about the seasons, with genres focusing on the six seasons (*sadritu*) and the four rainy months (*chaumasa*), or on the twelve-month cycle of changing weather conditions and the different seasons (*barahmasa*).[23] Texts like the *Viṣṇudharmottarapurāṇa*, likely written around the sixth century AD, that bring together information from separate treatises about poetry, music and astronomy present the seasons as part of a predictable cycle that returns and can be anticipated by human actors – without allowing for the fact that there were considerable variations over the short and the long term: summers tormented and springs bloomed.[24]

Many good examples abound from this period of efforts to influence and control the weather to ensure clement conditions and plentiful agricultural yields. For example, the *Mahāpratisarā-Mahāvidyārājñī* recommends attaching a particular spell to the top of a flagstaff, promising that doing so 'calms all types of winds, cold-spells, untimely clouds, lightning and thunderbolts', as well as offering protection from 'crop destroyers' such as 'hosts of stinging insects, flies, locusts and worms'.[25] This is just one of a host of texts that focuses on rain-making rituals and protection from adverse weather and its consequences that can be found in Sanskrit, Chinese and Tibetan sources spread across multiple centuries.[26]

The idea that nature could be understood, controlled and shaped by humans belied the reality of unusual and severe weather conditions – or of other dramatic interventions such as natural disasters. Tsunamis were regular occurrences in the Mediterranean, for example, including ones of extraordinary force and violence. The port of Caesarea in the eastern Mediterranean in what is now Israel had been home to a mega-harbour built using fast-drying hydraulic cement made from ash from

Mount Vesuvius imported from Italy, while cobble and rubble that were laid down as foundations for piers had been brought from Asia Minor, Cyprus and Greece. The harbour was hit four times by major tidal waves – in AD 115, 551, 749 and 1202 – with each doing substantial damage.[27]

Other earthquakes and tsunamis were devastating, as an account of an episode in AD 365 makes clear. 'Shortly after dawn,' wrote one late Roman author, 'the solidity of the whole earth was made to shake and shudder, and the sea was driven away, its waves were rolled back and it disappeared,' stranding many ships on dry land and enabling people to walk about and pick up fish with their hands. 'As if insulted', the sea then roared back through the 'teeming shoals and dashed itself violently' on to islands and the mainland, flattening buildings, killing thousands and changing the face of the earth.[28] While some have questioned whether this description involves considerable dramatic licence on the part of an author representing a localised disaster as a universal cataclysm, it is true that the coastline of Crete was raised by as much as nine metres, likely as a result of multiple seismic events rather than a single episode.[29]

Volcanic eruptions could be devastating too, as the eruption of the Ilopango Tierra Blanca Joven volcano in what is now El Salvador shows. The force of the eruption, dated to AD c.431, was astonishing, with modelling of deposits suggesting the ejection of fifty-eight cubic kilometres of bulk rock to a column height of almost thirty kilometres.[30] The ash from the eruption would have altered ocean chemistry in the Pacific, generating a significant increase in marine productivity and bringing about changes to atmospheric carbon dioxide and oxygen levels in the years that followed. It is likely that the eruption resulted in a global cooling of around 0.5 °C, with evidence from Antarctic ice cores implying that the effects were more pronounced in the southern hemisphere. Everything within eighty kilometres of the vent would have been destroyed, including vegetation, which would have taken decades to recover.[31]

It is not easy to establish the effect on the climate of Central America because of the current paucity of high-resolution sampling – something that may be resolved in the future. Tree-ring data from central China point to this period as one of drought, raising the questions of whether this was linked to the volcanic eruption, to a cooling episode related to solar activity or both; there is the question too of what role, if any, climate pressure played in a new wave of consolidation and migration by pastoralist groups in this period, with some scholars proposing that the pressure on vegetation on the steppes served as a catalyst to movements westwards.[32] The most notable of these were the Huns, who surged into the Europe under a formidable new leader, Attila, in the decade after the eruption.[33]

In such circumstances, taking the long view was important. 'From time to time,' wrote the Greek historian Polybius, 'as a result of floods, plagues, failures of crops or similar causes, there occurs a catastrophic destruction of the human race, in which all knowledge of art and social institutions is lost.' This was to be expected, for 'such disasters, tradition tells us, have often befallen mankind and must reasonably be expected to recur'. No matter how bad things seemed, however, the human population would recover and renew 'as if from seeds'.[34]

This perspective was admirable and, on balance, was broadly correct. However, some disasters were worse than others, and could on occasion seem apocalyptic in the scale of the devastation they inflicted. In the first half of the sixth century, a sequence of climate-related phenomena brought about change on a profound scale around the world. Crucial to this was a series of large eruptions that took place in the 530s and 540s. Tree-ring data from the Altai mountains, the Austrian Alps and bipolar ice cores point to precipitous falls in temperatures in this period, which were amplified by reduced solar activity and oceans that were already cooling.[35]

Rather than starting with a single, massive explosion, the signal that can be detected from ice-core data may relate to at least two major and possibly multiple eruptions of North American, Icelandic or Kamchatkan volcanoes in c.536 and around 540.[36] This was then followed by tropical eruptions in the decade that followed which produced long-lasting effects, especially in terms of sulphate deposition.[37] Despite the assertions of many scholars to the contrary, this did not include Ilopango, whose eruption a century earlier was one of the ten largest of the last 7,000 years but whose dating has long relied on inaccurate reading of radiocarbon data.[38] The clusters of eruptions in a 'machine-gun' effect may have been part of the reason why the after-effects were so severe – with the timings and latitudes of the explosions being of particular significance.[39] Some scholars have suggested that at least one of the tropical eruptions was of an undersea volcano, which would explain the presence of warm-water marine microorganisms dating to this period that have been found in ice cores.[40] Taken together, volcanic activity produced a dramatic and substantial drop in temperature: one scholar has argued that 'the 530s and 540s were not just frosty. They were the coldest decades in the Holocene.'[41]

The ejections threw enormous quantities of debris into the atmosphere that formed a 'dust veil' around the earth and were much remarked on by contemporary commentators. Writing about the years 536–7, the Byzantine author Procopius noted that this was a time of terrible portents, with the sun giving out so little light and heat that it resembled the moon. John of Lydus recorded that the sun was all but obscured by clouds which dimmed its light to the extent that it could barely be seen for almost a year. The *Nan Shi*, written in Tang dynasty China, meanwhile reported that in the mid-530s 'yellow dust' fell 'like snow', presumably a reference to sulphuric deposits and ash falling back to earth after being ejected into the atmosphere.[42]

In fact, though, the cooling effect may not have been the most important result of the volcanic activity. It is certainly the case that crop yields are sensitive to temperature change, especially at high altitudes and latitudes.[43] But what mattered more was not so much the onset of cold conditions as the effect of the dust veil. Terrestrial photosynthesis and agricultural production are closely linked to solar radiation, with the result that blocks on the sun's light and warmth inevitably present major challenges for plant life, including that grown as human food sources.[44]

Some estimate a population decline of 70 per cent or more in some areas north of the Yellow River, although the evidence for such claims is not enormously robust.[45] More striking is the emphasis placed by some accounts of this period on famine and food shortage, from the *Annals of Ulster* noting the 'failure of bread' to the author of an early Japanese chronicle which reports King Senka in the mid-530s lamenting that 'food is the basis of the empire. Yellow gold and ten thousand strings of cash cannot cure hunger. What avails a thousand boxes of pearls to him who is starving of cold?'[46] Nevertheless, the point here is obvious: material wealth was all very well, but it did not put food in people's mouths.

Data from bristlecone pines from the San Francisco Peaks in Arizona and from precipitation reconstructions in New Mexico show that the

540s was a period of extreme cold in the North American south-west. It has recently been argued that this was the cause of the rapid population dispersal that took place at precisely this time, and moreover that the warm and wet conditions that returned later in the sixth century drove a process of major socio-economic reorganisation on the Colorado plateau. A generation-long crisis that affected a vast area brought about a sophisticated response that saw agricultural knowledge and technology spread across communities which adopted similar patterns of sedentary lifestyles, including the widespread domestication of turkeys. It also catalysed the evolution of complex societies, spurred the emergence of villages and ultimately led to the development of hierarchies that served as foundational elements of Ancestral Pueblo culture.[47]

The dramatic change in conditions also had a major impact in Scandinavia, where this period saw the greatest change in settlement patterns in Sweden for 6,000 years as villages were abandoned in their scores. In Denmark, a sharp rise in the number of prestige objects at sacrificial sites has been linked to expressions of desperation and appeals to the divine during the traumatic downturn of the mid-sixth century.[48] Indeed, the cold weather, and specifically the diminution of the sun's strength, was such that it transformed ideas, influenced beliefs and set down models that were warnings for the future. As some leading scholars have argued, the dust veil provided the model of the *fimbulvetr*, a great winter in which Norse myths held that 'snow will drift from all directions', great frosts and keen winds will be everywhere and 'the sun will do no good'.[49] This was the prelude to the Ragnarök, the final battle at the end of the world.[50] Cinemagoers who have helped make *Thor: Ragnarok* one of the highest-grossing films of all time, taking almost 1 billion dollars at the box office, may not realise that they have a set of volcanic eruptions a millennium and a half ago to thank for their thrills.

The century of change that followed the eruptions has been the focus of considerable attention from scholars, who have often taken the volcanic explosions and associated climate impact to be the start of a sequence that includes the 'transformation of the eastern Roman Empire and collapse of the Sasanian Empire, movements out of the Asian steppe and Arabian peninsula, spread of Slavic-speaking peoples and political upheavals in China'.[51] As we shall see, they also had a bearing on major transformations in the Americas, in Africa and across South Asia, as

well as paving the way for the rise of Islam and the creation of a vast
Arab empire in the wake of the death of the Prophet Muḥammad.

The hundred years that followed the eruptions of the mid-sixth century certainly saw socio-economic, political and cultural transformations
of extraordinary magnitude. However, it is more useful to assess the
volcanic activity and its impacts not as being the cause of a series of
epic social, political and economic revolutions, but rather as having
aggravated existing problems, and exposed fractures that then produced
radical change. For example, food shortages were a product of pressures
of demographics as much as of low yields. Had King Senka's realm had
a smaller population, feeding it would not have been as difficult as he
claimed.

The vulnerability of cities was not just linked to climate, as we
have already seen, for they could also collapse quickly as a result of
flooding, the shift of trade routes, external pressure or a combination of
all three – as the case of Pātaliputra shows. Pātaliputra had once been
described as the greatest city in India, with buildings of such beauty
that 'no human hands of this world could accomplish'; it was a city that
was not only gorgeous, but would 'endure for 5,000 years'; by AD 600,
however, it was in ruins, likely the result of catastrophic flooding in the
wake of sustained rainfall that affected the flow of the Ganges, although
precisely when this happened is not clear.[52]

Certainly, by the start of the seventh century, all the trading cities
of north-west India had fallen into decline. They remained that way
200 years later when the entire region was described as being short of
water, filled with bandits and a likely graveyard for invading armies
that would 'die of starvation and hunger'. While such testimonies pose
important questions about their aims and audience, the fact that the
ties that had held the Gupta empire together dissolved quickly in the
course of the late sixth century tells its own story about the fragility of
states that seemed to be well organised, stratified and resourceful.[53]

It was perhaps no coincidence, therefore, that the beneficiaries of
this age of upheaval were societies, peoples and cultures that were better
able to adapt or take advantage of opportunities – such as Angles and
Saxons, who translated roles as mercenaries in Roman Britain into
political dominance, much as Slavs in the Balkans, Langobards in Italy,
Berbers in North Africa, Arabs on desert fringes in Palestine, Syria and
Iraq, Avars in the Pontic steppe and Huns in South Asia found their

bargaining positions improved to the point that they too transitioned from being marginalised groups to becoming significant and even dominant in their own right.[54]

Naturally, there were different trends in different regions, but the general picture showed the vulnerability of hierarchical, urbanised societies. In some parts of Europe, for example, notably the north-west, the impact of the downturn of conditions was mitigated by the fact that the climate had become steadily colder and wetter in the previous decades: this had led to innovations in plough types better able to cope with heavier soils and also to the adoption of more diverse forms of food production that depended less on agricultural cultivation and more on pastoralism. As some scholars have also pointed out, recurrent barbarian invasions and attacks had broken down extended supply networks and encouraged communities to become both more localised and more self-sufficient. This provided obvious buffers against the kinds of difficulties faced by urban populations that were poorly placed to develop coping strategies in the face of pressure on food supplies.[55]

Dealing with shortages was a source of concern in many parts of a world that went through a seismic set of shocks in the mid-sixth century. In Gupta-era India, for example, scholars like Varāhamihira wrote about the dangers that accompanied dust veils in the sky, warning that if these lasted for three or four days, 'food grains and liquids [and] juicy substances will be destroyed'; if it lasted for longer, 'there will be mutinies in the armies of kings'.[56] Jain texts like the *Titthogālī*, produced around the same time, provide apocalyptic visions of a world dominated by floods, drought, famine and persecution and offer insights into this tumultuous period, even though, as some scholars note, it is not entirely clear how much we can rely on such accounts as historical testimonies and to what extent they are echoes and responses to other works in circulation at the time – in this case, the *Mahābhārata* in particular.[57]

The anxieties about food shortages and political dislocation were topics of great interest to Chinese historians and especially to those seeking to influence the decisions of emperors. Wu Jing, a court official in the Tang dynasty writing in the early eighth century, was scathing about how the Emperor Wen (r. 581–604) oversaw famine following a great drought and 'did not permit the distribution of supplies to relieve the people'. This was madness, wrote Wu Jing: when he died, Emperor Wen had enough granary supplies to last 'for fifty or sixty years'.[58] The fact that Emperor Wen was the founder of the Sui dynasty – from whom the Tang seized power – means that we should take Wu's comments with a pinch of salt. But his aim here was plain: in times of environmental catastrophe, the role of the ruler was to plan for shortages and to distribute emergency aid when needed. There was no point stockpiling too much; but, equally, enough should be set aside for when problems arose. If this advice was not taken, it was inevitable that there would be 'peril and ruin'. Needless to say, those who could help the Emperor 'preserve the realm' had vitally important roles to play. What mattered more was the lesson that had been learned: the stability of empires depended on their resilience through times of climate stress. Famine did not just kill the poor or drive anger towards the ruler; it brought regimes down.[59]

The effects of turbulence were also severe in Teotihuacan in the Valley of Mexico, which went through a period of intense social, political and religious upheaval in the middle of the sixth century that included the desecration and smashing of idols and artefacts of the deities related to rainfall, as well as systematic destruction of temples and palace complexes. This is usually interpreted as a case of popular dissatisfaction with the response of the elite to solving problems and in particular with the pressures of declining rainfall levels – and furthermore is usually correlated with the volcanic eruptions of the mid-sixth century.[60] Certainly, the fact that building work stopped elsewhere in Central America, most notably across the territories controlled by the Maya around 540, suggests that the scale of disruption caused by volcanic eruptions was both immediate and spectacular.[61]

By c.600, Teotihuacan had become a shadow of its former self, while parallel state structures elsewhere like Monte Albán and Río Viejo entered periods of decline too, as populations drifted away from large cities to smaller urban settlements, and thus to local elites with lower ambitions, less expansive geographic connections and more limited resources. The archaeological record attests to a major diminution in distribution networks of goods like obsidian, a volcanic glass that was prized for its utility and rarity and for its role in ritual. New valley centres emerged at locations like Lambityeco, Jalieza and El Palmillo, but none came close to the scale of the large cities that had dominated Mesoamerican cultures in the past.[62]

In South America, the eruptions marked a new phase in a devastating series of environmental and climatic changes made all the more difficult by increasingly frequent and severe El Niño events in the sixth century characterised by alternating patterns of severe flooding followed by long periods of drought. There are indications that an El Niño event had begun to cause problems of aridity and depletion of marine supplies as warm, nutrient-poor water displaced the rich cold waters off the coast of South America, with radiocarbon dating of clamshells suggesting that this had begun even before the eruptions – although dating of marine data needs to be handled with caution.[63]

Across the regions usually described as part of the Moche culture, the results during the sixth century were dramatic, as irrigation systems were overwhelmed, soil damaged and food production underwent precipitous decline.[64] Windblown sands proved almost impossible to deal with in some locations on the north coast and in the Cerro Blanco complex, where the city was overwhelmed – and further affected by one or more major pluvial events.[65] Competition for water and food was such that communities took steps to protect themselves by building fortifications, for example in the Jequetepeque and Zaña valleys.[66] As in Teotihuacan, reactions against the elites shaped artistic and religious responses, with the reversion to archaic iconographic motifs best understood as a turn to ancestors for help at a time of turmoil.[67] Some cultures, however, adapted better than others, with Wari and Tiwanaku expanding from the central highlands of the Andes around the end of the sixth century and building empires – or at least establishing cultural supremacy – that covered much of the coastal region and interior of what is now Peru, Bolivia and northern Chile.[68]

This was a time of change in many parts of the world. For example, the Christian Aksumite kingdom that dominated the Horn of Africa and what is now Ethiopia, Eritrea and eastern Sudan had intensified contact with the state of Ḥimyar across the Red Sea at the start of the sixth century, sponsoring an expansion of Christian churches in the Arabian peninsula and interfering in politics across the region by installing rulers who were sympathetic to if not directly controlled by Aksum.[69]

By the 550s, however, things in Aksum looked bleak: grand buildings fell into disrepair and in so far as they were used at all seem to have become ad hoc shelters for what one scholar has called 'squatters'.[70]

Quarries were abruptly abandoned, pointing to the cessation of major construction projects, while the reduction in the quality of the coinage, both in metallic terms and those of production, provides another indicator of a state that had been vibrant, expansive and ambitious entering a state of terminal decline.[71]

Scholars have long noted the depopulation of Aksum in the middle part of the sixth century, often linking it to problems of oversettlement and the unsustainable depletion of resources, in particular of ivory, timber and water.[72] Some have also underlined the precipitous collapse in trade across the Red Sea following the capture of Ḥimyar first by the Persians and not long afterwards by the Arabs, thus depriving Aksum of its crucial overseas markets and depressing the economy further still.[73] Aksum began to fall into a state of decay, with non-ecclesiastical buildings and monuments no longer maintained, and the political centre – to the extent that it still existed – either moving to Kubar, a town whose location remains unknown, or possibly to a mobile capital that moved as the ruler progressed through his territories.[74]

Climate conditions may have played a decisive role in the depressed circumstances in which Aksum found itself, and indeed may explain many of the extraordinary changes that took place in this period in Africa, Europe and Asia. As we have seen, the dust veil of the volcanic eruptions put pressure on food production and in many cases contributed to environmental, material and social stress. It had another effect that was perhaps even more fundamental in shaping the revolutions of the sixth century: exceptionally cool weather created unusual opportunities for rat survival and flea reproduction, which in turn resulted in an outbreak of plague. The need to support crop failures with greater shipments of food across the Mediterranean may also have provided a surge in connectivity that resulted in an intensification of networks that enabled the rapid spread of disease. And finally the suppression of sunlight may have played an important role too, namely inducing a deficiency in vitamin D which is crucial to the human immune system and especially to its ability to respond to bacterial infections. These factors now dovetailed to create the perfect conditions for a major pandemic.[75]

The first written reports of plague come from the Egyptian port of Pelusium, a key node linking the Mediterranean with the Red Sea and

Indian Ocean. From the summer of 541, disease took hold, spreading to Gaza and across North Africa, before reaching Constantinople a few months later and then spreading 'over the whole world', according to the Byzantine author Procopius. The disease came close to annihilating the entire human race, noted the author, who claimed that at its peak plague was killing 10,000 people a day in the imperial capital. Bodies were left in heaps, as there were not enough people to bury them.[76] The Justinianic plague, as it has become known, is purported to have killed millions – tens of millions in fact, if not more, with mortality rates of perhaps 50 per cent of the population of the Mediterranean as a whole.[77]

As more recent experience has shown, even small rises in mortality levels can create major economic compression and in some cases can overwhelm governments. Disruption does not only come from large numbers of deaths, in other words, but also from dislocation of trade, transport and communication networks.[78] Nevertheless, estimates about the extent of the Justinianic plague have come under sustained attack in recent years, with scholars arguing that a range of evidence including pollen data, papyri, inscriptions and coin minting do not point to major social, demographic or economic change. According to these scholars, the impact of the disease was simply 'inconsequential'.[79]

Others have pointed out that even if there were high levels of mortality, population losses were not universal, but rather were highly localised and of marginal consequence.[80] Such highly sceptical assessments are certainly eye-catching. More sanguine assessment suggests that while it is certainly true that estimates of mortality levels involve guesswork and leaps of faith, there are good reasons to suspect that the plague did enormous damage. One of the most compelling is that multiple related but distinct plague strains from this period can be identified in France, Spain, Bavaria and even in Cambridgeshire: the fact that the disease took hold in isolated, rural locations is telling – and suggests that if plague spread in marginal, lightly populated settings, then it must have been immensely damaging in densely inhabited cities and towns.[81] Furthermore, the imperial authorities in Constantinople issued laws in the early 540s that sought to impose controls on wages, with some workers 'demanding prices double or triple what was formerly customary', which fits naturally alongside what we know about later effects of high-mortality pandemics on the labour force – namely a sharp contraction in the number of workers following large numbers of deaths. Other emergency measures taken in Constantinople likewise point strongly to a state desperately trying to shore up its finances at a time of sudden and extreme pressure.[82]

Furthermore, there are several indicators of panic and breakdown. One comes from the ending of public services such as organised rubbish collection in Roman provinces in the Levant.[83] It has been suggested too that in the Negev Highlands, the decline of viticulture and of luxury 'Gaza wine' which provided the economic basis for settlement in this region resulted not from later Islamic conquests, but from challenging climate conditions and more importantly from population contraction caused by plague and from shortages of the manpower required to plant, harvest and process the wine.[84]

A similar problem seems to have been responsible for the collapse of the Marib dam in what is now Yemen towards the end of the sixth century, which caused a flood that 'overwhelmed the lands, gardens, edifices and homes until the inhabitants of the country had perished and its citizens been wiped out'. According to one later author, the dam's failure resulted from the lack of maintenance in general and from the reduction in the size of the workforce needed to prevent its deterioration.[85]

Genetic studies of populations in the eastern Mediterranean and the Near East reveal mutations that make them more susceptible to autoinflammatory disease and as a result offer heightened resistance to *Yersinia pestis*, the bacterium that causes plague. That genomic reconstructions are consistent with the Justinianic plague, with the Black Death of the 1340s and regular occurrences thereafter, not only suggests that the outbreak of the sixth century left a genetic imprint but that this was conferred as a result of widespread exposure to infection by those who survived the pandemic.[86]

A very dramatic decline in the black rat population of Europe around this time may be the result of the pandemic, although it may also be that cooling temperatures or the breakdown of communication and trade networks in western Europe and later in the Balkans influenced the population size of these and other pests; or, of course, a combination of some or all of these factors may explain the dominant genetic impact of a second wave of rat introduction to temperate Europe starting in the eighth to tenth centuries.[87] Either way, plague did not only affect humans

– but changed the distribution and size of rat colonies across many parts of Europe too.

The devastation that the pandemic wreaked may have spread to the Congo rainforest in West Africa, which saw catastrophic population collapse that was broadly coincidental with the Justinianic plague. Communities that had been present for many centuries disappeared both rapidly and entirely, save for a few scattered communities.[88] Although genomic evidence points to a Central Asian origin for the plague pathogen, groundbreaking new research is showing that there were deaths from plague in multiple locations in Europe years and perhaps even decades before the pandemic took hold in the 540s.[89] It may be that plague spread in Africa before it did so in the Mediterranean and elsewhere, though more research will be needed to determine if that was the case, or if plague was indeed a cause of catastrophic demographic decline in Congo or elsewhere in Africa.[90]

What matters more, then, than the original source of the plague strain are the routes it took and the timing of the transition from small clusters of plague victims to a pandemic that produced mass casualties. It may be no coincidence, for example, that the disease was first tracked in North Africa near the mouth of the Nile and that evidence from the Aksumite kingdom also shows significant demographic contraction in precisely this period. Some sources talk of plague arriving 'from the land of Kush' as well as from Ḥimyar, which suggests transmission through the Red Sea from sources in Central Asia or China via maritime or land trade routes. That plague outbreaks took place in Persia only after it had struck further west, and that epidemic disease in Chinese societies only struck decades later, suggests that it was the disease climate and wider climate conditions in the Mediterranean and Red Sea that provided the ideal incubator for the disease to establish itself.[91]

As one recent commentator has noted, however devastating the 'twin catastrophes' of severe climatic change and plague were, they did not lead to the collapse of the Roman empire. Indeed, they did not even bring down Justinian's regime – or topple the Emperor himself, though he may well have caught the disease and survived it, according to an ingenious reading of coins issued in this period showing the ruler with a classic boil on his neck filled with infected pus.[92]

While it is important to note that there is considerable evidence for continuity in many regions in the Mediterranean economy and notably in the Sasanian empire, which seems to have been unequivocally expanding throughout the sixth century, most indicators suggest major transformations resulted from the pandemic.[93] Of course, significant population decline would have had implications for centralised governments, resulting in lower productivity, reduced tax receipts and diminished ability to fund large-scale projects.[94] This of course affected the operation of the state; but ironically, as the evidence of Justinian's attempts to curb wage inflation implies, many do better in a shrinking economy as they are able to negotiate or acquire better rewards and status for themselves.[95] As we have seen before, 'decline and fall' means different things for different people.

The question marks about the effects of plague are part of a broad and sometimes bad-tempered debate between contemporary scholars – especially in the case of the Roman economy.[96] What is perhaps clearer is that the middle decades of the sixth century saw a series of major social and political changes in many parts of the world in which climate and disease environments were factors that in some cases aggravated or catalysed existing socio-economic trends.

For example, detailed work on resettlement patterns in sixth-century Scandinavia and the Baltic reveals widespread instances of land being abandoned and of shifts from grain production to animal husbandry. These were responses to a complex set of circumstances that included adaptation to a colder climate which began with the eruptions of the 530s and persisted for more than a century in the northern hemisphere. They were also responses to the impact of plague and the stagnation and collapse of trading networks that accompanied the waning of Roman authority in northern Europe – circumstances that fundamentally changed the dynamics of markets, altered economic and cultural horizons and transformed land ownership and labour relations.[97] Over time, this brought about a rise of large-scale land ownership and the creation of substantial estates that led to the emergence of a new 'super elite'. If this did not provide the basis for the Viking Age, it was certainly a precondition of its emergence.[98]

According to some scholars, the traumas of the mid-sixth century also help explain psychological, emotional and religious changes in Europe and beyond. For example, it has been noted that the conceptual development of the Virgin Mary as the protector of Constantinople evolved from this period, with some questioning if this was connected, directly or indirectly, with plague.[99] Others have noted that disease, bad weather, military setbacks and earthquakes in the mid-sixth century prompted heightened concern about the imminent end of the world – which continued and even strengthened in later decades even as it was patently obvious that such fears remained unfulfilled and unfounded.[100]

Gregory of Tours wrote of floods that were so bad that it was impossible to sow grain, and recounts earthquakes, possible meteorite strikes, hailstorms and fires that were 'stirred up by God'. His warnings that 'There will be plagues, famines and earthquakes in various places, and false Christs and false prophets will rise up and they will make signs and prodigies in the sky' reflect an obvious interest not only in severe climate events but in how to interpret them.[101] For Gregory, it seemed clear: these events marked 'the beginnings of the pains' that Christ warned would presage the end of time.[102]

Apocalyptic prophecies that foretold the end of time were not just a fundamental part of the Judaic and Christian tradition but a recurring one, too. Even before the challenges of the mid-sixth century, fears about the end of the world seem to have been rising with some intensity. Gloomy ideas transmitted along trade routes, particularly within the eastern Roman empire. As we shall see, they fell on particularly fertile ground in Arabia and the Middle East. But it was perhaps no coincidence that pagan communities as far away as Scandinavia began to offer precious metals such as gold and silver as part of sacrificial offerings in desperate efforts to ward off what seemed to be imminent and final catastrophe.[103]

Such cases can seem highly persuasive at first glance, but it is worth stressing that this source material is complex and needs to be placed within the context of many other works with varied and different conceptual formulations at this time which do not talk of the impending apocalypse.[104] Nevertheless, there is compelling evidence from many periods and regions that bouts of existential angst and crisis could be important in recalibrating, re-evaluating or strengthening ideas about the divine.[105] Some have noted that Buddhism made inroads in Korea and Japan precisely when the dust veil created by volcanic eruptions was at its height.[106] The expansion of Christianity into Central Asia and the conversion of steppe peoples, meanwhile, was specifically linked by contemporary accounts to the time of plague and to remedies that had proved effective: nomads who had had 'the symbol of the cross tattooed in black on their foreheads' were asked by envoys from Constantinople sent to form an alliance why this had been done; there had been a plague, they replied, 'and some Christians among them had suggested doing this [to bring divine protection] and from that time their country had been safe'.[107]

Recent work on palaeoclimatological proxies and historical records from central Italy in the sixth century reveal not only a period of very wet conditions in the later sixth century that result from a negative phase of the North Atlantic Oscillation, but also a sharp increase in reports of water miracles in accounts of saints' lives. As well as showing how climate events were reflected and interpreted by contemporary authors, these also represent – intentionally or otherwise – a means of increasing the already growing authority of bishops, monks and the Church more broadly.[108]

Traumatic single events and periods of climate shock not only shaped beliefs; in other words, they also delivered opportunities for the expansion of institutions – such as the Christian Church – and for the consolidation of authority and power for those who could offer explanations for, and protection against, future catastrophe. That proved to be significant in the early part of the seventh century when another sharp, swift downturn in temperatures across the steppes of Central Asia played a part in another set of revolutions that once again had far-reaching consequences.

II

The Golden Age of Empire
(c.600–c.900)

The orderly government of kingdoms is based upon religious beliefs, good manners, law and order.

al-Iṣṭakhrī, *Kitāb Masālik al-mamālik* (tenth century)

The first twenty-five years of the seventh century were a period of extraordinary political instability across Europe, Africa and much of Asia. The Roman and Persian empires engaged in a phase of almost continuous warfare that saw many of the greatest cities across the Middle East and North Africa change hands – and both empires embarking on military campaigns that not only brought their rival to their knees but threatened to destroy them. The swings in fortune were not just rapid but momentous: in 626, the Persian army was massed by the walls of Constantinople, with many citizens cowering inside the great city fearing the worst; just a few months later, after mounting one of the most spectacular rearguard actions in history, the Emperor Heraclius had advanced eastwards at lightning speed and crushed a large Persian force at Nineveh in what is now Iraq.[1]

Within a few weeks, the Roman army reached Dastagerd, home of the favourite palace of the Persian shah Khusro II, capturing an enormous quantity of booty that included textiles and fabrics, sugar, spices, silver and all sorts of valuable equipment, and recovering 300 military standards that had been taken in battles in the past. The confrontation between Rome and Persia had been couched on both sides in ominous, apocalyptic terms as a fight between light and darkness.[2] News of the advance and the overthrow of Shah Khusro by factions at his court reached Constantinople and was framed in biblical terms: 'the arrogant Khusro, opponent of God' had fallen and had been 'cast down to the depths of the earth, and his memory is utterly exterminated from the earth'. Such was the lot of the man who had spoken unjustly, arrogantly and with contempt 'against our Lord Jesus Christ the true God and his undefiled Mother, our blessed Lady, Mother of God and our ever-Virgin Mary'.[3]

It was no coincidence, then, that before returning to the imperial capital, Heraclius went to Jerusalem to stage a victory procession and the solemn restoration of the True Cross that had been looted by Shah Khusro. It was a day of great celebration, accompanied by 'the sound of weeping and wailing'; tears of joy 'flowed from the awesome fervour of the emotion' of onlookers who were so overcome that they were not able to sing hymns of thanks.[4]

The key to the Roman triumph had been an alliance made with the powerful ruler of the Türk empire which in the mid-sixth century had established itself across much of the steppe that ran from the northern lip of the Black Sea in the west through to the plains of Mongolia in the east. In 626, Heraclius had reached a comprehensive agreement with Ton Yabghu, *qaghan* of the western half of the lands controlled by the Türks at the time. Heraclius had been compelled to concede astonishing terms to cement this agreement. These were confirmed in an extraordinary meeting between the Emperor and the Qaghan during which the former removed a crown from his own head and placed it on that of the Türk ruler, who was then presented with elaborate gifts for him and his retinue, and shown a picture of the Emperor's daughter who was offered up in marriage.[5] A marriage tie between a member of the imperial family and a barbarian nomad ruler was unprecedented; it demonstrated both the weakness and desperation of the Emperor's position and the importance of the support of the Türks against the Persians.[6]

Winning over the Türks as allies was crucial in turning the tide in favour of the Romans and against the Persians. The nomad grouping had risen in prominence and power around the same time as the climate crisis of the mid-sixth century. Quite how and why it did so is not entirely clear, although to judge from other expansions of nomadic states, the correlation of higher-than-normal levels of rainfall and a consequential improvement in the carrying capacity of pastureland on the one hand and an increase in the sizes of flocks and herds on the other may have been significant.[7] This seems to have been the case on this occasion – but it also mattered that the Türks had been able to reinforce their superiority through the extraction of tribute from sedentary states, through control of trade routes and networks and through the sale of animals, particularly horses.[8]

By around 550, the Türks had been able to establish a dominant position on the steppes, either absorbing their nomad rivals or displacing them eastwards into the Korean peninsula, westwards towards and beyond the Danube or southwards into India.[9] These included the Avars, who had streamed into the Balkans and were later sought out by Khusro II in the 620s as partners for the assault on Constantinople in a sign of the military capabilities of nomadic forces, their organisational ability and the important role they played in the politics of this period.[10]

Something similar was happening in East Asia at precisely the same time as the dramatic events in Persia and beyond were playing out. In 618, an aristocrat named Li Yuan led an uprising capitalising on antagonism towards the Sui dynasty, whose expensive infrastructure projects, such as the extension of what ultimately became the Grand Canal that links Beijing with Hangzhou, along with poorly thought through campaigns against Goguryeo in Korea, absorbed financial resources, demanded significant manpower and proved highly unpopular. The period of uncertainty that followed presented an opportunity for Illig, the Türk *qaghan* in the east, and notionally the supreme leader of the entire Türk confederation, who was keen to take steps against the Sui, whose wall-building efforts (which became part of the Great Wall system) closed off the best pasture and agricultural lands to the nomads.

In the 620s, the Türks were plundering northern China with abandon. In 626, the *qaghan* and his forces reached the Wei River, threatening to strike at Chang'an (modern Xian), the capital city of the newly established Tang dynasty under the Emperor Gaozu – Li's chosen regnal name. Perhaps the Türks considered an all-out assault on the city, one of the largest in the world at the time. More likely, the invading force were seeking to extract a hefty ransom in exchange for withdrawing; this was duly paid and the invaders retreated.[11]

This was the high point for the Türks in the east. Within four years, the vast confederation had disintegrated, crushed by a huge Tang army that advanced into the grasslands of the steppe and inflicted a devastating defeat.[12] Having played a crucial part in the defeat of the Persians, the Türks in the west also experienced a rapid decline, falling into difficulties in the 630s before disintegrating and paving the way for a reorganisation of peoples on the steppes north of the Black and Caspian seas. The suddenness of the collapse of the western Türk empire, mirroring that of the east, was astounding. One moment, both had been receiving tributes by the cartload; the next, they had disappeared into irrelevance.[13]

The dramatic changes in fortunes have often been ascribed to internal divisions within the Türk world. In the west, the murder of Ton Yabghu, the dominant figure on the steppes, is usually seen as instrumental in prompting a period of instability, internal competition and decline.[14] In the east too, friction within the leadership was seen as a factor in the precipitous reversal on the steppes, as contemporaries noted. A report produced for the Tang Emperor in 629 by Zhang Gongjin, the ambitious commander of the northern city of Daizhou, specifically drew attention to the rivalries as a source of opportunity. In this report, Zhang argued that the nomads were unusually vulnerable to a counter-attack and urged the Emperor to take advantage of favourable circumstances. These included their rivals' over-reliance on non-Türks, disaffection among subordinate tribes and opposition to Türk overlordship by some local populations. Another reason for their weakness, said Zhang, was a sudden period of harsh climate conditions – and the fragilities that were likely to arise as a result.[15]

New research on the tree rings from Asia, Europe, the Americas and Tasmania as well as on ice-core evidence from the Arctic and Antarctic suggests that a major eruption took place in 626, with the lack of Antarctic signal being highly suggestive of a volcano located in northern latitudes. The fact that the eruption's sulphate anomaly in the record in northern Greenland was the largest of any in the last 2,000 years (with the exception of Laki in the later eighteenth century, which we will assess later) suggests that it was responsible for a sudden drop in temperatures in the Altai mountains in Central Asia (as well as elsewhere) in the years that followed, with reconstructions indicating a cooling of as much as –3.4 °C at its peak.[16]

As well as cooling temperatures, dust veils had an impact on vegetation growth. The effects would not have been uniform: regions and locations with finely balanced ecosystems or ones near the limits of their exploitation only need small modifications to produce a cascading effect. The steppe belts were a perfect example, especially following benign climatic periods that had enabled increases in flock and herd sizes. Moreover, the effects could be exacerbated by changes to sea surface temperature in the tropical Pacific and the North Atlantic that shaped rainfall patterns in Central Asia and may have been a further aggravating factor.[17]

This helps explain the many references in the Chinese sources to unusually harsh weather, including heavy snowfall and droughts that caused food shortages and famine across large areas of central and even southern China. While it is not possible to quantify precisely what this meant for agricultural production, perhaps more important here was the impact on the steppes and on livestock in particular: modest increases in snowfall can impact mortality levels as a result of greater difficulties in finding suitable food sources.[18] These problems are made dramatically worse when they follow drought, partly because animals have been weakened by malnutrition.[19] In the first century AD, for example, the Book of the Later Han states that 'two thirds of the people and domestic animals' of the Xiongnu nomads – predecessors of the Türks – died following a short but sustained period of drought.[20]

It was not just sedentary states, therefore, that could prove to be fragile when faced with a set of circumstances that ratcheted up pressure and risk. In modern times, unusually cold periods have resulted in precipitous losses of livestock, numbering in their millions.[21] Such disasters have often triggered social consequences stemming from economic distress, including increased poverty and mass migration.[22] In the case of the Türks, it is not hard to see how the build-up of challenges identified by Zhang Gongjin combined to create a cocktail of crises. Livestock mortality, food shortages and harsh climate conditions helped unweave the ties that bound tribal groupings together – undermining the authority of the leader and causing the confederation to splinter.

The breakdown of this system could happen rapidly, in other words. As one contemporary report put it, 'the Türks rise and fall solely according to their sheep and horses'. A decline in livestock numbers, especially if sudden and dramatic, could have very serious consequences for the fortunes of individual nomad groups within the wider confederation, with the impact felt most keenly by those of the lowest status and means.[23] As such, shocks, climatic or otherwise, could suddenly make what seemed to be resilient and powerful structures brittle, providing opportunities for subordinate groups and ambitious individuals to undermine what has been aptly described as an 'inherent instability of the system of alliances and hierarchies that made a centralized leadership relatively weak'.[24]

When a massive Tang army marched to exploit the situation in the winter of 629–30, the Türks fell back into the grasslands but were caught off guard by a surprise attack that led to heavy casualties, many prisoners being taken including the *qaghan*, who spent his last years in captivity before reportedly dying of a broken heart. Just as this served as the breaking of the Türk empire in East Asia, it was the making of the Tang emperors, who were to rule much of what is now China for the next three centuries. After seizing the throne in 626, Emperor Taizong had taken the traditional titles of 'Son of Heaven' and 'August Emperor'. The dramatic change on the steppes significantly enhanced his legitimacy and strengthened his hold on power. It also emboldened the emperor to add to his titles. 'I am the Son of Heaven of the Great Tang,' he announced in a ceremony in front of the imperial palace in Chang'an. From now on, he announced, '[I] will also attend to the affairs of the Heavenly Qaghan.' New seals were created to stamp all correspondence with rulers and peoples to the north and the west of territories directly under his control. He had not just defeated the nomads, in other words; he now claimed to be their overlord.[25]

To the west too, the western Türks experienced a time of turbulence, characterised by uprisings, revolts and assassinations of senior figures, including the supreme leader, Ton Yabghu.[26] Local rulers in locations like Turfan, Kucha and Samarkand refused to pay tribute as they had done in the past, and opted to look instead to the Tang court for protection. The scale of this upheaval was remarkable. But so too was its speed; when the famous Chinese monk and traveller Xuanzang visited the western Türks in 630 and described the *qaghan* as wearing green silk robes and being surrounded by an entourage dressed in embroidered finery and lavish textiles, he would have assumed he was witnessing an empire at its apogee; in fact, he was present months before its collapse.[27]

Xuanzang was in fact surprised by his reception, not because of how alien and strange it seemed to him, but because the ordered and structured ceremonial element that he witnessed was similar to that of the world he came from. Symbolic displays of power, paying obeisance to the ruler while kneeling, the grasping of feet and touching the ground with one's head were all customs that were widely practised not only on the steppes but at courts from one end of Eurasia to the other, linking the Mediterranean to the Middle East, East Asia and the steppes of Central Asia. This was an interconnected world, in other words, far removed from crude formulations put forward by writers at the time (and in the modern world) of sedentary states and a chaotic nomadic world. Rituals and symbols were not uniform, of course, but the boundaries of these worlds were porous, flexible and adaptive.[28]

The upheavals in the steppe world opened doors for the Tang, who expanded into Korea as well as westwards, and deftly cemented their gains by reshaping ceremonies to include foreigners and honour their status; this served to incorporate and celebrate new members of the empire rather than treating them as subordinates. As it happens, this aped how nomads themselves often dealt with the absorption of new peoples (and stands in stark contrast to the European age of empire in Africa and the Americas in particular, as well as in Asia).[29]

This was important too at the other end of the Eurasian land mass where the tumultuous changes on the steppe, combined with the epic confrontation between Rome and Persia, opened the door for the dawn of a new age. Extraordinary resources of time, manpower and money had been poured into a war that had gone on for some two decades. The fighting not only had an impact on cities and markets in the Mediterranean, Levant and Persia, but depressed the economies of networks and regions that were connected to them, such as in southern Arabia.[30] The effects were felt further afield too, with some scholars linking a sudden and sharp decline in much coveted garnet in Europe at this time, which came from South Asian sources, with the dislocation of trade in general and in the Red Sea in particular.[31]

One of those to offer explanations for the seismic changes was a young merchant named Muḥammad, who had gathered a small but loyal following around himself. His warnings of imminent doomsday struck an obvious chord in the context of more challenging economic conditions along trade routes in the Ḥijāz, the continuous and ominous fighting to the north and apocalyptic propaganda in both Rome and Persia – although it is clear that there was substantial resistance to his radical message of the need for the purity, unity and loyalty that would win God's favour, as well as to the deviation from other religions that were well established in Arabia.

The decisive moment for the Prophet Muḥammad and his companions came in 628, when in a masterstroke of political accommodation he reached agreement with the ruling elite in Mecca, in which he decreed that the direction of prayer should not be towards Jerusalem but towards Mecca itself, while the Kaʿba, previously a pagan shrine, was designated the focal point within the city.[32] This paved the way for reconciliation between different factions in Arabia and laid the basis for a common identity that would provide an umbrella under which all the different peoples of the region could gather. The agreement, enshrined in the Treaty of al-Ḥudaybiya, proved to be turning point not only for Mecca, the surrounding region and Muḥammad himself; it was one of the major turning points in world history.

As it happened, the timing could not have been better. For one thing, in addition to the military setbacks, Persia was also experiencing a virulent outbreak of plague, with many being killed in the capital city, Ctesiphon – with the re-emergence of the disease perhaps linked to the eruption of 626 and the associated ecological and pathogenic conditions that followed.[33]

Muḥammad, his movement and followers emerged into a world that had been scarred by warfare and disease, by economic contraction and by a collapsing global order of sedentary states that were exhausted, and of nomadic confederations whose ties were dissolving. His own thinking had been apocalyptic too, as he warned that the time was fast approaching when the sun would be extinguished, the stars would fall, the seas would boil, the skies would be torn in two and graves would spill open, throwing out their cadavers. The Final Judgement was nigh, warned Muḥammad: the virtuous would go to paradise; the wicked to hell.[34]

As it happened, Muḥammad and his followers emerged to be the prime beneficiaries, creating a vast new empire in the process. But it might not have turned out that way. Around 630, with Rome and Persia teetering, economically and militarily exhausted, the scene was set for the nomads to expand their dominance from the steppes to create one of the greatest empires in history. The Türks held the fate of the Roman empire in their hands; Persia was teetering on the brink from their crushing defeat and the ravages of disease; in East Asia, the Tang were vassals in all but name.

The timing, then, of the collapse of the Türks stalled what might have been the creation of an empire even greater in extent than that of the Mongols, who were able to do what the Türks did not in the thirteenth century. With a different set of climatological and ecological developments precisely in the late 620s and early 630s, it might have been the Altaic languages of the steppes and the devout Buddhism of Ton Yabghu that spread quickly in all directions across the whole of Asia, and perhaps even across Europe and North Africa. The bulk of those prizes now went instead to Muḥammad and his followers – and to the Arabic language and to Islam.

The 630s and the decades that followed saw the explosive expansion of Arab armies from the Ḥijaz across the Middle East through North Africa into southern Europe and deep into Central Asia. The result was a wide spread not only of Muslim beliefs and teachings but of Arabic culture. This included the Arabic language, of course; but ideas about music, poetry and fashion spread too, as well as about art and architecture.

Arab elites in the Iberian peninsula harked back to homelands, building palaces and monuments, and commissioned historical accounts and complex genealogies of early Arab leaders with the result that Spain became 'more Arab than Arabia'.[35] One poet spoke to the single palm tree in his garden, lamenting that both were a 'stranger in the west, far from our Orient home' – recognition that ecosystems and people alike did not find it easy to adapt to new locations.[36]

The creation of an enormous new world that bonded together the core territories of the Persian empire stretching from the Caucasus and Asia Minor in the west as far as the Himalayas in the east along with some of the richest and most important parts of the Roman empire – including the breadbasket of Egypt, the harbours of the Levant and a heavy influence in the Mediterranean – also underpinned environmental and ecological changes that had far-reaching consequences.

This was partly the result of the spread of irrigation techniques and agricultural technologies across lands conquered and controlled by the Arabs; but the diffusion of crops and cultivars was no less important in knitting together a world that was not only culturally joined together but also linked through tastes for specific foods, flavours and recipes.[37] Hypotheses about the extent, nature and footprint of this 'Islamic Green Revolution' are hotly contested, with some pointing out that foods and crop species that are often linked with the spread of Islam had been grown long beforehand in other parts of the world.[38] Evidently, a measured evaluation is dependent on assessing archaeobotanical evidence, a discipline which is still an emerging area of scholarly research.[39]

What was perhaps more important than the availability of new crops was the fact that the new Arab empire provided two things that had been in short supply in preceding decades: peace and manpower. While the politics of succession at the core of the Arab leadership were bloody and brutal affairs, with three of the first four successors of Muḥammad being assassinated, elsewhere the new masters were careful to leave communities and populations largely alone and unscathed – as long as they paid their taxes. Some complained, stating that financial obligations trebled under the new overlords, although it is not clear whether such claims can be taken at face value.[40] For the most part, however, transitions appear to have been smooth and uncontested.

Relying on officials and administrators who were familiar with market practices, agricultural yields and financial expectations was a key part of this success. In Roman and Persian territories, provincial bureaucrats were retained, something that presumably left the local population sanguine as long as their property, rights and beliefs were left untouched.[41] In Central Asia, the Arabs inherited a system which had been developed to good effect by the Türks and which relied on local urban elites who used their positions to protect their own interests.[42] This too was largely left intact by the new masters.[43]

Remarkably, although there was regular internal friction within the world that was built by Muḥammad, his heirs and successors – who held the title of caliph – competition and threats from beyond were minimal for centuries. While the battle of Talas in 751 soon came to be regarded by commentators as marking the end of Muslim expansion eastwards and the conquests of the Tang dynasty to the west, it was the mountain ranges of the Pamirs and Himalayas that acted as the most effective insulation to keep the state (the caliphate) from any serious pressure from the east. In the west, likewise, there was little by way of challenge. Expeditionary forces that moved north from Spain into France found limited fruits worth picking; the only place where there was pushback and rivalry was in Asia Minor, where the eastern Roman – or Byzantine – empire hung on for dear life and occasionally mounted a show of bravado. However, as archaeological evidence from Anatolia shows, from the middle of the seventh century onwards, regular raiding, coupled with the rising costs of military defences, led not only to long-term economic decline but to lowered living standards and falling life expectancy which created a vicious circle downwards that took centuries to correct.[44]

In contrast, in the lands under the control of the Arabs, a boom time followed, with cities expanding, swelled by rural workers migrating from the countryside. This did not just enhance civic life; it also fuelled greater economic and demographic growth both within cities but also between them as old and new networks began to throng with people, goods and ideas. These developments sparked change in other regions as new connections and opportunities emerged.

For example, the lands to the north – steppe belts fanning out across thousands of kilometres – flourished in parallel, with the Khazars thriving as a result of growing demand for animals and their by-products as well as from trade across the territories they controlled directly or indirectly through networks of subordinate tribes. To the south-east, meanwhile, Indian Ocean trade networks began to blossom too in conjunction with stable and prosperous regimes both in the Middle East and in East Asia. One prime beneficiary was the Śrīvijaya state which grew from Palembang on the island of Sumatra to control the Melaka Straits – a key artery for east–west trade.

By the seventh century, Śrīvijaya was a leading centre for Buddhist scholars and scholarship, and purportedly home to a thousand monks.[45] Its political and economic ambitions now grew rapidly, as attested by inscriptions in Old Malay that talk of conquest of other coastal towns, often at the cost of considerable bloodshed.[46] Scholars debate the nature of the Śrīvijaya and whether it is best understood as an empire, a kingdom or a chiefdom, partly because of the complexity of the relationship between the ruler (the *datu*) and the *orang laut* – the coastal peoples who were co-opted and incentivised to expand the reach of the state.[47]

What is less in doubt is that levels of exchange across the Indonesian archipelago, South and South-East Asia began to intensify as horizons narrowed, contacts rose and goods moved in increasing volumes through local, regional and long-distance maritime connections linking the coastlines and peoples of the Mediterranean, Red Sea, Indian Ocean, South China Sea and Pacific. From the eighth and ninth centuries, new states mushroomed across this whole region, such as at Pagan in what is now Burma, Angkor (Cambodia), Đại Việt and Champa (Vietnam) and the Chola (India), which shared similar characteristics of bureaucratic centralisation, cultural integration, theocratic kingship and territorial expansion at the expense of smaller, less robust polities as a result of forest clearing and the extension of ecological frontiers or both. The simultaneous trajectories of so many complex states were 'strange parallels', to use the term coined by one modern historian, interdependent regional powerhouses whose successes fuelled each other's continued expansion and success.[48]

The Arab conquests opened up new geographic, ecological and cultural frontiers. Drawing heavily on Greek scholars like Hippocrates and Aristotle who wrote about how the environment affected personal character, Arab writers considered the peoples that they now came into contact with thanks to their widened horizons. Slavs and Franks in western Europe, wrote al-Mas'ūdī, 'have large bodies, grotesque natures, difficult characters, limited intelligence and heavy tongues'. This was because they lived in damp and cold regions which saw near endless rain and snow, and which also explained why their skin colour was so white that it sometimes passed for blue. The lack of warmth made clear why their religious beliefs could be considered limp and wishy-washy. People living further north were even more stupid, grotesque and brutish – and because of the cold were so fat that their faces were round, making their eyes look small. Black people, meanwhile, had the 'bravery of lions and the slyness of foxes', but because of the heat had short concentration spans, although they were also very taken with dance and rhythm. It was simple, said al-Jāḥiẓ: climate shaped 'character, personal qualities, manners, speech, desires, ambitions and appearances'.[49]

The conquest of North Africa also opened up new links beyond the Sahara desert, the final frontier faced by the great Muslim empire. A key reason for this was the development of the trans-Saharan gold trade and access to the rich deposits located at the headwaters of the

Niger and Senegal rivers. Although the first written records of this trade come from Arabic texts, this region's integration into intercontinental exchange networks had occurred well before the expansion of the caliphate, with gold from West Africa underpinning Roman minting of coins in Carthage at least a century earlier.[50]

Interest in West Africa was partly shaped by reports of large reserves of resources, manpower and gold, which encouraged travellers, scholars and writers to find out more, with the results sometimes blending myth with reality. For example, many Arab authors believed that the populations of the world descended from the three sons of Noah – including those descended from the youngest, Ham, who had been cursed by his father for seeing him naked. As punishment, wrote Ibn Qutayba, God 'changed his colour and the colour of his descendants'.[51] Ham's progeny, who had been turned black, had then crossed into Africa where some settled in the east, some in the west.[52]

In fact, the story of the 'Curse of Ham' that appears in the Book of Genesis says nothing about skin colour, nor about population migrations or dispersals in Africa. Arab authors were adopting long-established Jewish and Christian tropes about blackness that became the subject of considerable discussion and speculation that extended into hypotheses about reproductive sciences.[53] The idea that black skin colour was connected to divine punishment caught hold. It did not just prove toxic, but became what one scholar has called the 'single greatest justification for Black slavery' that was to have profound consequences in later centuries.[54]

Arab geographers were nevertheless intrigued by the peoples living in sub-Saharan Africa, by their customs and behaviour – though many also paid attention to the wealth and the gold mines in the kingdoms of West Africa, and especially of Ghana, 'whose king is very powerful'.[55] Locations which were home to gold mines were strictly patrolled and access was forbidden; once extracted, gold was made into bricks and sold in towns like Sijilmāsa, on the northern edge of the Sahara, a city so rich as a result of its role as emporium that it was home to four enormous mosques and a main street so long that it took half a day to walk from one end to the other.[56]

Commentators were both curious and dismissive of religious and cultural practices – evidently in order to justify their own superiority, to emphasise the 'civilising' influence of Islam and also because development of the slave trade went hand in hand with demeaning racial caricatures. Many peoples in Africa, wrote al Iṣṭak̲h̲rī, did not deserve lengthy descriptions: 'the orderly government of kingdoms is based upon religious beliefs, good manners, law and order, and the organisation of settled life directed by sound policy. These people lack all these qualities.'[57] It was hardly surprising, therefore, that many learned to be cautious when dealing with Arab traders: some insisted on negotiating at a distance and in using signs to reach agreements – the result, wrote one author, of 'their extreme wariness', which he equated to the 'wariness of domestic animals faced by beasts of prey'.[58] It was not hard to conclude that this trepidation was well judged given unbalanced views about race and the threat of enslavement.

By the ninth century, Muslim Arab and Berber merchants were active in the city of Gao, situated on the banks of the Niger River, involved in the trade not only of gold but of textiles and ivory too.[59] Gao's location as a crossroads was important, even if, as some scholars suggest, it should be understood as referring to a sprawl of settlements rather than to a single one.[60] The growth in population density and building across sites such as Gadei, Old Gai and Gao Saney may have been a result not only of new opportunities opened up by trans-Saharan trade, but also of variability of rainfall and flood patterns and in particular of the resultant shifts of sediment and fish stocks.[61]

Similar processes of rising connectivity can be found in the Americas around this time. This was a period of transition in the American south-west, as people of the Puebloan cultures moved from pit houses into larger, more elaborate and more permanent structures.[62] It was also a point where political powers began to concentrate under the leadership of single chiefdoms in the Mississippian cultures of the midwest, east and south-east of North America, which in turn led to competition that drove consolidation and expansion.[63]

In all these cases, the causes of change were complex, but the principle of a chain reaction seems to have been particularly important in the way entities galvanised each other. Certainly, demographic expansion was a factor, which was closely linked to the ability to grow and store food, and to cope with weather shocks or with long-term climatic changes. Likewise, stability was dependent on the emergence of organisational sophistication that had to adapt and expand as the powers and the claims of the centre grew – something that presented growing pains of its own.

This was always easier said than done, especially when it came to the relationship between population growth, consumption patterns and sustainability. A decline in both the quantity and size of land crabs in the Caribbean from the first to the sixth centuries resulted from human activity and over-exploitation, while precipitous falls in marine life, including reef fish, as a result of overharvesting likewise shows that food availability which supports increasing numbers of people can become quickly exhausted.[64] Coping with this equation normally required adaptation, new lifeways, migration – or starvation.[65]

Urban settlements too could be victims of their own success, as the case of Panjikent neatly shows. A magnificent jewel of a city in what is now Tajikistan, Panjikent was a booming centre along the Silk Roads, serving as a key point of exchange for Sogdian and other merchants, some of whom were engaged in long-distance trade in luxury goods. The city blossomed as trade grew, expanding rapidly in the fifth century and becoming home to magnificent palaces whose murals depicting scenes from the Indian epic *Mahābhārata*, from the Persian classic *Shāh-nāma* and from ancient Greek literature still take the breath away today.[66] Panjikent's glory came at the expense of the natural environment. Over time, the surrounding area was subject to major deforestation, while the demands made on the land both for agricultural production to feed the population and to make bricks for construction projects resulted in severe soil erosion.

Those presented challenges in any circumstances. But when there were even modest changes to precipitation levels and to water supply from springs, they not only became problematic but insurmountable. While the abandonment of what had once been a key node along the Silk Roads may have owed something to the Arab conquest of this region and to the diversion of trade routes, there were obvious consequences for cities that lived beyond their means.[67]

Perhaps not surprisingly, then, in the world of the caliphate, pressures came primarily from within. Competition between rival factions at the court could sometimes spill into dynastic upheaval and, in the worst cases, secession. A case in point was the overthrow of the Umayyad dynasty in the middle of the eighth century by the 'Abbāsids after almost a hundred years in power – although members of the former managed to hold on to power in Spain, establishing an independent state centred on Córdoba.[68]

Revolutions like these could spark reform and change that rebooted the socio-economic and political system, primarily through a purge of the elite and their replacement with a new class and caste. As new leaders sought to win hearts and minds, they would often spend lavishly not only on their immediate client base but on public works and on patronage that could appeal more widely. Rather than an Islamic golden age, there was a series of golden ages or golden waves, overseen by rulers with different resources at their disposal and different aims.

For example, considerable effort went into gathering and translating Sanskrit texts about science into Arabic, to bringing physicians from India to instruct about medicine and to becoming better informed about other parts of the world. This was partly motivated by a genuine sense of curiosity, although, as has been recently argued, it also owed something to emulating earlier, pre-Islamic Sasanian models where rulers were champions of scholarship. There were benefits too, of providing support to and privileges for non-Muslim subjects: inclusiveness and tolerance were not just indicators of open-mindedness; they were smart politics in a world studded with minorities.[69]

Skills in subjects like astronomy and astrology were important too in providing additional tools for the legitimisation and validation of political decision-making, which had heightened significance after a change of dynastic leadership.[70] The transfer of information, knowledge and technical expertise between India, China, the Middle East, Constantinople and beyond was a hallmark of the rivalry between interconnected and interlocking empires.[71]

The explosion of literacy from the mid-eighth century, which is sometimes linked to the introduction of paper to the Middle East from China, was above all else a sign of rising prosperity. Wealth and leisure time helped generate interest in book production, as well as funding for scribes and for scholars to research, think and write. Some writers were voracious readers – like al-Jāḥiẓ, who reportedly rented bookshops so he could read volumes all night; according to one source, he was killed by a stack of volumes that collapsed on him when he was elderly and frail.[72] Others were workaholic compilers of information, like Ibn al-Nadīm, a scholar who lived and worked in Baghdad in the tenth century, who set out to make lists of the 'books of all peoples, Arab and foreign ... dealing with the various sciences', recording who 'composed them and the categories of their authors, together with their relationships and records of their times of birth, length of life and times of death as well as the location of their cities', while also listing their virtues and faults.[73]

Naturally, the abundance of wealth that enabled this rich cultural world was equated with divine benediction. Both were closely linked with benign climate, and, above all, the availability of food. After all, this had been one of the messages given by God to Muḥammad in a cave near Mecca: those who were obedient would experience 'gardens of perpetual bliss', promised one verse that was later codified to create the text known as the Qur'ān.[74] God 'has promised to believers, men and women, gardens under which rivers flow, to dwell therein and beautiful mansions of everlasting happiness'.[75] Paradise would be a place where all one's needs would be catered for, from fruits to whatever else one desired.[76]

This was of course all familiar in other Abrahamic faiths – Judaism and Christianity, where God has a central role as provider of a benign environment in general and of enough food in particular. God had provided manna from heaven, a special 'fine, flake-like thing' that had to be collected before it was melted by the sun's rays, and was either eaten raw with a taste like honey, or baked into cakes that would not 'stink or get maggots in it'.[77] When Jesus had gone fishing in the Sea of Galilee, he had managed to fill the net to bursting – after the efforts of his disciples had proved fruitless.[78] Perhaps the most famous example of divine intervention to provide food, however, comes with the feeding of the 5,000, the only miracle that appears in all four gospels.[79] In Judaism, Christianity and Islam, God was not only the creator of the natural world, he was also the ultimate provider of resources, able to guarantee provisions when needed and to provide benign climatic conditions at will.

Indeed, in Islam God's control over the natural world was absolute to the extent that animals were purported to be particularly alert on Fridays, as they realised that this was the day that the end of world would take place. Muslims were warned not to slap creatures in the face or to mistreat them as 'all animals hymn the praises of God'.[80] Just as the Bible underscores the idea that God perfected all creation, the Qur'ān speaks of the order, beauty and harmony of nature as being divinely ordained. Moreover, when taken as a whole, nature has intrinsic and inherent value, not simply being useful for human beings. As some contemporary scholars contend, the earliest teachings of Islam argue that biodiversity and rich ecosystems stem from God's creation and therefore deserve respect and maintenance.[81]

Naturally, the bounty of plenty was not just the result of a benign God but also the sign of a good ruler. The Caliph al-Mutawakkil (r. 847–61) was praised by one contemporary poet who noted that it was not God alone who deserved thanks, as he wrote in rapture:

Thanks to you, vast tracts of land have become fertile.
For sure, the world cannot be barren as long as you are its
protector.[82]

Such praise went hand in hand with an expansion of irrigation and of agriculture. In Central Asia, for example, the number of oasis settlements and the area that they covered had already started to rise before the Arab conquests. They continued to do so afterwards, even though it is clear that the Arab moves into this region met with considerable resistance and deliberate destruction, at least in some locations.[83] The development of crop growing must certainly have been facilitated by the fact that, around 650, climatic conditions shifted across Central Asia, the Tibetan plateau, the steppes of Mongolia and eastern China, becoming both cooler and more humid, with additional rainfall beneficial to plant cultivation as well as to vegetation cover that enhanced livestock rearing.[84]

In this bustling, vibrant world of widening horizons, rising populations and growing wealth, 'high cuisine' was an important way of expressing cultural identities, especially for the urban bourgeoisie that in time extended beyond just the rich and powerful.[85] Although some wryly noted that Baghdad was 'a joyful land for the affluent' but 'for the destitute, an abode of anguish and sorrow', there was great demand for the best ingredients, as well as information about where to find and how to prepare them.[86]

Do not be fooled into thinking that the wealthy have different food to normal people, warned the author of one Baghdadi cookbook. The difference in flavour and experience was down to the meticulous cleanliness of the ingredients – as well as of the pots in which they were cooked. It was important, therefore, to have good hygiene practices in the kitchen, including having separate knives and chopping boards for meat and for vegetables and to use good-quality materials for handwashing.[87] Personal tastes could sometimes be quixotic, such as those of the Caliph al-Wāthiq who had a penchant for aubergines and once ate forty in a single sitting.[88] Some gourmands were dismissive about vegetarianism, claiming that dishes made only of vegetables were a fraud, often being presented to look like meat but tasting nothing like it. 'Trifling vegetable dishes are not for me,' wrote one poet in protest; 'where is my kebāb? Where are the fried dishes? The succulent roasts and the spiced meat? Bring them on!'[89]

While metropolitan concerns preoccupied most contemporary commentators in all these regions – they paid particular attention to the corridors of power – during times of demographic growth, urbanisation and rising patterns of consumption, maintaining agricultural sustainability was essential. This in turn was dependent on multiple factors that varied significantly not just from region to region but sometimes even across modest distances. Climate was of course one element; but so were distances to market, the availability of labour forces to work the land, crop specialisation, soil suitability and optimal yields.

What mattered most of all, however, was water management, and not just in terms of assessing agricultural production. Pre-industrial states were heavily dependent on their agrarian economies and above all on land tax. Trade was important, certainly, and rightly receives much attention from modern scholars. However, revenues from land were considerably higher than they were from long- or short-distance trade.[90] Not surprisingly, considerable effort went into assessing land quality, ownership and the revenues it would produce. Muslim rulers conducted regular land surveys, therefore, just as their Sasanian predecessors had done before them.[91] In addition, state officials were careful to make records of hydraulics, irrigation systems and water resources, in order to try to maximise returns that could be collected by the crown.[92]

Such evaluations naturally depended on the fact that rainfall should be consistent and – equally importantly – that demand made of the land and of the available water sources did not rise. When the balance tipped in the wrong direction, problems followed. Around the start of the ninth century, for example, a set of challenges emerged that brought with them major socio-economic and political consequences. To judge from tree-ring records from the Altai mountains in Central Asia, salinity levels in the Aral Sea and observations about the North Atlantic Oscillation shifting strongly towards positive mode, a long-term trend away from colder, more humid conditions towards warmer and drier ones for many parts of Eurasia began to emerge around c.800. The challenges this posed can be seen from the decline in the number of oasis settlements, which fell by almost 70 per cent over the decades that followed.[93]

This was not a phenomenon unique to Central Asia, as seems clear from the abandonment of towns, villages and agricultural land in western Iran as well as in the floodplain of the Euphrates – one of the most productive areas not only in the Middle East but in the world.[94] In addition to the climatic shift was ecological stress caused by rapidly increasing salinisation of soil. Wheat yields in the rich alluvial plains of Mesopotamia fell precipitously in the years c.820–50; that those of barley rose provides one indication of the imbalance in salinity levels, since the latter is much more robust than the former in soil that has become degraded.[95] The importation of coerced labourers from Africa to remove the salt layers by hand to keep fields cultivable provides another.[96]

These problems were compounded by the fact that around 800 wealthy elites had started successfully to build up landholdings, using status, connections and political pressure to influence taxation systems and water allocations.[97] This brought about short-term gains at the expense of long-term sustainability – both economic and environmental: in the early ninth century, the Abbāsid state was able to collect around 125 million dirhams per year in revenue from land tax in Iraq; a century later, the figure had fallen by almost 80 per cent.[98]

Such collapses in available funds created problems of their own, not least pressure on the leadership to explain why they could not live up to the largesse and glories of their predecessors, ranging from their magnificent building works to their territorial expansion to the lavish spending that could often seem to be a distant memory. One response was to criticise profligacy, over-the-top lifestyles and irresponsible outgoings. According to Alī ibn 'Īsā who was writing in the early tenth century, almost a third of the caliphate's annual budget was spent on the harem, on eunuchs and on senior officials at the court.[99] With expenditure like this, it was not surprising that shocks could exacerbate grievances and open the door for protest, instability and sometimes worse.

In Baghdad in 919–20, for example, there were riots as a result of food shortages that were only pacified after grain reserves were distributed at subsidised prices.[100] Two decades later, a famine struck that was so bad that 'houses, vineyards, and gardens were sold for joints of meat and cakes of bread. And men used to pick out the grains of barley from the dung of horses and asses and eat them.'[101] A few years later, lack of food in the Jazira was so terrible that 'many went mad, attacking one another mercilessly and savagely devouring each other'.[102]

Whether one can rely on such accounts is unclear, even in cases where they seem to correlate with climate data that suggest unusual weather patterns or events, such as droughts, floods or periods when it was hotter or colder than normal.[103] Of course, food shortages were not always related to climate or even to productivity as they were heavily influenced by market expectations. Some tried to mitigate these risks by buying up grain, such as the Fatimid authorities in Egypt – although, tellingly, this practice was abandoned because too often grain stocks had been bought above market prices and could not be sold.[104] Equally, it was inevitable that speculators would seek to capitalise on high prices and try to drive them higher – an activity that the Caliph al-Hākim dealt with by threatening punishment. 'I will cut off the head of whoever possesses any grain whatsoever,' he announced during a crop failure that followed a poor Nile flood in 1007; 'I will burn his house and confiscate his wealth.'[105]

As the Caliph's reaction makes clear, such episodes provided opportunities for the disaffected, the ambitious and the brave to chance their arm in attempts to improve their position. Economic problems that could flow from harvest failure, shock weather patterns or military instability could be exploited by rivals to the Caliph, his court and interest groups eager to capitalise on fundamental underlying structural problems flowing from the concentration of too much power in too few hands. Peripheral regions were often the breeding ground of dissent, the source of revolt and a springboard for newcomers to rise to power.

In the case of the caliphate, one of the key sources of social, economic and political problems was that reforms introduced to provide the state with a stable and secure income ultimately served to strengthen the hand of the rich and powerful further still. The creation of the iqtā' system during the mid-tenth century devolved tax-collection rights in exchange for services; apart from being open to abuse, it meant that the aims of the state and of the elite were not aligned, with the latter seeking to benefit at the expense of the former, and indeed of tenants and farmers who would be squeezed as hard as possible for rents. This aggravated existing causes of widespread dissatisfaction and rising inequality, further stretching a system that was already inefficient, open to corruption and inclined to favour large landholders at the expense of the weak.[106] Perhaps climatic change fanned problems whose embers were already glowing, but it was the failure to keep the rapacious magnate class under control that was the ultimate cause of the regular crises.

It was a similar story in Tang dynasty China. Many scholars have recently sought to associate climatic change with the fall of the ruling family and the implosion of the empire, arguing that droughts, floods, locust invasions and famine spread over fifty years, along with harsh winter and weak summer monsoons inspired waves of rebellion.[107] These cannot have helped, but the root causes of instability, revolt and ultimately the overthrow of the last Tang Emperor in 907 lay in what one historian called the 'violently anti-aristocratic feeling' that was manifested by 'hatred of the old families and hatred of high officials'.[108]

These feelings owed much to the success of the wealthy in feathering their own nests at the expense of the wider population and in managing not only to protect these gains but to keep augmenting them. Tang dynasty China had become a closed shop, observed Zheng Qiao, a commentator writing in the twelfth century. Appointments were not made on merit, but owed everything to family background, with officials (who were themselves all drawn from the same narrow social group) consulting genealogies and marriage ties, records which they kept up to date and relied on as their guide to show not only who was who, but who should be allowed to be part of the great and the good.[109]

By the 870s, wrote one chronicler just over a hundred years later, 'waste and extravagance [at the court] became more extreme by the day', while corruption and dishonesty were rife among bureaucrats. Successive years of drought had caused difficulties, but the real problem had been that the effects 'were not reported truthfully' to the authorities, so action that would have averted suffering was not taken. The result was that 'the common people wandered about in search of food' and, with nobody to help them, soon 'gathered into bandit gangs, moving about like swarms of bees'. Worse, because peace had reigned for so long, government soldiers had little experience and were often defeated by bands of marauders.[110]

Rebellions in the 870s and 880s critically weakened the state, damaged the credibility of the imperial dynasty and strengthened the hands of provincial warlords whose own machinations led to a splintering of territories controlled by the Tang after the murder of Emperor Daizong – who was given the name Ai (哀) or 'lamentable' after his death. As one historian has put it, the period that followed, known as that of the Five Dynasties and Ten Kingdoms, was the logical culmination of increasing regional development at the expense of the centre and had been decades in the making, if not more than a century, depending on one's perspective.[111]

Identifying the deep roots of problems that were affected and exacerbated rather than caused by shifting climatic conditions is also helpful in understanding the fate of the Maya culture in Central America – whose 'collapse' has become one of the most celebrated and commonly cited examples of the dangers of climate stress and environmental degradation and a clarion warning about the troubles of the present and the future. According to some, for example, climate factors brought about pressures that were so severe that catastrophe was unavoidable. 'There was nothing [the Maya] could do or could have done,' in the words of a group of leading scholars; 'in the end, the food and water ran out and they all died.'[112]

This is something of an exaggeration. The Maya did not all die – with most estimates suggesting a population today around the 9 million mark, as well as a cultural footprint that is not only recognisable but extremely important in Guatemala, parts of Mexico and elsewhere in Central America.[113] Moreover, while there was undeniably a period of profound change, some Maya polities continued to survive and flourish, with the last only falling to the Spanish in 1697 – almost 800 years after a decline that was supposedly apocalyptic.[114]

The Maya culture of Central America had blossomed for many centuries, reaching a symbolic apogee with the construction in AD 747 of Temple IV at Tikal, which had the distinction of being one of the tallest buildings in the Americas before the arrival of the Europeans at the end of the fifteenth century.[115] Tikal had long been an important urban centre, one of a series of cities, polities and settlements knitted together by extensive trade networks that facilitated the movement of luxury goods such as obsidian.[116] These networks were also closely linked to the retention of political authority by individuals and family groups that preserved some degree of autonomy from each other.[117] Long-distance reciprocal contact was maintained, including with Teotihuacan over a thousand kilometres away, which was a source of cultural reference points and even the inspiration for the reproduction of major buildings in Tikal.[118]

The distribution of power, status and wealth was not even either between cities or within them: people living in locations that were better connected and close to principal trade routes were more prosperous than those on the periphery, while those ruled over by autocratic elites saw high levels of wealth inequality.[119] A mesh of political alliances was held together by officials and diplomats who shuttled between locations and frequently appear on carved monuments, murals and polychrome vessels.

Recent excavations of a small plaza compound at El Palmar in the eastern part of the Yucatán peninsula have revealed a particularly fine burial chamber of one such individual, a diplomat named Ajpach' Waal, who held the title of *lakam* (standard-bearer) and was involved in negotiating terms between El Palmar and the ruling dynasty of Copán, located 350 kilometres to the south.[120] Ajpach' Waal had acquired jade and pyrite inlays to his teeth during puberty or as a young adult, as was common for individuals of high status, especially those in the Lowlands, as a means of displaying social distinction.[121] Pioneering research into his osteobiography – the study of his skeletal remains – reveals that the inlay had fallen out of his right canine; the fact that dental plaque had formed around it and hardened into calculus suggests that he was not able to afford a replacement, which in turn raises the question of whether his status and that of El Palmar suffered during his lifetime in the mid-eighth century.[122]

Despite regional ups and downs, long-term demographic growth over many centuries was substantial, supporting city states with populations numbering in the tens of thousands.[123] Some estimate that total population across Maya lands peaked at around 10 million, if not higher.[124] Tikal's importance had risen under the leadership of Jasaw Chan K'awiil and his son, Yik'in Chan K'awiil, who oversaw a sequence of military victories and political expansion from the very end of the seventh century.[125] This was a factor in the size of the city's population tripling or perhaps even quadrupling in the course of the next hundred years or so.[126]

Urbanisation required significant investment if settlements were to be sustainable – and able to expand. This included the creation of irrigation networks, as well as development of sophisticated engineering solutions to maximise water capture that ranged from dam building to sand-filtration technologies to cleanse water of sediments and impurities.[127] Another crucial element was tackling erosion, which was partly done through sophisticated forestry and soil management.[128] This allowed the expansion of food production based on maize, beans, squashes, sweet potato and other root crops and enabled it to meet the rising demands of densely distributed and rising populations.[129]

Speleothem data from Yok Balum Cave in Belize, along with lake-sediment cores in northern Yucatán and titanium content from the Cariaco Basin in the southern Caribbean all point to a protracted period of drought in the mid-ninth century that lasted for several decades.[130] Average rainfall levels dropped by around 50 per cent and occasionally by as much as 70 per cent, with these shifts likely the result of changes in solar activity, volcanic eruptions or both.[131] This exacerbated problems caused by rapid deforestation that has been linked to the dual need to expand agricultural land and to obtain wood to fuel fires required in order to bake calcium carbonate from shells or stone to create quicklime – a substance that can be used as a binder, for whitewashing and for construction.[132] Cutting down trees amplified drought conditions since less water evaporates from cleared land, which has a negative impact on cloud formation and on rainfall.[133]

It helped then that the Maya cultivated a wide range of plants, but what is more, that many of these were drought resistant.[134] Although it is the lack of rainfall that has attracted a great deal of attention among historians seeking to link climatic change with apparent civilisational collapse, from an ecological perspective what may have mattered more was not rainfall deficiencies but rising temperatures in the summer and their impact on maize: as modern research has shown, maize yields decline by as much as 1 per cent for each day that air temperatures are greater than 30 °C.[135] This was important in Maya culture, not only from the point of view of food availability, but also because of the symbolic relationship between political legitimacy and maize. Harvest depletion raised questions about the competence and special status of rulers, which in turn provided obvious incentives to maintain authority and acquire prestige through military conquest.[136]

Rising levels of violence in the eighth and ninth centuries therefore created instability and put pressure on trade networks. However, it also created opportunities for the accumulation and appropriation of greater levels of wealth and resources by soldiers and officials. This widened the demand for luxury goods such as ceramic vessels, and drove the evolution of new styles.[137] It also increased demands on rulers whose kingship rested on elaborate rituals such as bloodletting, sometimes involving venomous stingray spines, which 'opened a conduit between the natural and supernatural worlds'.[138] Such rituals were designed to allow elites to assert and maintain their distance from commoners – as was the case in Monte Albán culture in the Central Valleys of Oaxaca, in what is now Mexico, where rulers erected carved stones to show off their genealogy and so preserve a gap between themselves and commoners. In Monte Albán, increasingly challenging times led to rising levels of elite competition that likewise played a role in driving deurbanisation, depopulation and the decline of a functioning state apparatus.[139]

Another problem may also have come from materials used in decorative preferences and techniques employed by the Maya. Recent scholarship has drawn attention to the cinnabar that was used to produce blood-red colours in dyes and paint used for claywares and architectural ornamentation as well as in ritual activities. High in mercury, cinnabar would have been washed off buildings and artefacts during rainstorms, leaching into reservoirs on which the city depended. Over time, pollutants rose to dangerous levels, made worse during dry periods when cyanobacteria became even more concentrated and toxic than normal. The pollutants brought the sustainability of cities into question and undermined the authority of their rulers and elites; moreover, the effects of chronic mercury exposure on obesity and cognitive function may even have played a role in undermining social structures that were already under pressure.[140]

As scholars of Mesoamerica often stress, the transition from a highly connected patchwork of densely populated settlements was neither sudden nor consistent. In Oaxaca, for example, many aspects of domestic life continued uninterrupted for many centuries until the Spanish conquest.[141] In Maya lands, different urban centres faced different challenges and different experiences, in both the short and the long term: while some sites experienced severe depopulation and the breakdown of bureaucratic systems, others did not.[142]

What collapsed was not the Maya world as a whole, but rather the skeleton that held it up – the mesh of networks that linked locations together, facilitated the exchange of goods and ideas, and provided a canvas for alliances, rivalries and competitive kingship. Faced with drought, crop shortages and loss of mechanisms of control, rulers found it impossible to meet expectations and to maintain authority.[143] This was the reason why grand monuments and palaces stopped being built and why they began to be occupied not by rulers and officials but by what appear to be squatters – ad hoc groups who used the buildings for cover from the elements to cook, eat and sleep.[144]

Changes in climate, even pronounced and challenging ones, were not unusual in Mesoamerica (or indeed in other regions around the world), especially for cultures that had survived and flourished for decades, centuries or longer. Crucial, then, was the balance between environmental pressure on the one hand and social and economic fragility on the other – something worth thinking about in the present day. Hyperconnectivity such as globalisation can provide obvious benefits in terms of exchange of all kind; but these can mask the vulnerabilities and shocks that can accelerate if one or more link in the chain becomes uncoupled, intentionally or otherwise. Factories closing, ports being congested or natural resources being cut off not only create inflation, but in the worst case, can lead to socio-economic pressure, revolution and state failure.

It is also worth remembering that demographic contraction, the abandonment and decline of cities and the reduction in travel, communication and exchange might disappoint archaeologists and historians, who revel in studying grand monuments, in assessing power structures and kinship patterns. For most people, the retreat of geographic, socio-political and ecological frontiers did not necessarily equate with the catastrophe or even the decline that is often associated with the end of supposedly 'golden ages'. Likewise, transformations of human societies evidently had an impact on flora and fauna as usage of the land changed in response to demand: deurbanisation and city failure meant local ecologies were no longer used the same way or in the same scale, which meant a different set of winners and losers for animals and plants alike. And it is worth remembering that the climate pendulum can swing both ways. Long-term shifts in rainfall distribution, in aridity or in temperature do not always result in the unravelling of connections. Sometimes, they can galvanise new opportunities and spur the creation of new worlds.

12

The Medieval Warm Period
(c.900–c.1250)

They thought that the Day of Judgement had arrived.

Ibn al-Athīr on the people of Mosul (thirteenth century)

In 1965, the historian Hubert Lamb noted that evidence had been 'accumulating in many fields of investigation pointing to a notably warm climate in many parts of the world, that lasted a few centuries around AD 1000–1200'. His proposed name for this period, the Medieval Warm Epoch, has since been modified and is now usually referred to by scholars as the Medieval Climate Anomaly or the Medieval Warm Period.[1] Some too have pushed at the dating boundaries, arguing that the period of long-term generally warm climate conditions in fact dates from c.800 and lasted until 1200 or even 1250.[2]

From a European perspective, this was a time of favourable weather patterns which ensured that atmospheric circulation in the North Atlantic was a reliable source of warm, dry air which in turn resulted in a reduction in the number both of wet and cold summers and of bitterly cold winters. This created ideal conditions for agriculture, generating good harvests, inflicting few shocks and, perhaps best of all, providing a climatic context that was reliable and stable.[3] Of course, temperatures and rainfall levels do not remain constant or benign; nor does it mean that there was coherence or consistency across Europe as a whole – something that is in any event difficult to assess given that considerably less attention has been paid to the south and the east of the continent than to the north and the west.[4]

The benign climate conditions that favoured Europe had different consequences for other parts of the world, although it is important to note important regional variations. For example, while rainfall in Iran, Armenia and Palestine was significantly below average for much of this period, there is little evidence that this was true of northern Syria or of western or central Anatolia.[5] Climate impacts varied from region to region, even over relatively small areas – as was the case in what is now Bulgaria, central Greece and western Asia Minor, which do not show uniform precipitation patterns in the early Middle Ages.[6]

Tree-ring data from Central Asia, combined with reconstructions of the Aral Sea salinity, point to the climate being cold and dry, especially from around 900 onwards.[7] This chimes with evidence from the northern parts of China, though the rising cultivation of citrus trees and subtropical plants in Henan province to the south provides a reminder that experiences were not only different but could be sharply contrasting.[8] More detailed and more recent surveys drawing on peat cellulose, stalagmites, ice cores and tree rings show that many parts of what is now China were generally warmer and wetter than average during this long period.[9] Taken as a whole, however, six of the ten warmest decades globally of the last millennium were clustered in the period 950–1250.[10]

These changes have been linked by some scholars to large-scale shifts in the El Niño–Southern Oscillation (ENSO), the Atlantic Multidecadal Oscillation (AMO) and the North Atlantic Oscillation (NAO).[11] Others however have suggested that they were driven by high levels of solar irradiance and low levels of tropical volcanism.[12] The lack of the latter seems to be the more important, given that while reconstructions of radioactive beryllium (^{10}Be) and carbon (^{14}C) isotopes show that while the main phase of the Medieval Climate Anomaly corresponds with relatively high solar magnetic activity levels, these were not unusual compared to other periods. As such, reduced volcanic activity and variability of the ocean-atmosphere system appear to have been the main motor of what was a major global climate reorganisation.[13]

The warming of the tropical Indian and western Pacific oceans induced a broad range of changes to atmospheric pressure in the North

Atlantic and shifts in monsoon rainfall patterns in Africa and South Asia, and affected aridity levels in the Eurasian subtropics.[14] Much of the Pacific Rim, including San Francisco Bay and coastal Peru, was affected by drier conditions than normal, albeit at different stages between 800 and 1250.[15] There were long and unusually regular periods of hydrologic drought in California and Patagonia, as well as in the Upper Colorado Basin, although peatlands in north-eastern America do not seem to have been affected.[16]

That warming also brought change in Africa, although establishing the precise impact is more difficult because so little work has been done on data gathering and on climate reconstructions – a reflection of the way that this continent's histories have been marginalised more generally.[17] Nevertheless, information gathered from over forty sites shows uneven and highly varied patterns, with marked cooling in the Benguela Current Upwelling System that dominates the south-western coast of the continent, mirrored by warming across southern Mauritania, the Gulf of Guinea and the Congo River basin.[18]

How these shifts in long-term weather patterns affected different regions is not easy to determine: although socio-economic changes that began in the early Middle Ages may correlate with climate contexts that themselves were not uniform, constant or immediate, that is not the same as asserting that one causes the other. Even where connections seem both plausible and compelling, identifying the precise relationship between environmental and societal changes requires caution.

For example, in the southern Mediterranean, changes were modest. The eastern parts of the Roman empire that had survived the Arab attacks had regrouped. Characterised by centralised control of Constantinople and a bureaucratic, military and religious apparatus which sustained integration as well as a common Roman identity now went through a period of contraction where cities became smaller and networks less vibrant. Exchanges between the Middle East, Levant, North Africa and Spain were based on a similar model of political structure that provided protection of property rights, administered justice and collected taxes. The difference was one of scale, with cities like Damascus, Cordoba, Fustat and many others being both larger and more numerous than those in the Eastern Roman – or Byzantine world – that was centred on the Aegean, the Balkans and Greece.[19]

Both stood in sharp contrast with western and northern Europe. There, the decline and fall of Rome brought about fragmentation and atomisation. There were brief spells of consolidation, most notably during the reign of the great ruler Charlemagne who succeeded in uniting much of what is now France, the Low Countries, Germany and northern Italy into a single realm. The high water mark came with Charlemagne's coronation as Emperor of the Romans by Pope Leo III on Christmas Day in 800. This was a brief and exceptional moment in what was otherwise a period of almost no long-distance trade and a narrowing of horizons that lasted for hundreds of years.

Rather than the romantic figure of Charlemagne, who became a totemic symbol for those keen to provide a reference point for ideas about European unity, the regional emporia of Comacchio and Torcello in Italy, Verdun in France and Birka in Scandinavia were better examples of reality. Curiously, each was set in locations that were ecologically marginal and geographically peripheral; but more importantly, each was a trading zone that was an internal market, where commerce took place between local producers and local consumers rather than with those of other regions.[20]

This in turn provided a setting for a very different form of social and economic development. In the absence of a sophisticated bureaucracy, a new baronial class emerged in western Europe that was able to establish authority over both the labour force and over productive land. The magnates faced challenges from each other, of course, as well as from other competitors – namely the Church, which built up extensive landholdings and sought to protect and maximise its own socio-economic and political position. Together, these represented a superclass of dominant individuals, families and landowners who sought to concentrate power in their own hands, and retain it within kinship groups closely related by marriage or in institutional ownership that kept assets away from the king.[21]

Ideas about the 'feudal revolution' that were commonplace a few decades ago have been replaced by much more sophisticated interpretations, many of which emphasise the variety and importance of other participants in early medieval society in western Europe – such as guilds, urban groups, parishes, regional assemblies and universities.[22] Some commentators have also stressed that the initial weakness of property rights spurred innovations over time in the formalisation and consolidation of power as aristocrats built up assets and status that ultimately proved transformational.[23] The evolving role of the Church, as the recipient of endowments and distributor of alms, patronage and influence, was also a significant factor in social, institutional and ecological change.[24]

The consequent transformation of both human societies and the natural environment in the early Middle Ages was so profound that some scholars have talked of this period as being the time of 'the most significant agricultural expansion since the Neolithic'.[25] The role of new technologies in driving yields and production has long been emphasised by medieval historians. Particular attention has been paid to the importance of horse collars and to the development of heavy ploughs that were much more effective in turning the heavy clay soils of northern Europe. These improved weed control, enhanced drainage and had the twin effect of boosting yields and requiring less work by farmers, thereby freeing up time and resources that could be allocated to other activities.[26] The breaking up of larger estates into smaller units resulted in important social change, altering perceptions of the land and landscape for the peasantry as well as for the emerging baronial class.[27]

What was most important about the uplift in agricultural productivity was its relationship to urbanisation: by improving per capita income and spurring the development of transport and commercial networks, the changes in the agrarian economies of western and northern Europe set off a chain reaction that led to the growth in size and number of towns. This in turn encouraged specialisation and cultural experimentation and resulted in further migration from the country to urban settlements – and drove cycles of dynamic growth.[28]

The rise of towns and cities – and the regular arrival of newcomers and strangers – had profound effects on urban cultures as well as on those who lived in the countryside, with new customs, ideas, fashions and tastes producing an explosive transformation of Europe.[29] As long-distance trade networks developed, intellectual, cultural and geographic horizons expanded too, typified by concepts like pilgrimage to Jerusalem, a journey that was long, expensive and often risky, but that gave considerable kudos to those who were able to visit the sites where Jesus Christ had lived, died and risen from the dead.

Ideas about nature also evolved, as did concern about competition for resources that varied from region to region; these ideas were evidently influenced by changing lifestyles, as well as by engagement with flora and fauna.[30] Concepts of the natural world could be ambiguous: as one leading historian has pointed out, growing one's own food in early medieval Italy 'was both a necessity and a luxury'.[31]

For some scholars, however, climate shifts provide an invaluable key to help understand the early Middle Ages. It has been suggested, for example, that warming was at least a contributing factor in the move of the Bulgars into the Middle Volga region from around AD 800, as it was in the establishment around this time of the Volga Bulgarian state, which opened up trade routes fanning out to Scandinavia, Byzantium, the Middle East and Central Asia.[32] These networks spread goods as well as ideas and religions: when the Arab envoy Ibn Faḍlān visited the Volga Bulgars just over a century later, he reported that the ruler presided over a court complete with elaborate rituals, and with expensive materials and goods on display that had come from Constantinople and Baghdad – although he was unimpressed by the incomplete understanding of Islamic teaching.[33]

Journeys to the east were mirrored by those heading both south and west. New evidence based on mammal faeces, pollen samples and charcoal analysis suggests settlement of the Azores by Scandinavian peoples – which has been linked to anomalous winds and warmer temperatures in the northern hemisphere.[34] The expansion of Scandinavian peoples across the North Atlantic into the Faroe Islands, Iceland and Greenland in the ninth century has also been strongly linked to the retreat of the polar ice cap that allowed ice-free sailing, to the northerly migration of fish stocks and to the emergence of favourable growing conditions on land.[35] Such colonisation was not easy, for it involved not only a jump into the unknown but also leaving families and friends behind. A major eruption of the Hallmundarhraun volcano in Iceland around c.900 must have been unsettling and inspired at least some settlers to build an enormous boat-shaped structure in a 1,600-metre-long lava cave in the interior, with animals including sheep, cattle, horses and pigs apparently sacrificed as burned offerings to appease pagan gods.[36]

This was just one of dozens of eruptions in the period c.850–c.1250, with some – such as the Eldgjá volcano in 934 – proving particularly damaging to flora and fauna alike.[37]

Perhaps not surprisingly, it was mainly men who set out in the first waves of colonisation. Although Y-chromosomal and mitochondrial DNA (mtDNA) evidence suggests that Norse women were involved in expansions to the Shetlands, Orkneys and northern parts of Scotland, genomic data reveals that in Iceland most settlers were lone men who brought enslaved women from the British Isles to satisfy their sexual desires through rape and coercion.[38] Some made it as far as North America to the L'Anse aux Meadows site in Newfoundland, although the settlement proved unsuccessful.[39] Advances in dating techniques allow us to know that a Scandinavian community had become established by 1021 – when trees were cut down with metal blades used by Vikings, but not by indigenous peoples.[40]

Trade between new colonies in the North Atlantic and Scandinavia was primarily centred around low-bulk, high-value prestige goods that could be sold back home for rich rewards, most notably walrus hides and ivory tusks.[41] That produced problems of its own, with Iceland evolving into a society that was dominated by those who had arrived first and built up the best and largest landholdings; by the mid-tenth century, it was 'over-chieftained' – one reason why the national assembly, the Althing, was established, with the earliest laws concerned with protecting landholdings from new arrivals and landowners from each other. There were creative efforts to correct the manpower shortage, with texts like the *Íslendingabók* setting out how Erik the Red was able to tempt others to journey to Greenland with an astutely judged sales pitch that spoke of verdant and abundant land and of limitless opportunities that this auspiciously named island offered.[42]

The expansion from Scandinavia into the North Atlantic was part of a wider intensification not only of regional trade and knowledge networks, but of long-distance trade too – most notably to the east and to the south that in turn brought enormous volumes of silver coinage into circulation first in the Nordic and Baltic lands and then elsewhere.[43] These activities, coupled with human interventions in the landscape such as building new settlements, farming and rearing domesticated animals, changed ecosystems – as did the hunting of animals for food and trade.[14] One outcome was a world of plenty – and of equality: while evidence from Roman sites suggests men ate 50 per cent more protein than women, in Scandinavian societies in this period women and girls had access to the same food sources as men and boys and had better health outcomes as a result. Some scholars have even suggested that this may help explain why there are such high levels of female autonomy and gender equality in contemporary Scandinavia.[45]

The impact of human activities in geographically constrained locations with finite resources could be severe: within decades of the first settlers arriving in Iceland it is possible to detect soil erosion and deforestation that continued for several hundred years before long-term conservation practices were adopted.[46] Genomic analysis, combined with radiocarbon dating and use of written sources, shows that the local population of walrus was wiped out soon after the Norse settlement in Iceland, with overhunting the most obvious cause of extinction – even if warming climate and volcanism may have been additional stress factors.[47]

The pressure on managing resources led to shifts in consumption patterns, such as in Iceland with deliberate moves away from pigs and cattle to sheep, whose rising numbers may be explained also by attempts to produce wool for domestic use and for export.[48] Isotope data shows that adaptation in Greenland led to terrestrial meat-based diets being replaced over time by reliance on marine protein sources.[49]

In other words coping strategies were needed not only at times of climate shifts, but also to manage the consequences. The favourable conditions and material incentives that opened up routes between Scandinavia and the North Atlantic led to different sets of questions that were part of a constant negotiation and renegotiation between human agency and the natural environment: the arrival of any animal or plant species could have major ecological consequences. It just so happened that humans brought cascades of change in their wake as a result of interventions in the landscape: the need for settlement, food, water and other resources meant that human impact on ecosystems was profound. Plants and animals that were brought by humans, deliberately or otherwise, were part of a 'natural' set of changes that had anthropogenic causes – such as weeds, seeds and parasites that were carried in the guts of pigs, or plants carried for food, for planting or simply as insulation or packaging that then took root in a new ecological setting.[50] These outcomes, sometimes referred to as human ecodynamics, are particularly clear with the occupation of previously uninhabited locations that are then occupied and transformed by human settlement.[51]

In this sense, islands offer unique and important insights into understanding how landscapes, flora and fauna were affected – and also how quickly. Remarkably, then, the examples of Iceland, Greenland and the Faroes have obvious parallels with similar colonisation experiences of islands in the South Pacific which took place at the same time and were likewise closely related to changes in global climatic patterns.

Starting around AD 800, there were abrupt reductions in zonal sea surface temperatures in the West Pacific Warm Pool and a push of rainbands to the north. Moisture levels in the archipelagos of Vanuatu, Samoa, Tonga and Fiji began to fall, resulting in the driest period of the last two millennia. The inhabitants of these islands who had settled there almost 1,500 years earlier had not previously sought to explore islands further north – or, if they had, they had done so impermanently, leaving little or no sign of settlement. That now changed as weather patterns went through a fundamental shift, bringing winds that facilitated exploration of and journeys to Polynesia's margins.[52] Much of the long-distance travel to and from the islands was undertaken by double-hulled canoes and, as computer modelling has shown, discovery and settlement of new locations was deliberate and systematic rather than a matter of chance.[53]

Radiocarbon analyses show waves of migration first to the Cook Islands and then beyond into East Polynesia, the vast area between Hawai'i, Rapa Nui and New Zealand. Successive groups of islands were colonised not only by humans but also by livestock that they intentionally brought with them, such as pigs, and those that they presumably did not, such as rats.[54] As in the North Atlantic, settlement led to transformation of vegetation as land was cleared for crop cultivation, for fuel sources to burn or both, with appreciable soil erosion as a result. Leaf wax biomarker hydrogen isotopes confirm that these islands saw enhanced rainfall levels over a long period of time, confirming the reason for their attraction as alternatives to islands where life was becoming more difficult.[55]

That may also explain social change during this period in other Pacific island groups: increasing aridity clearly put a premium on capturing rainfall as and when it fell. As such, the investment in the building of canals, aqueducts and terraces was not only a natural response but an important one.[56] Projects like this required high-levels of co-operation, in terms of energy requirements to build and maintain such structures, and also to share equitably the benefits and proceeds. This in turn provided a context for transitions from small, dispersed communities into larger groups that over time became stratified and hierarchical.[57]

Changing lifeways were also evident in other parts of the Southern Hemisphere during the period of the Medieval Climate Anomaly, although warming started later than in the north.[58] In addition to paleotemperature reconstructions, archaeological evidence suggests that human population levels in continental Australia rose rapidly around 1000. While patterns of demographic increase may reflect migration patterns, studies of other periods suggest that major changes in population size of hunter-gatherers in Australia were triggered by high-amplitude environmental changes.[59]

Aboriginal hunter-gatherers used mobility and technologies to manage their exposure to risks in resource availability.[60] In this instance, however, it appears that higher levels of water and food availability encouraged a shift to sedentary lifeways.[61] Benign conditions meant that more people could safely live together rather than having to spread out to find resources – one reason for the striking reduction in archaeological signatures in lower-latitude settings in this period.[62] To give one example of the scale of change, a major pluvial event in southern central Australia c.1050–1100 led to Lake Callabonna being filled by around ten to twelve times more water than at its highest-known historical filling. Local populations were being presented, in other words, with new options of how to live, co-operate and flourish.[63]

Societal and environmental transitions are also evident at sites in the Caribbean starting c.900, where centuries of aridity gave way to significantly higher levels of moisture. This was a time too of rising sea levels, which forced islanders to move inland to locations such as Coralie on Grand Turk Island in the Turks and Caicos Islands.[64] The abundance of rainfall favoured agricultural surplus; but there were other outcomes too – not least rising levels of interaction between island groups and between the islands and South America. This can be attested by major transformations of pottery styles, as in the Antilles, which are often interpreted as a sign of sharply increased interactions between islands, and also as a response to the loss of confidence in traditional gods whose preservation of previously arid conditions was now seen as hostile, if not cruel.[65] On top of this was the introduction of a suite of new species of plants and animals into and across many parts of the Caribbean, which served as catalysts for change. For one thing, they reduced the previous heavy dependence of local populations on fish, crab and bird stocks, whose numbers began to recover as a result; for another, island forests were cut down to make way for flora and fauna that were introduced for the first time to the island.[66] Changes in climate, in other words, brought about changes in interaction, in belief systems and even in diets.

This happened elsewhere too around the same time. One such case was the American Bottom, the region stretching from the confluence of the Mississippi and Missouri rivers with the Kaskaskia. This region was home to dispersed, isolated single-home farmsteads and small settlements, of which Cahokia was one of the most prominent. From c.900, Cahokia exploded in size and importance, drawing in migrants from outside the wider region, in what is often called a Big Bang moment, sparking a cultural efflorescence in pottery and other arts.[67] Centred on a single, enormous 6.5-metre-high platform, Cahokia benefited from being set within a vast floodplain that offered ideal conditions for growing crops, while permanent lakes and swamps contained plentiful fish stocks that were central to Cahokian diets, and access to rivers provided water and served as trade and transport arteries.[68] Being able to expand exponentially during times of demographic growth offered possibilities, in other words, that more geographically constrained locations did not.

Temperature profiles at six sites in the North Atlantic from 200 BC–2000 AD

Source: Cronin et al, 2010

The Medieval Climate Anomaly was not congruous, and was pock-marked by regional as well as periodic differences. There were hiccups and one-off episodes too that caused disruption. Most obvious were significant volcanic episodes – such as the highly explosive 'Millennium' eruption of Changbaishan, situated on what is now the border between China and North Korea, which took place in c.940–50. Another is a recently identified but previously unknown tropical eruption that may have been part of a 'forgotten cluster' of explosions at the start of the twelfth century that has been linked to crop failure and poor weather across Europe; it obscured the moon, whose light was so diminished, according to one chronicler, that at nightfall 'it was completely extinguished withal, that neither light, nor orb, nor anything at all of it was seen' – a reference to a dust veil made up of ejected materials.[69]

Then there were natural disasters that were epic in proportion and consequences. On 19 July 1048, the Yellow River breached its northern bank, causing a cataclysmic flood and gouging a new channel 700 kilometres in length up to the Gulf of Bohai. Myriads drowned, 'turning into food for dish and turtles', in the words of contemporaries. Famine followed that was so bad that 'fathers and sons ate each other.' At least one million people were displaced, producing social, economic, political, environmental and humanitarian shocks that some scholars argue took eighty years to play out.[70]

The case of the catastrophe of 1048 illustrates the importance of stability for new farming techniques to take hold and for demographic growth and societal shifts that also followed. For example, levels of rainfall that were considerably higher than they had been in the past help explain transformations in Ancestral Pueblo cultures in the Four Corners region of the American south-west. Spread across where the states of Utah, Colorado, Arizona and New Mexico meet today, rain-fed farming which supported cultivation of maize began around a thousand years ago; this was especially important since maize was a food source that made up 60 per cent of the Puebloan diet, if not more.[71]

Higher food production was of course a bonus in its own right. But no less important was the fact that greater availability was connected with the familiar processes of elite control of resources and ritual, the creation and extension of regional trade networks and political, social and economic centralisation – as well as population expansion.[72] This was typified by the rise of the Great House system in Chaco Canyon in the San Juan Basin in north-western New Mexico, where these structures served as residences of the powerful, as administrative centres, as places with a ceremonial purpose – or all three.[73]

The growing complexity of these communities helped spur the extension of procurement networks from Chaco Canyon from around c.900, with marine shells from the Gulf of California, copper from west Mexico and cacao from Mesoamerica being found in increasing numbers as horizons and elite consumption demands expanded.[74] Scarlet and military macaws whose long and colourful feathers were particularly prized as indicators of prestige, wealth and spiritual status also began to be imported in large numbers from Central America. In Ancestral Pueblo communities too, they played key roles in repetitive rituals that helped enable group cohesion and solidarity, while underscoring the hierarchies that cemented and augmented power and status of those in the highest social groups.[75]

Processes of centralisation and expansion put pressure on resources, both locally and also beyond. There were ways to alleviate these by ecological interventions. In locations in the Andes, such as near Cuzco, for example, extensive planting of the fast-growing *Alnus acuminata*, which does well in poor, degraded soil, bears witness to sophisticated large-scale agro-forestry that was designed to reduce erosion or perhaps to ensure a fast-growing supply of timber. This may apply also in the Ancash highlands where a complex set of hydraulic systems was created at this time, apparently as a result of co-operation between rural communities.[76] The aim, evidently, was to re-landscape in order to mitigate against risk, and to ensure long-term sustainability of natural resources.

Key, then, was ecological awareness; but so too was the development and improvement of new technologies. Sometimes, this was spurred by the arrival of new, colonising powers. In the case of Andean societies, for example, the Wari brought about a range of new ideas as they expanded the territories and peoples under their control from around 600. The most important of these was the introduction of terraced agriculture in the high sierra, something that was supplemented by construction of advanced canal systems that were capable of discharging 400 litres per second. This was important in locations where there were significant demands on water supply: to give a sense of scale, the economic capital of the Tiwanaku, allies of the Wari, was located at Chen Chen. It has been estimated that the people who lived in and around this conurbation required 250 million litres per year to sustain its population and its agricultural production. This necessitated a delicate balance not only with the requirements of fields further upstream but with other groups such as the Tumilaca.[77]

In such delicate contexts, then, population expansion, competition for water, overexploitation and of course subtle shifts in weather patterns could all present problems that could develop into existential threats. Just how serious those could be can be seen with the sack of Chen Chen and the collapse of Tiwanaku in the tenth century.[78] While some scholars have linked the fall of Chen Chen and the contraction to the return of drought conditions, the sudden and dramatic dispersals of people from cultural centres point tellingly to the failure to adapt to changing circumstances by populations who were living beyond their means and capabilities.[79]

It was no coincidence, therefore, that the Chimú - effectively the heirs of Wari – invested time and energy in major infrastructure projects. These included the construction of the largest canal system in the western hemisphere along the hyperarid coast of Peru that connected the Chicama and Moche valleys. The aim was to ensure water supply to Chen Chen and the surrounding area, which both began to flourish once again. This time, in a clear sign that lessons from the past had been learned, the canal was extended further still as the city's populations or needs grew.[80] Steps like this were important at any time; during long periods of El Niño-related droughts, however, they were essential.[81]

Other locations in South America experienced different climatic changes from c.800. The steep, forested slopes of the mid-elevation Andes in northern Peru, saw unusually high levels of rainfall in this period. Conditions were so wet in this region that it seems that maize was abandoned in favour of other crops, such as potatoes, squash and beans.[82] The Chachapoya culture emerged around this time, which was also a moment of the expansion of Wari influence in this region. Key to the success of the latter was the introduction of widespread consumption of beer laced with hallucinogenic substances that expanded the numbers of those able to partake in experiences of euphoria and heightened awareness.[83]

Similar patterns could be seen in the eastern Amazon, where changes to the South American monsoon system brought a substantial shift from humid to drier conditions. Here too there were wholesale changes in diet, as well as in living patterns, marked by the rise of scattered settlements known as the Guarita phase.[84] In the southern highlands of Brazil, coastal communities not only saw an expansion of trade networks and of ceramic distribution, but underwent a demographic boom between c.800 and 1100.[85]

Such expansions excite archaeologists and historians; whether they produced such enviable outcomes for all involved is not so clear. In Japan, for example, a notably warm period starting in the late twelfth century was marked by demographic growth and higher population densities. This came at a cost, however: sharply falling life expectancies. Part of the reason for this was regular conflict and warfare that themselves reflect competition for resources, which in Japan's case was closely connected to the rise of a warrior class.[86] However, there was another, more important explanation: people living close to each other spread disease, often made worse by domesticated animals, rodents and parasites. Poor drainage and unhygienic conditions created a disease environment that could be dangerous, even deadly. New research on skeletal remains indicates that life expectancy fell as both the number and size of towns and cities rose.[87]

Curiously, there were long-term beneficiaries of epidemic disease. Some scholars argue that frequent outbreaks of smallpox – whose spread accelerated thanks to the intensification of trade as well as because of rising proximities and numbers of hosts – created 'endemic afflictions of childhood', thereby allowing immunity to build up over time.[88] Those who survived, in other words, built up resistance that proved helpful later in life. It is intriguing, then, that while phylogenetic analyses, ancient DNA (aDNA) and palaeopathological studies around smallpox are not (yet) able to shed light on smallpox in particular, there are indications that this disease spread widely across the Indian Ocean worlds during the early Middle Ages: the great polymath al-Bīrūnī writes about smallpox being blown by winds from Sri Lanka, while the emergence around the same time of the cult of the Hindu goddess Shitala as a figure of worship in Bengal – with a particular association with this disease – suggests a growing awareness of, and concern with, the transmission of infectious disease.[89]

A 'cluster of writing' about smallpox along the Silk Roads from the seventh to eleventh centuries during a time of intensified commercial, political and cultural contacts provides another illustration of how increased contact spreads disease, even if it says nothing about infection levels or mortality rates.[90] The dissemination of ideas about therapeutics, healing techniques and medical knowledge was a crucial characteristic of the regions that were connected by land and sea in Asia.[91] In the past, as well as the present, inter-connections of trade did not just help move goods, people and ideas; these networks also spread disease and death.

As it happened, there had long been extensive connections between the Indian subcontinent, South-East Asia, China and beyond: evidence from the Gupta dynasties in India attests to local, regional and long-distance contacts, diplomatic, commercial and otherwise in the fifth and sixth centuries.[92] There are clues too of maritime connections: for example, communities from southern India are recorded as living in Chinese ports in the sixth century; Tamil inscriptions of a trade guild in Takaupa in what is now Thailand point to trade relations between one side of the Bay of Bengal and the other in the ninth century; a set of copper plates from the mid-ninth century, meanwhile, inscribed in old Malayalam as well as in Arabic, Middle Persian and Judaeo-Persian show the level of cosmopolitanism in the Indian Ocean in this period. Nevertheless, while connections existed, they were relatively modest in scale.[93]

That changed dramatically from c.900, as connections, ties and exchanges rose quickly and dramatically. Climatic shifts certainly seem to have been a key part of the knitting together of ties. Cave speleothem deposits and other indicators reveal that there were regular monsoons and consistent rainfall for a three hundred years time-frame, with a short interlude in the years c.1030–70 which is usually attributed to the Oort Solar Minimum, a period when the sun's activities were significantly reduced.[94] As in other regions, what mattered as much as shifts in temperature and precipitation was the stability of climate patterns: it is certainly true that higher levels of rainfall made a major difference in particular boosting rice production across many parts of Asia – which helped improve calorie intake, freed up labour and contributed to demographic growth.[95] The low incidence of sudden or long-term shocks, however, was no less important in providing a platform for what has been described as the 'jump start' of a new era in the history of Asia: namely large, centralised states that rose in parallel.[96]

* * *

A set of empires began to grow up across South, South East and East
Asia in the early Middle Ages. One good example is provided by the
Song dynasty in China which from the mid-tenth century was able to
consolidate large parts of the lands that had disintegrated following the
collapse of the Tang at the start of the tenth century, and establishing
relations with neighbouring states like the Dali kingdom in what is now
south-west China's Yunnan province in the process.[97] The success of the
Song owed much to mundane, bureaucratic reforms. Instrumental in
the economic boom, for example, was the widening of participation
by merchants in political decision-making. Likewise, the investment
in and improvements in literacy and education brought rewards, as did
greater sharing of knowledge through rising book production. These all
influenced imperial policy towards external trade in an age of change.[98]
Benign and stable climatic conditions were important; but monetary
policies, urbanisation, the rise of vibrant market towns and the networks
that bound these together both domestically and internationally were
what helped transform the Song into an imperial power.[99]

The success was mirrored by the rise of dynasties elsewhere which
emulated, influenced and competed with each other. Strikingly, they
rose as a group, fuelling each other directly and indirectly. The success
of Chola dynasty in India, of Pagan in what is now Burma, of Angkor
in Cambodia, of Śrivijaya in the Indonesian archipelago, of Đại Việt
in what is now Vietnam all took place more or less contemporaneously
as part of a rapid broadening of geographic, commercial and cultural
horizons across the Indian Ocean and across large parts of Asia.

Such was the intensity of these connections by the tenth century, that
instead of finding whisps of evidence that show long-distance exchange,
we find envoys were being received in Kaifeng in China from the Arab
lands, from Chola India, from Śrivijaya Sumatra and from the Champa in
central-south Vietnam.[100] Dining tables in the Middle East were adorned
with Chinese ceramics of such high quality, wrote one contemporary, that
'you can see water sparkle through them, even though they are ceramic'.[101]

Much of this contact was driven by trade, as well as by curiosity and
information gathering, in some cases across distances of thousands of
kilometres: the Mahārāja of the Śrivijaya, wrote the Arab geographer
al-Masʿūdī, 'commands an empire without limits ... the fastest vehicle
could not in two years make the tour of the islands which are under

his domination. The lands of this king produce all kinds of spices and aromatics, and no other sovereign in the world extracts as much wealth from his country.'[102]

Central to the creation of what developed into a global trading network that linked many parts of Asia, North Africa and Europe was the Chola dynasty in southern India. The origins of the Chola are not clear; either they supplanted the Pallavas who had held sway over much of this region before the ninth century or they were the beneficiaries of good timing as a stable agrarian economy emerged as a result of improvements in rainfall patterns.[103] The Cholas were formidable rulers; but they were lucky ones too.[104]

From the late tenth century, territories under Chola control underwent a series of physical transformations. The most striking of these was the transition from small shrines and places of worship into grandiose religious institutions. In addition to the resources that went into building enormous temples such as at Gangaikondacholapuram in what is now Tamil Nadu, hundreds of musicians, dancers, actors, goldsmiths, text reciters and more were employed to emphasise the power, wealth and splendour of the rulers and the divine protection they enjoyed.[105]

Chola rulers were keen to profit from maritime trade, with one ruler urging merchants to be generous to foreign visitors, to offer them hospitality and gifts as a way of ensuring preferential treatment.[106] Inscriptions dating to the second and third decades of the eleventh century suggest that Chola rulers were extending their interests more deeply into South-East Asia, presumably in order to gain control of regional and long-distance trade. In a case of expand or be damned, doing so was important to prevent access being blocked by Śrivijaya through the straits of Melaka and suffer higher prices – or throttled supply as a consequence.[107]

The Chola kingdom evolved rapidly in response to the opening up to other regions and parts of the world. Monetisation was one key element in a series of steps taken that strengthened bureaucratic control over the agrarian economy. This had as its parallel the extension of cultural norms in the form of the spread of the cult of Śiva across Chola territories, as a means of promoting a common identity.[108] Like other post-Gupta dynasties in South Asia, such as the Paramāras, Cāllukyas and Rāṣṭrakūṭas, the Chola leaders also looked to connect their genealogies

with the ancient past, tracing links to the sacred texts of the Purāṇas as
a way of emphasising their authority – most notably in a set of famous
copper plates discovered in 1905 at the shrine of the Vāṭaraṇyeśvara
temple in what is now northern Tamil Nadu.[109]

Crucial too were the steps taken by successive rulers and administrators
to plan for unexpected circumstances: with heavy dependence on
monsoon rains, major investment went into the construction of tanks to
trap and store water.[110] Adaptation and innovation were important too.
The relocation of the capital first from Uraiyur to Thanjavur and then
to Gangaikondacholapuram provides one example; the implementation
of a fundamental reform of land tax to distinguish between the ways
crops were watered, presumably to optimise central revenues for the
benefit of the crown, provides another.[111]

It was a similar story in Angkor where massive storage facilities were
constructed to help cater for a huge influx of inhabitants that some
estimate rose to as much as 750,000 at its peak, bringing about the
'largest low density urban complex in the pre-industrial world'.[112] The
city, whose central area and periphery covered more than 1,000 square
kilometres, was studded with magnificent temples, ritual dance halls
and palace complexes that owed a heavy debt to southern India, to
both Hindu and Buddhist influences and to a blend between the two.[113]
Angkor was characterised by large-scale hydraulic systems that helped
provide water to inhabitants and mitigated against sudden collapses in
precipitation levels.[114] This was essential not only for social and political
stability, but also for ensuring food supplies for a population whose
sheer size meant there were obvious risks to even modest climatic
shocks, such as droughts or floods.[115]

In other cases, such as Pagan and Đại Việt, water supplies were less of
a problem thanks to the abundance of the river systems. Nevertheless,
it has been suggested that higher-than-usual rainfall levels across
several centuries helped reduce mortality rates from droughts in the
highlands. This encouraged migration which in turn produced a boost
in manpower and capacity in towns and regions that further galvanised
productivity both locally and further afield.[116]

Each of these states had the good fortune to expand without obvious
competition – or to absorb rivals with ease. In due course, however, they
came into contact with each other and so competed for resources and
for status. Market pressures could drive prices down, while rivalry over

control of goods and products, as well as over transportation nodes, could lead to antagonism and in some cases military confrontation. Burmese royal interest in ports that served as transit points between the Indian Ocean and China sharpened ill-feeling locally in the eleventh century, while the struggle to control ports led to multiple conflicts between Angkor and Champa not long afterwards.[117]

The interlocking worlds of the Indian Ocean were not a new phenomenon, for even in antiquity there were close connections that linked the coasts and interior of many parts of Asia with Africa, the Mediterranean and Europe. What was different was the scale of activity: the rise of states and kingdoms that were imperial in scope and ambition created velocities of commercial and cultural exchange that were remarkable in both scale and volume. This was clear even as early as the ninth and tenth centuries, when ships carrying tens of thousands of pieces of ceramics made in China, bronzes from India, polished mirrors from Java, glassware from Egypt flitted back and forth – and occasionally ended up on the sea floor.[118]

This spurred interest in other parts of the world too. One Chinese Emperor wanted to know how elephants and rhinos were caught. 'To capture elephants,' one envoy replied, 'we use decoy elephants to get so near them that we can catch them with a big lasso. To catch a rhinoceros, a man with a bow and arrow climbs a big tree, where he watches for the animal until he can shoot and kill it. The young rhinoceros are not shot as they can be caught.'[119]

We know of rhinoceros horns, elephant tusks and rattan mats, as well as silks, textiles and umbrellas, being brought to Song China by traders from closer to home, such as from Champa, which also acted as a base for middlemen for goods from further afield, such as Javanese fabrics and aromatics from the Middle East. While it is tempting to look at exotica and the array of different kinds of things being exchanged, it is noteworthy that closer contacts also drove trends to standardisation – with a case in point being the ceramic industries of Đại Việt, Angkor and Java that over time adapted to mirror Chinese ceramic design.[120]

Then there were the cultures of the Toutswemogala and Mapungubwe kingdoms in the Limpopo–Shashi basin of what is now Botswana that grew and thrived from the eighth to the twelfth centuries.[121] Both lay to the south of Great Zimbabwe, which likewise flourished in southern Africa from the eleventh century and was centred on a settlement surrounded by massive sets of walls made up of almost 1 million pieces of stone, and with soapstone figures of birds with human lips mounted on perimeter columns.[122] These walls acted above all as statements of authority, rather than as defensive structures, built as enclosures for the most powerful members of a community numbering many thousands of people. The Hill Complex, the oldest of the enclosures, was the spiritual centre and the location of royal rituals that included appeasement of ancestors and sacrifices to the gods for clement weather. Finds of goods and luxury objects from Persia, Syria and China show that southern Africa was also involved in the rise of long-distance trade networks during this period.[123]

So were other societies, cultures, peoples and locations in Africa. The island of Kilwa played a central role in the Indian Ocean connections, linking the Swahili coast and East Africa in particular not only with maritime routes but also with southern Africa. Kilwa's prominence was partly the result of its location at the southern end of the route down which ships could sail from the north with monsoon winds, which made it an export centre for ivory, wood and above all gold from mines in what is now Mozambique; it enjoyed a rich trade in imported goods from as far away as China, attested by ceramic records dating from the eleventh century onwards.[124] Kilwa was 'one of the most beautiful towns in the world', according to Ibn Baṭṭūṭa, who nevertheless notes that its Muslim sultan and inhabitants were in a state of constant conflict with those living on the mainland.[125]

Kilwa was just one town among many dotted along the East African coast, extending from Somalia to Mozambique and northern Madagascar, which were typically home to mosques, tombs made of coral and houses of elites who were closely and directly involved with trade between the hinterland and the Indian Ocean.[126] Glass-bead assemblages provide particularly important insights into connections with the Gulf and South Asia. So too do large numbers of coins from Fatimid Egypt dating to the tenth and eleventh centuries that have been found in East Africa and Madagascar, which not only provide evidence of a north–south trade axis, but suggest Madagascar and the Comoros Islands as the likely sources of exceptionally pure rock crystal used for the magnificent ewers produced in North Africa in this period.[127]

While individual towns differed greatly in size and social structure, they were part of a wider, mutually interdependent web that fed regional trade, bound together by linguistic similarity, by Islam which had spread effectively by the twelfth century, and by rituals that helped build and deepen relationships between different rulers and peoples – of which feasting was one of the most important.[128]

The continuing development of these long-distance networks brought challenges as well as opportunities. Locations like Unguja Ukuu on Zanzibar became victims of their own success: deposits which built up from agricultural activities that supported growing populations, combined with the dumping of food remains and general waste, helped amplify sediment levels that resulted in the decline and eventual abandonment of one of the key trading hubs in East Africa.[129]

Elsewhere, the widening of connections led to states seeking to maintain monopolies over trade to protect existing interests and assert the authority of the political centre – one reason why the authorities in Song dynasty China interfered regularly in foreign trade, issuing bans, announcing reforms and following the movement of goods and traders carefully. This spilt over into other sectors as people became richer and, in some cases, more adventurous: the rise of a literati class who commented on courtly and official life, looked at history as a mirror for the present and created intellectual networks of their own was a major development, a by-product of a widening of society that was partly a result of increased international contacts.[130]

All around the world, the period from c.800 to c.1200 was one of profound change – a time when the effects of a climate reorganisation were enhanced by a lack of major volcanic activity. Migration to new regions, the development of new hydraulic technologies and the expansion of agricultural production were just some of the strategies that were adopted and modified to cope with a new set of questions that were different in different regions.

In some cases, ecological exchange had significant consequences. For example, the adoption of drought-resistant, early-maturing Champa rice by Chinese farmers following a drought in the Yangtze and Hui river valleys at the start of the eleventh century is thought to have been instrumental in providing a greater level of insurance against weather shocks in the future; the new strain not only helped guarantee calorie intakes and enabled more mouths to be fed, but in doing so also created conditions for political stability.[131] Perhaps not surprisingly, then, this was also the time when a set of other interventions in the natural landscape were carried out in East Asia, including the creation of large irrigation works, the construction of dams and deforestation to open up new areas for agricultural exploitation.[132]

Economic expansion, agrarian production and population growth were not only themes across many parts of Asia, Africa and the Americas but in Europe too where population levels also shot up, even if estimates of a tripling or even a quadrupling between c.800 and 1200 may be on the high side.[133] Some historians have noted that this set off a race between humans and their ecological and environmental settings, with the former battling to keep pace with the needs of ever-rising numbers of people that would almost inevitably end in crisis.[134]

What mattered above all, though, was not headcount in total terms, but population distribution and density. Cities like Hangzhou and Kaifeng are often thought to have been home to a million people, and while such estimates are both conveniently round and notably large, there is little escaping the fact that they were home to very substantial numbers of inhabitants.[135] As in other major cities in other regions and in other periods of history, their practical needs – for food, water and fuel, not to mention luxury goods – were substantial and not only put a strain on ecological resources but demanded supply chains that were efficient and robust. This was a problem for locations like Pagan, later known as Arimaddana-pura (or 'the City that Tramples on Enemies'), which had to contend with flooding and regular river sedimentation.[136]

At Angkor, a city that was home to 3,000 temples including the magnificent Angkor Wat complex, an elaborate system of reservoirs and canals that trapped and distributed fresh water helped to supply inhabitants as well as to irrigate fields locally. Schemes such as this not only depended on a great deal of manpower for construction, but also needed constant maintenance to prevent sand and debris from blocking the canals.[137] This meant heavy labour costs and required supervision and co-ordination, as well as technical skills. The water system was designed to reduce risk of one kind, but in doing so introduced those of another kind: any lapse in or challenge to centralised authority could quickly escalate to create problems that could leave the livelihood of the inhabitants and the viability of the city in doubt.

The ease with which urban settlements could be threatened by relatively modest fluctuations in weather conditions is made clear in the middle of the twelfth century in the Levant, where any delay to the rains provoked concern and distress and where their arrival was greeted with joyful celebration. As Nūr al-Dīn, a well-known and important leader at the time, reached Baalbek, 'it happened by the predestined decree and celestial mercy that the heavens opened their fountains with rains, dews, outpourings, and heavy showers lasting from Tuesday until the following Tuesday. The water courses overflowed, the pools of Hawran were filled, the mills turned, and the crops and plants that had been withered were restored to fresh green shoots.' This fortuitous timing generated political capital for Nūr al-Dīn, with people declaring that the rains were the result of 'his blessed influence, his justice, and his upright conduct'.[138]

The people of Mosul were less fortunate at the end of the 1170s when drought and famine were so severe that citizens demanded that the sale of wine be prohibited, judging that the lack of rain was an act of divine punishment for the impiety of those who dared drink alcohol. To make matters worse, dust storms blown by strong winds had made the skies 'so dark that a man could hardly see anyone he was with', causing people to pray at all times of the day, begging God for forgiveness. 'They thought that the Day of Judgement had arrived.'[139]

The swings in fortune are captured in a treatise written by the physician ʿAbd al-Laṭīf al-Baghdādī which offers a snapshot of life in Egypt at the start of the thirteenth century. While it starts with accounts of abundance – including a memorable recipe for a giant pie made with three roast lambs and some ninety chickens and other birds that would be ideal for a large picnic for family and friends – it goes on to provide harrowing details of sudden food shortages and of the spread of disease. So many died, he writes, that skulls were 'stacked up on top of each other in layers … to an onlooker, they resembled a newly cut crop of watermelons, heaped together at the harvest'. When ʿAbd al-Laṭīf saw them a few days later, 'the sun had scorched the flesh off them and they had turned white'; they now looked 'like ostrich eggs piled up'.[140] It did not take much for times of plenty to give way to horrors such as this in cities that were densely packed and where a sudden shortage of supply could not be quickly addressed.

The dry conditions were a source of bafflement for some commentators, especially those who read about the past history of the region. William of Tyre, a prominent archbishop and chronicler at the time of the Crusades, was puzzled when he read an account written by

Gaius Solinus several hundred years before about different parts of the then Roman world. 'I am surprised at Solinus' statement that Judea is famous for its waters,' wrote William in the mid-twelfth century, given that the conditions at the time of writing were arid and that households had to depend on rainfall and on their wits to survive. 'I cannot account for this,' he added, 'except to conclude that either he was not telling the truth, or that the face of the earth has changed since that time.'[141]

As it happens, evidence from lake sediment, speleothems, pollen data and tree-ring samples shows that conditions were becoming drier and colder in Anatolia, Syria and the Balkans in the second half of the twelfth century.[142] Rising aridity also seems to have been the signature in many parts of Central Asia in the same period, with evidence of progressive settlement contraction in the decades leading up to 1200 – a period that also saw long and damaging droughts on the Mongolian plains.[143] Perhaps not surprisingly, this was a time of considerable turbulence and instability among nomadic groups, whose intense competition for dwindling resources points to the gravity of the situation as it unravelled.[144]

A dramatic and seemingly sudden demographic downturn in and around Jenne-Jeno, a city on the floodplain of the Niger and Bani rivers, in c.1200 also hints at a possible role played by climatic factors. The previous decades had seen an extraordinary blossoming of the city's art and architecture which came to an abrupt end at this time, with other large urban settlements in the Niger floodplain, such as Dia and the Akumbu communities in what is now Mali, likewise undergoing major contractions.[145] In the absence of environmental data, hypotheses about climate factors is highly speculative – and perhaps even dubious, given that Jenne-Jeno's sister city of Jenne, home to the famous Great Mosque that was located just a few kilometres away, does not appear to have experienced the same negative fortunes at the same time or afterwards.[146]

In fact, population collapse could equally result from disease outbreaks or from political infighting and instability. For example, while some scholars have argued that the decline of Cahokia in North America was a form of ecocide caused by the exhaustion of natural resources (wood in particular), and other scholars have suggested that evidence of demographic decline shown by faecal stool samples is correlated to changes in summer precipitation around 1200 which affected maize yields and made life more difficult. Others still have noted that more mundane explanations may be the most plausible: leaders within Cahokia society tried to gain too much power, prompting local rebellion and infighting between settlements that spilt over into bitter conflict. The need for groups that had previously co-operated to protect themselves from each other explains why towns became fortified with defensive walls, palisades and moats – rather than changing precipitation levels.[147] Naturally, resource scarcity caused by changes to seasonal rains may have exacerbated the situation; but climatic change was likely to have been a contributing factor rather than the direct cause of decline.

Such distinctions are important, not least because of the contemporary focus on searching for lessons from societies that were apparently ecologically or environmentally unsustainable and sowed the seeds of their own demise. In the case of Angkor, for example, it has been convincingly argued that what proved difficult to overcome were not so much the decades-long droughts and intense monsoons that are indicated by archaeological evidence as well as by tree-ring and other data, but rather the variability between the two. Multi-year arid periods followed by long phases of very heavy rainfall made planning difficult and undermined the complex socio-economic systems required to manage the needs of the city and its inhabitants.[148]

As historians have also noted, a new phase in the strengthening of maritime connections, especially in the Mekong Delta, may have persuaded elites to relocate closer to the benefits and rewards of trade. This left the hydraulic networks that had made Angkor into one of the greatest cities in Asia overwhelmed by a combination of a lack of expertise, available labour, investment, leadership and oversight. Heavy expenditure on temple building, maintenance and public services was also a strain: Khmer rulers had embarked on a 'building orgy', with Jayavarman VII (r. 1181–1218) overseeing a construction programme 'never equalled by any other monarch in any other country'; this included a network of more than a hundred hospitals, each of which had a walled compound, a stone-lined pool and a 'library' building, as well as a sanctuary, ritual platform and storeroom for herbs, spices and precious items sent out three times a year from the king's warehouse. All of these were completed within four years of him taking the throne and evidently required enormous funding.[149]

Such irrational exuberance did not need much to turn the good times into bad ones. But by the early fourteenth century, a tipping point had been reached. This was a difficult period in Khmer history. There was widespread desecration of Jayavarman's temples, while artefacts were destroyed. Ritual burials of Buddha icons were carried out with idols carefully placed in beds of sand to protect them.[150] The costs of living rose too fast, and as they did so confidence in the old way of doing things faded too. The hard realities of keeping up appearances was what brought about decline, rather than catastrophic collapse brought on by climatic change.[151]

Similarly, while the decline of the Pagan kingdom in South Asia has multiple intertwining explanations, the most compelling reason for a trajectory that corresponded to that of Angkor was that it too was a victim of its own success. Successive rulers in the eleventh and twelfth centuries had proved capable of fusing ideas that bound together diverse peoples and cultures, and finding ways to express these in magnificent monumental architecture, as well as through art, literature and language and through more mundane steps such as the standardisation of weights and the monetisation of the economy.[152] Pagan's rulers invested considerable resources in religious patronage of Buddhist monasteries and institutions as a trade-off for support and confirmation of authority. Constructions such as the Ananda, Shwesandaw and Thatbyinnyu temples – to name three of the 2,000 that were built between the tenth and thirteenth centuries – were intended to serve as statements of power, not just as links to the Buddhist cycle of reincarnation: royal donations were supplemented by support from elites who hoped thereby to gain status as well as to earn merit that might ensure for them a better next life.[153]

Inevitably, over time this tipped the balance away from secular authority in favour of monastic establishments. As more land, its produce and tax revenues passed into the hands of religious devotees, the less authority the crown had over territories it nominally controlled – and the less able it was to make investments for the future, whether in trade, in military capabilities or simply to retain the loyalty of magnates.[154] Pagan had benefited from the springboard provided by benign conditions that favoured agricultural exploitation, enabling a sufficient and steady food supply that encouraged a process of urbanisation that in turn led to specialisation, to a glorious cultural flowering and to a place within vibrant local, regional and even intercontinental trade networks. In due course, that took a toll from which recovery proved impossible: it would be both an oversimplification and an exaggeration to argue that Pagan and Angkor declined because of such excesses. Nevertheless, it is hard not to acknowledge the impact of the depletion of financial resources and of what amounted to an effective transfer in royal wealth to religious foundations. But, taken together, it may well be that the most important way to understand the decline of these great cities in South and South-East Asia is as victims of their own success.[155]

Shifting consumption patterns, combined with an inability or unwillingness to adapt, also explain the failure of Norse communities in Greenland. A cooling process from the early fourteenth century saw an extension of Arctic pack ice, significantly reduced rainfall, shorter growing seasons and more demanding conditions for domestic animals. Each and all were challenging. But what may have proved more telling was that Norse Greenlanders did not adopt Inuit Arctic-adapted technologies, such as the toggle harpoon and the kayak – perhaps out of pride or because of an 'intense cultural conservatism' that impelled them to looked down on Inuit peoples and keep separate from them. Osteological and DNA evidence suggests that intermixing between settlers and Inuit was essentially non-existent.[156]

In fact, even the unwillingness to adopt tools that had proved their worth may have been less important than developments elsewhere that had nothing to do with climate or adaptation and a lot to do with market forces: Greenland's main value was a source of exports of furs and walrus hides and teeth. From the early 1300s, however, new trading networks linking Novgorod and Russian cities into the White Sea opened up new sources of animal skins and furs that provided competition for supplies from the North Atlantic, while hemp ropes offered alternatives that were cheaper and more plentiful than those made from thick and sturdy walrus hides. The market for walrus tusks, meanwhile, was depressed on the one hand by rising supplies of elephant ivory from Africa, and on the other by changing tastes and cultural shifts that included a turn away from ivory in religious art production.[157] To top it all off, as climate conditions became more challenging and economic realities more difficult, life also became more precarious for the Norse communities, who were subject to ever more attacks by Inuit (whom the the settlers called Skrælings) driven by opportunism and by competition for resources and location; these raids were devastating and demoralising.[158]

Nevertheless, just as there had been a reconfiguration of global climates starting around 800, patterns now begun to change around four centuries later, albeit neither suddenly nor consistently. The impacts started to be felt in the eastern Mediterranean and Central Asia as early as the late twelfth century, and appear to have been matched by changes that took effect in the Pacific Rim around the same time. Wetter conditions coincided with a new wave of migrations and settlements from Polynesia to Hawai'i, Aotearoa (New Zealand) and Rapa Nui (Easter Island).[159] This was also the moment of a single contact between Polynesian and South American population groups – either as a result of a group or groups of the former journeying across the entire expanse of the Pacific and returning successfully, or as a result of groups of the latter setting off and reaching Polynesian islands c.1200.[160]

The causes of population dispersals across the Pacific were complex. While a shift in climate was likely an important factor, the custom of primogeniture – which consolidated power for the benefit of a single family member – may have encouraged searches for new opportunities.[161] Whatever the case, new settlements brought about ecological change to islands that had previously not been inhabited, including deforestation and the collapse of bird and mammal species caused by human interventions, such as hunting or land clearing. On top of that was the damage done by domesticated animals that were brought intentionally by new settlers and by rodents like the Pacific rat that were transported alongside human travellers – perhaps inadvertently, but perhaps not.[162]

By the late thirteenth century, life on many Pacific islands was becoming precarious. Evidence of elevated numbers and growing strength of storms, falling sea levels and rapid reduction of food sources in coastal areas transformed many island communities. The harvesting and trading of pearl oysters across the South Pacific by inhabitants of the Cook Islands came to an end, while fishing seems to have declined sharply on Easter Island. Coastal embayments that had provided comfortable living spaces became brackish lakes or swamps in the Solomon Islands, for example. In New Zealand, the drop in temperatures meant that it was no longer possible to grow the sweet potatoes that had been brought by the first human settlers who had arrived c.1250, leading to a reliance on the roots of the bracken fern instead.[163]

When we think about the Middle Ages, we tend to focus only on western Europe and think in terms of an age of kings and barons, peasants and priests, of the rise of institutions such as the Church or guilds. Taking a more global view of this period raises important questions about the exploitation of natural resources, about the role of technological change in raising agricultural yields or about the adoption of new crops or of new strains such as Champa rice. It also raises questions about the role played by climatic shifts and about the distinction between changes in weather conditions and the challenges posed by their variations.

Taken as a whole, however, the period c.800–c.1250 was one of profound intensification of connections – within interlocking and interdependent worlds in Asia, Africa and Europe, or within communities in the Americas which emerged as central if independent hubs that had little or no interaction. It was hardly the case that this long time-frame provided conditions of peace and harmony that enabled relentless economic and demographic growth: in fact, one could argue the opposite was the case, as empires and dynasties rose and fell, states fought, were subsumed by or conquered each other or faded into obsolescence as new competitors emerged who could supply goods more quickly and more cheaply.

Nevertheless, as with other periods, it is hard to escape the fact that the fundamentals of ecological equilibrium and environmental sustainability underpinned the cultural, political, socio-economic, diplomatic and military histories of individual kingdoms, states or regions. Reliable food and water supplies were central at all times, but especially during periods of demographic expansion. Societies had to contend with finite natural resources; when those became exhausted or came under stress because of over-exploitation, because of shifts in rainfall patterns or because of conflict, disease or the failure of infrastructure such as river defences, disaster soon followed. That provides some food for thought for the present and future – as well as the past.

Cave paintings from Lascaux in the Dordogne, southwest France.
The depiction of animals became increasingly common around 40,000 years ago.

Mohenjo-Daro, one of the main Indus Valley settlements. Planned around a
grid system, the city was one of the largest in the world around 2500 BC.

Female figurine from the Indus Valley. Women were often portrayed as deities in many civilisations, including across Mesopotamia.

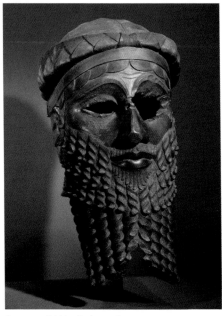

Sargon of Akkad. An early empire builder, he had 'neither rival nor equal', according to one source.

Tortoise shell that was used in Shang dynasty China to predict the future – including the weather.

Detail of a painted casket from the tomb of King Tutankhamun. Horse-drawn chariots were not just symbols of power, but methods of control.

Scene from the Ramagrama stupa at the temple complex at Sanchi, Madhya Pradesh. It depicts a visit by the great leader Ashoka to collect relics belonging to the Buddha. Buddhism was one of many religions that offered explanations about engagement with the natural world.

Gold adornment from Issyk Kul, modern Kazakhstan, of winged horses, c.4th century BC. Objects such as these showed status and wealth – and the importance of horses in the steppes of Central Asia and beyond.

Roman-era mosaic from southern France. The Romans idealised the countryside and control over nature.

Pyramid of the Sun in Teotihuacán in the heart of Mexico. The city was given its name (literally 'City of Gods') only much later.

Portrait vessel from the Moche era of South America (c.1–800 AD).

Aztec gods Cipactonal and his wife Oxomoco devising the calendar. The peoples of Central America – like others – set great store by their mathematical and astronomical calculations.

The Emperor Justinian (r. 527–65). One of the greatest of the Roman emperors – but one whose reign coincided with a catastrophic pandemic.

Above left: Temple V at Tikal, in what is now Guatemala. The building marks an apogee of Maya culture in Central America. *Above right:* A ceramic pot of the Aztec rain god Tláloc. Offerings were given to deities who brought rains, water and fertility across Mesoamerica.

The magnificent temple complex of Angkor Wat at Angkor in modern Cambodia. The city's huge population depended on sophisticated water management systems.

Monks Mound at Cahokia, one of the largest and most important
pre-Columbian settlements in North America. Cahokia's success owed
much to the rich soils of the American Bottom region.

The eruption of Mount Samalas in 1257 was one of the largest in the last
millennium and had global consequences. It was so violent that it created
Mount Rinjani on Lombok, Indonesia, with its distinctive Segara Anak caldera.

Peter Bruegel the Elder's *Hunters in the Snow*. Painted in the mid-sixteenth century, this oil painting is often closely associated with ideas about the Little Ice Age (c.1550–1800).

Above left: Conditions on slave ships were appalling and were all about maximising profits for those with capital to invest – and no concerns about the fates of their fellow human beings. *Above right:* A sample of a potato plant from 1660. Potatoes have played an extraordinary role in global history, and not only in re-shaping diets around the world.

Opposite: A page from Thomas Jefferson's meteorological observations. The American statesman took multiple temperature readings every day for decades – including on the day of the signing of the Declaration of Independence on 4 July 1776.

ly.	hour.	thermom.		day	h. m.	°
	9 - 0 A.m	81½		9	5 - 30 A.m.	75
	7 - P.m	82.		9		77½
2.	6. A.m.	78.			6 - 30 P.m.	81½
	9 - 40' A.m.	78			9 - 45	78.
	9. P.m.	74		10.	8. A.m.	75.
3.	5 - 30' A.m.	71½			9 - 15.	76½
	1 - 30. P.m.	76			2 - 0. P.m.	80.
	8 - 10.	74.			4 - 45'	82.
4.	6. A.m.	68.			6 - 30	81½
	9.	72¼			9 - 30.	78.
	1. P.m.	76		11.	5 - 30. A.m.	74.
	9.	73½			8.	76½
5.	6. A.m.	71½			9 - 40. P.m.	75.
	9.	72		12.	7. a.m.	72.
	9. P.m.	74.			9.	72.
6.	5. A.m.	74.			8 - 50. P.m.	72.
	9.	75.		13.	5 - 30. a.m.	71½
	4. P.m.	77.			11.	74
	10.	74.			2. P.m.	76
7.	6. A.m.	71.			6 - 45.	76
	10.	73.			7 - 25	76
	1. P.m.	74.			9 -	75
	3 - 20'.	75			rain	
	9 - 30.	74		14.	6 - 50. A.m.	73.
8.	5 - 35' A.m	75			rain	
	9.	77½			9 - 30.	72
	2. P.m.	80.			rain.	
	5.	81.			1. P.	71½
	8 - 15'	80			rain	
	9 - 30	79			5 - 35	70
					5. - - -	

Cast-brass plaques currently in the British Museum. Part of what has become known as 'the Benin Bronzes', these were looted during a reprisal raid by British forces on the Oba (ruler) of Benin in 1897.

Guano was both highly prized as a fertiliser and a source of profits that some could only dream of. This image from the mid-nineteenth century shows bird droppings being collected along the coast of Peru.

Below left: Eunice Foote's seminal article on the effects of the Sun's rays (1856). Foote proposed that atmospheric water vapour and carbon dioxide could combine to create what later became known as the 'greenhouse effect'. *Below right:* Svante Arrhenius likewise worked on the influences of human activities on climate change, including assessing carbon dioxide concentrations in the earth's atmosphere.

Much of the Soviet Union's new infrastructure in the twentieth century was built by coerced labour. Here prisoners are being made to work in sub-zero temperatures in the 1930s to build a canal linking the White and Baltic seas.

Below left: Rubber plantations grew at an astonishing rate in South-East Asia. Hundreds of thousands of acres of forest were cleared in the twentieth century to meet global demand. *Below right:* A poster from Mao's ill-fated 'Wipe Out the Four Pests' campaign of the late 1950s. Efforts to eliminate mosquitoes, flies, sparrows and rats had unforeseen consequences.

A nuclear bomb test in the South Pacific. The number of these tests likely had an effect on global climate patterns in the mid-twentieth century.

Dagmar Wilson and Coretta Scott King leading a protest march for peace at the United Nations in 1963. Women were instrumental in raising awareness about environmental destruction in the post-war period.

'Let us harvest the rich virgin lands!' urges a poster highlighting Nikita Khrushchev's overhaul of Soviet agriculture of the 1950s and early 1960s. The scheme did not go well.

Poster for *When the Wind Blows* (1986) – a bleak and haunting animated film that told the story of a couple in rural England trying to survive after a nuclear attack.

Weather control became a key part of military and political thinking in the twentieth century. Many were convinced that it was a matter of time before the climate could be manipulated at will.

President George H. W. Bush at the Earth Summit in Rio de Janeiro in 1992. 'Our children will judge us by the actions we take from this day forward,' he said. 'Let us not disappoint them.'

Newly cleared forests ready planted with oil palm trees; palm oil is used in everything from lipstick to soap to ice-cream. Clearances have significant effects on biodiversity as well as soil chemistry.

Air pollution is a major health hazard. In South-East Asia, 99.9 per cent
of the population live in areas that exceed World Health Organization
guidelines. In some parts of India, concentrations of fine particulate matter
are twenty-three times levels considered safe.

Transitions to energy sources that are carbon neutral are a key part of
the challenge of the twenty-first century. Scientists estimate that the
chances of staying within the target of keeping global warming below
1.5°C agreed at the Paris Accords in 2015 are effectively zero.

13

Disease and the Formation of a New World
(c.1250–c.1450)

Time itself has slipped through our fingers.

Petrarch, *De rebus familiaribus* (1348)

The twelfth century saw repeated series of fractures within and beyond the Chinese world. The efflorescence that followed the collapse of the Tang 200 years earlier had seen the emergence of the Song dynasty as well as a host of other states of varying size and capabilities across East, South and South-East Asia. Kingdoms such as Đại Việt in what is now northern Vietnam, Dali in modern Yunnan province and Goryeo (Koryŏ) in the Korean peninsula had been part of a tapestry of states that rose together and benefited from increasing levels of trade as well as from cultural, political and sometimes military competition.[1]

The mesh of relationships was often uneven, complex and unstable. In 1127, the city of Kaifeng was sacked by Jürchen nomads in one of the most dramatic moments in history. At the time, Kaifeng was one of the largest cities in the world – perhaps even the largest measured by population size – and also one of the most splendid, lovingly described by Meng Yuanlao in a treatise known as *Dreams of Splendour of an Eastern Capital* which records the friendliness of the inhabitants as well as the richness of civic life before the city's fall. 'Peace stretched on day after day; people were many and all things were in abundance,' writes Meng. 'Youths with trailing locks practiced naught but drumming and dancing, the aged with white speckled [hair] recognized neither shield nor spear. Season and festival followed one upon the other, each with its own sights to enjoy.'[2]

Now devastation followed. The Emperor Huizong and his family were captured and deported to Jürchen homelands in Manchuria. 'Once I lived in heaven above, in pearl palaces and jade towers,' wrote the Empress Zhu, who was raped by her captors as she was marched northwards into the cold; 'now I live among grass and brambles, my blue robe soaked in tears. I hate the drift of snow.'[3] It was a disaster, lamented the poet Li Qingzhao:

And you should've been more cautious,
Better educated by the past.
The ancient bamboo books of history
Were there for you to study.
But you didn't see ...[4]

Disruption extended into the steppes, where peoples like the Naiman, Kereit and Qonggirad were obliged to submit to and make contributions in kind to the Jürchen overlords. The Mongols were treated so badly that, in the words of one leading scholar, they 'almost ceased to exist'.[5]

That changed in the late twelfth century, thanks to another Mongol leader who proved rather more resilient and successful. There were many reasons for the rise of Temüjin: his charisma, determination and leadership skills were crucial, as was his strategic decision-making. However, he owed his rise to his marriage to Börte, the daughter of a Qonggirad chief, which gave status to the Mongols and to Temüjin personally. To keep Temüjin in his place, Börte was kidnapped by the dominant Merkit tribe, which likely resulted in her rape; the birth of her firstborn son, Jochi, soon followed. Temüjin's response seems to have been astute – he brought up Jochi as his own, and later placed him in charge of all Merkit territories and beyond.[6]

Temüjin consolidated his power steadily in the 1180s and 1190s, perhaps helped by a period of extremely dry conditions that made life on the steppes challenging and offered opportunities for those looking to profit from the weakening of both the authority and the resources of more powerful groups.[7] By 1203, he had become master of the Orkhon Valley, home to hundreds of *balbal* stone statues and to a famous set of inscriptions proclaiming that whoever controlled this area would be the master of all the steppe nomad tribes. For nomadic peoples, this was the centre of the world; by taking control, Temüjin could benefit not only from prestige but from *sülde* – the life-affirming source of unity and power that provided a mandate for further expansion.[8]

This was important, for steppe empires required expansion to survive: growth was not only useful but essential in supplying the rewards to keep social and political structures in place. The fluidity of tribal groups and confederations meant that leaders needed to engage in the constant acquisition of material rewards – in the form either of the payment of tribute or of spoils of conquest – in order to keep interest groups, supporters, rivals and potential rivals happy.[9]

By 1206, Temüjin was master of the steppes, expanding his authority through military force, dynastic ties and relentless expansion through the elimination of rivals. That year, he gathered an assembly of the Mongol elites and representatives of all the other groups that had been defeated by or had submitted to him and was proclaimed the supreme ruler with the title Činggis Khan or universal ruler.[10] The consequences of not recognising his overlordship were then made clear: those who had defied him would be hunted down one by one and measured 'against the linchpin of a cart'. Those who were not murdered would be captured, enslaved and shared out among Mongol supporters as prizes.[11]

Achieving a dominant position on the steppe was one thing; using that platform to create the largest land empire in human history represented an achievement of a different magnitude altogether. In the decades that followed, states and dynasties in what is now China were picked off one by one, as the Western Xia, Jin, Dali and Song fell in turn leading up to the fall of the house of Ch'oe in Korea in 1258; this left Japan as the final remaining objective in East Asia. To the west, Mongols had also spread westwards like wildfire, forcing the submission of Central Asia before Činggis Khan's death in 1227, and reaching central Europe barely a decade later.

There were many explanations for these successes – including the selective use of extreme violence as a tool of control and intimidation, the organisational abilities of highly mobile scouting troops in identifying new targets, logistical brilliance in information gathering, adoption of new techniques and tactics (such as the use of artillery) and supreme innovation on the battlefield.[12]

But the fuel for success may have been provided by an extraordinary pluvial period between 1211 and 1225 in Mongolia that was the wettest in more than 1,110 years. These conditions significantly increased the carrying capacity of the landscape, producing more abundant grasses and dramatically expanding the amount of available pastureland. This provided the basis for enormous increases in the size of livestock herds, above all of horses. Činggis Khan, his followers and successors may well have been tactically brilliant, yet it was their good fortune that the Mongols not only could draw on vast resources to defeat their enemies and expand their empire, but were able to do so at exactly the right moment.[13]

The significance of this accident of timing can be underlined by noting a downturn in conditions in the later 1220s caused by another 'machine-gun effect' of multiple volcanic eruptions such as Mounts Reykanes and Reykjaneshryggur in Iceland, and Mounts Zao and Aso on the Japanese islands of Honshū and Kyūshū which erupted between 1226 and 1231 and left acid spikes on northern hemisphere ice cores. These were likely responsible for drops in temperature that are attested by tree rings in Scandinavia and the Qinghai–Tibetan plateau, perhaps exacerbated by changes to El Niño patterns in the Pacific.[14]

This period was one of severe food shortage in Korea and of catastrophic suffering in Japan, which experienced one of the worst famines in its history in the early 1230s. This was followed by an epidemic that was referred to as the 'barbarian sickness' (*ibyō*) and by floods that left dead bodies lining the riverbanks in Kyoto. The story was the same throughout Japan, wrote the nobleman Kageyukōji Tsunemitsu in his diary: put simply, it was a time of 'sad lamentation'. Although mortality rates across the Japanese archipelago would not have been uniform, data from censuses and from taxation and land records suggest population falls of around 20 per cent in the course of four years of woe.[15]

Around the same time in Novgorod in what is now Russia, one chronicler recorded that crops were ruined by frost, leading to so many starving to death that it was impossible to bury the bodies. 'The wrath of God was such that the people not only ate the dead but even the living would kill and eat each other; and they ate horseflesh, dogs, cats, and other animals. Whatever one could find, he would eat – moss, pine, elm, linden bark and leaves.' While the impact on the Mongols is not clear, the fact that this was the time when they expanded across swathes of territory, overwhelmed the Xia and destroyed what was left of the Jin may well indicate that these developments are best understood as knockout blows being delivered to already weakened states that were now teetering on the abyss. The prizes taken by the Mongols were in many cases low-hanging, fading fruit rather than glittering, extraordinary glories.[16]

The administrative and bureaucratic structures that the Mongols put in place as they brought more and more territory, people and cities under their sway were nevertheless impressive in their scale and their efficiency. In many cases, these were extensions or adaptations of systems that the Mongols had been exposed to and were then introduced across the empire – such as the service for bringing post and information both reliably and quickly to and from the centre. Usually referred to as the *yām*, this was a variant of the Chinese *yi* (or *li*) system of communication which had been introduced by Uighur and Kitan advisers enjoying prominent roles at the Mongol court.[17]

Another key part of the Mongols' success was the incorporation of local elites – something that runs contrary to the common assumptions today of the conquerors as brutal, bloody and ultra-violent.[18] Such is this reputation, indeed, that some scientists have claimed that the Mongol invasions killed so many people that this had an impact on the global carbon cycle and lowered atmospheric CO_2 content – a proposition that gained attention in the press around the world, generating headlines such as 'Why Genghis Khan was good for the planet'.[19]

In fact, growing evidence shows that, although the Mongols were beneficiaries of an ideal set of climatic conditions for their own needs, the world they were stepping into – at least in Central Asia – was one of urban and demographic contraction, at least partially caused by ecological and hydrological pressure as long periods of aridity forced changes of lifestyle and in some cases abandonment of settlements.[20]

While there can be little doubt about instances of extreme violence on the part of the Mongols, of rather more significance was the creation of the Pax Mongolica – a term intended to explain the essential stability of Mongol overlordship after the explosive period of expansion. Rather than devastate and destroy the world, the Mongols oversaw an intensification of trade across Eurasia, which sits at odds with the widespread depiction even today of Činggis Khan and his successors as bloodthirsty and barbaric.[21] A 'bridge of justice' should be built across the Mongol lands, one khan was told, 'for where there is justice, the world is prosperous'.[22]

The interconnected world created by the Mongol conquests facilitated commercial exchange — something the leadership was actively keen to promote.[23] The Mongols were prolific builders of cities, most notably Karakorum, which was founded in 1220 and was the recipient of heavy investment over the decades that followed, including an opulent and spectacular palace.[24] Many new towns were founded on the banks of the major rivers of Central Asia and on the shorelines of the Black Sea and the Sea of Azov.[25]

As recent experience of pandemic has reminded us, greater volumes of trade, higher levels of travel and closer links between people living in different regions accelerate the transmission of disease. It was no coincidence, for example, that the increase in population size in Song dynasty China and the expansion of territory under centralised control also saw an increase in the number and severity of outbreaks of epidemic disease.[26] While leprosy is well attested in Europe, the rash of leper houses that began to be founded in the eleventh century would most naturally seem connected with the rise of socio-economic interactions across the continent in precisely that period.[27]

The rise of the Mongol empire too seems to have been closely linked to the expansion of pathogens and disease. Multiple accounts talk of large numbers of casualties in sieges mounted by Mongol forces in the course of the thirteenth century, including the assaults on Kaifeng in 1232 and on Baghdad in 1258 which were seminal moments in the establishment of control in East and West Asia respectively. In the case of the latter, so many had died from 'pestilence' that there were not enough people to bury the dead, with the result that dead bodies were simply flung into the River Tigris. The fact that the Mongol ruler, Hülegü, moved camp at least five times in 1257 while preparing the assault on the city may reflect efforts to relocate in order to contain or stay ahead of outbreaks of disease.[28] Literary sources as well as network analysis of authors who trained in Damascus in the thirteenth century reveal how epidemics affected other parts of the Middle East at this time.[29]

The significance of this evidence has long been overlooked, as has material relating to demographic contraction and political instability in East Africa and in Ethiopia in particular.[30] In recent years, a handful of pioneering scholars have been questioning whether plague was responsible for some or all of these developments. For example, the site of Akrokrowa in Ghana was abandoned around the middle of the fourteenth century, as were multiple associated sites, with new settlement patterns being adopted that suggest a sudden shock or shocks traumatic enough to alter people's way of life. This took place precisely at the time when the Black Death was ravaging other parts of Africa as well as the Middle East and Europe.[31]

Groundbreaking research has produced startling new avenues of investigation into how the Mongols may have unwittingly created a perfect environment for plague to be spread along trade and travel routes and for it to be taken outside natural foci where it exists endemically, such as in the Tien Shan mountain range of Central Asia. One way in which this happened was the expansion of Mongol food tastes, as well as through their fur and leather requirements.

As Monica Green has suggested, marmots were prime suspects as vectors of disease transmission: textbooks written by leading dietary physicians extolled the utility of marmot meat as a source of protein, and also remarked positively on the warmth of marmot skin as well as its resistance to rain and water. It seems that the spread of the Mongols and of their dietary and fashion preferences may have played a role in helping disseminate pathogens from their natural habitat, where they could enter the human body through ingestion of meat or simply through inhalation.[32] It may be that the higher prevalence of plague was related to greater exploitation of marmots from their habitats; or it could also be that changing environmental conditions provided the impetus for the migration of marmots and other hosts to locations that brought them into closer contact with human populations.[33]

Extraordinary breakthroughs have also led to a full mapping out of the genome sequence of *Yersinia pestis* aDNA and to the phylogenetic tree of *Yersinia pestis*.[34] This shows the sudden divergence of *Y. pestis* into four branches shortly before the Black Death in a moment that geneticists have called the Big Bang. The causes, context and date of this divergence are not clear, although some have suggested a spillover event may date to the first two decades of the thirteenth century within a context of Mongol contacts with the Turfan Uighurs or the Qara Khitai.[35]

Another line of investigation has focused on the massive eruption of the Samalas volcano on Lombok in what is now Indonesia that took place in the late spring or early summer of 1257.[36] The eruption was one of the largest of the last millennium, if not the largest.[37]

Its violence led to thick tephra covering Lombok, with pyroclastic density currents up to fifty metres thick burying around half the island.[38] The eruption appears to have had an important climatic impact, including the draining of the stratospheric beryllium (^{10}Be) reservoir for over a decade, as well as resulting in an El Niño-like event in its immediate aftermath or enhancing pre-existing weather patterns.[39]

The most important of these was an exceptional period of high solar irradiance which had peaked in the late 1240s – and reached levels that have not been matched for almost eight hundred years. The effects of the Samalas eruption were global, producing intense monsoons in Asia, severe drought in the western flank of the Americas and unusual and unpredictable conditions in England where extremely strong storms in some years and 'intolerable heat' in others had an impact on harvests and on wheat prices. These uncertainties fanned baronial opposition to the crown in England, where King Henry III was faced with considerable political pressure in 1258 as a result of a severe food crisis that stemmed in part from bad weather, but also from the panicked response to rising prices. People fled from the countryside to towns and cities in search of food. As London became swamped by refugees, it was not long before bodies could be found 'in all directions, swollen and livid through hunger, lying by fives and sixes in pigsties, on dunghills, and in the muddy streets, their bodies woefully and mortally wasted'. Poor harvests put pressure on supplies, but the responses made a bad situation much worse.[40]

The middle decades of the thirteenth century were a time of intense volcanic activity. As it happens, eruptions of Zao in Japan in 1227 and of Rekjanes in Iceland in the same year and again in 1231, left a heavier signature on the dendrochronological evidence than Samalas, presumably because of the timing of the latter.[41] Two further eruptions that have recently been identified but whose locations remain unknown can be dated to 1241 and 1269. All these eruptions resulted in the mass ejection of aerosols into the atmosphere.[42] 1258 and 1259 proved to be two of the coldest summers in the northern hemisphere of the last millennium – with conditions in western Europe, Siberia and Japan proving to be particularly harsh.[43]

Volcanic eruptions were one part of the story. The behaviour of the sun was another. The period of enhanced irradiance gave way to a Grand Solar Minimum, a time of sharply reduced solar magnetic activity – first christened the Wolf Minimum in 1980.[44] A number of studies have demonstrated there were sharp changes in atmospheric radiocarbon and cooler global temperatures starting around 1260 and lasting for eighty years until the middle of the fourteenth century.[45] As astronomers note, solar minima have limited impact on temperature, with differentials typically no more than 0.3 °C.[46] What was significant, then, was that this reduced solar activity not only coincided with yet another series of major volcanic eruptions in 1269, 1276 and 1286 but also with the weakening of the El Niño–Southern Oscillation, the Asian monsoon and the North Atlantic Oscillation. The results were striking, with higher levels of moisture to the Pacific coast of South America as well to North America, ending a long period of drought in each, while also resulting in sudden reductions in monsoon rainfall in South and South-East Asia from the 1280s, which played a role in harvest failure, hunger and famine.[47]

In north-western Europe, weather became more unsettled, heightening the risk of food pressure and increasing the possibility of the spread of disease – for humans and for livestock. These pressures were compounded by the strong demographic growth that had characterised the early Middle Ages and by the resultant development and exploitation of fragile ecosystems. A case in point comes from a belt of sandy terrain stretching from Breckland in England across continental Europe through North Sea coastal regions as far as Russia.[48] Mismanagement or overuse of the top layers of loose quartz could result in heightened vulnerability to floods and to uncontrolled sand drifts that could destroy intensively cultivated arable land very quickly.[49] As one leading scholar has put it, the prospect of supply-side food shortages came at precisely the moment when societies found it hard to cope with them.[50]

Accounts from Scotland, England and Ireland paint a picture of conditions that were difficult and unpredictable, with repeated indications of crops failing and outbreaks of famine in the second part of the thirteenth century. One report of 'a great lack of productivity of the land and unfruitfulness of the sea, as well as turbulence of the air, as a result of which many people fell ill and many animals died' was typical of many that show the difficulties in trying to contend with unusual storm patterns, insect infestations, excessive rain or unusual cold.[51]

The falls in revenue at international fairs such as those held at St Giles Winchester and St Ives, which collapsed by at least 75 per cent between 1285 and the 1340s show this was a time of disruption and contraction. The dramatic decline in the value of English overseas trade handled by foreigners, and the halving of rental values of property in Cheapside in London across three decades at the start of the fourteenth century, point to the scale of the economic repercussions in this period.[52] So too do cases of the spread of sheep scab in 1279 which halved national wool output and led to supply shortages in locations that were dependent on high-quality raw materials for their own production – such as in Flanders, where riots broke out as a consequence.[53]

It has been suggested that rising numbers of cases in Scotland in the 1290s of scab – an intense reaction to the *Psoroptes ovis* mite – along with severe winters, poor summers and failed harvests played a role in fomenting dissent against King Edward I of England, especially after his treasurer, Hugh de Cressingham, sought to raise taxes in 1296–7; this provided a rallying point for anti-English sentiment, spurring revolts led by William Wallace that dominated much of the following decade.[54]

In East Asia, the century that followed 1260 saw regular drops in mean temperature, with harsh conditions in the 1270s, 1310s and 1350s attested by multiple written and climate sources from Korea, Japan and China.[55] For many parts of China in particular, the period 1300–60 was one of almost constant natural disaster, including frequent supertyphoons that devastated the coastline, regular flooding of the Yangtze and Yellow River deltas and a series of epidemics in the middle of the century that claimed hundreds of thousands of lives. In 1308 alone, according to the Yuan chronicle, bitter snowstorms on the Mongolian steppe killed huge numbers of livestock and forced almost a million people to migrate southwards – where they had to survive on handouts from the court. Famine and disease killed almost half the population in the Yue region of the Yangtze Delta in the same year, while so many perished in Jiangzhe province that 'Dead bodies lay upon each other. Fathers sold their sons, and husbands sold their wives, the sound of their crying was so loud that the Earth trembled.'[56]

This presented challenges for the imperial court and the Emperor in particular, whose authority – linked to having a mandate from heaven – was compromised by such regular and severe setbacks. Indeed, in 1297, the Yuan Emperor changed his reign title to Dade (大德, or 'Great virtue') in an attempt to end the downturn of fortunes.[57] Imperial power was not just affected in its abstract terms: failed harvests meant reduced revenues, which were then further stretched by distribution of food and disaster relief. This in turn impaired the Emperor's ability to dominate the aristocracy, which relied on regular disbursements from the centre. In 1307, Emperor Wuzong was barely able to pay half of what he had promised because 'the exchequer's coffers ... [were] empty'. Four years later, the amounts paid were only a third of that figure; and those in turn had halved by 1317.[58] In other words, beneficiaries received less than a tenth of what they had been getting a decade earlier.

It was perhaps not surprising then that shortage, suffering and fear brought about social and political turmoil. Many parts of the Middle East had seen an unusually large number of freak weather events in the decades after 1260, including damaging wind storms – such as one in 1319–20 that destroyed many houses in Damascus, leaving people trapped indoors by fallen trees. The same storm struck Aleppo, accompanied by dust, hail, thunder and lightning, and uprooted oak and olive trees as well as vines. As is true of all such events, the poor were hit hardest, first through the loss of homes and livelihood, and then by the limited or non-existent resources that might enable them to start from scratch; needless to say, one additional outcome was the inflationary effect on prices that pushed commodities, goods and food to become even more unaffordable.[59]

In this context, it was no surprise that social unrest was a consistent feature of this period. Its effects were particularly clear at the start of the fourteenth century in a phase of rapid climate change that has become known as the Dantean Anomaly, after the contemporary poet, author and philosopher Dante Alighieri. Reconstructions of weather conditions in 1302–7 point to several years of dry summers, coupled with large-scale urban fires across Europe which led to steps being taken to mitigate future risks; these included the city of Siena buying the port of Talamone to help ensure food security, as well as investments by several cities in more resilient infrastructure, such as wells that were new and deeper than those that had been dug in the past.[60]

Nevertheless, the decade that followed saw problems that have led to this period being described as experiencing 'arguably the single worst agrarian and food crisis in northern and central Europe in the last two millennia'. The problem, as so often, was not a single failure of crops but back-to-back harvest shortfalls. In 1315, 1316 and 1317, torrential rains resulted in grain yields that were respectively around 40, 60 and 10 per cent of their usual levels.[61] Temperatures were also conspicuously cold, with chroniclers in northern Italy reporting that wine stored in barrels froze solid, as did wells and springs, while many trees died of cold. For this reason, and presumably as a result of high consumption, the price of wood rose sharply.[62] Scholars believe that 10–15 per cent of Europe's population died of hunger and hunger-related disease in these years in perhaps 'the single worst subsistence crisis, in terms of relative mortality, in recorded European history'.[63]

These shocks produced civil unrest across many parts of northern Europe, with a mob of men, women and children gathering and rampaging across France, attacking castles, officials of the crown, priests and lepers before targeting Jews in particular across Languedoc in 1320. Anti-semitism had a long history in Europe, most notably during the passage of elements of the First Crusade who committed atrocities in Cologne, Nuremberg and elsewhere as they made their way east towards Constantinople and Jerusalem in the mid-1090s.[64] On this occasion too, it would seem that Jews were attacked because they were perceived to be responsible for the misfortune of others, or because they were vulnerable targets that were perceived to be wealthy. As noted in the Introduction, the pogroms of the early 1320s were part of a pattern that saw the probability of persecution rise by between 1 and 1.5 per cent according to decreases in growing season temperatures in the previous five-year period. In other words, the worse the weather conditions, the greater the likelihood that minorities would be attacked.[65]

This was mirrored elsewhere: in Egypt, for example, churches and monasteries were looted and sacked in 1321, while Christians were accused of trying to set fire to mosques, with some arrested and tortured into 'confessing' their apparent crimes.[66] As in Yuan dynasty China, climate shocks presented political challenges. In the case of Muslim Egypt, Nile failures tended to result in crackdowns on minorities (including Christians), as well as enforcement of decrees against prostitution, beer drinking or the wearing of either immodest or over-luxurious clothes.[67] Such instances were also typically accompanied by the sultan ordering the construction of new mosques, the repair of existing structures, the distribution of money to urban inhabitants and the grant of enhanced powers to religious authorities in order to buy their acquiescence.[68]

Others dealt with the situation in ways that were perhaps intentionally more vague. One decree issued by Pope John XXII in 1320 urged that 'sorcerers, the infectors of God's flock, flee from the midst of the House of God'. Those who 'make sacrifice to demons or adore them' or who have made 'binding pacts with them' should be identified and punished accordingly. Such pronouncements, of course, fit within a wider context of the Church hierarchy exerting control, and using accusations of heresy or 'evil deeds' to consolidate and concentrate power. Nevertheless, it is difficult not to be struck by the similarities with other times of need, of pressures on food supplies and of harsh weather – most notably, as we shall see, with systematic persecutions in the late seventeenth century, predominantly of women.[69]

Traumatic though the second and third decades of the fourteenth century were, however, one of the most striking features was the resilience of urban populations in coping and surviving. Cities like Florence, Pisa and Lucca were hit repeatedly by food shortages, by inward flows of people from the countryside and by hefty rises in wheat prices not only between 1314 and 1322 but on multiple occasions before the 1340s.[70] The inhabitants of Cairo meanwhile were understandably delighted by the visit of Mansa Musa, ruler of the Malian empire in West Africa, who spent so much gold on his way to Mecca in 1324 that its value fell, according to one contemporary.[71] Whether this was really the case, or whether deflationary pressures were a reflection of economic recovery following shock is not clear; but the fact that life returned to normal is an indication that it was possible to come out the other side of even the most difficult experiences.[72]

Ironically, then, the most significant environmental, ecological and epidemiological development in the first decades of the fourteenth century, at least in Europe, was not the impact of harvest failure but the spread of bovine pestilence that swept across the continent; this probably arrived from the Far East and was likely linked, once again, to the enhanced connectivity brought about by the creation of the Mongol empire across much of Eurasia. By 1315, the disease was ravaging central Europe; over the next five years, it devastated herds across much of what is now Germany, France, the Low Countries and Denmark before reaching the British Isles in 1319. In England and Wales, the disease

killed almost two-thirds of the bovine population, with total losses in
some regions balanced out by other locations which saw few or even
no deaths. Nevertheless, as owners grappled with what was happening,
many panicked, trying to get whatever price they could before their
herds were struck – with obvious implications both for value per head
of cattle and oxen and for the rural economy as a whole.[73]

Clearly the loss of assets was devastating to rural communities that
were already suffering from the harvest shortfalls of previous years. But
so too were consequences that played out over the longer term. One was
the fact that bovines such as oxen had important roles in agricultural
production, most notably in field ploughing, so their reduced numbers
meant that either future crop yields would be lower or more human
effort (and therefore calories) would have to be expended to produce
the same return. Whether the pestilence in question was a form of
anthrax, foot and mouth or rinderpest, the cold weather of the 1310s
may well have had a role in reducing resistance to pathogens: colder
weather, combined with smaller quantities of fodder, logically resulted
in animals having to expend more energy to maintain their body
temperature, making them more prone to disease – especially in the
case of cows, whose immune systems tend to become weaker during
the nine months of the year when they are pregnant or lactating.[74]

The consequences of this perfect storm were dramatic. Although
many historians have pointed to evidence of a robust recovery in
places like Dresden, Hamburg and many parts of England and a
return to normal by the mid- to late 1320s, others have questioned
whether the decline or loss of protein sources in diets over more than
a decade may have had a role in compromising immune systems as a
result of malnourishment in adolescents – exacerbating susceptibility
and vulnerability to disease later in life. According to this ingenious
suggestion, one of the causes of the mass suffering of the Black Death
in the 1340s lay in the weather conditions and disease environment
of three decades earlier.[75] Demonstrating this with certainty is not
straightforward, and requires detailed data sets that allow mortality
profiles to be tested empirically, as well as skeletal evidence that might
support ideas about malnourishment; what is clearer is that when bad
weather, crop failure and disease struck in the 1340s, it did so with a
vengeance.

* * *

The Mongol expansion across Eurasia had brought about an extensive ecological reorganisation of the steppes as well as beyond. This was primarily driven by the increased demands for pasturage for livestock – enhanced by political control that gave the ability to enforce rights to pasturage. It is telling that some accounts praise individual administrators for convincing Mongol leaders not to convert newly conquered lands to pasture; others report that where once cities had flourished, mulberry trees had grown, vines had borne fruit and fields had produced good harvests, now the landscape consisted of rich pastures that were home to cattle and horses, an abundance of hay and very little grain. Optimal conditions of damp and mild weather encouraged grass growth; however, they also produced expansion in the populations of rodents such as marmots, susliks and gerbils, which thrive in grassland.[76]

An extreme drought episode between 1336 and 1339 triggered a set of chain reactions that proved devastating. The lack of rainfall, attested by the dendrochronological records, would have led to a sharp decline in vegetation levels which would in turn not only put pressure on the rodent population through rising mortality levels because of shortages of food and water supplies, but also increase susceptibility to fleas carrying pathogens. A sharp spike in burials in two East Syriac Christian cemeteries in the Chu Valley in the Issyk-Kul region of Kazakhstan, including a rash of inscriptions to 'death through pestilence' in 1338–9, provides persuasive evidence of a zoonotic crossover from rodents to humans, with fleas seeking new hosts to spread the bacterium.[77]

As one leading scholar has pointed out, it was not easy for plague to spread: not only were plague foci far removed geographically, physically and geologically from places with high population densities, but the disease had to move across multiple animal species and over long distances – including across diverse climatic zones – to spread effectively. As such, while the role of the rat has usually been the focus of scholarly attention, it is worth noting the multiplicity of vectors that likely combined to help spread plague from the late 1330s. Among these are commensals – that is to say, animals that 'share our tables', living in, around and near human settlements, eating food scraps – and wider biological systems or groups such as wild rodents, including rats, hares and rabbits, and ruminants, including cattle, camels, goats and sheep.[78]

Other spreaders of plague include predator birds that feed on rodents, including large birds of prey and migratory birds, and also carnivores and scavengers such as wolves, foxes, jackals and hyenas that were common across Eurasia and would eat infected carrion, thus spreading the disease. That *Y. pestis* has been shown to survive in flea faeces in both soil and plant likewise shows that there were multiple routes for the transmission of plague, while the disease's ability to persist within rat colonies for a hundred years without the need for importing new infection underlines how prevalent the disease can be once it is established. Some cities ravaged by the Black Death – such as Constantinople – experienced more than 230 outbreaks in the 500 years after the 1340s, averaging around one outbreak every two years.[79]

The most powerful spreaders of plague, however, were humans. The hyperconnected world created by Mongol expansions enabled high levels of economic and cultural exchange, as well as rapid communication of information along trade routes. These links were well suited to the shipping of materials such as carpets and clothes, which were ideal for the spread of fleas, ticks and lice and therefore infection, as were the traders and the animals who transported them. The long-distance movements of foodstuffs in general, and of millet in particular, likewise provided perfect conditions for 'tag-along rodents', the parasites that accompanied them and the bacterium itself.[80]

That the connections facilitating the exchange of goods, ideas and information and at the same time providing the sinews of Mongol power across much of Eurasia were the very same that spread disease and death should not surprise us: in more recent times too we have learned that thinking about globalising forces solely in socio-economic and geopolitical terms makes it all too easy to forget that air, sea, rail and road links are the ways in which pandemics spread – and do so rapidly.

In the case of the Black Death that took hold of Europe, North Africa and the Middle East in the 1340s, the shipment of grain seems to have played a key role in the catastrophes that followed. Of particular importance were worsening growing conditions in Italy since the late thirteenth century that had led to increasing demand for exports from the Mongol Golden Horde; the Horde occupied prime territory that was ideally located for shipping through ports on the Black Sea. This produced an economic boom for the Mongols, while simultaneously creating dependency of populations in southern Europe on regular shipments from afar for their calorie provision. So important was the grain that flowed towards Europe that control of the sea lanes and the goods that moved along them became a source of competition and confrontation between Italian maritime republics. The competition between Venice and Genoa was so intense that it led to a war that lasted for most of the 1290s before the latter emerged triumphant.[81]

Demographic growth in Europe in the later thirteenth and early fourteenth centuries also posed a problem. Urban populations could be struck by sudden and acute crises – as was the case in 1329–30 in Italy where major uprisings followed harvest shortfall and soaring prices. In Florence, all-out anarchy was only narrowly averted through quick action by the authorities in importing grain and selling bread to citizens at subsidised prices. Those living in the countryside were no better off; indeed, quite the opposite was the case. Repeated subdivision and subletting of land into smallholdings served as a barrier to economies of scale and provided limited incentive for innovation. In times of shortage, therefore, families had few safety nets – and often fled towards cities in search of relief from hunger and suffering. This in turn meant there were even more mouths to feed in one location, which often made a bad situation worse by driving prices even higher.[82]

From the 1330s, unsettled conditions became common – and a source of concern. In 1338, for example, one chronicler reported that locusts descended on central Europe like a snowstorm, so blinding as to block out the rays of the sun, and so cacophonous that people could not hear each other speak. The insects destroyed everything: 'everywhere they landed, no sowing, no fruits, no hay nor any other of such crops of the earth remained, but they digested all', while many 'pigs, dogs and other animals' died as a result of the infestation. 'All the men of the church and the people of Prague then made processions with holy relics and banners, and begged God's mercy to eliminate such injustice.'[83] Locusts whose eggs 'were as large as bats' descended on many parts of Italy the following year, devastating fields and causing chronic food shortage as well as disease.[84]

In this context, the collapse of relations between the Mongols and the Italian maritime republics at the start of the 1340s was particularly poorly timed, coming at a moment when 'all parts of Eurasia were experiencing exceptional levels of environmental stress'.[85] These challenges were not uniform either chronologically or geographically, with tree-ring reconstructions in northern Scandinavia and central Sweden indicative of harsher conditions than those in Novgorod and southern Finland around the same time.[86]

This took place around the same time as the accession of Janībek Khan as the new ruler of the Golden Horde which controlled access to the Black Sea in 1342. This sparked a new round of competition between Genoa and Venice, both of which were keen to have their previous trade concessions confirmed by the new ruler. The antagonism between the two city states was such that as well as seeking to preserve their own advantages, both also sought to undermine their rival, arguing that their commercial privileges be withdrawn. As it happened, both were left unhappy. Following the murder of one of Janibek's high-ranking subjects by a western trader, the ruler of the Golden Horde decided to act against both states. He not only ordered the imprisonment of 'Latins' living in his territories, but also banned trade outright – including grain shipments.[87]

The effect was twofold: first, the blockade of shipping put further pressure on food supplies in the mid-1340s at a time of rising scarcity in Europe and in Italy in particular when harvests were badly affected by higher-than-usual levels of rainfall and also a period of cooler weather than normal.[88] Second, the embargo on grain movements inadvertently prevented the transmission of pandemic disease that would have been carried unwittingly in the cargoes of the ships and perhaps on the bodies of those who sailed them. As such, when the Golden Horde agreed peace terms with Venice and Genoa in 1347, it did not just unlock trade, it unleashed death.[89]

It has long been assumed that plague had little or no impact in China. This is now being challenged by close reading of complex sources whose significance has long been overlooked by scholars.[90] As for Europe, the Middle East, North Africa and likely parts of sub-Saharan Africa, the evidence points to wide-scale disaster: an estimated 40–60 per cent of the population perished of disease in what remains the 'highest [mortality] of any large-scale catastrophe known to humankind, save for the impact of smallpox and measles on indigenous peoples in first-contact events of the early modern period'.[91] So great was the demographic and economic collapse that metal production and atmospheric lead dropped to undetectable levels in ice-core analysis – the only time they have ever done so in at least the last 2,000 years.[92]

'The year of 1348 has left us alone and helpless,' wrote the Florentine poet Petrarch, whose son was to succumb to the disease in an outbreak a few years later. 'Where are our dear friends now? Where are the beloved faces? Where are the affectionate words, the relaxed and enjoyable conversations?' The pandemic was one 'without equal for centuries', so devastating that it had 'trampled and destroyed the entire world'. Those who have lived through pandemics more recently will recognise Petrarch's wistful observation too that 'time itself has slipped through our fingers'.[93]

The Black Death did not spread evenly across countries and states in Europe, with space–time mapping suggesting that the highest levels of virulence in France were in the warm, dry south, likely propagating from the port of Marseille, while what is now Belgium and the Netherlands were spared the worst ravages of the plague, at least in its first waves.[94] Limited coverage of plague in Polish chronicles, along with payments of tithes to the papacy that remained constant throughout the 1340s and 1350s, have led some to suggest that mortality levels were low or even negligible – and to question the impact in other regions too.[95] Some places were saved by steps taken by the authorities to force urban populations to self-isolate. In Milan, for example, immediate action to implement preventive measures led to only three households being infected – sparing the rest of the population as a result.[96] Cultural, ecological, economic and climatic factors explain why the Black Death proved devastating in many parts of Europe but left others unscathed.[97]

The first accounts of the arrival of plague in North Africa describe a disease 'worse than any seen before in the Islamic world'. It arrived in Alexandria in 1347 on a ship full of corpses drifting into the harbour, manned by a handful of surviving crew who themselves all died shortly afterwards. The pandemic spread through the city like wildfire, killing hundreds every day before reaching other towns and cities in Egypt – including Cairo, which recorded 7,000 daily deaths at its peak.[98]

Part of the problem was that cities were ideal crucibles for the spread of disease, because people lived in close proximity to each other, because of inadequate sanitation standards and because of the density of rodent populations that helped spread the plague bacterium. It is instructive, for example, that basic widening of streets, repair of buildings and the move away from wood for housebuilding in the United States in the twentieth century had an immediate effect on the size of rat colonies, producing a decline of almost 80 per cent in just two years: in other words, the layout of cities had and has a direct relationship with disease environments and as such provides an important insight into outcomes during pandemics. While the idea of filthy houses and poor hygiene standards dominate the popular imagination (and are often taken for granted by medieval historians), many cities took seriously the challenges presented by rising populations and infrastructure needs and invested heavily in creating preventative health programmes and devising ways to deal with domestic waste disposal, burials and artisanal pollution, not to mention maintaining wells, canals, bridges and roads that were of benefit to citizens individually and collectively.[99]

That the Black Death overcame many of these measures – which included procedures to contain or stop the spread of infectious disease – speaks volumes about the robustness of the pathogen and its ability to find new hosts. The decline in the number of craft workshops and textile makers in Alexandria in the 1340s and the decades that followed suggests that around half the population died as disease swept through the city.[100] In the case of Cairo, meanwhile, changes to the city were profound, reminding us that demographic disaster has an impact on the built environment and on the way that buildings are used, maintained and constructed.[101]

Socio-economic background was a factor in Cairene mortality rates during the pandemic, with those from poor households more likely to die than those with access to medical care, better levels of nutrition, more space and better domestic hygiene levels. It was the poor who suffered most during the Black Death, then, with particularly high fatalities among those who sought refuge in cities to escape trauma or simply to find food in the face of rising prices. Those drawn from the countryside to Cairo died in large numbers, as did newcomers to Dijon in France, who were more likely to die than those who already lived in the city or who had moved there before the outbreak of disease.[102]

The movement of people in the face of adversity must have played a part in the acceleration of the spread of the plague, just as in later periods pilgrims travelling to and from holy places were crucial in spreading disease.[103] Changes to land use also seem to have been an important reason why the pandemic spread in many parts of Europe: the rapid urbanisation and demographic growth of previous centuries brought ecological pressure not just in terms of exploitation of agricultural resources, but also in land-use practices and in particular the adoption of the open-field system. Even modest changes could have important outcomes. For example, in modern land-management practice, hedges are often removed to create efficiencies for crop planting. This comes at the expense of numbers and diversity of birds, and also causes an increase in vermin populations. While little is known about how hedges were maintained during the fourteenth century, if they were at all, such issues may have played a role in helping the Black Death maintain a lethal velocity, as it kept finding more animals and humans to infect and kill.[104]

On top of all this, the cocktail of catastrophe was enhanced by the fact that the 1340s was one of the most turbulent decades in global military history, not only in Europe but in the Middle East and North Africa. This was a period that saw bitter warfare between England and France, and between Scotland and England; relations between Florence and Pisa and between Genoa and Venice were not just fractious but violent, marked by repeated military confrontations; the Byzantine empire was disrupted by a brutal and long-running civil war, to say nothing of the rising threat of the Ottomans whose raids and attacks were a constant feature in this period; to the east, competition between rival descendants of Činggis Khan saw the Golden Horde to the north of the Black and Caspian Seas take on the Il-Khanate masters of Persia and the Levant, who were also involved in a protracted confrontation with Mamluk Egypt. In normal circumstances, the demands made on manpower, resources and food consumption for military forces in each of these cases would have caused strain. In a period when multiple proxies point to more unsettled, more variable and less predictable weather and therefore to harvest uncertainty, these combined to create a deadly context which ensured that a virulent pathogen circulated commercial disruption, economic compression and political turbulence. Each served to magnify the problems caused by the others, producing a downward spiral of which large-scale mortality was just one feature.[105]

Other features included major cultural changes as a result of concern about health and life expectancy, the loss of loved ones and attitudes to spending money rather than saving it, made worse by the repeated waves of plague that served as brakes on recovery of all kinds and as a near constant source of anxiety for those who had avoided death and suffering previously.[106] Waves of horror accompanied those of plague that fanned out across Germany in the late 1350s, just as life seemed to be getting back to normal.[107]

People in Scotland had heard reports of 'the foul death of England', and 'declared that it had befallen them through the revenging hand of God'. The disease struck the Scots in 1349 just as they were gathering 'with the intention of invading the whole realm of England'.[108] It was the first of four devastating waves that swept away huge numbers of people – with one-third of those who had survived the first outbreak dying in 1362–3, a third of the remainder perishing in 1380 and yet more in 1401. And of course, while naturally drawing most attention, plague was just one disease among many. In 1402, for example, dysentery spread across eastern Scotland, causing many to die 'with the overflowing of the bowels'.[109]

Cycles of plague outbreak repeatedly scarred Egypt, with deaths in 1403, 1407, 1412, 1430 and 1460 running into the thousands in each case, and sometimes the many tens of thousands. 'Most of the cities of Upper Egypt have been obliterated,' wrote al-Maqrīzī of 1412; 'Cairo and its outskirts have lost half their wealth. And two-thirds of the population of Egypt has been wiped out by famine and plague.'[110] Plague recurred repeatedly in cities in the Ottoman empire such as Salonica, Aleppo, Damascus and Trebizond for the best part of 500 years – which helped build up ideas of the Ottoman world as 'a sickened land and peoples' or 'the sick man of Europe'.[111]

The heavily Eurocentric slant of medieval history in general and of accounts of the Black Death in particular can make it all too easy to ignore the impact that plague had outside European cities and societies. As one leading historian has stressed, current knowledge about the plague in East Asia, South Asia, Central Asia, the Middle East and North Africa 'is at best fragmentary and disconnected'.[112] This is problematic, not least since genetic evidence reveals that plague found in sub-Saharan Africa today descends from the strains that caused the Black Death and suggests that the disease took hold and may be responsible for the abandonment of settlements and apparent population losses of as much as 50 per cent in West Africa. It may further explain cultural shifts, such as the sudden prominence in Ethiopia of St Sebastian and St Roch – saints closely associated with plague in Europe.[113]

The limited research done on its consequences beyond Europe mean that it is not easy to assess the damage in locations that have been poorly studied, especially in this period. And even in regions where written, archaeological and secondary material is more abundant, there can be a temptation to overlook how damaging the pandemic was in collapsing connectivities established in the aftermath of the Mongol conquests. Cities in territory controlled by the Golden Horde, for example, were hit badly by plague – such as Caffa in the 1340s, one likely entry point of the disease to Europe, or Saray in 1364.[114]

The decimation of the labour force through disease must have affected agricultural production across the steppes, as well as having an impact on harvesting and on rearing of livestock. These pressures clearly had implications not only for trade but for supply to cities that depended on crops, meat and dairy products as well as the textiles and fabrics that were sourced from nomad populations. It has been convincingly argued that the creation of a new mint and new coinage types and the later introduction of new lightweight silver coinage were responses to acute economic pressures that arose from the spread of pandemic disease in the Eurasian steppes.[115] Another indication of stress was the soaring price of peoples enslaved by the Mongols and sold on to Italian buyers who were desperate to find labour in the face of demographic collapse in the 1350s. The fact that many were sold by members of their own family shows how desperate conditions were around the Black Sea at this time.[116]

As one scholar puts it, the Golden Horde seems to have been 'an orderly, stable, well-functioning political system' until 1359; at that point 'the only word one can use to describe it ... is "anarchy"'.[117] Competition for power and control became intense, reflecting the fact that there were fewer rewards and benefits available than in the past, and that resources were dwindling at speed. If vicious infighting provides one indication of growing crisis, the rapid expansion of other powers into Mongol territory provides another: in the second half of the fourteenth century, Lithuanian forces pushed south dramatically, taking swathes of lands and cities, and by 1400 extending control right up to the shore of the Black Sea.[118]

This was a rare case of opportunism. More common was the collapse of long-distance travel and trade, and a narrowing of horizons as societies turned inwards. Even before the pandemic, there had been signs of change across parts of Central Asia. In the 1330s, a wave of conversions to Islam marked a move away from the cosmopolitanism and laissez-faire policies of inclusivity which may not have characterised Mongol society but were certainly an important part of it.[119] This now accelerated in the period after the outbreak of the Black Death as the steppe world went through a period of socio-economic contraction. The disappearance of languages such as Syriac, Volga Bulgarian (related to modern Chuvash) and Standard Turkic (linked to modern Kazan Turkic) in the second half of the fourteenth century is most logically explained as a sign of a retreat to a single common identity at a time of pressure and duress.[120]

In Egypt, the Black Death had a devastating impact on the labour force. The problem centred on irrigation: vast amounts of manpower were required to maintain the canal system. Several accounts from the late Middle Ages put the number of those required to work on regional, large-scale irrigation in Upper and Lower Egypt well above 100,000 people, with many more involved in local community schemes – with perhaps around half the male population between the ages of fifteen and fifty participating on a seasonal basis to clear canals of silt, remove blockages and shore up dykes. Sharp population loss through disease had obvious implications for villages, let alone for more substantial networks of water distribution. In some cases, irrigation canals did not just fall into disrepair but became unusable.[121]

Not surprisingly, in all areas and regions that had been struck by the plague, landowners and the previous beneficiaries of the rich surpluses tried to retain their status and to extract the rewards of agricultural production. In England, small emerging elites in the countryside started to build larger houses, set back from the roadway to mark rising status linked to increasing stratification in village life, which was in turn accentuated by the wealthier inhabitants taking advantage of opportunities to enclose and take ownership of open land and fields.[122]

Unlike in Europe, where the aftermath of the Black Death saw the start of a social revolution that stemmed from the successes of the peasantry and the labour force in negotiating better bargaining positions that translated into better living conditions, in Egypt elites were able to protect their own interests with considerable success. This had long-term consequences that were economically regressive, throttling opportunities for social mobility, stifling innovation and creating long-term stagnation. Ironically, the principal cause of this divergence lay in the fact that the Mamluk state in North Africa was more centralised and better administered than fragmented societies in most parts of Europe. In some contexts, this was an advantage; here, it proved to be an Achilles heel that prevented reform and held back progress.[123]

The experience of plague brought about a series of important changes – including biological ones. As it happened, many of those who survived the plague may have done so because they were carrying a genetic mutation that served to offer a layer of protection to the immune system. Ironically, this mutation, known as ERAP2, is now associated with several autoimmune disorders, such as Crohn's disease, lupus and rheumatoid arthritis: in other words, alleles that protected against Black Death confer raised risk for diseases for people alive today.[124] Evidence from burial sites in London suggests that the onset of the menarche (first menstrual cycles) occurred at a younger age for adolescent women – something that can be best understood as a sign of better living conditions after the pandemic.[125]

There were cultural shifts too. For example, there was what some scholars refer to as a 'nuptial frenzy' as those who had survived the plague rushed to get married – resulting in a period of bumper fertility.[126] Great efforts were made to catalogue experiences of plague, including medical tracts explaining the symptoms and effects of the disease, as well as liturgical poetry written by Sephardic Jews recording the suffering – and also the violence – that had accompanied the Black Death.[127] Over time, writings by and for Jews came to suggest Jewish resistance to plague – usually attributed to higher levels of sanitary behaviour and prudence.[128]

The aftermath of plague also saw new fashions, diets and attitudes towards luxury goods, partly as a result of the emergence of competitive consumption, partly because smaller populations meant goods cost less, and partly because of the euphoria of having dodged death's call.[129] Protein availability not only soared in Europe because there were fewer people to feed, but as farmers and herders were forced to find new markets for their livestock – opening up what one historian has called an international meat trade.[130] It has been argued too that the deaths of so many copyists during the pandemic led to the fall in the price of paper – something that spurred literacy levels and perhaps even helped prompt the print revolution of Johannes Gutenberg and others.[131]

Systems to detect plague were also devised and developed, with varying degrees of success. While the Ottoman empire struggled to deal with regular outbreaks, public health authorities in Milan monitored plague cases with extreme vigilance following a devastating outbreak in northern Italy in the 1360s which Petrarch described as not only equal to but worse than that of the 1340s – and which killed the poet's beloved son, Giovanni. A network of early-warning stations reported new cases, with particular attention paid to main travel routes, mountain passes and the periphery of Lombardy. In the first half of the sixteenth century, surveillance systems failed, largely as a result of a long period of inter-city warfare and turbulence which put pressure on expenditure and diverted attention elsewhere. The results were dramatic. Anxiety about plague was not misplaced, as is clear from the death of half the population of Milan in 1524; it took years for the city to recover. Fear of plague bred suspicions, with some refusing to believe reports of new cases or to suspect skulduggery by nurses, gravediggers and others keen to continue collecting 'high, hazard-pay wages' and to exaggerate the situation in order to enrich themselves.[132]

Recovery from the demographic contractions of the mid-fourteenth century took a long time. By 1550, the population of north Italy stood at around five and a half million – a rise of more than 40 per cent from a century earlier. Nevertheless, this was still a long way short of where things had been on the eve of the Black Death.[133] The fact that the setback had not been reversed 200 years later speaks volumes about the scale of correction caused by the perfect storm that struck at the height of the Middle Ages.

There were counter-intuitive benefits for those who survived: the large numbers of deaths meant there were more 'carts, wagons, horses, oxen, mules, boats, ships, barns, and granaries' available per capita – in what amounted to a 'plague bonus'.[134] Ironically, climatic variability, harvest failure, dependence on long-distance trade, intense warfare and a shifting disease environment created conditions that decimated communities across three continents – but often also served as a catalyst to long-term growth. This was especially true in Europe, and above all the north of the continent. For some, the plague did not so much have a silver-lining as serve to change the world.[135]

The unusual constellation of factors that created the Black Death helps explain why seemingly similar periods did not produce similar results. In the mid-fifteenth century, for example, another phase of reduced solar activity known as the Spörer Minimum produced a period of unusual cold in north-western and central Europe, with conditions in the 1430s being particularly hostile.[136] In addition to wine freezing solid in bottles, crops failed in 1432, 1433, 1434, 1436, 1437 and 1438 in England, Germany, France, the Low Countries and elsewhere – leading to price rises not only for food but also for firewood in Paris, Cologne and Augsburg. Manorial accounts from southern England suggest sharp declines in fertility rates of ewes, and mortality rates in sheep multiple times greater in 1438 than in normal years.[137]

Large-scale eruptions in the 1450s of an as yet unidentified volcano and of the Kuwae underwater caldera in the South Pacific, which was more explosive than the massive Tambora eruption of 1815, resulted in lakes in the Alps freezing solid, so that they could be crossed on horseback. We learn of trees in Ireland not growing leaves till May and of celestial phenomena that were much commented on by chroniclers across Europe who noted that the sun had lost its power, that it shone weakly with a blue colour and that fog and haze hung in the air – like the dust veil of the sixth century.[138] The volcanic events left an indelible impression on oral traditions in Oceania, where stories were told centuries later that explained that the eruption of Kuwae was the work of a powerful sorcerer who had been tricked into committing incest with his own mother and had unleashed the cataclysm as revenge.[139]

Extreme weather events elsewhere could and did pose serious challenges. For example, a severe storm in northwestern Europe in January 1362 was so strong that a monk was blown out of a window, falling beams killed people taking shelter in churches and many famous buildings were damaged. As one scholar reminds us, while little consideration has been given to the impact of major storms in the pre-modern era, it is worth noting that an unusually strong storm in October 1987 brought down an estimated 15 million trees in Britain.[140] Storms in central Europe in December 1999 brought down 400,000 telephone and power supply lines, along with 138 million cubic metres of trees.[141] Even though much depends on tree type, wind direction and speed, tree anchorage (including root system, soil and drainage), tree density and height, as well as forestry techniques, the scale of damage that can result from such weather events is enormous and can impact all aspects of the natural environment, from animal and plant habitats to reshaping landscapes – as well as producing socio-economic shifts that result from adapting to major upheaval.[142]

Curiously, then, the centuries that followed the creation of the Mongol empire, the intensification of commercial and cultural ties across Asia, Europe and Africa and the sweeping pandemic that proved so destructive were the prelude to a long and slow period of consolidation in which states and peoples that had been on the geographic periphery of long-established political centres and of the networks that linked them together found themselves able to take advantage of new opportunities not only to expand but to explore new worlds. The results were to change the nature of the world we live in – quite literally.

14

On the Expansion of Ecological Horizons
(c.1400–c.1500)

> The famine was so great that the old Mexicans sold themselves.
>
> *Annals of Chimalpahin* (early seventeenth century)

The aftermath of the Black Death saw the creation of a set of new states, new worlds and new interactions. In Europe, the next century witnessed the twin trends of military and bureaucratic innovation as states fought each other almost incessantly. New technologies and tactics could not fail to emerge during the Hundred Years' War between the kings of England and France which had started shortly before the pandemic, while the need to raise taxes and armies provided a motor that drove both the centralisation of power and the professionalisation of those who ran kingdoms, controlled budgets and chose how to deploy expenditure.[1]

It was a time of change too for those able to take advantage of new opportunities. The dukes of Burgundy, for example, oversaw a rapid political and territorial expansion from the late fourteenth century, forging canny alliances with the kings of England as well as compromises with magnates and cities across the Low Countries – like Bruges and Ghent – that saw the Burgundian state become the commercial and culture powerhouse of the western half of Europe.[2] It was matched in the east by the Grand Duchy of Lithuania, whose military successes even before the great pandemic were followed by momentous victories

such as Kulikovo in 1380, by the conquest of territory as far south as the
Black Sea and by the invasion of Crimea in 1398.[3]

To the south-east too, things were changing rapidly, with a Serbian
empire taking shape under Stjepan Dušan in the 1340s, helped first by
the death by plague of his brother-in-law, the powerful Mladen Šubić,
which opened up a pathway to the conquest of Bosnia and Dalmatia,
and then by two decades of civil war in the Byzantine empire which
left the latter poorly placed to defend its northern cities and territories.
New opportunities also emerged for the Ottomans, who established
a foothold in mainland Europe in the early 1350s, which served as a
springboard for future gains across the Balkans – which opened up
following the battles of Maritsa in 1371 and Kosovo in 1389. The latter
acquired totemic status in later Serbian historiography.[4]

These successes paved the way for the successful assault on
Constantinople in 1453, and eventually took the sultan's armies deep into
the heart of Europe. Success followed success as the Ottomans advanced
deep into eastern and central Europe. By 1529, troops were poised for an
assault on Vienna. The city was saved by the rain: torrential rainfall led
to Ottoman heavy cannon becoming stuck in the mud. Had the sultan's
forces moved two months earlier – or if the weather had stayed benign –
Vienna might have fallen and the history of Europe would have been
dramatically different.[5]

As it was, the Ottoman expansion into Europe paid dividends of a
different sort, for the presence of an existential threat to Christian states
across the continent concentrated hearts and minds: Ottoman military
expeditions not only resulted in fewer violent engagements between
kingdoms in Europe but also shortened conflicts by more than 50 per
cent from the early sixteenth century at least until around 1600. This
was remarkable given that the Ottoman surge into Europe coincided
with the Reformation and the splintering of Protestant and Catholic
communities. The military prowess of the Ottomans and the momentum
behind the expansion of Islam and of Turks challenged ideas about the
role of the Church and the moral rectitude of leaders, while the threat
of further conquests provided grounds for co-operation among those
most at risk. Ironically, then, the Ottomans were instrumental in driving
the cohesion that enabled Protestantism to establish itself in a way that
previous reformist movements had failed to do, a process that had a
profound impact on the religion, politics and economy of Europe.[6]

The Ottomans expanded elsewhere too, taking on the Safavids of Persia whose ruler, Shah Ismail, was renowned for his moustache, for being 'as brave as a game cock' and for being an excellent archer. Ismail was a man who spent most of his time 'hunting, or in the company of rosy-cheeked youths, quaffing goblets of purple wine, and listening to the strains of music and song'.[7] A lightning campaign in 1514 against the Safavids resulted in a crushing defeat and led to the fall of Tabriz, a city famous for its magnificent library, its miniature painters and its expert carpet makers. This not only removed the threat to the Ottomans' eastern flank, but also freed up resources to enable an attack on Egypt, which duly fell with a whimper in 1517. The result was that the Ottomans now controlled a global empire that spanned three continents – Europe, Africa and Asia.[8]

The Ottoman encounter with Europe was part of an Age of Discovery that historians typically link with European voyages across the Atlantic and to Asia at the end of the fifteenth and start of the sixteenth centuries. This says much about how study of the past prioritises European experiences instead of providing a wider, more balanced and more meaningful context. In fact, the Ottoman field of vision expanded rapidly in this period to the east, as well as into Europe. The conquest of North Africa opened up maritime routes stretching not only into the Red Sea but into the Indian Ocean World – a web of vibrant connections that had meshed South-East and South Asia, the Gulf and East Africa, as well as reaching beyond to China, Japan and Oceania for centuries.[9]

This engagement was accompanied by rapid and almost continuous expansion along the north-western coast of the Indian Ocean, including the occupation of Aden, Mocha and Basra – which opened up new gateways to the east – as well as conquest of the coasts of Sudan and Eritrea.[10] Although control of many of these new regions was often precarious and dependent on co-operation with local rulers, these advances went hand in hand with sharply increasing interest in and the accumulation of information about peoples and places as far afield as southern India, Sri Lanka and Melaka.[11]

Horizons were being recast elsewhere, such as the Solomonic dynasty in Christian Ethiopia, whose extension and consolidation of authority and power locally continued apace. By the start of the fifteenth century, its *nägäst* – or kings – ruled over the largest political entity in the Horn of Africa in this period, and began to dispatch diplomatic missions to Venice, Rome, Valencia, Lisbon and other cities. These were not intended to seek European technology, solicit military support or show submission, as has long been assumed by western scholars, but rather were confident displays of an ambitious and successful power that saw itself as at least the equal of contemporary societies elsewhere.[12]

Nowhere was this clearer than in the 1440s when Pope Eugene IV invited 'Prester John, illustrious emperor of the Ethiopians' to attend the Council of Florence, a landmark event that was intended to resolve schisms that had ruptured the Christian Church and that in some cases dated back not just centuries, but a full millennium. The initial response, composed in Gəʿəz, the ancient language of the Ethiopian Church, to emphasise independence and autonomy, was curt to the point of being dismissive. Although a delegation led by a certain Petros duly met the Pope at the Council in 1441, his message, likely delivered in Amharic, then translated into Arabic and then into Latin, left little to the imagination: the Ethiopian Church had not 'crumbled' as others had in the face of intimidation, heresy or poor leadership. That was no thanks to the papacy, he went on, pointing to 'the negligence of your predecessor Roman pontiffs' who had shown no concern for 'the sheep of Christ', let alone deigned to visit over the past 800 years, even though the first followers of Christ had done so. These were emphatic words that reflected the self-confidence of a state that could draw on the deep past to frame expressions of its power.[13]

This was a time of discovery for others too. Long before the age of European navigation across the Atlantic, sailors in Oceania were making regular return journeys that in some cases covered thousands of kilometres in double-hulled sailing canoes, leapfrogging across what one leading scholar has called a 'sea of islands' that connected tapestries of peoples, ecologies and resources. Rising levels of exchange in the Indian Ocean world, including along the coast of East Africa, also characterised a period of intensifying contacts from around the middle of the fourteenth century.[14]

In East Asia, constant environmental pressure from the 1320s onwards had produced severe famine, epidemic disease and drought so bad that land in Anhui province in China split 'with cracks like the back of a tortoise'.[15] Uprisings gathered pace against the Yuan dynasty, whose emperors, one poor peasant named Zhu Yuanzhang complained, were 'weak and failed to do their jobs'. Galvanising support among those who could barely survive because 'food was scarce, with grasses and bark serving as our nourishment', by the 1350s Zhu had become a leading force in a revolt that was gaining pace. By 1368, he had deposed the Yuan and become the master of the empire himself – and the father of a new dynasty which he named Ming, or 'Brightness', while taking for his own imperial name Hongwu, or 'Magnificent in War'.[16] This provides another example that has caught the eye of historians seeking to correlate environmental change with the fall of dynasties in China.[17]

Whereas the Yuan and the Song dynasty before them had sought to promote and prioritise long-distance maritime trade, the new Ming rulers took a very different approach. In 1371, just three years after the Hongwu Emperor seized power, a ban was imposed on seaborne commercial traffic, forbidding those living in coastal regions from setting sail in any capacity whatsoever. Foreign trade and contact with people beyond China were to be reserved for the imperial court and its officials alone.[18]

This did not mean that Ming dynasty China cut off its engagement with the world beyond. In fact, in the first half of the fifteenth century, considerable energy and expense went into equipping fleets under the command of Zheng He that were intended to be statements of Ming power. Reports written by Ma Huan, who accompanied several of these missions, provide rich detail about the rituals, wealth and interactions of interlocking political and commercial centres along the Malabar and Coromandel coasts of India and also about what is now Sri Lanka. In the latter, he wrote, people flocked to scour the sandy beaches whenever it rained to search for precious stones, which were thought to be nothing else than 'the crystallised tears of the Buddha'.[19]

Like other senior commanders on these expeditions, Zheng He was Muslim – likely a consideration for appointment for voyages that set out into regions that were experiencing a wave of Islamicisation. Muslim gravestones had become increasingly common across South-East Asia in the course of the fourteenth century, perhaps spread by refugees coming from Champa during a time of upheaval to places like Surabaya, Gresik and Cirebon in Java, each of which became what one scholar has called 'early centres of Islamic propaganda'.[20]

The Yongle Emperor, who dispatched Zheng He, issued orders in 1407 to guarantee the safety of Muslims within Ming dynasty China itself, proclaiming that 'no official, military or civilian personnel shall despise, insult or bully [Muslims] and whoever disobeys this order by doing so shall bear the consequences'. While this points to the extent and speed of the spread of Islam in the fourteenth and early fifteenth centuries, it can also be understood to demonstrate the challenges posed by the arrival of newcomers, new ideas or both.[21]

Although the fleets also reached East Africa, South-East Asia was a particular focus of the expeditions' attention. Ma Huan also wrote at length about the Majapahit kingdom, which was centred on the island of Java and knitted much of South-East Asia together. It had experienced a golden age of cultural brilliance in the fourteenth century under the Hindu King Hayam Wuruk that is richly evoked in a text known as the *Nagarakṛtāgama*; this records the building of temples and palaces and the rituals undertaken to honour ancestors and to mark phases of the moon, as well as painting a vivid picture of the arts, crafts and handiworks being carried out by skilled artisans making goods for their customers. In addition to diplomatic exchanges with Ming China, the Majapahit kingdom built up contacts with dozens of cities and states stretching from southern Thailand through the Malay peninsula, across Sumatra through the Sulu archipelago and into the Pacific Ocean. These kings may not have been quite the vassals that the author of the *Nagarakṛtāgama* seems to claim, with the payment of tribute not necessarily a reliable indicator of submission. What is less in doubt, however, is that this was a time of widening horizons and perspectives.[22]

It was a similar story in South Asia several decades later, where Bābur, the founder of the Mughal dynasty, in 1505 marched south into India for the first time, seeking to expand his authority, to acquire new riches and to undergo fresh experiences. 'I had never seen a hot climate or any of Hindustan before,' he wrote, 'a new world came into view – different plants, different trees, different animals and birds, different tribes and people, different manners and customs. It was astonishing, truly astonishing.'[23]

This did not mean that he was entranced by what he saw: India, wrote Bābur, 'is a place of little charm. There is no beauty in its people, no graceful social intercourse, no poetic talent or understanding, no etiquette, nobility, or manliness. The arts and crafts have no harmony or symmetry. There are no good horses, meat, grapes, melons or other fruit. There is no ice, cold water, good food or bread in the markets. There are no baths and no madrasas. There are no candles, torches or candlesticks.'[24] As others found in this period, encounters with new worlds were often uncomfortable, prompting unfavourable parallels with familiar places that were left behind.

It is perhaps not surprising, then, that the global climate records for the second half of the fourteenth century and the first decades of the fifteenth are conspicuously quiet: it seems that this was a long period of stability with few episodes of upheaval or major shocks. While demographic losses to the Black Death and the associated lowered usage of the land might explain why this was a time of continuity in Asia, Europe and Africa, the fact that similar patterns can be identified in places unaffected by the devastating plague pandemic provides an insight into the importance of multi-decade spells of stable conditions.[25]

It is striking, for example, that in Polynesia, chiefs from the Tongan island of Tongapatu began to centralise authority from the early fifteenth century, sparking social change that resulted in commoners reduced to tributary serfs and a small group of elite leaders emerging under the Tu'i Tonga, the supreme ruler. As new forms of monumental architecture developed in this period, such as mounds, altars and a massive complex of stone tombs, other islands across the region saw transformations of their material culture and rising levels of imports – most notably of stone. Oral traditions likewise commemorate the emergence of new chiefly lineages as well as the adoption of new rituals around this time.[26] Although the way to understand these developments is subject to debate, they are most logically explained as an expansion of Tongan culture, if not direct political overlordship.[27]

How these changes related to developments in Hawai'i, which had been first settled around AD 1200, is not clear: however, it seems that while voyages between these islands and central Polynesia had been relatively common, they stopped around the same time as the Tongan expansion. This may have been entirely coincidental; whatever the case, the Hawai'ian archipelago became isolated from the rest of the world and remained so until the late eighteenth century when an expedition under James Cook reintroduced the islands into global exchange networks.[28]

The case of Hawai'i was unusual, for the patterns elsewhere were ones of integration and consolidation. By the early fifteenth century, for example, a new empire was being forged by the Mexica (often referred to as Aztecs), who emerged as the dominant power in central Mexico through a combination of military success and skilful political diplomacy.[29] The world of the Aztecs rested heavily on control of water resources, evidenced by dense networks of canals not only in the principal city of Tenochtitlán which was located on an island in the great lake of Texcoco, but also right across central Mexico. These waterways greatly facilitated movement of goods and people over long distances.[30] Water was a core part of the identity of the peoples of this rising empire, recorded in pictographic records, in song and in belief systems.[31]

Management of water resources was also of crucial importance to the long-term viability of the Aztec world as a whole, as well as to Tenochtitlán itself, whose population perhaps reached 200,000 at its peak. Major investments in terracing and irrigation were vital for supporting such large numbers and densities.[32] These were also responses to threats posed by ecological conditions in the Valley of Mexico, which was subject to the twin risks of early frosts and late rains.[33]

Although Aztec rulers tried to ensure that reserves were kept in royal granaries to protect against weather shocks and food shortages, there were occasions when not even these proved enough. The so-called Famine of One Rabbit in the middle of the fifteenth century during the reign of Moctezuma I (c.1440–c.1469) was the result of almost a decade of climate turbulence, including bad frosts, droughts, locust plagues and crop failures, which overwhelmed food stocks. This led to starvation, disease and large-scale migration from cities including Tenochtitlán, Texcoco, Chalco and Xochimilco.[34]

In 1454, according to the *Annals of Chimalpahin*:

> the people died of thirst. From Chalco came foxes, ferocious beasts, lizards, etc., and they devoured the people. The famine was so great that the old Mexicans sold themselves; they took refuge in the woods where they lived unhappy and feeble. For four years there was nothing to eat in the country, so that the older Mexicans sold themselves and two divisions, it is said, delivered themselves into servitude ... They put themselves into holes and died in any place they found and the lizards devoured them for there was no one to bury them.[35]

Some scholars have argued that the famine was instrumental in prompting a further wave of Aztec expansion, which might explain why priority was given to the annexation of Totonacapan, which had not been so badly affected by the disasters of the 1450s.[36] Whether or not linking the Aztec conquests to the traumatic experiences which also included children being sold into slavery is plausible, it is certainly striking that there were major upgrades to dykes, canals and aqueducts in the aftermath of the disaster. Not surprisingly, considerable effort went into increasing the amount of land that could be cultivated in order to ensure there was a bigger buffer in the future. The intensification of agricultural production was an obvious and logical response to a catastrophic sequence of events.[37]

How these difficult conditions played out in other parts of Central America are more difficult to evaluate. Drought in Yucatán appears to have played a role in prompting civil conflict and political collapse between c.1441 and 1461 at Mayapan, the largest and most important city in the peninsula following the demise of Chichen Itza four centuries earlier. While lack of rainfall posed its own problems, the compression of trade links, reductions in mobility and rising levels of violence were what drove change: even then, however, the re-emergence of a resilient network of small states shows that the failure of some created opportunities for others.[38]

In South America, the age of discovery saw the Chimú bring more territory and peoples under their control from the start of the thirteenth century. The heart of the Chimú state was the city of Chen Chen in the Moche Valley, whose population of 30,000–40,000 drew in agricultural produce from the surrounding region and beyond. Chen Chen was home to palaces, three-sided constructions called *audiencias* that provided living and working quarters for the Chimú nobility and their families, and to plazas and temples. The city was located close to the Pacific Ocean, alongside wetlands, fields, desert and mountains.[39]

By around 1310, the Chimú had moved into the Jequetepeque Valley to the north and had taken the strategically important location of Farfán by force before consolidating control over the Casma Valley to the south, where a new provincial capital was built at Manchan, at the intersection of major routes running north–south and also to the east.[40] Like many societies in the past, human and animal sacrifices were regular features in the Chimú world, given as offerings to the gods in a range of contexts. Around 1450, a mass sacrifice was made of more than

140 children and 200 camelids (llamas and alpacas) at Huanchaquito–Las Llamas on the western coast of Peru, just north of Chen Chen. Stratigraphic evidence from the site suggests that the sacrifice was made following heavy rains or perhaps a flooding event which has recently been connected to a weather event, likely associated with the El Niño–Southern Oscillation which disrupts the marine food chain in northern and central Peru and brings about coastal flooding. It seems plausible, if not likely, that the offering of young boys and girls was made in an attempt to appease the adverse weather and to bring about a return of more clement, hospitable conditions.[41]

Sacrifices that were intended to respond to climatic and natural disasters can also be found elsewhere in South America from around the same period. An archaeological expedition to the summit of the Misti volcano led to the astonishing and spectacular discovery of the bodies of several children ritually killed and buried alongside dozens of figurines made of gold, copper, silver and spondylus shell – as well as ceramic and wooden vessels. The pre-Columbian Andes peoples held belief systems that articulated ideas about the landscape and about ancestors and ethnic identity, and involved a sacrifice ritual known as the *capacocha* ceremony in which prestigious goods and human sacrifices were presented as offerings to the gods. Important moments, such as major disasters like bad weather that affected crops, were taken as signs of angered deities that could be calmed only by a major offering. The context for and date of this particular ritual sacrifice is not clear, although some have linked it to a major eruption of the Misti volcano and to the climatic dislocation that likely followed.[42]

The Inca too proved highly successful as empire builders from around AD 1400, succeeding in creating the largest empire in the history of the Americas before the European colonisations. They did so through a combination of alliance-building which included copious consumption of *chicha* (a maize-based alcoholic drink), intimidation and selective use of force – eventually overcoming the Chimú themselves around 1470.[43] The rise of the Inca owed much to the warm weather conditions which opened up the potential for farming at higher altitudes, not least around their capital city of Cuzco – as well as to the fact that climatic patterns remained consistent for long periods. Investment in the construction of agricultural terracing, which was so important in highland societies, was not only a hallmark of Inca success but underpinned its resilience;

the same was true of the development of agroforestry techniques and the cultivation of a suite of crops that produced exceptionally high levels of seed protein and oil content.[44]

The Inca also developed regional exchange networks which were of central importance to ideological, cultural, economic and political control through the monopoly of acquisition and redistribution of long-distance trade goods, metals, ceramics and luxury objects. This sometimes took place through co-opting newly incorporated local elites who were able to benefit from Inca overlordship, sometimes through a light touch that left sacred sites undisturbed (such as the Yumbo site of Palmitopamba), but also sometimes through direct control.[45]

Perhaps the most important – and enduring – feature of Inca rule and empire was the enormous programme of road building to create a royal highway (*qhapaq ñan*) using labour tribute; covering an estimated 30,000 kilometres of interconnected coastal and highway routes, it was punctuated by way stations (*tampukuna*) and relay-runner posts (*wasi*).[46] The aim was to improve the speed and lower the cost of moving caravans, personnel and information between the provinces and the capital, thereby improving levels of control as well as stimulating trade.[47] The increase in the vibrancy of exchange networks and in activity more generally can be identified by the sharp rise in mite abundances from the mid-fifteenth century that are suggestive of the increased density of large herbivores, above all llamas, that were being used as part of trade caravans in the Patacancha Valley and beyond.[48]

The importance of this road network was not limited to the period of the Inca empire: its role in shaping subsequent economic development after the Spanish conquest of much of South America and later has been such that it boosted living standards, including average hourly wages, nutrition and educational achievement – with a particularly pronounced effect on women. Land near the roads rose in value as a result of the ease of accessibility and lower costs of travel and trade. This not only had long-term consequences in the Inca world and in the period of colonial rule – but beyond as well: locations close to the 'Royal road' of the Incas have today high levels of property rights and protection, more schools, greater literacy levels, and better levels of child nutrition than those further away.[49]

This has parallels with the web of the land-based Silk Roads that criss-crossed the spine of Asia. Again, locations close to primary routes developed significantly higher levels of economic activity than those situated at distance, as well as greater rates of intergroup marriage that can be identified across many centuries – right up to today.[50] The identification of such long-term impacts is revealing, instructive and, as we shall see, crucial for understanding even broader patterns of inequality which are global in their nature and which allow us to see just how far the world of the past has shaped the world of today.

Taken as a whole, the fourteenth and fifteenth centuries were a time of transition, a series of new ages of expansion and discovery – at least for those who were able to impose themselves on others. Of course, things looked rather different for cultures and societies that found their customs and lifestyles swept away – whether by conquest or by accommodation. Cahokia had been home to a thriving culture, based on agricultural exploitation of the rich, fertile soils of the American Bottom region. By around 1400, however, it had fallen into decline. Many explanations have been suggested for the dispersal of population and the abandonment of the central site located near the modern city of St Louis that was based around more than 100 burial mounds. These hypotheses include climatic change that impacted rainfall levels and flood frequency, soil erosion and deforestation due to overexploitation, and unsustainable levels of pollution caused by poor sewage clearance.[51] Each and all of these use archaeological and environmental data (including faecal matter, pollen records and tree ring evidence) to build a picture of ecocide – that is to say, a picture of resources being depleted to the point that collapse was inevitable.

In the case of Cahokia, what is perhaps most important is that use of the political centre declined over many decades, centuries even, suggesting that there was not a sudden point of no return or a conflation of factors that caused dramatic change.[52] Rather, whatever looks like a decisive moment of no return around 1400 was in fact the result of a long and drawn-out process of socio-economic and cultural changes. The most significant of these was the decentralisation of power, the emergence of local elites within Cahokia with distinct residential areas and their own mound plazas, and the resulting rise of intergroup identities, rivalries and antagonisms. While ecological exhaustion and climatic shifts cannot have helped, the key to long-term abandonment may have been tensions and conflicts between different groups that created incentives to seek new opportunities elsewhere.[53]

The same seems to have been the case in southern Brazil where previous robust trade networks sagged and subsided, just as they did in southern Africa where repeated droughts starting in the late thirteenth century seem to have been at least one of the factors in the migration of the Hima and Tutsu peoples towards Lake Victoria and of the Luo towards the Nile. There were more large-scale movements of people a century later, at a time when the Nile in Cairo was so low that it could be crossed on foot.[54]

In other cases, change came through the arrival of new overlords. An obvious case in point is Timur, the most prominent of the successors of the Turco-Mongols, masters of much of Asia, who brought swathes of territories, cities and peoples under his dominion in the late fourteenth and early fifteenth centuries. Like all empires, Timur's centralised resources drew labour, funds and expertise from all corners to support and adorn the political heart, in this case enabling the creation of architectural jewels – like the magnificent Bibi-Xonim mosque, the Ulugh Beg madrasa and the mausoleum of Timur himself, all located in Samarkand in what is now Uzbekistan – to say nothing of the art, literature and scientific scholarship that went hand in hand with imperial patronage.[55]

Conquest, expansion and change of dynasty did not always create centripetal forces, for they could also could send shock waves that resulted in fragmentation. This was the case with Timur's brutal sack of Delhi in 1398, which served as a catalyst for the splintering of a sultanate that had itself built up a position of control over much of the Indian subcontinent. Its collapse sparked a blossoming of city states and the intensification of spheres of regional interaction as tax revenues that previously flowed north were now reallocated locally. Delhi's loss was Gujarat's and Orissa's gain; the Malabar and Coromandel coasts benefited in the same way.[56]

Likewise, the recoil of the Mongols after the fall of the Yuan dynasty in China opened opportunities for others. In the 1390s in Korea, General Yi Seong-gye imposed a major land reform after seizing power that was centred on trying to distribute land more equitably than in the past. As had been the case at Pagan, one of the problems had been the concentration of land ownership in the hands of wealthy families, as well as the creation of large-scale endowments by Buddhist temples and minorities. 'In ancient times,' noted one of the new ruler's senior advisers, 'all the land belonged to the state, and the states then granted land to the people ... There was no one who did not receive land and there was no one who did not cultivate land.' Yi – who took the name of Taejo when he seized power – dramatically burned all the existing land registration records and set about a wholesale redistribution of land and favours.[57]

While some of the motivation naturally came from the desire to weaken rivals and potential rivals, another aim was to boost food production and to make more intensive use of the land. One of Taejo's grandsons, King Sejong 'the Great', took considerable personal interest in this cause, ordering the compilation of a farm manual *Straight Talk on Farming*, which stressed the importance of sowing at the right time, the correct use of effective fertiliser and the need to weed regularly, and explained how to plough during the autumn months. 'Farmers are the roots of all the nations of the world,' states the preface to the text. Moreover, it goes on, 'because all food and clothing are produced by farmers, our government must give them foremost priority'.[58]

The Lê dynasty in Đại Việt adopted a similar view after taking power in 1428 – an event celebrated with a 'Great Proclamation on the Defeat of the Ming' which declared that 'Now our Great Việt is truly a cultured country', one with its own customs, landscape and people.[59] As in Korea, a new code set out measures to protect peasant landholdings as well as public lands, with 'eighty strokes of the heavy stick' ordered as punishment for transgressors. Steps such as these were not simply acts of benign rulers – indeed it could be argued that they represented the opposite, seeking to ensure that all taxes passed directly to the crown rather than to those who might be able to conceal their assets or resist the demands of the fisc. It is noteworthy, then, that eighty strokes of the heavy stick could be administered to anyone who 'cultivates public rice land but does not make grain payment within the time limit'.[60]

As one leading scholar of the European Renaissance makes clear, by the end of the fifteenth century, perspectives had widened dramatically. Paintings were being created in Bruges that showed rugs from the Ottoman empire, metalwork from Islamic Spain and porcelains and silks from China. Pictures like the Arnolfini portrait by van Eyck showed not only the scale of trade networks that linked continents together, but also the competitive urge to acquire – itself a precondition for growth and for the presentation of status. This famous painting might be viewed as the record of a pair of individuals, but it also represents a 'celebration of ownership', not only of the picture itself but of the many lavish and expensive objects depicted.[61]

This proved to be important from an ecological and environmental perspective. The connections that had linked Asia, Africa and Europe together had not been driven by royal policy or command, except of course during phases of conquest and expansion. While it is true that such periods could lead to increases in regional and interregional trade, the principal outcomes more usually resulted in nets being cast by the political centre to drag resources towards it, rather than in the facilitation of exports. There were exceptions, such as in Song dynasty China and even during the early decades of the Mongol empire. By and large, however, the driving force behind increased levels of exchange came from merchants, from middlemen, from shipowners and from those in oasis settlements and towns whose populations were well supported by their hinterlands and well located for benefiting from the sale and resale of commodities and goods that were rare, expensive and desirable.[62]

These were the seeds of what some historians have called the capitalist world economy that gathered strength in the eighteenth century and beyond.[63] Between 1500 and 1800, global trade expanded at an unprecedented rate, growing on average by 1 per cent a year: by 1800, twenty-three times as many goods were being carried on ships as had been the case 300 years earlier – a reflection of course of increased levels of exchange, but more pertinently of the impulse to acquire, demonstrate and accentuate status, and above all of the deployment of merchant capital to respond to and feed these desires.[64]

That such demands placed strains on ecosystems had been understood even in antiquity. The point was being articulated ever more forcefully and regularly by 1500. Long before that, depletion of forests in Germany had become so acute that protection was the rule and special permission was required to clear woods. In the 1230s, Archbishop Eberhard of Salzburg expressly forbade the transformation of forests owned by the Church into arable land, so that trees could grow again.[65] A few decades later, King Henry VII of Germany declared that 'harm had come to him' and his realm as a result of 'the destruction of the forest of the kingdom' and its reuse as cultivated land – which he regarded as a calamity.[66]

By the early fifteenth century, Venice was taking steps to try to ensure control of wood supplies from the Italian mainland, recognition of the vulnerability that states could experience as a result of overconsumption of sustainable resources. With no forests of its own, the maritime city state was vulnerable to shortages – as well as price inflation. The latter was expensive; the former was hard to overcome. During an unusually cold period in the 1530s, for example, the authorities complained that the apparatus of government, as well as industry, had ground to a standstill: 'Our mint cannot function, nor can many artisans such as glassmakers, dyers and metalworkers.' This was a dire state of affairs, and a precarious one too.

Henry VIII of England had complained of precisely such a problem halfway through his reign when he passed an act that acknowledged the 'great and manifest likelihood of scarcity' as a result of 'the great decaye of Tymber and Woodes'. As one commentator put it a century later, many cut down wood but 'few or none at all doth plant or preserve'. There used to be ten times as much wood in England as there was at the time of writing, he said. This was dangerous, he added bluntly: 'so it may be conceived, no wood, no kingdome'.[67]

Such concerns were shared in many parts of the world – and not surprisingly. Wealth and power were closely linked to agricultural production, which gave rulers a strong incentive to support land clearances and the transformation of woodland into fields that could support settlements, expand cultivation and produce cash crops that were taxed as a share of each harvest.[68]

There were exceptions, such as King Krishnadevaraya Tuluva (r. 1509–29), whose poem on statecraft, *Āmuktamālyada*, advised kings to preserve and protect forests. Deforestation that cleared land to allow more agriculture should be discouraged, not least since it impinged upon the 'tribal people, who roam about in the forest and hill areas'.[69] Such concern with the rights of indigenous peoples was to prove extremely unusual in relation to other developments that were taking place at the same time in other parts of the world.

Among those expanding geographic and cultural horizons in the fifteenth century were the Portuguese. In popular history, much of the credit for voyages down the western coast of Africa and to the various islands and island groups in the Atlantic Ocean such as Madeira, the Canaries and the Azores is given to Prince Henry, whose nickname 'the Navigator' provides a clue to the central role given to this ruler during a time when Portuguese engagement with new worlds rose rapidly.

Henry should better be described as a co-ordinator of the interests of financiers, merchants, navigators and cartographers who were the driving force behind the development of sugar plantations in Madeira and of the use first of Muslim and then of African forced labour. By 1444, Lançarote Passanha had founded the Lagos Company with the intention of formalising the enslavement of fellow human beings; soon ships were sailing south likewise motivated by the profits to be obtained from the import of malagueta, also known as 'grains of paradise', used as a substitute for Indian pepper, and by the discovery of gold markets at Elmina in what is now Ghana.[70]

These voyages, and perhaps most importantly the profits they generated, helped propel more journeys over longer distances. By 1475, Portuguese ships had crossed the equator going south, and two decades later had reached the southern tip of Africa. The crown benefited from expeditions that returned with lucrative cargoes – whether of human beings, foodstuffs or gold – taking a fifth of the value of goods in tax. According to some chroniclers, such as Fernão Lopes de Castanheda, who was writing several decades later, Portugal's rulers in the 1480s and 1490s were also keen champions of these maritime journeys because of their desire to wrest control of the spice, medicine and jewel trades that passed through the Red Sea from Egypt and Venice.[71]

One of those who took part in these adventures was Christopher Columbus, whose strong religious convictions and desire to use the proceeds of trade with Asia to finance a reconquest of the Holy Land from the Muslims inspired an ambitious plan to sail not south but west across the Atlantic Ocean in the hope of reaching the rich markets of China, Japan, India and South-East Asia. When he failed to win backers in Portugal for his project, he turned to Ferdinand and Isabella of the unified kingdom of Castile and Aragon, eventually setting sail in August 1492 from Palos de la Frontera in what is now southern Spain.[72]

Columbus provided an excitable account of the people and places he discovered after successfully reaching the islands of the Caribbean. Apart from the 'gold and metals' that he claimed to have found, he stressed that he had come upon a natural environment that was ideal for development. The island which he named Hispaniola was home to 'great farms, groves and fields, most fertile both for cultivation and for pasturage'. Another island, which he named Juana (Cuba), likewise was 'exceedingly fertile' and blessed with many safe harbours.[73] Particular attention was paid to the kinds of things grown by locals, including cotton and foodstuffs like kidney beans and maize; the latter 'has a very good taste when cooked, either roasted or ground and made into gruel'.[74] 'Your Highnesses must believe me', he wrote, 'that these islands are the most fertile, and temperate and flat and good in the whole world.'[75]

Reports of the ecological richness of these discoveries were matched by the accounts of those who accompanied the men who followed Columbus across the Atlantic. The Venetian Zuan Caboto (who is better known as John Cabot) set out from Bristol five years after Columbus' first voyage in search of the spice islands of South-East Asia, and instead landed on the north-eastern coast of what is now Canada. Although the rich fishing waters of the North Atlantic had been well known for centuries, the abundance of fish was almost overwhelming. The seas were 'swarming with fish', wrote the Duke of Milan's envoy in London to his patron. Cod were so numerous and easy to catch, it was said, that rather than using nets, weighted baskets simply needed to be lowered over the side of fishing boats and then hauled aboard. The fact that the first recorded cargo of cod brought back to Europe by Hugh Elyot in 1502 had a value equivalent to the annual income of a prosperous landed estate showed that there were fortunes to be made from the natural wealth of parts of the world that Europeans now found themselves able to access with ease and regularity.[76]

Alongside such excitement came considerable disappointment and confusion. Having read all the works of the great philosophers and poets, wrote the Jesuit priest and naturalist José de Acosta in the late sixteenth century, 'I was convinced that when I reached the equator I would not be able to bear the dreadful heat.' Instead, he went on, 'I felt such cold that at times I went out into the sun to keep warm.' All this made him doubt his education. 'I laughed and jeered at Aristotle's meteorological theories and his philosophy.' It was clear that the Greek scholar had no idea what he was talking about.[77] This troubled others too. It was obvious that 'the ancients, guided by inference', had got things badly wrong, noted Bernabé Cobo; those who now live in the Americas know from experience that 'it is the opposite' of what the great scholars of antiquity had assumed when it came to hypotheses about the world, its temperatures and its climate.[78]

The harshness of conditions in the Americas prompted much speculation. Some suggested that South America was full of holes and caverns of water, and that the strength of the sun caused cooling vapour to escape; others argued that heavy rains were responsible for counteracting heat and helped create a cold climate as a result. Others still threw their hands in the air. The American tropics were on the same latitudinal parallel as those of Africa, yet were much colder. 'What the causes of these differences may be, I do not know,' wrote one commentator.[79]

The realities differed sharply from expectations that the new climates would damage or destroy newcomers. Early reports of the 'New World' had emphasised that ecologies were very different, including – in the words of the English natural philosopher Richard Eden – 'Contagious ayre and extreeme heate'. So alien was the natural environment, he added, that it 'altereth the formes and qualities of [European] thynges', such as wheat and cattle.[80] Some worried that those who travelled to the Americas would be overwhelmed by the 'change of air, diet and drinking of water' that would lead to 'sore sicknesses and grievous diseases'.[81]

This in turn helped sow the seeds of an ecological hierarchy that was to find an echo in ideas that developed about race. Tobacco grown in England was 'better for the constitution of our bodies' than that grown on American soil, stated the herbalist John Gerard, author of *Great Herball, or Generall Historie of Plantes*.[82] In the minds of at least some Europeans, things grown in the Americas were smaller, more dangerous and worse than those grown back home. Animal life, wrote the Comte de Buffon in the eighteenth century, 'is less active, less varied, and even less vigorous' in the Americas; not only was the number of species smaller, animals were generally 'inferior in size to those of the old continent'.[83]

These were ideas that hardened as they became established, and influenced thinking for centuries. Writing around the same time as de Buffon, the Dutch philosopher Cornelius de Pauw claimed that 'Europeans who pass into America degenerate, as do the animals; a proof that the climate is unfavourable to the improvement of either man or animal.' The climate and natural environment of North America were such that they prevented reason itself. 'America has not yet produced one good poet, one able mathematician, one man of genius in a single art or a single science,' wrote Guillaume Raynal, one of de Pauw's contemporaries.[84] De Buffon himself was equally scathing about the peoples living in North America: the indigenous population and the waves of immigrants who had settled had certainly formed a distinct 'new people', he wrote, but it was not an impressive one, defined by 'their ignorance, and the little progress the most civilised among them had made in the arts'.[85]

Such views were not shared by all. Alexander Hamilton was dismissive of 'men admired as profound philosophers' who could be so foolish as to have 'gravely asserted that all animals, and with them the human species, degenerate in America, [claiming] that even dogs cease to bark after having breathed awhile in our atmosphere'. Such comments betrayed the 'arrogant pretensions' of their authors and of Europeans in general. Moreover, they could be easily disproved by studying scientific data that few had even bothered to gather, let alone think about.[86] Quite right, agreed Thomas Jefferson; it had taken centuries for the Greeks to produce 'a Homer, the Romans a Virgil, the French a Racine and Voltaire, the English a Shakespeare and Milton'. America was a young nation and needed more time to produce poets of similar genius.[87]

The creation in the minds of Europeans of the Americas as a 'New World' of course paid no attention to the indigenous populations who had settled and lived in these continents for tens of thousands of years before the transatlantic crossings. In the conceptualisations of early European travellers and the waves of settlers who followed them, the Americas were virgin lands, ripe for exploitation. To all intents and purposes, the ways that land had been used by local populations were irrelevant – as were the local populations themselves.

It was certainly true that the first waves of those who sailed across the Atlantic from Europe, who became known as *conquistadors* or conquerors, were motivated by the rapid rewards of movable wealth – particularly the gold, silver and jewels that were pillaged from the Mexica (or Aztec) in Central America and the Inca in South America and were piled up on the quayside in Seville in such volumes as to be compared to wheat being brought in from the harvest. But after these initial surges of pillage and looting, the Americas became primarily important as landscapes whose ecologies could be transformed and exploited.

The energy – and cruelty – with which that process took place has its own important story, as we shall see, one centred on human bondage, on slavery and on grotesque ideas about race. The terrible treatment of indigenous peoples in the Americas played a central role in these abuses, but so too did that of peoples from Africa who were torn from their families, transported across the Atlantic in horrific conditions against their will and forced to work for their 'owners' and 'masters', helping to create great fortunes for them in the process.

In some ways, then, the disappointment of many of those who visited or settled in the Americas with the environmental conditions that they found raises a question that can all too easily be overlooked – namely that of why Europeans sought to expand across the Atlantic in the first place rather than stay closer to home. After all, Portuguese and Spanish cultivators had successfully set about creating plantations in places like Madeira, the Azores and São Tomé in the second half of the fifteenth century and early sixteenth; why then did they not turn their attention to doing the same in West Africa itself? It was after all a region blessed with climate conditions that were similar to or better than those on the other side of the Atlantic, with land and soil well suited to growing crops such as sugar, rice and cotton that proved so lucrative in the Caribbean and the Americas. West Africa was also home to a labour force whose size is self-evident from the slave trade itself. Not only that, West Africa was closer to European markets and as such meant transport was quicker, less risky than crossing the Atlantic and therefore much cheaper thanks to lower transportation costs.

As one later English official put it, 'everything that thrives in ye West Indies' would thrive in West Africa. As others also noted, 'the shortness of distance, and the safe passage between England and the Africa coast, compared to the whole [transatlantic] voyage' made the idea of plantations in West Africa infinitely preferable to long-distance connections – and would likely produce three times as much sugar to boot.[88]

Although some have questioned whether soil fertility and acidity, as well as rainfall distribution patterns, were entirely appropriate for cash crops, the fact that these were successfully grown from the mid-nineteenth century suggests that there were other reasons why they were not grown at the time of European maritime expansion. São Tomé and other such locations could have been a 'stepping stone' to similar developments elsewhere along the coast of Africa; instead, the latter became a source of labour rather than a place for intensive agricultural investment.[89]

Historians who have focused on the disease environment in tropical Africa have noted that Europeans were at a major epidemiological disadvantage because of the lack of immunity to malaria and yellow fever that local populations had built up over millennia. From the outset, Europeans in West Africa were confronted by a 'striking angel with a flaming sword of deadly fevers', with the result that visitors and settlers 'tended to sizzle and die'.[90] In contrast, the biological odds swung heavily in their favour in the New World where, ironically, indigenous peoples were the ones who suffered as a result of diseases that were brought by new arrivals and the animals that were brought with them.[91]

This was all true enough, but what was more significant was the fact that the political systems in states in West Africa were highly developed to the extent that it was nigh on impossible to contemplate colonisation – at least for several centuries. In fact, until the nineteenth century, Europeans were rarely able to penetrate 'beyond a cannon shot from the coast'. The kingdoms of Kongo, Benin and Oyo and others were perfectly able to resist incursions, and were not much threatened by military pressure exerted by small numbers of people a very long way from home, whose settlements amounted to a handful of coastal fortresses and whose commercial activities depended not on coercion but on negotiation.[92]

As one scholar has put it, Europeans were convinced of the potential of Africa. The problem was that they 'failed completely' to gain access to, let alone control of, gold mines, and were unable to develop the plantations that they wanted. 'African resistance', therefore, 'resulted in Europeans taking slaves away in ships as a second-best alternative to working slaves on African plantations or mines'. Although things would change dramatically in the centuries that followed, the reality at the time of the European engagement with the Americas was as clear as it was emphatic: 'the slave trade was a symptom of African strength, not weakness'.[93]

There were many reasons why the Americas were colonised, including misinformation, poorly informed and elevated expectations, competition between settlers and of course the push towards further discoveries in what was to Europeans at least a New World that offered excitement and opportunities, albeit ones that were often disappointed. The primary reason, though, was mundane and easy to overlook. While there were many and complex motivations for the different ages of discovery that characterised all parts of the world in the fifteenth century and beyond, the least convincing was the notion of an idealistic gathering of information about new peoples and new parts of the world for the sake of learning and knowledge. What drove the next cycle in global history was the pursuit of profit. It reshaped political power, transformed ecologies and ultimately changed the climate itself.

15

The Fusion of the Old and the New Worlds
(c.1500–c.1700)

Another great pestilence struck this land, bringing death and
destruction to the native population.

Diego Muñoz Camargo (late sixteenth century)

The establishment of routes across the Atlantic was one element of a much
wider web of connections that tied most of the world's great landmasses
together. Just five years after Columbus had set out westwards, Vasco
da Gama had rounded the southern tip of Africa, headed north as far as
Malindi and then made his way across the Indian Ocean, reaching the
coast of what is now Kerala in southern India. Others followed swiftly
in his footsteps, sailing not only along the coastline of Africa or even to
the Red Sea, Gulf and South Asia but beyond, to Sri Lanka, South-East
and East Asia. Within a matter of decades, ships were criss-crossing
the Pacific, creating what might be reasonably described as a globalised
commercial system.

This went hand in hand with a dramatic ecological transformation
that was also global in its nature, scope and impact. While attention is
often paid to political frontiers being reset by the forces of colonialism
and by the social and economic consequences of new and even deeper
contact and integrations at local, regional and intercontinental level,
the reshaping of the natural environment had impacts that are difficult
to exaggerate. Flora and fauna were transplanted into new settings,
sometimes intentionally and sometimes not. Ecological boundaries

that were demarcated by climatic conditions became boundaries for human interaction. Public imagination and conceptualisation of the natural world were reshaped in such dramatic ways as to distort how we think about history as the Old World expropriated and refashioned the New World in its image.

For example, in today's world, we think of tomatoes as being synonymous with Greek and Italian cuisine, or perhaps with that of Spain – home to tomato-throwing festivals such as 'La Tomatina' which takes place every year near Valencia. We relate paprika to goulash, the Hungarian national dish; we associate pineapples with tropical Africa or South-East Asia, chillies with India, peanuts with the satay sauces of Thailand and Malaysia, and potatoes with Sunday roasts that are the hallmarks of the British family weekend experience. Yet none of these foods is native to Europe, Africa or Asia; all are from the Americas.[1]

At stake was not the greater availability of different crops and their diffusion to different parts of the world, but a full-scale global ecological revolution. The process was that of the harnessing, exploitation and consumption of resources that changed landscapes, transformed ecosystems and shaped human settlement patterns as a result of the deployment of the labour that was needed to sow, reap, extract and ship foods, minerals, materials and goods to those able and willing to pay for them. That in turn provided motors of economic growth and social change that were disproportionately concentrated in centres of power that were able to build up and maintain control over key resources, to militarise and secure trade routes and to protect themselves from competitors and rivals both at home and abroad.

The result was the creation of global empires, above all centred in Europe, that were directly related to the relentless rolling back of commodity frontiers – the expansion of geographic boundaries to appropriate and exploit ever greater natural resources. The search for more and more materials, whether in the form of minerals or crops, brought about the wholesale transformation of landscapes, as seams of precious metals were mined and forests and flatlands turned into fields and plantations. This had ecological costs, but also had obvious economic consequences as booms in production of precious minerals or cash crops that generated the greatest profits often gave way to gluts in supply that could not be easily absorbed, which in turn led to downturns in prices.

High levels of availability accompanied by falling costs had benefits of their own, with widening access serving as a driver of social change and creating conditions well suited for an 'industrious revolution' of populations in Europe; these populations gained access to ever-increasing volumes of goods at ever-decreasing prices that initiated a cycle of rising levels of disposable wealth and greater participation in the celebration of ownership so richly shown in the Arnolfini portrait of the last chapter.[2] The consequences were significant in the Old World, but they came with a heavy price in the New, where the benefits were not shared equally or equitably either socially or ecologically.

Sugar, once a luxury product for the elite alone, serves as a perfect example of this process. Sugar plantations were created in the Atlantic on Madeira, the Canary Islands and São Tomé before being introduced by the Flemish and Dutch to Brazil and to one island of the Caribbean after another; it spread further, as production later took off in Louisiana and above all in Cuba, which by the nineteenth century was the most productive sugar island in the world. By that time Java in what is now Indonesia had become an important centre for production, as had Taiwan – with centres then being developed in Mauritius and Réunion in the Indian Ocean and Fiji in the Pacific; then vast plantations in Brazil were created to grow sugar for ethanol in industrial quantities as a form of 'sustainable' fuel.[3]

Sugar was one commodity among many – with cotton, coffee, cacao, rubber, timber, furs and many others following similar trajectories of massive expansion of exploitation that in turn depended on four interrelated factors. First, the forcible occupation of land, often through recourse to violence, with the most productive terrain seized, secured and then turned over to the production of cash crops; second, the availability of a labour force to engage in plantation, cultivation, harvesting or extraction of resources; third, a logistical infrastructure that enabled the transportation of goods and an institutional framework that protected property rights at source, during transit and on arrival; and finally, the existence of markets with growing appetites and discretionary spending power.

Some scholars have argued that the goods and products that flowed back across the Atlantic in particular fuelled the industrial revolution and the rise of Europe.[4] Indeed, Columbus' crossing and the opening up of maritime trade routes to Asia shortly afterwards have long made an impression on economists. 'The discovery of America,' wrote

Adam Smith in 1776, 'and that of a passage to the East Indies by the Cape of Good Hope' by Vasco da Gama just six years later were 'the two greatest and most important events recorded in the history of mankind'. These new routes united 'the most distant parts of the world … enabling them to relieve one another's wants, to increase one another's enjoyments, and to encourage one another's industry'. Smith was no rosy-eyed globalist who saw the benefits of wealth being evenly distributed: it was just that 'At the particular time when these discoveries were made, the superiority of force happened to be so great on the side of the Europeans.' They took advantage of this in ways that were grotesque, committing 'with impunity every sort of injustice' in remote lands.[5]

Smith was principally concerned with the political and economic transformations which followed in the wake of the opening up of new trade routes that eventually linked all the world's continents together. By the time that he was writing in the 1770s, Europeans had spent the best part of three centuries fanning out in all directions, driven above all by the search for resources, goods and produce that could be extracted by force or acquired cheaply but that commanded higher prices elsewhere. The process of colonisation later became closely wrapped up in the promotion of values and ideas about religion, about race, about interpretation of the past and even about scientific exploration. But the motor for expansion from the 1490s was the deployment of mercantile capital and interests.

It was no surprise or coincidence, therefore, that the colonial expansions that brought misery and injustice to so many, as Smith had noted, brought different results in the countries that benefited most from the opportunities offered by a new global trading system. Those who became wealthy through commerce were determined to protect their assets, both by minimising tax demands from the crown and by limiting the powers of the sovereign to interfere in trade generally or to target its beneficiaries – entrepreneurs, risk-takers and investors. Ironically, enslavement, murder and dispossession from their land by Europeans in many corners of the world were mirrored by the entrenchment and enhancement of rights back at home: parliaments, law courts and institutions that were independent of the king were set up, often acting specifically to limit the powers of the sovereign while also defending the interests of elites who were determined to retain their independence and to use their fortunes to demand and fashion further political, social and economic reforms.[6]

Naturally, the aim of these reforms was to develop further still the sinews for economic growth and further institutional improvements. They were not uniform, with interest groups in northern Europe – in England and the Netherlands – proving far more effective than their counterparts in the south in Spain and Portugal, especially after 1650. Among other things, the relative skill of parliamentarians in preventing extraordinary taxes, curtailing warfare with neighbours and borrowing prudently led to significantly lower interest rates, which served to provide a competitive advantage in global trade, but also served to ensure lower levels of risk, thus favouring long-term investments.[7]

From a European perspective, this led to what economists refer to as the Little Divergence, a time when northern Europe pulled away from the south in terms of socio-economic and political performance, which led to waves of democratisation that although slow were regular when viewed over the long term.[8] But the path towards higher participation in the political process, and away from absolute monarchies, can be traced back to the initial bursts when the 'superiority of force' enabled Europeans to gain at the expense of those living in other parts of the world.

The first waves of those who crossed the Atlantic focused on movable wealth, and above all precious metals, jewels and fine objects. Attention turned quickly to the conquest of political systems that offered the greatest and most lucrative spoils. Chief among these were the Aztecs of Central America and the Inca kingdom in the Andes, both of which were effectively dismantled by relatively small numbers of men who had left Spain looking for fame and fortune. The invaders' success owed something to the military technologies that had been honed by centuries of warfare in Europe, particularly the use of firearms and other weapons, although identifying where and how these were decisive is not straightforward. Horses also gave the newcomers a decisive edge in communication and transportation as well as in military confrontations. The ability of individual Europeans to forge local alliances and to exploit differences with rivals among the ruling elites both within and outside their realms however also proved crucial in enabling modest numbers of outsiders to bring down what had appeared to be robust, resilient and sophisticated state structures.[9]

Precious metals flooded out of the Americas, with some vessels packed so full that they even used gold for ballast.[10] Gold and silver flooded back to Seville, where they were stacked up like wheat, often in such large volumes that the customs house (the Casa de Contratación) 'could not accommodate it all and it overflowed onto the patio'.[11] Such booty was part of a one-off, exceptional transfer of wealth that reflected large-scale appropriation from leadership structures that had themselves succeeded in centralising power and drawing in riches from all parts of their realms, and sometimes beyond it too through the collection of tribute. After this initial explosion, however, a different model was required to extract revenue, whether through crops or through mining, such as from Potosí, which accounted for more than half of global silver production for more than a century: one that was based on mass labour.[12]

Columbus himself had not only noted the possibility of relying on local populations in the Caribbean and the Americas as a workforce but had hinted at the ability to make them do so through compulsion: 'they are fit to be ordered about and made to work', he wrote, and to 'plant and do everything else that may be needed'.[13] The enslavement of the indigenous peoples of the Americas is often mentioned only in passing by historians of this period, if indeed it is mentioned at all. However, those who do discuss the way the local populations were forced to work suggest that hundreds of thousands and perhaps as many as a million people were put into bondage and made to work for Europeans who had crossed the Atlantic.[14] This was despite the fact that Queen Isabel had been horrified by the enslavement of indigenous peoples, had ordered them to be freed and returned to their homeland and had even, in 1501, declared that they were free vassals who should be treated well and should pay tribute to the crown.[15]

This view was not only shared by clerics such as Bartolomé de las Casas, who argued that God had created all humans to be equal by their nature and in their recognition of the Almighty, but also repeatedly enshrined in law, such as the Laws of Burgos in 1512, Pope Paul III's papal bull of 1537 and further legislation issued by the King of Spain in the 1540s.[16] As happened in many other cases during the creation of colonial empires, there was a profound disconnect between theory and practice, between the declarations of law on one side of the world and the hard realities on the ground in another.

The arrival of the first Europeans, along with their requirements and expectations, met with fierce resistance from indigenous peoples. The first settlement in the New World, La Navidad on the island of Hispaniola, which had been built using the remains of a Spanish ship, the *Santa María*, was destroyed by the Taíno population. In some cases, resistance took passive forms of protest against the appropriation of land, including the refusal to plant their annual crops – a ploy described by the contemporary commentator Gonzalo Fernández de Oviedo as an 'evil scheme' that cost the lives of Spanish settlers and local populations alike. The Spaniards took desperate steps to survive, eating all the domesticated animals they could find, including those they had brought with them, before turning to lizards, salamanders and snakes, 'of which there were many types, but none poisonous'. Any living creature they could find 'ended up in the fire, boiled or roasted'.[17]

Experiences such as these resulted in a range of measures designed to reduce food insecurity. One was the introduction of species that were familiar to Europeans as reliable sources of protein – pigs being the most important, above all because of their prolific reproduction rates, short gestation, large litters and ability to eat almost anything. This was one part of a major ecological upheaval that was accelerated by the introduction of sheep, goats and cattle – a process captured by one eyewitness in the early seventeenth century, who noted that there were 'immense flocks of cattle, such as horses, mules, oxen, cows, sheep and goats, which have pastures always fresh in every season', spreading out in all directions 'as far as the eye can see'.[18]

This of course had dramatic effects on native species of animals of all kinds; but it also had a major impact on topsoil and plant life. For example, in some cases, overgrazing brought about soil erosion which dramatically reduced the quality of the land; that in turn reduced the quality of forage that was available, which led to a decline in the average weight of animal stocks, affecting animal reproduction rates and thus impacting herd and flock sizes, which reduced available protein and textile sources. Deforestation likewise brought about a host of consequences, including deterioration of the water regime that in some places made land unsuitable for long-term settlement and encouraged the spread of arid-zone species. Anthropogenic change, inadvertent and otherwise, had drastic impacts in the Americas.[19]

The impact of the biological revolution was so widespread that some historians have termed it a form of 'ecological imperialism'. Native biota were usurped and transformed by the arrival of new peoples with new habits, lifestyles and demands – and by that of their domesticated animals, whose 'eating habits, trampling hooves, and droppings, and the seeds of the weedish plants they carried with them ... altered forever the soil and flora' of regions that were colonised.[20]

The Europeans brought with them a 'portmanteau biota', in part by design, in part by mistake. In addition to domesticated crops and animals, which spread to, bred with and affected domestic species of flora and fauna, there were weeds, seeds and pests that also had profound effects on all aspects of the peoples, places and living organisms they came into contact with.[21] So too did pathogens. Local populations had had no contact with or experience of a range of diseases brought by the Europeans, leaving them immunologically defenceless.

Smallpox and measles proved devastating for the inhabitants of Central America, with waves in the 1520s leading to catastrophic loss of life, with some accounts describing the number of rotting bodies left unburied as almost overwhelming.[22] The 'cocoliztli' pandemic of the 1540s was likewise disastrous, which is clear not only from written accounts but from chemical analysis of a cemetery in southern Mexico that points to ruinous population loss.[23] Subsequent study of these bodies shows that they had died after being infected with a subspecies of salmonella known as *S. paratyphi C*, a bacterial cause of enteric fever, that was perhaps brought across the Atlantic by European carriers who were either asymptomatic or survived the crossing. The fact that contemporary accounts suggest that both indigenous peoples and Europeans died during the outbreak provides an indication of its virulence.[24]

This was followed by a further outbreak of disease in 1576 that was devastating. 'Another great pestilence struck this land,' wrote Diego Muñoz Camargo, 'bringing death and destruction to the native population.' So many died, he added, that 'the native population was almost extinct'. The pandemic was accompanied by ominous celestial signs – namely 'three circles in the sun, resembling bleeding or exploding suns, in which the colours merged'.[25]

The apocalyptic scale of suffering was reported by others too, such as the Franciscan friar Juan de Torquemada, who wrote that 'a great mortality and pestilence' that spread through Spanish-controlled territory in Central America 'was so fierce that it ruined and destroyed almost the entire land. The place we know as New Spain was left more or less empty.' So many died, he added, that 'nobody had the health or stamina to help the sick or to bury the dead. Large trenches were dug in the cities and towns, and from morning till sunset, priests did nothing but carry the dead bodies and toss them into the ditches without any of the solemn rites usually reserved for the dead, because there was simply not enough time to do so. At night they covered the trenches with soil.' According to Torquemada, more than 2 million died as a result.[26] Modern research indicates that this estimate was not far off the mark. One study suggests that 80 per cent of the indigenous population died in the pandemic of the 1540s and that there was a mortality rate of 45 per cent three decades later.[27]

Some Europeans believed that indigenous peoples were being punished by God because of their belief systems which were not so much pagan as diabolical and demonic.[28] The response, therefore, was to enforce hard-line religious practices on local populations. As it happened, this also provided an opportunity for individual missionary priests to appeal for greater resources to help minister to their flocks – in a classic case of not letting a good crisis go to waste.[29]

It was not disease alone that proved devastating, however, but rather a cascading set of factors that became closely intertwined, as forced labour, slavery, resettlement and malnutrition combined to lower resistance and increase vulnerability to illness and suffering. 'I saw ... cruelty on a scale no living being has ever seen or expects to see,' wrote the Spanish friar Bartolomé de las Casas of his experiences in the earliest days of European settlement, in a horrified report designed to inform those back home of what was happening in the New World.[30] These thoughts were echoed by Jerónimo Mendieta later in the sixteenth century. The indigenous population 'was falling each and every day, consumed by inhumanity and cruelty'. The greed of the Spanish, he said, was such that not only were the settlers 'pitilessly watching them die like flies', but were 'exploiting them for the short time while they yet survive, for it will not be long before none of them remain'. Vast regions were being devastated by greed, leaving no records of the countless inhabitants who had once populated them.[31] Adverse climatic conditions made life more difficult for new arrivals and presumably for indigenous populations as well.[32]

Expeditions into what is now New Mexico in the 1540s that were inspired by excitable and exaggerated rumours of rich cities waiting to be conquered were accompanied by the use of violence against Ancestral Puebloans by the Spaniards, including rape and torture on a scale that was so horrifying that they became the subject of a royal commission set up to investigate abuses. Contemporary sources claimed that antagonisms were sharpened by competition for shelter between new arrivals and local peoples, which resulted not only in displacement of the latter, but in the burning of their homes and pillaging of sources of fuel 'to provide insurance against the cold, which was extreme'.[33]

In the eyes of some modern historians, the fact that the invading forces were woefully ill-equipped for the conditions and lacked warm clothes was decisive, with food shortages and the looting of blankets and turkeys – whose feathers were used to make cloaks – proving to be sources of friction and confrontation. The most infamous of these incidents was in 1599 when Spanish soldiers and their native auxiliaries massacred almost a thousand men, women and children at Acoma Pueblo, spurred, at least in part, by 'the struggle for warmth [which], even more than food, likely created the worst hardships'.[34]

Exceptionally cold weather in the south-east and south-west of North America played a part in this, with freezing winters and heavy snows, including in parts of California in the 1540s, while alternating patterns of droughts and storms in the last decades of the sixteenth century made life difficult for settlers in Virginia, which one eyewitness spoke of as a land of 'sterility and death'.[35] In South Carolina, outposts were abandoned under pressure from a combination of food shortages, the animosity of Guale Indians and inclement weather; as one contemporary recorded, it did 'nothing but rain all the time' when grain had to ripen. 'So here we are lost, old, weary and full of sickness.'[36]

Life became more difficult still in the late sixteenth century in the Americas – and elsewhere too, as we shall see – following a series of eruptions of Colima in Mexico in 1586 and of Nevado del Ruiz and Huaynaputina in the Andes in 1595 and 1600 respectively, as well another major volcanic event at an unknown location around 1592. While none of these was particularly large in its own right, they had a similar effect to chain-reaction explosions in the sixth century, leading to a rapid phase of cooling that brought temperatures down sharply across Eurasia and North America. In 1601, the temperatures around the northern hemisphere were some 1.8 °C below their long-term average, making the summer the coldest of the last 2,000 years – and part of the coldest decade of the same period – if tree-ring data is to be believed.[37]

The late seventeenth century was a time when there was one disaster after another in North America, with conditions so bad that many Spaniards questioned whether it was even worth holding on to the territories that had been taken. Spanish Florida was 'a wasteland', warned one commentator, that 'renders little harvest after much labour'. The entire enterprise of supporting colonies in North America was misguided, Alonso Suárez de Toledo told King Philip II of Spain; 'to maintain Florida is merely to incur expense because it is and has been entirely unprofitable nor can it sustain its own population. Everything must be brought from outside.'[38] As if to underline the scale of the challenges, the first English colony – at Roanoke in what is now North Carolina – failed, with its inhabitants reported to have been massacred, to have starved to death, dispersed or fled.[39] Perhaps not surprisingly, Suárez de Toledo reported to Philip II that whatever problems the Spanish had to face in North America, the English were not one of them.[40]

The cold did not affect all regions the same way, nor all peoples, with many different responses and coping strategies. Neutral Iroquois in eastern America shifted from cold-sensitive protein-rich beans to a diet that involved more deer hunting both for protein sources and for the animals' hides, although this in turn led to rising competition for hunting territories.[41] Other Iroquois groups migrated, relying on a decentralised clan and kinship system to ensure a balanced distribution of resources during times of scarcity. Then there were the Algonquin societies in the Chesapeake and Potomac, which responded in a different way, concentrating power and authority in the hands of a hereditary elite as part of a hierarchical social, economic and political system that marked a radical departure from previous systems of organisation.[42] Changing settlement patterns in the south-west and in eastern woodlands which included greater population densities at some sites and the construction of fortified villages reflected the emergence not only of hierarchies but of violence and of higher levels of competition for resources.[43] These were just some of the changes that are also captured in a range of oral histories of different peoples that were passed down through generations right across North America that recall a set of climatic and environmental crises and enshrine them as lessons of resilience and triumph over adversity.[44]

The histories of indigenous populations were not framed by their interactions with Europeans and new settlers, as there were complex relations not only between local groups but within them too.[45] However, the pattern was rather different in regions that were conquered or dominated by arrivals from Europe, where the scale of demographic collapse was astonishing: while precise figures are not easy to come by, some estimates suggest that the Valley of Mexico was home to 1.5 million people around the year 1500, but to just 325,000 seventy years later; by the middle of the seventeenth century, this had fallen again to as little as 70,000. In other words, the population around 1650 was just 5 per cent of what it had been before the arrival of the Europeans.[46]

The scale of loss was such that indigenous historians like Hernando de Alvarado Tezozomoc and the great Don Domingo Chimalpahin took to writing down accounts of their ancestors for fear that there would soon be no one left to recall the deeds and events of the ages before the transatlantic crossings began.[47] More recent commentators have focused on other consequences of the large fall in population size, most strikingly arguing that the loss of life on such a scale induced dramatic environmental and climatic change: one study has suggested that the demographic decline resulted in a massive reduction in land use, with tens of millions of hectares of land that had previously been cultivated reverting to prior states, impacting CO_2 levels and global surface air temperatures in the sixteenth and seventeenth centuries.[48]

The hypothesis that the 'Great Dying' of the indigenous populations of the Americas which was caused by violence, malnutrition and disease itself led to climatic change is ingenious, but it is also speculative and problematic. For one thing, it relies on assumptions about population sizes before 1492, as well as about those of a century and more later, which are very difficult to assess with a real degree of certainty – to the extent that some models concede a margin of error in calculations of almost 100 per cent.[49] For another, it is not clear that reforestation did occur in the Americas, and if it did, where it did and what its precise impact was; furthermore, even if forests did grow back up and even if plant biomass rose as a result, it is also worth noting that the sixteenth and seventeenth centuries were a time of rapid deforestation in Asia, which would (presumably) have offset some of the changes in the Americas.[50]

Perhaps the most obvious problem, however, is that the climate evidence from ice-core data points to a sharp decline in global atmospheric carbon dioxide in the 1590s, and not in the 1540s and/or 1570s, two decades that were periods of catastrophic mortality rates caused by pandemic disease, at least in many parts of Central America. The suddenness of this signature suggests that the causes of the shift had less to do with population collapse and changes to the use of land, and more to do with other, non-human factors – including volcanic eruptions that we know took place at precisely this time.

What is perhaps less contentious is the broader impact of population loss on the fate of the Americas, on Europe and on Africa. The ecological dividends of the New World had been a key factor in the early transatlantic trade, most notably sugar to start with, soon followed by tobacco which was thought to have almost miraculous medicinal qualities. Settlement and colonisation happened not by accident but by design – and both were closely linked to exploitation of new landscapes, and to the wealth that these could generate.[51]

Key, however, was availability of a workforce that could plant, grow, harvest and process the fruits of the land; this was especially important for cash crops that were labour-intensive and year-round processes. From the outset, the new arrivals from Europe saw forced labour and slavery as the obvious solution to producing sugar: this required the digging of trenches, planting of new cuttings, weeding, cutting and then crushing and processing within twenty-four hours before the juice began to ferment, through a process of boiling and refining that was often compared to the fires of hell.[52]

In Brazil, tens of thousands of people were enslaved and set to work on sugar plantations, sometimes through slaving expeditions into the interior of Bahia that also brought Europeans into conflict with each other over who could seize the lion's share of indigenous inhabitants who could be set to work to create profits for strangers from thousands of kilometres away. The large-scale population loss among those living in the Americas therefore presented a major problem. Outbreaks of dysentery in the 1550s, bubonic plague a decade later and repeated bursts of smallpox soon afterwards killed many of those working for planters in Bahia, compounded by starvation resulting from the failure to plant enough food crops under the pressure of falling population numbers.[53] 'The number of people who have died here in Bahia in the past twenty years seems unbelievable,' wrote one Jesuit missionary. 'No one ever imagined that so many people could ever be expended, far less in so short a time.'[54]

Population losses were not uniform throughout the Americas, nor even within individual regions. Those living in coastal estuaries, by rivers or near water sources that favoured waterborne disease were far more at risk than those living in the interior, where heavily forested homelands supported higher numbers of game animals as well as soils that produced surpluses. This explains why the Spanish faced regular if not constant hostility from indigenous populations on the Atlantic coast of North America for more than a century after the initial settlements, which suggests that demographic collapses were highly localised. As one leading scholar has put it, when it comes to coastal communities 'massive depopulation did not occur'.[55]

European settlers had been preoccupied with boosting the labour force even before the first waves of demographic disaster through disease and other factors. Crucial to this had been the development of the 'plantation complex' where enormous quantities of sugar as well as other cash crops could be cultivated and produced, and where Italians and Portuguese had learned from experience with agricultural production on the offshore islands in the Atlantic such as the Canaries.[56] Access to the slave markets of West Africa had played an important part in the economic and ecological exploitation of these locations; it now proved crucial in the search for profits across the ocean.[57]

By the early sixteenth century, large numbers of Africans were being transported from one side of the Atlantic to the other. So many were being made to work on the sugar plantations in Hispaniola that 'the land appears a copy or image of Ethiopia', wrote one Spanish writer.[58] However, there was some trepidation among new settlers about the shipments of large numbers of enslaved, unfree or indentured people. In theory at least, one concern was that Islam would be transported to the New World. 'In a new land like this one where faith is only recently being sowed,' stated one royal decree in 1543, 'it is necessary not to allow to spread there the sect of Mahomet or any other.'[59] These echoed long-standing fears that had led to Africans taken to Spain and Portugal before the crossings of Columbus and others being forcibly baptised and given religious instruction designed, if not to turn 'idolators' into good Christians, then at least to reduce the threats they posed.[60]

More of a concern than the possible spread of Islam was the numerical inferiority that was soon a source of deep anxiety for early European settlers in the Americas. As early as 1503, some like Nicolás de Ovando, Governor of Hispaniola, were arguing in favour of a complete ban on the trafficking of Africans across the Atlantic because they fled, joined the indigenous population and taught them 'bad customs'.[61] These worries were rooted in fear of uprisings, and were framed by claims that Muslim Africans were effective at inciting rebellions and at encouraging indigenous populations to revolt too. Such fears were not ill founded, with the first major revolts breaking out in the 1520s prompting the repeated issue of decrees by the Spanish crown that specifically forbade the sale of enslaved Muslims from Africa without a special licence.[62]

One alternative was to try to attract people to migrate voluntarily or through forms of incentivised migration. In England, these ideas had been brewing since the start of the Tudor period, primarily in relation to Ireland – which figured in the English imagination as a wilderness ripe to be civilised by settlers who would tame the 'weeds of Irish society' and turn Ireland into 'our new world'.[63]

Reports about the natural environment on the other side of the Atlantic capitalised on such rosy portrayals of virgin territories waiting to be transformed. Accounts such as that by Thomas Harriot which paid close attention to the commodities and produce that grew and could be grown in the fields and forests of North America were read carefully in Europe and in England in particular, where Queen Elizabeth I took a keen and personal interest in the opportunities these new lands offered both in their own terms and as a potential source of profits that would benefit the country and protect it against Catholic rivals and enemies closer to home.[64]

Initially, the numbers of those who crossed the Atlantic voluntarily from northern Europe were modest. In the fifty years that followed the first English attempts to settle in the Americas in the 1580s, colonists established themselves in Virginia, New England, Bermuda, Barbados and the Leeward Islands. Communities were so sparse, however, that some have suggested that there were more people from the British

Isles being enslaved in North Africa by the Moors, usually taken captive at sea, than the sum total of those who had made their way to the New World.[65]

However, the flow of settlers rose steadily throughout the seventeenth century. In many cases, the rewards of agricultural production were reaped not through coercion of a labour force controlled by fear and violence, but through co-operation and trade. However, it was not lost on some early settlers that opportunities had opened up because of the decimation of local populations through disease. In a charter of 1620 that set out the claim for what became known as New England, King James I declared that 'within these late Yeares there hath by God's visitation reigned a wonderfull plague' that resulted in 'the utter Destruction, Deuastacion, and Depopulation of that whole Territorye'.[66] The large numbers killed in this outbreak, perhaps caused by leptospirosis complicated by Weil syndrome resulting from contamination of soil and fresh water by rats brought from Europe, helped smooth the way for colonisation.[67] Ironically, however, it also paved the way for co-operation locally as competition for the best resources became less intense, and as those who survived were incentivised both to trade and to seek accommodations while being forced to regroup and rebuild strength.[68]

For Europeans, such experiences must have seemed providential, but they were also a reminder of the need to build up labour forces that could actually take advantage of opportunities opening up as if by magic. While much of the commentary about the New World, its climate and nature was whimsical, anecdotal and plain wrong, there was also a concerted effort to explain the potential rewards on offer. For example, in the 1630s, John Winthrop sent a letter back to London reporting that while Boston was characterised by 'sharp and longe' winters, 'the natives, they are near all dead of the smallpox, so the Lord hath cleared our title to what we possess'. If that was not enough of an attraction, men like the minister Francis Higginson were careful to let others know that there was 'hardly a more healthful place to be found in the world that agreeth better with our English bodies' than Massachusetts. 'Many that hath been weak and sickly in old England', he went on, 'by coming hither have been thoroughly healed and grown healthful and strong.'[69]

The climate of New England was 'most apt' for those born and bred in England, and its lands 'most fitt for the generation and habitation of our English nation', wrote Thomas Morton, whose book about his experiences in North America turned him into something of a celebrity when it was published in 1637. The 'English and Scotch' had another advantage over others, wrote Philip Vincent around the same time, namely 'a facultie that God hath given the Brittish Ilanders to beget and bring forth more children, than any other nation of the world'.[70]

The focus on being 'good breeders', as Vincent put it, was significant for it points to the problems of labour shortage and the attempts to overcome them. While having more children was an obvious solution, such population growth took time, and also required accounting for high levels of infant mortality that kept population numbers low. A quicker fix came from indenture, a contract of servitude whereby an individual agreed to work for a designated master for a set period of time, in return for passage to an agreed location on the other side of the Atlantic.[71] Efforts went into extolling the prodigious resources and rewards that were dangled to lure workers overseas. There were 'extravagant encomiums', wrote one disbelieving cynic, that promised there were islands in the Caribbean where 'Gold is more plentiful than Ice, Silver than Snow, Pearls than Hailstones.'[72]

In the case of England and to some extent France, the availability of a labour force and its migration to the other side of the Atlantic were part of a wider story of socio-economic changes in the sixteenth century that saw improvements in agriculture and gains in production which resulted in shortages of work for the peasantry. Merchant capital was quick to connect this surplus to the surging demand for workers in the New World, exploiting indenture contracts to offer opportunity, transport and work for a fixed period, normally of four to five years, followed by a cash payment at the end.[73] Some, like Francis Bacon writing in the early seventeenth century, were unimpressed by the quality of those agreeing or sometimes being forced to go to work on plantations in the New World, which were being filled with 'the scum of people, and wicked condemned men'. For others, early colonists in places such as Virginia and Barbados were 'loose vagrant people' whose interests centred on 'whoreing, thieving or other debauchery'.[74]

Perhaps not surprisingly, however, not everyone saw things the same way. In addition to developing a labour market in the colonies, the dispatch of cohorts of unskilled workers would rid England, as well as Scotland and Ireland, of potential troublemakers at home, at the same time strengthening the ties of dependence with the mother country. There were 'many thousands of idle persons' in the British Isles, wrote Richard Hakluyt in the early 1600s, who have 'no way to be sett on worke'. They are 'either mutinous and seeke alteration in the state, or at least very burdensome to the common wealthe and often fall to pilferage and thevinge and other lewdness'. If they were not sent off abroad, they were bound to end up on the gallows.[75]

Indenture contracts for the poor (whether mutinous or otherwise) offered a credit mechanism that gave those who set off overseas the ability to borrow against future returns based on their labour. Their arrival, as members of the House of Lords put it in 1640, was 'hailed with delight by planters who wanted cheap labour' on the other side of the world.[76] It was a solution that suited worker and owner alike.

Planters often saw little distinction between indentured servants who came from Europe and those shipped from Africa, with the former often being referred to as 'white slaves'.[77] In some cases, enslaved African people were heavily outnumbered by indentured Europeans: by 1645, around 24,000 people had settled on Barbados, which was fast becoming a crucial engine in the growth of the plantation economy. Some three-quarters of the population were white, with many either previously or still indentured servants.[78] Africans were much more desirable – something reflected in prices that were almost double those for white indentured workers, although this differential also reflected attitudes to investing in those whose freedom had been taken away for good rather than reverting to them after a defined period and whose labour therefore 'belonged' to their 'owner' as long as they were alive.[79]

Scholars have pointed out that many Africans in the New World were highly regarded and valued for their knowledge and expertise in particular areas. Those from the regions of Wolof, Fula and Mandinga in West Africa which had well-established equestrian and cattle-raising traditions were relied on for livestock management – skills that continued to be valued well into the seventeenth century. Experience with mining and aptitude in pearl fishing were also sought after in enslaved African workers. This finds a parallel in the way that Spanish settlers not only valued but made use of the smelting and technological skills of indigenous populations when it came to metal working in Central America; another parallel lies in the help given by Arawak peoples in South America whose agricultural knowledge and acumen were so invaluable that promises were given to them that they would be left as 'free people', commitments that were soon broken.[80]

The heavy reliance on those who were enslaved, indentured or otherwise unfree characterised Europe's engagement with the new worlds that they had stumbled into in the Americas. The extraction of resources of all kinds, from pearls to precious metals, from crops like sugar to tobacco, required manpower. Political crisis in England spurred an intensification of those demands. Protracted and bloody civil war in the 1640s resulted not only in the emergence of a new leader in Oliver Cromwell but in a major change of direction and ambition in foreign affairs. Cromwell's determination to stamp out royalists and rivals led to interventions in Ireland that resulted in more than 3 million hectares of land being confiscated from Catholic owners and handed out to his supporters, while a fleet sent to Barbados in 1651 paved the way for the conquest of Jamaica shortly afterwards as part of a wider effort to weaken Spain by taking its prized possessions in the West Indies.[81]

England's imperial ambitions transformed engagement with islands in the Caribbean as well as beyond. As early as the sixteenth century, indigenous Taíno populations as well as peoples across Central America had been decimated by overwork, malnutrition and epidemic disease.[82] Ideas for solving labour shortages were soon being sought in Africa. Supply was not only about numbers though: Spaniards in New Spain and other colonies often suggested that one African was as productive as four indigenous workers – with sums paid for enslaved people often reflecting high differentials accordingly. Perceptions about physical strength may or may not have corresponded to reality. What was more important, as one commentator noted in the 1650s, was that a coerced African labourer 'begins at three in the morning and works until eleven at night', compared to indigenous workers, who work 'from eight or nine in the morning until six in the evening'.[83]

The English decision to invest in global affairs in a direct and meaningful way through conquests of islands in the Caribbean brought about a significant acceleration in transatlantic slavery. Part of the motivation was a major yellow fever outbreak in Barbados that began in 1647 and spread not only through the Caribbean but across Central America too; this brought into sharp focus the need to find ways to exploit lucrative honeypots that could help generate valuable resources as well as revenue.

Life expectancies for Europeans who came to places like Jamaica were dismal, especially in urban areas, and, what is more, declined rather than improved in the century and a half after that island's capture by the English in 1655.[84] Yellow fever proved especially lethal to white settlers, more so than to enslaved people shipped from Africa, though the latter likely did not have immunity to this disease despite many historians' claims to the contrary.[85] Yellow fever was just one of a suite of diseases that proved devastating, alongside typhus, smallpox, influenza and malaria. Taken together, these all posed considerable threats not only to the indigenous populations, but to those from further afield – including enslaved people and the indentured and unfree from Africa, as well as European settlers of all kinds. Although the disease environment for each was different, some required particular conditions to flourish: malaria and yellow fever, two of the most dangerous and virulent diseases, were both spread by mosquitoes that needed hot, wet conditions in which to breed and spread the parasite and virus to human populations.[86]

The Atlantic coastal regions of South, Central and North America – as well as the islands of the Caribbean – provided perfect breeding grounds for the mosquitoes and for these two deadly illnesses. This ecological environment, which was perfectly suited to a particular insect species and as a result ill suited to most humans, shaped political, economic and military fortunes not only across this region but in Europe too. Yellow fever and malaria, diseases transmitted in the Americas by females of the *Anopheles quadrimaculatus* mosquito, were prices to pay for the so-called Columbian Exchange – the transfer of people, foods, ideas and pathogens that followed Columbus' voyage of 1492. Millions died as a result of the epidemic diseases which were largely responsible for shaping the boundaries of what one scholar has called 'mosquito empires' formed in the centuries that followed: where mosquitoes led, only those who were rash, brave or obliged followed.[87] Europeans were not the only colonisers, in other words, who reshaped the world; so too did insects.

Until the middle of the seventeenth century, the scale of transatlantic slavery from Africa was relatively small, averaging around 2,700 souls per year – a number that can all too easily mask the horror of the loss of dignity and liberty and the traumatic experience of being settled involuntarily in poor conditions, thousands of kilometres away from home.[88] The numbers being sent across the ocean against their will now rose dramatically in what has been described as a moment of profound transformation, nothing less in fact than the 'Africanisation of the Caribbean'.[89] Ecological opportunities provided by soils that lent themselves to cash crops, the efficient deployment of capital in the search for profits, the establishment of triangular trade system, the development of supply chains and the pricing of human lives all came together to create opportunities for some and misery for countless others.

16

On the Exploitation of Nature and People
(c.1650–c.1750)

For God's sake [capture] it without delay.

William Beckford (1758)

The English engagement with the Caribbean established the principle of conquest and colonisation as a state enterprise. It did not take long for new institutions to be set up to formalise investment overseas. The most important of these was the Royal African Company (RAC), founded in 1660. While this had obvious parallels with the East India Company (EIC) and the Levant Company which had both been founded a few decades earlier, it was not long before the RAC branched out beyond metals and agricultural crops into slavery. Indeed, in 1663, a charter was granted to the RAC by King Charles II for 'the buying and selling, bartering and exchanging of, for, and with any negro slaves, goods, wares and merchandises whatsoever to be vended or found' in West Africa.[1]

These developments had a dramatic impact on the ecologies of the Caribbean, the Americas and Africa itself. Shipments of enslaved people rose sharply, rising almost tenfold in the course of the seventeenth century, and reaching over 80,000 per year by the middle of the eighteenth century. By 1680, enslaved people made up around half of Europe's trade with Africa; a century later, in the 1780s – the peak decade for transatlantic slavery when more than 900,000 souls were sent from the coast of Africa – it accounted for over 90 per cent.[2]

The demand for enslaved people was driven by the vast profits that could be generated from tobacco, cotton, indigo and sugar – all of which required large amounts of labour. Returns on capital were substantial and rapid, with the enslaved paying back the cost of acquisition within eighteen months, 'with god's blessing', as one visitor to the Caribbean put it.[3] The incentive to make money drove technological and political change as well. Ships were developed that were fortified and specially designed to discourage uprisings of those being transported across the Atlantic.[4]

Shipowners in England seized the opportunity to invest in building bigger, better and faster vessels – and also a lot more of them. Tonnage of English shipping grew by seven times in just over a hundred years after 1570.[5] While the Spanish and Portuguese dominated the first century of transatlantic trade, the English and later the British built up a dominant position to become by far the biggest carriers of enslaved people, transporting around 2.5 million in the eighteenth century, around 40 per cent of the total.[6] The Dutch moved quickly too, quadrupling tonnage between 1500 and 1700 as they too became so involved in the slave trade that some contemporaries contended that they 'managed the whole trade' of the colonies in the western Atlantic, furnishing islands like Barbados 'with Negroes, coppers, stills, and all other things appertaining to the engines for making sugar'.[7]

Advances like these were crucial to the development of merchant shipping fleets, and to the evolution of military forces whose mastery of the seas would later prove the bedrock of imperial power and global empires – most notably the Royal Navy. The demand for seamen far outpaced population growth in England and then Britain as a whole, and helped change perspectives. Once an isolated corner of northern Europe, the British Isles now transformed into a global hub whose population became exposed to peoples, languages, commodities and climates in other corners of the world.[8]

Long-distance trade was part of what is often referred to as a triangular system through which textiles and manufactured goods were shipped from Europe to be exchanged in Africa for enslaved people who were then transported across the Atlantic, before vessels travelled home carrying sugar, tobacco and cotton. Such trade also played an important role in the creation of insurance markets, which in turn relied on the accurate pricing of risk. This in turn was linked to mathematical models and the careful evaluation of factors ranging from the competence and experience of a ship's captain to the route being sailed to the value of the cargo being transported – whether human or otherwise. This too had long-term consequences for the emergence of financial markets and the creation of centres of commercial power that established cities like London and Amsterdam as economic powerhouses.[9]

Such efficiencies had awful consequences, most notably in driving the price of enslaved people downwards. In just over a decade after 1664, average market prices in the West Indies fell by 25–30 per cent; over the same period, supply more than doubled. In other words, coerced labourers became cheaper and cheaper, which in turn stimulated demand further still.[10] In time, that changed attitudes, hardening the hearts and minds of enslavers, who came to treat fellow humans as though they were simply disposable. It was 'cheaper to work slaves to the utmost', said one planter in the middle of the eighteenth century, 'and by the little fare and hard usage, to wear them out before they become useless and unable to do service; and then to buy new ones to fill up their places'. In one of history's strange twists and turns, John Newton – a notorious trader of West African men, women and children who had never said a word in public against slavery – was the author of 'Amazing Grace', a hymn that became closely associated with the civil rights movement in the twentieth century and with those struggling for justice thanks to lyrics that include the euphoric lines 'My chains are gone, / I've been set free. / My God, my Saviour has ransomed me.'[11]

As one leading scholar has put it, enslaved people were 'systematically worked to death', with the Caribbean best described as a 'slaughterhouse'.[12] The ships that sailed across the Atlantic were designed to maximise profits by cramming as many humans as possible into the smallest amount of space, in horrific conditions with poor ventilation, shocking sanitation and perfect for the spread of disease. On one crossing, 200 captives died as smallpox took hold – with signs of the disease evidently missed before loading, when each individual would or should have been carefully inspected to ensure they were in good health.[13]

From the point of view of the merchant, death at sea was a poor use of capital. In the seventeenth century, it is estimated that 30 per cent of those who left Africa on Spanish ships and 20 per cent of those on

British ones never made it to the other side of the Atlantic. It was not long before these figures began to improve, driven by more rigorous inspections at point of sale and by better care during the crossing – though it would be hard to describe conditions as anything other than inhuman. The British again were pioneers in this regard, bringing mortality rates down to 10 per cent by the second half of the eighteenth century.[14] Captives were typically taken on deck every day to breathe fresh air and made to exercise, with the threat of a lashing from a cat-o'-nine-tails concentrating the minds of those who were reluctant.[15] This allowed the disposal of excrement and the smoking of the quarters with tar, tobacco and sulphur, before they were washed down with vinegar – steps thought to be helpful, but which offered little help against illness.[16] Nevertheless, these steps were an aspect of the efforts made to reduce loss of life, or, to put it more accurately, to improve profitability.

The relentless process of improvements had other important knock-on effects. Apart from the technological advances that boosted efficiency and provided greater returns in investment, there was a host of other advances that flowed from the maritime world. These included increasing sophistication of logistics required to provision and supply ships as well as ports and settlements that were often many thousands of kilometres apart and that required first-class administrators and bureaucrats who learned not only from their mistakes but from each other. This in turn helped professionalise the economy and for that matter government apparatus. It is hard to measure exactly what uplift these provided; but it stands to reason that they were extremely important.[17]

Those lessons were not shared well in places where personal responsibility was high, accountability low and temptation made irresistible by lack of oversight: in fact, as recent research has shown, the economics of Spanish trans-Pacific trade between Manila and Acapulco from 1570 to 1815 that transported silver as well as cochineal, sweet potato, tobacco and chocolate was compromised by overloading of vessels, endemic corruption of officials and poor decision-making related to bribery of ships' captains. These led to many late departures that took chances with the monsoon winds and ocean currents and dramatically raised the chances of shipwreck. As all the galleons were owned by the crown, losses of vessels such as the *San José* in 1694 which carried a cargo equal to 2 per cent of the Spanish empire's entire GDP had catastrophic consequences for Spain's economy.[18]

The relationship between the state and mercantile interests was different in northern Europe, though not completely clear-cut: after all, Queen Elizabeth I herself was an early investor in the slave trade, persuaded by Sir John Hawkins that 'Negroes were very good merchandise in Hispaniola and that store of Negroes might easily be had upon the coast of Guinea' to boost the royal coffers.[19] The overloading of slave ships went hand in hand not only with the search for gain but also with the dehumanisation of those being shipped. In recent years, considerable attention has been paid to questions around race, racism, slavery and the way that humans have sought to justify treatment of those of different skin colour, religious beliefs or sexualities, attention that has helped revolutionise the way we think about these issues.[20]

At the time, attitudes built on ideas and influences from the classical world, of which Europeans now claimed to be the true heirs. Herodotus, for example, opined that 'soft lands tend to breed soft men', a reference not to political leadership but to geographic bounty: 'it is impossible for one and the same country to produce wonderful crops and good fighting men'.[21] Being tough meant coming from land that required hard work; places that offered easy lives produced lazy men.

Montesquieu was heavily influenced by what he read, tweaking views that were not only widespread but repeated almost unquestioningly in texts by authors like Pliny, Strabo and Vitruvius. People 'are more vigorous in cold climates', wrote Montesquieu in *L'Esprit des lois* in 1748, exhibiting 'more confidence in oneself' as well as 'more courage' and a better sense of 'one's superiority' than those living in warmer conditions. 'The peoples in hot countries', he went on, were 'like old men': timid. Those living in Asia were 'indolent of mind and incapable of any exertion or effort'.[22]

Sweeping statements like this reveal much about finding justifications and explanations for how the cold climate of northern Europe enabled the rise of European global empires. They were also heavily imbued with ideas about race and about the right of European men to dominate other peoples and cultures in distant countries and continents. Those who lived in hot climates, said Montesquieu, were greatly in need of a wise legislator, unlike those who lived in Europe, who were, of course, ideally placed to provide such wisdom to others. Cold air restricts the body's fibres, he wrote, meaning that people lived like 'young, brave men'. In contrast to the lure of luxury experienced by those in other parts of the world, those from cold countries had 'very little sensibility for pleasure'. Indeed, Muscovites were so tough and so inured to comfort that they needed to be flayed to even feel alive.[23]

As Montesquieu realised, elaborate intellectual somersaults were needed to explain the apparently self-evident truth that those in northern climes had superior temperaments, were racially superior and deserved to rule over others. After all, if this had been the case, why was it that northern Europe had been home to barbarian tribes at the time of the Roman empire, and had produced so little by way of culture, scholarship and philosophy in comparison with the Mediterranean region? As always, Montesquieu had the answer: the Germanic tribes 'rose and quit the forests to subvert that great empire'. In the end, in other words, the cold, calm north had won out.[24]

Montesquieu and others were well aware that others had built empires in hot climates, so again manoeuvred to distinguish between the achievements of those with 'a certain vigour of body and mind' who were capable of making 'long, painful, great and intrepid actions', like Europeans, and those who lived close to or near the equator. Those who inhabited 'hot climates' suffered from cowardice which 'has almost always made them slaves'. Empires that had grown up in Mexico and Peru before the arrival of Columbus were fundamentally inferior, he argued. They were 'despotic', he said – presumably unlike their European counterparts.[25]

This was nothing compared to how those from Africa were written about, treated and reacted to – which makes for painful and difficult reading. Africans who were sold for shipment across the Atlantic were inspected by buyers as if they were animals rather than humans, even being compared to the purchase of 'black cattle'.[26] In the middle of the eighteenth century, one Lord Chancellor used a High Court judgement to confirm his opinion that 'a Negro slave ... is as much property as anything else'.[27]

Similar views were expressed in the notorious incident of the *Zong*, a ship whose captain had overloaded his vessel and made navigational errors that resulted in water supplies running low. As shortages and sickness took hold, over 130 Africans, many bound and shackled, were thrown overboard to drown. The ship's owners, a syndicate based in Liverpool, went to court to claim on their insurance for losses in a case heard by the Chief Justice, Lord Mansfield – a towering figure in maritime law. Mansfield had ruled that the fate of those who drowned 'was the same as if Horses had been thrown overboard'. This was not a case of murder, in other words, but one of dealing with excess farm animals.[28]

Those who benefited directly or indirectly from the slave trade thought little about the source of the wealth that built grand country houses, funded great art collections and improved the quality of life in comfortable surroundings at home.[29] Britain is a country, wrote one of Lord Mansfield's contemporaries, 'whose grandeur is founded on commerce' – rather than on the blood, sweat, tears and lives of enslaved black men, women and children toiling on sugar and tobacco plantations in the Caribbean and Virginia.[30]

Others did not so much turn a blind eye as convert reality into fantasy. 'Let an intelligent and educated man compare the deplorable state' of those who had been sold into slavery 'in Africa with the pleasant and easy life which they enjoy in the colonies', one delegate told the French Colonial Assembly. They should be thankful that they were 'sheltered by all the necessities of life, surrounded with an ease unknown in the greater part of the countries of Europe', with access to good healthcare, surrounded by loved ones and 'in peace with their children'. It would be hard to paint a more misleading picture. Rather more accurate were the views of the coerced labourers themselves, who cursed the way that Europeans exploited everything they came into contact with: 'The devil was in the Englishman,' said one enslaved African in Barbados; 'he makes the negro work, he makes the horse work, the ass work, the wood work, the water work and the wind work.'[31]

Some admitted to being moved by the plight, sorrow and despair of those whose lives were spent making money for others, against their will. 'My Heart hath been ready to bleed for those poor Wretches,' admitted one surgeon on a slave ship.[32] Others spoke of their sadness at seeing the separation of 'fathers from sons, husbands from wives, brothers from brothers', declaring that it was impossible not to be moved by such scenes: 'what heart, however hard it might be, would not be pierced with piteous feeling to see that company?'[33] It was dreadful to see the 'Barbarity daily exercised on the Bodies of the miserable Negros; The piercing Cryes and dolefull Lamentations that every Day enters one's Ears both in Town and Country.'[34]

Such sentiments rarely translated into more prolonged reflection, and were rather the passing thoughts of those briefly moved by a sense of compassion. It did not take long to become 'cruel and hard-hearted', and some even encouraged their children to become just that when they were young. 'The first Play-Thing put into their hands is commonly a Whip with which they exercise themselves upon a Post, in Imitation of what they daily see perform'd on the naked Bodies of those miserable Creatures, till they are come to an Age that will allow them Strength to do it themselves.'[35]

Ideas of African barbarism also took hold – dispensing with the history of West Africa and its contemporary sophistication. Dutch writers in the seventeenth century had written with admiration of Benin City, for example, describing its 'very broad streets, each about 120 feet wide', and the royal quarters that were 'easily as big as the town of Haarlem and enclosed by a remarkable wall', while the palace of the Oba (ruler) was 'so large that you can feel no End', so vast in fact that walking its length was exhausting.[36]

These accounts were replaced by representations of Africa as dangerous, barbaric and cruel. If Africans were not sold, wrote the same ship's surgeon who claimed to be upset by the 'Abuse and Death' of enslaved people, other Africans 'will surely starve them to death'.[37] Similar views were expressed to a parliamentary committee by Richard Miles, a slave trader who had 'bought some hundreds – some thousands of slaves' and claimed that if he had not done so, many would have been 'sacrificed at the burials of great men'. Another slaver confirmed that buying those who had been enslaved helped save their lives, 'preserving them from being sacrificed, or otherwise put to death'. In such warped logic, the slave trade became one of salvation rather than of suffering and sacrifice.[38]

It was not just that coerced labour from Africa was all about lining the pockets of merchants, investors and landowners – as in Barbados, which saw land prices multiply by almost twenty times in a few years and went from being a sparsely populated, nondescript island to 'one of the richest Spots on earth under the Sun' following the development of sugar plantations in the mid-seventeenth century.[39] The fortunes that were made for white plantation owners were built on the fact that Africans who were shipped across the Atlantic were far more productive, especially in tropical climates, than those who controlled and traded their lives. Nowhere was this clearer than in West Africa itself, a place described by one scholar as a place for most white men 'to get away from or to die in'.[40] In the early eighteenth century, the Royal African Company lost half those it sent from Europe to West Africa within a year. Just one out of every ten Englishmen who went to work in this region came home alive before the middle of the nineteenth century.[41] Such was the gamble of trying to strike it rich.

The main cause of such high mortality levels was the disease environment. The lack of innate immunities of European populations stood in sharp contrast to those that had been developed over thousands of years by African populations to cope with malaria, a single-cell parasite that is transmitted to humans by mosquitoes. The two most prevalent strains of the disease are *vivax* malaria, a mild form that is rarely fatal, and *falciparum* malaria, which is far more virulent and deadly, with the latter requiring higher temperatures than the former to become infectious. Classic symptoms include fever, chills and nausea and, in serious cases, impaired consciousness, coma and death. Repeated infection results in deteriorating health, including decreased ability to resist other diseases. Populations in sub-Saharan Africa have developed a vast range of immunities to malaria – including the sickle-cell trait, a blood-cell disorder that can reduce the likelihood of developing cerebral malaria after a *falciparum* infection by 90 per cent. Populations that do not present a high frequency of sickle-cell traits have other forms of innate resistance, including the HbC allele or high levels of antimalarial antibodies.[42]

Falciparum malaria had established itself in the Caribbean and South America through multiple independent introductions that were linked to the slave trade – attested by genetic and archaeological evidence, with mitochondrial DNA showing a clear link between African and South American haplotypes.[43] Settlers from Portugal and Spain may also have brought the disease with them, as malaria was endemic in both countries at the time of the conquest.[44] Strains may also have entered the New World from Asia and the western Pacific either before Columbus' crossing or as a result of the subsequent creation of trans-Pacific trade – which would help explain the high genetic diversity of *vivax* parasite in the Americas.[45]

However, malaria does not seem to have taken hold in the colonies of North America until the 1680s, when it did so suddenly and with a vengeance. A series of epidemics hit most southern US colonies, with high mortality in Virginia and South Carolina, where Charleston was particularly badly affected. *Falciparum* malaria then became established, leaving long-term impacts on the white population, such as in Christ Parish in South Carolina, where records show that 86 per cent of the population died before the age of twenty, and more than half before the age of five.[46]

The spread of malarial disease is dependent on three variables, namely the parasite itself, mosquitoes – of which several species prefer to feed on humans rather than on any other vertebrate – and climate, itself closely linked to the habitats of mosquitoes, which require enough water and hot enough temperatures to reproduce, survive and spread. The early 1680s in particular were a time of turbulent and unusual climate conditions, with major El Niño events in 1681 and 1683–4 and again in 1686–8.[47] The correlation has led some scholars to ask whether these weather anomalies played a significant role in allowing *falciparum* malaria to cross a threshold to establish itself successfully in North America.[48]

The appearance of malaria was to have profound consequences that lasted centuries – plaguing the south of what became the United States long into the twentieth century, reducing life expectancy, impacting productivity and being both an indicator and a motor of poverty in the American south.[49] As it was, the second half of the seventeenth century saw a wider shift by plantation owners in states like Virginia; they had previously had a preference for indentured English workers who spoke the same language as the owners, had similar customs and were familiar with English agricultural practices. Demand was difficult to satisfy even before the arrival of malaria – at which point those who could choose where to work actively avoided the American south if they could. Combined with this, however, was a dramatically enhanced need for a labour force that resulted from so-called landgrab in the decades after 1650 in which millions more hectares that were ideal for cultivation were seized – further stimulating appetite for the cheapest, most effective workers. Hardening racial prejudice played a role, with some like the Reverend Godwyn Morgan declaring in 1680 that the terms 'Negroe' and 'Slave' had 'by custom grown Homogenous and Convertible'. As if to underline what this meant to him and to many of his contemporaries, he added that 'Negro and Christian, Englishman and Heathen, are by the like corrupt Custom and Partiality made Opposites.'[50]

As far as colonial settlers were concerned, therefore, African manpower was the answer to multiple problems. As malaria kicked in, decimating the existing population, the desire for more coerced labourers rose sharply. Buyers were not keen to buy enslaved people from just anywhere, but had strong preferences for acquiring those from malaria-ridden regions in Africa, whose populations had high levels of resistance to the disease. This resulted in what one scholar has called a 'malaria premium': higher prices for people most likely to survive infection – and therefore prove to be better 'investments'. Captives from the Gold Coast were considered robust, strong and little affected by disease, while those from the Niger Delta were thought to be less hardy.[51]

It was not so much disingenuous as disgusting then that the depiction of blackness in general and of enslaved peoples expressly shipped because of their genetic advantages should become associated with inferiority, when the opposite was the case. Those brought across the Atlantic in subhuman conditions and subjected to lives of misery were not only the backbone of fortunes and status for their 'owners', but were doing work that the latter were poorly suited to do themselves. Far from being 'inferior', those put into chains were not only stronger but genetically better suited to the Americas than those who exploited them.

In a further surreal twist, states set about codifying the deprivation of rights of fellow human beings – both to justify the principles of enslavement and racism and to enshrine these ideas in law. Slave codes began to be passed by one state after another that formalised the status of slavery as a lifelong condition, as having a racial basis and as being an inherited condition, with coerced labourers being defined as the property of their 'owner': 'buying' another human meant controlling and benefiting from the fruits not only of their labour but that of their descendants too.[52]

This was first set in law in Virginia in 1662 whereby the legal status of 'all children born in this country shall be held in bond or free only according to the condition of the mother'. This was known as *partus sequitur ventrem* – or 'the offspring follows the womb', which among other things meant that the children of women who had been raped by owners of coerced labourers or other white men were condemned to lives of servitude, thus concealing the acts of sexual violence that had led to their conception.[53]

Controlling fertility became an important part of the lives of enslaved women, and in due course led to advances in the development of herbal remedies by black women to control the spacing of childbirth, to act as contraception and to track stages of pregnancy.[54] This was itself partly a response to the fact that an estimated 54 per cent of pregnancies among enslaved women resulted in stillbirths, infant mortality and early childhood mortality.[55] Appalling experiments on women of African descent in the first half of the nineteenth century, almost all without consent, proved foundational for modern gynaecology, obstetrics and healthcare for women – debts that are likewise little known and even less acknowledged.[56] This does not mean that the benefits have been shared equally: today in the United States the disparity between deaths of infants born to white and black mothers is actually greater than it was in the first half of the nineteenth century, while pregnancy-related mortality is three to four times higher among black women than among white women.[57]

Then there is the Mason–Dixon line which is often seen as marking the boundary between southern slave states and northern free states. This was not only a political and commercial frontier but an epidemiological one too, with *falciparum* malaria flourishing to the south – precisely where slavery was not only prevalent and defended as a principle, but eventually a fundamental cause of civil war in the United States in the 1860s.[58] The role that climatic factors had in creating the rich soils of the American south has already been noted in this book, along with the role these play in shaping US presidential elections. However, this can be developed further, with recent research suggesting that the counties with large populations of enslaved people at the time of the abolition of slavery in the 1860s are more likely today not only to vote Republican, but to oppose affirmative action and express racial resentment and sentiments towards black people.[59]

The major expansion of the exploitation of human labour deepened the ecological transformation of large swathes of land in the New World which diverted the benefits to those in the Old, much of it delivered by the manpower and lives of those who had been transported from Africa. This included the importation of coffee, tea and chocolate, of silver and raw materials that were hacked out of the ground and from the sides of mountains, and of cod from Newfoundland that found its way on to European tables by the boatload.[60] The benefits were manifold. For example, indigo production soared in the 1740s, rising from just over 2,200 kilos in 1746 to more than 62,000 just twelve months later.

Such jumps owed much to the consolidation of estates that allowed efficiencies of scale, but they owed more to the huge increases in the size of the coerced labour force working the land, such as in Jamaica where numbers almost doubled between 1740 and 1774 – a figure that excludes many tens of thousands who arrived on the island but did not survive. The sheer numbers of enslaved people being transported across the Atlantic, combined with good climatic and economic conditions, meant that prices fell – driving a permanent cycle of demand for coerced labour, and rising production.[61]

So valuable did this trade become that it was a key part in the struggle for the mastery of Europe, as well as in the creation of and competition between fledgling global empires. Naval victories by the British against the French in West Africa and the Caribbean not only sapped the finances of the latter, but made it more difficult to fund efforts against the former either locally or elsewhere. During the Seven Years' War of 1756–63, the conquest of the West Indies loomed large. One plan involved the British swapping control of Guadeloupe, which they had taken from the French, for France's territories in Canada, so great was Canada's perceived strategic value. This was preposterous, William Burke wrote in riposte to discussions in Parliament. Canada, he told MPs, 'produced no Commodity except Furs and Skins', and offered 'little Returns to make the English merchant'.[62] Guadeloupe, in contrast, produced sugar, cotton, tobacco and more besides that was worth about forty times the wealth of the trade of Canada each year.[63] Despite these entreaties, the French managed to get back Guadeloupe as well as Martinique and St Lucia – but at the price of all its possessions in North America, except for the territory of Louisiana west of the Mississippi River.

The preoccupation with controlling resources was intense. 'The negroes and stock' of Martinique, an island controlled by the French, 'are worth above four millions sterling', William Beckford had told the prime minister, William Pitt the Elder, in 1758. 'For God's sake', he said, capture it 'without delay'.[64] A prominent enslaver, Beckford had his own interests to promote – a pursuit made easier by the fact that he had the ear of the Prime Minister, who took to saying that Beckford 'had done more to support government than any minister in England', thereby surprising other parliamentarians who viewed him 'as a wild, incoherent, superficial buffoon'.[65] Beckford was motivated, above all,

by concerns that were shared by many involved in the sugar industry –
namely, that taxes would be levied to help support British war efforts
at a time when 'the low price of sugar, high freights, high insurance,
and hoops and staves' threatened to eat into their profits. Pitt obliged,
overruling his own Chancellor of the Exchequer to abandon a special
tax on sugar, earning the goodwill and support not only of Beckford but
of many others in the colonies. It paid to have friends in high places.[66]

The volumes of sugar being imported back to Europe naturally
brought prices down rapidly, enabling consumption by an increasing
share of society; this in turn encouraged people to earn more money
that could be spent on luxuries that were previously unavailable,
unaffordable or both. This was a core part of an 'industrious revolution'
that drove a rise in productivity and an increase in levels of income.[67]
Some scholars have linked this to a further set of other social responses,
such as growing interest and investment in new ideas, that stemmed
in part from rising wages and the gains that could be obtained from
innovation and technology.[68]

As such, the increase in goods, materials and foods being drawn
back to Europe from the Americas in general was important from the
perspective not just of ecological dividends being recirculated from one
part of the world to another, but of powering a much wider division
between what are often referred to today as the global north and the
global south. Those in Europe did not just advance because they were able
to dominate other parts of the world, or even because they could adapt
ecosystems for their own benefit and reap the rewards; those rewards
themselves brought further gains that themselves proved crucial. What
mattered most was not so much the quantities of goods brought back
to Europe as the range and novelty of what became available during the
Age of Discovery. Elites secured the bulk of the profits available from
transatlantic trade, although these were not always substantial. What
made a difference, then, were the gains from variety, as life became
sweeter and more stimulating and, furthermore, became available to
rich and poor alike.[69]

The case of sugar again proves instructive. In 1700 England imported
around 7,000 metric tons; a century later, this had risen to 150,000
tons. On a per capita basis, consumption of sugar in England increased
twentyfold between 1663 and 1775.[70] Apart from having the obvious
effect of dramatically increasing the availability of what had previously

been a luxury product, this consumption of sugar also provided a calorie boost for the growing working class in Europe. It has been argued that the increase in sugar imports between 1600 and 1850 increased English welfare and living standards by an astonishing 8 per cent.[71]

Sugar availability was significant in other ways too. For example, the innovation of adding sweetener to hot drinks such as coffee and tea produced not so much a spike in demand from China as a tsunami, as consumption rose by 400 times in the course of the eighteenth century. This in turn galvanised the East India Company and others to invest in direct trade with East Asia, and in time to begin production of tea in India. The ecological transplantation of natural commodities to new locations was paralleled by that of coffee, native to Ethiopia and Yemen, which was then introduced to plantations in Java in South-East Asia and Surinam in South America by the Dutch soon after 1700, to Martinique and Saint-Domingue in the Caribbean by the French and to Jamaica by the English – by 1800 Jamaica was producing a third of the world's coffee.[72]

In some cases, there were unexpected side-effects of the transplantation of plants and the opening up of new markets. High taxes on tea and tobacco imports to Britain produced a thriving smuggling industry, with some estimates suggesting that between 50 and 90 per cent of all the latter were brought into the country surreptitiously.[73] While difficult to measure accurately, it stands to reason that there were gains arising from the improved navigational skills of sailors who had to sneak in to unwatched bays and from a new industry whose job it was to stay one step ahead of government tax authorities – perhaps coincidentally an area where Britain has long led the world: even in 2021, British tax havens in the Cayman Islands, Bermuda and the British Virgin Islands are ranked as the three top enablers of corporate tax abuse on the planet.[74] Engagement with the New World led to new possibilities and new tastes – and demanded new ways of doing things that had long-lasting consequences.

Near insatiable demand for tea which rose dramatically in the course of the nineteenth century, meanwhile, led to the production of opium in Bengal which was typically sold at auction in Calcutta in exchange for tea sold in Canton (Guangzhou) in order to avoid bans imposed by Qing dynasty emperors.[75] Trade eventually became so important to the British Exchequer that the government used military force to gain favourable trading terms, control of China's customs system, access to ports such as Shanghai and even direct sovereignty over Hong Kong – part of what has later become known in China as the 'Century of Humiliation', and a fundamental part of contemporary conceptualisations of world affairs in Beijing and beyond.[76] Again, it is hard to overstate the impact of new global connections, of new patterns in supply of and demand for commodities and of the attractions of a financial world that enabled merchant capital not only to reshape ecologies but to transform global politics.

As with sugar, increasing availability meant lower prices and greater affordability. The impact on social interaction was profound. There is 'no bourgeois household where you are not offered coffee', wrote one Parisian in the eighteenth century, 'no shopkeeper, no cook, no chambermaid who does not breakfast on coffee with milk in the morning. In public markets and in certain streets and alleys in the capital, women have set themselves up selling what they call *café au lait* to the populace.'[77]

These interactions were not just about improvements to the quality of life, but were important to the exchange of ideas, to innovation and to collaboration. It was no coincidence, for example, that tea and coffee houses played key roles in the development of the stock exchange, of the insurance industry and of political debate and in the dissemination of print media during the Age of Enlightenment – a term that suggests a time of exceptional brilliance in Europe without acknowledging the debts that were owed to the planters, pickers and diggers who were an essential part of the story.[78]

Sugar, coffee and tea were just some of the commodities that characterised a time of intensive and deepening global exchange; attention could also be paid to individual goods like tobacco, timber or furs that played a role individually and collectively in the acceleration of pre-industrial capitalism, which was centred on geographic, ecological and environmental inequality and exploitation.[79] Commodities could see explosive growth in production, as was true of cotton, with exports from the United States to Great Britain rising ninety-three times between 1791 and 1800 and then rising another seven times before 1820.[80] This owed much to the development of the Whitney gin, a device that separated cotton fibres from their seeds, and the huge

British demand in terms of domestic consumption and for export, not to mention the almost doubling in size of the United States after the Louisiana Purchase, which was followed in 1845 by the annexation of Texas. The annexation delivered land that not only could be turned over to cotton but was ideally suited to be offered as rewards to political allies and to become part of what one eminent scholar has called the 'military–cotton complex'.[81] As with the other phases of the expansion across the Atlantic, the impetus of mercantile interests and landowners, combined with a targeted and relentless search for a return on capital, proved not so much crucial as the driving force.[82]

Often, there were steep environmental prices to pay. Overwork of the land damaged soils, caused a decline in fertility and resulted in erosion. Even by the start of the eighteenth century, some were commenting that fields had turned into 'barren, rocky gullies, runaway land, waste land ... much worn out and not so fertile as it was.' According to one clergyman in Barbados in the 1730s, 'the face of the earth appeared, as it were, a dry crust, burnt up and gaping', while a contemporary noted that 'excessive drought, the number of people running off, and the miserable condition and poverty' of an island that had been the source of colossal fortunes only a century before now meant that famine and disaster seemed inevitable.[83] According to Eric Williams, one of the Caribbean's most influential historians, the ecological degradation was a blessing in disguise, helping to reduce profitability and therefore easing the way to the abolition of slavery. Commercial realities rather than compassionate sensibilities were what made the difference.[84]

Plantation agriculture in the Caribbean opened up other environmental hazards too. These included destructive landslides that were dangerous to workers and damaging to ecosystems. Rampant deforestation did not only have implications for soil run-off but magnified the threat posed by hurricanes because of the removal of trees that served as natural protection for animal and plant life alike.[85] Colonial legacies of ecological intervention remain an important factor in climate and biotic vulnerability to this day, to judge from some studies.[86]

The secret to the rise of Europe – and especially of northern Europe – was the ability to access, develop and control millions of 'ghost acres' of land that were either blessed with natural resources or that could be repurposed to produce crops and goods that were necessary and desirable thousands of kilometres away. This had the twin effect of diverting calories, energy and raw materials to Europe and so supporting economic and demographic growth that would otherwise have not been possible, and also of liberating land for other uses.[87]

Much of this was poorly understood and taken for granted, with buyers and sellers knowing and caring little about where goods, products, components or resources came from. A good example comes from hat fashion in Europe in the second half of the seventeenth century. King James I is sometimes given credit for starting the craze when he ordered twenty beaver hats when he became king of England in 1603. Sweden's successes in the Thirty Years' War influenced the popularity of tall 'cavalier's hats', with hat styles then influenced by the latest French fashions. This led to surging demands for beaver fur, which has the ability to hold its shape through felting, and for vicuña, an alternative from the Andes. In short order, this led to beaver populations in most parts of Europe being all but hunted to extinction, and to commodity frontiers being expanded elsewhere, most notably in what is now Canada and Peru. Tens of millions of beaver pelts were imported to Europe from North America, with England alone re-exporting more than twenty million between 1700–1770.[88]

Other components for these hats – such as mastic gum that was used as a stiffening agent and ostrich feathers for decoration – led to surging pressures on plant and animal life in West Africa, Sudan and the Levant where these items were in seemingly inexhaustible supply. All were parts of a process that has been termed 'Invisible Globalization' – that is to say, the centralisation of resources and the expansion of supply chains alongside patterns of overconsumption and depletion. All went unnoticed by consumers.[89]

This was true of no end of other commodities, goods and products. Animals were hunted to the edge of extinction and sometimes beyond. Demand for ivory piano keys and billiard balls in the nineteenth century in the drawing rooms of the burgeoning middle classes in Europe and North America decimated African elephants; beavers were trapped in vast numbers not only for their fur but for the castoreum that could be extracted from their anal scent glands and was used in popular treatments for fever, headaches, spasms, epilepsy and mental illness. Demand for products from whaling, including whalebone and oil made from blubber that was boiled down and used in street lighting, in miners' lamps and in lubricants for guns, watches, sewing machines and typewriters, led to the collapse of whale populations in the Atlantic and the opening up of new whaling stations as far away as the Falklands and the Pacific in the search for new supplies.[90]

Rubber booms saw another surge in the plantation agricultural economy and the intensification of monoculture; demand was spurred both by the growing use of this material and by the development of vulcanisation which stabilises rubber and allows it to be used across a much wider range of temperatures. The booms were concentrated in the Amazon forests of Brazil, where latex was collected from rubber trees, until Sir Henry Wickham smuggled tens of thousands of seeds back to Kew Gardens in London. These were then transplanted to South and South-East Asia, where they were used to establish rubber plantations that not only became central to local, regional and global economies, but led to mass migration through the familiar practice of indentured labour.[91]

Developments such as these fuelled the Great Divergence – the shift by which the western world caught up with, overtook and dominated vast parts of the world. To most Europeans, this was a story of entitlement as much as one of enlightenment, a natural outcome of inherent sophistication and progress which those living in other parts of the world could not and never would match, a result even of racial superiority. Underpinning all these ideas, however, were the cold realities of resource exploitation, of the creation and improvement of supply chains and of turning blind eyes to abuses, of not asking too many questions about who lost while others gained.

Such questions were hardly new. After all, many societies had experienced problems of over-consumption, exhaustion of natural resources or strains on ecosystems that were excessive and unsustainable. Centralised states had at their roots the principle of drawing in resources of all kinds – whether minerals or metals, animals or foodstuffs, tax revenues or human labour – from the periphery to the core. What was different in the centuries that followed Columbus, however, was the global scale on which these cycles operated. Exploitation of landscapes no longer just meant pushing geographic frontiers beyond existing territories, it meant an expansion across oceans and continents.

This in turn was enabled by advances that are easy to dismiss as dull or to overlook altogether, in matters like the logistics of shipping, including docking facilities that enabled rapid turnaround of vessels and the efficient onward distribution of goods; or like the calculation of pricing when purchasing goods at source, which in turn owed much to information gathering and analysis, taking in factors such as the size of harvest, the value of commodities or weather factors.

Calculations like these were extensive and constant, and not just for Europeans involved with trade of all kinds. For example, traders in Angola deliberately restricted the number of captives in the market place at any one time in order to drive prices up. Then there was disease, illness and malnutrition that also led to extensive loss of life: an estimated 6 million enslaved men, women and children who never even made it on to a ship, let alone across the ocean, lost their lives as a result of the transatlantic slave trade.[92]

Trading seasons for trafficking captives from Africa to the New World meanwhile were dictated by rainfall: it was difficult for merchants to move heavy cargoes during downpours that directly affected the ability to reach coastal ports from the interior. Navigational dangers increased during the tropical rainy seasons in Africa, the Guianas and north-east Brazil, while estuarial sandbars off the coast of West Africa and reefs off that of the Americas posed additional risks that rose and fell depending on season and on seasonal variations.

This was not the full extent of the relationship between slavery and climatic conditions. In fact, Atlantic rainfall in Africa not only impacted harvests, but shaped the distribution of crops, determined growing seasons and directed the farming calendar. This was important, of course, in its own right for domestic agricultural production, but it was significant too for the wider trade in terms of selling cash crops to Europe and for buying foodstuffs for enslaved Africans. As such, there were optimal times for sellers to seek to profit, with exports highest in September–October, immediately after the yam harvest, and lowest during yam planting in May–June. These rhythms suited merchants, as they also fitted around supply constraints in Atlantic Africa and allowed provisioning before the ocean crossing. As one agent of the Royal African Company put it, the 'best time for Negroes to arrive is between December & June, being a healthy time & affording plenty of provisions, and ye rest of ye Year being ye reverse'. Such realities underpinned the transatlantic slave trade.[93]

The role of climatic and weather conditions was closely linked in other ways too. For example, higher temperatures had an impact on rainfall levels and on agricultural production. Rises of just 1 °C resulted in measurable and significant reductions in exports of coerced labour – a phenomenon best explained by the rising costs of enslavement since provisions needed to be obtained for those who had been enslaved and for those who enslaved them. These trends were particularly marked in drier regions which were more sensitive to changes in temperature and rainfall and where agricultural productivity varied accordingly. Furthermore, warmer conditions resulted in higher mortality levels, both as a result of an increased disease burden and because of food scarcities, with again the results being most striking in regions that were least ecologically resilient and those where pathogens flourished most easily.[94]

As with the sources of goods that made life in Europe more interesting, more exciting and more varied, these rhythms were difficult to identify or even to see. In some ways, this is not a surprise, for invisibility characterised the way that history itself was written. The arrival of the pilgrims aboard the *Mayflower* in Plymouth Bay in 1620, for example, was seen for centuries as a seminal moment in the telling of the origins of the United States, while the landing of the *White Lion*, the first known arrival of indentured Africans at Old Point Comfort in Hampton, Virginia, a year earlier was relegated to footnotes, or not mentioned at all. The foundation of many of the most illustrious seats of learning in the United States, including Princeton, Yale, Georgetown and Harvard, owed much to the profits of slavery and in the case of University of Virginia to coerced labour itself.[95] US politics likewise owe much to the fruits of slavery, with more than 1,700 members of Congress between 1789 and the 1920s 'owning' or having once 'owned' other human beings.[96]

Not everyone was blind to the new world that was being opened up by long-distance trade. 'I never more will drink Sugar in my Tea,' wrote Aaron Thomas, ship's purser who saw for himself the backbreaking way sugar was made and the conditions it was made in; 'it is nothing but Negroe's blood'.[97] Others too were clear-sighted about just how important the exploitation of humans and nature was. 'If the negro trade was lost,' one businessman of the mid-eighteenth century observed, 'the Colonies must be lost.'[98] There was a choice to be made, in other words: wealth and glory; or compassion and fewer rewards.

Daniel Defoe, later to find fame as author of *Robinson Crusoe*, set out the equation in similar terms. 'No African trade, no negroes,' he argued; then 'no negroes, no sugars, gingers, indicoes etc; no sugar etc no islands, no islands, no continent, no continent, no trade.'[99] He could have added 'no empire, no rise of the west'. Defoe was no critic of the slave trade, in fact quite the opposite, this comment coming not from a critique but from an impassioned defence of the importance of British commerce: the slave trade was vital to trade in general; in Defoe's eyes, that justified the violence done to African men and women.[100] Some, like the Quakers, took increasingly principled stands, with Joseph Woods in 1784 questioning whether it was 'better a thousand poor unoffending people should be degraded and destroyed [than] the inhabitants of Europe should pay a higher price for their rum, rice and sugar?' Those who buy any such goods, observed a contemporary, 'become partaker in the guilt of it'.[101]

Few took such principled views, preferring to look the other way – or to benefit directly from trade. The 'discovery' of the Americas was of course nothing of the sort; indeed, despite his place in historical narratives, Columbus was not even the first European to cross the Atlantic. And yet the new arrivals reshaped and renamed the world they now stepped into and that they regarded as theirs by right. New names reflecting the triumph of Europe were layered over those of indigenous peoples. Continents, countries and states came to bear the names of their conquerors and new masters, from America to Colombia, named for Amerigo Vespucci and Columbus, from Pennsylvania, named after William Penn, to Venezuela – literally 'Little Venice', because the houses that local populations lived in on stilts surrounded by water reminded the Spanish of the Italian city.

Yet the idea that the Europeans embarked on an Age of Discovery, a golden age that is associated with a sense of curiosity and exploration, of adventure and search for knowledge obscures the fact that the driving force behind the expansion of horizons was merchant capital and the thirst for financial rewards. There were scholars, certainly, whose minds were inspired and opened by new possibilities; but, as in any day and age, scientific research has always been closely linked to the availability of resources, of funds and of time that allows the flapping of intellectual wings in place of the planting and harvesting of crops.

This model shaped the world of today in the Americas as well as in Africa. For example, regions in the Americas that were well suited to cash crops that required intensive farming and benefited from large pools of labour became places with high levels of inequality and limited distribution of rights among the population. Locations better suited to less intensive forms of agricultural production on the other hand – regions that favoured the growing of wheat, for example – proved to be more egalitarian, with better distribution of rights among the population. The short explanation for this is that crops that required lower labour input produced smaller profits and therefore meant there was less to fight over, and more reason to co-operate. This is one of the reasons why it is possible to identify a strong relationship between a country's socio-economic development and its distance from the equator.[102]

What mattered, of course, was the combination of temperature, rainfall and soil quality that in turn together determines where particular crops could – and could not – be grown. But the disease environments of tropical locations were of key importance too. In an alternative reality where African sailors had crossed the Atlantic first and taken advantage of conditions the way that the Europeans managed to do, the last 500 years of the world's history would have been very different indeed.

Whether this different scenario would have meant no transatlantic slavery, however, is another matter. For one thing, the ability of Europeans to ship millions of people across the ocean owed much to the existence of established markets for enslaved people in Africa that they could tap into easily and quickly – even if the insatiable appetites became overwhelming almost immediately. By 1516, some African leaders were already demanding a stop to the enslavement of local populations, begging for it even, for it was already clear that it brought negative and damaging results locally. African slavery was not only a transatlantic issue. Around two million of the estimated seven million people enslaved or sold by the Ottomans came from sub-Saharan Africa. Some caravans are attested to have conveyed thousands of captives, while other records suggest annual sales to port cities in the Red Sea of as many as 10,000 people per year.[103]

Over the long term, the impacts of slavery on Africa were catastrophic. By 1800, the population of the continent was half of what it would have been had the slave trade not existed.[104] That of course meant that manpower that would and could have been involved either in ecological transformations of Africa or in other productive activities was deployed elsewhere and for the benefit of others.[105] Indeed, even the end of the slave trade did not immediately result in more positive outcomes; in some parts of Africa, for example, prisoners of war had previously been sold into captivity. With this option closed off, many were put to death.[106]

But there were more profound effects that changed social and political development in many parts of the continent too. Supplying captives to traders on the coast demanded constant raiding to secure near endless numbers of captives – which created a vicious and circular world of enslaving or being enslaved. That in itself had other consequences, not least demand for weapons and above all guns – areas in which Europeans had an advantage by the time of enhanced contact with Africa, an advantage maintained and furthered in part thanks to the sheer number of conflicts in Europe and between Europeans that incentivised the development of improvements in reliability of firearms.[107] The demand for guns, itself in part a function of the need both to defend against raiding and to use in raids, became a motor of intensification of the slave trade in its own right, as well as propelling the emergence of highly centralised states dominated by military elites such as Oyo, Dahomey and Asante, each of whose fortunes were closely connected to the European expansion.[108]

While some of these states consolidated, others splintered, with the Jolof confederation splitting into a series of smaller kingdoms under the pressures and demands of slave traders that were so great that some rulers in West Africa complained that it was not only prisoners of war, members of minorities and those of low status who were being sold and shipped across the seas, but high-ranking figures and even members of royal families.[109]

In this age of mounting violence and insecurity, it was perhaps not surprising that ties between villages weakened, communities became introverted and levels of trust plummeted dramatically.[110] Societies atomised and disintegrated, leading to fragmentation that encouraged differentiation along ethnic and other grounds. Research has suggested that these breakdowns developed into long-term issues which are still prevalent today in many parts of West Africa, and explain low levels of co-operation, low levels of trust and poor economic performance. Regions that provided large numbers of captives to be shipped across the Atlantic are worse off now thanks to the historical effects of slavery. In other words, it is not just that peoples and places in Africa paid a heavy price centuries ago; they continue to do so to this day.[111]

Slave raiding had a significant impact on age and gender balances, with demand being heavily skewed in favour of securing male captives of working age whose labour was most valuable across the Atlantic. The instructions given to Cesar Lawson, a slave-ship captain, by a member of the merchant consortium in Liverpool were typical. The vessel under Lawson's command was 'allowed to carry 400 Negroes, and we request that they be all males, if possible to get them, at any rate buy as few females as in your power'. It was important, Lawson was told, to 'be very particular, select those that are well formed and strong; and do not buy any above 24 years of Age'. The captain's patrons further instructed him to keep the ship 'very clean' and to maintain discipline on board. Lawson was also told to ensure that no officers or crew 'abuse or insult them in any respect'. In the world of slavery, this passed for a rare expression of kindness.[112]

Investigation of the impact of gender both in locations where enslaved people were taken from and in those they were exported to is complicated by the fact that the proportion of women and children taken across the Atlantic varied greatly by region and changed over time. Price differentials between men and women generally were much smaller in the Americas than in Africa, partly because planters soon recognised that enslaved African women had a high work rate.[113] In fact, on some estates, women made up the majority of field hands. Gender shaped work roles beyond the fields too, with men being used as mechanics, blacksmiths, carpenters, coopers, masons and so on, while women were made to cook, nurse and act as midwives – jobs that brought less prestige and fewer material rewards than the men's, which in turn affected models and expectations both within slave populations and beyond.[114]

Rape and sexual fantasies were particularly virulent among enslavers and white settlers, with some scholars arguing that ideas about abuse and the use of violence became more prominent, not less so, over time.[115] Women who chose to acquiesce, either in the hope of material rewards or to self-convince that they had choices that did not exist in reality, had little chance of seeing their offspring being freed, although again patterns were different in different locations with manumission of sexual partners and children more common in the Spanish West Indies than in most other regions.[116]

There was, of course, considerable variation in the ways that enslaved people were treated, dependent not only on the personality of the 'owner' but on the size of plantation, its location and what work a coerced labourer was made to do: those working in enslaver homes typically benefited from better food, clothing and housing than those forced to work in the fields, and endured less exposure to disease, heat exhaustion, frostbite, snakebite and more. Hierarchies were created among enslaved social groups, in other words, based on proximity to those who exploited unfree labour.[117]

The effects of the slave trade in Africa were also significant. For example, scholars have noted that the prevalence of polygyny (men having more than one wife) is strikingly high today in regions that were sources of large numbers of enslaved men, such as Guinea, Togo and Benin – where the numbers with multiple wives are three times higher than in countries in East Africa. This has been linked to skewed sex ratios caused by the strong preference for male coerced labour in the Caribbean and Brazil, the main destinations for the exports of human captives, mirrored by the demand for enslaved women from East Africa who were sold into slavery as household staff and concubines in the Middle East and India.[118]

There is evidence to suggest that these patterns were reduced in districts that received intensive Christian missions during the later colonial period.[119] This is important, as models show that the banning of polygyny reduces fertility, increases savings, produces significant uplifts in GDP per capita and reduces gender inequality – raising separate questions about the impact of slavery and the period that followed its abolition in the early nineteenth century.[120] In regions where polygyny persists, on the other hand, incidences of HIV infection rates, violence and child mortality rates are high.[121] However, there are more progressive outcomes too, including higher levels of women's participation in the labour force, an increased likelihood that women will vote and broader-minded views about women as political leaders.[122]

Such then were the profound and many impacts that resulted from the 'discovery' of the Americas that followed Columbus' crossing of the Atlantic in 1492. Global ecosystems were knitted together through transformations of landscapes, the emergence of new tastes driven by widening demand that were in turn propelled by consumption patterns and spending power, above all in coastal regions of western and northern Europe. All this came at the expense of indigenous populations in the Americas and of the millions shipped from Africa who ironically were better suited and more resilient in climate and disease environments than the Europeans who dominated, controlled and 'owned' them.

The sense of entitlement extended even further. The philosopher John Locke, for example, writing in the late seventeenth century, proposed that every person had an equal right to use natural resources, as these were provided by the 'spontaneous hand of nature'. If, however, one worked the land oneself, that land should become one's property. For Locke, human intervention changed the status of land itself from 'the common state of nature' to something that 'excludes the common right of other men'. Labour, in other words, was the foundation of private property.[123]

Ideas like this naturally had resonance as European settlers fanned out around the world, especially when they moved into lands that had not been cultivated or were lived on by nomadic and tribal peoples. As the Reverend Samuel Stoddard, a Puritan minister in Boston, put it in 1722, it was entirely reasonable to take land from the indigenous peoples as they 'made no use of it, but for hunting'. In this particular case, the local population had at least been paid a nominal sum for some of their land; they were lucky, said Stoddard, given that the land had no value in their hands. For the settlers, however, it would be rendered valuable after 'our improvements' had been made to it.[124]

It was a similar story in other parts of the world, where Dutch, British and other jurists and philosophers relied heavily on humanist ideas that private property related only to land that an owner cultivated and transformed. Anything else was *terra nullius* – virgin land that could be claimed and taken over.[125] One good example comes from Assam in India, where British authorities declared that land that was not planted with cash crops was deemed to be 'waste land'. This meant that it was deemed unowned and capable of being claimed or given out as a grant to settlers, usually for free.[126] Another example can be found in legislation in the United States, such as the Homestead Act (1862), which entitled any adult citizen or intended citizen to claim 160 acres of surveyed government land as long as they improved and 'cultivated' it, and allowed land to be taken from indigenous peoples if it was deemed suitable for agriculture or grazing – by newcomers.[127]

The story is not just one of human exploitation of nature, nor even one of humans' exploitation of each other. Insects played a role in shaping social, economic and political change, as did pathogens, precipitation patterns and soil conditions not only in the New World but beyond. One thing, however, was clear: the intensification of exchange had dramatic consequences. It seemed inevitable, therefore, that the emergence and development of new technologies was likely to produce a familiar pattern of exploitation of all kinds, of inequality and of unsustainability. The question was who would do well, and who would pay the price.

17

The Little Ice Age
(c.1550–c.1800)

There was great scarcity in the cities and villages of India … men took to eating one another.

Abū al-Faẓl, *Ain-I Akbari* (late sixteenth century)

In April 1939, a committee on glaciers at the American Geophysical Union delivered a report which stated that glaciers had advanced over the course of the previous century by a 'far greater extent and volume' than they had done in the past, and suggested that 'We are living in an epoch of renewed but moderate glaciation – a "little ice age".'[1] Although the committee took the long view of the duration of this epoch, suggesting a time frame of 'about 4,000 years' for global cooling, the idea of the Little Ice Age caught the imagination of historians, who typically use the label to refer to the period from the sixteenth to the nineteenth century.[2]

While the problems of trying to identify even vague start and finish points given the vast range of global, regional and local variations are obvious, many scholars have argued that it is possible to paint a broad picture of changing climatic conditions that spanned several centuries, which is attested by tree-ring data, morphostratigraphic evidence and glacier evidence.[3] Three periods of low solar activity are often noted as being key factors in climatic changes in this period – namely the Spörer Minimum (1420–1550), the Maunder Minimum (1645–1715) and the Dalton Minimum (1790–1830).[4] The Little Ice Age is said to be characterised by significantly lower average temperatures in Europe, shown for example by canals freezing in the Low Countries and by the timing of Baltic ice that blocked ports such as Tallinn and Stockholm.[5] Some reconstructions have indicated that temperatures in Sweden and Switzerland fell by as much as 2 °C and 5 °C below normal respectively.[6]

According to some assessments, the hundred years or so from the late sixteenth to the late seventeenth centuries were unique in being the only significant period of cooling in both the northern and southern hemispheres known in history.[7] Others have suggested that the Little Ice Age as a whole corresponded to a time of weaker Asian and East Asian monsoons, greater humidity in Central Asia and higher rainfall in the Indo-Pacific.[8] The concept of the Little Ice Age as a global phenomenon has taken root and become popular with historians and general readers alike.[9]

This is not entirely surprising, as the period overlaps with a time of profound social, economic, political and ecological change. The seventeenth century in particular has been described as a 'General Crisis', a time of tumult across much of Europe, Asia, Africa and the Americas, with more wars taking place around the world in the 1640s than in any other era until the 1940s.[10] As one recent magisterial book put it, climatic extremes combined with poor harvests, food shortages and famine, disease and conflict to produce a 'fatal synergy' that had dramatic impacts on human societies around the world during the period between c.1600 and c.1700.[11]

A host of transformations have been associated with the challenges posed by harsher climates. For example, it has been suggested that adaptations to the challenges presented by the Little Ice Age shaped not only the way people lived but the way they built and lived in houses and interacted with each other. In Sweden, a shift towards tiled stoves – which use less wood and retain more heat than open fireplaces – has been linked to the cooling of northern Europe and to the practicalities of needing to maximise indoor temperatures and comfort. The move away from large, multifunctional halls towards corridors and smaller rooms brought with it ideas about privacy and intimacy in personal relationships – as well as about individuality and exclusion for those who were not invited into small rooms to listen to, take part in or overhear conversations.[12]

The effects of the Little Ice Age apparently impacted European art, with paintings dating to the period 1550–1849 showing a higher proportion of clouds and darkness than earlier or later periods.[13] Paintings such as Bruegel's *Hunters in the Snow*, produced in 1565, have become iconic representations of a phase in history that is supposedly distinctive and identifiable for its coldness.[14] Social behaviour and tastes during the early modern period ranging from fashion to the growing popularity of beer – which has been associated with the cooling temperatures that affected vineyard production, forced the cultivation of wine to the south and both reduced its availability and raised its prices – have likewise been linked to the Little Ice Age.[15]

The persecution of women through accusations of and executions for witchcraft has also been linked to climatic change, adverse weather conditions generally and hailstorms, with women disproportionately being made scapegoats for poor harvests, food shortages and high prices of grain.[16] Although the estimates of excitable commentators in the eighteenth century who claimed that 9 million witches had been executed in the previous 200 years can be dismissed out of hand, there can be little doubt that both perceptions and realities of the suffering of women were extraordinarily high in the second half of the sixteenth and the first half of the seventeenth centuries.[17] The scale of persecution was astonishing. Some cases saw hundreds tried and sentenced to death.[18] Prosecutions for 'witchcraft' in Catholic and Protestant parts of Germany were similar until around 1600, when the numbers of women accused and tried rose sharply in the former – becoming twice as common than in the latter.[19] Regardless of ratio, this should all be seen within the wider context of women being blamed for societal problems, not least in the modern day for ecological stress, economic shocks and disease, including HIV/AIDS.[20]

Some historians have suggested that cooler weather may have led to high levels of clinical depression in the late sixteenth century, with the 'fashionable malady' of melancholy prevalent across royal households and intellectuals circles in England, France and Spain (in England it was known as the Elizabethan malady). Cases of the 'winter blues' were 'well suited to an age of long winters and raining summers', fuelling a rash of literary works focusing on despair – of which Montaigne's essay 'Of Sorrow' is one obvious example. Challenging climate apparently led to an escalation in cases of suicide – although this had the mirror effect, at least in popular imagination, of implying that suicide itself was the cause of bad weather.[21]

The correlation between ecological catastrophe and the expansion of glaciers in the European Alps has led some to connect concern with climatic change with the contemporary search for explanations and scapegoats.[22] Individuals like the Swiss scientist and politician Renward Cysat were blunter when it came to speculating on the causes of challenges and change. In recent years, he wrote c.1600, the weather has taken 'such a peculiar and astounding course and undergone such extraordinary alterations'. He had little doubt about what was to blame. 'The years have shown themselves to be more rigorous and severe' than in the past, he declared, 'because of our sins'. This, he said, should serve 'as a warning to future generations'.[23]

Sustained periods of cold weather have a marked effect on agricultural production: a drop of just 1 °C over a period of a few years can result in the kilocalories from solar radiation falling by around 10 per cent per square centimetre. In temperate zones, diminished sunlight as a result of lower levels of solar irradiation or cover thrown up by volcanic dust can sharply reduce growing periods of crops, pastures and forests. Moreover, cold temperatures often had associated effects – including changes to rainfall levels, and negative alterations to microbial activity in the soil that in turn affected the decomposition of organic material and thus the fertility of the land.[24]

Climate enhances vulnerability to disease – a factor that has been stressed by historians, who have pointed out that colder temperatures significantly raise the risk of the spread of disease, largely as a result of weakened immune systems that are closely connected to malnourishment. It has been pointed out that in the case of China, for example, the six centuries following 1370 saw almost 6,000 epidemics, with a cold climate raising the risk of outbreak by 35–40 per cent.[25]

Indeed, specific events that took place during the period of the Little Ice Age have been explained as being understandable primarily through the lens of unusually cold conditions and of changing climate in general. For example, heightened storm activity in the later sixteenth century, which included the incidence of severe storms rising by 400 per cent, has been cited as the decisive factor in the demise of the Spanish Armada in 1588 in conditions where wind speeds approached hurricane strength. As one author has noted, great gales destroyed more of the Spanish fleet than the combined guns of English warships.[26] It has been suggested that Sweden's status as a European great power was first compromised and then shattered by a cluster of cold winters in the 1690s and early 1700s, as Swedish armies were 'broken by cold and disease', leading to the loss of the country's Baltic possessions and the occupation of Finland by Russian forces.[27]

While the idea of the Little Ice Age may seem an attractive one, a way of explaining cultural and behavioural change, the context of disease transitions shaping the outcomes of military confrontations, it is not without its problems. For one thing, it is simply not the case that temperatures dropped simultaneously in different parts of the world for sustained periods, and certainly not over the course of many centuries. As the Intergovernmental Panel on Climate Change (IPCC) puts it, evidence does not support the hypothesis of a Little Ice Age as globally synchronous.[28]

There were long periods even in the seventeenth century when climate conditions were entirely within their normal range in the northern hemisphere, including during decades of multiple crises, such as major wars and devastating outbreaks of plague.[29] Even where temperatures dropped below average, such as in northern Europe in the 1590s, there was no such drop in other parts of the continent, such as southern Italy or the eastern Mediterranean, where tree-ring data is not in any way unusual.[30]

Marine-core data, sea-ice incidence and ice-core isotope records from this period are marked not by the presence of a persistent multi-century signal, but rather by its absence.[31] Even the advance of glaciers from c.1400 to c.1800 that are supposed to be a key marker of a new cold phase are in fact neither unusual nor markedly greater than advances made a century or two earlier.[32] As two leading scholars put it diplomatically, evidence for a Little Ice Age that is climatically distinct and clearly cooler than periods before or after 'remains elusive' even for northern Europe and the North Atlantic.[33]

Likewise, although it is tempting to suppose that solar minima and a reduction in irradiance and in the numbers of sunspots had an impact on global, regional or local climatic conditions, more careful evaluation shows that the effects were modest, if not entirely negligible. 'Grand solar minima' such as the Maunder Minimum produced a measurable difference in temperature of around 0.3 °C.[34] Indeed, studies on the likely outcome of future solar minima suggest that this is at the top end of the scale for temperature change, with a range of 0.09–0.26 °C a more appropriate outcome – which has obvious implications for any hypothesis about a Little Ice Age.[35]

Moreover, even cases where the Little Ice Age seems to have influenced change can look different when provided with wider context. It is easy to focus on Bruegel's *Hunters in the Snow*, for example, while overlooking *The Harvesters*, painted by the same artist in the same year, which evokes a warm summer, bountiful harvest and golden light. While it is true that some landscape painters in northern Europe produced a large number of winter pictures, such as Hendrick Avercamp, many others did not. Rather than acting as a guide to colder climate, it may be that paintings depicting snowy scenes represent the contemporary tastes of buyers and artists alike.[36] These tastes changed: while icy landscapes were popular in the late sixteenth century, during the chilliest decades of the Maunder Minimum a hundred years later, 'Dutch customers preferred bright, sunny scenes.'[37]

Similarly, the correlation of witchcraft trials and cold temperatures seems convincing at face value. After all, one papal bull issued in the 1480s specifically declares that evildoers 'raise and stir up hailstorms and tempests, and cause lightning to blast both men and beasts', noting that many reports had been received in Rome of occasions where 'the produce of the earth' including fruit, wheat and other crops had been destroyed by sorcerers.[38] However, some scholars have argued that rather than reflecting deteriorating climate conditions, the persecutions of the late sixteenth and early seventeenth centuries should be understood in the context of rising instances of syphilis and resultant mental illness that presented as disturbing; it may equally and also be seen as a case of men seeking to secure medicine and healthcare as a male preserve, or as a result of an economic downturn that might be related to poor yields. Likewise, other explanations might explain rising anxieties that helped spur animosities and violence, such as slowdowns in trade, increased demands by the state or breakdown between rival political factions.[39] Recent research into mass psychogenic illness also raises the question of whether the 'witch craze' of this period was a case of more and more people being caught up in accusations and counter-accusations in a self-fuelling and self-fulfilling cycle.[40]

Or there is the case of the defeat of King Charles XII at the battle of Poltava in 1709 and subsequent setbacks that proved so damaging for

Sweden. While cold weather certainly did not help, it was more telling that the Swedes had not only overextended their resources by investing in long-distance and expensive warfare, but had overreached into Russian territory, leaving their supply lines exposed and vulnerable to harassment and disruption. When combined with a set of strategically naive mistakes on the battlefield that exposed the King, his leading generals and his men to unnecessary risks, it was hardly a surprise that things turned out badly over the following years. Perhaps, in fact, the main surprise was that the fallout was not even worse.[41]

Identifying exactly what effect cooler temperatures have on trends, fashions and even individual events is not easy and requires careful judgement. So too does evaluating climatic change during a long period where there was considerable variation and fluctuation both geographically and chronologically. There were some decades – notably the 1590s, the 1680s and the 1810s – when there is evidence of substantial disruption to regular climatic patterns that saw cooler temperatures across multiple locations as a result of volcanic activity, of particularly strong El Niño events or of the two reinforcing each other. Equally, however, it is important not to treat a period of over 500 years as one that presents any consistency across either time or space.

Temperatures do seem to have dropped sharply in the early fifteenth century. Ironically, it appears that this was a result not of a weakening of climatic and solar patterns, but rather of their strengthening: the Atlantic Meridional Overturning Circulation (AMOC) usually pushes water from the tropics to higher latitudes where it then meets colder Arctic water, causing it to lose heat, become more dense, sink to the bottom of the ocean and become part of global circulation patterns. In the late fourteenth century, however, abnormally strong transfers of warm water were circulated northwards as a result of the major strengthening of AMOC, in turn causing a period of extreme sea-ice export that released enormous volumes of cold water into the North Atlantic, affecting salinity levels and leading to the decline of the AMOC.[42] The outcome was an age of intensive cooling which was reinforced by a sequence of major volcanic eruptions that served to magnify reductions of solar irradiance persisting from c.1400 until the early seventeenth century.[43]

This serves as a useful reminder of how important it is to separate out different phases, causes and outcomes in a long period that is often treated as a single phenomenon. In fact, there are important trends that do give a coherence to the centuries leading up to the start of the industrial revolution. Clearly the linking together of the world's major landmasses through the opening up of sea routes across the Atlantic and the Pacific resulted in new exchanges of goods, foodstuffs and ideas as well as in the large-scale movements of people, both through choice and through enslavement. The ecological changes that this produced were substantial, as we have already seen in the case of the Americas and the Caribbean, as were the new pathways that opened up for the spread of disease, such as the transfer of malaria and yellow fever from the Old World to the New, and of syphilis in the other direction.

The spread of these and other diseases was facilitated by intense competition for resources and power, not least in Europe itself, where states not only jostled for position in their own continent but sought to outmanoeuvre their rivals' advantages in others. One result was a dramatic rise in military confrontation, with a war between the major powers in Europe under way in 95 per cent of all years between 1500 and 1700. This itself was partly the result of the development of new technologies and their constant improvement, most obviously in the field of gunpowder. The spread of cannon brought about major changes to defensive tactics and fortifications as it became clear that town and city walls could be quickly destroyed. Among the innovations were massive earthen bulwarks, often covered with brick, which could withstand fire. The fact that they could be built in numerous locations and away from urban settlements (albeit at considerable cost) transformed warfare, changing battlefield requirements, not the least of which was the need for more men and better training. All this meant higher levels of professionalisation, more equipment – and more money.[44] These were all fundamental parts of a military revolution which not only meant that 1600–1700 was the 'century of the soldier', as one contemporary commentator has so aptly put it, but also played a key part in the rise of west.[45]

Military innovation was in turn closely connected to wider social, economic and institutional transformations. The decades and centuries that followed Columbus' crossing of the Atlantic saw successive waves of centralisation, above all in Europe. There was a wide range of reasons for this, but constant war was perhaps the most significant. The need to put more and more men in the field required ever-greater revenues, which inspired an almost insatiable appetite for taxation. Between

1500 and 1780, therefore, the combined tax take of major European powers increased by a factor of twenty. On average, 80 per cent of government expenditure was devoted to military costs, with spending often exceeding tax revenues during times of war. In turn, this brought about a financial revolution in government debt finance.[46]

It was not just the military who became better equipped, better trained and better run as a result. One effect of the drawing of more resources into political centres was an uplift in the quantity and quality of bureaucrats whose job it was to assess, collect and allocate funds. States that invested time and energy in creating institutions, educating administrators and appointing on merit, and in applying methods that rooted out corruption, did better than those that did not. Indeed, failure to reform in some cases led not only to poor outcomes but to states being 'partitioned out of existence', as was the case with Poland. Capacity building was closely linked to the creation of institutions that were run by salaried officials appointed and judged by performance and competence – at least in theory. The development of these institutions was of crucial importance in establishing constraints on the autonomy and autocratic powers of rulers, and they thereby served as staging posts towards higher levels of accountability and increased levels of political participation.[47] Indeed, some have argued that the relative weakness of the crown may even have been 'an essential precondition for the eventual success of England's overseas enterprise' – and empire.[48]

The building of an efficient state apparatus was not a rapid process, but rather one that evolved over the course of many decades and centuries. There were nonetheless rapid accelerations that had knock-on effects not only on the role of the state, but also on urbanisation, disease and health and the natural environment. In France, for example, public revenues rose by three times in the course of just forty years between 1600 and 1640 in the face of demand from the centre. While this is striking from a purely economic point of view, no less dramatic was the effect it had on population migration. As monetary (and other) resources were sucked into the capital, so too were people, eager to take advantage of opportunities there or because those that had existed locally had vanished. Paris accounted for 60 per cent of all French population growth during the first half of the seventeenth century. The speed and scale of urbanisation meanwhile acted like a shot in the arm

for the city's merchant class – who suddenly found themselves with
hosts of new customers – and like a mirage for those elsewhere whose
markets melted away before their eyes.[49]

The centralisation of the state and the collection of resources at its
disposal were particularly pronounced in parts of northern and western
Europe that were connected to Atlantic trade, demonstrating the
significance of the maritime routes to Africa, the Americas and Asia
which accelerated growth after 1500.[50] Urbanisation, however, produced
further developments through unexpected side-effects, the most notable
of which was disease. Europe's cities were 'major death traps' that were
unhygienic and ideal for the spread of illnesses. London recorded twice as
many burials as baptisms in many years during the seventeenth century.

Perhaps counter-intuitively, the rise of urban populations led to
higher wages, with large numbers of people packed together driving
demand and stimulating supply within cities and in the countryside.
The rate of urbanisation shocked some, like King James I, who claimed
in 1616 that the entire population was being crammed into the capital.
Soon, he said, 'England will only be London, and the whole country be
left waste with everyone living miserably in our houses, and dwelling
all in the city.' This was hardly exceptional, he added; it simply made
London like Italy – and especially Naples.[51] London, wrote John Evelyn
in the 1660s, was the 'suburbs of Hell', afflicted by a 'fuliginous and
filthy vapour' that hurt the lungs and ensured that 'Catharrs, Phthisicks,
Coughs and Consumptions rage more in this one City than in the
whole Earth besides'.[52]

Paris was no better, a city that one contemporary complained was
'always dirty' and where visitors would get covered in 'black onctious
oyl, that wher it sticks no art can wash it off'. Paris, he added, was
so grotesque as to merit the nickname 'vagina populorum'.[53] While
some complained, the reality was that large-scale urbanisation was
a prerequisite for an acceleration of production and consumption
patterns, a catalyst for engagement with other parts of the world and in
due course a motor of the industrial revolution itself.[54]

As we shall see, demand for goods and foods stimulated global trade
and came with ecological and environmental prices that were charged to
the parts of the world they originated from, not only unseen but usually
unpaid by those living hundreds and often thousands of kilometres
away. Key to that story was the role of cities relative to their hinterlands

and their ability to absorb large and growing populations. While cities drove demand, fostered innovation and drew in manpower, they were also co-dependent on their hinterlands and on agricultural production. Large urban settlements could be uniquely vulnerable to weather shocks, although risk could be mitigated in cities that had access to the sea or river systems and well-established long-distance connections.[55]

Clearly, changes to the production and availability of crops, fodder and lumber – bulky items that were heavy and expensive to transport – as a result of weather shocks or long-term climatic change had immediate and direct impacts on both urban and rural populations. Reduced availability of food brought about demographic declines through increased mortality and disease, through lower fertility levels and through people leaving cities. During the seventeenth century, for example, the Balkans and Anatolia lost half their population, with declines of 80 per cent in some regions and large numbers of rural villages simply abandoned.[56]

Food shortage had other consequences that did not just alter the relationship between town and country, but changed the character of urban life – even without catastrophic incidence of famine: reduced agricultural productivity means that workers in the rural economy have less to sell, which means that they earn less. This then affects the prices of goods and services made or sold in towns and cities, as demand falls. Employment prospects in the latter diminish as a result, making life in cities less profitable. As such, assessing temperature yield ratios and their correlation to historical wheat prices helps show that climate is a fundamental motor in powering growth, and that it is a crucial factor in elucidating economic growth and explaining why it was so uneven across Europe in the early modern period.[57]

This unevenness again shows that there was no 'one size fits all' during the Little Ice Age. For example, from the winter of 1539, a long spell of high pressure over western and central Europe was mirrored by low-pressure systems over the Atlantic and western Russia. One Russian chronicle noted that 'Springtime was cold and during all summer there were floods, and the rye did not grow and was frozen in spring, and all the meadows at the banks of the rivers and lakes were flooded,' while 'in the autumn there was a lot of rain, and the sun was not seen for 2 weeks' in November. This produced a different set of problems to continental Europe, where Lake Constance dropped so low that the lake floor could be seen, while water transport became difficult and, in some cases, impossible. Wildfires broke out in Portugal and Germany; town fires were more frequent in Germany in 1540 than any other year since 1000. As well as animals suffering from heatstroke, water shortages led to tree stress, poor vine and cereal harvests and a collapse of water power for mills – all of which led to sky-rocketing prices.[58]

High prices were closely related to a range of factors, including hoarding practices by unscrupulous traders who sought to take advantage of demand and panic; in addition, ineffective measures were adopted to alleviate suffering as meagre resources were ineptly distributed or food was not brought in from other sources elsewhere. Desperate scenes could result. In India just after the accession of Akbar to the Mughal throne in 1556, there was 'great scarcity in the cities and villages of India … men took to eating one another; some would join together and carry off a solitary man and make him their food'.[59]

Some scholars have suggested that the Atlantic Meridional Overturning Circulation (AMOC) underwent a period of significant change around the 1560s, provoking the Grindelwald Fluctuation – a reference to a marked increase in Alpine glaciers that starts at this time. This period also saw significant falls in precipitation in the Mediterranean and a rise in north-western Europe, especially during spring and summer months, which shortened growing seasons by six weeks.[60] The effects may have been felt further afield, with Shona oral traditions and other evidence from sub-Saharan Africa pointing to severe droughts, locust infestations, famine and epidemic disease in northern Zimbabwe and on the Mozambique coast in the 1560s.[61] Average temperatures in China went into a long cycle of cooling, with only three years between 1569 and 1644 warmer than normal.[62]

It was not impossible or even difficult to adapt to changed circumstances. In northern Europe, for example, many farmers switched to growing barley, oats and rye rather than wheat, which is more sensitive to cold.[63] Nevertheless, other circumstances made significant differences to experiences. For example, in 1569–73 a succession of long and severe winters brought heavy rain and disastrous flooding throughout much of Europe. Lakes, rivers and the Baltic Sea from Denmark to Finland froze as a result of temperatures that were not unusually low, but stayed low for weeks at a time. The effect on harvests was considerable, exacerbated by successive failures, leading to grain prices in central Europe reaching the highest point that they would reach until 1877. Landlocked regions suffered more than those by the coast, while cities with access to the sea were less badly affected than those in the interior.[64] This was a time of considerable disruption that affected south-east Europe and the Middle East too, where one chronicler described food shortages and famine in 1574–5 as causing the worst suffering for hundreds of years.[65]

Plagues of locusts that swarmed into Italy threatened to make things worse still until Pope Pius V held a trial in St Peter's Square, attended by locusts that appeared 'in such numbers that the sun was obscured', and excommunicated them. With that, 'they soon disappeared, never to be seen again'.[66] Not surprisingly, some feared that famine and high prices would lead to unrest – as in southern Germany, where conditions were particularly bad.[67] Perhaps anxieties were so acute because climate had been so stable for many decades, and preparedness to cope with crises was declining. As such, cold and harvest failure seemed like 'a bolt of thunder after a long period of economic sunshine'.[68]

The most dramatic effects, though, were felt in the Low Countries, where tensions had been boiling because of resentment against Spanish authority, aggravated by religious animosity between Protestants and Catholics. There had been problems of food availability with low harvests even before unusually poor weather made things worse, exacerbated by war between Denmark and Sweden which made alternative supply difficult and led to a tripling of the price of wheat in some parts of Flanders over the course of six months in 1565 alone. Government ministers had already been warning of unrest when religious mobs took each other on in scenes of escalating violence that saw churches set on fire, their contents and interiors destroyed by Protestant mobs as part of the *Beeldenstorm* (literally 'image storm') that set the tone for poisonous relations in the Low Countries and that did not improve when thousands were killed by storm surges and gales in the early 1570s. The culmination of this time of upheaval was the brutal sack of Antwerp in 1576 by Spanish soldiers who had been frustrated by weather conditions during a failed siege at Leiden and had not been paid. The ferocity of the attack on Antwerp was instrumental in uniting the provinces of the northern Netherlands and led to the Union of Utrecht, which effectively created the Dutch Republic.[69]

In this instance, a wide range of factors had reinforced each other, had served to compound and deepen existing crises and had produced outcomes that were to have long-lasting significance. A similar chain reaction came with a sequence of large-scale eruptions of low-latitude volcanoes between 1580 and 1600 at Billy Mitchell in Melanesia (1580), Kelut and Ruang in Java (1586 and 1593), Ruiz in Colombia (1595) and Huaynaputina in Peru (1596), the latter being the largest eruption ever recorded in South America.[70] Indeed, its effects were so pronounced that some scholars have questioned whether there was a contemporaneous eruption elsewhere that might explain the sheer size of the signature in ice cores and tree rings in the northern hemisphere.[71] Either way, the result was the severest short-term cooling of the last 600 years – if not more.[72]

The climatic upheavals were dramatic – and global. In Scandinavia, a series of long and severe winters began in 1587 with a year without a summer – a fabled *fimbulvetr* of Norse mythology.[73] In the autumn of 1589, for example, much of Italy suffered from very heavy rainfall which caused flooding in Campania, Tuscany and Rome, and affected sowing of crops. A bad situation turned to crisis in the 1590s as harvest yields fell by two-thirds, leading to reserves becoming exhausted. Cities like Naples, Bologna and Mantua responded by expelling anyone who might be considered a burden – foreigners, students, the poor – in desperate attempts to preserve food supplies, to avoid overcrowding and to alleviate the risk of theft and potential revolts.[74] Cities saw not only sharp rises in mortality, contributing to a drop in population in and around Bologna of almost a fifth, but also a collapse in births of 44 per cent.[75]

Towns and regions such as the Po Valley saw the worst famine and the steepest demographic collapse since the Black Death. The disaster provoked long-term change. For one thing, it prompted farmers to accelerate a shift to more intensive production techniques as well as to sacrifice food variety in favour of greater calorie availability through a widespread shift to maize.[76]

For another, it helped draw parts of Europe more closely together, linking the Baltic region into southern European trade networks following the desperate search for emergency supplies by city states and local rulers during a time of crisis. Cities like Genoa, Venice and Livorno established strong links with the Baltic region, importing not only grain but also other foodstuffs, including beans, salted meat and fish, as well as metals, textiles and hides. Much of this trade was brokered and shipped by the Dutch, whose investment in the 1590s paid dividends at the time but above all a few decades later after relations with Spain had been stabilised. The returns on investment of 100 per cent were to prove vital in the acceleration of northern Europe in the early modern period in general and of Dutch capitalism in particular.[77] Such were the curious by-products of disasters and response efforts.

Spain was already in an economic tailspin as a result of persistent overstretch in wars in continental Europe and slowdowns in extractions from its empire overseas. It was hit hard too by almost a decade of high rainfall and inclement conditions, followed by a decade of drought. As well as resulting in a death toll that some estimate was as high as 600,000 – a demographic shock from which recovery took more than a century – the effect on livestock was particularly severe with the loss of a third of the country's sheep, which in turn had knock-on effects for the wool industry. All this was made worse by a major increase in taxation intended to help the crown to make good on a debt burden that had left it crippled and prone to taking poor political and strategic decisions. As we have seen, this coincided with discussions about whether the imperial project was worth pursuing or should be abandoned as too costly. Spain, noted one author in 1621, was like a patient suffering from a critical illness; sometimes 'you have to cut off an arm' to ensure they would live.[78]

Harvest failures in the 1590s are attested to in Scotland, Sweden and Austria, the latter experiencing several major uprisings as the situation turned bleak. It was said that in Ireland, starvation left the population unable to walk and resembling ghosts. In England, things were grim too, with famine, great floods and in the 1590s and 1600s frosts that were so strong that the Thames froze over and was able to host 'frost fairs'. Particularly strong storm conditions in this period went hand in hand with sand drift that had considerable long-term ecological impacts on the Atlantic coast of western Europe.[79] Some scholars have pointed out the attention paid to celestial events and climate phenomena in Shakespeare's writings in the 1590s and 1600s, and have suggested that climate was an important framing device in tragedies and comedies written in this period.[80]

Climatic stress not only characterised South-East Africa in the late sixteenth century but was a factor in political change. As one Portuguese missionary put it, the region was afflicted by 'chastisements' from the late 1580s including locust plagues that 'devoured all the crops, gardens and palm groves' leaving them bare for the next two years, famine and a severe outbreak of smallpox. This served to weaken the Zambezi Tonga chiefdoms, provoking social instability, sharpening local rivalries and opening the door for Portuguese intervention in this region. Initially, this took the form of military assistance provided by traders who had hitherto been confined to coastal regions; it was quickly turned into concessions over land, access to mineral resources and a role in local and dynastic politics that ultimately served as a springboard for colonial expansion – and for the economic impoverishment of Zimbabwe culture states.[81]

This was a time of environmental and ecological change in other parts of Africa, most notably the southern frontier of the Sahara which expanded by some 300 kilometres to the south. Disease environments shifted too as the habitat of the tsetse fly increased, prompting population movements and altering agricultural capacity. Dating precisely when and how this happened is important, particularly if the transitions are to be linked with the collapse of the Songhai empire and the conquest of Timbuktu in 1591 by Moroccan forces.[82] After all, Moroccan operations to the south to gain control over oases and trade networks had begun many decades earlier and were driven by the desire to access gold and salt on the one hand, and powered by rising ambitions and capabilities on the other.[83] Nevertheless, the fact that Timbuktu did not regain its prominence, its wealth or its size is to be explained, at least in part, by the onset of desertification – which also drove outflows of people from regions that were most exposed to changes in weather patterns around the end of the sixteenth century.[84]

In China, floods, unusually cold temperatures and food shortages in the 1580s produced the worst famine in a hundred years. The effects were made worse by outbreaks of epidemic disease that likely included hydrocephalic meningitis that caught hold in densely populated cities – cities, as one contemporary observer complained, that were home to houses 'so closely crowded together that there is no spare space and in the markets there is much excrement and filth'. These were perfect conditions, in other words, for 'malarious fevers, diarrhoeas and epidemics [to] follow each other without stopping'.[85] The Jesuit missionary Matteo Ricci noted that 'all the rivers in northern China' froze over in the late 1590s – a period of notably lower average yearly temperatures than normal.[86]

South-East and South Asia saw significant falls in precipitation, with rain levels well below normal. Spanish tax lists from the Philippines point to a population loss of 25 per cent between 1591 and 1608.[87] In the early 1590s, reports Abū al-Faẓl, the author of the *Akbarnāma*, 'those learned in stars announced death and scarcity', which allowed the Mughal Emperor Akbar to take emergency action to make sure that supplies were available to all.[88] This worked – to start with. Lack of rain over several years eventually took a terrible toll, with one Turkish source speaking of 'fearful famine' and 'general distress' made worse by epidemic disease. As large numbers died, leaving cities, towns and villages empty apart from 'streets [that] were blocked with dead bodies', some who were desperate resorted to cannibalism.[89]

It had been a similar story in the Ottoman empire, where famine-relief efforts at the start of the 1590s initially proved successful – notwithstanding the demands of supplying cities that were often ecologically constrained by the limitations on what their hinterlands could supply and the distribution of populations in the interiors of the Balkans, Anatolia and parts of the Middle East that were logistically difficult and expensive to support. The situation now became bleak as drought took hold and turned into the longest that the eastern Mediterranean had experienced in 600 years and by far the worst in the history of the Ottoman empire.

Rampant inflation ravaged the economy and drove prices up, leading to chaos as the government responded by increasing taxation and debasing the coinage in an attempt at fiscal stimulus. As people abandoned the countryside for cities, discontent spread – as did disease, most notably plague. Banditry was rife as strongmen took power into their own hands, leading to breakdowns of law and order, including direct challenges to central authority such as by a figure known as Karayazıcı ('Black Scribe'), who managed to bring regions and towns in what is now south-east Turkey under his control. Although this and other uprisings were eventually put down, some scholars have argued that the events of the 1590s and 1610s brought about fundamental changes in the Ottoman empire. Chief among these were a sharp reduction in sultanic authority and the transformation in the political and economic fortunes of the Janissary Corps, which henceforth became the major power broker in the internal workings of the empire.[90]

The late 1590s and the first decade of the 1600s was a traumatic period in Russian history, known as the Time of Troubles. It too was characterised by political upheaval and by chronic shortages as a result of successive years of harvest failure. One contemporary was shocked by what he saw. 'I swear to God,' wrote Conrad Bussow, 'it is the honest truth that I saw people lying dead in the street with my own eyes; in the summer people ate grass and in winter they ate hay, like cows.' The dead could be seen with straw and manure in their mouths, having evidently been trying to eat anything to stay alive, while some had even resorted to eating human waste, he wrote. 'Children were chopped up and boiled by their parents, parents by their children, guests by their hosts and hosts by their guests.' People flocked from the countryside to the city in search of food and handouts. While Bussow's assurance that 500,000 people died in Moscow alone is exaggerated, mortality rates were clearly very high indeed – with perhaps as much as a third of the population succumbing.[91]

Extensive though the global climatic rearrangement was, however, it is important to underline that in many cases the unsettled and unusual weather conditions served to accentuate existing vulnerabilities rather than act as primary causes of disaster. In many of the cases of food shortage in the late sixteenth and early seventeenth centuries, for example, problems stemmed from cities that had expanded to a size that left them exposed to risks that were themselves aggravated by the attempts of those living in the countryside to try to find safety – and things to eat. Pushing communities over the edge that were living on land that was at the limits of its carrying capacity did not require dramatic, extreme or apocalyptic weather events; rather, as many of the accounts from different parts of the world make clear, the problem was not the cold or rain or drought, but rather that it stayed cold, wet or dry for longer than people were used to – and that adaptation proved difficult, especially at speed.

Perhaps understandably, animosities rose against those who benefited from shortages and price rises – or at least those who were thought to be benefiting. In Ottoman Turkey speculators were fiercely denounced for hoarding grain and taking advantage of the suffering of others.[92] In Russia, many were stirred up by 'the devil of greed, enabled by God to punish the entire country' by holding on to grain or buying it cheaply, only to resell it at higher prices.[93] In England, some blamed foreigners, with one good example captured by a poem written on the wall of a Dutch churchyard in London in 1593, which declared that foreign merchants were responsible for high prices, before posting a dark warning to them to look after 'your goods, your children, and your dearest wives' because 'like Jews you eat us up as bread'.[94]

But hoarding and price gouging were symptoms of crisis, not its causes. In many of the cases that marked the late sixteenth and early seventeenth centuries, there were deep-rooted underlying factors that helped turn challenging situations into catastrophes. In Russia, for example, the combined impacts of inflation, increasing tax demands by the state, heavy costs of war and ineffective leadership by tsars who were divisive, indecisive or both produced intense competition within the elite – factors that were aggravated by a rising population that almost doubled during the sixteenth century.[95] The ability of monasteries to build up property and acquire tax exemptions resulted in artificial prices and a distorted market, which persisted until the start of the eighteenth century when Peter the Great sought to limit monasteries' rights to own land, assigned monks a fixed income and attempted to reduce the number of clergy as a whole.[96]

In this sense, climatic stress served to bring ongoing issues to the boil. The Ottoman experiences in the late sixteenth and early seventeenth centuries owed much to the fact that the state was engaged in military operations against the Austrians and the Persians simultaneously – at a time when the costs of warfare were increasing both in real terms and in terms of the numbers of men needed on the front lines and in support positions. This produced a draw on resources that removed margins for error – not least when it came to harvest collection.[97]

Poor leadership combined with military overstretch also brought down empires in this period – not climate. The collapse of the Tungoo empire, which was marked by the sack of many of its principal cities in 1599, is a case in point. The empire was one of the largest, if not the largest, in South-East Asian history. It had grown out of a series of spectacular military successes, shrewd diplomatic compromises and alliances to cover much of what is now Burma, Thailand, Cambodia and Vietnam. It may be that breakdowns in regional and global silver bullion markets put strain on the economy, although evidence to this effect is scant. It may be that weak monsoons prompted food shortages, rising prices and perhaps a plague of rats that infested the capital in 1596. But the more prosaic answer, as one scholar has put it, is that the empire 'overheated' and that it was a victim of its own success – a polite way of saying that its structure was weak, its operation unviable and its ability to dominate the periphery from the centre poor to non-existent.[98]

The reason then that all the streets of the capital, Pegu, 'especially those which lead to the temples, are scattered with skulls and with the corpses of these wretched Peguans' was partly due to famine, as one visitor to the city observed the year after its sack by the combined armies of the rivals of the Tungoo dynasty. Many lay dead because they had been killed in street-to-street fighting as anarchy took hold; others had been killed by command of the King, who had turned on many of his own subjects, suspecting them of treason. As a result, many dead bodies had apparently been thrown in the river, which is hardly navigable any more even in a small boat because of the multitude of corpses'. The collapse of the empire, said another European commentator, meant that 'the former seats of great and powerful lords became the abodes of tigers and other wild beasts', with an eerie calm now hanging over a once great city that was described as 'the greatest silence on Earth that human thoughts can imagine'.[99]

The many crises of this period could and did present opportunities. In England, for example, the severe conditions of the late 1590s led to the formulation of legislation that included the Act for the Relief of the Poor of 1598, which was largely repeated in 1601. This passed on to individual parishes the responsibility for looking after the poor and the associated costs, a move that proved both revolutionary and remarkably long-lasting. Even though such ideas drew on those that dated back long before this period, the principle that households should be taxed to support their poor neighbours was one that had crucial significance for attitudes to welfare, to communities, to money and to the creation of a 'robust safety net' which was to come into play when all other options had failed.[100]

Others seized opportunities for change in different ways. Christian priests and missionaries in North America took advantage of unusual weather to proselytise the indigenous populations, exploiting precipitation shortfalls to undermine existing beliefs. Franciscans working among the Timucua of Florida in the early seventeenth century were advised to ask questions such as 'Have you made rain?'; or 'Have you conjured a rain storm or a thunderstorm with superstitions?' Local populations were told not to perform traditional rain ceremonies, as they should 'be aware that it will not rain unless God our Lord is served'.[101]

Rainmaking contests were a key part of interaction with indigenous peoples in the Americas and an important way for Europeans to try to win hearts, minds and credibility. One of the questions most often asked of Christians was about the ability to modify the weather through appealing to God. 'If we believe in your God, will it snow?' Montagnais Indians in Quebec asked French Jesuits in the early seventeenth century. 'It will snow,' came the reply, soliciting the follow-up questions of 'will the snow be deep ... shall we find moose ... shall we kill some?' The Jesuit replied, 'Yes; for as God knows all things, as he can do all things, and as he is very good, he will not fail to help you, if you have recourse to him, if you receive the Faith, and if you render him obedience.' This seemed to impress the listeners. 'We will think about what thou hast told us,' they said, before heading off to the woods and promptly putting what they had heard to one side.[102]

Changing religious beliefs and reconfigurations of the engagement with the natural world were not only connected to European interaction with local populations. In the late 1620s, a new wave of unusual weather brought about a combination of El Niño and a weak Azores anticyclone that had global impacts lasting several years. The Indian monsoon failed in four successive years from 1628 to 1631, followed by unusually high rainfall in 1632. The impact in India and Burma was catastrophic, with some suggesting that as many as 3 million people died in Gujarat alone. Demographic devastation was financially crippling, with the reduced revenue from some provinces meaning that administrative expenditures could not be covered for decades.[103] This came at a pivotal time in South-East Asia, just as the Dutch East India Company was establishing itself as a political as well as an economic force by taking over most of the lucrative spice trade, crippling the economies of port states like Demak, Japara and Surabaya in the process. These then fell to Sultan Agung, the ruler of Mataram, centred near modern Yogyakarta on the island of Java, who oversaw the creation of a distinctive court culture that blended elements from Islam with Hinduism and Buddhism.[104]

In the 1630s, however, greater emphasis was placed on Islamic features, with Agung abandoning the Javanese calendar and its anchor around the lunisolar Śaka year for the Islamic lunar year, sponsoring patronage of Qur'ānic teaching and making great play of the importance of pilgrimages. While this may have derived from Agung's own evolving beliefs, it is striking that it followed on from a series of uprisings centred on a group of villages in central Java and above all on the holy site of Tembayat.[105] It is tempting, therefore, to associate the suffering of the early 1630s with the need to win over starving and restive populations with conciliatory measures; after all, turbulence of all kinds can be a powerful motivation for seeking explanations and finding solutions.[106]

The effects of these patterns were global, with an estimated 30 million perishing from hunger and related causes around the world before more normal conditions returned in the mid-1630s. The flooding of Potosí in South America forced the suspension of mining activities, while cataclysmic flooding in Mexico between 1629 and 1634 left fields saturated, washed away nutrients and made it almost impossible to travel, let alone transport goods and foods.[107] In Europe, 1628 was another 'year without a summer', with snow falling in Switzerland twenty-three times in the Bernese Oberland, crop failures and a notable rise in the number of witch hunts and accusations of sorcery in the Alps and Rhine basin.[108]

Far more damaging, however, were war and disease which decimated populations, transformed societies and had consequences that lasted for centuries. Wars of religion between Protestants and Catholics overwhelmed much of continental Europe in what became known as the Thirty Years' War (1618–48), a time of furious, sustained conflict and suffering. The 1630s onwards saw the start of fierce fighting between Spain and France that lasted for more than twenty years, while England was ravaged by civil war for most of the following decade. The 1640s were brutal, a time of constant unrest, suffering and turbulence within as well as between states. There were revolts and uprisings 'everywhere in the world, for example in France, England, Germany, Poland, Muscovy and the Ottoman empire', noted the Swedish diplomat Johan Adler Salvius, who was at a loss to understand the rise in violence. Perhaps 'this can be explained by some general configuration of the stars in the sky', he suggested – musings that were shared by others around the same time.[109]

To make matters worse, plague took hold, spreading from northern France in 1623 to England, the Low Countries and Germany before reaching Italy in 1629–30. Mortality rates were particularly high in major cities in northern Italy with around 30 to 35 per cent of the population of that region – some 2 million people – dying from plague.[110] The effects of war and disease were even more severe in the territories of the Holy Roman empire, with demographic declines of 35 to 40 per cent.[111] The dual effects of conflict and plague were so great that they impacted inequality into the eighteenth and perhaps even into the nineteenth century. Had Germany not had these experiences, it would probably have followed trends elsewhere and become much more unequal. While this made Germany more egalitarian, it also meant it was on the wrong side of economic divergence in Europe that left it having to play catch-up from the mid-nineteenth century onwards.[112]

The outcome was equally bad in Italy, especially when further waves of plague struck in the 1650s and led to around a million deaths in the kingdom of Naples – or somewhere between 30 and 43 per cent of the population. This was dramatically worse than north-western Europe, where plague intensity led to the deaths of around 8–10 per cent of the population in England and marginally higher in France. Plague did enormous damage in Italy because it struck so quickly and because its decimation of the urban workforce brought about shocks to production capabilities – and because deaths among elites in cities like Venice and Genoa who organised and invested in long-distance trade had material impacts on short- and long-term economic fortunes.

But perhaps most important of all was timing: plague struck at the worst possible moment, at a time when manufacturers were facing intense and rising competition from countries in northern Europe. In this sense, the disparity in mortality rates between north and south was crucial, as it allowed the former not only not to suffer but left it in a position to capitalise on its emerging advantages.[113] This was a crucial staging post in the Little Divergence – the point where northern Europe caught up with, then overtook and pulled away from the south. By 1650, English wages were 10 per cent higher than those in Italy; by 1800, they had risen to a staggering 150 per cent higher.[114]

These changes were not due only to warfare and disease. Spain's transatlantic trade had contracted sharply from the start of the seventeenth century, leading to a contagious financial crisis that brought Genoa and its banking system to its knees. Part of the reason for this was the maturing of economies in the New World that saw greater levels of internal trade combined with an increased ability of local officials, landowners and merchants to build up their own resources. Where around 30,000 tons of goods were shipped to Spain in the 1610s, less than half that quantity was being sent across the Atlantic thirty years later. Other economies were also experiencing transitions that were in some cases accelerated by wars that had substantial impacts on supply and demand. Industry in Venice contracted by half in the first decades of the seventeenth century, as did exports from Tuscany, while English sales of woollen cloth in continental Europe suffered a similar level of collapse.[115]

The great upheavals and crises struck many as extraordinarily severe and unusual. 'God Almighty has a quarrel lately with all mankind, and

given the reins to the ill spirit to compass the whole earth,' wrote James
Howell, a historian, in the middle of the seventeenth century. For more
than a decade 'there have the strangest revolutions and horridest things
happened, not only in Europe but all the world over, that have befallen
mankind, I dare boldly say, since Adam fell, in so short a revolution of
time'. It would seem, said Howell, 'the whole world is off the hinges'.[116]

Particularly harsh weather was part of the story of disruption. Rather
than taking place every five years or so, the El Niño–Southern Oscillation
(ENSO) took place twice as frequently, including in 1638, 1639, 1641, 1642
and a further eight times in the next twenty years. The western United
States and Canadian Rockies experienced prolonged periods of drought,
as did the Valley of Mexico, where processions were held invoking the
intercession of the Virgin Mary to bring rain. Challenging conditions
were the norm in many places around the world, from the coldest winter
ever recorded in Scandinavia to the failure of rice harvests in South-East
Asia, chronic food shortages in India and Central Asia and chaos in many
parts of East Asia where the monsoon was extremely weak in this period.[117]
The unusual conditions were aggravated by the eruptions of Komagatake
in 1640 and above all of Parker Peak in the Philippines a year later, which
extended drought conditions by at least three years.[118]

   This came at a particularly testing time in China, where the fall of
the Ming dynasty has often been linked to poor weather generally and
to climatic change in particular.[119] It is not hard to see why. In 1640,
poor harvests were accompanied by locust swarms, food shortages,
sky-high prices and outbreaks of disease. The next three years (1641–
4) then produced the worst drought in 500 years.[120] Zeng Yuwang, an
eyewitness who wrote about the suffering that followed, leaves little to
the imagination. First came a plague of locusts who were so many that
they piled up a foot high in people's compounds and forced Zeng to
cover his face with a fan so that he could breathe when he was out of the
house. 'I could barely stand up under the weight of the locusts clinging
to my fan and to my clothes and hat.'

   In 1642, he wrote, famine struck, causing huge loss of life. 'All along
our route home,' he wrote, 'we saw corpses scattered about in the fields,
and innumerable children abandoned by the side of the road ... To
be walking properly attired down an unending road of corpses was
a nightmare the likes of which I had never known. Approaching the

village, we saw six or seven people stripping bark off elm trees' trying to feed themselves.[121]

In Suzhou, the engine room of Ming dynasty economic development where anything and anything could be bought in the early seventeenth century, where shops lined the streets and merchants houses with their delightful gardens characterised the prosperity of the city and its inhabitants, things looked bleak indeed. 'Most of the residences in the city are empty and they are falling into ruins,' recounted Ye Shaoyuan, a scholar who wrote extensively about the traumas of these times. 'Fertile farms and beautiful estates are for sale but there is no one to buy them. Formerly the city of Suzhou was prosperous and its people tended to be extravagant. It is natural that after a period of prosperity a period of depression should follow; but I never dreamed that I should have to witness these misfortunes in the days of my life.'[122]

Anxieties were heightened by reports of rebel armies taking up arms, the result of years of uprisings and unrest gathering steam under disgruntled Ming officials like Li Zicheng and Zhang Xianzhong, who captured the imagination – and gained the support – of the peasantry by promising land reform and an overhaul of the taxation system, as well as by being successful bandit leaders who were able to deliver rewards to their supporters.[123]

By the end of the 1630s, the rebel army numbered in the tens of thousands – if not more – and was not only capturing cities and taking control of provinces but was soon bearing down on Beijing itself. Worrying omens spelled trouble in a state where disorder was understood as a sign of divine disfavour. Further harbingers of doom were provided by an earthquake in early 1644, by strange-coloured pears appearing on trees, by ominous winds that rose shortly before the start of the annual veneration of Confucius and by not a single baby being born in the capital, Beijing.[124]

As rebel forces closed in on the capital in the spring of 1644 and then gained entry through the city gates, the Chongzhen Emperor apologised to his relatives that they should be part of his unlucky family, instructed his wife to commit suicide, told the younger members to disguise themselves and flee, and then hung himself from a tree. He reportedly left a letter noting that his 'inadequate virtues and weak flesh have invited punishment from Heaven'. Although his officials deserved the blame, he went on, 'I must die but am ashamed to face my ancestors.

Therefore I take off my crown and cover my face with my hair.' And with that the Ming dynasty that had ruled China for almost 300 years came to an end.[125]

While the climatic misfortunes of the 1640s cannot have helped, given the starvation and disease, and also given the loss of prestige, credibility and authority of the Emperor personally, the reality was that the fall of the Ming had been a long time coming and had deep roots. These included population growth that almost tripled from about 70 million around 1400 to more than 200 million by 1644, a rise in military spending in the late sixteenth century which meant that 76 per cent of all government expenditure went towards provisioning and equipping troops, and using them in costly engagements, particularly on the northern frontiers, and an eye-wateringly profligate waste of resources on the 'confusions of pleasure'.[126]

Lavish levels of spending on the palace and the court – which included maintaining 3,000 court ladies and an apparent 20,000 eunuchs – went hand in hand with high levels of corruption, endemic bribery being a product of low salaries and regular opportunity and temptation. Eunuchs collected imperial rents, administered tax, oversaw government storehouses and ran the secret police. This came at a high cost in regular payments, combined with inefficiencies and inflated costs as a result of regular kickbacks pocketed by eunuchs, and also at the price of the Emperor being shielded from reality as these officials grew more powerful. As one popular ditty addressed to the Emperor from the 1630s put it:

You're getting on, your ears are deaf, your eyes are gone
Can't see people, can't hear words ...
How can you be so high? Come down to earth.[127]

The problems in Ming China were made worse by a cycle of desperate measures, poor decision-making and bad luck. Cuts to the postal service of some 30 per cent in 1629 were imposed in order to save money but ended up rendering efforts to deal with uprisings more difficult as information supply and orders slowed down. The ranks of the poor, hungry and disgruntled were swelled by officials who had lost their jobs in similar cost-cutting exercises. Growing banditry added to the sense of confusion and chaos, as well as to ideas about incompetence at the highest levels of government. Major changes to patterns of global commerce, over which the Ming dynasty had little control and even less understanding, were a factor too: China had been the world's great silver sink, sucking in bullion from the Americas via Manila. But from the start of the seventeenth century, trade competition from other locations in South and South-East Asia had risen sharply; Japanese decisions to cut off trade with Macao shut off an important market; while the capture of Melaka by the Dutch likewise had a major impact on Chinese silver imports.[128]

The collapse of the silver mines of Potosí in 1630, the reduced number of Chinese trading ships and the loss of galleons sailing across the Pacific because of poor seamanship and strong typhoons also proved damaging. Among those wrecked in 1638, for example, was the flagship *Nuestra Señora de la Concepción*, the largest vessel built at that time.[129] Losses such as these brought about change, since they represented major setbacks for the Spanish crown, which monopolised trans-Pacific trade. The year 1640 marked a turning point not only as the end of a period of rapid growth and high profits dating back to the 1580s, but because it saw a fundamental curtailment of shipping from Acapulco, the principal point of embarkation in the Americas to Manila, the entry point to Asia. In order to try to control the bribery that had made ships leave port overloaded, late or both, it was decided that the number of vessels travelling the route each year across the Pacific should be reduced – to one.[130] For China, it was like the oxygen supply being cut off to a patient that was already sick.

This was all made worse by dislocation in the provinces which played a major part in the collapse of central tax collection, while falling levels of agricultural production created pressures of their own: some estimates suggest the amount of land being cultivated in China fell from almost 200 million acres at the start of the seventeenth century to less than a third of that by the time of the collapse of the Ming.[131] China was trapped in a death spiral, in other words, locked in above all by the failure to adapt to or deal with mounting problems. But it means the collapse had roots that extended far deeper and lasted for far longer than climate crises in the build-up to 1644 or even in the years that preceded the fall of the Ming.

In the circumstances, then, the turbulent climate in the late 1630s and early 1640s clearly did not help; but this was one factor among a very wide range of others that dovetailed and came together to bring about the end of a dynasty. It was a similar story in many other parts of the world, as political challenges, economic contraction and demographic loss through warfare, disease and famine proved so deadly at roughly the same time. It did not take much to upset precarious balances that relied on fine margins of error.

However, it was not the same story in every part of the world. Tokugawa Japan, for example, and the Low Countries were able to navigate very similar climatic conditions that proved so devastating in China without the catastrophic problems of epidemic disease, famine and the overthrow of the ruling elite. The secret to this success lay in the mundane: the competence of officials and administrators in recognising problems, in anticipating the challenges they would bring and in planning ahead accordingly.[132] Climate was an aggravating factor, in other words, where problems already existed.

It is important to steer away from the irresistible temptation of the historian to identify watershed moments, to pinpoint events that can be described as turning points. The fall of the Ming dynasty clearly had repercussions and a high level of symbolic resonance both at the time and looking back with hindsight. However, the reality is that such transitions had much more ambiguous meanings for most contemporaries. Who sat on the throne was presumably less important than taxation demands that emanated from regional and imperial political and administrative centres. In that sense, the significance of the transition to the Qing was more obviously felt in the arrival of new hairstyles, new customs and new pastimes that were introduced at the court and then fanned outwards: frontier products like fur became the markers of elite Chinese fashion, while mushrooms from Mongolia, freshwater pearls from Manchuria and exotic foods from South-East Asia and Oceania put a raft of commodities under pressure as they were exploited for the benefit of metropolitan markets.[133]

Nevertheless, the challenges of the 'general crisis' of the seventeenth century produced important responses. The most significant of these was about putting plans in place to mitigate risk and to reduce the chances of being exposed to a range of problems being shaken into a cocktail at the same time. There were many ways to go about doing this: improving communication and efficiency; putting time, effort and resources into agricultural sciences; and opening up or taking over new lands that brought with them solutions to one of the oldest problems in human history: how to enable cities to expand and grow beyond the extent of their own hinterland. As we have seen, parts of western Europe had already done just that in the New World. Others too were about to try to do the same.

18

Concerning Great and Little Divergences
(c.1600–c.1800)

I am astonished at my own moderation.

Robert Clive (1772)

The opening up of trade routes linking Europe, Africa, the Americas and Asia brought about a series of commercial, social, political, biological and ecological revolutions. The demand for resources and goods, the search for profits and the acceleration of exchange brought more and more people into ever closer contact with each other, knitting the world together in the process by linking overlapping regional networks into a globalised economy.

Taking advantage of opportunities demanded investment in skills, the development of information-gathering and the systemisation of knowledge. For example, shipping routes were standardised by the Dutch in a formal set of sailors' orders, with fleets sailing from harbour twice and, from the 1630s, three times per year for collective security – while they also sought optimal wind conditions: in a world where time was money, setting sail in adverse conditions was not just risky but costly. On average, ships sailing into easterly winds covered 218 kilometres per day; those sailing into westerlies covered just 167 kilometres – almost 25 per cent less. Winds were part of unpredictable cycles of weather patterns. For example, westerly winds were common in the 1730s and the early 1750s, but much less so in the 1740s and after 1755.[1]

Knowledge that had commercial and strategic value was kept secret. Ships' captains who were among the first waves to travel from Europe to the Americas and Asia were under orders to make maps, and to keep them secret from potential rivals. The Venetians were desperate to acquire accounts of the voyage of the Portuguese Pedro Cabral around the Cape of Good Hope to India in 1500, though this was not easy 'because the king has placed a death penalty on anyone who gives it out'. Shrewd intelligence solved the problem within three months, with one leading Venetian writing back home to say that he had been able to get hold of maps not only of the routes to India but beyond as well.[2]

Other materials were also kept secret, such as a text compiled in the seventeenth century by the botanist Georg Rumphius, who was sent out to South-East Asia by the Dutch East India Company to gather knowledge about plants. His catalogue, entitled *Het Amboinsche kruidboek*, ran to over 7,000 pages, with descriptions of more than a thousand plant species from all over the Indonesian archipelago. But this was no altruistic contribution to modern scholarship. In fact, it was considered so sensitive and commercially important that it was not allowed to be published for several decades and did not appear until 1741, some forty years after Rumphius' death.[3]

This was part of a wider set of new connections that had developed almost as soon as European sailors had reached the Americas and rounded the tip of southern Africa to gain access to the worlds of the Indian Ocean and beyond in the 1490s in a rapid, deliberate and systematic exploration of oceans, coastlines and peoples. Within the space of a few short decades, Spanish and Portuguese explorers and mariners had reached the Americas, India and the Philippines, had sailed around the Cape of Good Hope, had come into contact with Andean, Mesoamerican, Asian and Oceanic peoples and had circumnavigated the globe. Scholars and scientists chronicled and recorded the peoples, geographies and natural histories that had been encountered, compiling encyclopaedic works of extraordinary scope and scale, such as the thirty volumes produced by Francisco Hernández, who had set out for the New World accompanied by painters and engravers to help capture what he saw and discovered.[4]

Such ventures were championed and paid for by Iberian rulers who, rather than sharing with others what scholars found out, saw the chance to help understand and fashion new empires taking shape not in adjacent territories to ones that they already controlled – which was the abiding model of political expansion – but in lands far from home. The Spanish and Portuguese models differed in many ways. While the latter tended to rely on military control of specific enclaves, the former focused on establishing seats of learning in new locations. Almost all major cities in the Spanish empire were endowed with hospitals, printing presses and even universities – with the first set up in Santo Domingo in what is now the Dominican Republic in 1538, followed by Michoacán in Mexico the following year and Lima just over a decade later.[5]

There was a further purpose in supporting learning overseas, as the Dominican monk Tommaso Campanella stressed c.1600: 'This learning will make the one who possesses it master of the sea, the land and the people, and will illustrate the empire more than any other thing imaginable for making a King great,' he wrote. 'Because God Himself wants His works to be known and He gives them to he who knows them.' Empires were a manifestation of divine benediction, in other words, and therefore understanding and controlling them was a matter not just of interest but of duty.[6]

Such rosy ideas about the value of scholarship should of course be set within the context of an age of intolerance and prejudice that characterised both the Inquisition and the Reformation in Europe in the late fifteenth century and beyond. As violence erupted between Catholics and Protestants, as Jews were subjected to terrible persecution and countless innocent victims across Europe were accused of sorcery, heresy or both, science and learning became battlegrounds too. Rulers issued decrees ordering the handing over of Lutheran works on pain of fines and imprisonment; popes issued lists of forbidden books, which included texts deemed inappropriate because of where they were published or because their publishers had issued other works that were heretical.[7]

It should also be stressed that European engagement and growing familiarity with new parts of the world, new peoples and new cultures did not result in relationships based on mutual respect – indeed, far from it. As we have seen, attitudes to men, women and children from Africa developed into condescension, brutality and oppression, producing painful wounds that still scar the contemporary world; the treatment of indigenous peoples in the Americas, likewise, was above all one based on entitlement, manipulation and inequality that also has legacies today.

Such ideas took time to foment, develop and then harden – but harden they did as views about race as well as about other parts of the world became set in stone. It took several leaps of faith to underline the European sense of superiority when it came to climates that were hostile because of disease environments or uncomfortably warm. Some had their own theories about who might flourish and why. 'The Lustier and Fresher' one was, speculated Dr John Fryer, the harder it was to adapt to living in India. 'But to Old Men and Women', on the other hand, 'it seems more suitable.' Others were gloomier: one English doctor in the late seventeenth century mourned that 'we are here, as Exotick Plants ... not agreeable to the Soil'.[8]

As far as Asia was concerned, supposed 'exoticness' was no barrier to the growing conviction that Europeans were uniquely privileged among all the peoples of the world. This was self-evident, wrote Voltaire. Europeans had proved their 'great superiority of genius and courage over eastern nations'. Preconceptions and prejudices became routine, both about others and about Europeans themselves. 'Europe may be the smallest part of the world,' wrote one German scholar in the late eighteenth century, 'but it is the best.' There could be no doubt, claimed a contemporary, that Europe had attained 'the ultimate peak of perfection'.[9]

Those living elsewhere were worthy of contempt – an attitude reinforced by a flood of travel literature written by a new class of scholar gentlemen who set off around the world in pursuit of scientific knowledge and learning, snootily complaining about the 'hard and dangerous work' of investigating peoples and places that usually fell short of expectations. Timbuktu, for example, was not a glorious desert city that was home to wealth and wisdom, but a profound disappointment for one European visitor – 'nothing but a mass of ill-looking houses, built of earth. Nothing was to be seen in all directions but immense plains of quicksand of a yellowish white colour.'[10]

Racist tropes developed into sneering caricatures of other peoples' cultures. Persians were barely able to use knives, forks, plates or napkins, wrote one traveller in the early nineteenth century. Mongols and Tibetans were good natured, open and trusting, and showed none of the 'deception, cowardice and baseness of the Hindus', said the philosopher Georg Hegel in a series of lectures delivered in Berlin around the same time.[11]

In European imagination, the Ottoman empire became synonymous with plague and illness, to the point where (as we have seen) it became routinely referred to as 'the sick man of Europe', both in reference to a supposed sclerotic political system and literally as a sick land unable to deal with ill-health in a rational manner.[12] Sickness, laziness and perversion were all rolled into perceptions of the Turks. Constantinople, the Ottoman capital, 'has been not only the hot bed of every sort of Eastern vice', stated British Prime Minister David Lloyd George in 1919, but 'the source from which the poison of corruption and intrigue has spread far and wide'.[13] Such views were commonplace.

They also became central parts of reframing history in a way that celebrated 'western civilisation' at the expense of all others. 'A single shelf of a good European library', wrote the celebrated historian Thomas Macaulay in 1836, 'was worth the whole native literature of India and Arabia.' As for education in India, he opined, it was essential for the British to do their 'best to form a class who may be interpreters between us and the millions whom we govern – a class of persons Indian in blood and colour, but English in tastes, in opinions, in morals and in intellect'. India needed non-Indians to run it; this could be done only by creating and training a new 'class' who had been disconnected from their Indianness and made to think like Europeans. English was 'better worth knowing than Sanskrit or Arabic', wrote Macaulay; moreover, it was much easier to learn than Greek. Given that intelligent English youths could master Herodotus and Sophocles easily, how hard could it be for 'a Hindoo to read Hume and Milton'?[14]

Entitlement over other people was mirrored by entitlement over nature. As we have seen, islands in the Caribbean as well as enslaved people shipped from Africa were brutally exploited to extract crops that could be sold for high profits, most notably sugar, coffee, cotton and other cash crops. The human costs of these enterprises were catastrophic; in many cases, the ecological consequences were also dramatic – described by one scholar as 'reverse tonsures, shaved bald but for little tufts of forest left on their mountain crowns'. Exhaustion of the soil was exacerbated by deforestation which exposed the land to rain and heat that sapped the nutrients further still. This was neither sustainable nor clever business: it did not take long for it to become necessary for firewood to be imported to provide fuel for sugar furnaces, with wood eventually even being imported to the Caribbean from as far away as England.[15]

Such ecological degradation was initially limited to a relatively small number of locations – above all in north-eastern Brazil and the Caribbean. Taken as a whole, European exploitation of territories in the Americas was surprisingly constrained. For one thing, the aim of establishing colonies was above all not to increase production but to control trade. Indeed, in some cases, severe restrictions were introduced to limit exports in order to maximise the financial return to the crown. For example, concessions giving monopoly on brazilwood, a red dyewood that was much in demand for European textile makers and was so valuable that it gave its name to a vast country, were auctioned every three years; strict limits were set for volumes that could be exported, with the King of Portugal ordering the death sentence in 1605 for anyone caught smuggling. Control of the diamond trade was also enforced, with physical removal of everyone who was not involved in mining – and especially priests, who were notorious for their involvement in contraband trade.[16]

In fact, one of the most noteworthy features of the New World when looked at as a whole was how poorly it was exploited by European settlers. Spanish olive and wine producers and Portuguese salt manufacturers successfully lobbied to block production in the Americas in order to keep prices of their own domestic goods high. Before the third quarter of the eighteenth century, Spain received around 150,000 animal hides from the Americas; changes to trade policies and taxes in 1778 altered the dynamics of trade, resulting in 800,000 hides being shipped across the Atlantic the following year – and 1.4 million being sent in 1783. This was part of a tenfold increase in colonial trade between 1778 and 1796. In fact, rather than exploiting 'virgin lands', trade from the colonies in the Americas often ran at a fraction of its potential for some 300 years after the first crossing by Columbus.[17]

One result was that there were widespread environmental changes in North, Central and South America in the centuries that followed 1492 that were not caused by human activity – but quite the opposite. This was a time of major reforestation and wildlife gains in many regions. The major reductions in human populations as a result of disease, hunger, warfare and dislocation meant that land that had been cultivated became overgrown, to the benefit of wild plants and the detriment of foodstuffs like fruits and nuts. Likewise, animals that had been heavily exploited for food or status recovered – such as birds whose feathers had been valued as symbols of prestige and currency, or apex predators like jaguars who gained from expanded habitats with the thinning out of the men who had hunted them and of settlements that had displaced them. Animals whose fortunes and numbers had risen with human success, like turkeys, dogs and llamas, suffered decline as their owners, breeders and keepers died out. But one of the most curious outcomes of the European settlement of the Americas – outside the locations given over to the cultivation of a small number of valuable crops – was the significant wildlife gains across vast territories. Taken as a whole, the Americas were more heavily forested in 1800 than they had been 300 years earlier.[18]

Dramatic ecological change was not just connected to the interventions of Europeans. The peoples of the indigenous world that linked what is now California, Nevada, Utah, Arizona, New Mexico, Sonora and Baja California with the Colorado Delta and on to the east coast had long been part of regions that had been defined by geographic boundaries made up of rivers, mountains and deserts, of economies that interlocked or were distinct as a result of tools, commodities and livestock, of food systems that were based on similar plant and animal resources and water availability, and of transport and communication networks. Over time, these relationships and balances began to change, both as a by-product of what was happening far to the east and above all because of the introduction of horses, which prompted dynamic change across swathes of North America. As raiding strategies evolved, so too did new confederacies emerge under the Haudenosaunees, Cherokees, Comanches, Navajo, Apaches and others, in response to pressures posed by European settlers, to the need to defend or attack territory controlled by other indigenous peoples, and to the urge to enshrine the power of elites within what some scholars have termed 'regimes of violence'.[19]

The opening up of new lands was not limited to the Americas. In India in the late sixteenth century, for example, the expansion of the territorial boundaries of the Mughal empire brought about more liberal ideas that were connected to the incorporation of more minorities and the expedience of inclusivity. The Emperor Akbar was instrumental not only in welcoming Jains to his court, but in being influenced by their philosophies and practices. This included vegetarianism, which Akbar adopted with enthusiasm. 'It is not right that a man should make his stomach the grave of animals,' he declared, adding that he would ban the eating of meat if it were practicable. In any event, he said, 'from my earliest years, whenever I ordered animal food to be cooked for me, I found it rather tasteless and cared little for it. I took this feeling to indicate a necessity for protecting animals, and I refrained from animal food.'[20]

This should not be confused with conservationism or environmentally sustainable practices. In fact, Mughal rulers were motivated by seeking military victory, imposing public order, increasing land revenues and annexing new territories. The expansion of the empire into Bengal in the late sixteenth century was a perfect example of these aims intertwining as ambitious local rulers were forced into submission over the course of a long series of military campaigns. Local populations – many of whom were devotees of forest goddesses – were displaced as new settlers set about securing land grants before cutting down the jungle and clearing land to make way for rice cultivation. This came at a heavy price for forest dwellers, who were not inhabitants of peripheral lands that were wild and savage, but rather played hugely significant roles in the collection and cultivation of spices, resins and other non-timber products which required knowledge and skills that were difficult to acquire.[21]

Most of these 'pioneers' were Muslims, and built small rural mosques to support new settlements and communities as wild wetlands, marshlands and forests were transformed into domesticated lands. The effect was to open up new regions which eventually produced vast amounts of foods, commodities and tax revenues that helped sustain and fuel the Mughal empire in the seventeenth century.[22]

The principle was the same under the Qing dynasty which took power in the middle of the seventeenth century in much of what is now China. The new rulers were able to restore order both speedily and effectively after the chaos that had accompanied the fall of their Ming predecessors, which of course speaks volumes about the structural problems most to blame for the calamities that had proved so devastating in the 1640s. As administrative reforms kicked in, central coffers swelled as agricultural production rose sharply. Over the next two centuries, the amount of land under cultivation rose from around 100 million to 200 million acres – equivalent to a rate of 500,000 more acres per year being turned over to the growing of crops.[23]

This came with significant ecological costs. Intensive deforestation and uplift in farming resulted in high levels of erosion and soil degradation that badly affected the Yellow River and the Grand Canal. The results proved to be not just environmentally damaging but economically catastrophic: the costs of maintaining the canal systems rose fivefold in the century after the 1730s, eventually consuming as much as 20 per cent of all government spending by the 1820s.[24] This was a poor use of capital and manpower under any circumstances; but such inefficiency left Qing dynasty China badly placed to deal with the global competition that emerged in the nineteenth century.

The Qing oversaw a period of territorial expansion in all directions – to the west, to the north and to the south. By the end of the seventeenth century, the dynasty had established control over large parts of Mongolia and Xinjiang as well as Taiwan, meanwhile extending influence into Tibet.[25] By 1759, the Qianlong Emperor (r. 1735–96) was declaring that all rival states and kingdoms had been subdued and that he had brought about 'eternal peace and security on the borders'.[26]

Many of the lands that the Qing moved into did not offer manifold natural bonanzas. While expanding into the interior brought kudos, taking over steppe lands, mountains and deserts – all of which featured prominently in the newly conquered regions – did not offer obvious gains. In any event, if, when and where they did, they came at a price: goods needed to reach consumers and customers; exploiting regions with low population densities of their own that were far from towns and cities meant that transportation of high-volume, bulk items – like crops – was expensive and even financially unviable. As a result, attention was almost exclusively focused on exotic and luxury commodities, such as furs and freshwater pearls, that were light, high value and offered the best returns.[27]

In other parts of the world, cities rose up along new trade routes. For example, Lima, Panama, Havana, Buenos Aires, Rio de Janeiro, Manila and beyond all grew in response to silver mines and shipments from South and Central America.[28] The expansions of the seventeenth and eighteenth centuries meant that many parts of the world cross-fertilised, galvanised and stimulated each other. But this did not happen in China: no new cities were founded; there was limited resettlement of populations by choice or by force; and the gains of expansion were primarily existential rather than material. That was a poor equation. High costs, low production and low wages all placed drags on growth, as well as on expenditure. That all served to set Qing China on a trajectory that contrasted sharply with Europe, where acute anxieties had led to considerable efforts being made to improve yields, promote resilience and reduce the risk posed by climate shocks.

The traumas of the mid-seventeenth century had prompted considerable thought on how to get better returns from the soil. For example, Walter Blith was a passionate advocate of land management, urging farmers to study the subject more carefully, offering advice that he promised, in the manner of a modern marketing guru, would lead to a doubling or trebling of crop growths, if not a fivefold and perhaps even a tenfold return on expectations. This was just one example of burgeoning interest in agronomy and in the efforts to protect against future shortage.[29]

The devastation of the Thirty Years' War, not least in the form of heavy loss of life through war, disease and hunger, perhaps played a crucial role in focusing minds. It was probably no coincidence, for example, that in the seventeenth century, young scholars like Ludwig von Seckendorff not only spent time thinking about soil conditions, fertility of the land and jurisdiction over the land, but considered these to be essential elements for all rulers to be informed about in order to govern their states successfully. This was not just about mitigation of potential threats; it was also closely connected to questions about labour-force size and potential gains from higher levels of productivity. 'A country in which subsistence and commerce flourish can never have too many inhabitants,' wrote Johann von Justi, an influential scholar in the eighteenth century. Finding ways to support higher population densities was important, he argued, in enabling rulers to generate higher incomes.[30]

Such lessons were not lost on those who understood how science, politics and economics were all closely linked. 'Agriculture', wrote Frederick the Great, the pre-eminent figure in eastern Europe in the mid-eighteenth century, 'is the first of all the arts, without which there would be no merchants, kings, poets, philosophers.' As he observed to his correspondent, the French author and polymath Voltaire, 'making domain lands cultivable interests me more than murdering people'.[31]

Such views can easily mask widespread anxieties about production and about supply being able to keep up with demand. London alone had grown by seven times in the hundred years after the reign of Queen Elizabeth I (1558–1603), noted William Petty. What would happen if it grew by the same amount again, he wondered; would there be enough 'bread and drink-corn … fruits, garden stuff, hay, timber, coal?'[32] One way to solve the problem, was to spawn new colonies from which resources could be extracted – which could prove something of a learning curve: England's experiences in Ireland, for example, were to provide a template in North America and in Canada in particular in relation to displacing existing populations from their lands, to the use of force within racial and religious frameworks and to ideas about supposed superiority and entitlements.[33]

Another method of boosting productivity was to work the land more intensively, through higher labour levels or through innovation. Qing China, for example, imported vast quantities of fertiliser primarily made of rapeseed, cottonseed and soybean cakes amounting to perhaps 3 billion kilos per year by 1750. This was part of a revolution in land management and also in consumption, welfare and health. Life expectancy and living standards not only matched but were superior to those almost everywhere in Europe other than south-east England, where levels were about the same. This was particularly impressive given that the population of Qing China more than doubled in the course of the eighteenth century.[34]

One explanation for how demographic growth happened was the process known as the Columbian Exchange which linked the Americas to the other continents of the world. Although this term is often used by historians to describe the new connections made in the decades and centuries after 1492, it is unsatisfactory – above all since it implies symmetry in the two-way relationship between the Americas and the rest of the world. For one thing, the movement of people was almost exclusively in one direction: as we have seen, millions were shipped against their will from Africa to the Americas, while millions more settlers in due course crossed the Atlantic from Europe; the number of those heading the other way, as visiting dignitaries or as coerced labourers, was extremely low.[35]

To be sure, silver flows from the Americas were truly transformational; but taken in broader terms, the New World was the recipient of a great deal more from the Old than the other way round. In addition to the involuntary and voluntary movements of people, countless horses, sheep, pigs, cows, chickens, geese, cats, rats and more were shipped across the Atlantic, while only squirrels, turkeys and guinea pigs came in the opposite direction. Cacao, corn, tobacco, pineapples, beans and chilli peppers were exported from the Americas, but many plants and crops were imported from Europe, Africa and Asia, including citrus fruits, rice and bananas – with all three becoming staple parts of the diet.[36]

Old World diets had more of an effect on those living in the New, at least initially: indeed, by the 1580s, indigenous populations in 190 communities which were surveyed about high mortality rates and sickness levels explained that a century earlier, before the arrival of the Spanish, people had eaten less food, consumed less salt and drunk less alcohol – while also claiming that hygiene levels had been higher. Although it was noted that the tried-and-tested remedies of local healers had been more effective than medical ideas brought by new settlers and that bans on polygamy had limited birth rates, the fact remained that the arrival of Europeans had heralded the end of an age of moderation and replaced it with one of excess.[37]

Nevertheless, while relatively few in number, there were foods that had a major impact after they were exported from the Americas. Manioc (cassava) and maize were first brought east across the Atlantic by Portuguese traders in the sixteenth century and became much prized by local farmers in West Africa above all for their robustness against drought. Manioc was drought-resistant as well as high-yielding, and, better still, not only resistant to locusts but also easy to store. It was labour-intensive, requiring soaking, grating and squeezing before being left to dry and then ground into flour; this process was vital in order to remove metabolites that have the ability to produce highly toxic hydrogen cyanide when degraded by plant enzymes. Failure to do so properly causes paralysis and poisoning.[38]

Maize too took well to African soils, providing two crops in the time it usually took for a single sorghum harvest, and moreover was both easy to store and difficult to spoil. It became such a crucial source of diets in Benin, Ghana, Nigeria, Togo, Cameroon and Angola that it is often seen today as an indigenous, traditional crop rather than a relatively recent import.[39] Both of these crops were of crucial importance in increasing the number of calories available in regions that were sources of large numbers of enslaved peoples who were sold. Ironically, therefore, they played a role in the slave trade by enabling more people to survive famine, by increasing the dependence on women in agricultural production and by enabling stores of food to be built up that reduced labour demands locally. This created pools of 'surplus' workforces that made men available for sale to traders.

Even more important, however, than maize and manioc was another crop, whose impact on global history is hard to overestimate – not least because it too mitigated risks posed by weather shocks and climatic shifts. Indeed, one could reasonably argue that it did as much to change human lives as some of the most significant medical discoveries in history or as much to change economic outcomes as events like the industrial revolution. As well as providing security against famine and disease, it changed health outcomes, drove urbanisation and reduced conflict. The humble potato changed the world.

Potatoes provide more calories, vitamins and nutrients per acre of land sown than other staple crops. The skin of a medium-sized potato contains 45 per cent of the recommended daily intake of vitamin C – unlike wheat, oats, barley, rice and maize which contain none at all; a medium potato also provides significant amounts of the recommended daily intake of vitamin B6, as well thiamin, riboflavin, niacin, magnesium, iron and zinc. In addition, potatoes need less land and yield well over ten times more calories than crops such as wheat. An acre of potatoes typically produces three times more energy than an acre of oats, wheat or barley.[40]

Potatoes can grow on land that is not suitable for the cultivation of wheat or rice – and, what is more, can be grown in between the growing seasons of other crops and on land left fallow between periods of grain cultivation. As well as being robust in poor weather conditions, potatoes provide diversification that mitigates the threat of famine. Relatively easy to store, they also allow for surpluses to be built up as contingencies during winter months for both human and animal consumption. The latter is important too: as fodder for livestock, potatoes also help feed larger numbers of pigs and cattle, which increases opportunities for meat consumption, as well as boosting quantities of manure which in turn further helps with crop production.[41]

Cultivated in the Andes for thousands of years, potatoes were not immediately appreciated in Europe after being brought back in the sixteenth century, partly because the potato belongs to the poisonous nightshade family and partly because of associations made between the lumpy, discoloured appearance of the crop and leprosy. Slowly adopted at first, potatoes started to be planted in Spain, Italy, England and Germany by around 1600, by which time they were also being introduced by sailors to ports in Africa, Asia and Oceania – although widespread cultivation in Europe, India and China did not take off until the late seventeenth and early eighteenth centuries.[42]

By this time, the potato was widely considered to have almost miraculous properties. Peter the Great had toured Europe in the late 1690s and sent a bag of potatoes back to Russia with instructions that they should be distributed to farmers in different districts. A few decades later, the Senate issued directives to grow 'earth apples which in England are called potatoes' as part of an effort not only to encourage cultivation but to counter grain failures that had led to hunger and disease in Finland and Siberia in the 1760s. It took time to catch on, partly because some were convinced that the potato was 'that forbidden fruit' which Adam and Eve had eaten in the Garden of Eden (presumably because of the name *pomme de terre*); 'whoever eats it disobeys God', said some traditionalists in Russia, 'violates the holy testament and will never inherit the Kingdom of Heaven'.[43]

Others were more pragmatic. 'The food produced by a field of potatoes', wrote Adam Smith in *The Wealth of Nations*, was 'much superior to what is produced by a field of wheat'. Smith was convinced that 'No food can afford a more decisive proof of its nourishing quality, or of its being peculiarly suitable to the health of the human constitution.' The strongest men in London and the most beautiful women in the dominions, he said, were those who were from the 'lowest rank of people in Ireland, who are generally fed with this root'.[44]

The adoption and dissemination of potatoes had remarkable effects in many different spheres. Military records of more than 13,000 French soldiers born between 1658 and 1770 suggest that adult height went up by an average of over one centimetre as a result of nutritional investments made during early life that can be directly attributed to consumption of more calories in general and of potatoes in particular.[45]

More remarkable still is the fact that the adoption of potatoes in Europe increased agricultural productivity, reducing the requirement for land in the process and thereby watering down competition for its control. Although it stands to reason that the availability of cheap food can reduce the cost of maintaining large armies and therefore make warfare more attractive, lower land values typically lessen the probability of violent conflict because incentives to arm are lower, as are the probabilities of winning a war. Statistical modelling suggests that the introduction of potatoes dramatically decreased conflict not only in Europe but in other continents and regions where it was established.[46] The introduction of sweet potatoes, first to the southern part of China and then to the north, served to reduce peasant rebellions primarily by providing insurance against rainfall shocks that led to failures of other crops.[47]

By mitigating risks of adverse weather conditions, whether temporary or part of wider, long-term shifts, potatoes opened up new opportunities for larger populations by reducing food shortages, increasing calorific intakes and improving health and life expectancies. Furthermore, they also enabled higher population densities to be supported, which in turn drove the rise and growing size of cities in the seventeenth and eighteenth centuries. It has been estimated, for example, that potatoes account for around 25 per cent of the increase in population sizes in this period – and for an even greater percentage of the increase in urbanisation.[48] This was crucial because of the primary role that cities played (and play) in fostering commercial exchange, stimulating demand and spurring innovation.

If food in general played an important role in propelling the growth of cities around the world, then so too did high wages – which are fundamental to explaining the phenomenon that has been mentioned before in this book and that historians call the Great Divergence, the phase when Europe caught up with and overtook what had long been much larger and more established economies in Asia, most notably in India and China.

There were many reasons why this happened. Some scholars who have investigated urban growth in Europe have looked at the importance of the continent having been part of the Roman empire and having access to Roman road infrastructure a thousand years or more later.[49] Others have looked at the effects of Protestantism and at the profound changes which its rise brought about for education in Europe, where countries in the north and west pulled away from those in the south and east. Martin Luther had been a passionate advocate of schooling for both boys and girls, insisting that all should be able to read the Scriptures. It was disgraceful, he argued in *A Sermon on Keeping Children in School* in 1530, that some parents were so selfish, stupid and unChristian as to leave their children ignorant or to withdraw them from school. In doing so, he said, they were serving the devil. Such admonitions had long-term impacts, with Protestant cities and lands adopting policies of education for all children and requiring the building of new schools. Strikingly, while all countries with Protestant majorities had nearly universal literacy by 1900, none with Catholic ones did, with many a long way behind.[50] This pattern was replicated between 1910 and 1938 in the United States, where educational outcomes were better in areas with high proportions of Protestants in their populations.[51]

The rise of institutions of all kinds in Europe was important, as were the introduction of the printing press and the expansion of book production which opened up ideas and knowledge to more and more people. The creation of universities was significant too, not least because they ensured increasing numbers trained in law, which had a positive effect on economic transactions by reducing the uncertainties of trade – a development that increased levels of trust as well as improving volumes and speeds of exchanges.[52]

Lowering transactional risks helped make the use of capital cheaper, which in turn helped stimulate competition. Interest rates in Europe, especially in Great Britain and the Netherlands, were much lower than they were in China: Adam Smith noted in the 1770s that while 3–4.5 per cent could be considered normal in London, current rates in China were said to be as high as 12 per cent.[53] He was not quite right. In fact, loans in China often ran at 2 per cent per month, or 24 per cent annually; in some periods, rates were more like 50 per cent per annum.[54] Although there were regional variations across South, South-East and East Asia and fluctuations too depending on the period, it is clear that the costs of taking loans and of doing business were extremely high when compared to Europe – and that these contributed to both dampening economic growth and suffocating innovation.[55]

Both economic growth and innovation were fuelled by competition, something that was readily available in Europe where political fragmentation created opportunities to test and refine ideas, to sharpen reactions and to drive change. This manifested itself in the near constant warfare which served to improve military technologies, as has been touched on already, and which proved essential when it came to the age of colonisations of Africa, Asia and Oceania in the nineteenth century. But competition fuelled scientific discovery and advance too, although these should be seen in the context of gender divergences that often are not mentioned: in northern Europe and in Britain in particular, male scholars and entrepreneurs drove innovation, creating and fuelling networks that shared and celebrated inventions, as well as providing capital for those with commercial application.[56] Women were locked out of these networks, intentionally and otherwise, and had to publish their work anonymously or not at all.[57]

The circle of men, science, money and opportunity was particularly effective in Britain, where the key elements that drove progress and ultimately resulted in the industrial revolution were high levels of pay and high productive output of the labour force. The workforce in Britain were far more productive and also better paid than anywhere else in Europe, which meant that new inventions were adopted earlier, faster and on a larger scale than anywhere else in the world. What mattered most was not schooling or literacy, but mechanical competence – something that was in uniquely high supply in Britain where skilled craftsmen introduced a constant stream of small improvements to machinery.[58] Examples included the spinning jenny, a basic and cheap hand-powered cotton-spinning machine, changes in which metals were used for gears in early textile machines and standardisation of engine size that enabled inventories of spare parts to be kept on hand that in turn saved time and money in the event of breakdown or failure.[59]

As a result, agricultural workers easily outstripped their peers in terms of time and speed. Whereas it took around seven days to reap a hectare of wheat in England at the end of the eighteenth century, the same process took more than sixteen days in France. These efficiencies enabled higher wages and a slew of further benefits for English workers, including higher calorie consumption than their French peers. It has even been suggested that because English workers were better paid, protein from meat was more readily affordable, which is particularly important for brain development in babies and infants. It seems that workers may not only have been richer and more productive than in other countries, but smarter too.[60]

'The working manufacturing people of England eat the fat, and drink the sweet, live better, and fare better, than the working poor of any other nation in Europe,' wrote Daniel Defoe in 1726; 'they make better wages of their work, and spend more of the money upon their backs and bellies, than in any other country.' Defoe could have added that, as well as providing leisure time, high wages underpinned the acquisition of skills, ranging from literacy to numeracy to craft competence, all of which were in high demand in an economy that was advancing rapidly.[61]

It was a very different story in China, where pressures to innovate and to compete were conspicuous by their absence. For one thing, only a tiny sliver of the rural population was engaged in wage labour – perhaps as little as 1–2 per cent during the late Ming dynasty.[62] Moreover, for all the territorial gains made by the Qing in the late seventeenth and eighteenth centuries, the Pax Manchurica that was so triumphantly proclaimed as a symbol of heavenly order proved to be a badly exposed Achilles heel. The push towards natural barriers such as mountains and deserts removed neighbours that were threats or competitors, and forced parallel empires out of view. The military had to contend with threats that were modest and containable from Korea, from South-East and Inner Asia and from Cossack adventurers. None prompted revolutions in military technology or tactics, and what is more, none led to administrative, social or economic reform, or to investment in innovation, productivity or industry, which were so instrumental in the rise of the west. Europe's divisions and fractured political history since the time of the Roman empire made for intense competition. Warfare and fracture had long been sources of weakness within Europe; now they became a strength. As Europe galloped forward into the future, in other words, China remained stuck in the past.[63]

Higher levels of resilience against unpredictable weather conditions helped societies deal with food shortages and upheavals in which climate had been a factor in the past. There were still one-off events that could have dramatic consequences. For example, a particularly severe storm in southern England in December 1703, described by Daniel Defoe as 'the greatest, the longest in duration, the widest in extent, of all the Tempests and storms that history gives any account of since the Beginning of time', led to thirteen Royal Navy ships being lost at sea along with 8,000 seamen. This jeopardised English naval supremacy at a testing time, coming as it did during the War of Spanish Succession, which amounted to a battle for control of the heart of the continent of Europe.[64]

Sequences of unusually frequent and severe storms in the Atlantic in the 1760s and 1770s were instrumental, meanwhile, in reshaping trade networks and shifting political alliances. A disastrous hurricane season in 1766 which saw six major storms hit the Caribbean basin brought about severe food shortages that prompted desperate appeals for help from Spanish islands being placed in newspapers in cities in the US colonies. American support brought warnings from the British government, which regarded the sale of flour and other goods as trading with the enemy and issued official demands to cease trade forthwith.[65]

Another brutal hurricane season in 1772 which resulted in terrible destruction across the Caribbean again, damaging houses, shops, hospitals and civic infrastructure in Havana followed by crop failures, presented what one scholar has called 'a onetime windfall' for merchants in Philadelphia. Five of the six richest men in the city, at that time the leading financial centre on the eastern seaboard, were already actively involved in trade with Spain, and attention now began to turn to asking what Britain's European rivalries had to do with the colonies in North America – a question that was compounded by growing concern about taxation demands and the use of unconstitutional powers.[66] This had resulted in a gathering in October 1774 that brought together representatives of twelve of the thirteen colonies to meet and issue the Declaration of Colonial Rights that proposed a boycott of British goods and sent a list of grievances to London, along with a petition to King George III.[67]

Americans began to look at the opportunities presented closer to home, both in the Caribbean and in Spanish Louisiana, where New Orleans had also experienced severe food shortages. Representatives of the newly formed Continental Congress were established in French Caribbean seaports, while channels of communication with Spanish islands were opened to discuss the possibility of trading tobacco from North America for gunpowder. In October 1775, the Continental Congress passed a resolution that authorised exports to foreign ports in exchange for weapons, ammunition, gunpowder and cash – all of which were essential given the rising tension with Britain and the adoption of a boycott of goods a few months earlier. Leading figures in North America, such as John Adams, one of the Founding Fathers and signatories of the Declaration of Independence, were convinced that there 'will be little Difficulty in Trading with France and Spain, a great deal in dealing with Portugal and some with Holland'.[68]

The realisation that the environmental and economic vulnerabilities of the American south and of the Caribbean represented commercial and political opportunities for the colonists and weak spots for the British was an important, if little commented on, part of the story of the weeks before, during and after the Declaration of Independence on 4 July 1776. Embargoes on British shipping effectively cut off Jamaica and other British colonies, creating shortages and suffering that were made worse by a violent early-season hurricane that struck in June of that year. In the French islands in the Caribbean, in contrast, there was a 'festival-like atmosphere' as traders cut deals, booked future agreements and filled warehouses to the rafters with goods.[69] There were plenty of good reasons why the French and later the Spanish entered the War of Independence on the side of the Patriots in the second half of the 1770s; but one factor was certainly that all benefited from closer integration of commercial ties in the western Atlantic and Caribbean, which owed much to the efforts to mitigate risk from adverse weather shocks.

Not all were so fortunate when it came to finding solutions to these problems. Climate conditions in India in the second half of the eighteenth century were challenging, consistently resulting in lower-than-usual agricultural productivity and higher-than-average grain prices. Regular warfare between potentates in South Asia pushed the price of agricultural input higher still as did rising rent burdens, made worse by the deterioration of central authority. This had the unfortunate effect of raising nominal wages while driving up the cost of living, serving to make India less competitive at a time when trends in Britain and elsewhere in Europe were heading in the other direction. In 1750, India had been responsible for around a quarter of the world's industrial output; by 1900, it accounted for just 2 per cent.[70]

This precipitous decline can also be explained, at least in part, by colonial policies that included a lack of investment, tariff protection that supported foreign-made goods, and the extraction of resources and labour by the East India Company that established a dominant position in much of the subcontinent by the middle of the eighteenth century.[71] This combination proved deadly in the late 1760s when a weak monsoon resulted in a poor harvest in Bengal and Bihar in 1768 which was followed a year later by even lower levels of rainfall, prompting the superintendent of Bishunpur to note that 'the fields of rice are become like fields of dried straw'.[72] The East India Company administrators in the region took little interest in the calamities that were unfolding. Their primary concern was to collect revenues in order to meet financial targets that helped enrich shareholders – so attention was focused on gathering obligations, which by the start of 1770 were reported as being on track and as expected. Although it was noted in January that a slight remission of land tax might be necessary, at the start of February officials stated that they had not 'yet found any failure in the revenue or stated payments'.[73]

By this time, the price of rice had already gone up to ten times its normal rate, and there were reports that starvation levels were so bad that people were selling their children to try to raise money. Malnutrition and disease took hold, followed by death on an almost unimaginable scale. More than 76,000 people died in Calcutta alone between July and September 1770; as many as one in five Bengalis died that year – numbering more than a million. 'The only remedy', wrote one contemporary, 'is the mercy of God.'[74]

Some East India Company officials tried to put emergency measures in place, setting up distribution centres to give out food to those who were suffering, although this proved to be a losing battle. Others, however, seized the chance to enrich themselves, buying up stocks of rice that forced prices higher still and then selling them on at huge mark-ups. A number made 'great fortunes' as a result, with one employee of previously modest means earning the modern equivalent of millions of pounds from the disaster which he promptly sent back home. Although some, like Horace Walpole, were horrified by reports coming from India, investors were delighted: the East India Company share price soared, convincing the directors to issue a dividend of 12.5 per cent – the highest that the company had ever paid out.[75]

There were to be consequences in due course, not least because the famine provoked a collapse in land values in and remittances from Bengal, which led to a sell-off that, coupled with unsold stockpiles of tea and high company debts, precipitated a downturn in fortunes that was as dramatic and rapid as the collapse of major investment banks in the financial crisis of 2008. In need of a state bail-out, senior EIC figures were hauled into Parliament to explain and justify what had happened. The fact of the matter, said Robert Clive, who had taken control of the fisc of Bengal just over a decade earlier and had enriched himself mightily, was that he could have helped himself to even more. 'I stand astonished at my own moderation,' he told nonplussed MPs.[76]

That the motivation of greed turned a difficult situation into a catastrophe was to have obvious parallels with a similar situation in Bengal in the twentieth century as millions again became exposed to the lethal combination of poor weather, bad harvest, incompetent administration and the consequences of predatory speculation. In these cases, it was human decision-making that played a significant and even a decisive role in making a difficult situation worse. The result was mass mortality on an epic scale.

Sometimes, however, it was coincidence of timing of natural phenomena that exacerbated problems. A case in point comes with the eruptions of the Laki fissure in Iceland in 1783. Between 8 June and 25 October that year, Laki produced ten stratospheric eruption episodes, with eruptions continuing until the following February. Local impacts were dire, as 80 per cent of the sheep, almost the same proportion of horses and half the cattle in Iceland died, primarily as a result of grazing lands becoming poisoned. Around one in five of the human population perished as famine and 'haze hardships' (Móðuharðindin) gripped the island for several years.[77] The effects were felt more widely too, with unusually frequent episodes of respiratory illnesses across many parts of Europe, along with multiple accounts of acute headaches and asthma attacks and high mortality rates in England in the summer of 1783.[78] The summer months saw 'a constant fog over all Europe', observed Benjamin Franklin, 'and a great part of North America'; he added that 'the rays of the sun seemed to have little effect towards dissipating it', and suggested that it was probably the result of major volcanic activity in Iceland.[79]

The cloud of haze was watched as it spread eastwards, reaching Berlin, Padua and Rome by mid-June, St Petersburg, Moscow and Tripoli in Lebanon by the end of the month, and then the Altai mountains in Central and East Asia by the start of July.[80] The summer was unusually warm across many parts of Europe: model simulations suggest that the abnormally high temperatures were the result of high pressure over northern Europe causing cold polar air to bypass western and central Europe, as well as the Middle East. The anomaly also brought about lower-than-normal rainfall in the Sahel and North Africa, and in many parts of China and South Asia. It is not impossible that, without the eruptions, conditions might have been even more challenging across the northern hemisphere.[81]

The period that followed saw significant temperature drops in the northern hemisphere, where the years 1784–6 were the coldest in the second half of the eighteenth century. Measurements from New Brunswick show that the winter of 1783–4 was the coldest in 250 years, while harbours and channels in Chesapeake Bay were closed for the longest period in recorded history as a result of build-up of ice. The Mississippi River was filled with ice fragments at New Orleans for a week in February 1784. North-west Alaska saw some of the coldest conditions of the previous millennium, while unusually low temperatures were recorded from Paris to Prague, from Lebanon to Japan; droughts and frosts in Mexico in 1784–5 were also part of the disturbed weather patterns in this period.[82]

It has been argued that these conditions were caused by a protracted El Niño event in 1782–4 that was coupled with a negative North Atlantic Oscillation phase.[83] However, recent research has shown that large high-latitude eruptions can prime the Pacific to create climatic perturbations – and that effects are particularly severe when El Niño–Southern Oscillation or La Niña episodes are already developing, magnifying the effects in the process.[84] The timing and scale of the Laki eruption was therefore not only important in its own right, but as an amplifier of other emerging climatic trends that were developing at this time.

While most parts of the world coped with the challenges posed by this period of unsettled and unusual weather well, some did not. That said much about how important underlying factors were in the context of climatic impacts, and shows that turbulence should be usually be seen as a catalyst of change rather than its primary driver. For example, the early 1780s saw famines in several locations across India, with death tolls running into the millions and leaving some regions entirely depopulated.[85] While many reports commented on the lack of rain and the effects this had on food production, non-climatic factors, most notably hostilities and warfare in southern and western India and the resulting reduction in the area of land under cultivation because of the diversion of manpower were significant. So too was the fact that crops were deliberately burned on several occasions as intentional acts of aggravation. All this heightened risk by reducing margins of error to the point that populations were left highly exposed to weather and food shocks.[86]

There were other occasions where something similar happened. For example, in Japan, a poor harvest in 1782 was followed by four years of failure that were made worse by the eruption of Mount Asama, which covered with volcanic ash a large part of the major food-producing Kanto plain surrounding the shogun's administrative capital at Edo. Some domains reported no harvest at all.[87] Ash made its way into the waterways, adding to silt build-up that was already a labour-intensive chore for villagers to deal with; now it proved deadly.[88]

As people were reduced to eating horses, cats and grass, hunger and disease took hold, resulting in a million deaths in the space of little more than four years. This led to a chronic shortage of labour, as was made clear by the fact that in the twenty years after the so-called Tenmei famine wages rose by 500 per cent while rice prices essentially remained stable.[89] The eruptions and poor weather were the straw that broke the camel's back in a finely balanced relationship that had led agricultural exploitation to develop to the edge of its ecological margin. Too much depended on harvests delivering anticipated yields and on waterways working as expected. Relatively modest alterations to both left vast populations at risk of hunger, disease – and death.

In Egypt, climate shock in the late eighteenth century triggered a set of socio-economic and political changes that had been brewing for years, decades even. Since the 1760s, local elites in the Ottoman province had grown increasingly assertive in their relationship with the central authorities, successfully building up their landholdings, not least through control over tax collection – where the incentives were to collect as much as possible from the population and pass on as little as possible to the state. Indeed, by the 1770s, they had established power bases that were not so much independent of the Ottoman empire as competing with it, sending private militias into Ottoman Syria in attempts to boost their influence and their fortunes further.[90]

The high-pressure anomaly of 1783 and the effects of the Laki eruptions deepened these existing problems. The failure of the Nile floods, which reduced irrigation by almost 20 per cent, sparked an intense period of drought, famine and poor yields. One chronicler wrote that the 'grain became scarce ... the price of wheat was on the loose ... and the poor suffered greatly from hunger'. Others commented on the lawlessness, violence and banditry which rose as shortages were accompanied by outbreaks of disease. The rich seized the opportunity to further their dominance, no doubt in part driven by their anxiety to protect their falling incomes during a time of uncertainty. This in turn reinforced determination in Constantinople to cut rapacious magnates down to size and restore control over the empire's wealthiest province.[91]

Plans were drawn up for a military occupation of Egypt that was designed to weaken the elites, depose the Mamluk emirs and impose tighter supervision over the provincial administration. A powerful force dispatched in 1786 under the command of Gazi Hasan Pasha, an experienced and trusted officer, met with immediate success. Before it could achieve its full objectives, however, it was recalled the following year when Catherine the Great ordered Russian forces to begin an offensive in the Black Sea and annexed Crimea. Ottoman efforts to push back the Russians came to nothing, as did the best-laid plans in Egypt, where as one leading scholar puts it, 'defiant rural leaders returned and quickly picked up where they had left off' – embedding a system of 'crony politics that would shape Egypt for at least the next 150 years'.[92]

Some locations, in contrast, benefited from the climatic upheavals. Evidence from the region between the Niger Bend and the various parts of the Voltas in Burkina Faso and Ghana suggests a period of wet conditions, as does the fact that Lake Chad and Lake Ngami in what is now Botswana rose to very high levels in the 1780s. Of course, heavy rains present problems of their own, causing waterlogging of fields and rough seas, as seems to have happened on the Guinea coast in the early 1780s, a decade which tree-ring data from South Africa indicates was the wettest decade for 200 years.[93]

These upheavals were soon followed in 1789–93 by what has been referred to as a 'mega-Niño', an episode so strong that it has only occurred a handful of times in the last two millennia.[94] Recent scholarship that draws on climatological records from the western Pacific suggests that the event was in fact a very strong La Niña event from 1788–90 which flipped to a strong El Niño event that lasted from 1791 to 1794.[95] This was a global reorganisation of weather systems that had global consequences, some of them long-lasting, some dramatic and some both.

Cold and wet conditions made settlement of south-east Australia by the British precarious. Officials appointed to set up a new administration in Sydney commented on the inclement, tempestuous weather that made life difficult. It was almost impossible to produce bricks because of the driving rain and howling gales, while newly created roads were all but impassable. 'The rain came down in torrents, filling up every trench and cavity which had been dug about the settlement,' wrote one observer, 'causing much damage to the miserable mud tenements' that had been built by convicts shipped out to help build what looked like an unpromising new world.[96]

The loss at sea of a crucial provisioning ship not only devastated morale further still, but led to the introduction both of drastic rationing and of martial law. Finally, in September 1790, conditions seemed to be improving – only for the rains to stop and be replaced by heatwaves which were compared to the 'blast of a heated oven'. It was again all but impossible to grow food, because of the lack of precipitation. By November 1791, water restrictions were introduced, with the only source of natural water a small freshwater stream that ran into Sydney Cove. The project of colonising Australia was looking like an act of folly – until benign and stable conditions finally settled around the middle of 1794. But it had been touch and go whether it was even worth trying to make a success of it.[97]

As it happened, the rainy conditions of the late 1780s in Oceania and the Pacific helped preserve a totemic moment in history. Captain William Bligh and a handful of sailors had been cast into an open boat after mutineers took control of HMS *Bounty*, a Royal Navy vessel that had been sent into the South Pacific on a botanical mission. In the diary he kept during the long trip at sea that followed, Bligh commented not only on the cold but also on the regular rainfall that allowed stocks of water to be maintained. This was vital in helping Bligh eventually reach safety – and therefore in the story of the mutiny becoming one of the best-known episodes in history. In a normal year, notes one scholar, survival would likely have been impossible, and the story would have been one of 'the disappearance of Captain Bligh and the *Bounty*', or just another unremarkable case of a ship lost at sea, in other words.[98]

The effects of the climate upheavals in other parts of the world affected rather more people than the escapades of a handful of British sailors. In India, so many died in a famine that struck in 1791 that it became known in folklore in Bijapur as the *Doji bara* or the 'skull famine', because so many bones of unburied victims could be found lying by the side of roads; census figures from the Madras Presidency, from Andhra Pradesh, from Hyderabad, the Deccan and elsewhere give a frightening sense of the scale of mortality, once more measuring in the millions.[99]

Prolonged drought in Natal and Zululand that started in 1789 brought about the Mahlatule famine (literally 'the time we were forced to eat grass'), the most severe episode known before the mid-nineteenth century, which nonetheless served as a springboard for the reorganisation and consolidation of political control that strengthened and accelerated the rise of the Zulu kingdom.[100] In North America, a period of mild, dry winters helped spur migrations of indigenous peoples including the Cree, Blackfeet and Gros Ventre, perhaps because higher than normal temperatures expanded the pasture available for horses: this led to increases in herd size and created competition for the best grazing lands. But there were challenges too that resulted from smallpox outbreaks and from shifts in the bison population which scattered in the face of climate factors and diminished as a result of epidemic bovine disease.[101]

On the eastern seaboard, meanwhile, pronounced warming enlarged the tropical range of the *Aedes aegypti* mosquito, resulting in outbreaks of yellow fever as far north as Philadelphia. This sparked an important division between political parties in the newly independent United States, with the newly formed Federalists and Republicans blaming each other for the outbreak. The former party argued that weak immigration policy had enabled the disease to be brought from Haiti, while the latter blamed Philadelphia's poor hygiene, bad governance and worse morals. Among other things, the dispute put pressure on elected officials to act swiftly to show their competence, leading to new approaches to investment in civic infrastructure, in emergency relief and in improving the quality of medical care.[102]

In Europe, change was in the air. A cold and hard winter in France in 1787–8 had been followed by a late and wet spring that led to grain prices rising by 50 per cent – an upsurge that proved crushing for the peasantry and had knock-on effects for the wider economy: as food costs soared, sales of wine fell sharply as financial resources were used to buy bread. By the summer of 1788, unrest was building across the country, with small-scale uprisings and shows of dissent in regions that had been badly hit, such as Provence, Picardy, Brittany, Franche-Comté,

Languedoc and Poitou. Bad storms that brought hailstones large enough to kill hares and knock partridges out of trees did further damage to vines and to wheatfields. Things were so bad that it was estimated that one in five people in Paris alone were dependent on charitable relief to survive.[103]

The French King, Louis XVI, took steps to try to calm the situation, sending out calls for the people of his kingdom to draw up *cahiers* of complaints and grievances and to present them in Paris. In just three months in early 1789, over 25,000 documents were prepared – a clear indication of the scale of the problems that were brewing. Steps were taken in the winter of 1788 to address food shortages by offering trade privileges for American flour imports, including exemptions from freight duties, as well as agreed premium prices – concessions that were extended in April 1789. As one historian has put it, 'the urban merchant's windfall was the frontier settler's despair': grain being shipped across the Atlantic was a godsend for growers, traders and shippers, but bad news for local populations who were faced with shortfalls and higher prices – if not both.[104]

The fact that the late 1780s had seen a surge in migrations to frontier communities in New York, Vermont and Pennsylvania put further pressure on grain supplies, made worse by a bitterly cold winter and warnings of poor harvests. As one newspaper in Vermont put it at the start of June 1789, 'The great want of that necessary article BREAD ... will be more severely felt the next season than it is the present.'[105] By this time, acts of resistance and defiance had become commonplace in France. These ranged from petty breakdowns such as poaching to more violent protests such as the Réveillon riots in April 1789 which served as a sign of things to come. Leaflets were circulated such as 'What No One Has Yet Said' which declared that the authorities were failing in their primary duty of ensuring there was enough bread for all. Another, 'Four Cries of a Patriot of the Nation', put it more bluntly. Citizens needed to be armed, immediately; and aristocrats should be banished. What was the point of 'preaching peace and liberty to men dying of hunger? What use would a wise constitution be to a people of skeletons?'[106]

The French revolution set fire to Europe and much of the world beyond. In the years that followed the storming of the Bastille on 14 July 1789, France re-emerged not as an egalitarian kingdom but as a putative empire, with ambitions in Egypt and India, to say nothing of a series of spectacular military victories that made Napoleon Bonaparte the master of much of continental Europe. There were reverberations beyond too – including revolution in 1791 in Saint-Domingue (Haiti) which was followed by independence.

France's failure to retake the island, its most profitable colony, owed much to the organisational skills and charismatic leadership of Toussaint Louverture and to the determination of Haiti's inhabitants to pursue 'liberté, égalité and fraternité' for themselves. It owed something too, however, to the yellow fever that devastated a force sent by France to reimpose authority. The failure to retake a colony that was the most valuable real estate in the world at the time, that produced much of the world's sugar and coffee and that was responsible for 40 per cent of France's foreign trade was devastating, and led to the panic sale of Louisiana to the United States to raise much-needed funds to pay for ongoing and constant military operations in Europe.[107] This changed the political landscape of global geopolitics: it shifted the gaze of the newly independent US to the south and to the west. In doing so, it opened the doors to the conquest of a large part of the continent of North America.

Ironically, France's turmoil provided Britain with several shots in the arm. The violence of revolution as well as the upheavals that started to engulf Europe soon afterwards, even before the rise of Napoleon, served to make Britain in general and London in particular a magnet for talent from all over. Skills, capital and connections descended on London from the Low Countries, Germany and of course France itself. As one historian has elegantly put it, new arrivals were like bees bringing pollen with them to produce new blooms and, in doing so, to germinate Britain's industrial take-off.[108] Such were the unforeseen consequences of major shifts in politics, ideas and climatic change.

19

Industry, Extraction and the Natural World
(c.1800–c.1870)

Fog everywhere. Fog up the river where it flows among green airs and meadows; fog down the river, where it rolls defiled among the tiers of shipping, and the waterside pollutions of a great (and dirty) city.

Charles Dickens, Bleak House (1853)

Engagement with new parts of the world spurred some to take a deeper interest in what could be learned from geography, from history and from science. Writing in the middle of the eighteenth century, the philosopher David Hume considered what the authors Horace, Juvenal and Diodorus Siculus had said about the weather and climate in Rome as well as elsewhere in the empire. It would have been preferable, he notes, 'had the ancients known the use of thermometers'. Nevertheless, comparing the accounts with the present day, it is reasonable to conclude that 'the winters are now much more temperate at Rome than formerly'.[1]

In Hume's own time, he wrote, the Tiber froze about as often as the River Nile – never, in other words. Likewise, Ovid's description of the Black Sea freezing every year either spoke of a very different climate, or reveals that Ovid was lying. There was only one explanation, concluded Hume. 'Plainly', he said, human activity was responsible for causing the planet to warm. This must have been done, he went on, primarily through deforestation and the cutting down of trees 'which formerly threw a shade upon the earth, and kept the rays of the sun from penetrating it'.[2]

The question of man-made climate change was one that preoccupied many settlers in the colonies in North America, including several of the Founding Fathers. In the 1760s, Benjamin Franklin wrote to Ezra Stiles, later President of Yale, stating that temperatures were becoming milder as a result of deforestation. 'When a Country is clear'd of Woods,' he said, 'the Sun acts more strongly on the Face of the Earth.' Solar warmth 'melts great Snows sooner than they could be melted if they were shaded by the Trees'. While a 'regular and steady Course of Observations' would be needed across multiple years and taking measurements from several parts of the country to confirm it, Franklin was persuaded that real changes were happening – and that human activity was responsible.[3]

Franklin reported in the same letter that he had recently been in England where he had visited Cambridge to compare notes with John Hadley, the Professor of Chemistry. Contacts such as these were a by-product of fast-expanding global trade networks and a new age of information-gathering from around the world as Europeans began to translate voyages of discovery into extensive commercial contacts and into regional dominance and colonialism. While trade, political and military priorities had led the way, science and scientists had sometimes followed hand in hand and had sometimes done so soon afterwards, supported by rising levels of wealth that funded scholarships and academic institutions and encouraged individual curiosity.

In 1768, for example, Captain James Cook was commissioned by the Admiralty to undertake a voyage to the Pacific Ocean, with the aim of tracking the transit of Venus across the sun the following year. Missions like this were commendable, Samuel Johnson had written two decades earlier, because unlike expeditions which set off with 'intent like merchants' to trade, or had military ambitions, they were motivated solely by the joy of knowledge for knowledge's sake.[4] Such sentiments sounded reassuring, noble even. But they masked the fact that those leading the expeditions, like Cook, often had other motives too: in this case Cook was also given secret orders to search the South Pacific for a southern continent whose existence had been much speculated on and whose discovery was considered a matter of signal strategic importance to Britain and its global interests.[5]

The question of the weather, of climate and of changing conditions dominated the minds of some. Thomas Jefferson had an obsessive interest, starting a diary on 1 July 1776 just as he was drafting the Declaration of Independence, and recording two temperature readings per day for the next fifty years. Indeed, on the morning of 4 July, the day that America's independence from Britain was declared, Jefferson visited Sparhawk's stationery store in Philadelphia to buy himself a new thermometer.[6] We know from his diary that as the Declaration itself was being delivered to Congress, he was busy recording that the ambient temperature was 72.5 °F.[7] As the United States was being born, one of its principal architects was thinking about humidity and atmospheric pressure. He was presumably not entirely happy, at least about the tools at his disposal, for the day after independence he went back to Sparhawk's to buy a barometer so that his findings could be even more accurate.[8]

One of Jefferson's pet theories was about the changing climate in North America in the late eighteenth century. Summarising in a book his 'data for estimating the climate of Virginia', he set out observations about sudden temperature changes, frosts and the impact on plant and animal life. This led him to conclude that 'A change in our climate ... is taking place very sensibly. Both heats and colds are become much more moderate within the memory even of the middle-aged. Snows are less frequent and less deep,' and while 'The elderly inform me the earth used to be covered with snow for about three months in every year,' it no longer was – and nor did rivers freeze often, as they used to.[9]

Jefferson's views chimed with the prevailing wisdom among scholars in North America that the climate was changing rapidly. Hugh Williamson of Harvard had written almost twenty years earlier that 'our winters are not so intensely cold, nor our summers so disagreeably warm'. This was due to land-use change from forest to open fields, which gave the earth a hard smooth surface, and rather as 'the face of a looking glass or any polished metal will reflect more light and heat' the result was a warming of the land and of the temperature. This was good news for the future, he said: 'clearing the country will mitigate the cold of our winters [and] it will also increase the heat of our summers'. As soon as trees had been felled, 'we shall seldom be visited by frosts or snows'.[10] Climate change was 'so rapid and constant', agreed Samuel Williams in 1794, that 'it is the subject of common observation and experience'. Not only that, he added, but it could be observed 'in every part of the United States'. There was no scope for this to be 'a matter of doubt'; rather, it was a matter of fact.[11]

That was all nonsense, retorted Noah Webster, who is best known for his famous dictionary. Webster took issue both with such statements and with the supporting evidence. 'Mr Jefferson seems to have no authority for his opinions' other than 'the observations of elderly and middle-aged people', he said, adding that there was plenty of evidence to suggest that climates had not changed. The claims by men like Williams that temperatures had risen by ten to twelve degrees in the last century and a half were implausible and should lead any rational commentator to conclude that such views depended on 'insurmountable difficulties' and were quite unreliable.[12]

This was less a case of climate-change denial than a demand for rigorous scholarship to back up claims properly. As it was, competing views had already developed that set out entirely the opposite hypothesis – namely that the earth was cooling. One pioneer was the Comte de Buffon, who looked into questions like the locations of oceans and continents, sea-level changes and the formation of mountains. Widely read in the second half of the eighteenth century, Buffon proposed not only that the earth had been getting colder since its creation, but that it would keep on doing so. Draining marshes, deforestation and urbanisation had certainly helped raise temperatures, but ultimately little would stop an inevitable freeze.[13]

For some, the worry was less about warming and cooling than about the problems of rising populations and the pressure of food shortages, a topic that had generated considerable discussion since the 1770s when the Bengal famine, outbreaks of wheat infestations by the Hessian fly, hurricanes in the Caribbean, the American War of Independence and a run of bad harvests in Britain and Ireland all raised fears about the impact on the poor, about the viability of the colonies and about the potential for disaster in the future.

In 1798, Thomas Malthus published a gloomy tract *On the Principle of Population*. The power of population, he wrote, 'is so superior to the power of the earth to produce subsistence for man, that premature death must in some shape or other visit the human race'. The more people were alive, the greater the difficulty in producing enough food for them all. Fortunately, he went on, such are 'the vices of mankind' that humans were often the best source of population control, most notably through fighting wars that brought death to many and served as a cap on the number of those consuming resources. Inevitably, however, this would not always work – for 'Man cannot live in the midst of plenty.' As a result, the spectre of 'gigantic inevitable famine' loomed large to resolve competition between the total available 'food of the world' and the number of people living in it.[14]

Such concerns led Sir Joseph Banks, President of the Royal Society, to investigate ways of making warm climate plants and crops frost resistant. This was a task that gained additional urgency in the light of scientists' conviction that the world was going through a pronounced phase of climatic change, although it was a matter of disagreement whether the problem was one of cooling or warming. It was 'unquestionable', wrote the Scottish chemist John Leslie in 1804, 'that the climate, over the whole of Europe, has assumed a milder character'. It was clear, he argued, that 'our earth must grow continually warmer' thanks to the sun's rays. There could be little doubting, he added, that the climate of central and northern Europe 'has gradually become milder', for natural reasons. This had nothing to do with human activities. If it did, then the role was a marginal and peripheral one: anthropogenic activities 'have no influence whatever in altering the average of temperature'.[15]

Scholars like Harvard Professor Samuel Williams were not so sure: Williams suggested that 'the heat of the earth has been gradually increasing' as a result of colonisation and human-inspired ecological change in New England. If this was true, argued others like the influential Buffon, then it was an anomaly that would pass: the clear trend was towards cooling that would eventually cause the planet to freeze.[16]

Hypotheses and disagreements like these were in part a reflection of the awareness and realities of a rapidly changing world. The first decades of the nineteenth century saw a series of profound technological, political, socio-economic and ecological shifts that remoulded geographies, accelerated exchanges of goods and people and transformed landscapes in ways that were both dramatic and rapid. This was an age of scientific discovery and of the dissemination of information, an age of the creation, expansion and enhancement of transport, trade and communication networks, a time when improvements in communication and transportation were developed and rolled out, a time when productivity shot up as a set of industrial and scientific revolutions bore fruit.

These changes had the greatest impact in Europe, a world that in the late eighteenth century was 'shaped by bereavement, peopled with orphans and widows', where half of all children died before the age of ten and only one in ten people reached the age of sixty. Harvest failure, famine and epidemic disease were commonplace, all aggravated by squalid conditions in towns and cities where mortality rates were so high that there was constant demand for people to migrate from the countryside.[17]

Part of the stimulus for change came from the military revolutions that transformed battlefield tactics as well as from the demand for manpower by states that became increasingly centralised as a result. In the early eighteenth century, the number of casualties in even major battles in Europe amounted to a few hundred, and rarely much more than that. With soldiers carrying enough ammunition for about fifteen or twenty minutes of fire, and at a slow rate of discharge, army sizes were modest, as were levels of training. By the end of the century, however, firefights could last many hours – and the number of dead and wounded typically ran into the tens of thousands.[18] Around 1.7 million men (and a few women) who served in the French armies between 1798 and 1815 died, with a large proportion not killed in battle but dying as a result of injuries, infection or disease.[19]

Britain was able to mass-produce weapons and ammunition – helped by control of the saltpetre of Bengal and Bihar, by far the richest in the world, which produced vast quantities of nitrates, the essential ingredient in gunpowder. In the years 1808–11, during the height of the Napoleonic Wars, the British were able to supply 336,000 muskets, 100,000 pistols and 60 million cartridges to help Spanish guerrillas opposed to Napoleon, to say nothing of the production of weapons, cannon and ordnance for use by British forces in their own campaigns.[20]

The Napoleonic Wars had another curious effect too: the demand for manpower to serve in Britain's armed forces against the French produced labour shortages as men were recruited to fight, with an estimated 350,000 men under arms at the peak of the conflict. Towns and regions where recruitment was heaviest prompted adoption of, investment in and improvement of labour-saving technologies – such as threshing machines. This produced long-term socio-economic benefits, even after peace returned to Europe in 1815.[21]

The enhanced role of the state created demands for higher levels of political participation, a mood that reached crisis point at St Peter's Field in Manchester in August 1819, when a crowd of perhaps 60,000 gathered to protest against the lack of parliamentary representation; this was a time when voting was a right confined to the elites, with just over 10 per cent of adult men (and no women) being allowed to vote in elections, which in some cases were not even held regularly. Magistrates called in troops, who brutally broke up the protests with a cavalry charge that resulted in deaths and many casualties. Soon known as the Peterloo Massacre, the event became notorious for the use of force against unarmed demonstrators.[22]

The demand for reform was driven by a number of factors, including economic stagnation and unemployment stemming from a conclusion to almost two decades of constant warfare after the defeat of Napoleon at the battle of Waterloo four years earlier. But depressed climatic conditions also played a role in disrupting harvests, creating price shocks that led to the cost of grain doubling and to poverty deepening across much of Europe and elsewhere around this time. In the winter of 1816–17, noted a leading newspaper in Manchester not long afterwards, workers in large manufacturing towns were without work and desperately short of food. Parishes gave what they could, but this bore no comparison to what was needed. As one influential newspaper at the time put it, Britain was 'a nation supplicating for bread – a people sinking for want of food'.[23]

This stemmed in part from measures adopted by the government following the end of the Napoleonic Wars which served to benefit landowners and the wealthy. One was the abolition of wartime income tax; another was the introduction of the Corn Laws that imposed a ban on the import of grain, which inevitably drove prices upwards and dragged people deeper into poverty.[24]

Another cause of problems, though, lay on the other side of the world. This was Mount Tambora in what is now Indonesia, whose eruption on the evening of 5 April 1815 is the largest of the last ten thousand years. The effects locally were devastating as tens of cubic kilometres of magma were spewed out and ejected as much as forty-three kilometres into the atmosphere, while the explosions were heard 2,000 kilometres away. A tsunami fanned out, with some reports of waves as high as four metres devastating multiple islands, including Java. As many as 120,000 people lost their lives in South-East Asia as a result of the famine and disease that followed.[25]

In the three years before Tambora erupted, global temperatures had already become distinctly cooler, partly because of the eruptions of Mount Soufrière in the Caribbean and of Mount Mayon in what is now the Philippines in 1812 and 1814 respectively. If these magnified the effects of Tambora, it did not help that 1816 corresponded to an unusually weak maximum in the sunspot cycle, a phenomenon that is known to affect sea surface temperatures. The impacts around the world were so great that 1816 has become popularly known as 'the year without a summer'.[26]

The consequences were dramatic. In July 1816, *The Times* of London warned that dangers lay ahead: 'Should the present wet weather continue', the paper noted, harvests were likely to fail 'and the effects of such a calamity at such a time cannot be otherwise than ruinous to the farmers and even to the people at large.' It was a similar story in many parts of Europe, with 'melancholy accounts' being received 'from all parts of the Continent of the unusual wetness of the season; property in consequence swept away by inundation and irretrievable injuries done to the vine yards and corn crops. In several provinces of Holland, the rich grass lands are all under water, and scarcity and high prices are naturally apprehended and dreaded. In France the interior of the country has suffered greatly from the floods and heavy rains.' Mortality rates in some parts of Europe increased, most notably in Switzerland and Tuscany.[27]

It is perhaps no coincidence that a group of English writers that included Percy and Mary Shelley and Lord Byron, who were spending time in Geneva in the summer of 1816, made repeated reference to dark storms, unusual skies, violent winds and rains in their writings. Indeed, on one evening in June, the group came up with the idea of a ghost-story contest in order to entertain themselves during that long, cold summer; thus it was that Mary Shelley had the idea for *Frankenstein*, one of the most famous novels of all time, and one in which celestial anomalies, lightning strikes, thunder and storms feature prominently.[28]

A collapse in cereal crops in New England led not only to severe food shortages and price surges, but also to large-scale livestock death thanks to lack of animal feed. 'Never were such hard times,' wrote Thomas Jefferson, as people found themselves in a state of 'unparalleled distress'. It was likely, perhaps even inevitable, he wrote, that there would be local insurrections, uprisings and breakdowns of law and order as a result. Newspapers compared the situation to the biblical 'famine of Egypt', though that analogy did not allow for the scale of the financial crisis, the emotional upheaval or the abandonment of towns. One historian has even argued that Tambora was a 'primary cause of the United States' first major economic depression'.[29]

The eruption devasted other regions, such as the Indian subcontinent, where a shift in monsoon rains, a failure of trade winds and a three-year depression of the thermal cycle of South Asia not only led to major reductions in crop yields and of maritime trade, but also served to change the microbial ecology of the Bay of Bengal. In 1817, unusually early and heavy rains brought about a surge in cases of cholera that caused death on an almost unimaginable scale. The bodies of the dead and dying were gathered together, wrote one eyewitness, with funeral pyres burning non-stop to incinerate the rich and the poor, and other corpses picked over by vultures or jackals. It was 'a scene of woe which completely baffles the power of description to portray'.[30]

More than a million were estimated to have died as a result of what one report of 1820 argued was the 'distempered' state of the weather since 1815. It seems that climatic factors had indeed played a decisive role: changes in water temperature and salinity supported zooplankton that serves as cholera's main aquatic host, while unusual and unseasonal flooding served as the source of the bacterium's nutrients, at the same time delivering the pathogen into the water system of coastal regions. This was almost uniquely dangerous in Bengal because of the low-lying land in the river delta.[31]

While assessing how many people died involves wide margins of error, another indication of the devastation it caused can be seen in the widespread panic within affected communities, with large numbers of towns being depopulated as people fled. Some took to traditional methods of salvation, turning for protection to deities such as Kali and Ola Bibi, whose cult grew quickly around this time. Although climate had a role in the prevalence of cholera, diet, sanitation and hygiene were even more important factors, since cholera was above all a disease of poverty.[32] By the early 1820s, cholera had spread by land and sea through South-East Asia to China and Japan and westwards to Persia and Russia, and then on to Europe, where it took hold at the start of the 1830s.[33]

Disease, poverty and limited employment prospects all played a role in driving waves of migration from Europe. The end of the Napoleonic Wars meant that the job market was inundated with 200,000 demobilised soldiers at a time when massive government contracts to provide supplies to the military of everything from uniforms to musket balls to ropes to canvas for ships were scaled back or abandoned altogether.[34] Appetites for better futures were ravenous: a British government scheme for settlement in South Africa was set up in a bid to attract 4,000 people willing to move continent to settle in the Albany district of the eastern Cape; more than 80,000 submitted applications.[35]

This was not to say that all greeted the arrival of newcomers with glee. George Washington, the first President of the United States, was scornful of the quality of those who reached the shores of North America in search of better lives. They were nothing more than 'banditti who will bid defiance to all authority', 'worthless fellows' and a bunch of 'savages'.[36] From the point of view of those facing a protracted depression in Europe, the prospect of moving to start a new life elsewhere was an increasingly attractive one, especially in lands where new opportunities were opening up and infrastructure and facilities were constantly being improved, such as in the United States, where more and more steamboats plying the great rivers created new networks, cheaper transportation and the prospect not only of wealth but of freedom too.[37]

The numbers of those leaving the British Isles accelerated sharply. Between 1790 and 1815, around 180,000 emigrated from England, Scotland and Wales.[38] In the three decades that followed, numbers swelled dramatically, followed by a series of surges in the second half of the nineteenth century. In the seventy years after 1850, some 45 million people migrated from the Old to the New World.[39] This proved crucial for the development of the Americas as a whole, with new arrivals not only serving as fresh pools of labour, but bringing with them ideas, knowledge, cultures, genes, institutions and languages that helped promote rapid socio-economic and political development.[40] It spurred change in Europe too, with the mass exodus reducing the size of the workforce, thereby driving up wages and offering further rewards for innovation, mechanisation and industrialisation.[41]

Reports were sent back home that talked not only of opportunities in new lands, but of freedoms. As Joseph Hollingworth, a recent arrival in North America in the 1820s, put it in a letter to his relatives in Huddersfield, 'in this country there are no Lords, nor Dukes, nor Counts, nor Marquises, nor Earls, nor Royal Family to support nor no King'. Not only that, there was no sign of poverty. 'I have never seen in this Country a Beggar such as I used Daily to see in England,' and, perhaps better still, no sign of a tax collector 'taking the last penny out of the poor Mans Pocket'. It was a wonder too that the President of the United States began speeches by addressing not 'My Lords and Gentlemen' but his 'Fellow Citizens'. This was a place of dreams, remarked Hollingworth, and turning to what he called 'Old English poetry', described his new homeland as 'A land where tyranny is no more / Where we can all be free.'[42]

Underpinning such notions of freedom was a variety of expansive ideas about nature, about ecological transformation, about 'improvement' of virgin lands – and about the displacement of those who already lived there. Indigenous populations, commonly lumped together and referred to as 'Indians', were routinely dismissed as 'citizens of an inferior order' – like Jews, Gypsies, enslaved people and 'free negroes'. Native peoples were 'the very filth of civilized society', opined some, worthy to be 'left to the rapacity of noxious vermin'; it was only a matter of time before they would end in 'total extermination'. In any event, it seemed obvious to many that the lands where they hunted, farmed and subsisted should be taken over by settlers. This would involve displacing existing communities, since, as one commentator claimed, 'if any thing is certain', it is that 'savage and civilised man cannot live together'.[43]

Views like this prompted discussions about mass deportations and in due course led to government policy and legislation that expelled Chickasaw, Choctaw, Muscogee-Creek, Cherokee and other nations, pushing them west into what survivors called 'the Land of Death'. It has 'long been the policy of Government', said President Jackson in his State of the Union address in 1829, to introduce indigenous peoples to 'the arts of civilisation'. Such efforts had failed entirely, he said, as was clear from the fact that they had 'retained their savage habits'. The best solution, therefore, set out in the 1830 Indian Removal Act, was to encourage migration to the west – which in practice led to forcible deportation.[44] It was a 'great pity', wrote former US President Thomas Jefferson, 'and indeed a scandal that we let that race of men disappear without preserving scarcely any trace of their history', although he too argued that deportation would free up land for white labourers.[45]

It was a similar story elsewhere. In Canada, First Nations were pushed into reserves away from the best land, which was taken over by new settlers. In Australia, the 1830s and 1840s saw Europeans move inland into territories that some described with giddy excitement as being 'as green and fresh as Eden', forcing out the Wurundjeri, Boonwurrung and Wathaurong people who had managed and lived on grasslands for centuries, cutting off their access to water holes and in some cases even accusing them of trespass.[46] New Zealand was presented as a wilderness ready to be tamed and turned into a bucolic idyll by hard work and perseverance, with little or no reference to those who already lived there. New worlds were waiting to be transformed. All they needed were people – or, more precisely, the right kind of people. Europeans, in other words.

Mass migration not only dramatically changed demographics and population distribution but also reshaped the natural world. The number of settlers in Australia grew from 1,000 to 12,000 between 1790 and 1810, before climbing to 1.25 million fifty years later – a rise of more than a hundredfold. The population of Ontario rose by a multiple of twenty-three, from around 60,000 to 1.4 million, in roughly the same period – while those of Ohio, Indiana, Illinois, Michigan and Wisconsin did much the same, from just over 250,000 to 7 million in total. Similar shifts could be found in Alabama, Mississippi, Arkansas plus Missouri, Florida, Louisiana and Texas, which together grew from 150,000 to more than 4.6 million – and in the original thirteen colonies, plus Vermont and Maine, where numbers rose from 3.8 million to 15.9 million between 1791 and 1861.[47] In 1830, Chicago consisted of 'about half a dozen houses'; sixty years later, it had a population of 1.1 million.[48]

This pattern of expansion was not limited to the Americas and to the west. It was mirrored in the steppe region of the European part of the Russian empire, where the population increased more than eightfold between 1700 and 1800, almost trebled again before 1850 and then trebled again before 1914 – rising from around 380,000 to more than 25 million. Moreover, these numbers did not include the seasonal migrant labourers who came to work on farms each year. But it was not only population movements that explain the dramatic rise in numbers; so too did the high fertility levels among new settlers.[49] The abolition of serfdom in the 1860s did much to help loosen ties between the rural peasantry and the land, sending waves of those seeking new opportunities to lands that changed from being exploited as pastoral, partly nomadic economies to settled arable farming.[50]

Colonial expansion did not repeat the pattern in Qing China, where there was little enthusiasm for long-distance migration into Xinjiang,

Inner Mongolia and Manchuria – regions that offered few of the attractions or rewards promised by reshaping agricultural systems. It did not help that the lands conquered by the Qing in the late seventeenth and eighteenth centuries were remote and difficult to reach, or that there were no sea and river routes allowing the transport of bulk materials in both directions to be carried out easily and relatively cheaply. What mattered more, however, was that these lands afforded little by way of commodity frontiers that might render their exploitation attractive in the first place, and no obvious upside of converting challenging terrains into large arable landholdings that could enrich new owners and support large settlements locally. Regressive policies adopted by the Qing, in particular preventing recipients of land grants from buying or selling land, and the practice of tying labour to landholdings, created barriers that offered few incentives and even fewer opportunities.[51]

This acted as a brake on Chinese social, economic and even political development during the heady nineteenth century that saw vast empires take shape in other parts of the world. Past climatic and geological serendipity which created coal reserves were also to prove important at the start of an age of fossil fuels that in some ways is still responsible for the way the world is today. The exploitation of coal, combined with advances in technology that paved the way for the industrial revolution, helped to transform productivity in Europe. Britain in particular was blessed with coal-fields and a scientific community that developed, refined and improved the methods by which enhanced energy resources could be put to use. That included improving coal extraction, which lowered costs further still. The impact was astonishing. By 1850, some 18 million people in Britain used as much energy as 300 million in China.[52] This was a reflection of multiple factors. Most important of these, according to some historians, was rising demand – which in turn reflected new and evolving ways in which energy could be used.[53]

The scale of such demand was impressive. In Britain, coal production doubled between 1815 and 1830.[54] In this sense, the distribution and location of coal deposits proved extremely providential for Britain. Much turned on the location of coal-fields. While China matched Europe in its living standards, its sophisticated and commercialised agriculture, its vibrant scientific community and its advanced print culture, its coal-fields were a long way from population centres in general, and from the densely populated Yangtze Delta that was the heartland of manufacturing and production in particular.[55]

Coal-fields in Britain – above all in Northumberland and Durham – were far closer to towns and cities, where there were high energy demands. Indeed, the availability of coal spurred the growth of cities which could attract cheap labour and were either connected to new canal systems or located on the coast. Manchester and Birmingham were two obvious beneficiaries; so were Glasgow and Liverpool, whose population rose by nineteen times during the course of the eighteenth century.[56]

Part of the success of provincial cities was because transportation costs for large volumes of bulky coal were high: the cost of coal in Newcastle, for example, was one-eighth of what it was in London. Coal was important not only as a source of cheap energy, however, but also because it spurred immediate and major gains in productivity, most notably thanks to the steam engine and the railway, which combined to connect locations together and to lower the costs of transport and of exchange while increasing their speed.[57] It helped too that there was an enhanced ability and interest in supporting research and the development of new technologies which ensured ever greater efficiency thanks to profits from overseas trade that created a pool of capital looking for returns.[58]

The bondage of other human beings was crucial in this regard: what mattered above all was not the purchase and sale of enslaved peoples, but rather the fruits of their labour in the form of sugar, tobacco, coffee and cotton. As new research shows, not only would Britain have been substantially poorer and more agricultural in the absence of slave wealth, but it also benefited from the proceeds of slavery being invested in other businesses and technology. In other words, it could be argued those who toiled in servitude provided the fuel that accelerated the industrial revolution in Britain.[59]

Taken together, pools of capital, new ideas and technologies helped drive urbanisation and the growth of London in particular, which stimulated the growth of the coal trade and in doing so helped seed towns close to mines that needed manpower and capital, offered financial rewards to investors and encouraged consumption in new locations too. This powered a housebuilding boom that was accompanied by changes in living habits and architectural styles, requiring 'an entirely new style of house' as heating with coal replaced burning of wood.[60]

In China, in contrast, regions with lower resources and worse soils were the ones that saw the most rapid demographic growth – which exacerbated strains on resources rather than alleviating or solving them.[61] This too marked a different trajectory to Europe, where the creation of colonies overseas had created networks of extraction that funnelled resources from one continent to another. As we have seen, some commodities that were highly prized were cash crops, like sugar, cotton and tobacco; but there were real needs for more mundane materials too, including a number that were bulky and expensive to move. For example, by 1650, perhaps as much as 200,000 hectares of forest land had been cut down in Europe – around 40 per cent of the entire area. Almost the same again was cleared between 1750 and 1850. This was a process of land being repurposed for other uses, and also a reflection of unsustainable consumption patterns. The answer was to look to sources abroad.

One key area was the Baltic, which had long serviced the timber needs of western Europe. Mature trees needed for ships and large buildings took a hundred and twenty years to grow; moreover, substantial volumes were required: a single galleon required two thousand oaks, or around twenty hectares of forest. These needs had been instrumental in opening up trade into the Baltic and stimulating the success of the towns of the Hanseatic league that were dotted around the coast of the North and Baltic seas.[62] With industrialisation taking off, demand now rose sharply: imports of wood climbed from 2.5 million cubic metres in 1850 to 15.5 million cubic metres seventy years later, with wood-pulp imports rising even more steeply in the same period.[63]

Fundamental to the extraction of resources were radical ideas about nature, about land and about the right to remodel the environment in whatever way they wished and thought best. The natural world became something to be tamed and defeated, a notion fuelled by the conviction that human ingenuity, hard work and new tools could now shape and repurpose ecologies better and faster than ever before. Many scholars tie these attitudes directly to Europe and European religious, cultural and philosophical sensibilities. Hegel was dismissive about the way that East Asians engaged with nature and suggested that their cosmological frameworks prevented them from thinking in abstract terms or freely. His view of Africans as exemplifying 'natural man in his completely wild and untamed states' also captured the supremacist sentiment that it was for Europeans to inherit the earth, and that other races were not only inferior but unworthy and incapable of doing so. For Hegel, then, the urge to 'do violence' to nature was an aggressive statement that reflected emerging mainstream ideas, bundling whiteness, power and entitlement into a toxic framework that set Europeans at the apex of humankind and of all living animals and plants.[64]

Nature became something to be not only exploited but defeated as standing in the way of human progress. As one American engineer modestly put it when proposing the construction of a canal between the Black and Caspian seas that would supposedly double the size of the latter, change rainfall patterns and improve the soil fertility of the steppes, schemes such as this would represent 'a great triumph of a nation over Nature' which 'would be far the greatest conquest in the annals of human material progress'.[65] His scheme would restore desert lands to 'their primaeval condition, as the abode of countless millions of men and beasts'. This was important, noted one contemporary commentator, because 'the world is none too large for its present population'. Stopping the advance of nature was essential; anyone who could help do so 'will be a benefactor of his race'.[66]

Not everyone was convinced that human activity was positive and some instead worried about sustainability and about the long-term damage that was done to the environment. Alexander von Humboldt was concerned about the combination of deforestation and the increase in irrigated agriculture that turned plains into deserts. 'By the felling of trees that cover the tops and sides of mountains,' he noted, 'men in every climate prepare at once two calamities for future generations; the want of fuel and a scarcity of water.'[67]

Humboldt was hardly alone in his concern about or his awareness of the link between deforestation and aridity, which was known, as one leading historian puts it, by 'every literate person' in the nineteenth century.[68] New regions brought wealth to some and disappointment to others, observed one English visitor to Australia. Though 'Anglo-Saxon energy at last triumphs over every obstacle', this victory came at a cost: 'Nature, as if offended, withdraws her beauty from the land; the pasture gradually loses its freshness, some of the rivers and lakes run low, others become wholly dry.' Wild animals 'are no more to be found'.[69]

* * *

Anxiety about the effects of deforestation became part of mainstream scholarship – and policy. In Russia, measures to promote the conservation of forests were put in place as early as 1802, with the Ministry of State Domains establishing a forestry corps to oversee protection.[70] In due course, efforts were made to gather information about pre-existing territories and about the large swathes of land in Siberia and Central Asia which were brought under imperial control from the middle of the century. Russian scientists and landowners became increasingly concerned about rising aridity and intense and worryingly regular droughts. Many had read the works being written in the United States and in Europe and picked up on the theme of deforestation as a cause of changing climate. Cutting down trees had exposed the land in southern Russia to the easterly winds, reported a survey in the early 1840s, noting that this 'must be the main cause of the disastrous impact of droughts which have been intensifying recently'.[71]

Although the Valuev Commission, which published its findings in 1873, stated that the climate had become 'more severe and drier' thanks to land being cleared, not everyone was convinced either that this was the case or that human activity could influence climate in the first place. Senior military officers who surveyed the empire's provinces complained that ideas about changing climate were more often than not based on anecdotal evidence and on comments made by members of the local population that were of doubtful reliability.[72] Nevertheless, the common view was that the climate on the steppes was changing, and for the worse; to help understand how and why, networks of weather-measuring stations were set up across the empire, to try to build a coherent picture that relied on data rather than opinions.[73]

Similar concerns were being raised elsewhere. In Mexico, the polymath Michel Chevalier considered how to develop the economy of Mexico after French intervention in the 1860s. One of the principal problems, he argued, was that a place that had once been a veritable Eden had been turned into a 'barren and desolate wasteland' by overexploitation by the Spanish. Deforestation had been catastrophic, he wrote, not only because it led to spells of aridity and changing patterns of precipitation, but because the land had been starved of nutrients, depleting its productivity. This naturally had an impact on the diet and poverty of the local population – and led in turn to falls in productivity and competitiveness, to economic distress and to political instability. The answer, for Chevalier, was to look at 'the extent to which the country could be reforested'.[74]

Protecting forests and indeed replanting trees became a central part of British colonial policy – starting with India and the Charter of Indian Forestry, which annexed all forests that were not privately owned and declared them to be state property. Similar measures soon followed in Australia, Canada and Africa, where 'large tracts of the country' were said to be drying up as a result of too many trees being felled. Despite the claims of some scholars, the motivations of the authorities in taking control of forests had little to do with conservation: in fact, what was at stake was that colonial authorities insisted on exploiting timber resources that were vital to the extension of political and economic control. The consequences for peoples who lived in the forests – and who had done so for many generations – were disastrous.[75]

Though some raised concerns, the reality was that deforestation continued at a staggering rate in the nineteenth century and beyond. Between 1850 and 1920, around 152 million hectares of the world's tropical forests were converted to grasslands, almost two-thirds of which (some 94 million hectares) were in sub-Saharan Africa and South and South-East Asia – that is to say, in the heartlands of colonial expansion.[76] Ironically, when it came to rationalising change, the standard narrative was that local populations were poor guardians of nature and were primitive in their approach to agriculture and that the development of new landscapes was not only to their benefit but beyond their capabilities. Such claims, of course, were untrue.[77]

From 1750 to 1900, around 600,000 to 800,000 hectares of the world's most fertile arable land were opened up for exploitation. Newly settled regions in the Americas, Australia, New Zealand and southern Africa became important sources of wool, meat and grain and among the largest producing regions in the world. This was not just the result of those hunting for land and scouting for prospects, but also a function of the predatory use of claims to legal possession and land title and by insistence on the importance of 'improvement' of land and nature, which gave incomers the 'right' to take control of territory.[78] In many cases, such as in India, colonial administrations simply asserted – and enshrined that assertion in law – that all uncultivated land belonged to the state. This too was an aspect of the widely held assumptions that indigenous peoples were ignorant and careless and adopted policies that ruined forests. The British saw themselves as guardians of the environment who needed to protect the natural world from the predations and practices of peoples who had lived there for centuries and even millennia.[79]

In time, ideas like this were carried further still – namely, not only taking control of land but pushing people off it altogether. National parks established at Yellowstone in the United States, Banff in Canada and Tongariro in New Zealand in the 1870s and 1880s were based on the idea that, to protect nature, human beings needed to be excluded entirely, even if that meant forcible displacement.[80] In some cases, this led to violent protest, as in German East Africa where orders to protect forests granted the right to expel inhabitants from newly created reserves.[81]

Colonial expansion cemented the power of the global north by giving access to the best land all over the world, controlling its usage, monopolising the fruits of its production and enshrining the reality of poverty and limited freedoms for those excluded from its resources and from land 'ownership'.[82] Even today, wildlife conservation – whether to do with animals or plants – often involves benefactors with deep pockets or well-funded and well-resourced charities seeking to 'preserve' the natural world by keeping human beings out of conservation areas. In a curiously neo-colonial twist, the wealthy of the developed world protect nature from being spoiled and save it by ring-fencing it, often literally, from indigenous populations: the creation of the Messok Dja protected area in the Republic of Congo by the World Wildlife Fund, without the consent of local Baka communities, or the eviction of more than 70,000 Maasai from their lands in northern Tanzania to create a game reserve are just two examples among many.[83] In fact, creating national parks and protected areas does not necessarily benefit wildlife – and certainly does not do so in predictable, uniform ways.[84]

Ironically, of course, for all the concern about the effects of anthropogenic change on forests and the impact on soil erosion (and therefore on yields), the demand for goods and commodities was not merely ravenous but ecologically catastrophic. As we have seen, animals were hunted for their pelts to the brink of extinction and sometimes beyond in North America; in southern Africa, ivory was the driver of the expansion of resource frontiers. In the late 1870s, British and Boer hunters advanced north into modern Zimbabwe, northern Botswana and eastern Zambia in search of elephants. Export figures show the shocking scale of slaughter, with thousands of elephants killed each year in the second half of the nineteenth century.[85]

Ivory was highly desirable in Victorian-era Britain and the United States, as well as elsewhere, used for fashion accessories ranging from collar studs, hairbrushes and vanity sets to sewing cases, toothpicks and napkin rings.[86] Major demand came from piano manufacturers, as these instruments became popular in working-class bars and music halls and also as status symbols for the burgeoning middle classes, whether in British homes or newly settled farming communities on the Great Plains. The rise of billiards as a social pastime also increased demand; the ivory used for billiard balls had to come from soft young specimens – and from only part of the tusk at that.[87]

There were attempts to slow the trade down, if not stop it altogether, with Khama, the Tswana King, trying to introduce controls over hunting by sharply raising levies for elephants shot in his territory. This had little effect on consumers living a long way away, whose ideas about nature and the majesty of wild animals and preconceptions about the continent of Africa proved a heady cocktail that glamorised hunting and hunters; indeed big-game hunters like R. G. Cumming became household names and even outsold Charles Dickens when they wrote their memoirs filled with derring-do tales about how supposedly brave (white) men were able to push large animals to the brink of extinction thanks to their proficiency with the rifle.[88]

These were all new developments in the opening up of 'ghost acres' – that is, the colonial powers' exploitation of land, resources and commodities in other parts of the world. The British were by far the best, most organised and most determined at spawning clones in other continents, such as South Africa, North America and Australia. In each, political, legal and religious institutions were created that imitated those at home and were controlled by those who spoke the same language and had strong family ties with the mother country. The growth of the Anglosphere was explosive: the number of English-speakers rose sixteenfold between 1790 and 1930, from 12 to 200 million. It was not that the Spanish, Russians, Chinese or others who expanded into new lands or developed extractive, centralising policies in the same period did not meet with success; but as one leading historian puts it, 'it was the Anglophones who bred like rabbits'. It was the British who met with conspicuous success in creating infrastructure networks that sent resources, commodities and goods in one direction and people in the other. Despite what conventional wisdom might say, it was only in the nineteenth century that Britain became great.[89]

Of course, success came at the expense of other people, as local populations were displaced or coerced in the Americas, in Africa, in Asia and in Australia either by Europeans directly or by their descendants. Ironically, the push towards independence in the United States was not predominantly caused by the rejection of Britain, or by revulsion towards British rule and British identity; rather it was more because leading Americans felt that they were treated as not being British enough and were accorded second-class status, above all in the lack of representation in the political process in London.[90]

On the face of it, the United States was fiercely republican in character, demeanour and self-identification; in behaviour and practice, however, it proved to be an expansive, militaristic and extractive power in its own right: the Louisiana Purchase in 1803 was followed by the occupation of Florida in 1810, by the expansion of geographic horizons to the west in the decades that followed and in the 1840s by the seizure of about half of Mexico. The gains that came thereby were distributed to elites and to commercial interests, at the expense of those who were conquered, displaced or subjugated.

The processes of acquiring territory and of opening up new commodity frontiers set in motion a chain of other changes, including investment in transportation connections and a rapid period of urbanisation. In the late 1770s, the three principal settlements in Kentucky were home to a combined population of 280 people. By 1782, some 8,000 European settlers had joined them; by 1790, the number had risen to 73,000.[91] Development of bigger and faster steamships brought shipping costs down and prices with it – though not in uniform ways. For example, in the 1820s, the cost of sending freight by river from Philadelphia via New Orleans was a third of what it cost to transport it overland; a few decades later, transatlantic shipping had become so efficient and cheap that it cost less to transport flour from North America to Liverpool than from Dublin across the Irish Sea.[92]

The rise of coal-powered steamships, combined with increases in their size, speed and reliability in difficult seas, stimulated the creation of a global network of refuelling stations and in doing so galvanised the development of new ports and coastal cities where goods could be loaded and unloaded and which became important in their own right.[93] The opening up of the Suez canal in the late 1860s cut shipping times dramatically, with obvious implications for prices; the massive expansion of railways in Britain and Europe, the United States and Canada and in other continents during the nineteenth century was a similar story.

Cheaper, quicker, more reliable connections did not just boost economic exchange, but brought about dramatic social and cultural revolution too: provincial cities blossomed as they became integrated into networks giving access to ideas about art, music and literature that had previously been the preserve of the wealthy. Literacy levels in France rose by 20 per cent in the 1830s and about the same again in both the 1840s and 1850s; museums opened in the second half of the nineteenth century at great pace in the major cities in Europe, seeding discussions about the past and the present across ever wider sections of society. Not everyone approved: with operators like Thomas Cook capitalising on a booming tourism industry, some complained that visitor numbers spoiled the experience for others. English tourists, ran one complaint, were 'seemingly everywhere; there is no lemon tree without an English lady smelling its perfume, no picture gallery without at least sixty Englishmen, each with a guidebook in their hands checking everything is where it should be'. All these changes transformed connectivity, shrinking distances between regions and widening cultural horizons.[94]

So too did technological breakthroughs: the invention of a concentrated meat extract by Justus von Liebig not only proved commercially lucrative (and spawned many imitators), but also played an important role in the 'meatification' of metropolitan diets, as did mechanical refrigeration, which made the transport of meat highly profitable and much more efficient from the late 1870s.[95] Meat and protein increasingly became a part of working-class diets in London in the later nineteenth and early twentieth centuries, bringing health benefits to adult populations and aiding brain development in young people.[96] Meat consumption likewise grew in parts of Asia, such as in Japan, where the Meiji Restoration of 1868 prompted new attitudes to beef in particular, and in China. China was scouted by the beef baron William Vestey, whose family business had built up a global network that shipped millions of carcasses. He thought China offered enormous long-term potential – although one study in 1912 poured cold water on the idea of exporting to Asia because the inhabitants, 'where they are not entirely vegetarian, are too poor to buy imported meat'.[97]

There were clear winners and losers in the acceleration and deepening of globalisation that characterised the nineteenth century. The British were prime beneficiaries, above all in terms of rising living standards and availability of goods and commodities. For example, by the 1890s, Britain absorbed 60 per cent of the meat and as much as 40 per cent of the wheat that was traded globally.[98] There was good news too for low-income households, with the repeal of the Corn Laws and the rise of imports from America ensuring that the price of a loaf of bread halved between 1840 and 1880.[99]

Constant improvement of machinery expanded production, improved efficiency and applied pressure on costs. Wheat exports from the United States rose from 5 million hectolitres to 100 million in the thirty years after 1840. The introduction of mechanical reapers doubled productivity, while steam-powered grain elevators meant that 500,000 bushels of grain could be processed in ten hours at a cost of five cents per bushel.[100] Mechanisation reduced the amount of labour required to yield a hectare of wheat from 150 hours to just nine. Breeding techniques doubled milk-fat production per cow and resulted in draught horses becoming 50 per cent bigger (and therefore stronger) in the United States between 1860 and 1890.[101]

The fruits of these gains flowed to those who had capital to invest and to those who could take advantage of mass production – large-scale arable farmers in the United States and livestock owners in Australia and South America, for example, or shareholders in railway companies that paid handsome dividends. They brought despair to those who were squeezed out, however, such as cereal farmers in Britain who could not compete with the flood of imports, and to labourers who were forced to leave the countryside to look for work in towns that had low levels of hygiene and a high incidence of poverty and disease – a way of life epitomised in works such as *Bleak House* by Charles Dickens.

There were similarly few benefits in parts of the world that were either left behind or received little by way of hard infrastructure in the form of roads, schools, hospitals and railways, and little by way of soft investment in institutions, education and local capacity building. States that were notionally free from colonial rule, such as in South America, behaved as classic extractive satellite states, exporting raw materials and relying on imports for domestic consumption. The changes to the global economy brought mixed blessings for India and South Asia. Between 1810 and 1860, India lost much of its domestic textile market to Britain as prices were driven downwards, which had dramatic consequences because of the relative rise in the price of grain in the same period.[102]

So while Europeans revelled in plentiful and cheap food, others were not so fortunate: as many as 16 million died of hunger in India in the decades between 1875 and the start of the First World War – a prolonged catastrophe that colonial administrators treated as a fact of life and one that had positive side-effects, which included forcing indebted smallholders off the land while serving as a welcome check on the growth of India's population size.[103] Huge numbers died in famines, often during times when massive consignments of wheat from India continued to be exported, especially to Britain. This mattered little, according to one blunt official writing to the Viceroy in the second half of the nineteenth century: 'still they reproduce themselves with sufficient rapidity to overcrowd every employment that is opened to them.'[104]

Problems came too from opportunistic attempts to turn profits. By the end of the 1850s, for example, the United States was the biggest producer of cotton in the world, exporting around 3.5 million bales annually – much of it grown by coerced labour on plantations in the Deep South. The Union blockade of Confederate ports during the civil war dramatically strangled trade, with just 10,000 bales exported in 1861–2, a drop of almost 99 per cent.[105] Although slavery had been banned in Britain by the Abolition Act of 1833, the British economy relied heavily on the textile industry and as such not only benefited from cotton imports but relied on them to keep its factories working. The supply failure raised concerns about shortages of materials and about threats to public order. There were riots in several cotton towns in England, as well as elsewhere in Europe. 'No crisis in modern times has been so anxiously watched,' declared *The Times* in the early 1860s in reference to the US civil war, 'nor has any European war or revolution so seriously threatened the interests of England.'[106]

Anxieties were heightened by the rise of radical ideas that were themselves closely linked to long-term socio-economic problems. Rapid and large-scale urbanisation was transforming the literal and the political landscapes of Europe, which had already been hit by a series of revolutions in 1848. These upheavals were framed by Karl

Marx and Friedrich Engels as expressions of the class struggle by the oppressed against those who controlled the means of production; in fact they owed rather more to serious food shortfalls that had triggered famine and hunger riots, most notably in Ireland, Flanders and Silesia in the mid-1840s. The violence that spread across many parts of the continent resulted in governments cancelling investment programmes and so negatively impacting mining and metal production from the spring of 1847. This contributed to the groundswell of protest and the demands for reform, freedom and greater rights that peaked the following year.[107]

Attention quickly turned to India, therefore, as an alternative source of cotton – perhaps not surprisingly given the repeated attempts to stimulate production, most of which had faltered thanks to the low quality of Indian cotton and to poor transportation networks that added to costs. The invention and widespread adoption of the Whitney gin had prompted a revolution of its own in the United States: in 1801, a cotton picker typically averaged around 28 pounds per day; by the end of the 1820s, this had risen more than fourfold, to just over 132 pounds a day, almost trebling again just over a decade or so later to 341 pounds. The coerced labourers who were forced to work at these staggering rates of productivity benefited factory owners and factory workers (albeit to a much lesser extent) in cities like Manchester on the other side of the Atlantic, as efficiency in weaving mills improved by six to ten times between 1820 and 1860. Ironically, the gains made slavery even more lucrative, ensured that enslavers in the American south became richer still and reinforced their determination to hold on to the source of their wealth – forced labour.[108]

The pressure on supplies caused prices of Indian cotton to rise by almost five times in the early 1860s. A surge in land clearances followed as producers sought to capitalise on the high prices. More than a million hectares were turned over to cotton growing from other subsistence crops. Railway construction dissected Bihar, a province in the heart of the Indian subcontinent, where the town of Khangaon quickly became what one contemporary called the 'largest cotton outpost of the British empire'.[109]

Others too tried to grasp the opportunity, including in Central Asia where some Russians in this period hoped the local populations would be 'our Negroes' – an aspiration that leaves little to the imagination.

During the US civil war, cotton production in Central Asia boomed, rising more than fourfold in the period 1861–4.[110] Then there was Lower Egypt where 40 per cent of all fertile land was converted to cotton cultivation, as was much of the substantial personal landholdings of the Ottoman Viceroy, Sa'id Pasha.[111]

All these opportunistic moves paid immediate dividends. However, they came at a high cost. For one thing, the settlement of the American civil war reintroduced large supplies to the market, with improved availability placing downward pressure on prices. For another, the expansion of cotton growing led to the introduction of agricultural slavery in other parts of the world, most notably in the Nile Delta, which became home to large numbers of enslaved people imported from East Africa: efforts to end slavery in one continent brought about just the opposite in another.[112]

The attraction of high returns also led to financial strains and overstretch among those who borrowed money for seeds, tools, food and labour. When prices turned, many found themselves overextended and with debts that they struggled to pay. In Egypt, this produced waves of land abandonment, bankruptcies, a growing body of landless labourers and the polarisation of social relations in the countryside as inequalities rose. Much the same happened in India, where some have argued that indebtedness, displacement and desperation – combined with the tightening suffocation of colonial demands – provide the backdrop to the Deccan riots and to the death of millions from starvation in the 1870s.

Minimal efforts went into investing in irrigation because of mistaken beliefs about the richness of the black soil of Berar and partly because of the incompetence of colonial officials who prioritised cotton over grain. The failure of wages to keep up with prices resulted in malnutrition and higher susceptibility to disease as well as starvation. Famine cycles in the 1890s struck again and again, ruining countless families and enriching those who hoarded food – and even continued to export it during years of acute crisis.[113] That starvation rates were particularly high in Berar tells its own story about attempts to cash in on short-term opportunity.[114] Such were the consequences that flowed from a conjunction of the search for profits, the unsustainable exploitation of the landscape and nature taking revenge when pushed beyond its limits.

20

The Age of Turbulence
(c.1870–c.1920)

A gallant body of Englishmen have fallen victim to their efforts to bring West Africa within the outer fringes at least of civilisation.

Daily Telegraph (18 January 1897)

If the human consequences of disaster were severe and shocking, so too were the ecological implications of rapid transformations of landscapes that were motivated by the chase for a fast buck. For example, forests were hacked down to create cotton plantations. This had obvious knock-on effects for flora and fauna alike, exacerbated by the practice of paying bounties for tigers, panthers, wolves, bears and hyenas – apex predators whose disappearance produced major environmental change. Worse, land that was cleared often proved to be unsuitable or was exploited so badly that yields dried up, as (literally) did rivers, lakes and water resources as a result of soil erosion or because clearances induced changes to regional rainfall patterns.[1]

The story of cotton was one that was repeated time and again, particularly from the second half of the nineteenth century onwards when global markets became more integrated, transport networks improved and information-sharing accelerated. For example, the demand for rubber shot up as a result of Charles Goodyear's pioneering work in the 1830s on vulcanisation, the chemical process that improves elasticity, hardness and resilience, and then again by the development of the pneumatic tyre, perfected by John Dunlop in 1888.[2]

Rubber production rose in leaps and bounds from around 2,500 metric tons in 1851 to over 20,000 tons thirty years later. That was just the start of it. By 1900, some 50,000 tons was being produced; by the eve of the First World War, it was closer to 120,000 tons. During this time, rubber became 'the most important, most market-sensitive, most sought-after new commodity in the world', much in demand for bicycles and cars, as well as for a seemingly limitless new range of goods and products, from shoe soles to electrical insulation.[3]

This versatility too spurred a rush to alter the use of land and take advantage of the demands – and profits – on offer. The rate of transformation was not so much bewildering as epic. Vast amounts of rainforest in South-East Asia were cut down and replanted with *Hevea* seeds from the Amazon. In the late 1890s, around 145 hectares had been planted in Melaka and Selangor, in the Malay peninsula. By 1910, that had risen to around 220,000 hectares, with plantations being established in other states too, such as Perak. Four years later, more than 445,000 hectares had been cleared for rubber-plant cultivation. By 1914, more rubber was being exported from the Malay peninsula than all of South America combined. The Dutch East Indies followed the same pattern, with large-scale planting in the same period resulting in rubber trees covering 245,000 hectares in 1914 and almost double that amount five years later. By the 1930s, millions of hectares of the world's richest rainforests had been replaced by rubber trees by entrepreneurs who recognised that cheap land could be the basis of great fortunes, especially once cars began to be mass-produced and demand for rubber for tyres soared even more. The excitement of being a pioneer and of the rewards that came with it was palpable: as one new planter put it, he felt he had the 'morale of a conquistador', a phrase that summed up the prevailing attitudes as well as the process of extraction. The Americas had been denuded; now it was the turn of other parts of the world.[4]

Ecological reconfigurations had always followed the transplantation of crops, flora and fauna. Highly integrated global markets that were linked by increasingly efficient and cheap transport networks enabled this to happen faster and more deeply than at any point in human history. Exploration of unknown lands had the same outcome, usually motivated by the search for new resources. Central to this was a philosophical framework which asserted that uncharted parts of the world were not only wild and lived in by savages, but home to endless sources of wealth.[5] As David Livingstone put it, European expansion brought three elements that would be welcomed by all: 'Christianity, commerce and civilisation'.[6] The dovetailing of evangelical ideas about racial superiority, religious virtue and capitalism was a core element of the way that Europeans, and the British above all, saw both themselves – and the rest of the world.[7]

Rubber was just one commodity among many. Some were plant-based, like cocoa, coffee and tobacco, which were all introduced into new regions around the world. It was a process that depended in many locations on the availability of quinine, the first effective treatment of malaria, which itself came from the bark of the cinchona tree native to the Andes which spread rapidly from the 1820s.[8]

The exploitation of resources that accelerated in the nineteenth century was not limited to crops and plants. Minerals and metals too were in high demand as new technologies were developed and rapidly scaled up, while manufacturing accelerated too. A good example was tin, which played a key role in a range of industries, including textile production, mechanical engineering and military armaments. Its main use was for tin cans for preserved food – a function vital in conserving rural food surpluses and transporting them to cities, and therefore crucial to urbanisation, industrialisation and globalisation.[9] Tin production in Europe quickly became exhausted, leading to the search for supply from elsewhere, most notably South-East Asia. As with rubber, volumes grew sharply. Automation meant that daily output of cans per worker rose from 60 in 1847 to 1,000 less than thirty years later; production then went up by a factor of almost four between 1874 and 1914. In the following decades, demand was all but insatiable: in 1962, for example, the United States alone produced almost 50 billion cans, corresponding to more than 250 cans per person per year.[10]

Another highly prized commodity was whale oil, obtained from rendered blubber, which was a prime source of lamp fuel. Around 162,000 Greenland whales were killed between 1500 and 1800. Although whaling had long been part of the economy of the Arctic Sea and the North Atlantic, the industrial revolution brought about a surge in demand because the low viscosity of whale oil made it a valuable source of lubricants for use in machinery: it did not dry out, congeal or corrode metals.[11]

The United States became the epicentre of the global whaling industry, home to more than three-quarters of the worldwide fleet by the 1840s, in large part because stocks of bowhead and right whales in the Atlantic had been so overexploited by the start of the nineteenth century, forcing mariners to look elsewhere for their kills.[12] This served to open up the heart of the Pacific basin, integrating the littoral of South America in particular into global commercial networks.[13]

The global hunt for new commodities extended to other resources that were in abundant supply in the region, most notably guano, the excrement of wild birds and bats whose prodigious value as a fertiliser had been well known for many centuries before the European arrival in the Americas. Birds congregated and bred in flocks on the coast of what is now Peru, wrote one observer, flocks so enormous as to be incredible 'to anyone who has not seen them'. Inca kings had made it illegal to interfere with the birds during breeding season on pain of death, so prized was the guano – a word derived from the Quechua for 'shit' – which was spread over fields each year to help crops grow.[14]

Its value as a fertiliser rose dramatically during the age of scientific advancement, as investment in plant sciences and concern about increasing food production to feed growing populations directed attention to improving crop yields. Experiments conducted by Alexander von Humboldt in the early nineteenth century on samples that he brought back from his travels to the Pacific coast of South America revealed extraordinarily high concentrations of nitrogen-rich uric acid in guano that astonished the scientific community and sparked a global hunt to find sources before exporting them back to Europe and the United States.[15]

The British Isles, the Caribbean, Canada and South-West Africa were stripped of their supplies. But it was the coast of what became the independent state of Peru in the 1820s that proved the greatest source of guano, as well as the most lucrative. Between 1840 and 1879, almost 13 million metric tons of guano were exported from Peru's islands and coastline, with an estimated value of £100–£150 million. Most of this was shipped to Britain, though some was sold to farmers, landowners and planters in the Low Countries, Germany and the United States – as well as to islands in the Caribbean, to Mauritius and Réunion in the Indian Ocean and even to China.[16]

Guano's properties were so prodigious that it took on a role in popular culture. Joseph Conrad's *Lord Jim* includes a character excitedly talking about his plans to make a fortune by buying a cheap steamer and sailing to 'a guano island' that he had discovered, which despite lacking safe anchorage and despite lying in the path of frequent hurricanes promised to deliver untold riches.[17] One story printed in a San Francisco newspaper in the 1850s claimed that when the hatches were left open on a ship transporting guano it caused the timbers of the vessel to sprout, the masts to grow and a cherry table to grow fruit; indeed so potent was its fertilising power that cockroaches grew to enormous sizes, so large 'that they could get up the anchor and make sail on the brig'. According to the sailor who reported this story, 'there is nothing like guano to make things grow'.[18]

Other prizes lay in wait too. Huge saltpetre deposits in the Atacama desert quickly caught the eye of prospectors. Like guano, saltpetre nitrates were valuable as a fertiliser. They had another quality too: their ability to burn at an explosive rate. They were therefore of enormous interest both to miners and to armies whose artilleries became increasingly powerful in the course of the nineteenth century, especially after Alfred Nobel pioneered ways to transport and detonate nitroglycerine in the form of 'dynamite'. This set off a race to find and secure sources that were ideally both plentiful and cheap. It did not take long for attention to turn to the Atacama desert, one of the driest places on earth where rich quantities of sodium nitrate ($NaNO_3$) had built up over the course of many millions of years.[19]

As happened elsewhere, the appetite for extracting resources from South America had consequences. For one thing, it led to the flexing of muscles by those who benefited from the natural wealth to ensure that they were able to continue to do so. Blockades of Buenos Aires by British and French warships in the mid-1840s were paralleled around the same time by the advance of the United States at Mexico's expense as the former took control of more than a million square kilometres of the latter's territory: the 'storming of the Halls of Montezuma', a reference to the assault on Chapultepec castle in Mexico City, was so important that it features in the official hymn of the US Marine Corps.[20]

US ambitions soon widened further still. With the Guano Act passed by Congress in 1856, US citizens were given the right to lay claim to islands and territories on behalf of the government.[21] Several of these are still in US hands, including the Midway Atoll, famous first as a staging post for telegraph communication and flights between the United States and Asia and then as the location of the battle on which the Second World War in the Pacific turned.[22]

For another, the race to acquire natural resources led to a spending spree at both government and individual level in Peru, based on the sudden bonanza brought by guano and nitrate sales and on the belief that this would never end. New railways, roads, hospitals and city sewerage were testimony to this newfound wealth, as were the fine wines, cigars and fashions that were brought from all over the world. Reality eventually caught up with the overenthusiasm, overspending and overborrowing as Peru lurched from a costly civil war in the 1850s into hostilities with Spain in the 1860s and then, at the end of the 1870s, into the War of the Pacific (1879–84), which has been described as 'one of the largest armed conflicts ever fought in the Americas'.[23]

The adoption of new technologies and the identification of new ingredients or materials that were suddenly in high demand often brought unusual spillover consequences. The discovery by James Lind, a British naval officer, that the consumption of citrus fruits prevented scurvy led to the Royal Navy buying vast quantities of lemons – enough to supply 1.6 million gallons of lemon juice in the years 1795–1814 alone.[24] This caused production to soar, especially in Sicily, which was exporting 2.5 million cases of lemons a year to New York by the mid-1880s, with citrus cultivation on the island yielding more than sixty times the profit per hectare of any other crop. When combined with the weak rule of law, low levels of inter-personal trust, high levels of local poverty and a series of land reforms introduced in the first part of the nineteenth century, these created the perfect circumstances for the rise of groups that could exploit high margins and also benefit from conditions of uncertainty. Lind's solution to scurvy helped sow the seeds for the rise of the Sicilian Mafia.[25]

Thanks to the craze for rubber, guano and nitrates, the deepening of trade connections with South America also had far-reaching effects. In 1842, an unusual blight appeared in potato crops in Philadelphia and New York, as well as in many parts of Europe. Ireland was severely affected, in part because the population was dependent on potatoes as a source of food and calories and as a form of payment, but also because of pre-existing poverty closely correlated with pressure on the land, poor agricultural practices made worse by landlord absenteeism, and inadequate government policies. The results were dramatic, with as many as one million men, women and children dying from hunger and disease in appalling conditions. For many, the choice was simple: migrate or die, or as one leading scholar has expressed it, 'the next world or the new world'.[26]

Recent sequencing of DNA from strains of the blight suggests that it likely had an Andean origin and had been transmitted around the world among new potato varieties exported from Peru that owed much to the country's growing accessibility and its place in global commercial networks.[27] Ireland's population fell by almost 40 per cent between the mid-1840s and early 1850s: in addition to those who died, well over a million people emigrated, with most heading across the Atlantic to North America.[28] Not surprisingly, the profound suffering left scars that lasted for generations and influenced ideas about British rule, about sectarianism and about rebellion.[29] The rapid creation of large Irish communities in the US, meanwhile, became a part of mainstream political discourse in the United States in the mid-nineteenth century, as many questioned whether large influxes of migrants might not only impact prosperity but perhaps even damage the nation's institutions.[30]

That so many new arrivals in the late 1840s were less educated than those who had reached the US from Ireland beforehand was reflected in sharply lowered literacy levels and in their children's poor educational and vocational outcomes.[31] Nevertheless, as the Irish population grew, it retained pride in its roots – as is clear from the fact that today almost 40 million Americans claim Irish descent – influencing cultural events such as St Patrick's Day as well as the results of US presidential and congressional elections for more than a century and a half.[32] These dramatic demographic trends re-shaped Ireland and the United States, and can be traced back to a crop disease in South America that had piggy-backed on intensive global trade networks. Moving plants from one side of the world to another brought the risks of doing the same for pathogens.

The exodus that followed the Irish potato famine was part of a much wider set of migrations closely connected to the exploitation of foods, minerals and natural resources. As slavery in the Americas had shown, crops like sugar, tobacco, coffee, cocoa and cotton counted for little without labour to plant, cultivate, harvest and process them. The advances achieved in the nineteenth century were part of an interrelated and complex series of technological developments, deepening trade ties, rising aspirations and knowledge-sharing that above all required vast amounts of labour, almost all of which was in the wrong place. Clearing forests, reworking land and establishing new plantations therefore meant bringing in workers from locations near and far, So while the nineteenth century can be seen in terms of the reshaping of the natural environment, it should be stressed that this relied on massive changes in the human environment too. Just as crops were transplanted around the world, so too were people; strikingly, much of the movement of the latter was involuntary.

The second half of the nineteenth century and early years of the twentieth are often characterised by wistful ideas of impoverished Europeans packing up their few belongings and heading across the Atlantic where new lives awaited those willing to work hard and where new opportunities awaited, especially after the end of the civil war and the abolition of slavery. While migration to North America in the second half of the nineteenth century was substantial, the decades that followed c.1850 saw even greater mass movements of people in and around Asia. Almost 30 million Indians migrated around the Indian Ocean and South Pacific, often at the instigation of the colonial authorities, in many cases involving debt obligations whereby loans were provided in return for labour. Nearly 4 million from India moved to Malaysia; more than double that number to Ceylon (modern Sri Lanka); more than 15 million moved to Burma and around 1 million to other parts of South-East Asia, to islands in the Indian and Pacific oceans and to Africa.[33]

Movements on this scale were paralleled by the migrations of some 50 million Chinese, who relocated to South-East Asia, to the Dutch East Indies, Thailand, French Indo-China, Australia and New Zealand, as well as to Pacific and Indian Ocean islands. All this caused a major shift in patterns of global demographics. The Americas and South-East and North Asia experienced massive population growth, with numbers increasing by a factor of four or more between 1850 and 1950. While the trajectories ran in parallel, there were some fundamental differences, not the least of which was that density of settlement rose far more sharply in South-East Asia than in the United States as around the same number of migrants moved into less than half the space.[34] By 1900, there were more Indians on Mauritius than Africans, and more Indians in Natal than there were Europeans. Forty per cent of the population of Hawai'i were of Japanese origin, and almost 20 per cent were Chinese.[35]

These enormous movements of people were above all driven by the need for workers on plantations, in mines or engaged in processing materials, minerals and metals. Again, much of the movement was involuntary. Populations across Polynesia were devastated by raids that aimed to enslave and transport adult men against their will and force them to participate in collecting guano in Peru and preparing it for export.[36] Indian and Chinese 'coolie' labourers were shipped across the Atlantic under false pretences or against their will as indentured labour to help build railways, roads and port facilities, or to work in hospitals or as domestic staff, or to carry out backbreaking work such as picking crops.[37]

Forced sexual exploitation and rape of women were rife, underpinned by western tropes about people living in Asia being prone to laziness, pleasure-seeking and promiscuity. Among other things, this also reinforced the idea that not only should women 'serve' their masters, but that their doing so should not come at great cost. 'Some of the China women are very good-looking,' observed Robert Hart, Inspector General of the Imperial Maritime Customs in China. 'You can make one your absolute possession for 50 to 100 dollars and support her at a cost of only 2 or 3 dollars per month.'[38] This was just one among countless examples that shows how ideas about race and entitlement combined to reinforce the belief that it was not just humans holding sway over the flora and fauna of the world, but Europeans doing so over all other peoples too.

In this sense, the abolition of slavery in Britain, Europe and North America was symbolically and commercially significant, but it brought little by way of changes in attitude or behaviour. Colonial administrators found it easy to convince themselves of their own virtues. 'There was hard work,' wrote one official posted to India, 'but it seemed worth doing; it was always concerned with other people's lives.' Yet physical abuse was routine – and, when administered by Europeans, either went unnoticed, unpunished, or all too easily excused. In India, labourers with low work rates, who complained or did not meet their targets were beaten and in some cases even killed.[39] So great were the losses in Congo to maltreatment, torture, starvation and disease as the local population were forced to gather rubber during a boom that lasted around forty years from the 1880s until the early 1920s that every ten kilos of rubber exported in this period came at the price of one human life. This appalling lack of regard meant that the population in the Congo fell by 70 per cent in the course of these decades.[40]

While Europeans and their descendants who had fanned out across the world's continents in a series of increasingly large waves were the main beneficiaries of the fruits of the land and the fruits of human labour, they did not benefit evenly: in fact, the proceeds went to those with capital, rather than to those in or linked to Europe – including to those from and living in Asia, Africa, the Americas and Oceania. It was capitalism's 'ruthless terrorism', in Karl Marx's phrase, that spurred economic growth, but it also came at a heavy price.[41]

The relentless demand for commodities of course did not just put pressure on ecologies that were transformed as forests were cut down, new crops introduced and planted or hillsides split open in search of minerals, for major interventions were also needed to support, house and feed large workforces that themselves put strains on the local environment. One farm in southern Cambodia, for example, had to supply between three and five metric tons of food per day for the workforce of 3,000 who toiled on a neighbouring rubber plantation, while also maintaining 500 cows, 800 pigs and herds of sheep and goats.[42] Such projects inevitably had trickle-down effects on local economies, but the cold reality was that the lion's share of rewards, both direct and indirect, went to make the rich even richer.

It did not help that investments could be precarious because the race to create supply could sometimes overwhelm demand – pushing prices down and turning glittering El Dorados into bleak wastelands. At the height of the rubber boom Iquitos in the Peruvian Amazon rainforest fleetingly became home to some of the most expensive real estate in the world, with prices for property reputedly surpassing those in cities like New York, only to fade as new sources of rubber came on tap in other parts of the world. Or there was Irkutsk in Siberia, which by the end of the nineteenth century was home to an opera house, several cathedrals and no fewer than thirty-four schools; it was also a repository of hope and optimism about the future fuelled by rapid industrialisation, major investment, not least in railway connections, and rising living standards. 'Siberia is going to be another America,' remarked one contemporary visitor, a prediction that remains unfulfilled.[43]

Then there were the by-products of extraction that condemned local communities, as well as local wildlife and plant life, to generations of side-effects, as well as to widespread ecocide. The tailings from metal industries, for example, caused rivers to silt up from Tasmania to Perak in Malaya by increasing sediment supply to local stream systems.[44] Aquatic environments around the world were poisoned by metal pollution resulting from efforts to extract copper, zinc, lead, tin and more from the landscape.[45] Highly toxic sodium cyanide, used to concentrate gold, which was first developed in the late nineteenth century and was almost immediately employed in the Witwatersrand gold fields in South Africa, poisoned the landscape.[46] Even today, around 100 million kilos of cyanide is used annually in North America, of which more than three-quarters is assigned to gold mining, while some scientists estimate that gold mining alone generates around 1 billion metric tons of waste rock and tailings each year.[47]

The conviction that transformations of the natural environment were certain to bring about positive long-term change proved to be spectacularly ill-placed – throwing ideas about cultural and ethnic supremacy into sharp relief. The expansion of British political power in India sparked an ambitious set of new ideas to transform what looked like basic and inefficient networks of canals, small dams and terraced fields fed by water channels. These were dramatically expanded with technologies that were thought to be far superior and would prevent the devastating famines that seemed to be regular occurrences – as in 1837–8 when some 800,000 people died in the North-Western Provinces, Punjab and Rajasthan owing to the failure of monsoon rains.[48] If the problem was obvious, so too was the solution: 'all Indian famines are caused by drought', in the words of a report by the Indian Famine Commission set up in the second half of the nineteenth century after more than 5 million people had died from hunger and disease; to stop this happening, 'the first place must unquestionably be assigned to works of irrigation'.[49]

The problem, according to one British engineer, was that in 'one year a portion of the whole crop ... is destroyed by the overflowing of the rivers', while 'in another the crop is destroyed by a failure of the rains'. The answer was to open up all of India to new irrigation schemes, which would be 'a work of such magnitude ... that nothing approaching it has ever been seen in the world'.[50] The excitement of those who built and commented on the works that followed was no less breathless and self-congratulatory: a dam built in 1852 that regulated the flow of the Godavari River in Dowleswaram was described as 'the noblest feat of engineering skills which has yet been accomplished in British India'.[51]

Similar sentiments were expressed about the 700-mile-long Ganges canal which was opened two years later. One senior official who had conceived the project talked of his pride that 'a few hundred Christians in the heart of a foreign country' and surrounded by heathens had been able to create 'a work of civilisation for the benefit of these unenlightened multitudes'. The dark night of 'false religion, tyranny and war' and 'the bitter consequences of evil' had given way to the bright day 'of Christianity, good government and peace'.[52]

In the next three decades or so, more than 65,000 kilometres of new canals were built, irrigating some 5.4 million hectares – with the costs largely passed on to the Indians themselves.[53] Railways were constructed that offered returns to private investors that were guaranteed by the state, which proved a recipe for a huge programme of track-laying that turned India into the largest railway network in Asia and the fourth largest in the world. This required vast amounts of steel and iron to make the tracks, engines, wagons and carriages, as well as wood, coke and coal for sleepers and fuel. The subcontinent shrank to 'one 20th of its former dimensions', stated one British engineer, who also noted that 'the power of steam' had done much to deliver social progress by undermining 'one of the most rigid and exclusive caste systems in the world' by linking regions together, enabling the mobility of workforces and modernising the subcontinent.[54]

The scale of such changes was breathtaking. In Burma, the land devoted to rice growing rose twelvefold, turning the Irrawaddy Delta into one of the world's largest areas of rice production. The delta's population soared from 1.5 million to over 4 million in the period 1852–1900. This was mirrored by Dutch activities in the East Indies and those of the French on the Mekong; in the latter case, massive removals of mud and soil made the canalisation of the river one of the largest earth-moving exercises in human history.[55]

It would be tempting to assume that the development of infrastructure was intended first and foremost to help the local population, an objective born from a sense of magnanimity, duty and philanthropy. Of course the primary purpose was to support the concept and reality of empire. For example, investments in irrigation systems served the dual purpose of boosting agricultural yields (and therefore tax receipts) and mitigating the effects of famine, which could prove costly: in the 1870s, for example, almost £10 million was spent on famine relief – the equivalent of around £200 million today.[56]

While the expansion of the railways likewise brought opportunities for India and Indians, the increase in velocities of exchange benefited the colonial administration by boosting revenues, at the same time promoting British industries that exported raw materials in large quantities. Apart from the fact that one motive for building railways was to connect ports with the interior and thus enhance both the quantities and speed of exports and imports, there was a further important motivation. As the Governor General of India, Lord Dalhousie, told Parliament in 1853, an extensive rail network 'would enable the government to bring the main bulk of its military strength to bear upon any given point in as many days as it would now require months'.[57] The railways encouraged trade, linked regions together, spurred urbanisation, impacted land values and helped widen literacy; but above all they strengthened Britain's political, military and economic stranglehold on the subcontinent – just as they did for Russia in Siberia, for the United States in the midwest and beyond and for China in Manchuria.[58]

While colonisers helped themselves to the best of the rewards, there were local beneficiaries too, of course. Local farmers keen to improve their lot provided support, stimulus and even labour for agricultural improvement.[59] Substantial opportunities became available to those able and willing to move into regions as they were opened up through new transport links and overhauled by large-scale irrigation projects. A good example is the western Punjab, where the scale and pace of ecological transformation was dramatic. One million people moved into nine 'canal colonies' built between 1885 and 1940 for a workforce that tended to 13 million hectares of land which had previously been little more than scrub and desert but had now been turned into lush green fields by new irrigation works. This transformed Punjab into what one leading scholar has called 'the engine of agrarian growth in India'; the region was home to almost half of all canal-irrigated land in British India and the source of high levels of revenue and of manpower, both of which helped reinforce colonial rule over the subcontinent as a whole.[60] While farmers and those who worked this newly available land clearly gained from the process, control over nature provided resources for the British to strengthen their grip on India.[61]

This was not how colonial administrators themselves saw it – or at least not how they wrote about it. One major report issued in 1918 stated that the investment in India's agricultural development had been so profound and so successful that 'the terrible calamities which from time to time depopulated wide stretches of country need no longer be feared'. The interventions in the natural environment had eliminated risks of climatic stress: failure of the rains had once meant not just 'privation and hardship' but also 'wholesale starvation and loss of life'. This was not the case any longer, since the problems had all been 'scientifically studied and a system worked out'. Nature itself had been tamed.[62]

Such self-confidence went hand in hand with the certainty that knowledge and enlightenment brought from far away were greatly superior to anything that had existed or had developed locally. In a volume written by a British official in East Africa, modestly entitled *The Book of Civilisation*, the author claimed that 'Folk are poorly fed in Africa because most Africans are not good farmers, and particularly they're not good farmers because they do not know how to keep the soil fertile.' This was infuriating, he said, adding almost with a sense of despair: 'When cattle in Africa are used for their proper purposes – to provide milk for the children and meat for [them]selves, and manure for [their] fields and draw carts and ploughs – then, but not sooner will the people thrive.'[63]

Although views of the supposed inferiority of indigenous practices were typical and widespread, they were not uniform. One British agriculturist who was asked by the Secretary of State for India to assess the condition of Indian farming concluded firmly that he did 'not share the opinions which have been expressed as to Indian Agriculture being, as a whole, primitive and backward', but rather believed 'that in many parts there is little or nothing that can be improved'. In so far as it was even possible to generalise in these matters, the reality was that it was not 'inherent bad systems of cultivation' that were problematic, he wrote, but the absence of infrastructure in some districts in the subcontinent. Moreover, rather than adopt new practices from Britain, Europe or elsewhere, far better results would be achieved 'by the transference of a better indigenous method from one part [of India], where it is practised, to another where it is not'.[64]

Disdain for farming practices in Africa – and for African peoples in general – was likewise not shared by all. As one French botanist wrote, 'the indigenous cultivator is not the contemptible lay-about ... that is so often represented' by many contemporary commentators; rather, local practices not only got the best out of the land but did so in ways the European efforts had conspicuously failed to do, often at great financial as well as ecological cost.[65] British officials, wrote Archibald Church, a member of the Parliamentary Commission of Enquiry into the East African colonies, 'would do well to make a careful study of native methods of agriculture before they arrogantly assume that the methods of the European are far superior to those of the African'.[66]

Such views reflected rare cases of open-mindedness and did little to change opinions or to affect policies and practices in relation to cultivation and extraction that often made things worse, not better, replacing one set of problems with another. For example, biodiversity suffered as habitats that were home to animals and plants were cleared to make way for monoculture production.[67] In India, malaria was such a 'serious menace' as it spread along canal networks that some questioned whether the building programme would deliver any positive outcomes at all.[68] The digging of new canals often brought disputes about property rights, while poor drainage sometimes not only meant irrigation was inefficient but gave rise to unsanitary conditions that enabled disease to catch and spread.[69]

Large-scale irrigation was also problematic because, although initial results always looked promising, the threats from waterlogging and soil salinisation were no different in the modern world from what they had been in Mesopotamia and elsewhere thousands of years earlier. So it proved in Punjab, where the cold reality soon became clear, that canalisation was not the panacea that had been hoped for. By the 1940s, yields in many parts of the Punjab had fallen by 75 per cent or more; at least one million acres of former cultivated land had been made barren by unsustainable irrigation practices.[70] This was bad enough on its own terms; but even worse given the infrastructure of housing, schooling, healthcare, roads and more that had been provided for the large populations that had moved into these regions in the expectation that they would have rosy long-term futures, rather than precarious ones.[71]

Then there were the dangers of overdependence on a single or principal crop, which could produce economic downturns – and major shocks. While cocoa had been grown in South and Central America, as well as the Caribbean, from the start of the twentieth century, a massive development of plantations in West Africa eventually led to this region becoming the primary global producer of cocoa, as yields rose almost twenty times from around 40,000 tons a year to some 700,000 in the five decades after 1885. The glut led to a collapse in prices that had obvious beneficial consequences for consumers who were able to obtain this calorific bonus at increasingly low prices, but also consequences for workers and investors that were distinctly less rosy. The problem was exacerbated by the biological vulnerability that put the reliance on cocoa into sharp focus: in the 1940s, the 'swollen shoot' pathogen devastated plantations in West Africa, destroying livelihoods and stability with it, leading to riots in Accra and desperate attempts to stop the spread of the disease that eventually led to more than 100 million trees being cut down.[72]

In many other parts of Africa, human reshaping of the environment with the establishment of plantations on the one hand and of conservation zones on the other had a series of profound impacts. These ranged from forcible resettlement to competition for water, land and manpower. Settlement patterns shifted as railways were built into the interior resulting in the enlargement of local towns and markets that put ecological pressures on surrounding regions.[73] Matching impacts were effected by livestock herds that stripped vegetation from new regions where they had been forced to move, and by tribal and intertribal rivalries that were often fanned by decisions made by colonial authorities to support (and arm) one particular social group over another.[74]

Above all, however, the colonies were where resources could be claimed and used by the colonisers. Commodities like palm oil, rubber, cotton, timber, coffee, cocoa and metals were all needed for industrial production, for no end of manufactured goods, ranging from soap to clothing, from vehicle tyres to engine belts for use in factories. Soon after his appointment as Special Commissioner to Uganda, one of the first questions Harry Johnston asked was: 'what resources does the country possess for the development of a profitable trade?' He was trying to find out, of course, what had value that could be sent abroad to consumers back home.[75]

The world was a rich paradise, whose fruits were to be enjoyed by those who had inherited the keys to the garden of Eden. Nature could be tamed, just as local populations could be 'civilised'; scientific knowledge could be applied to impose and maintain control. All this powered a conviction that the world offered unlimited riches that could be developed and tapped for ever.

For example, after a tour of East Africa in 1907, Winston Churchill did not mention the people of the region, commenting instead on its fertility and natural beauty, the coolness of the air, richness of the soil and abundance of running water. Africa, he went on, 'must one day become the great centre of tropical production, play a most important part in the economic development of the whole world'. Africa, in other words, would be remodelled to suit the needs of people beyond, not those of its own people.[76] Theodore Roosevelt, who followed a very similar itinerary to that of Churchill a year and a half later just after standing down as US president, put it more bluntly. East Africa, he said, was 'a white man's country, a country that should be filled with white settlers'.[77]

This was an expression of ideas and convictions which had been developing since the first crossing of the Atlantic and which were intoxicatingly strong in northern Europe and in those who had emigrated from northern Europe to other corners of the world. Ideas about Anglo-Saxon tribes from the distant past fed into fetishised mythologises of history which held that the Germanic peoples had succeeded in overthrowing Rome thanks to their profound devotion to liberty and their determination to throw off the shackles of authoritarianism. These concepts which had been forcefully articulated by Montesquieu in the eighteenth century, had proved highly attractive in Britain and Germany and were particularly seductive in the United States. The British success against Qing dynasty China in the First Opium War (1839–42) was hailed by the *New York Herald*, for example, as 'not a victory for British imperialism, but a triumph for the Anglo-Saxons'.[78] Such views were all the more striking at a time when many in the US were critical of British imperialism.

Some went further still. For example, in the 1850s, one congressman could be found musing that one day 'the two great branches of the Anglo-Saxon stock, the one pressing from the Bay of Bengal, and the other from the golden gulf of California, would meet in some beautiful group of sunny isles in the Pacific ocean, and together clasp their united hands in love and peace around the globe'.[79] If some in the United States could not hesitate to express their gratitude that the Atlantic seaboard 'was settled by colonies of the Anglo-Saxon race', others left little to the imagination about what this meant: 'history warrants us in believing in races of men, as well as of inferior animals', said Edward Everett, a man whose glittering career saw him serve as Governor of Massachusetts, US senator, representative of the United States to the United Kingdom and President of Harvard. 'The Anglo-Saxon race', he went on, 'from which we Americans trace our descent is surpassed by none other that ever existed.'[80] He meant some Americans, of course – those with pale skin.

The British were hardly immune from such high self-regard. Indeed, some even argued that Britain itself had evolved as a result. 'England has outgrown the continent of Europe,' said Benjamin Disraeli in 1866. She was 'no longer a mere European power; she is the metropolis of a great maritime empire, extending to the boundaries of the farthest ocean ... she is really a more Asiatic power than a European'. This all stemmed from the engagement with other continents, peoples and ecologies and reflected the awareness, if not the conviction, that the sources of Britain's wealth lay overseas.[81]

British rule was 'steadfastly progressive', *The Times* claimed a month after the end of the First World War, 'the most beneficent in design and execution known in the history of mankind'. These views were shared by many, such as Gertrude Bell, who wrote to her father not long afterwards, reflecting that 'truly we are a remarkable people. We save from destruction remnants of oppressed nations, laboriously and expensively giving them sanitary accommodation, teaching their children, respecting their faiths' – even if few showed much gratitude for it.[82]

Sir Michael O'Dwyer, who gained notoriety as the Lieutenant Governor of Punjab after the Jallianwala Bagh massacre of 1919 in Amritsar when British troops opened fire on peaceful protesters complaining about the arrest of pro-independence leaders, was even more emphatic. 'The British Empire in India is the greatest achievement of our race,' he wrote. 'It has been built up by the blood, the brains and the energy of our ancestors.' In any event, he insisted, 'INDIA NEVER HAD – unity, security, peace, justice, communications, public health – until the British came.'[83]

Such self-congratulation hid the mechanics of how control was ruthlessly exerted to keep local populations in place – and apart. One historian has noted that Indian regiments were 'organised to kill each other should the British desire it', a scheme attested by comments made by Sir Charles Wood, Secretary of State for India in the second half of the nineteenth century. Wood had shown his colours when he opposed support for relief in Ireland during the terrible famine of the mid-1840s, claiming that this was 'a calamity' which had been 'sent by Providence' and would help stimulate better farming practices in the long run; famine, mass starvation and death on an apocalyptic scale were a price worth paying.[84] His view of India and Indians was no more enlightened. The reason for keeping ethnic groups apart in the army, he wrote, was simple: 'If one regiment mutinies, I should like to have the next so alien that it would be ready to fire into it, so that Sikh might fire into Hindu, Gurkha into either, without any scruple in case of need.'[85]

The overwhelming dominance of colonisers offered almost unlimited and irresistible temptations.[86] The Belgian King Leopold II simply asserted that the Congo River basin was his own – sparking a set of diplomatic crises not in West Africa but in Europe, where some objected that they had been deprived by Leopold's action from undertaking their own exploitation. Eventually, Leopold's 'ownership' of territories five times the size of France was endorsed by European politicians at a conference in Berlin in 1884–5, where the primary concern was to ensure that disputes between colonial powers should be settled peacefully – in order to avoid a war close to home. This mirrored attitudes to resources, to the environment and to human life outside the settler world: all were fruits that were ripe for the taking.[87]

Africa was one of the last to be plucked, a function of long-standing ideas about the continent and its many different peoples and cultures that were deeply rooted in prejudice and racism. It was significant, though, that as had been the case with the establishment of transatlantic slavery in the sixteenth century, Europeans proved singularly poor at making inroads beyond the coasts and their immediate hinterland. This speaks of the resilience and adaptability of local political leadership, and shows that their consolidation of authority was a logical response to the threats posed by non-native traders as well as to the opportunities that they offered.

By the late nineteenth century, the accelerations brought by the industrial revolution tipped the balance decisively in favour of European outsiders, most obviously because of the improvements in weaponry which meant that more could be done more quickly by fewer men. For example, in 1897, James Phillips, the Deputy Commissioner and Consul for the Niger Coast Protectorate, set off to overthrow the Oba (king) of Benin, accompanied by half a dozen officials, two businessmen and a train of porters. Their ambush – and Phillips' death – became a cause célèbre in London. 'A deplorable catastrophe' had taken place, said the *Daily Telegraph*, which lamented that 'a gallant body of Englishmen have fallen victim to their efforts to bring West Africa within the outer fringes at least of civilisation'. The paper did not ask why this gallant body's interest and motivations might not have been welcome.[88]

Reprisals were demanded throughout the press to avenge the apparent compromising of honour, and to teach the Oba 'the wholesome lesson that not only is English faith to be trusted and English friendship to be sought after, but that English justice, when outraged, is something to be feared'.[89] The expedition to 'avenge' Phillips brought about one of the more shameful episodes of destruction, murder and theft of African cultural and political heritage, not least of the 'Benin bronzes' which today stand in some of the world's leading museums, and which have been central to calls for the return of looted artworks to the places and the peoples from whom they were taken.[90]

At the time, though, teaching the Oba and the people of Benin a lesson in how they should understand their position in the world seemed perfectly natural. After being detained, the Oba was informed that he was no longer the ruler of his own lands; 'the white man', he was told, 'is the only man who is King in this country'. The local population were to be shown 'how a country of black men ought to be governed', by a British judge sent to oversee a 'trial' of the Oba to consider his role in the death of Phillips. If any evidence was found connecting the leader to the murder, the judge told him, 'I will hang you here.'[91]

The seemingly unstoppable rise of the west caused soul-searching and even sorrow. In Vietnam, for example, some called for a 'new learning' that moved away from traditional models to linguistic, economic and political innovations. 'In ancient times', wrote one Vietnamese author in 1904, 'Asia was the source of civilisation.' Vietnam had possessed 'fertile land, a moderate climate, plenty of rice, silk worms, and forestry products and a seacoast longer than most'. But now 'we no longer have forestry products and other resources. We no longer control hundreds of goods and benefits of our country, such as cloth, crepe, velvet, silk, shoes, sandals, handkerchiefs, spectacles, umbrellas,' or dozens of other items, all painstakingly listed.[92]

Others among the colonised communities saw things in a similar way, but called for action. 'Asia has sat too long in darkness and in the shadow of death,' railed a group of radical Indian revolutionaries in exile in 1909, calling for an Asian parliament to be established made up of 'educated Indians, Osmanlis, Egyptians, Japanese, Chinese, Arabs, Armenians, Parsis, Persians, Siamese, and others' to help rise up against the domination of European imperial powers.[93] In other cases, passionate anti-colonialist calls for change brought fleeting successes – as in much of the Russian empire after the abdication of the Tsar in 1917, when extensive rights of self-determination were given to minorities in an initial blossoming of optimism and creativity, before crushing persecution set in once again. A burst of poetry, creative writing and visual arts lit up Russian Central Asia before being suffocated as the Soviet Union began to centralise and to deprive of oxygen everyone and anyone who might threaten the communist dream. In such scenarios, suspicions fell quickly and decisively on those who were ethnically distinct or lived in peripheral, frontier regions – or both.[94]

Anti-colonial attitudes hardened, often reinforced by strong ecological agendas that pointed out how lands were being stripped, overworked and degraded for the benefit of people living thousands of kilometres away. 'India's salvation', wrote Mahatma Gandhi in 1909, 'consists of unlearning what she has learnt during the past fifty years.' Modernity had brought unhappiness and oppression. 'The railways, telegraphs, hospitals, lawyers, doctors, and such like have all to go,' he went on. 'Machinery is the main symbol of modern civilisation; it represents a great sin.'[95] Urbanisation, he predicted later, was sure to lead to 'a slow but sure death for [India's] villages and villagers'.[96]

True happiness could only come from living 'the simple peasant life' and rejecting the technology, the consumption and the ravages of colonialism. 'The British Empire today', he wrote bitterly soon after the end of the First World War, 'represents Satanism … repression and terrorism.' It 'will stoop to any atrocity' to boost itself, he added; it was a source of evil and 'the height of tyranny'.[97]

Others too were struck by the pace and deleterious effects of modern life. Vsevolod Timonov was one Russian scholar who was sharply critical of the way 'Nature's harmony' was being ruined in the search for 'immediate profits'. Human activities and lifestyles meant that factories 'spew their foul-smelling gases into the atmosphere, greatly impairing the enjoyment of nature'. The result, he concluded, was that the 'climate is being ruined'.[98] Another Russian author, Leo Tolstoy, renounced alcohol, became a vegetarian and gave up handling money as part of a search for a simple, self-sufficient and notionally traditional lifestyle – which in turn inspired Gandhi to create a utopian community in South Africa based on the same principles and ideals. Tolstoy had been so influential, in fact, that Gandhi wrote to tell him of his development, to thank him for his guidance and to reveal that the ashram had been christened 'Tolstoy Farm' in his honour.[99] For Gandhi, the idea that thousands of men worked in factories and mines in conditions 'worse than [those] of beasts' and women laboured 'for the sake of a pittance' was despicable; it could hardly be described as 'civilisation', he said.[100]

The hypothesis that humanity was dangerous – to itself and the natural world – gained currency in the nineteenth century, with Ralph Waldo Emerson urging limits to how the wilderness was treated, and George Perkins Marsh warning that man wages 'an almost indiscriminate warfare upon all the forms of animal and vegetable existence around him'. Human intervention was both damaging and relentless: as 'man … advances in civilization, he gradually eradicates or transforms every spontaneous product of the soil he occupies'.[101]

The logical culmination of such thinking came in a famous essay written in the 1960s, in which Lynn White argued that Judaeo-Christian society had introduced a world view that had made people less likely to care for the environment because it promoted humans as superior to and separate from nature – with the result that they became the primary causes of ecological crisis.[102] This reveals much about Eurocentric ideas from the Enlightenment onwards which characterise 'eastern' belief systems as more environmentally friendly, and sees people living in Asia in particular as being imbued with noble, almost mystical attitudes to life, spirituality and nature; this itself is a form of orientalism. In fact, religious tradition was not the problem nor the source of the problem; greed and personal gain were.

Although it would be wrong to suggest that landscapes had been untouched, forests left intact and ecologies undisturbed until Europeans arrived to disrupt a golden age, the reality was that extraction during the colonial era, facilitated by technologies and labour forces that made land clearance quicker and more extensive than ever before, took place on scales that were unprecedented.[103] Moreover, as recent scholarship has emphasised, agricultural policies (in particular in relation to woodlands) were closely linked to population resettlement, the consolidation of authority and social engineering, and therefore to the extension of political control, especially in ecologically and geographically peripheral regions. As one scholar has put it, environmental and colonial rule went hand in hand.[104]

Nowhere was this clearer than in North America, where indigenous peoples were pushed off their lands by force and sometime by deception. Moving the 'Indians' to 'suitable lands west of the Mississippi', stated James Barbour, the Secretary of War of the United States, would 'produce the happiest benefits upon the Indian race' – without mentioning the benefits that the lands they vacated would bring to their new occupiers. Some tribes, such as the Senecas, had themselves to blame, according to officials who said that the suffering that followed forced auctions of property and migration under duress was 'their own fault' for not choosing better substitute territories to move to.[105]

Other politicians urged brutal treatment in order to claim the best lands for crops and grazing for themselves. At the start of the Black Hawk War of the 1830s, one editorial urged John Reynolds, the Governor of Illinois, to 'carry on a war of extermination until there shall be no Indian (with his scalp on) left to the north part of Illinois'.[106] The pattern of dispossession was one that extended for decades. At noon on 22 April 1889, to cite one example, a gunshot signalled the 'right' of settlers to claim land in Oklahoma that had been assigned to the Creek and Seminole peoples as part of previous settlements and pressured migrations. By the end of the day, 2 million acres had been claimed in plots of 160 acres; by sundown, new cities with populations of as many as 10,000 had been founded, with street grids planned, lots staked out and thoughts already turning to the structure of civic government.[107]

The furious pace of change offered tantalising prospects of fortunes to be made and of social mobility. It also captured the imagination of writers like Anton Chekhov, whose play *The Cherry Orchard* describes perfectly the ambiguities of the old being replaced by the new, and the challenges of adaptation, as Madame Ranevskaya, a blue-blooded landowner, struggles to come to terms with a changing world epitomised by Lopakhin, a self-made man whose father had been a peasant and grandfather a serf. The play ends with Lopakhin buying the estate where he had grown up, accompanied by the sound of the orchard being chopped down – with the repurposing of nature to suit modern needs and tastes a symbolic nod to the changing times and the 'progress' of the modern era.[108]

Such advancement came at a price. Marx expressed his concern about the relentless determination to 'exploit the earth's surface, the bowels of the earth, the air', which was driven by the desire to profit at nature's expense.[109] The rush to exploit the land brought about ravenous demand for returns on investment, and led to ecological exhaustion. The development of Alabama was paid for by capital gathered in faraway locations, from

Wall Street to cities across Europe. The black prairie lands, with their dark, fertile soils high in calcium deposits left over from previous periods of climatic change, were particularly attractive. Few cared that the switch to monoculture and heavy exploitation produced very different results to the way that the Creek people, for example, had managed their lands through sophisticated models of biodiversity that combined multicropping with careful rotation to ensure long-term sustainability.[110]

It was true that colonialism did not always lead to degradation. Imperial authorities in Japan invested considerable resources in massive tree-planting activities in Korea, planting more than a million trees in the first decades of the twentieth century and protecting forests by law – although there was also heavy investment in megaprojects such as dams, ports and other monuments that echo European policies and practices. Japanese motivations here were centred not on conservationism or attitudes to forests, but rather on the centrality of wood in the pre-industrial history of Japan, on the conceptualisation of the country as a 'luxuriantly green realm'. The reason to preserve forests was not about protecting nature, but rather the promotion of national identity, statecraft and rulership at a notional level, combined with securing access to unlimited stocks of wood for fuel and building. Taken together, this all demonstrated the prescience of the emperor and the imperial court.[111]

Taken as a whole the story of the nineteenth century – and particularly the second half – was one of the world becoming smaller and smaller, with peoples and places being brought closer together as a result of the development of railways, bigger, faster and more reliable ships and new technologies such as telegraph systems that allowed information to be shared quickly. It was not only an age of new infrastructures that accelerated globalisation but one that saw the rise of cities, which sprang up like mushrooms and became swollen by inflows of inhabitants. It was perhaps not surprising that many of the largest cities were located on the coast, with good harbour access that allowed ships to load cheaply and efficiently. Ports were funnels for commodities, natural resources and minerals extracted from colonial hinterlands, as well as valves for manufactured goods that flowed in the opposite direction, out from industrialised Europe into new markets.[112]

There was growing awareness that the intensification of exchange and of consumption came at a price – though, needless to say, that meant little to investors in railway financing across North America, Russia and Africa who barely thought about the ecological costs that came with deforestation or with the imposition of livestock, crops and mines. The construction of a railway line in Equatorial Africa proved to be one of the deadliest in history, resulting in tens of thousands of deaths from overwork, disease, malnutrition and physical abuse – but providing guaranteed profits for the French company that built it.[113]

Those with more of a conscience, open mind, or both, paid greater attention to the ways in which environmental and climatic change had significant consequences. Phenomena like the Carrington Event in 1859, a massive solar storm and coronal mass ejection that knocked out telegraph stations around the world, or the dramatic eruption of Krakatoa in 1883, which had an explosive power that was an estimated four times greater than the most powerful thermonuclear device ever detonated – or forty times the combined force of all the conventional explosives used during the Second World War – did much to focus attention on weather patterns and to spur research into how they were changing and the extent to which this was the result of human activities and interventions in the landscape.[114]

Scientists had been trying to gain a better understanding of climatic shifts for some time. In 1801, for example, the renowned astronomer Sir William Herschel wrote a paper in which he suggested that sunspots were a crucial factor in weather conditions and also in harvest yields, wheat prices and economic cycles. Others followed in the ensuing decades, including Arthur Schuster, who observed that good vintages of wine seemed to be correlated with the eleven-year solar cycle.[115] William Jevons went further in the 1870s, linking the sun's activities in the eighteenth and nineteenth centuries with prices, commercial crises and a pattern of 'unquestionable collapse' of economies.[116]

Others investigated atmospheric conditions. The French scientist and mathematician Joseph Fourier was one of the first to look at this methodically, suggesting that gases in the atmosphere served as a barrier that trapped heat and kept the earth warm – and he questioned whether 'the progress of human society' around the globe could be responsible for 'remarkable changes', including 'variations in mean temperatures'.[117] Another pioneer was Eunice Foote, who by the 1850s, realised how significant her findings were for science and for the future of the planet. Foote conducted experiments in which she heated combinations of gases in glass cylinders and observed that carbon dioxide and water vapour not only became hotter faster than other combinations, but also were 'many times as long in cooling' as cylinders containing different gases when removed from the source of the heat. 'An atmosphere of that gas would give to our earth a high temperature.'[118]

By the turn of the twentieth century, further research by John Tyndall, Samuel Pierpont Langley, T. C. Chamberlain and, most famously, Svante Arrhenius had established the principles of the greenhouse effect – although this research was driven not by any concern about rising levels of CO_2 but rather by efforts to investigate the world's geological and geophysical past. Arrhenius was not worried about global warming – indeed, quite the opposite: he wondered whether heating the earth might be beneficial and lead to a 'more equable' environment, more abundant plant life and higher food production.[119] Despite these advances, climate science was for decades pushed to one side as a fringe subject, a pursuit for amateurs, crackpots and oddball academics rather than one worthy of serious investigation.[120]

If such events brought questions about solar cycles, volcanic activity and weather patterns into public consciousness, so too did an upsurge of popular science writing and early science fiction, much of which focused on anthropogenic change, on ways in which human behaviour altered the climate for good or for ill. As early as the 1820s, authors like Faddei Bulgarin could be found pondering a world where Russian scientists had successfully warmed its Arctic coastline and turned it into a maritime idyll. A few decades later, Byron Brooks and Lysander Richards separately came up with the idea of the Sahara desert being transformed into fields through manipulation of the climate. British writers speculated about the horrors of alterations in the Gulf Stream patterns that would surely result from the construction of the Panama canal – and would bring down the empire as a result of a fundamentally changing climate.[121]

Or there was George Griffith, in whose book *The Great Weather Syndicate* the hero, Arthur Arkwright, develops a machine that changes the climate and promises to make him the 'master of the fate of the world' by enabling him and his 'syndicate' to sell bespoke weather conditions to those willing and able to pay for them.[122] Another author who wrote on a similar theme was Jules Verne, whose *Sans dessus dessous* was a biting satire of a plan by reckless investors to tilt the axis of the earth, equalising global climates in order to eliminate seasons and differences in the length of the day and night, but also so as to melt the Arctic and gain access to enormous coal reserves.[123]

What Verne and others could not have known was that sea levels were not only rising globally in the late nineteenth and early twentieth centuries but were beginning to do so at an accelerated rate.[124] Recent research pinpoints a starting date of 1863 for the sea-level rises, which do not seem to have been associated with mass loss of the Greenland Ice Sheet or with anthropogenic factors; instead, they were more likely linked to ocean circulation changes resulting from the Atlantic Meridional Overturning Circulation (AMOC), the strength of the Florida Current and the strength and position of the Gulf Stream, as well as to changing patterns in atmospheric winds, buoyancy fluxes and pressure.[125]

Works conjuring up dystopian (and utopian) futures that were framed by climatic change point to a growing interest in the ways that human activities had an impact on the natural world and the environment. This was not entirely surprising, especially in places that had industrialised early and intensively: air pollution in cities in Britain was so severe that it raised infant mortality by as much as 8 per cent, while industrial coal had a measurable and highly detrimental effect on urban life expectancy during the later nineteenth century.[126] In this period, atmospheric pollution from coal-fired industrialisation was fifty times higher than today and had a strong negative impact on health outcomes and on adult heights.[127]

City living was always noisy, bustling and chaotic. What made the nineteenth century different, however, was the scale and speed of the cities' growth. The population of Manchester quadrupled between 1801 and 1851, as people were drawn by the promise of work.[128] Such rapid expansion of urban areas, combined with poor sanitation and heavy levels of pollution that were emitted mainly from the burning of fossil fuels, shocked visitors. Friedrich Engels was struck by the 'foul air' of the city that caused 'blood-spitting, hard, noisy breathing, pains in the chest, coughs [and] sleeplessness', and where women and girls not only had to make dresses 'in almost total exclusion from fresh air' but worked in such poor light that many had trouble with their eyesight and some even went blind. This, along with the capitalist principles behind the factories that belched thick smoke and the work which was 'monotonous and demeaning', was nothing less than 'the barbarous exploitation of the workers'.[129] These conditions were mirrored all over the north-west, which in the words of a contemporary of Engels was full of 'little Manchesters'.[130]

While wealth and status have always provided one barrier to disease, they did not always offer sufficient protection. In 1841, William Harrison died just a month after being sworn in as the ninth President of the United States. Although his death has long been attributed to pneumonia, and to his decision not to wear a hat, overcoat and gloves on a bitterly cold inauguration day, it seems that what actually killed him was enteric fever, most likely caused by poor sanitation in Washington, DC, and specifically by the White House water supply being located close to marshy public ground where sewage flowed freely in the absence of a city sewer system. It seems too that two of his successors – Presidents Polk and Taylor – suffered from illness during their presidencies that may have stemmed from the same source.[131]

One reaction to the growth of cities, to pollution, to disease and to the degradation of the environment was a surge in literature, most notably in romantic poetry, which extolled the countryside and family life as an undisturbed idyll, unspoiled by human intervention. These were all a sharp contrast to the reality of industrialisation, which according to Wordsworth meant that mothers had 'no nice arts / Of needlework' to look forward to, while young boys had the 'short holiday of childhood' curtailed as they were pushed into factories as workers and 'prisoners'.[132]

It did not take much to grasp, in other words, that modernity came at a cost – and that those costs were not evenly distributed globally. Air pollution, for example, was largely restricted to places that industrialised early and quickly, above all cities in northern Europe and North America; water and riverine pollution, on the other hand, was devastating in regions where severe deforestation and land clearing had taken place. The large-scale movement of people had been instigated both by urbanisation and by the distribution of labour, whether by force, as a response to demand or as a reflection of the search for opportunity. As a result, the concentration and dispersal of people shifted more dramatically and more quickly than at any point in human history. As they did so, they brought consequences whose legacies are very much part of the challenges we face in the present day. Perhaps the most obvious and resonant was the impact of epidemic disease.

Urbanisation brought more and more people into closer contact, which not only accelerated the rhythms of commercial and cultural exchange but promoted the spread of infectious diseases. These did not only affect humans: in the autumn of 1872, for example, equine influenza took hold across North America, appearing first near Toronto and reaching the Gulf coast by the end of the year and the west coast by the spring. Although the disease rarely proved fatal (with mortality rates of around 2 per cent), it had the effect of leaving horses unable to work through exhaustion; this caused major economic disruption, in no small part because of the central role played by the animals in hauling 'coal, bales and boxes' in cities. One result was a sharp rise in the cost of goods and services, which some believed was the result of opportunistic price gouging. Another was the vulnerability of cities to fire, with one enormous blaze in Boston attributed to the lack of horses for pulling fire-tending equipment into place.[133]

Epizootic disease elsewhere had unexpected outcomes, most notably in East and southern Africa where an outbreak of rinderpest in the 1890s killed millions of cattle as well as sheep, goats and other livestock; this precipitated famine that resulted in the deaths of two-thirds of the Maasai from starvation, weakened local social and political structures, caused widespread violence and ultimately opened the door to the expansion of colonial authority at the expense of local elites.[134]

In other places, too, misfortune for some brought opportunity for others. An outbreak of measles in Fiji in 1875 led to an increase in demand for Indian labour, not least because of high immunity levels acquired during infancy in South Asian populations.[135] Pacific islander plantation workers in Queensland and elsewhere in Australia experienced unusually high mortality rates from measles, most notably during the first year, which shows how epidemiological boundaries shifted as migrant labour moved – or was moved – in order to transform and refashion landscapes for growing labour-intensive crops.[136]

As one prominent historian has put it, the intensification of intercontinental connections in the nineteenth century brought about 'the unification of the globe by disease', with trade routes, migration corridors and troop movements serving to create a 'common market of bacilli' criss-crossing the world and linking it together.[137] Epidemic and pandemic disease were part of the price to pay as human interactions and activities spread pathogens from one continent to another. Ports like

Bombay, Cape Town, Singapore, Hong Kong and Calcutta were crucial not only for the loading and unloading of goods but as transmission points in an interconnected global disease network.[138]

Other nodes were important too, of which Jeddah was perhaps the most significant. Since the time of the Prophet Muḥammad, Muslims had been expected to undertake the hajj, a pilgrimage to Mecca that involved the performance of a series of rituals. The journey became synonymous with outbreaks of cholera that intensified with the rise of steamer ships and the opening of the Suez canal. Cholera broke out in Mecca on average every three years in the nineteenth century, with eruptions in 1831, 1865 and 1893 each resulting in tens of thousands of deaths.[139]

Such events had parallels. As it was expressed at a sanitary conference in 1866, pilgrimage was 'the most powerful of all the causes which conduce to the development and to the propagation of epidemics of cholera'.[140] As well as Mecca, 'places where congregations of Hindoo pilgrims take place' were notorious for the rise in infection, which fitted into colonial tropes of the pilgrim as 'filthy, superstitious, inherently irrational and resistant to sanitation measures' introduced by the state.[141]

The Hindu festivals that are held in northern India at alternating locations on four sacred rivers – Hardwar on the Ganges, Nashik on the Godavari, Ujjain on the Shipra River and Prayag at the confluence of the Ganges, the Yamuna and the mythical Sarasvati – were a particular concern. This was especially true of the Kumbh Mela, which takes place roughly every twelve years and attracts myriads of pilgrims: the gathering in 2019 was estimated to have been attended by 50 million people.[142] These festivals became a deadly source of infection and dissemination, for example in 1867, when torrential rains saturated the ground, resulting in a major epidemic which spread despite attempts by the colonial authorities to screen pilgrims and to quarantine those showing symptoms. Disease spread despite the construction of stone kilns to burn excrement, as well as trench latrines that were then covered by dry earth – efforts that were initially so successful that one observer noted that 'not a single Anglo-Saxon, a race far more sensitive to foul smells than Natives, perceived [any smell]'. Events in 1879, 1891 and 1901 also brought serious outbreaks.[143]

Even such disasters could have silver linings. For one thing, large-scale disease encouraged scientific research and investigation by scholars and doctors such as John Snow, Filippo Pacini and Robert Koch to identify the causes of cholera, to devise treatments and eventually to identify the cholera bacillus itself.[144] For another, suffering brought about shifts in ideas about civic responsibility, with some communities like the Parsis in India taking pride in founding hospitals to tend to the sick. Ten major international conferences between 1851 and 1900 which saw researchers from all over the world share their findings and hypotheses represented high levels of collaboration.[145]

Other cases provided further evidence of the downsides of increasingly integrated transport links and high population densities. In the summer of 1889, the first cases of what appears to have been a highly virulent strain of influenza were detected in Bukhara in Central Asia. Not long afterwards, cases were being detected across the Russian empire, leading to factories and schools being closed in St Petersburg where almost 20 per cent of the population became infected. The disease spread quickly to Germany and by the end of the year was ravaging cities on the east coast of North America, before affecting populations from New Orleans to San Francisco – as well as reaching Constantinople, where half the population fell ill. Within four months, the virus had circumnavigated the globe, killing about 1 million people.[146] A new hypothesis has suggested that the pandemic might in fact have been the result of a coronavirus, rather than influenza; whichever it was, the ways in which human populations lived and travelled provided ideal routes for the pathogen to spread around the world.[147]

Concerns about the spread of disease, political fragility and economic dislocation prompted rallying calls for international co-operation. At a conference on plague prevention in 1911, for example, the Qing Viceroy of Manchuria, Xi Liang, urged delegates to work out how to develop safeguards that 'will result in the saving of human lives, not only of this country but of others also, should this terrible disease unfortunately break out elsewhere'. Efforts to implement strategies to identify infectious disease and prevent its spread became a major focus for policy planners and non-profit organisations in other parts of the world, as well as for international collaboration: the ill-fated League of Nations set up an agency in 1924 that was dedicated to health organisation, while the Rockefeller Foundation invested more money in China in pursuit of disease prevention than any country outside the United States.[148]

This attention was partly a response to the most celebrated pandemic of the last century, the Spanish flu of 1918–20, which was likewise able to piggyback on a particular set of conditions that were as advantageous for the virus as they were disastrous for the people it infected. Estimates of the numbers who died from Spanish flu are wildly divergent, with some suggesting a figure of around 17 million and others suggesting that it may have been as high as 100 million.[149] The mass demobilisation of troops following the end of the First World War, combined with low calorie consumption as a result of four years of conflict in Europe, Asia and Africa which reduced natural immunity, helped spread the disease around the world.[150]

Poor hygiene levels in hospitals treating infected patients made matters worse still, and it has been suggested that a large number of fatalities – perhaps even the majority – resulted from secondary bacterial pneumonia caused by common upper respiratory-tract bacteria.[151] Location mattered too: mortality rates in cities were high, partly because populations living in close proximity provide obvious vectors for transmissions. But death rates were also conspicuously high in cities that burned more coal; those that did experienced tens of thousands more fatalities than those that did not.[152] Urbanised, industrialised cities proved to be more fragile and vulnerable, in other words, than places that were less connected.

This did not just matter in the short term, for there were long-term consequences, as recent research on the Spanish flu has shown. The disease had profound effects on infant mortality in Brazil, for example, where it also significantly altered gender ratios.[153] Evidence from the United States surveying outcomes over many decades shows that the cohort of children *in utero* in 1918–20 had increased physical impairments, lower incomes and reduced educational attainments compared to those born before or after the pandemic.[154] No less strikingly, civic responses to the Spanish flu were closely linked to voting patterns and political extremism in Germany: areas most affected by the pandemic, and where per capita spending by local authorities was low in the following decades, saw markedly higher numbers voting for Hitler and the National Socialists in the early 1930s.[155]

Spanish flu caused the highest number of known deaths in history from a single pandemic. It has long been suspected that a key factor in its potency was the use of chlorine gas on the battlefields of Flanders and northern France, which either caused or accelerated the mutation of a virus that had originated in Asia and been brought to Europe by Allied troops.[156] It seems, however, that a series of climate anomalies in the form of torrential rains and depressed temperatures – especially during the autumns of 1917 and 1918 – also played an important role. Historians have noted that the adverse weather had a material bearing on the outcome of battles during the First World War, such as flooding of trenches, cessation of operations because of waterlogging and substantially increased casualty numbers. Heavy rains and cold were the result of an unusual spike in cold, wet marine air from the North Atlantic – the highest concentrations in a century. These also provided ideal conditions for the survival and replication of the virus in Europe, both by enhancing infectivity in pools of water and perhaps also by interrupting the migration of birds and ducks whose infected droppings contaminated water sources and helped magnify the spread of disease.[157]

The Spanish flu was a product of the suffering of the First World War that had itself left as many as 20 million dead; a result of military hostilities in Europe, Africa and Asia and also because of chronic food-supply crises – above all in the German, Ottoman and Russian empires. Russia had been the source of around a third of global wheat supply before 1914, so the closure of the Bosporus inevitably put a strain on economies around the world, exacerbated by challenging climate conditions: at the start of the war, unfavourable weather led to shortfalls from major producers such as Canada and Australia, which had both suffered from drought, the latter yielding barely a quarter of its usual output; the crop in South America was unexpectedly poor, while an early winter freeze in the United States produced transportation delays west of the Great Lakes. This prompted panic in London, where cables were sent to India urging the Viceroy to boost exports to maintain supply – and to keep prices down, as farmers and speculators in the United States in particular were already moving in to take advantage of the squeeze.[158]

The demands of the military for manpower made matters worse, with declines in grain production in Russia closely correlated with the military draft. The thinning out of the labour force brought about declines in the regions where crops were grown, and commensurate falls in yields. Peasant farms additionally reacted to conscription and to the uncertainties generated by wartime by refusing to sell grain, which deepened food crises and ultimately resulted in shortages and queues for bread. All this helped lay the ground for revolution, not least by highlighting the ineptitude of the tsarist government.[159]

In some cases, most notably Central Asia – a region that was poorly integrated into the Russian empire – ineptitude was closely linked to abuse of power by local officials responsible for overseeing conscription, which in turn led to popular dissatisfaction and then to a series of open revolts.[160] While these are often seen by modern scholars as the seeds that grew into the Russian revolution, a more useful and compelling way to understand the uprisings is as part of wider contemporaneous resistance movements in South Asia, North Africa, West Africa and Ireland and beyond that aimed to throw off colonial rule.[161]

One reason for this was the insatiable demands made on colonial societies by those who controlled them and drew on resources in ever greater volumes in increasingly interventionist ways, with mistreatment of local populations proving even more antagonistic than before the war.[162] In France, more than 350,000 hectares of forest were cut down during the First World War in order to provide wood for trench building, camps, fuel and more – which had obvious implications for ecological systems in France and for the pressures on its colonies to make up six decades' worth of tree harvests.[163]

African copper had already electrified European cities by providing millions of kilometres of cabling to help power everything from elegant soirées to factories churning out products. It now helped deliver death, destruction and ecological damage on an unprecedented scale: copper was sucked in from fields that had been opened up and exploited in Central Africa as well as elsewhere to create casings for the 1.45 billion artillery shells fired in western Europe alone, not to mention the copper and lead required for countless rounds of rifle cartridges and small-arms bullets. Their use brought about soil contamination in the killing fields of northern Europe that is still detectable today.[164]

These resources were all secured by colonial officials who thought it was their right to exploit and extract what was needed to undertake a war between European powers based on hypercompetition between small groups within the ruling elites. Millions of men were enlisted from India, from Africa and from Indo-China, not to mention those who volunteered to serve in the armed forces of Australia, New Zealand and Canada. The pressure on the colonies to serve their masters acted as a catalyst for revolutionary and anti-colonial movements which had already begun to communicate and co-operate through networks ironically flourishing in precisely the cities and ports that had fostered globalisation and cemented the economic power of the industrialised world.[165]

As the US President Woodrow Wilson put it in 1917 when trying to justify staying out of the war, the United States was 'the only one of the great White nations that is free from war today', and it would be 'a crime against civilisation' for that to change. In any event, as he told one key adviser, it was essential for the US to stay out of the conflict so that when peace eventually came it could help rebuild 'the nations ravaged by war'; this was vital, he said, so that 'white civilisation and its domination of the world' should continue unchallenged.[166]

Wilson was hardly alone. Max Weber, whose belief that agricultural productivity was low in less developed countries because peasant farmers were 'irrational' proved highly influential with economists and policy planners in the 1950s and 1960s, wrote that 'a dross of African and Asiatic savages and all the world's rabble of thieves and lumpens' stood armed to fight on the western front in 1917 – without mentioning that many of them were about to give up their lives thanks to the whims and preconceptions of a set of emperors and leaders obsessed by ideas about the balance of power in Europe.'[167] Even in the 1930s, the Governor General of the Netherlands East Indies said bluntly that 'We have ruled here for 300 years with the whip and the club, and we shall still be doing it for another 300 years.'[168]

In the context of nature, the environment and climate, such views were important indications of a taxonomy and ranking that placed white Europeans at the apex of humanity; but they also represented a conceptualisation of the natural world as being at the disposal of those who saw themselves as best suited and most entitled to use and control the fruits of the land, agricultural and mineral alike, while also being their most capable guardians. The First World War is often seen as marking the end of empires and the beginning of the end of colonialism. Much depends on the eye of the beholder, for when it came to nature, the environment and anthropogenic climate change, the century that followed witnessed a profound acceleration and intensification in the way that resources were extracted, consumed and used. The consequences of the twentieth century were to shape the fate of the twenty-first – and of the future beyond.

21

Fashioning New Utopias
(c.1920–c.1950)

It is imperative to preserve the German landscape, for it is, and always was, the ultimate foundation of the power and strength of the German people.

Adolf Hitler (1936)

Judged as a whole, the century that followed the start of the First World War was a sequence of catastrophes unparalleled both in human history and in that of the natural world. Millions of people were killed as a result of wars that were conceived, or rather misconceived, in courts, chanceries, presidential offices or revolutionary camps. Millions were killed as a result of persecution and hatred founded on grotesque preconceptions about race, religion or ethnicity. Millions died as a result of famine, deliberate starvation or lack of access to basic healthcare. The suffering of the last hundred years or so has been by far the greatest in recorded history in terms of its scale and its horror.

The heavy human costs were matched by those inflicted on the natural world. The last century was one of profound ecological change that ranged from the impact of mass urbanisation to the large-scale extraction of resources required to support, service and power populations living in cities of unprecedented size. It has been a period during which the consequences of how we live have been poorly understood or little thought about – with the result that environmental and climatic changes of the present and future are being and will be shaped by what has already happened in the past, rather than simply by decisions made today.

Much of this has been driven by higher levels of exchange within and between different regions and continents and by an intensification of globalisation, whose roots long predate the modern era. New technologies provided the basis for a sharp acceleration from the eighteenth century onwards, with mechanisation, industrialisation and improvements in speed and cost of transportation by rail and by sea all providing gains that helped tie the world together. All depended on minerals like coal, iron, copper and steel, which in turn attracted attention to places holding reserves that could be exploited.

These often lay in locations that were difficult to reach or were not under the control of those most motivated to gain access to them. Oil provides one obvious example. The value of petroleum became obvious in the later part of the nineteenth century as an energy source that was easier to extract and cheaper to transport and store than coal, as well as having a higher energy density. Oil – and gas – did not replace coal, but rather served as a supplement long into the twentieth century, only displacing the latter from the 1970s as coal-fired steam engines gave way to petrol-powered combustion engines, as oil and gas heating took over from coal ovens and as hydrocarbons became a primary source of electricity production.[1]

Oil strikes in Pennsylvania and then by the Caspian Sea provided the spark for a scramble to establish further commercially viable wells in these and other locations around the world, with attention falling on Burma and the East Indies and then on Persia following a massive series of discoveries from 1907. These each provided the seeds for personal fortunes and for the rise of giant oil companies, including Standard Oil, Burmah, Shell and BP, and other operators such as the Nobel Brothers, which ultimately served as a source of wealth for Alfred Nobel and the prizes that bear his name.[2]

Production rose dramatically in response to soaring demand that was itself fuelled by innovations such as the motor car. In 1900, there were just over four thousand vehicles on the roads in the United States. Twenty years later, that had risen to almost two million.[3] Production was dominated by Model T Fords, whose success was based on streamlined, standardised design and a moving assembly line that allowed costs to be driven down and affordability up.[4]

By this time, ensuring reliable supplies had become a major area of concern for countries in Europe, whose geological and climatic history had left a continent more or less bereft of crude oil, with a few modest exceptions. Britain was particularly vulnerable, as leading politicians and officials realised. 'The only big potential supply', wrote Sir Maurice Hankey as the First World War was still raging, 'is the Persian and Mesopotamian supply.' As such, gaining 'control of these oil supplies', the Secretary to the War Cabinet went on, should be considered nothing less than 'a first-class war aim'.[5]

Oil shaped the history of the Middle East, and arguably much besides. There were other enormous resources, of course, not least in Venezuela, whose coastline is still home to the world's largest reserves.[6] But taken together it was not hard to see why the oil of Iran, Iraq, Saudi Arabia and elsewhere in the Gulf was being referred to in the 1940s by US government envoys as 'the greatest single prize in history'.[7] The lack of access to oil sources played a crucial role in the strategic unravelling of the Second World War, responsible both for the ill-fated German attempt to push south into the Caucasus to gain access to oilfields and for Japan's attack on Pearl Harbor in 1941, which was motivated at least in part by Tokyo's acute awareness of Japan's energy limitations and by its desire to have a free hand in securing oil and other minerals in resource-rich locations across South-East Asia.[8]

Oil and gas have been central to global geopolitics over the eight decades that have followed, shaping the rise of petrostates, helping pay for high-profile football teams and World Cup tournaments, Formula One races and eye-poppingly lucrative golf tournaments, while also being used as a source of oligarch wealth and latterly as a weapon of control for Russia in its relations with Europe and the west following the invasion of Ukraine in 2022.

Anxiety over energy prices has been a crucial factor in the United States' domestic politics as well as in its foreign, defence and economic affairs, whether relating to pipeline construction in North America, to permits for drilling new wells or to developing methods for unlocking shale which has helped turn the US from an importer into an exporter of all oil products, including crude and refined petroleum products.[9] The energy needs of India and China, both of which are heavily dependent on oil and gas imports, are not merely an important part of the long-term future of the world's two most populous countries, but are matters of such fundamental vulnerability as to determine their policies towards oil- and gas-producing states, as well as towards multinational companies.[10]

Oil was just one example of the competition for natural resources. Another was gold, a metal that exists on earth as a result of extraterrestrial impacts over the course of billions of years, and whose distribution in commercially viable quantities results from the haphazard points of collisions of comets, meteors and asteroids.[11] Gold has been highly prized since antiquity, a marker of wealth and status, and in the modern world is also valued for its low conductivity, which makes it useful for circuit boards and wiring in cars and household appliances, such as microwaves, refrigerators and stoves.

Gold mining is often both inefficient and highly polluting: twenty tons of soil and rock are typically required to produce enough gold to create a single ring – with much of the waste contaminated with the cyanide and mercury that are used to extract gold.[12] Mining activities in the Grasberg minerals district in Papua in Indonesia, home to the world's largest proven gold and second-largest copper reserves respectively, had already discharged more than a billion tons of metal-laden tailings into the local river systems by 2006 – with this quantity increasing significantly as production has risen sharply since.[13] The scramble for gold has been instrumental in encroachments on indigenous lands in recent years, such as in the Amazon, but previous gold rushes also brought about sizeable changes in population movements, such as in California, Colorado, Siberia and Australia. These movements in turn led to transformations of the landscape as land was cleared (usually by burning), as towns, cities and transport infrastructure were built to support miners, as ores were deposited in rivers creating long-term problems of pollution, and as livestock were introduced as sources of protein, usually after native animals had all been killed and eaten.[14]

Another example is copper, whose usage and value soared in the early twentieth century and have continued to rise ever since; it will perhaps even replace hydrocarbons as the most sought-after material as the world transitions, or seeks to transition, to more carbon-friendly technologies. Copper, according to Goldman Sachs, is the 'oil of the twenty-first century', although this label has been given to lithium, to rare earths more generally and to hydrogen, as well as to non-natural resources such as data.[15] The importance of copper can be seen in the fact that the average US-built car contains more than twenty kilos of copper, while a typical electric car requires six times the mineral inputs of a conventional vehicle.[16]

Demand for copper encouraged large-scale investment in regions and locations rich in reserves, not least the copper belts of Central Africa, where some reports in the early twentieth century suggested the presence of 'a quantity of minerals that one can practically consider inexhaustible' and of exceptionally high quality to boot. The extraction of copper and of other metals provided the impetus for railway construction in Africa, with lines connecting to port cities on the coast. It brought social change too, because manpower requirements meant that large numbers of young men were drawn into copper-producing regions with obvious consequences for gender roles in societies with high levels of male outward migration, for agricultural production which required higher levels of female participation and also for ecological systems: mining and processing resulted in enormous waste heaps, liquid-suspended tailings and the discharge of particulates into water systems, as well as substantial amounts of sulphur dioxide. All of this resulted in major damage to soil, animal and plant life and helped create virulent disease environments.[17]

The cutting down of forests for lumber for boilers, for poles to support mines and for railway sleepers did much the same, with the four largest mines in Central and southern Africa alone requiring almost 57,000 cubic metres of sawn timber per year. Furthermore, the demand for protein to feed workers helped devastate aquatic ecosystems, as fish stocks were heavily depleted by overfishing, which was facilitated by methods and techniques imported from Europe that did enormous damage.[18]

Copper demand was closely correlated with the global economy, which naturally meant that jobs and livelihoods were impacted by themes, trends and events that took place far away, out of sight. One example was the Great Depression, which had an effect on employment and settlement in Central Africa, leading to job losses as demand in other parts of the world collapsed. Another was the shock of the 1970s, when the oil crisis provoked not only sharp falls in demand for copper as economies contracted sharply and suddenly but a debt crisis in Zaire and Zambia, both of which were heavily dependent on copper revenues: the result was pressure to boost production at a time of falling prices to try to meet international obligations. Not surprisingly, this too had catastrophic environmental consequences.[19]

The list of minerals, metals and materials that were dragged towards the rich world of Europe and its settler offshoots, above all in the Americas, was and remains – a long one. Rich, developed economies depend on vast quantities of natural resources at the lowest possible prices. Europe could not have been rebuilt as quickly nor as cheaply following the catastrophic damage of the Second World War without access to plentiful supplies of copper and other minerals from Africa. Cheap oil and gas have been the lifeblood of economic growth, rising living standards and mass consumerism, with the benefits felt most keenly by societies and states which had industrialised early and which had built institutions that protected investment.

Uranium from the Shinkolobwe mine at Katanga in southern Congo was far richer than any other source in the world – producing uranium oxide whose quality assayed as high as 75 per cent (compared to 0.02 per cent from ores in the United States and Canada). So important was Shinkolobwe to the US nuclear programme, and consequently to the Cold War, that it was deemed to be essential to US national security. Congo as a whole 'offers natural resources of extreme importance to our domestic economy', noted the officials and engineers who worked on the Manhattan Project; they went to great lengths to conceal the origin of the uranium used in the first successful atomic bomb test, claiming it derived from Canadian sources. As a report of 1951 declared, uranium from Congo was of 'capital importance to the free world'.[20]

If peace, freedom and prosperity hinged on being able to secure a wide variety of resources it was notable how few were available domestically, especially in Europe. It was apparent too that, as the benefits of extraction flowed into the developed world, states that were resource rich often had little to show for their natural wealth. This was in part the result of what has become known as the resource trap, whereby countries with oil, gas and mineral resources have either become or remained authoritarian, with limited distribution of rights and equalities through the wider population. The tendency for commodity prices to fluctuate dramatically from year to year also means that revenues can be unpredictable, dragging economies through boom-and-bust cycles, which can result in debt crises that further affect institutional development. Resource-rich governments typically underspend on health, education and social services, and typically overspend on government salaries, on fuel and food subsidies and on large monuments.[21]

One problem was that investment was concentrated around regions and facilities that were important for extraction rather than for social development. For example, Congo, home to such rich reserves of copper and uranium, and also of cobalt, tin and gold, became home by 1960 to eight international airports, thirty major airports and another hundred lesser airports, thanks to the mining industry – but there was no parallel construction of hospitals, schools and infrastructure that benefited the local population. Bribery of local officials was an intrinsic part of gaining access to reserves with the least resistance, in the fastest time and at the lowest possible price.[22]

It was significant too, of course, that many places that were rich in resources – in Africa but also elsewhere – had to deal with the consequences of colonialism, which had left patchy and sometimes poor systems of government in place. These had themselves succeeded colonial social and political structures that reduced human capital, limited capacity and made governance difficult, not least because of the way that states and provinces were set up according to the vagaries of colonial control.[23] In many locations, such as in West Africa, pre-colonial pasts also played a significant role in influencing the period after independence.[24]

But there is another important element. Although the age of colonialism drifted away in notional terms as states became independent in the course of the twentieth century, the lure of accessing the fruits of the earth proved impossible to resist. Interference in the domestic affairs of independent states was something that has defined the last hundred years or so, with obvious examples ranging from US engagement in Iran and the Middle East to intelligence operations that supported and in some cases carried out political assassinations in places as diverse as Congo, Chile, Central America and South-East Asia.

The issue was not always about natural resources, as the context was often provided by the Cold War, or by geopolitical hypothesising about the need for friendly and reliable leaders and the removal of those who were hostile. But often there was a desire to maintain control over minerals, fuel sources and even vegetation, plants and food. For example, the introduction of the banana to the US in the latter part of the nineteenth century helped spur soaring demand for a tropical fruit that was nutritious and easy to transport, with the additional advantage of continuing to ripen after being picked. The popularity and profitability of the banana trade led to US fruit companies acquiring and transforming hundreds of thousands of hectares of land in Central America and the Caribbean into plantations. In doing so, they secured economic and political powers that were so overwhelming that the term 'banana republic' was coined as a shorthand for states that were highly dependent on a single crop and whose governments and officials were in hock to the major producers.

Such was the control over the levers of power wielded by companies such as the United Fruit Company (UFC) that a familiar record emerged by the early twentieth century of corporations successfully deposing governments and installing friendly political leaders who could be trusted to protect and further the interests of investors. This was what happened in Honduras, where President Dávila in 1911 was overthrown in a coup led by the memorably named Lee Christmas, an American mercenary, who had been hired by UFC. Lavish rewards followed, including land concessions, the rights to build ports, railways and buildings, and a cash 'reward' for helping engineer the revolution.[25]

During the Cold War, interventions and coups enjoyed support from, and were sometimes co-ordinated by, the CIA, as in Guatemala. UFC owned millions of hectares there by the Second World War, was the biggest landowner and largest employer in the country and generated annual profits twice those of Guatemalan government revenue.[26] Low salary levels, often paid only on a seasonal basis, were combined with high levels of systemic racism – with all non-white workers obliged to yield right of way to white people and to remove their hats while talking to them. Indigenous workers, said Sam Zemurray, President of UFC, were 'too ignorant' to translate dissent into any meaningful form of resistance. His confidence that UFC could afford to be complacent proved to be misjudged when major land reforms were introduced to help local farmers who were unable to access or use UFC's vast land bank – of which 85 per cent lay uncultivated.[27]

Concern about what this meant for UFC's business model, its valuation and its investors led to the US government being lobbied and issued with stark warnings that the country would become a Soviet satellite unless action was taken. This prompted the CIA to carry out Operation PBSUCCESS in 1954, which deposed President Árbenz Guzmán and installed a dictator in his place. This helped trigger a civil war that lasted for almost four decades, during which a series of military leaders were kept propped up by support, weapons and money disbursed from Washington.[28]

While important individually, these examples – and the many more that could be cited alongside them – found their significance in the role they played in the rapid modernisation, industrialisation and standardisation that drove change of all kinds in the twentieth century. Without raw materials, patterns of urbanisation could not have emerged as quickly as they did, nor those of long-distance travel, nor ultimately the creation of weapons of mass destruction. They had a climate context too, though, for the growth and spread of industry brought about rises in levels of manufacturing, production, exchange and consumption over the last hundred years – and also in emissions of carbon dioxide, greenhouse gases and aerosols that have formed a central part of anthropogenic climate change.

One of the most striking periods of accelerated warming took place between the 1890s and the 1940s. Although data for many key regions around the world is either limited or non-existent, a robust pattern emerges of a phase often referred to as the Early Twentieth-Century Warming. This period saw anomalies and failures in the Indian monsoon system in the first years of the century, pronounced Arctic warming and ice retreat in the 1920s and 1930s, heatwaves in North America that were broadly contemporaneous, drought in Australia in the late 1930s and 1940s and unusually cold winters in Europe in 1940–2. Despite the obvious limitations of the data, there was evidently a rise in mean surface air temperature that was particularly prominent over high latitudes in Europe, the Atlantic, the North Pacific and Canada.[29]

The global temperature rise points to a significant change in the energy budget of the earth's atmosphere, resulting not from a single cause, but from complex interaction between a number of factors. The lack of major tropical volcanic eruptions in the first half of the twentieth century was significant, for example, with no major ejection of rock, ash and particles into the atmosphere to induce cooling. Changes to solar irradiance, which appears to have increased marginally in this period, may also have played a role, although assessing this accurately is not easy. Decadal variability in the climate system, with the Atlantic Multidecadal Oscillation (AMOC), the Pacific Decadal Oscillation (PDO) or both, might also have been a cause of warming patterns.[30]

Reconstructions of hemispheric conditions point to another factor in this period of pronounced warming: increases in greenhouse gases. Indeed, these models suggest that a significant proportion of the global climatic change up to the middle of the twentieth century resulted from anthropogenic activities, with the fingerprint of human activities most notable because warming was stronger over land than over oceans, and moreover was enhanced in the Arctic – precisely what one would expect, in other words, from industrial production and rising emissions of greenhouse gases and of fossil-fuel combustion.[31]

These in turn arose from the changes in the ways people lived, worked and communicated. Of course, these changes did not take place uniformly around the world, but in some cases, they were dramatic. In the United States, for example, the size and number of cities grew rapidly as transport networks and innovations increased speed of travel, cut shipment costs and enabled production to take place further from customers. The mileage covered by national railroad networks rose almost sixfold from around 30,000 miles in 1860 to more than 160,000 thirty years later; the size of locomotives more than doubled in the same period, allowing greater loads to be pulled, while their speed increased substantially from around 12 mph to closer to 60. This was matched by the extraordinarily rapid rise of communication networks that helped co-ordinate trade flows, allowed for price convergence and stimulated economic growth: in 1860, the United States was covered by around 50,000 miles of telegraph wires; by 1890, that had risen to nearly 20 million miles.[32]

This helped spur changes in the nature of demographic distribution, not only by linking regions more closely together, but by altering the nature of towns, turning them from distribution points for agricultural products into industrial centres in their own right. Although it is significant that the number of cities in the United States more than doubled in the years 1880–1920, more telling were the changes in the sizes of cities and in demographic distribution as rapid and large-scale urbanisation took place in the same period: cities did not just become more numerous but acquired larger and larger populations, with sixty-eight cities having more than 100,000 inhabitants by 1920, compared to twenty just four decades earlier. While around a quarter of the people in the United States lived in cities in 1880, more than half did so forty years later.[33]

These dynamic transformations ensured considerable social change as the labour force moved out of agriculture into manufacturing. Innovation, efficiency and investment in new technologies and infrastructure turned the United States into an economic powerhouse, one where by 1900 labour productivity was twice as high as in Britain.[34] One reason for this was mechanisation and the rise of factories, which replaced artisanal and craft workshops for everything from textiles to ironmongery. The reason for the growth of cities was a complex one, linked to a range of factors including the development of the steam engine, immigration and domestic migration, and moreover in the United States such growth was neither geographically nor chronologically consistent.[35]

The transformation of the US paralleled similar processes in England, continental Europe and Japan, where railways, industrialisation, urbanisation and price and wage convergence likewise formed central themes of the later nineteenth and early twentieth centuries.[36] Dramatic though they were, those processes pale by comparison with what happened in Russia and the Soviet Union following the revolutions of 1917 and the seizure of power by Lenin and the Bolsheviks. The tenets of classical Marxism not only emphasised the importance of the urban proletariat in class struggle and the creation of a new political system, but also demanded development of the means of production. As such, in addition to being an authoritarian state, the Soviet Union was centred on top-down planning and the development of massive projects that came at profound human and ecological cost.

Long before the revolution, Lenin had argued that the key to turning Russia into a Communist model state was mechanisation and the provision of mass electrification, which he had written about at length while living in exile in Siberia in the late 1890s.[37] The nascent Soviet Union, he warned, would remain a petty, agrarian state based on peasant smallholdings unless dramatic action was taken. It was vital, therefore, to modernise the country and to do so rapidly. The success of the revolution itself rested on electrification: it might take years, but electrification would guarantee against 'any return to capitalism'. Put simply, he said in a report of December 1920, Communism 'is Soviet power plus the electrification of the whole country'.

Electrification was part of a wider modernisation that would rid the Soviet Union of 'the illiterates we have', because 'we need cultured, enlightened and educated working people'. For this reason, he argued, 'our best men, our economic experts' had been brought together to make a plan that 'indicates how many million barrels of cement and how many million bricks' would be needed to carry out an ambitious plan for the State Commission for Electrification of Russia – known by the acronym GOLERO.[38]

This was a statement of intent that presaged a series of monumental
construction projects conceived to fulfil ideological dreams and to
satisfy political dogma. Many caused massive ecological damage.
Vast tracts of forests were cleared, often wastefully and almost always
unnecessarily, with supply of wood far outstripping demand; copious
amounts of pesticides were used in agriculture, with rates deployed
often three to five times greater per hectare compared to Europe
and the United States; river after river was dammed from the 1920s
onwards to enable the construction of more and larger hydroelectrical
power stations to provide the power and water needed for the
production of the metals and resources essential to the creation of a
fully industrialised state.[39] Trofim Lysenko, an agronomist who found
his way into the corridors of power in the first decades of the Soviet
Union where he promoted views that were as controversial as they
were influential, liked to quote as his motto words attributed to the
plant breeder Ivan Michurin: 'We cannot wait for kindnesses from
nature; our task is to wrest them from her.'[40]

This was a view shared by Trotsky. 'Man has already made changes
in the map of nature that are not few or insignificant,' wrote the
revolutionary idealist. These were nothing compared to what could,
or should, be done. The time had come to cut down mountains and
move them, to divert rivers and to 'earnestly and repeatedly make
improvements in nature. In the end, [man] will have rebuilt the earth,
if not in his own image, at least according to his own taste. We have not
the slightest fear that this taste will be bad.'[41]

Electrification and resource allocation were essential too for the
transformation of cities, of which Moscow was the most notable
example. In June 1931, the Central Committee of the Communist Party
announced a major overhaul of Moscow's urban landscape, to include
the construction of a subway system, modern sewerage works, street
lights, parks and new buildings that could showcase the triumphs of
the working class and of the Soviet Union. The metro, noted Lazar
Kaganovich approvingly, 'is a symbol of the new society that is being
built', because 'into each step of the escalator penetrates the spirit
of new man, our socialist labour, there lie our blood, our love, our
struggle for a new man, for a socialist society'. He was not wrong, for
the subway, like many of the other works undertaken in Moscow and
elsewhere in the USSR, was built by political prisoners, class enemies,

intellectuals, Jews and others who fell foul of this new, self-serving, self-satisfied utopia.[42]

Everything – and everyone – was deemed to be secondary to the progress of the Soviet state. Coerced labour – derived from the gulag population that was swelling to enormous size thanks to denunciations, fear and vindictiveness – was deployed to build a canal linking the White Sea with the Baltic, a project that cost tens of thousands of lives, or to find and exploit gold and other metals in the Arctic north. Over a million prisoners were sent to Magadan, on the Sea of Okhotsk in Russia's Far East where temperatures as low as –37 °C have been recorded; labour was then distributed into penal colonies in Kolyma and Chukotka to claim the fruits of the land in much the same way as Orthodox bishoprics had once sent out missionaries to claim the souls of those yet to be enlightened by the Christian faith.[43]

Enforced collectivisation of farming under Stalin was supposed to improve yields by 20 per cent as a result of increased crop planting and the use of the latest agricultural technologies. The lack of co-operation from a peasantry resistant both to new methods and to coercion, coupled with requisitioning and repression, created circumstances that resulted in famine across Ukraine and southern Russia which claimed astonishing numbers of lives – perhaps as many as 8 million in 1932–3; life expectancy for men born in Ukraine in 1932 was thirty; for those born in 1933–5, it was just five.[44] The victims were disproportionately those of Ukrainian ethnicity, a reflection of bias in policymaking in the Soviet Union – a significant feature given the invasion of Ukraine by Russia in 2022.[45]

While there were many strands to repression in the Soviet Union, one constant was the conceptualisation of the natural world as something to be tamed, whose transformation represented the ingenuity and industriousness of the proletariat. Nature offered tools that could and should serve the construction of revolutionary society. The granite, marble, iron and steel needed to improve cities for the benefit of the urban proletariat were all hacked, hewn and mined from places that were often geographically remote and always ideologically light years away from where they were used – with little thought paid to the environmental costs of extraction. 'When we have seated the USSR on an automobile and the peasant on a tractor,' Stalin proclaimed in 1929, 'let the esteemed capitalists ... try to catch us up then. We will then be able to see which countries can then be "classified" among the backward and which among the advanced.'[46]

Central control and demand for implausibly high volumes of materials in impossibly short time frames led to exaggerations about production levels (which created a vicious circle by suggesting unrealistic results were possible) and to short cuts in connection with pollution, with the disposal of hazardous waste and with the health of miners and builders, whether forced to work or otherwise. For example, in 1928, Soviet authorities claimed that more than 73 million tons had been harvested – a figure likely overestimated by millions of tons. In any event, tens of millions of cubic metres of earth were moved annually by hand or by machinery, and millions of hectares repurposed every year.[47]

It would not be true to say that there were no concerns about the damage inflicted on the environment, or that no steps were taken to preserve and protect the natural world. As early as May 1918, just a few months after the Bolsheviks had seized power and as the former tsarist empire was dissolving into civil war, the new leadership passed legislation on the Basic Law of Forests. While the methods of enforcement were top-down, rather than bottom-up – put bluntly (and approvingly) by a contemporary forestry journal as aiming to 'centralise the forest economy of the country, dictate its will to the localities, and demand from them absolute economic obedience' – preservation was taken seriously in some areas by some people.[48]

There was concern among some agronomists that Soviet practices in forest preservation were resolutely antiquated. 'Remnants of tsarist practices are blocking the path to progress,' declared one article in the leading journal published in the USSR on woods and forests. While new approaches were transforming German attitudes to the landscape, the Soviet Union in the mid-1920s was characterised by 'obsolete concepts, technical backwardness, inertness, and a formulaic quality out of step with new German forestry ideas'.[49]

It did not help that there was bitter and near constant rivalry between the People's Commissariat of Agriculture and the Supreme Soviet of the National Economy about who had the right to set timber targets as well as policy in general.[50] There were disputes too about sustainability: a report in 1926 complained that 'fiscal goals overwhelmingly dominate our forest management', with too little thought being given to the effects which would leave a mark for 'fifty to eighty years'.[51] So what? retorted Fedor Syromolotov, a senior Communist figure who was a member of the Supreme Soviet of the National Economy; 'if all the forests' in regions like Smolensk, Tver, Novgorod and elsewhere 'were to be cut down, then these regions could be converted into areas suitable for the development of grain crops'. No one would suffer as a result, he went on; on the contrary, such a transition would be liberating.[52] Nevertheless, by the early 1930s nearly 3 million hectares of forests had been protected as reserves.[53]

By this time, concern about water led to the creation of an enormous forest preserve in the European part of Russia which became the largest in the world at the time. Cutting down trees within a twenty-kilometre belt along the Dniepr, Don, Volga, Ural and Dvina and multiple tributaries was made a criminal offence, with similar though smaller corridors alongside several other tributary rivers.[54]

Despite these efforts, Soviet leaders and thinkers had been clear from the outset that technology would ensure that the Soviet Union would succeed partly through the 'conquest' of nature. The White Sea canal was hailed by Maxim Gorkii, for example, as a victory of the people of the Soviet Union, not over capitalism or rivals abroad, but 'over the elements of the harsh nature of the north'. Gorkii chose not to mention that victory came at the price of tens of thousands of deaths of labourers who were made to work under the slogan 'We will instruct nature – and we will receive freedom'. Faith in technology, exaggerated support for decisions made by high-ranking party officials and the obvious consequences of dissent meant that there was little or no pushback about the pollution, increased salinity, loss of topsoil or other disastrous consequences of taking shovels and machines to change the environment to conform to the role that political tracts, textbooks and technocrats insisted it should play.[55]

Such problems were hardly confined to authoritarian states led by despots with effectively limitless powers. In Europe, the First World War had had a dramatic impact on consumption patterns. Before 1914, around 60 per cent of the energy value of the British diet came from imported foods. Although food shortages were rare and of relatively short duration during the war, there were significant shifts in what consumers ate, and were able to afford, with large increases in the consumption of sausages,

bacon, margarine and condensed milk, and falls in consumption of
sugar, butter, fruit and vegetables. Interventionist government policies
concerning agricultural production and food requisitioning were made
necessary by the war, not least by German U-boat activity leading to the
loss of around 2 million metric tons of merchant ships by the autumn
of 1916. These brought about significant changes in the use of the land,
in food availability and in attitudes to nutrition, which continued after
the war had ended.[56]

One consequence was an attempt to increase taxes from all parts
of the empire after 1918 to help fill a treasury that had been heavily
depleted by war. Another was the effort that went into boosting food
availability by opening up more and more farmland.[57] Yet another was
to try to create a circular economy by which people in Britain were
encouraged to buy from producers and farmers who were local or at
least based in the British empire: the Empire Marketing Board was
set up in the mid-1920s to urge customers to think about where their
food came from. 'Buying Empire goods means buying the produce
of your own Country and of the Empire Overseas, instead of the
produce of foreign countries' ran one early advertisement in a 1926
newspaper campaign headlined 'Message to the Shopping Public –
British First!' Customers were told, 'Every time you buy Canadian
salmon, Australian fruit, New Zealand lamb, South African wine,
Indian tea, you are dealing with the very people who go out of the
way to spend money on the goods made in your own country, and so
to create employment, pay wages and increase prosperity here.' Before
they buy, customers should always 'Ask – Is it British?' The royal family
were enrolled to support these efforts, George V and Queen Mary
expressing their personal sympathy and allowing it to be known that
their Christmas pudding in 1927 was made solely with ingredients
sourced from within the empire.[58]

Great effort therefore went into boosting agriculture in British
colonies overseas – with predictable results. Colonial administrators set
great store by the plough, identifying this piece of equipment as the
means by which Africa and Africans could be taught how to 'improve'
their lands and modernise, and what is more, in doing so to follow
European norms in terms of patriarchal nuclear families.[59] Needless
to say, rather than developing bountiful new fields and leading local
populations into a utopian state of rapture, the massive expansion of

ploughing did almost exactly the opposite: in one district in Uganda, the number of ploughs rose from just under 300 in 1923 to more than 15,000 in 1937. The results included high levels of erosion and gradual degradation of ecosystems that had previously regenerated under the light if time-consuming cultivation of the hoe. In West Africa, the experiment was so disastrous that British agronomists concluded that traditional methods of sowing and harvesting were far more effective than European methods after all.[60]

The introduction of unsustainable practices was hardly restricted to the colonies. Large swathes of the great plains of the United States had been converted to land for cash crops and for ranching in the latter part of the nineteenth century and in the early twentieth, spurred by the spread of railways, the availability of farm machinery and by high commodity prices following the outbreak of the First World War. All good things come to an end, however, and there were risks to agricultural booms – as farmers in Europe had learned the hard way in the 1870s: then, rising yields, innovation in fertiliser and soil chemistry and the availability of imports brought about a price correction that saw oversupply and a halving of prices two decades after their peak; the result was a spiral of falling land values and stagnation, most notably in Britain.[61]

Things turned out much worse in the United States as the First World War finally came to an end and, not long afterwards, so too did the American farming golden age. Encouraged by cheap credit, booming markets and expert advice, farmers had aggressively expanded wheat, corn, beef and pork production, betting on rising demand and ever higher prices. The Wall Street crash of 1929 dealt a shattering blow: prices plummeted, leaving crops worthless and making it impossible for many to meet crippling debt obligations. In 1932, half a million farmers across the midwest went on strike in protest, blocking roads with telegraph poles, deliberately letting milk and grain spoil and threatening to sell nothing to those living in cities. Franklin D. Roosevelt, the Governor of New York, had already observed what he called 'the dislocation of a proper balance between urban and rural life', which had inspired him to propose a new chapter in US politics. As he declared when opening his bid for the presidency, the time had come for 'something deeper and far more important for the future – in other words state planning'.[62]

The US farming crisis now went from bad to worse. The midwest experienced a decade of extreme heatwaves, drought and dust storms that devastated farmland, driving people into penury and hopelessness as banks foreclosed, inflicting further indignity and misery on farming communities – as described by John Steinbeck in *The Grapes of Wrath*. Dust storms lifted the topsoil that was already dry and exposed from lack of rain but also because the land had been worked almost to its limits in the preceding years. Although the 1930s were a time of warm temperature anomalies which deviated from the typical La Niña patterns that were part of the Early Twentieth-Century Warming phenomenon, reductions in vegetation and overexploitation of the land helped create a disaster.[63]

The result was a series of dust storms that one scholar has called 'the worst manmade environmental problem the United States has ever seen, whether measured in physical terms or by their human and economic impact'.[64] It was not hard to see why. Some storms were so intense that in the course of a single afternoon in April 1935 more than 300,000 metric tons of topsoil was estimated to have been whipped off the surface of the land – twice as much dirt as had been dug out to create the Panama canal over the course of seven years – leaving cities on the east coast covered in dust.[65]

Although the stark conditions which lasted for almost ten years eventually dissipated, the consequences were dramatic. By the 1940s, many areas in the Great Plains had lost more than three-quarters of their topsoil. There were significant population declines too as people moved away in search of better opportunities, or simply in order to survive. Not all areas were affected equally. However, it was not just farmland and farmers that suffered: counties which experienced high levels of erosion also saw higher levels of debt stress and bank weakness, which reduced the local supply of credit and impeded socio-economic recovery. The dust-bowl events did not just reduce land values at the time, but did so persistently and substantially, especially in the counties that were most severely affected.[66]

A report published in 1936, even before the disaster was over, had been highly critical in setting out the primary cause, which lay not in climatic anomalies and changes but in short-sighted and unsustainable human decision-making. Millions of acres of natural cover had been destroyed 'partly by over grazing, partly by excessive plowing'. Agricultural systems, techniques and expectations suited to humid conditions had been used on a semi-arid region, prompted by the high price of wheat during the First World War and the search for profits, made worse by innovations such as the tractor and the truck.[67] It did not help that the 1930s saw unusual weather patterns, including unusually dry springs and extreme heatwaves that brought record temperatures triggered by anomalous southward warm advection and continent-wide anticyclonic flows. But the dust bowl was an accident waiting to happen.[68]

The dust storms captured public imagination and that of the press. Columnists and science writers wondered whether the Plains would turn out to be the 'American Desert', echoing the conclusion of surveyors in the early nineteenth century that the High Plains were 'almost wholly unfit for cultivation' and might even serve as a barrier to the western expansion of the European population in North America.[69] One newspaper ran a column with the headline 'Erosion a World Problem', noting that 'abuse of the soil' by 'the unskilled farmer' had been devastating for many parts of the world, from Canada to Uganda, from Ceylon (Sri Lanka) to Australia. This mirrored articles that had appeared elsewhere in the world, such as in Britain, where there was considerable disquiet about the relationship between population growth, wheat and food production and degradation of the soil. Soil erosion, said Sir Daniel Hall in an issue of the *Journal of the Royal African Society* devoted to global conditions, was not a problem that was local or regional. It was a matter 'that affects the world, and in particular the British Empire'. As others writing in China, Japan and elsewhere observed, the problem truly was ubiquitous, and very serious to boot.[70]

These pessimistic and worrying accounts of the damage being done to the natural environment were part of a wider literature warning of the threat posed by humans in the hunt for resources and the pursuit of profit. The most influential and important work of this period appeared in 1939 with the title *The Rape of the Earth: A World Survey of Soil Erosion*, which not only compared land degradation with a disease, but put the blame for its spread squarely on the shoulders of Europeans and the 'European model' of trying to expand to the entire globe agricultural practices and methods that worked in one corner of it. 'Nature is in full revolt', wrote the authors, 'against the sudden incursion of an exotic civilisation into her ordered domains.' Land was being exploited 'beyond the limit of safety', they went on; the destruction of the earth's soils was 'proceeding at a rate and on a scale unprecedented in history.' Fertile regions were being turned into uninhabitable deserts.[71]

The authors, like others who produced similar tracts around this time, were hardly calling for anti-colonialism or for European settlers to withdraw from lands they had taken over, but rather the opposite: soil erosion was bound to make white territorial control more difficult given the problems it produced, so solving the challenge was essential if authority was to be maintained over potentially productive lands. It was 'increasingly clear', another author wrote around the time *Rape of the Earth* was published, that soil conditions in West Africa had become progressively worse 'as the influence of the white man has spread'. Before European colonisation, 'intertribal warfare and the ravages of disease' had taken their toll on 'man and animal alike', thus relieving pressure on the land. Now, peace brought by white settlers meant population growth, overuse of the soil, deforestation, overgrazing and other problems.[72]

By the late 1930s, some scientists were considering whether human activities might be producing other negative environmental impacts. Guy Callendar, a steam engineer, wrote a paper in which he noted that 'about 150,000 million tons of carbon dioxide' had been added 'to the air during the past half century'. The release of these emissions, he suggested, had resulted in the increase in the earth's land temperature, albeit by a small overall margin. Nevertheless, he concluded, this had obvious benefits, not least that the 'return of the deadly glaciers' that had plunged the world into Ice Ages in the past would now 'be delayed indefinitely'.[73]

The connection made by Callendar with emissions from fossil-fuel sources was not widely accepted by the scientific community, partly because of concerns about methodology and about the accuracy of the measurements he relied on.[74] There was – and indeed remains – vigorous discussion about the causes of the warming that characterised the first half of the twentieth century, and in particular about the nature, role and scale of putative anthropogenic factors.[75] Nevertheless, there were many others in the 1930s whose work pointed to warming patterns, in particular relating to the Arctic and changes that were both measurable and dramatic in Greenland and at Spitsbergen in Norway's Svalbard archipelago, where a meteorological station had been established in 1912.[76]

Amid concerns about climatic change, there were some who were thinking about conservation and preservation of the natural environment. In North America, for example, some rhapsodised about how 'tired, nerve-shaken, over-civilized people' were increasingly finding that 'going to the mountains is going home; that wildness is a necessity; and that mountain parks and reservations are useful not only as fountains of timber and irrigating rivers, but as fountains of life'.[77] National parks, wrote Stephen Mather, the first Director of the US Park Service, were 'not only show places and vacation lands but also vast schoolrooms' where people could learn 'to love more deeply this land in which they live'.[78]

Some of the most enthusiastic about linking national identity, happiness and nature were in Germany, where deeply held ideas about the protection of wildlife, of forests and of the 'right to wilderness' had taken root in the nineteenth century. These had fed into wider ideas not only about nature but about German identity itself: 'the love of nature is the root for the love of the fatherland', wrote Konrad Guenther in 1910; he later went on to argue that 'the chords of the German soul are tuned to nature'.[79] Nature, moreover, was a social equaliser: 'for only in nature is there no difference between poor and rich, high and low; only there does it cost nothing to gain a wealth of knowledge and happiness'.[80]

These idealistic views sharpened alongside poisonous ideas about racial 'purity'. 'Nature', wrote Hitler in *Mein Kampf*, 'knows no political boundaries. First, she puts living creatures on this globe and watches the free play of forces. She then confers the master's right on her favourite child, the strongest in courage and industry.'[81] Such comments delighted conservationists, who blended their views about the environment with Nazi political ideology. Conserving nature would help unite Germans, wrote Guenther, who was a vocal advocate of the notion that blond and blue-eyed people had a stronger attachment to nature than anyone else, and believed that making Germany ecologically pure would help with its racial cleansing too.[82] It was important, wrote Wilhelm Lienenkämper, not long after Hitler's government had passed a national conservation law, to set out the methodology for 'the protection of nature from a Nazi perspective'.[83]

'We will not only create a Germany of power', said Hitler, 'but also a Germany of beauty.'[84] This was why it was important, wrote Walther Schoenichen, a prominent biologist and later Director of the Department of Conservation, to eradicate 'foreign ways of building' inspired by the 'merchant's spirit' – a poorly concealed reference to Jewish capitalism. 'In the past the German soul always drew strength from German nature and the German landscape when it became necessary to overcome foreign influences and help Germanness break through again.'[85] Again, this seemed to chime with Hitler's own views. 'It is imperative to preserve the German landscape', the Führer claimed, 'for it is, and always was, the ultimate foundation of the power and strength of the German people.'[86]

In fact, things were not quite so straightforward. For all the words of encouragement, agronomists often squabbled with each other and had less influence on what was done than they wanted.[87] Moreover, the Nazis did little to protect nature: forestry expansion was minimal and often even counterproductive. While Hitler might have liked to talk about beauty and nature, he was more interested in overseeing massive building works in cities like Berlin and in the expansion of the autobahn network. He shrugged when asked about the impact on the landscape. Although beech forests should be protected if possible during motorway construction, they would ultimately 'have to yield to the demands of such a great technological project'. In fact, the costs of landscape protection amounted to 0.1 per cent of total construction expenditure – hardly the sign of high levels of environmental concern by the Nazi leadership.[88] That Hitler recommended physical training but did little exercise himself, walking gently downhill from his Alpine idyll before being driven back up the mountain by car, also suggests that his overall attitude to nature was one of ambivalence, if not downright indifference.[89]

There was little ambiguity, however, when it came to formulating ideas about the environment in the context of foreigners, of minorities or of plants. It was necessary, urged Reinhold Tüxen, head of the Central Office for Vegetation Mapping under the Nazis, to 'cleanse the German landscape of unharmonious foreign substances' – in this case, a small forest plant against which 'a war of extermination' needed to be fought. 'As with the fight against Bolshevism', he said, 'our entire occidental culture is at stake.'[90] This extended to new territories seized by the Germans during the Second World War. In an official edict issued in 1942, the Reich Commissioner for the Consolidation of German Nationality, Heinrich Himmler, set out 'Rules for the Design of the Landscape'. For territories captured in Poland, Himmler's order ran, 'it is not sufficient to settle all people in those areas and to eliminate foreign people. Instead, the area must be given a structure which corresponds to our type of being.'[91]

That 'type of being' had been in some German minds for a while. Long before the rise of Hitler and the Nazis, there had been suggestions that the defeat of Germany's neighbours would offer great opportunities for a major ecological transformation. Victory in eastern Europe during the First World War would open up the chance to replace 'clay huts and thatched and tiled roofs' with 'charming pictures of villages and little towns'. It would be possible to 'eliminate the urban-development sins of past decades'.[92]

Similar arguments accompanied the invasion of Poland in 1939. The east offered a chance, wrote the agronomist Heinrich Wiepking-Jürgensmann, to achieve 'a golden age for the German landscape and garden designer that will surpass everything that even the most enthusiastic among us had previously dreamed'.[93] It was a theme the same author returned to the following year: 'A German farmer is more fit for life, in the sense of having a higher calling, than a Polish baron, and every German worker possesses more creativity than the Polish intellectual elite. Four millennia of Germanic evolution point to an irrevocable chain of evidence.'[94]

Another leading figure, Alwin Seifert, who became an influential figure in the German green movement of the later twentieth century, bluntly spelled out what this meant: 'If the East is to become home for Germans from all over Germany, and if it is to flourish and become as beautiful as the rest of the Reich, then it is not enough just to cleanse the towns of past Polish mismanagement and construct clean and pleasant villages. The entire landscape must be Germanized.'[95] This was to be done by Polish workers, ordered Himmler, who needed to be kept separate from Germans, presumably for fear of 'contaminating' them.

Supposed racial superiority was often couched in terms of claimed abilities when it came to environmental stewardship. 'It is only man who can leave an imprint on a landscape,' stated the anonymous author of Der Untermensch ('The Subhuman') in 1942.

> Therefore, we find orderly fertility, planned harmony of fields, and well-thought-out villages on the one side of Germany; beyond we see zones of impenetrable thicket, steppe, and endless primeval

forests, through which silted up rivers are grinding their way
This badly used yet fertile lap of black soil could be a paradise,
a California of Europe, but in reality it is run down, coarsely
untended, imprinted with the seal of a cultural disgrace without
precedent, an eternal accusation against the subhuman being and
his rule.[96]

The Slavs, in other words, knew only chaos; it was up to the Germans
to transform and break nature into service.

What made this breathtaking, racist German self-confidence worse
was that it was a poor reflection of reality. Exploitation of agricultural
production during the war proved to be highly inefficient, characterised
by poor decision-making and by a set of priorities that not only did
not make the most of the opportunities available, especially in western
Europe, but proved actively counterproductive. German incompetence,
rather than sabotage or resistance, proved telling when officials failed to
generate the resources and outcomes that they could and even should
have managed.[97]

Poor planning lay at the heart of other German failures during the
Second World War, including provisioning of the Wehrmacht during
the invasion of the Soviet Union in 1941. While the cold Russian winters
have a prominent role in public imagination as an explanation for the
German army running to a standstill at Stalingrad and elsewhere, even
before the onset of winter it was clear that the invasion was compromised
by basic logistical problems in supply lines and the failure to deliver
food, fuel and armaments to where they were most needed.[98]

It is true there was an adverse climatic context for mass starvation in
Bengal in the early 1940s. 'Famine came, ghastly, staggering, horrible
beyond words,' wrote Jawaharlal Nehru, later the first Prime Minister
of India. 'Men and women and little children died in their thousands
daily for lack for food.' In some parts of the world, 'men were dying ...
killing each other in battle.' But in Malabar, Bijapur, Orissa and above
all in 'the rich and fertile province of Bengal', it was different: 'here
death had no purpose, no logic, no necessity; it was the result of man's
incompetence and callousness'.[99]

All in all, 2 to 3 million people died from starvation and associated
disease. Nehru was right that these deaths were in vain. As Amartya Sen
argued in a seminal study of the famine that was a key reason for his

winning the Nobel prize for economics, much of the stockpiling and the resultant price spikes during the famine were due to poor decision-making by the authorities which made a bad situation worse.[100] In fact, the most severe drought conditions predated food shortages by more than a year and, what is more, the months immediately before the famine took hold coincided with above-average precipitation levels. As such, political and military factors were primarily responsible for the high levels of mortality, rather than drought per se. Grain-import restrictions imposed by colonial officials, large inbound flows of refugees to the region, above all from Burma, high levels of pre-existing malnutrition and the prevalence of malaria all combined with deadly effect. Unlike most famines in India in the later nineteenth and twentieth centuries, the catastrophe in Bengal was mainly about human error, rather than about monsoon failure and problems of soil moisture.[101]

The horrors of the Second World War finally came to an end in 1945. Six years of bloodshed had led to tens of millions of deaths – with the numbers of those killed in combat dwarfed by those who had succumbed to starvation, to disease or to the gas chamber in the Holocaust. Men, women and children perished in numbers that owed as much to human cruelty as to human ingenuity in developing tools that could kill ever more efficiently and in ever greater numbers.

The single most powerful was a new type of explosive device of such devastating power that it arguably helped steer the world into a period of relative peace and prosperity – not least since the consequences of its use were so dreadful, so terrifying and so apocalyptic that they threatened global annihilation. Weapons of mass destruction thus seemed all but unusable in war situations, with two exceptions so far: Hiroshima and Nagasaki.

The development of these weapons had aroused concerns that they would do more than destroy the targets that they were dropped on. Hitler himself 'was unable to grasp the revolutionary nature of nuclear physics' and 'the idea quite obviously strained his intellectual capacity', according to his close confidant Albert Speer. Speer questioned leading German theoretical physicists about the detonation of a nuclear bomb, and pushed for an answer about 'whether a successful nuclear fission could be kept under control with absolute certainty or might continue as a chain reaction'. The fact that he could not get a straight answer troubled him – and troubled Hitler himself, who was 'plainly not delighted with the possibility that the earth under his rule might be transformed into a glowing star'.[102]

Those working on the Manhattan Project on the other side of the Atlantic also considered whether a nuclear detonation might result in a catastrophic ignition of the atmosphere, eventually producing a report which showed that this was unlikely to happen, at least with bombs of the current size. 'Whatever the temperature to which a section of the atmosphere may be heated,' argued three leading theoretical physicists, 'no self-propagating chain of nuclear reactions is likely to be started.'[103]

While only two nuclear devices have been used in warfare, more than 400 atmospheric explosions were detonated between 1945 and 1963, when the US, UK and the Soviet Union agreed the Partial Nuclear Test Ban Treaty, which prohibited atmospheric and underwater nuclear weapons tests – with a further sixty-three atmospheric detonations taking place after this date as China began and France continued testing. The total yield of the 504 atmospheric nuclear explosions between 1945 and 1980 (the date of the last test by China at Lop Nor) was 440 megatons.[104]

Although physicists were right that detonations would not ignite the atmosphere, they misunderstood what would happen to the radioactive dust produced by the explosions. Rather than spreading out evenly before falling back to earth several years later with depleted levels of radioactivity, it turned out that there were significant variations in the geographical distribution of radionuclide deposits. What is more, there were major differences in the descent of the fallout that were closely related to latitude, with northern climes being affected much more quickly than those to the south.[105]

In this context, then, it was noteworthy that the period immediately after the Second World War corresponded with the end of the Early Twentieth-Century Warming period. In the last months of the war, cities in Japan and Germany had been devastated by Allied bombing. In the former, an area of 461 square kilometres in sixty-nine cities including Hiroshima and Nagasaki had been bombed in US B-29 Superfortress raids, producing huge columns of smoke from urban fires. This was mirrored by what happened in Germany, where more than 80 per cent of the total bombing tonnage released by Allied aircraft on cities during the war was dropped in 1944–5. Much of this was done by incendiary bombs, which US Strategic Command found were four to five times as destructive as high explosives – and they destroyed 20 per cent of all residential properties in Germany, making 7,500,000 civilians homeless in the process. This produced smoke plumes and thermal currents that pilots described as 'a black hell'.[106]

Despite the very substantial amounts of soot released by the fires, it seems that this had little effect on global surface air temperatures.[107] More important seems to have been the repositioning of the boreal Hadley cell towards the equator in the 1940s and the cooling of the northern hemisphere.[108] It seems, however, that another key factor in the changes in climatic patterns and in the stagnation of global warming in the mid-twentieth century was the high level of nuclear activity in general and the atmospheric tests in particular.

The link between climate change and weapons tests was being suggested as early as the early 1950s, when a report by the US Atomic Energy Commission and the US Air Force looking into radioactive–biological hazards discussed the possible impact of radioactive debris on upper-atmosphere ionisation, and noted that the 'loading of the atmosphere with particulate matter … may affect the weather of the earth'.[109] As it was, public fears of weather modification had already risen following a lengthy period of drought in the American midwest, while unusually cold and wet summers in Japan likewise sparked widespread debate as well as papers written by Akio Arakawa and others that linked abnormal weather with nuclear detonations both individually and collectively.[110]

The effects of weather and climate disturbance formed part of Project Sunshine, a classified American programme led by the Atomic Energy Commission, the US Air Force and the RAND Corporation which reported in 1953 on the 'hazards of radioactive debris that, by one means or another, finds its way into a human being'.[111] The threat by the Eastman Kodak Company to hold the Atomic Energy Commission liable for damages after it turned out that contamination from tests done more than 1,000 miles away had ruined X-ray film had spurred attempts to understand the effects of releases of radioactive material – especially as bombs began to grow larger and as detonations took place at higher altitudes.[112]

A Congressional committee investigating the impact of nuclear weapons in the mid-1950s noted that 'many inventions of man – gunpowder, radio, airplanes, and television – have been blamed for changes in weather and climate'. It was 'only natural', therefore, 'that atomic and thermo-nuclear explosions, being amongst the most dramatic achievements of mankind, would come in for their share of the blame'.[113] While conceding that there had been an increase in 'unusual and undesirable weather' over the previous decade, the committee's report claimed that apparent abnormalities did not exceed the limits that could be explained by chance. Even so, it added, 'the failure to detect statistically significant changes' did not act as 'proof that physically significant changes have not been produced by explosions'. More work was needed to prove the case either way.[114]

This was a reflection of the scientific consensus at the time: as it had been with Guy Callendar's hypothesis about the effects of carbon emissions, it was a question of more data, more analysis, and avoiding rushing to conclusions.[115] Recent research, in part drawing on later studies of the possible effects of a nuclear war, has suggested that atmospheric explosions, especially of large hydrogen bombs and the fine dust released as a result, were the cause of the stagnation in global warming that took place from the middle of the twentieth century.[116] Ironically, however, what was at the forefront of the minds of military planners immediately after the Second World War was not whether the climate might be changed inadvertently by new weapons of mass destruction, but rather whether weapons could be developed that could change and control the climate.

22

Reshaping the Global Environment
(the Mid-Twentieth Century)

We had to bang on pots until the poor sparrows were exhausted. We did it for several days. There were many fewer sparrows after that.

Student, Chongqing Agricultural University,
on War on Pests (1958)

The idea of manipulating the weather is one that has long and deep roots. Rituals intended to encourage rains, to protect crops, to make harvests bounteous and to render climate conditions benign are known from societies stretching from the ancient past to the present day, from South Asia to the Americas. These rituals often involved a complex set of steps designed to win the favour of animate or inanimate deities, including the offering of foods or of sacrifices, and sometimes changes to behaviour, such as fasting, in order to earn the goodwill of supernatural powers.

By the nineteenth century, attempts to intervene in the weather had become rather more determined in their methods and aims. Interest was particularly strong in the United States, where several pioneers set about trying to conduct experiments to see if rains could be induced. One was James Espy, the first federally funded meteorologist, who conducted a series of experiments in the 1840s to generate artificial precipitation by setting fire to large tracts of forest in the hope that a giant column of heated air would create clouds and would thus generate rain. Although these experiments ended in failure, other scientists developed new ideas based on the supposed correlation between battles that involved heavy use of artillery shells and the rain storms that often seemed to follow soon afterwards.[1]

Despite the lack of results, the US government took a keen interest, with Congress allocating funds to the Agriculture Department in the second half of the nineteenth century for experiments to 'shock' rain from the sky through explosives – and further support when initial findings seemed promising. The follow-up proved less successful, with no rain being produced over the course of two months of incessant detonations in what observers described as full-scale imitations of a major battle. This was hardly a surprise, according to well-placed commentators: the premise was flawed, a waste of taxpayer dollars, and 'the silliest performance that human ingenuity could devise'. With that, the funding that had not yet been spent was returned to the Treasury.[2]

Rainmaking became something of an industry for confidence men in the early part of the twentieth century, who persuaded farmers to part with money in return for promising to induce rainfall that would result in bumper harvests. Men like Charles Hatfield were 'a charlatan's charlatan' – adept at staging compelling demonstrations of what seemed to be tried, tested and foolproof chemical products that made the most of the latest scientific discoveries.[3]

As is often the case, war brought about a spike in interest as well as heavy investment in new technologies that might offer strategic benefits. In Britain during the First World War, for example, the Advisory Committee for Aeronautics, overseen by the Department of Scientific and Industrial Research, ran a range of trials of innovations such as incendiary bullets for use against Zeppelin airships, model aircraft to help improve gunsights and hydraulic timing equipment for firing guns. Another project looked into producing artificial clouds.[4]

Interest in meteorology rose sharply as the value of being able to anticipate weather conditions on the battlefield became apparent. With the US entry into the war in 1917, the Signal Corps set about training a thousand men as military weather observers and forecasters – no easy job given the low regard for what one physicist called 'a guessing science'. It was not enough, noted the distinguished meteorologist Sir Napier Shaw, to rely on 'observations, map making, and forecasting'; what was needed, rather, was to develop specialist skills in mathematics and physics. The investigative field began to blossom in the 1920s, although it took time to do so: in the period 1919–23 in the US, while more than 600 PhDs were awarded in chemistry, 200 each in botany and physics and nearly a hundred in geological sciences, just two were awarded in meteorology.[5]

Advances in aeronautics and avionics led to attention being paid to dissipating fog – driven by the desire to clear landing strips for aircraft – with experiments in the 1920s using electronically charged sand, with follow-up trials on cloud formations that seemed to bear fruit. These were picked up by the national and international press, with the *New York Times* commenting that while weather modification was 'a subject in which quacks have revelled for more than a century', the lines of investigation being pursued by the military were not only promising but promised to 'change geography and history', indeed ultimately to transform 'the future of the human race'.[6]

The experiments being undertaken by the military ran out of steam in the 1920s, as did their funding; nevertheless, academic interest persisted both in seeking better understanding of weather formation and in efforts to influence it. This was helped by the establishment of the American Meteorological Society, set up to promote 'the advancement and diffusion of knowledge of meteorology, including climatology, and the development of its application to public health, agriculture, engineering, transportation by land and inland waterways, navigation of the air and oceans, and other forms of industry and commerce'.[7] Vapour transfer, ice-crystal process and the use of calcium chloride powder were just some of the ideas that were proposed and in some cases tested by scientists from Sweden to the US, from Germany to the Soviet Union. The Turkmenistan Institute of Rainfall in the USSR was reported to have been successful in 'producing rain from a cloudless sky' through the use of chemical reactions conducted on the ground or in the air.[8]

Research conducted during the Second World War into cloud-droplet size and crystal formation prompted a new wave of research into weather manipulation, with promising initial results. A team at General Electric's Research Laboratory in Schenectady, New York established that silver iodide and dry-ice particles could stimulate rainfall by providing nuclei around which moisture could cling, in a process known as cloud seeding.[9] The results were so exciting, and the implications so profound, that some leading figures in the technology community believed that 'control of the weather by man is a scientific possibility of the future', and that it might soon be feasible to 'neutralise a storm or detour it from its course' simply by 'pressing radio buttons'.[10]

For others the opportunity was not about how storms might be controlled but about how this budding new technology could be used as an offensive weapon. Vannevar Bush, who had overseen the US Office of Scientific Research Development during the war and played a key role in the Manhattan Project, wondered if it would be possible to create snow rather than rain: if so, he went on, being able to drop thousands of tons of snow on enemy troops would offer obvious strategic and military advantages.[11]

By 1947, a formal research programme had been established to look into 'the manipulation of gigantic natural forces for the benefit of mankind everywhere'. The aim of the programme – codenamed Project CIRRUS – was to boost the understanding of weather patterns in order to improve forecasting, as well as to attempt cloud modification using dry-ice granules, seeding with silver or dispensing water droplets into cumulus clouds in the hope of triggering a chain reaction to produce rain. A secret memo set out the objective of establishing the viability of weather modification from the strategical and tactical viewpoint. Not everyone was convinced, with one member of the military's Research and Development Board suggesting that money would be better spent on 'a lot of rattles and snake skins' which might be more likely to induce precipitation.[12]

Those involved with the project were more bullish, both about the experiments and about their implications. Dr Irving Langmuir, who had won the Nobel prize for chemistry in 1932 and was head of General Electric's Research Laboratory, had little doubt about the stakes involved. In August 1950, Langmuir featured on the cover of *Time* magazine above the headline: 'Can man learn to control the atmosphere he lives in?'[13] Rainmaking and weather control, he said in an interview in December that year, could be 'as powerful a war weapon as the atom bomb'. The effects of just thirty milligrams of silver iodide, Langmuir was quoted as saying, 'under optimum conditions equals that of one atomic bomb'.[14]

The potential of weather modification was immense, said Howard Orville, a former naval officer who had been appointed by President Eisenhower to serve as chair of the Advisory Committee on Weather Control. It 'may sound fantastic today', wrote Orville for *Collier's* magazine, but in the not too distant future it 'could well be a reality' that planes would be sent into the sky to break up tornados and hurricanes over Texas and elsewhere. Indeed, it might even be possible 'to influence all our weather to a degree that staggers the imagination'. What was needed was public support and funding for research. If both were given, 'we may be able eventually to make weather almost to order'.

Warming to his theme, Orville explained what this meant. For one thing, he wrote, 'Russia would be at a disadvantage,' since 'weather characteristically moves from west to east'. Being able to influence climate patterns did not just serve US citizens at home, but could be used as a weapon. Orville spelled it out: with control over the weather it would be possible to 'deluge an enemy with rain, or strike at his food supplies by withholding needed rain from his crops'.[15]

It was a topic at the forefront of much discussion in the 1950s as the Cold War took hold. The Soviet Union could be weakened by 'torrents of rain' caused by 'seeding clouds' that moved over its territory, suggested the science writer Frank Carey; alternatively, he noted, 'if an opposite effect is desired', it might be possible 'to cause destructive droughts which dry up food crops by "overseeding" those same clouds'. Fortunately, he added, echoing Orville, 'Russia could do little to retaliate because most weather moves from west to east.'[16] Just think, said the Texas senator – and future president – Lyndon Johnson, who had an even more ambitious vision: 'From space, the masters of infinity could have the power to control the earth's weather, to cause drought and flood, to change the tides and raise the level of the sea, to divert the Gulf Stream, and change temperate climates to frigid.'[17]

For all the bullish claims about the potential of climate control, the results of the experiments proved to be frustrating. While cloud seeding worked in practice as well as in principle, it was difficult, if not impossible, to produce quantities of rain that were economically viable – whether as a strategic weapon or to help farmers in the US and elsewhere.[18] Funding kept flowing, partly because of the worry that others might master weather-modification technology first. Were this to happen, the consequences would be dire, warned Orville. 'If an unfriendly nation solves the problem of weather control and gets into the position to control the large-scale weather patterns before we can,' he said, 'the results could be even more disastrous than nuclear warfare.' As the author of an article that appeared on the front page of the *New York Times* on New Year's Day 1958 put it, this could mean that 'New York City might be put under a few hundred feet of ice or a few hundred feet of water depending upon whether the temperature was raised or lowered.'[19] Professor Henry Houghton of MIT could only 'shudder to think of the consequences of a prior Russian discovery of a feasible method of weather control'; improvements to the climate of the Soviet Union 'could seriously weaken our economy and our ability to resist', regardless of whether such steps were taken with that aim in mind or were done inadvertently.[20]

These anxieties were heightened by the technological advances that the Soviet Union had made, not only in its development of nuclear weapons but also in the creation of a space programme that resulted in the launch of the Sputnik satellite in October 1957. Sputnik's launch took the United States by surprise and provoked shock verging on panic in some quarters, not just because it represented what some called a 'propaganda advantage' but because of what it revealed about Soviet ballistics and launch systems for its arsenal of warheads.[21] US scientists pored over pictures taken of the far side of the moon by Soviet spacecraft – and at the technologies used to do so.[22]

Not surprisingly, this raised concerns that the Soviets were making strides in weather modification too. Terrible drought caused by temperatures in 1946 that were much higher than usual, especially in June, resulted in harvest failure in many of the most agriculturally productive regions of the USSR.[23] This caused severe famine that lasted into 1948 and claimed more than a million lives – losses that came on top of around 24–27 million excess deaths during the Second World War.[24] One result was a top-down effort to create what one scholar has called the 'world's largest ecological engineering project': namely, a plan to change the climate of the Soviet Union.[25]

Announced on 20 October 1948, the 'Great Stalin Plan for the Transformation of Nature' was a considered attempt to counter the effects of anthropogenic climate change. That it was conceived and approved at all is remarkable, given Stalin's own certainty about human impact on the natural world. 'Geographical environment is unquestionably one of the constant and indispensable conditions of development of society,' he had written in the 1930s. But for the last

3,000 years 'geographic conditions in Europe have either not changed at all, or have changed so slightly' as to be irrelevant. Significant changes to the natural world, he went on, 'require millions of years'; this was unlike 'the system of human society' where 'a few hundred or a couple of thousand years are enough' to make a difference.[26]

The Great Plan now aimed to create 6 million hectares of new forests, a series of eight belts of woodland and the stabilisation of sands and soils, which would serve as a barrier to winds blowing from Central Asia and in so doing both cool southern Russia and induce rainfall in this region. A further set of decrees in 1950 expanded the scope of this major ecological transformation with the commissioning of projects to build dams and hydroelectric power stations, create new canal systems and plant more forests.[27]

The plan was not just ambitious but a centrepiece of propaganda within the Soviet Union: 'Which of the capitalist countries could take on a task of such a grandiose scale?' asked the head of the Ministry of Forest Management. 'None are able to cope with such a task. They are not interested in the people, but in the bags of money they protect. The robbery of their own and other people – this lies at the base of the programs of bourgeois countries.'[28]

The ambition and the triumphant overconfidence did not match outcomes. For one thing, central planning for what the transformation of nature would, could or should involve proved poor, as was so often the case with major projects in the Soviet Union, partly because of personality clashes but also because of endless bureaucratic infighting. For another, the assumptions about the science that underpinned both the aims and the methods were nowhere near as robust or advanced as had been expected.[29] As it was, days after Stalin died in March 1953, the project was shelved, the Ministry of Forest Management closed and the majority of workers employed in forest management were let go.[30] It was time for a new era, said Nikita Khrushchev in February 1956 when he came to review Stalin's rule in a famous speech at a closed session of the 20th Congress of the Communist Party in Moscow: Stalin never went anywhere, 'never met with workers and collective farmers', had little understanding of rural affairs and was completely out of his depth in rural areas. His last visit to a village was in 1928; instead, he relied on 'films that tarted up and made the real situation in the countryside look all nice and beautiful', with images of farmhouse tables groaning under the weight of turkeys and geese. No wonder his initiatives never ended in success.[31]

Stalin's other crimes, said Khrushchev, included the torture and execution of party members on false and trumped-up charges, his foreign policy mistakes and the dangers of his cult of personality – all of which were laid out to a shocked audience in a speech that lasted for hours. Stalin considered himself a genius, said Khrushchev repeatedly; and when a genius 'expresses his opinion, everyone has to repeat it and admire his wisdom' – especially if they had concentrated power in his hands as Stalin had. Stalin 'acted not through persuasion, explanation and patient cooperation with people, but by imposing his concepts and demanding absolute submission to his opinion'. The problems and failures of Soviet agriculture were laid firmly at Stalin's feet; these would now be remedied, said Khrushchev, and consigned to the past. It was time for new approaches that extended from the Soviets on the world stage to the treatment of the environment at home.[32]

Some of these initiatives were extremely ambitious. Considerable attention was paid in the US to the work being done by scholars based at multiple research institutes in the Soviet Union on cloud physics, with materials collated by US scientists and intelligence officers to show who was working on what project, where and with how much (or how little) success.[33] Soviet scientists had been looking into the melting of polar ice caps as well as contemplating 'large public works projects that would upset the entire wind circulation pattern of the Northern Hemisphere'. So advanced and ambitious was Soviet thinking that one engineer, Arkady Markin, had proposed an international project to design a dam across the Bering Straits that 'would pump the warmer waters from the Pacific into the colder Arctic Ocean' and sometimes in the opposite direction, 'cancelling out the Greenland, Labrador and other cold oceanic streams' and raising the temperature in New York, London, Berlin, Stockholm and Vladivostok by several degrees. Nevertheless, as Edward Teller, the 'father of the atomic bomb' put it, 'The Russians can conquer us without fighting through a growing scientific and technological preponderance.' The Soviet Union 'may advance so fast in science and leave us so far behind' that there would be 'nothing we can do about it ... What kind of world will it be where they have this new kind of control and we do not?'[34]

Concerns like these were carefully chosen to try to persuade Congress to provide funding for current and future research into geo-engineering schemes that were outrageously ambitious. These ranged from 'the large-scale use of coloured pigments over polar ice surfaces', which would 'melt the ice, and change the local climate', to a gigantic space mirror that would focus the sun's rays as a 'giant magnifying glass' and could 'prevent frosts over orchards, or melt Atlantic icebergs, open frozen harbours' or 'light entire cities or other areas safely at night'.[35]

Then there was the raw power of nuclear energy, the 'greatest of destructive forces', said President Eisenhower in his 'Atoms for Peace' speech at the United Nations in 1953, which could nonetheless 'be developed into a great boon for the benefit of all mankind' and could be used 'to serve the needs rather than the fears of the world – to make the deserts flourish, to warm the cold, to feed the hungry, to alleviate the misery of the world'.[36] Energy would become so cheap, suggested some senior officials, that there would be little point charging for it.[37]

Such were the new powers available to physicists and engineers, suggested Edward Teller, that the time had come to put right 'a slightly flawed planet'. Apart from the opportunity 'to remedy nature's oversights', the 'new and important discipline' of 'geographical engineering' had opened up limitless possibilities. 'We will change the earth's surface to suit us,' he said. New technologies, including nuclear explosives, 'will enable men to remould many landscapes'. These changes were not limited to reshaping landscapes; nuclear energy could bring about genetic mutations in fish that would create 'new breeds of higher food value'. Not only that: high-pressure explosions would be able to turn carbon into 'the hardest, rarest, and most beautiful material on earth: diamonds. Anyone visiting the earth in a hundred or two hundred years' time,' said Ralph Sanders, a contemporary military historian, would barely be able to recognise 'geographical landmarks' – let alone the other transformations that would have taken place.[38]

Scientists in the Soviet Union were making equally grandiose promises. The 'omnipotence of human genius', claimed Arkady Markin, meant there were no limits to the 'magnificent results' that could be achieved thanks to atomic explosions, which could help remodel nature by carving 'new canyons through mountain ranges' or creating new islands, as well as 'canals, reservoirs and seas'. There was surely more to come, wrote Nikolai Rusin and Lila Flit in *Man versus Climate*: 'we are merely on the threshold of the conquest of nature'.[39] Another Soviet author suggested several other schemes might soon be viable, including melting the North Polar ice-cap, the transformation of the climate of North Africa, the creation of the Chad and Congo seas, and the irrigation of the Sahara. These would open up 'millions and millions of fertile acres' that would 'yield two and even three crops a year for the benefit of mankind'. This would help the 'struggle of African peoples for national liberation'.[40] 'In a matter of decades,' the author predicted, gigantic dams and dykes, 'pumping stations capable of handling entire seas and other facilities as yet unknown' would put humans in full control of the world's 'heating system'. The only obstacle, he went on, was 'the continued existence of capitalism' – which served as a 'ball and chain hampering man in his progress towards a happier lot.' Fortunately, he concluded, 'nothing can stop the march of history ... and victory will come as surely as day follows night.' The ability to transform and shackle the earth, he wrote, 'was limitless'.[41]

There was little to show for all the promises of this brave new world. In fact, the claims for how science might help human progress – or produce devastating new weapons that were so awesome as to create and preserve a unipolar world – had already irritated some observers, who concluded that the breakthroughs were neither as imminent nor as dramatic as promised. One senior US official poured scorn on claims that 'one ounce of a particular biological material would be sufficient to kill 200,000,000 people': they were 'fantastic and have no basis in fact'. Statements made in the Merck Report prepared for the US Secretary of War in 1946 about the possibilities arising from biological warfare were dismissed as 'extravagant, inaccurate, and unduly spectacular in the light of present scientific knowledge' – ironic given that a report written around the same time had warned of the dangers of looking too far into the future and thinking that ' "Buck Rogers" warfare has arrived or that it is likely to arrive in the next ten years'.[42]

It was certainly true, noted a high-level report for the National Research Council of the United States Academy of Sciences published in the mid-1960s, that work over the previous decade had produced results that were 'intensely interesting' and provided encouragement 'for future efforts in weather modification'. However, it went on, the overall assessment was that 'the initiation of large-scale operational weather-modification programs would be premature'. There were 'fundamental problems' that needed answering before real progress could be made. All in all, therefore, said the report's authors, 'we must emphasise that the time-scale required for success' should be measured in decades.[43]

By this time, public opinion had become not just sceptical but in some cases actively hostile to experimentation that impacted the natural environment; this was coupled with concerns about the implications of what might happen in the event of a nuclear war – or if experiments were conducted of nuclear weapons being detonated on the surface of the moon, something that was being discussed seriously and regularly in some circles in the US.[44] Prominent public intellectuals like the mathematician John von Neumann warned that deliberate anthropogenic climate manipulation could have greater effects than a nuclear war. The title of his article, 'Can We Survive Technology?', gives an insight into growing fears about how human development was creating monsters it neither fully understood nor was able to control.[45]

If this was not a cause for alarm, the plethora of press reports about high levels of strontium-90 isotopes in milk certainly were. Even *Playboy* magazine published an editorial on 'The Contaminators', warning that children might 'die before their time … or after having spawned grotesque mutations'. In a stunt designed to provide reassurance, President Kennedy allowed himself to be filmed drinking a glass of milk and announced that he was serving it at every White House meal.[46]

Such thought as had been given to the implications of the reshaping of nature had downplayed the consequences. Radioactive fallout from nuclear explosions to open up new canals in South America and Africa or to create new deep-water harbours in Alaska 'would be negligible', claimed Teller. Underground detonations would have minimal impacts, with radioactivity trapped deep below the earth's surface.[47]

These promises and commitments proved to be wide of the mark. It had been clear from the many films made for the military and the public that, while much was made of nuclear energy's positive potential, the new power could cause terrible suffering. Films such as *Operation Plumbbob: Military Effects Studies* of 1957 showed the horrendous effects on pigs who were deliberately exposed to atomic tests. Films like *Atomic Tests in Nevada* (1955) opened with shots of St George, Utah in the still of the night. 'It's pre-dawn,' said the voice-over. 'Five in the morning. Pretty deserted at this hour. Everything is closed down. Everyone's asleep.' Before the town woke and people went to work, the sky turned blinding white. All got up and went on with their daily business. 'Nothing to get excited about,' the narrator told viewers, unconvincingly; the residents had seen this many times with previous tests. It was hard, however, for those watching not to feel that this was an age of fear rather than one of hope.[48]

Protests before a high-altitude test over Johnston Island in the Pacific, controversy over Project CHARIOT – a plan involving the detonation of a series of nuclear devices to create a new harbour – and the explosion of a thermonuclear device called 'Sedan' in Nevada which exposed more than 13 million Americans (or around 7 per cent of the population) to radiation, prompted a stricter set of guidelines.[49] These were set out in April 1963 by President John F. Kennedy in National Security Memorandum 235 'governing the conduct of large-scale scientific or technological experiments that might have significant or protracted effects on the physical or biological environment'. This was followed a few months later by the Partial Nuclear Test Ban Treaty, signed amid growing anti-nuclear protests, with public anxieties heightened by the Cuban missile crisis of 1962 and the possibilities of a Third World War. There had also been an accident at the Mayak nuclear facility in Chelyabinsk in the Soviet Union, when a storage tank exploded in September 1957, resulting in the worst nuclear accident in history at the time – and still ranked behind only the disasters at Chernobyl and Fukushima in terms of the release of radioactive materials.[50] As we shall see, there were other factors too that made a rapprochement between the US and USSR propitious at this time.[51]

It was not just the United States and the Soviet Union that had been looking into transforming the natural environment or at weather modification. Within a year of announcements that the US government was examining cloud-seeding programmes, research had begun in Australia, France and South Africa; twelve months later, at least twelve countries were conducting experiments in this area – and by the end of 1950 around thirty were doing so.[52]

Although these had all tailed off by the end of the decade in the face of minimal returns on investment, ambitions to change the physical environment did not. In India, independence was the moment for a major ecological transformation that would change and modernise Indian society. That, at least, was the hope of senior officials like Sardar Bahadur Sir Datar Singh, who had been vice-chair of the Imperial Council of Agricultural Research. 'For centuries, the Indian peasant has used the old plough and sickle; for centuries, he has suffered from inhuman toil and drudgery,' he wrote. 'The time to repaint the picture has come. The primitive peasant economy dating from five thousand years old Mohenjo Daro period must be replaced by scientific development and scientific technique, so that the peasant gets leisure for educational and cultural pursuits.'[53]

One way to do this was through concerted projects such as the 'Grow More Food' programme which led to an increase in domestic grain production between 1947 and 1951; another was for the government to take control of food grains, which it did in September 1948.[54] Another way to drag the country into the modern age was by commissioning a series of mega-projects. Before independence, India had just thirty dams over thirty metres in height; now, as one senior official put it, dams served as 'the symbols of the aspirations of new India, and the blessings that stream forth are the enduring gifts from this generation to posterity'. Dams, canals, power lines, roads and more were all seen not only as motors of progress but as ways of relieving dependence on the vagaries of nature – and of climate anomalies in particular. Dams would mean that India, and Indians, would be safe even 'in the worst monsoon'. As such, they represented a new future for a people that had not been able to build, nor even desire, projects before they 'became a nation'.[55]

Little was said about the human or ecological price of embracing modernity through major construction programmes. As many as 40 million people were displaced in India by dam-building projects and as many as 10 million more through other major construction programmes. Then there was the soil salination, flooding of woodland and fields, blocking or diversion of rivers and impacts on natural drainage – all of which had effects that became clear only later in the twentieth century. This was in addition to the spread of malaria, depletion of fish stocks, rampant corruption around water procurement and disaster caused by dam failure. As one scholar has noted, the beneficiaries were the rich landed classes, while those displaced suffered 'pauperization and increased marginalisation'.[56]

Dam-building was not just about prestige and vainglory, however: in the US in particular, dams were seen as crucial for preventing hunger – and for promoting peace. President Harry S. Truman envisaged enormous dams from the Yangtze Valley to the Danube, pooh-poohing those who were sceptical. 'These things can be done,' he told one adviser, 'and don't let anybody tell you different.' When dams are built, he said, and 'when millions and millions of people are no longer hungry and pushed and harassed, then the causes of war will be less by that much'.[57]

Truman did not have India in mind when he voiced support for dam-building, not least since he thought of it as a country 'jammed with poor people and cows wandering around streets, witch doctors and poor people sitting on hot coals and bathing in the Ganges' – or so he told Chester Bowles, a leading Democrat politician who had been Governor of Connecticut, when he expressed an interest in being appointed US ambassador to the country in 1951.[58]

Nevertheless, Truman and many of his successors as president proved enthusiastic promoters of the idea that these projects would be highly beneficial, encouraging initiatives for the Nile, Tigris, Jordan, Indus, Mekong, Irrawaddy and Ganges valleys. These were led by builders who took on the status of what one scholar has called the 'technocratic warrior elite', typified by men like Harvey Slocum, the chief designer of the 226-metre-high Bhakra Nangal dam in Himachal Pradesh in India whose reservoir held more than 9 billion cubic metres of water: Slocum modestly saw himself as part of a line that linked 'Napoleon, Eisenhower, Alexander the Great, all the great generals'.[59]

Dams also offered ways for neighbouring and rival states to work together – or to challenge each other. Following Partition and the creation of the separate states of India and Pakistan in 1947, the former shut down water supply to the latter in the Bari Doab, one of the major waterways of the Indus River. This led to violent clashes that threatened to develop into full-scale war and massive migration of Muslims and Hindus from one part of South Asia to another. The dispute escalated and was quickly brought before the United Nations, where Pakistan argued that 'the withholding of water' on which 'the survival of millions' of people depended constituted 'an international wrong and ... [is] contrary to the obligations of membership in the United Nations', while India claimed rights 'over the waters of rivers flowing through her territory', and insisted in any event that international law was ill-suited in this particular case because of the unique circumstances of Partition.[60]

The issue of Kashmir, wrote David Lilienthal, a senior figure in US public administration who had served as head of the Atomic Energy Commission, 'may not be solvable, now, short of war'. India 'must have more water or starve'; Pakistan, meanwhile, is naturally 'fearful of the future'. Tensions, however, could be reduced by common sense and engineering. Matters could be greatly improved if the 'whole Indus system' was 'developed as a unit – designed, built and operated' as a single entity, like the Tennessee Valley Authority in the US.[61] Efforts to broker an agreement over water rights were paralleled by US efforts to resolve the worsening conflict in

Vietnam two decades later, which President Lyndon Johnson proposed to alleviate in exactly the same way – this time, with the construction of a massive dam and irrigation works on the Mekong River.[62]

If the narrowness of perspective around identifying and solving problems is striking, so too were the limited successes achieved – and, in some cases, the cold reality of simply making matters worse. Rivalries between the US and the Soviet Union had intensified after the Second World War as the Cold War kicked in. The theatre for competition was global. Of particular concern to the US and to Europe was the Middle East both because of its geographic position and because of its enormous oil reserves. Iran attracted considerable attention as it was considered a 'weak post in the outer wall of the Free World's defence line', and therefore in need of shoring up. One obvious problem was that three-quarters of Iran's population were tenant farmers who owned no land and few assets; another was that credit was typically available at rates that were not so much extortionate as punitive, with money loaned at rates as high as 75 per cent.[63]

Considerable efforts were made, therefore, to try to help modernise the country, led by the Ford Foundation, the largest philanthropic organisation in the United States and one with impeccable connections at the very highest levels within the US government: Paul Hoffmann, the foundation's president in the 1950s who oversaw projects in Iran, had previously overseen the implementation of the Marshall Plan – the US initiative to provide foreign aid that helped rebuild western Europe after the Second World War. The Foundation introduced microfinance programmes, connected farmers with buying agents in Europe and established projects that boosted production methods, stimulated greater community participation in local affairs and improved creditworthiness. In addition, the US government poured in aid money, rising from around $27 million per year in the period 1946–53 to an average of more than $120 million a year after a CIA-engineered coup in 1953 that removed Mohammed Mossadegh from power; of course, there was also funding for a major dam, this time at Karaj, near Tehran.[64]

These initiatives met with almost total indifference from the Iranian political establishment, whose members stood to gain little from reforms, and might even lose from them. Perhaps worse still, for all the efforts to help modernise Iran, the end results were precisely the opposite of those that had been intended: long-term dependence on the west; high spending on military equipment, primarily from US sources, that while bringing rewards for American contractors served to concentrate economic and political power in the hands of the elite in Iran; and a pool of seething resentment among the majority of the Iranian population that proved ripe for exploitation either by the hard-line Tudeh party or by religious figures such as Ayatollah Khomeini, who had little trouble pointing to the inequalities in Iran – and showing that the United States was responsible for them.[65]

It was a similar story in Afghanistan, a country that had positioned itself well during the first half of the twentieth century and ended the Second World War with $100 million in reserves. Nevertheless, Afghanistan was 'a backward country', said the prime minister, Mohammed Daoud. 'We must do something about it or die as a nation.' He was particularly keen on building a dam. The resultant scheme in the Helmand valley was a disaster. Rather than crystallising Afghan dreams, it turned into a nightmare. As well as the water table being raised too high, salination quickly became a big problem. Vast sums were spent on construction salaries as well as on the wages of technicians during construction; costs were passed on to farmers, who were also subject to inefficient tax collection designed to repay a US loan for the project. Perhaps worse, it encouraged competition from the Soviet Union, which was keen not to lose influence in the region; this led to the construction of the Jalalabad dam and major development of the Amu Darya River, funded by a loan worth three times that of the US, and on terms that had little chance of being met. This all helped undermine the credibility of the king and the government – setting the scene for turbulence later in the twentieth century which has continued to the present day.[66]

Similar interventions in the landscape were attempted in other parts of the world, by the United States as well as by the Soviet Union, as part of the battle for hearts, minds and stomachs. Outcomes often did not match expectations. For example, with Latin America considered a 'fertile field for Communist agitation', efforts were made to introduce practices and models developed in the US. In Venezuela, these included new farming methods, such as the heavy use of synthetic fertiliser to increase yields, food-procurement plans and supply-chain optimisation. These were intended to stabilise the country's economy, calm union militancy and both stimulate and reward the Venezuelan middle classes with the conveniences of modern living, such as self-service supermarkets.

In fact, it stoked suspicion and resentment that the country was being transformed, in the words of one radio commentator, into 'an immense farm of economic slaves', underpinned by a plan for American industrialists to create a monopoly in Venezuelan food supply and distribution – and to turn the country into a colony in all but name.

By the 1960s, these fears, ill-founded or otherwise, resulted in the most visible face of US interests and activity, the oil tycoon and philanthropist Nelson Rockefeller, being branded 'public enemy no. 1' not by left-wing agitators but by Venezuela's Central Bank, in US businesses being firebombed and in protesters urging an end to the 'looting of our continent' by American investors.[67]

Competition between the United States and the Soviet Union had major political, economic and military consequences across all continents; but it had profound ecological implications too. In his inaugural address following his re-election as president in 1949, Harry S. Truman had noted that 'More than half the people of the world are living in conditions approaching misery.' This meant hunger and suffering. Put simply, he went on, 'Their food is inadequate.'[68] Truman had been taken with a briefing that assured him that the US had a 'potent weapon' in its arsenal: 'immense technological resources' that could not only 'capture the imagination of the peoples of other countries', but boost production of 'food, clothing and other consumer goods'.[69]

Some had warned Truman to tread carefully, with the National Security Council advising in December 1949 that while Asia was 'the source of important raw materials, many of which were of strategic value', the President should 'scrupulously avoid responsibility for raising Asiatic living standards'.[70] Trying to do so would cause no end of trouble. Far better, as one senior diplomat put it not long afterwards, to accept that 'the hands of the clock of history are set at different hours in different parts of the world'.[71] The spectre of the Soviet Union taking advantage of global poverty to promote left-wing causes and propel states towards revolution focused minds of policymakers in the United States: rural populations in Asia in particular 'were packed into what virtually constitutes an endless village', President Truman was told by his foreign aid adviser Isidor Lubin in the early 1950s. This was nothing less than 'a breeding ground for violent revolution'.[72]

This was an important strand in the evolution of American foreign policy. Soon after the end of the Second World War, former President

Herbert Hoover visited more than forty countries and territories to report back on the challenges they faced in terms of food supply and famine. The report prepared by his Famine Emergency Committee noted that 230 million people were at risk if wheat supplies could not be found for India. Within a few years, loans, aid and expertise were being given to dozens of countries around the world, with over 2,000 technical experts working in more than thirty-five different countries. At least one reason for these varied initiatives, many of which focused on agricultural production, was that overpopulation, resource exhaustion and hunger were all thought to dovetail into political instability and communist uprisings. They were therefore framed within the context of the Cold War and thus to US national security.[73]

But there was also a genuine desire to improve the lot of others in the world by sharing the American model of development that had been so successful, at least for the principal beneficiaries – namely, the descendants of European immigrants. Improvements and innovation had led to a golden age of agricultural productivity in the 1930s and 1940s, with developments of chemical fertilisers, pesticides, hybrid seeds, breeding programmes, mechanised farm tools, antibiotics and more resulting in an annual growth rate doubling or even tripling in the middle of the twentieth century.[74] What made this even more astonishing was that US crop output – which rose from some 70 million metric tons around 1910 to 100 million in 1965 and then nearly 140 million ten years later – was produced on smaller amounts of planted land, falling from 32 million hectares in 1954 to just 22 million in 1970.[75]

This bonanza had moved into national discussion and into government policy. The American Century, wrote the publisher Henry Luce in *Life* magazine early in 1941, meant that it was the 'destiny' of the United States to share 'with all peoples' not only the American way of living, but 'our magnificent industrial products'. Above all, he went on, 'It is the manifest duty of this country to undertake to feed all the people of the world who as a result of [the] worldwide collapse of civilization are hungry and destitute.'[76] President Truman put it in similar terms in his 1949 inaugural address. 'The United States is pre-eminent among the nations in the development of industrial and scientific techniques,' he said; 'our imponderable resources in technical knowledge are constantly growing and are inexhaustible. I believe that we should make available to peace-loving peoples the benefits of our store of technical knowledge in order to help them realize their aspirations for a better life.'[77]

Although Truman claimed that 'the old imperialism ... had no place in our plans', there was a neo-colonial zeal about these efforts. It was significant therefore that what started out as an attempt to support the developing world morphed into something rather different. The post-war period had led to challenges for American farmers and producers with export markets drying up as agriculture began to recover in parts of the world that had been ravaged by conflict. Federal granaries were filled to bursting point, with storage alone costing the US government around $1 million per day, and rising to half a billion dollars a year by the early 1950s. A 'Food for Peace' programme therefore solved several problems at once: feeding the developing world especially in Asia with the fruits of American fields that would make poverty 'obsolete'; opening the door on American values while closing it on those of the Soviet Union; reducing the threat of unrest and revolution; supporting American farmers; reducing stockpiles; and lowering expenditure.[78]

It sounded too good to be true, and it was. Dumping vast amounts of American surplus food depressed prices, hurting local farmers in the process, which then affected investment in agriculture and the countryside more generally. Although the Indian prime minister Jawaharlal Nehru had said in 1948 that 'everything else can wait, but not agriculture', the failure to boost domestic production sufficiently by the following year led to discussions with the US about imports.[79] Annual imports rose sharply and steadily. Between 1954 and 1960, annual imports of American wheat to India rose from 200,000 metric tons to more than 4 million. Nehru put a brave face on things when asked about hardship in the rural communities, saying that 'lower prices are much better than higher prices'.[80]

The reality, however, was that cheap grain weakened domestic production, increased dependence on imports, depleted foreign currency reserves and, rather than aiding socio-economic development, put pressure on an economy that began to show signs of stress and crisis in the late 1950s – just as the UN Food and Agriculture Organisation and some US officials had warned.[81] That was not all: after the signing in 1956 of the PL (Public Law) 480 programme which allowed food shipments to be made from the US to countries facing a deficit, discounted grain was paid for in rupees which were then held by the American government in India. The balance rose so high that by the

1970s the US had control over as much as a third of the money supply in India, a position of extraordinary potential and actual leverage – and an economic and political problem in its own right.[82]

As had happened in Venezuela, the dominance of the US position helped fuel precisely the opposite outcome to the one intended. A bad drought and poor harvest in 1957 were accompanied by Indian farmers being forced by the government to sell grain at discounted prices; inevitably, this led to a surge of support in several states for left-wing parties promising land reform (among other things) and, in the case of Kerala, communists leading the state government.[83]

This set off alarm bells, not least when Nehru made a well-publicised visit to the Soviet Union in 1955. The Indian Prime Minister had first been to Moscow in 1927 when he had been struck by similarities between the challenges facing the USSR and those confronting India. He later admitted to being impressed by the Five-Year Plans, an industrialisation programme which while scarred by 'all its defects and mistakes and ruthlessness' nevertheless served as 'a bright and heartening phenomenon in a dark and dismal world'.[84] On this visit, Nehru was taken to see farms as well as factories, and was then promised gifts of equipment, including hundreds of tractors, harvesters and other machines as part of an effort by Moscow to promote friendships by helping cultivate lands. This was just one example among many: development projects funded by Soviet aid and loans were put in place from Cambodia to Ethiopia, from Sudan to Nepal, many of which were linked to rural modernisation that turned barren parts of the countryside into productive fields or improved yields.[85] These were on top of wide-ranging efforts by the Soviet Union to influence anti-colonial movements around the world, often with the aim of undermining western influence in regions of strategic importance or with access to valuable natural resources.[86]

Taken as a whole, such approaches seemed a challenge to the visions of freedom and democracy that were being championed by the United States. In fact, while some major programmes really were statements of intent – most notably the Aswan High Dam, which was funded by the Soviet Union to the chagrin of a shocked US State Department – Soviet economists, agronomists and development experts frequently complained that they did not have the resources to make meaningful differences in countries they were assigned to, and drew comparisons specifically with the funds available to their American peers and rivals.[87]

That Soviet achievement did not match rhetoric was true also of agricultural and ecological transformation within the USSR in the decade after the death of Stalin in 1953. Despite a testy relationship with Stalin, especially in the latter's final years, Nikita Khrushchev had emerged as a decisive power broker in Soviet politics and eventually as the supreme leader. In the months after Stalin's death, he proposed and then began to implement a massive overhaul of the countryside named the 'Virgin Lands' programme which aimed to transform Soviet agriculture. Khrushchev planned to convert millions of hectares across the Soviet Union, primarily in Siberia and Kazakhstan, into mechanised, industrial-size farming lands that dramatically altered not only the character of the natural world but also the population's diets: in 1953, consumption of meat and milk was lower than it had been in 1914.[88]

This approach owed something to the Second World War and in particular to changing demographics. In some parts of the Soviet Union, the gender balance of women to men shifted to two and even three to one – both because more men than women were killed during the war, and because of mass urbanisation as men who returned from the front moved to cities for jobs and opportunities that offered better rewards and working conditions than could be found in the countryside.[89]

But much too was owed to Khrushchev's own experiences and views about agriculture. Soviet statistics, he asserted at the start of 1954, were distorted to the point of being totally incorrect. Grain production two years earlier had not been 130 million metric tons, as the authorities had claimed, but 30 per cent lower, at 92 tons. Moreover, he said, significantly less grain was being produced than before the Second World War. What was needed was the opening up of vast tracts of land – more than 40 million hectares – and the creation of giant farms that could profit from the benefits of scale.[90] This was intended to produce more grain for human consumption, but also to generate a major increase in animal feed that would therefore lead to a boost in production of protein, milk and eggs and an improvement in living standards.[91]

A key element in this was the planting of corn: Khrushchev had seen how famine had been avoided in 1949 in Ukraine where he had been based for more than a decade by the planting of this crop, and he now became an evangelist, seeing corn as the catalyst for a new revolutionary era in the Soviet Union. Corn, he said, 'has proven itself to possess boundless potential as feed' and was 'the decisive requirement' for boosting the output of protein and dairy produce.[92] Energetic central planning yielded immediate results, with planting of corn rising fivefold between 1953 and 1955 and then doubling again in the next seven years.[93]

The programme was not just about ploughing new lands and planting new crops. It was part of a much wider programme that included novel uses of genetics, soil chemistry, machines and engineering and accountancy to upgrade an entire sector in the search for abundance. Khrushchev was an ardent believer. 'This is all nonsense,' he told one official on a visit to Egypt who tried to explain tried-and-tested local methods of cultivation. 'You're wasting your time,' he went on; 'chemical agriculture is the answer.'[94] These new techniques and technologies would improve lifestyles. 'We want Soviet people to eat to their hearts' content,' Khrushchev proclaimed, 'and not just bread, but good bread, as well as sufficient meat, milk, butter, eggs, and fruits.'[95]

The transformation of the Soviet countryside was framed as the fulfilment of political destiny. 'In the competition with America, comrades, there is no doubt victory will be ours,' said Khrushchev. 'This is because our economy, based on the teachings of Marx and Lenin, develops without the bourgeoisie, the landowners, or the exploitation of man by man.'[96] A delegation that visited North America in 1955 had confirmed as much in a report prepared after its return to the Soviet Union. 'The enormous advantages of our socialist system became clearer to us' during the visit, it explained, especially given America's 'destruction of small farmers' – something firmly in line with Khrushchev's belief that in many parts of the world agriculture was 'concentrated in the hands of a numerically small group of capitalist farmers'. It was simple, declared the report: 'We enjoy enormous advantages over the USA.'[97]

This was not true. The Soviet Union was at a chronic climatological disadvantage when compared to the United States. The thermal conditions of the USSR were poorly suited to agriculture: 80 per cent of Soviet cropland was located in what would be classified as 'the least productive thermal zone' – four times as high as was the case in the US. Moreover, while around a third of US cropland was in the most favourable zone for agricultural production, the same was true for just 4 per cent of the equivalent in the USSR. Comparative rainfall levels were also very heavily weighted in favour of North America, where moist oceanic air helps support high continental moisture levels. According to some estimates, while 56 per cent of land suitable for cereal cultivation in the United States has an optimum combination of temperature and moisture, the same was the case for just 1.4 per cent of land in the Soviet Union.[98]

This was not the only challenge the Soviet Union faced when comparing its model with that of the United States and the capitalist west where the 'American way of life' rested on ideas (which had become realities) of superabundance, choice and low prices. Efficiency in supply chains, standardisation of quality, competition between retailers and wafer-thin margins had turned western supermarkets into extraordinarily successful enterprises – and effectively into shop windows for the rest of the world. Shelves groaning under the weight of goods and products made for compelling publicity and served as important tools in Cold War rivalries: while scholarly attention usually focuses on missiles, technology and ideology, the ways that nature was exploited provided a canvas that was easy to interpret and meant more to everyday citizens than high-flown discussions of theory.

The Americans knew it too: events like the 1957 Zagreb Trade Fair were perfect for what John Logan, head of the National Association of Food Chains in the US, called 'effective propaganda for the democratic way in Eastern Europe'. Another example was an exhibition in Sokolniki Park in Moscow in 1959 allowed by the Soviet authorities to mark a thaw in relations between the USSR and the United States. It was seized on by US planners as a chance to show off 'the freedom of choice and expression, and the unimpeded flow of diverse goods and ideas' that epitomised the free world – and that stood in contrast to those in the Soviet sphere. Behold an array of devices, said Vice-President Richard Nixon during a visit, such as automatic stoves and robot floor cleaners that would 'make easier the life of our housewives'. Although Khrushchev retaliated by telling Nixon off for a typically 'capitalist attitude toward women', it was painfully obvious that American ecologies and the US business environment produced results that were far beyond those of the Soviet Union. 'We don't have stores like this in Russia,' Deputy Premier Anastas Mikoyan said during a reciprocal visit to the USA in 1959.[99]

It was vital, Khrushchev said repeatedly in meetings in Moscow, 'to overtake and surpass' the United States, both in volume of production of foods and goods and in innovation in science and technology.[100] Colonial states were 'agog' at the abundance on display in the west; It was essential then to show that the socialist model could do more than keep up, that it could do better.[101]

It was not just competition with the outside world that drove Soviet socio-economic policy. So too did fears about the dangers of failing to reform, modernise and adapt. As Khrushchev later recalled, raising living standards was important for ensuring stability and continuity at home: poor housing, bland diets and shortages were all ingredients that might individually or collectively result in mass uprisings – as indeed proved to be the case in East Germany in 1953 and in Poland and Hungary in 1956. There were concerns too that a loosening of controls that had tightened inexorably under Stalin might create unintended consequences. 'We were afraid to lose control of the country,' wrote Khrushchev; 'we didn't want some tidal wave to come along that would sweep us away.'[102]

In other words, the transformation of Soviet agriculture was part of an extensive reorientation in the USSR and its satellite states.[103] In the late 1950s, for example, ambitious new decrees ordered mass housing construction to provide apartments that not only offered improvements in ways of living but created private spaces from which the state could be excluded, at least in theory.[104] Clothing brands like Moda Polska in Poland and magazines like *Sibylle* in East Germany were launched to promote ready-to-wear fashions and the latest styles – examples of a shift towards consumerism. They were also tacit admissions that the needs of citizens, and especially those of women, had been ignored because of the overwhelming focus on production. Efforts went into the manufacture of consumer goods that would 'alleviate women's work' in the home, while one leading newspaper began a new 'home and family' page aimed at female readers that focused on fashions, domestic advice and appliances for the home. The dramatic change from the tenets set out by Lenin and his acolytes was clear from Khrushchev's statement at the 22nd Party Congress in 1961 that 'personal ownership by the toiler of a large number of things ... is not at variance with the principles of communist construction.'[105]

If this sounded unconvincing, then the 'Virgin Lands' deserved an even less favourable verdict. Hundreds of thousands of new settlers were sent to Kazakhstan, putting pressure on infrastructure and creating social friction and ecological catastrophe that led to what has been called 'the worst case of ecocide' in the history of the Soviet Union – the reworking of the Aral Sea's tributaries into irrigation channels which was as ill-conceived as it has proved devastating.[106] The exposure of the dry seabed of the Aral Sea (once the world's fourth largest) had reached 87,000 square kilometres by 2010, with wind spreading 45 million metric tons of salty and contaminated dust each year and creating dust plumes that can reach 400 kilometres in length and 40 kilometres in width. This now directly affects some 5 million people, with one in two women suffering from serious gynaecological diseases as a result.[107] All in all, the transformation of the Aral Sea basin has been nothing less than a disaster.[108]

Khrushchev's master plan to reconfigure the natural environment ended in humiliation. In 1963, Moscow was forced to turn to Washington for grain imports, an admission that agricultural reforms had failed, and at the same time a damning indictment of what one US senator called 'the greatest weakness in all the Communist countries', namely 'their inability to produce food'. Other senators wanted to use the opportunity more aggressively. 'Food is a weapon in the cold war just as much as bullets or bombs,' said Strom Thurmond. Selling wheat to the Soviet Union would simply enable it to keep investing in military affairs; far better, he seemed to believe, to try to starve the USSR into democracy.[109]

The debacle cost Khrushchev his job. In October 1964, he was ousted as leader, with his colleagues and former supporters rounding on him for his 'hare-brained scheming' and in particular for his obsession with corn. As more generous recent scholarship has noted, Khrushchev was hardly to blame: the Soviet Union he had led and tried to modernise had been one ravaged by crises in agrarian production and more generally; moreover, condemning Khrushchev for driving forward a personal policy obscures the responsibility of others who had endorsed or tolerated plans to reshape the country on a monumental scale.[110]

If things did not go according to plan in the Soviet Union, at least they did not go as badly as in China. The country had been in turmoil following the Xinhai revolution of 1911–12, which deposed the last emperor, Puyi, and turned China into a republic. In the decades that followed, successive crises provoked by egregious settlements at the end of the First World War led to the rise of the Communist party; the result was civil war, which dramatically weakened resistance to Japan in the 1930s and during the Second World War – epitomised by the annexation of Manchuria in 1931 and the Nanjing massacre six years later in which somewhere between

40,000 and 300,000 people lost their lives. To make matters even worse, in February 1945, before the war in Europe had ended, Joseph Stalin made territorial demands, not least for control of the seaport of Port Arthur, as well as insisting on significant concessions over railway infrastructure – demands that were conceded by Churchill and Roosevelt without consulting the Chinese authorities.[111]

Stalin offered little by way of support or encouragement for the Chinese communists under the leadership of Mao Zedong – indeed quite the opposite. The Soviet leader contacted Mao in August 1945 to ask him to ensure that his troops avoided conflict with government forces. Mao took this to heart, issuing orders a few weeks later: 'We must stop [the] civil war and all parties must unite under the leadership of Chairman Chiang [Kai-shek] to build a modern China.' Part of Stalin's motivation was to try to ensure the withdrawal of US troops from East Asia; but there was also the lure of the industrial heartland of Manchuria, which in the autumn of that year was systematically looted by Red Army soldiers. As one eyewitness recalled, they 'stole everything in sight, broke up bathtubs and toilets with hammers, pulled electric wiring out of the plaster'; by the time they had finished, 'factories lay like raddled skeletons, picked clean of their machinery'. The value of this plunder was around $2 billion.[112]

With the fault lines of the Cold War already taking shape, a series of decisions were now made in Washington that had momentous significance. Unlike western Europe, which benefited from a massive injection of funds to help it rebuild, little or nothing was set aside to support China. Moreover, George Marshall, the architect of the regeneration of post-war Europe, was sent to China by President Truman with instructions not to back the government of Chiang Kai-shek, but to broker an agreement with Mao and the communists, persuaded by Stalin's apparent lack of interest in China, and by the soothing promises that 'Chinese democracy' under Mao would follow 'the American path'. Mao and his followers had little to lose and much to gain by making such commitments – especially when Washington imposed an arms embargo, which tied the hands of Chiang while those of his rivals were being untied and empowered with a massive injection of rifles, machine guns, aircraft and artillery from Moscow.[113]

The results were predictable and dramatic. Uncertainty and loss of confidence led to chronic price inflation, the cost of living being around 30,000 times higher in 1947 than it had been just over a decade earlier. Not surprisingly, this quickly spilled over into economic chaos and ensured the effective failure of the organs of bureaucracy as the government struggled to pay the salaries of officials, which in turn fuelled corruption and state failure. This played into the hands of the communists, who made a series of stunning gains in 1948 as one city after another fell into their hands; by the start of 1949, the Liberation Army had arrived at the west gates of Beijing.[114] A new world had been born.

What this new world would look like was far from certain. By the time the forces of Mao Zedong had established control over most of mainland China, there had already been considerable soul-searching among Chinese scholars, writers and strategists as they debated how and why China had been left behind by the countries of Europe, which had grown into global powers and suffocated China in the process during what became known as the Century of Humiliation. Many had radical ideas about what was needed to modernise the country and to turn it back into the behemoth that it had once been. One of the most vocal and authoritative was Sun Yat-sen, who was elected the first President of the Republic of China, albeit on a provisional basis, at the start of 1912. For Sun, the United States was the obvious model to replicate. 'In order to make sure of our success,' he wrote, addressing the American people, 'we intend to model our new government after yours.' There were many reasons for this; but 'above all', he wrote, it was 'because you are the champion of liberty and democracy'.[115]

He wrote a book about China's 'vital problems'. These included the British, who saw as allies and friends those who 'can render her services', though 'when her friends are too weak to be of any use to her, they must be sacrificed in her interests'. The British, he said, treat their friends no better than farmers treat silkworms – namely using them, and killing them.[116] Sun thought, wrote and spoke at length about what China needed. In 1924 he gave a speech setting out his views. Above all, China had to turn into a modern, industrial state. His vision of what this looked like was clear: a country studded with cities the size of London and New York, linked by 160,000 kilometres of railways, and connected by 1.5 million kilometres of new paved roads. This would transform a poorly connected agrarian economy into a powerhouse. About 100 million horsepower might be obtained, he suggested, from hydroelectric plants on the Yangtze and Yellow rivers This would be enough to 'supply railways, motor cars, fertilizer factories, and all kinds of manufacturing establishments'. It was time, in other words, for China to leapfrog several stages of development, to become an advanced country and reclaim its place in the world.[117]

Although Mao had been taken with the idea of China developing along the lines of the United States, declaring that 'Chinese democracy must follow the American path,' his ideas changed radically following the seizure and consolidation of power. They sharpened again during the Korean War when he committed hundreds of thousands of troops to support the forces of Kim Il-sung. The United States was 'thoroughly dark, thoroughly corrupt, thoroughly cruel', declared one leading newspaper, closely echoing Mao's thinking; it was a country that was the garden of 'Eden of a few millionaires, [but] the hell of countless millions of poor people'. Put simply, it was the 'source of … darkness, cruelty, decadence, corruption, debauchery [and] oppression of man by man'.[118]

Mao was after a different model. This was not to be found in the US after all, as a Central Directive issued in December 1950 made clear: 'Hate America. Despise America. And look down on America.'[119] Instead, Mao turned to Moscow for inspiration. 'Learn from the Soviet Union' was one motto adopted soon after a Treaty of Friendship had been agreed between Mao and Stalin in 1950; Russian would become compulsory in all Chinese schools; China was to follow not in American but in Soviet footsteps. As Mao confidently predicted, 'The Soviet Union's Today is our Tomorrow.'[120]

In this context, then, Khrushchev's savage denunciation of Stalin in 1956 had severe repercussions in China generally and for Mao personally. The seizure of power had come at a price of severe turbulence and violence which Mao had not only approved of but actively encouraged. He had set crude targets for identifying 'counter-revolutionaries', a deliberately broad categorisation that included the wealthy small landowners and teachers – and in some cases entire communities: for example, Shanghai, home to the largest foreign population of any metropolis in the world with the exception of New York, was 'a non-productive city, a parasitic city'. A cycle of violence had followed that was epic in scale and devastation, all encouraged by Mao. 'Killing counter-revolutionaries,' he said, 'is even more joyful than a good downpour.'[121]

It was not hard to see how the attacks on Stalin for being domineering, instigating violence and failing to introduce reforms that produced truly positive outcomes could be applied to Mao. At the 8th Party Congress in September 1956, a few months after Khrushchev's bombshell, the cult of personality was criticised while the importance of collective leadership was praised. Tellingly, 'Mao Zedong thought' was removed as a central tenet of Chinese communism in a sign of how vulnerable the leader's position had become.[122]

Highly sensitive to shifts in the direction of the wind, Mao reacted swiftly and decisively to the changes of tone and direction in Moscow. The Soviet model was not, in fact, one to follow, he said, berating those who 'copy everything and transplant it mechanically'. The USSR, moreover, had made many mistakes, especially in relation to the countryside and to rural populations – something which could not be said, Mao had the nerve to assert, about China. If that was bold, so too were his demands for what was now needed: a root-and-branch reform of the Communist party itself, which had clearly gone off the rails. Freedoms had been suppressed by those trying to take and preserve power for themselves, he said, in the process starving the people of their voice. It was important to 'let a hundred flowers bloom, let a hundred schools of thought contend'.[123] These were first-class survival instincts.

Mao needed to do more, though, than reposition his vision for China and reposition himself. Having already 'out-Khrushcheved Khrushchev' with his own brutal dismissal of Stalin, his policies and the weakness of those who had surrounded him, Mao had to come up with concrete proposals that were clear, simple and achievable. It did not take him long to find the ingredients for a perfect cocktail. These included a long-term plan, announced in 1956, for the 'chemicalisation, mechanization and electrification' of China to help boost agricultural production – just as Khrushchev was attempting in the USSR. As one formal report put it, 'the principle of ensuring food security via grain self-sufficiency needs to be the top priority of the development strategy, and this principle should never be slackened'.[124]

Mao distilled this for the masses with a single phrase: 'Man Must Conquer Nature' (*Ren ding sheng tian*). There was no reason to think that humans could not affect astronomy, geology and the natural world, he wrote, criticising Stalin's famous essay which had argued the opposite.[125] Nature could – and should – be forced to obey the will of its human masters. 'When we ask the mountain to bow its head, it has to do so! When we ask the river to yield the way, it must yield!' said Mao.[126] 'Man's ability to know and change nature is unlimited,' he remarked on another occasion – with one senior acolyte commenting with approval that 'No other world leader looks down with such disdain on great mountains and powerful rivers.'[127]

These words were chosen to galvanise people into action. Dams, hydropower plants and irrigation schemes were undertaken on a colossal scale – largely driven by Mao's sense of urgency that dramatic and rapid change was needed.[128] In 1958 alone, for example, almost 600 million cubic metres of earth were moved; the following year, twelve times more earth was moved in a single week than during the construction of the Panama canal. That, at least, was what the authorities claimed. If nothing else, the claim gives an indication of the resources that were poured into modernising China – and of the ambition that lay behind it all. Agricultural yields would rise, production would gallop ahead and within fifteen years, Mao predicted, China would have caught up with the symbol of western imperialism – Great Britain. In fact it was not long before he was arguing that this could be achieved sooner, perhaps in as little as two to three years.[129]

Those who doubted the logic, methods or principles found themselves demonised and disgraced. This had been the practice even before the massive projects of the late 1950s. For example, at the start of the decade, Huang Wanli had raised concerns about plans to build the Sanmenxia – or Three Gate Gorge – dam on the Yellow River, noting that the heavy load of silt carried by a waterway colloquially known as China's Sorrow would be likely to damage the dam and cause environmental damage, to say nothing of the forced resettling of almost half a million people. For his troubles, Huang was labelled a 'rightist', the standard designation of anyone deemed suspect, and his professional career ended overnight. This happened to many others, such as Chen Xing, who argued that dam construction would lead to waterlogged land and in turn to salinity and alkalisation. Those close to Mao, like his personal secretary Li Rui, who warned that the Great Leap Forward carried risks, found that their position did not protect them from being removed, not only from their positions but from public life.[130]

Many of those identified as 'rightists' found themselves placed in camps, often in remote parts of the country where they added to the labour force of many millions sent from the cities to cut down trees, dig canals, build roads and bridges and open up new lands. Vast areas of China were transformed as a result, with workers urged on with slogans to 'destroy the forests, open the wastelands', to 'squeeze land from the rock peaks, get grain from the rocks'.[131] Villages that had achieved outstanding results were held up as paragons to be emulated, examples of what could be achieved with hard work and devotion to the cause.

One such case, Dengjiabao in Gansu province, became the subject of posters and films, not least because of the positive story it told about gender roles. Efforts had long been made to encourage women to turn away from 'unproductive' handicrafts and to play a part in collective agriculture. With yields of grain more than doubling between 1952 and 1956, the local authorities put forward ambitious plans to create 'ten thousand Dengjiabaos'.[132]

The reality at Dengjiabao was rather different: as one official put it years later, 'most of it was exaggeration. It was bragging.' Moreover, after soaring briefly, yields soon returned to their previous levels – which threatened villagers' ability to feed themselves because of state requisitioning of grain.[133] If that told one important story, then the disastrous decision to connect the Great Leap Forward with changing ecosystems and food chains told another. Densely inhabited, the large cities in China had long suffered from public health problems, of which the spread of epidemic and pandemic disease was the most obvious. In the early 1950s, the authorities had claimed that germ warfare was being conducted by the US at a time of rising tensions through insects such as infected flies, mosquitoes, spiders, ants and more being deliberately dropped on urban populations.[134]

Inevitably, the rapid modernisation of China also now had to embrace an upgrade in hygiene standards, especially in cities – to reflect the utopian socialist ideal. Thus was launched a nationwide campaign to 'Wipe Out the Four Pests', egged on by Mao who insisted at the 8th Party Congress in 1958 that 'the whole people, including five-year-old children, must be mobilised to eliminate the four pests'. The pests in question were rats, flies, mosquitoes and sparrows. The inclusion of the latter in this list was based on the assumption that sparrows ruined harvests. 'We made ladders to knock down their nests and beat gongs in the evenings, when they were coming home to roost,' remembered one school pupil.[135] Between March and November 1959, almost 2 billion sparrows were killed across China.[136]

What hadn't been considered was that while sparrows did indeed eat grain, they also consumed vast numbers of insects. The extermination of so many of these birds – they became a rare sight for years afterwards – resulted in loss of grain due to insect infestation and so had a direct impact on harvest figures. This magnified the problems caused by rapid modernisation, urbanisation and industrialisation. The result of all these efforts, in the words of one leading historian, was 'the greatest human-created famine in history'. Between 1959 and 1961, the death toll from starvation beggars belief. While it is difficult to be precise about numbers, most historians estimate that somewhere between 35 and 50 million people perished because of food shortages.[137]

The impact on the environment was devastating too: as famine kicked in, animals of all kinds were slaughtered for food, from livestock to domestic pets to rodents; plant life suffered too, with anything edible consumed by those desperately trying to stay alive. Seeds and roots were pulled from the ground and eaten, while trees were stripped of their leaves and bark. The environment took decades to recover.[138] On top of this were the ecological consequences of the huge deforestation intended to create more lands for crops. In a rare moment of candour in 1966, Premier Zhou Enlai admitted his misgivings about the damage that had been done. 'I fear that we have made a mistake in harnessing and accumulating water and cutting down so much forest cover,' he said. 'Some mistakes can be remedied in a day or a year, but mistakes in the fields of water conservancy and forestry cannot be reversed for years.'[139]

In the course of the twentieth century, innovations across multiple scientific fields, the demand for food and energy and competition between political ideologies relentlessly drove forward the exploitation of natural resources. Some steadfastly refused to learn lessons that had caused such pain elsewhere. 'A revolution is a force more powerful than nature,' said Fidel Castro in the wake of Hurricane Flora, which struck Cuba in October 1963. Winds, storms, cyclones and hurricanes 'are nothing compared to the revolution', he said. This was correct, agreed *Revolución*, the newspaper that served as Castro's mouthpiece; 'Neither a hundred "Floras" nor a hundred imperialisms can defeat us,' declared the paper the day after the hurricane had passed. Quite right, noted Raúl Roa, Cuba's Foreign Minister. Referring to the great liberator of much of South America, he declared that 'like Bolívar, if Nature opposes us, we will fight against it'.[140]

It was important to take such fighting talk with a pinch of salt, to understand declarations of defiance as expressions of political determination in the face of adversity rather than as literal statements. Nevertheless, in the middle of the twentieth century, few stopped to think about sustainability – partly out of a sense of optimism that new technologies would bring about what Henry Wallace termed in 1942 the 'century of the common man'. Wallace, at the time Vice-President of the United States, was sure that new ideas, new inventions and new technology would be crucial in spreading freedoms and in particular in boosting agricultural yields. 'Modern science', he said in a speech heralded by one commentator as 'the Gettysburg Address of World War II', had 'potentialities of which we do not yet dream'. One thing that was possible to imagine, however, was that innovation 'made it technologically possible to see that all the people of the world get enough to eat'.[141]

By the 1960s, things looked rather less rosy. There were rare beacons of light and hope – such as Martin Luther King and the civil rights movement in the United States that aimed to give equality to all, or advances in rocket science that enabled humans to walk on the moon at the end of the decade. For many, however, it felt like storm clouds were gathering. The Cuban missile crisis provided a stark reminder of how high the stakes were and how fine the line was between a world of peace and one of destruction. War in Vietnam, which spilled into South-East Asia, reinforced these concerns.

In this context, it was perhaps no coincidence that a burgeoning counterculture did not just focus on freedoms and on resistance to consumerism, but had an important ecological message. Concerns were building in the corridors of power too that consumption of resources, choices of lifestyles and the search for economic growth all put pressure on and came at the expense of the natural environment. In some quarters, deepening anxieties led some to question whether there were circumstances in which human actions and decisions might even bring about the extinction of our own species. It was time for serious change.

23

The Sharpening of Anxieties
(c.1960–c.1990)

The magnitude of the first-order effects are so large, and the implications so serious, that we hope the scientific issues raised here will be vigorously and critically examined.

TTAPS paper on 'Nuclear Winter', *Science* (1983)

The Cold War played out in many different regions of the world in many different ways. One important, but often neglected, impact that it had was on attitudes to the environment. Concerns about the existential threat to life were crucial, however. As the American anthropologist Margaret Mead noted in the late 1960s, a generation was growing up fearful of global destruction, of cataclysmic nuclear war and of an impending apocalypse that was all but inevitable. Young people, wrote Mead, 'have never known a time when war did not threaten annihilation'.[1] While the slogan 'make love not war' is often used to illustrate the anti-war movement roused by US involvement in Vietnam, the emerging ideas of the counterculture were closely linked to ecological concerns.

Beat writers like Jack Kerouac had already begun to write about nature as an escape from city life and from the mundane realities of suburban America. In *The Dharma Bums* (1958), for example, Kerouac's narrator Japhy Ryder has a vision of a 'great rucksack revolution' in which 'thousands or even millions of young Americans' abandon bourgeois existence to head to the mountains to pray, to spend their time making children laugh and old people happy, motivated by the idea of 'giving visions of eternal freedom to everybody and to all living creatures'.[2] Drugs played a role too: LSD opened doors to higher meanings than were offered by a life of work, said Timothy Leary in New York in 1966. The purpose of existence, said Leary, was 'the glorification and the worship of God', which were after all 'ancient goals' that deserved endorsement today. As such, the right thing to do was 'turn on, tune in, drop out'.[3]

Black activists in the US, such as Coretta Scott King, were prominent in the anti-nuclear movement, connecting disarmament with the fight for racial equality.[4] Women also played a crucial role in bringing concerns to the forefront of public consciousness. On 1 November 1961, around 50,000 'concerned housewives' went on strike across the US, with protests in over sixty cities including New York, Chicago, Detroit, St Louis and Los Angeles that ranged from marches to picketing of nuclear facilities to lobbying officials, with the aim of banning atmospheric atomic weapons tests.[5] This was set out in an open letter written to President Kennedy by Dagmar Wilson, a children's book illustrator who was a prominent figure in the budding movement. 'American women today,' she wrote, 'want to see an end to nuclear testing by any country.'[6]

This was part of a much longer tradition in North America and in Britain, where women had been influential in the founding of conservation programmes to protect forests, waterways and wildlife and in the creation of animal rights movements prompted by concern about treatment of laboratory animals or about conditions in zoos and wildlife reserves.[7] By the mid-twentieth century, women's roles had become more restricted than they had been a century earlier, with men dominating public space while women were typically expected to be home-makers and to accept traditional female roles, for example by being teachers or librarians or giving advice about home economics.[8]

From the early 1960s, however, that started to change. Women's magazines such as *American Home*, *Redbook* and *Good Housekeeping* began to publish regular articles about pollution, falling environmental standards and the impact of these on family life. As one recent scholar has put it, 'the environmental cause seemed a natural extension of [middle-class women's] concerns as housewives and mothers'.[9] *Silent Spring* by Rachel Carson, published in 1962, provided a rallying cry therefore for those with deep concerns about ecological damage and about the degradation not only of the idyllic dreams of the American natural world but of the American way of life.

Carson mourned the time when 'all life seemed to live in harmony with its surroundings', characterised by 'fields of grain and hillsides of orchards', where foxes barked and 'deer silently crossed the fields, half hidden in the mists of the fall mornings' – all of which had been replaced by commercial farming that relied on industrial use of pesticides. It was as though 'some evil spell had settled on the community', marked by stillness – 'a spring without voices'. Chemicals had killed all the birds; they also lay behind 'sudden and unexplained deaths, not only among adults but even among children'. Farmers themselves 'spoke of much illness among their families'. It was unbearable: 'Everywhere was a shadow of death.'[10]

Pesticides, wrote Carson, 'have been found in fish in remote mountain lakes, in earthworms burrowing in soil, in the eggs of birds – and in man himself. For these chemicals are now stored in the bodies of the vast majority of human beings, regardless of age. They occur in the mother's milk, and probably in the tissues of the unborn child.' If this was not clear enough, one only needed to turn to the dedication, in which Carson thanked the polymath Albert Schweitzer and quoted his gloomy prediction that 'Man has lost the capacity to foresee and to forestall. He will end by destroying the earth.'[11]

Silent Spring was a runaway bestseller, its sales owing much to canny marketing by publishers who sent advance copies to women in prominent positions, to groups such as the League of Women Voters and to the heads of the Children's Bureau, the National Council of Jewish Women, the National Federation of Women's Clubs and more besides. *Silent Spring* also helped influence debate in contemporary politics, even prodding President Kennedy into commissioning a report on the use of pesticides.[12]

Activism around ecological issues, including the synthetic chemicals that were the focus of Carson's criticism, took place within a broader debate about private wealth and public poverty in which the Harvard academics Arthur Schlesinger Jr and J. K. Galbraith were perhaps the most prominent voices.[13] How was it possible, Galbraith argued in *The Affluent Society*, his own bestselling book that came out a few years before *Silent Spring*, that – while many were well off – cities could be 'badly paved, made hideous by litter, blighted buildings, billboards' and cabling that 'should long since have been put underground'? How was it possible that one could enjoy abundance, but have to do so 'amid the stench of decaying refuse'? Was this, he mused, 'the American genius?'[14]

Questions like these spilled into the mainstream, featuring prominently in publications like the *New York Times*, which complained in 1960 that despite the United States enjoying levels of 'private wealth never before seen on earth', the country's 'education is underfinanced. Streams are polluted. There remains a shortage of hospital beds. Slums proliferate, and there is a gap in middle-income housing. We could use more and better parks, streets, detention facilities, water supply. The very quality of American life is suffering from these lacks.'[15]

Well-placed senior figures in US politics wrestled with this problem. 'America stands today poised on a pinnacle of wealth and power,' wrote Stewart Udall, who served in both the Kennedy and Johnson administrations as Secretary of the Interior, 'yet we live in the land of vanishing beauty, of increasing ugliness, of shrinking open space, and of an overall environment that is diminished daily by pollution and noise and blight.'[16] Lyndon Johnson put it bluntly in May 1964, just a few months after he took over the presidency following Kennedy's assassination. 'We have always prided ourselves on being not only America the strong and America the free, but America the beautiful,' he said in a speech at the University of Michigan. 'Today that beauty is in danger. The water we drink, the food we eat, the very air we breathe, are threatened with pollution ... Green fields and dense forests are disappearing.' As a result, our 'communion with nature' was being eroded. This brought ominous risks, he went on, 'For once the battle is lost, once our natural splendor is destroyed, it can never be recaptured.'[17] Urgent action was required.

Anxieties about ecological devastation were compounded by coverage of the Vietnam War. That conflict had a significant environmental slant as a result of the deliberate use by the US military of herbicides and defoliants against woodlands, crops and rice fields to degrade the landscape, to deprive the Việt Nam cộng sản (or Vietcong) forces of food and so to weaken their capabilities and demoralise them. These tactics emerged from a classified research programme intended to explore the best ways to gain military advantages via interaction and interference with nature.[18]

Between 1962 and 1971, Operation RANCHHAND sprayed around 90 million litres of herbicides over South Vietnam – including Agents

Pink, White and Blue but most famously Agent Orange, the most extensively used.[19] In 1966, scientists protested to President Johnson, 'Even if it can be shown that the chemicals are not toxic to man, such tactics are barbarous because they are indiscriminate; they represent an attack on the entire population of the region where the crops are destroyed.'[20] A second petition, signed by 5,000 scientists including seventeen Nobel prize winners, was addressed to President Johnson, demanding that he end the use of herbicides in South-East Asia.[21] What the US was doing, said Yale professor and plant biologist Arthur Galston, who had spent several years studying the ecological and health effects of herbicidal war in Vietnam on plants and humans (including US servicemen), amounted to 'ecocide'.[22]

Then there were accidents that caused considerable damage and received widespread attention in the national and international press. In January 1969, a blowout in a well being drilled by Union Oil on Platform A off the coast of Santa Barbara in California led to oil being released into the ocean at a rate of around 40,000 litres per hour for eleven days. By the time it was stopped, almost 1.4 million litres had been spilled.[23] Activists who had formed a group within days which they named 'Get Oil Out!' helped generate media coverage for an event described a year later by one journalist as an 'ecological "shot heard around the world"'.[24]

Another shot came from a blaze on the River Cuyahoga in Cleveland, where oil and other pollutants caught fire in June 1969. Although the incident was small-scale and easily dealt with, it too captured the imagination, of the press above all. 'Some river!' exclaimed the author of an essay in *Time* magazine published in the aftermath. 'Chocolate-brown, oily, bubbling with subsurface gases, it oozes rather than flows.' The river 'receives the wastes of steel mills, chemical and meat-rendering plants, and other industries', ran a story in *National Geographic* in 1970, in an issue devoted to 'Our Ecological Crisis'.[25]

These disasters, coupled with other high-profile accidents such as the *Torrey Canyon* which spilled 160 million litres of crude oil after running aground off the western coast of England in 1967 fuelled calls for action. Carl Stokes, the first black mayor of a major US city, was instrumental, giving testimony in Congress and demanding the introduction of controls to deal with the 'rape of the Cuyahoga river' and its effect on the ecology of the Great Lakes.[26] President Nixon, who had been in office for just over a week when the Santa Barbara blowout took place, was clear that change was required. America needed to use 'the resources of the sea and of the land in a more effective way and with more concern for preserving' the natural world for the 'society we want for the future'. There could be no doubt, he went on, that environmentally damaging incidents had 'frankly touched the conscience of the American people'.[27]

Nixon was reading the tea leaves of a shift in public opinion which was so profound that the *New York Times* speculated that 'rising concern about the "environmental crisis"' was 'sweeping the nation's campuses with an intensity that may be on its way to eclipsing student discontent over the war in Vietnam'. Students were organising lobby groups, organising learn-ins, protesting against unnecessary packaging and in some cases even filing complaints against university authorities for their role in ecological damage and pollution. 'Environmental issues will obviously replace other major issues of today,' said one student at the University of Washington. Thank goodness, noted a young geologist at the University of Colorado, 'people are beginning to sense [the] urgency' of the crisis. Many students were already looking forward to 'D-Day' on 22 April 1970, reported the newspaper, to mark a nationwide campaign that some believed 'could be a bigger and more meaningful event than the anti-war demonstrations'.[28]

This proved to be an accurate prediction. Urged to create 'a huge grassroots protest' by Gaylord Nelson, a senator from Wisconsin who was one of the most powerful and long-standing advocates of stricter regulations on uses of the landscape and on water and air pollutants, students, scientists and conservationists worked together to publicise what was renamed 'Earth Day'. On 22 April 1970, around 20 million Americans demonstrated and staged teach-ins about the environmental crisis and propelled these concerns to the top of political agendas around the world.[29] As the famous broadcaster Alistair Cooke put it, Earth Day was 'the first mass reminder of our decaying and polluted planet'.[30]

President Nixon moved 'to place improved quality of life at the head of the national agenda for the new decade', noted the *New York Times*.[31] He was hardly the first to do so – his predecessor, Lyndon Johnson, signed more than 300 conservation measures into law, including the Clean Air Act of 1963, the Water Quality Act 1965 and the Wilderness

Act of 1964, the latter covering more than 9 million acres of national forests.[32] Nevertheless, Nixon now swung into action, signing the National Environmental Policy Act, revising the Clean Air Act and issuing an Executive Order to create the Environmental Protection Agency – all before the end of 1970.[33]

By this time, there had already been appeals at the United Nations pointing to the 'urgent need for intensified action at the national and international level, to limit and, where possible, to eliminate the impairment of the human environment'.[34] In the build-up to the first UN Conference on the Environment, held in Stockholm in June 1972, an unofficial report commissioned by Kurt Waldheim, the UN Secretary General, stressed the fragility of 'the planet's energy system' and the dangers of 'small but fateful changes which alter the balance of the seesaw'. What was needed, the authors urged, was 'a new capacity for global decision-making and global care ... a new commitment to global responsibilities'.[35] Put simply, 'Man must accept responsibility for the stewardship of the world.'[36]

'We share your concern at the rapid deterioration of flora and fauna,' said Indira Gandhi, the Prime Minister of India, in her address to the conference. 'Some of our own wildlife has been wiped out, miles of forests with beautiful trees, mute witnesses of history, have been destroyed.' However, while it was important, even essential, to 'deal with incipient environmental imbalances', it should not be forgotten that another species 'is also imperiled': man. 'In poverty,' said Gandhi, 'he is threatened by malnutrition and disease, in weakness by war, in richness by the pollution brought by his own prosperity.' Advanced countries had become rich through dominating others and exploiting their resources. 'We do not wish to impoverish the environment any further,' she went on, 'and yet we cannot for a moment forget the grim poverty of large numbers of people. Are not poverty and need the greatest polluters?' It was important to underline that 'we inhabit a divided world'.[37]

This harked back to commitments that had been made long before about trying to bring an end to poverty and hunger in many parts of the world. Although John F. Kennedy's inaugural address as president in 1961 is best known for his challenge to 'Ask not what your country can do for you – ask what you can do for your country,' his speech made promises that were as ambitious as they were wide-ranging. 'To those new states whom we welcome to the ranks of the free,' said the new President, 'we pledge our word that one form of colonial control shall not have passed away merely to be replaced by a far more iron tyranny.' The US would help those who needed help: 'To those people in the huts and villages of half the globe struggling to break the bonds of mass misery,' Kennedy added, 'we pledge our best efforts to help them help themselves, for whatever period is required – not because the communists may be doing it, not because we seek their votes, but because it is right. If a free society cannot help the many who are poor, it cannot save the few who are rich.'[38]

This echoed the promises made by President Truman more than a decade earlier. As we have seen, Truman was almost evangelical in his view of how the US could be a force for good in the world. 'America is faced with a solemn obligation,' he had said in a speech broadcast across the nation in April 1946. It was impossible 'to ignore the cry of hungry children. Surely we will not turn our backs on the millions of human beings begging for just a crust of bread … We would not be Americans if we did not wish to share our comparative plenty with suffering people. I am sure I speak for every American when I say the United States is determined to do everything in its power to relieve the famine of half the world.'[39]

At the heart of the problem of hunger, according to many policymakers and thinkers in the United States, was that population in the 'developing' world kept growing, 'not because birth rates are rising', as one CIA report written in the early 1970s explained, 'but because death rates, especially infant mortality, have fallen so sharply'.[40] What this meant had been a matter of considerable interest to general readers who had grappled with questions about the relationship between rising population, resource pressure and political instability that dated back to Malthus. These themes found new messengers in works such as Edward Murray East's *Mankind at the Crossroads* (1923), Warren Thompson's *Population and Peace in the Pacific* (1946) and above all William Vogt's *Road to Survival* (1948) and Fairfield Osborn's *Our Plundered Planet* (1948) – which paint ominous pictures of what the future had in store.[41]

These books proved highly influential, inducing Chester I. Barnard, the President of the Rockefeller Foundation, one of the wealthiest in the world, to ask his officers why programmes were being funded in Mexico to raise yields even though the earth's carrying capacity was already being stretched to its limits, if not beyond.[42] The logical conclusion of such perspectives was made clear in various reports, such as the 'World Food Problem' commissioned by the Rockefeller Foundation which underlined 'the conflict between population growth and unequally divided and inadequate resources', or the 'Food Crisis Report' prepared by the Ford Foundation in 1959 for the Indian government which asserted that 'the crux of the problem' facing India was the gap between the food supply and a rapidly rising population.[43]

The threat of a world facing a sky-rocketing population that would lead to mass starvation was not so much a problem to ponder in the years and decades ahead as one that was present and immediate. Americans had been told in 1946 that the situation in Europe and Asia was so dire that it was essential 'to save bread and to conserve oils and fat'. President Truman was emphatic in his message: 'Every slice of bread, every ounce of fat and oil saved by voluntary sacrifice, will keep starving people alive.' It was simple: 'Millions will surely die unless we eat less.'[44]

This was a message that was reinforced repeatedly in the 1950s and 1960s – although Paul and Anne Ehrlich's *The Population Bomb* of 1968 warned that the battle had already been lost. The chance 'to feed all of humanity is over', the authors declared in their hugely influential book. 'Hundreds of millions of people are going to starve to death.' The book's cover asserted that 'While you are reading these words, four people will have died from starvation. Most of them children.'[45]

The prospect of more and more people sharing the planet was one that added to anxieties about nuclear apocalypse and the transformation of the natural world into industrial heartlands that scarred the landscape and poisoned the soil, the water and the human population. Overpopulation now threatened to create yet another catastrophe. For some, it was obvious where blame should lie. 'The white race *is* the cancer of human history,' wrote Susan Sontag in the mid-1960s; 'it is the white race and it alone – its ideologies and inventions – which eradicates autonomous civilisations wherever it spreads, which has upset the ecological balance of the planet, which now threatens the very existence of life itself.' The Mongol hordes were far less frightening than the 'western man with his idealism, magnificent art, his sense of intellectual adventure [and] his world-devouring energies for conquest'.[46]

Others offered solutions. President Kennedy proposed 'the transfer of technology to food deficit nations' that would stimulate 'a scientific revolution which may well rival, in its social consequences, the industrial revolution'.[47] Such hopes were not entirely unfounded: research into high-yielding seeds that were responsive to high inputs of chemical fertiliser and irrigation had produced what is often referred to as the 'green revolution' – even if scholars today are often sceptical about how green this was, whether it amounted to a revolution and whether it was a legendary phenomenon rather than a real one.[48] Initial results showed that new dwarf wheat varieties, produced by cross-breeding short-strawed Japanese and Mexican strains, could produce four times as much as traditional varieties, thanks partly to improving resistance to rust and partly to resolving issues around late ripening. Similar approaches were then adopted to create new rice varieties, which were likewise early maturing, insensitive to day length and had short, stiff stems.[49] One of the most prominent pioneers, Norman Borlaug, won the Nobel peace prize in recognition of his work developing new cereal strains and then putting them 'into extensive production in order to feed the hungry people of the world'.[50]

These developments seemed to offer hope that the worst problems of rising global populations could be averted. In the late 1960s, the availability of what President Nixon called 'miracle seeds' promised a solution to the problems of global food supply.[51] The introduction of new hybrid crop varieties provided a substantial bonus, with recent studies suggesting that the calorific availability around the world was boosted by about 11–13 per cent.[52]

Ironically, in fact, the principal beneficiaries of substantially higher yields were in the countries that had already industrialised, that had developed new technologies and adopted them quickly.[53] Moreover, the surge in production generated a range of new and unexpected problems. Having been fearful of too little food being available, it was observed at a conference of leading experts, the concern was now about too much, with the Third World entering a period of surplus production: 'we worry more about idle arms than hungry mouths'.[54]

It was not wrong to be thinking of the socio-economic consequences. In South Asia, for example, improving fortunes and rising confidence precipitated a consolidation in landholdings as farmers with access to credit and know-how bought out smaller producers. In some parts of India and Pakistan, land values trebled as landlords moved in to exploit the opportunities on offer, aggravating social polarisation – especially in places where inequalities already existed.[55] As a UN study produced in 1974 expressed it, the green revolution compounded hierarchies rather than reforming them.[56]

The green revolution then was itself counterproductive, for one of the core aims had been not only to overcome nature but to soothe political instabilities. In practice, in many cases, it did precisely the opposite. In India in the early 1970s, for example, populists found support for their emphasis on the gap between the rich and the poor which was increasing as a result of improved harvests. In Sierra Leone, a country that is an integral part of the 'rice belt' of western Africa and one where rice is central to social life and economic activity, farmers who were already wealthy and integrated into rice markets could reap the rewards of new techniques, especially in lowland regions which did not favour traditional varieties of rice that were more dependent on rainfall.[57] In the Philippines, opportunities opened up for the consolidation of power by the regime and family of Ferdinand Marcos, which monopolised the supply of seeds, chemicals, loans and machinery – and controlled prices to boot.[58] The green revolution was supposed to offer 'technology as a substitute to both nature and politics, in the creation of abundance and peace'. Some scholars argue that it did neither.[59]

One reason for failure lay in the obvious fact that the green revolution was essentially seeking to overcome the ecological constraints in parts of the world that were poorly suited to large-scale agricultural production. Tropical locations that often saw intense rainfall were prone to erosion, soil run-off and the depletion of nutrients in lands whose cover had been stripped to convert to the growing of crops.[60] Overuse of synthetic pesticides and fertilisers led to the increase of heavy metals such as cadmium, lead and arsenic in the soil, whose pH levels rose as a result.[61] As one scholar has put it, 'the chief "success" of the idealised new agricultural technologies was to export the new high-yield but patently unsustainable agricultural systems of the industrial world throughout the tropics'.[62] The breathtaking scale of land degradation that results from agricultural practices associated with these initiatives is estimated to have affected nearly 2 billion hectares globally – as well as 2.6 billion people.[63]

The green revolution also had significant impacts both on ecosystems and on health. The crops introduced to India in the 1960s and later were water-intensive – with the irrigated agriculture that subsequently arose being the primary reason why the country is today experiencing water stress.[64] The planting of hybrid crops that would grow faster also led to a large decrease in the cultivation of sorghum, barley, groundnuts and more, which had repercussions for the ecological balance of plant and animal life.[65] It has been estimated that 100,000 varieties of indigenous rice were lost because of the introduction of hybrids from abroad.[66] Then there were the implications of farmers using pesticides without sufficient protection, a danger especially for women, who make up half the agricultural labour force in countries like India and who are particularly vulnerable if exposed to chemicals at a young age.[67]

Another problem came from massive population movements, both through government policies, through programmes funded by the World Bank and through the search for work and food. In Indonesia, for example, as many as 15–20 million settlers moved into the jungles of Sumatra, Kalimantan, Sulawesi and Irian Jaya, which not only changed local ecologies, but also brought widespread human rights abuses and the development of large-scale logging industries that today see around 750,000 hectares of forests cleared every year in this part of South-East Asia alone. Even by the mid-1980s, this was being called 'the World Bank's most irresponsible project'.[68]

The rise both in urbanisation across the less developed world and in the size of cities like Bombay, Manila and Mexico City brought new concerns about social and political stability and put pressure on city authorities to build infrastructure quickly enough to cope with soaring numbers of inhabitants – which in turn led to environmental degradation. There was scepticism in some places that production could keep up with demographics in the first place. When the Indian Minister of Food and Agriculture, C. Subramaniam, gave a talk at Stanford in 1968, Paul Ehrlich listened in disbelief. There was no way, wrote Ehrlich, that 'India is going to find a way to produce enough food to support some 12 million more people than they can feed today.' Millions would die, he predicted.[69]

For Ehrlich, 'Population control is the only answer.'[70] This was a policy that had been advocated since the 1950s by the World Bank, which had forecast that Indians could expect 40 per cent more income within three decades by adopting aggressive contraception programmes.[71] Borlaug was a passionate advocate too. Not enough people recognised 'the magnitude and the menace of the population monster', he insisted in December 1970. His work and that of others to develop new strains of rice simply bought time – 'breathing space' as he called it – perhaps thirty years.[72] Such dire warnings helped propel government interventions into reproductive health, which ranged from sex education and the promotion of family planning to programmes of large-scale forced sterilisation.[73]

Others did not see it that way, at least to start with. 'The more people there are, the earlier we can realize humanity's greatest ideal – communism,' ran one newspaper headline in China in the late 1950s. In the eyes of Mao Zedong, China's vast population was a positive asset, one that kept it insulated from dangers. 'We shouldn't be afraid of atomic missiles,' he told a shocked Nikita Khrushchev in Moscow in 1957. 'No matter what kind of war breaks out – conventional or thermonuclear – we'll win. As for China, if the imperialists unleash war on us, we may lose more than three hundred million people. So what? War is war. The years will pass and we'll get to work producing more babies than ever before.'[74] There was no need to put a cap on population growth or to advocate use of contraceptives, at least according to Mao. If China was home to half of the world's population in a hundred years' time, that would be fine as by then everyone would be 'college educated' and would 'naturally practice birth control'.[75]

Eventually, by the early 1970s, even China adopted measures designed to manage a population that was 50 per cent larger than it had been in 1949, introducing a family-planning campaign, raising the legal age of marriage to the mid-twenties in most regions and encouraging mothers to take more time between births.[76] By 1979, this had not produced the slowdown that was expected and a one-child policy was introduced.[77] While usually interpreted as a sign of central coercion by a Communist party determined to implement an effective modernisation programme, in fact almost all the principal ideas that underpinned the policy were borrowed from the west and from western science.[78] As is so often the case, this solution has created a new set of problems, not least that of a population that is not replacing itself and that is rapidly ageing, twin issues that are likely to have a profound impact on China's economy – and its social and political trajectories – in the future.

Many of these issues – and fears – seem remarkably familiar in today's world: the threat of nuclear cataclysm; ecological damage as a result of the pursuit of economic growth; unsustainable use of resources; demographic pressures from a rising global population on infrastructure and on food; over-reliance on new ideas and technologies that raise hopes but also open Pandora's box; regular statements by governments and intergovernmental organisations about the importance of co-operation coupled with an absence of real progress; and, most importantly of all, deep-rooted concerns about what the future holds. To these should be added concerns about climate change too.

Some scientists had become concerned that human behaviour was not just altering the natural world through pollution, pesticides and exhaustion of natural resources, but was also changing the climate itself. One paper published by Roger Revelle and Hans Suess in 1957 which looked at the increase of atmospheric carbon dioxide predicted that 'During the next few decades the rate of combustion of fossil fuels will continue to increase, if the fuel and power requirements of our worldwide industrial civilization continue to rise exponentially.' This would have ominous consequences, wrote the two scientists: 'human beings are now carrying out a large-scale geophysical experiment of a kind that could not have happened in the past nor be reproduced in the future'. The impact of the industrial revolution and the resultant addition of carbon dioxide through the burning of fossil fuels, especially since the middle of the nineteenth century, meant that 'Within a few centuries we are returning to the atmosphere and oceans the concentrated organic carbon stored in sedimentary rocks over hundreds of millions of years.' Revell and Suess did not venture to guess what the consequences would be, but they were certain they would be far-reaching.[79]

Other research around the same time began to produce alarming results. As we have already seen, interest in weather patterns, in climatic conditions and in short- and long-term changes had been a key part of military planning since the early twentieth century, if not before. At the height of the Cold War, as the Pentagon had put it, 'The Department of Defense has a vital interest in the environmental sciences since the military services must have an understanding of, and an ability to predict and even to control, the environment in which it is required to operate' – not least since guided-missile systems required as much information as possible about the lower and upper atmospheres, ionospheric conditions, geodesy and geomagnetism and more besides.[80]

Considerable energy and resources went into polar research, with major expeditions and bases set up in the Antarctic in the late 1940s with Operation HIGHJUMP and a decade later at Camp Century in Greenland under the authority of the US Army Corps of Engineers' Snow, Ice, and Permafrost Research Establishment (SIPRE), which included investigations into carbon dioxide monitoring not only in the present day but also in the past – as did monitoring stations elsewhere, such as at Mauna Loa in Hawai'i.[81]

Camp Century in particular proved a revelation: chosen as a site for Project ICEWORM, a proposed nuclear arsenal of 600 ballistic missiles, the scientists who were based there made a series of discoveries drilling into ice cores under the direction of Henri Bader, who had previously worked on carbon dioxide bubbles trapped in the ice sheets of Alaska. Using a drill rig brought in from Oklahoma, the team at Camp Century helped provide empirical evidence of previous climatic conditions dating back thousands and even tens of thousands of years – and in so doing helped provide perspective for long-run climate conditions that showed both the scale and the rate of change in the contemporary world compared to the past.[82]

Much work was also being done in the Soviet Union in this period on climatic change in general and on human impact in particular by scholars like Evgenii Fedorov.[83] One leading light was Mikhail Budyko, whose groundbreaking work on the relationship between the lower atmosphere and the earth's surface was picked up on with great interest outside the USSR by climatologists in the US and elsewhere.[84] One of Budyko's areas of investigation was the anthropogenic generation of heat in large quantities – and the question whether, where and how this might affect local, regional and potentially even global climates. Research revealed the polar ice–albedo feedback, whereby decreases in snow and ice in the polar region are intensified by solar heating that further decreases the snow and ice area and amplifies global warming.[85] This all suggested that human behaviour and activities were in the process of effecting 'major climatic change in the near future. As a result,' Budyko wrote in 1969, 'changes in climate that are natural are to be replaced in time by changes that are created and shaped by mankind.'[86]

Human impacts on climate became a subject of considerable discussion within the scientific community. One paper commissioned by the American Petroleum Institute, a trade association representing the oil and gas Industry, reported that prospects 'for the future must be of serious concern' given the amounts of carbon being released by the burning of fossil fuels.[87] Another paper, this one published in the journal *Science* in 1970, looked not only at fossil fuels but at the dust particles, particulates and air pollution they generated and above all at the cities which were home to 'by far the most pronounced and locally far-reaching effects of man's activities'. The 'noise' of natural climatic fluctuations could easily mask anthropogenic influences; what was urgently needed, therefore, was 'an adequate worldwide monitoring system to permit early assessment of these changes'.[88]

There were papers that considered when the present interglacial period would end, or examined historical global climate conditions, asserting that the world was currently in a cool cycle that would last for another 8,300 years.[89] Other studies written around the same time underlined the way ozone was being destroyed by nitrous oxide and could be further degraded by the advent of supersonic transport because of the high altitudes at which emissions would be released, or evaluated the effects of chlorofluorocarbons (CFCs) on the ozone layer. All served to concentrate minds on what the future might hold.[90]

Fears spilled over into the public domain. Even in the 1950s, the popular press carried articles that painted a bleak picture of a world about to enter a new ice age that might lead to a 'real estate boom in the Sahara'.[91] In the 1970s, such ideas were now taken up with gusto. Colder winters herald the dawn of a new ice age, promised the *Washington Post* in January 1970; 'Is mankind manufacturing a new ice age for itself?' wondered the *LA Times* four days later; a new ice age was predicted by the start of the twenty-first century, reported the *Boston Globe*, a few months afterwards – just a few examples among a vast array of similar coverage across print media.[92]

The gloomy, cold world that lay around the corner proved both persistent and irresistible. 'Tell-tale signs are everywhere,' said *Time* magazine in 1974; indications of global cooling were everywhere, and were already responsible for drought in sub-Saharan Africa, Central America, the Middle East and India. The changes were partly due to changes in the sun's behaviour – although 'Man, too, may be somewhat responsible for the cooling trend' thanks to 'dust and other particles being released into the atmosphere as a result of farming and fuel burning' that was 'blocking more and more sunlight from reaching and heating the surface of the earth'.[93] *Newsweek* followed a year later with the claim that while 'meteorologists disagree about the cause and extent of the cooling trend ... they are almost unanimous in the view that the trend will reduce agricultural productivity for the rest of the century'.[94]

This was not in fact the case at all: most scientists working on environmental and climatic change were wary of making sweeping statements about the present, let alone issuing catastrophising predictions about the future. Reports for the National Science Board in the US in 1974 noted apparent trends in cooling, but pointed out that 'major advances' were needed in order to establish whether, how and where emissions, pollutants or particulates were affecting current climates.[95] A report by the US National Academy of Sciences a year later suggested that 'a major climatic change would force economic and social adjustments on the worldwide scale' and therefore reiterated the need for more research; the NAS acknowledged that 'The climates of the earth have always been changing,' and it was not known 'How large these future changes will be'.[96] This was the line that Secretary of State Henry Kissinger took when addressing the General Assembly of the United Nations in April 1974: scientists needed urgently to investigate 'the possibility of climatic changes in the monsoon belt and perhaps throughout the world'.[97]

Nevertheless, some scholars expressed their opinion forcefully – and at the highest levels. The authors of one paper on global cooling wrote to President Nixon in 1972 to warn that the prospect of 'a global deterioration of climate, by order of magnitude larger than any hitherto experienced by civilized mankind, is a very real possibility and indeed may be due very soon'. This was circulated by the White House to several bodies 'for review and appropriate action'.[98]

What climatological changes meant for 'intelligence problems', national security and global affairs was set out in a CIA report prepared later that year. 'Climate has not been a prime consideration of intelligence analysis because, until recently, it has not caused any significant perturbations to the status of major nations.' This was because conditions in the twentieth century were the most benign for agriculture they had been for the best part of a millennium – right back to the eleventh century, according to the report's authors.[99]

Recent events had served to spook the spooks: drought in Burma, Costa Rica, Honduras and Pakistan, poor harvests in North Korea, the Philippines and the USSR, cold-damaged crops in Japan, the worst floods in a century in the Great Lakes of North America, unusually heavy rainfall in North Vietnam and poor weather, droughts and floods in the Soviet Union and China in the early 1970s had served to show that 'Climate is now a critical factor.' In particular, 'The politics of food will become the central issue of every government.'[100] The CIA noted a large and growing body of scientific work that indicated that 'A global climatic change was underway' and that this 'would create worldwide agricultural failures in the 1970s'.[101]

A separate, parallel CIA report looked at the implications in more detail. 'If climatologists who believe a cooling trend is underway prove to be right', its authors declared, there would 'almost certainly be an absolute shortage of food' around the world, led by drops in output in the Soviet Union and China because of shorter growing seasons, as well as more frequent monsoon failures in South and South-East Asia and in southern China.[102] 'Moreover, in periods when climate change is underway,' they continued, 'violent weather – unseasonal frosts, warm spells, large storms, floods, etc. – is thought to be more common.' All this would have 'an enormous impact' not only on the balance between food and population but on 'the world balance of power', largely because the US's geography made its ecologies less vulnerable to climatic change: this was likely to lead to anti-American sentiment on the part of countries which either were already dependent on the US or would become so.[103]

There was a strong chance of social and political upheaval, said the report's authors, as 'rural masses' became 'less docile' under pressure of famine and as the living standards of urban middle classes in sub-Saharan Africa, East Africa and India fell in response to food shortages and sharp price rises.[104] In the worst-case scenario, they warned, 'Massive migration backed by force would become a very live issue. Nuclear blackmail is not inconceivable.' As far as poor countries were concerned, it was likely that the population 'problem' would be solved 'in the most unpleasant fashion'. This meant mass starvation.[105] When George Will of the *Washington Post* got hold of a copy of this report, he summarised its findings bluntly: if the climate change occurs, he wrote, 'there will be megadeaths'.[106]

Some senior figures in the US administration believed that it was possible to leverage agricultural, climatological and economic power into gains. 'Food is a weapon,' said Nixon's secretary of agriculture, Earl Butz, and was 'now one of the principal tools in our negotiating kit'.[107]

The problem was that, despite having strong cards to play, the US had a weak hand when it came to energy. And this helped determine policy in ways that were far more influential than warnings about climate change, and furthermore went a lot further in encouraging lower consumption of fossil fuels, in promoting investment in renewable and clean sources of power and in provoking discussions about sustainability.

Ironically, then, the impulse towards lower consumption of fossil fuels came not as a result of warnings from scientists or from intelligence briefings, but as a result of events in October 1973 that took place thousands of kilometres away in the Middle East after a coalition of Arab states launched a surprise attack on Israel on the Jewish holy day of Yom Kippur – which gave its name to the ensuing war. To apply pressure on the US, the Arab nations that were part of OPEC (the Organisation of Petroleum Exporting Countries) cut oil production and then placed an embargo on oil shipments to the United States, as well as on those to other countries that were either supportive of and sympathetic to Israel or were thought to be. The result was a major energy crisis.

Within weeks, the US was facing shortages and what President Nixon in a televised address to the nation on 7 November called 'a serious national problem'. The United States had 'grown and prospered in recent years' by being able to rely on oil imports. The war in the Middle East had changed this, leaving the country facing 'the most acute shortages since World War II'. As a result, dramatic action was now needed by the American people: in the short run, that meant 'we must use less energy – that means less heat, less electricity, less gasoline'. In the long run, he went on, 'it means that we must develop new sources of energy' to provide for energy independence in the future.[108]

He spelt out a radical set of measures that would now come into effect. These included nationwide changes for every household. 'To be sure that there is enough oil to go around for the entire winter, all over the country,' said Nixon, 'it will be essential for all of us to live and work in lower temperatures.' That meant lowering 'the thermostat in your home by at least 6 degrees'; this had been endorsed by his personal doctor, who had told him that 'you really are more healthy' in a cool environment than in a warm one. Offices, factories and 'commercial establishments' were required to reduce temperatures by ten degrees – 'either by lowering the thermostat or curtailing working hours' If that was the only alternative.[109]

Sweeping changes were announced across the federal government. Energy consumption was being and would be further cut across 'every agency and every department in government'; state governors were encouraged to look at plans 'slightly altering the school year' (presumably to take advantage of longer hours of daylight during the summer) and at curbing 'unnecessary lighting' in communities; initiatives would be introduced to encourage greater use of public transport and car sharing; every government vehicle would have its speed capped at 50 miles per hour (except in emergencies) to conserve fuel; and a national highway speed limit would be set, based on combustion engines' optimal efficiency, to help save hundreds of thousands of barrels of oil a day. 'We must all co-operate to change,' said Nixon.[110]

The US should be inspired by the recent moon landing, as well as by the technological achievements of the Manhattan Project, the President went on, to develop 'the potential to meet our own energy needs without depending on any foreign energy supplies'. Announcing a new plan, which he christened Project Independence, Nixon pledged that by 1980 it would be possible for America to meet its energy needs from its own resources.[111]

Concerns about energy availability, rising prices, the environment and dependency on foreign exports with its implications for the economy and for national security all have resonance in today's world. So too do the promises and commitments by governments to solve problems with deep roots through a series of ambitious announcements that deliver little by way of action. It is true that the energy crisis of the early 1970s left a deep imprint on the United States, the most obvious of which is that the speed limit of 55mph which was signed into law a few weeks after the presidential address and remained in place for more than two decades – and in some parts of the USA, is still state policy. Early studies showed that these restrictions cost Americans almost 2 billion hours a year in lost time, but that lower speeds saved lives as well as fuel.[112] But few who take to the roads today, in electric vehicles or otherwise, link the speed they are allowed to travel with emergency measures intended to limit energy consumption.

One of the challenges, of course, was the power of the fossil-fuel lobby in the US which made moves towards cleaner energy difficult to achieve, as well as the resistance of voters to any increase in taxation to fund long-term programmes. The problems were epitomised during the presidency of Jimmy Carter (1976–80). 'The choices facing the members of Congress are not easy,' he said, when announcing his National Energy Plan in November 1977 in a nationwide address. 'With every passing month,' he said, 'our energy problems have grown worse. This summer we used more oil and gasoline than ever before in history,' with reliance on imports increasing rather than decreasing. This created inflationary pressures that were out of the hands of the US government, said Carter, who quoted his Secretary of Defense, Harold Brown, as saying that 'The present deficiency of assured energy sources is the single surest threat ... to our security and to that of our allies.' This was one reason why it was essential to create the new Department of Energy to oversee and co-ordinate responses to an extremely serious challenge.[113]

We need to 'cut back on consumption', said Carter, 'shift away from oil and gas to other sources of energy' and 'encourage production of energy here in the United States'. While supporting domestic fossil-fuel producers, he went on, the government would use tax incentives and penalties 'to hasten the shift from oil and gas to coal, to wind and solar power, to geothermal, methane, and other energy sources'. This was difficult, the President acknowledged, because America was facing 'long-range, future challenges', and politicians were elected for only short periods in office. Nevertheless, he concluded, 'I hope that, perhaps a hundred years from now, the change to inexhaustible energy sources will have been made.'[114]

Carter was not the first to suggest investment in research and development in clean energy. At the start of the 1970s, Nixon had introduced measures in Congress that included plans to improve energy efficiency as well as investment in technologies that enabled moves to cleaner fuels, including solar power, geothermal sources and hydrogen-based fusion reactors.[115] The money made available, however, was derisory and never likely to open up resources that were commercially viable. Carter, on the other hand, spoke with the zeal of an evangelist. 'The world has not prepared for the future,' he had said in a televised address to the nation soon after taking office. While advocating a boost in coal usage, he also set out ambitious goals to insulate homes and new buildings, and to develop and expand solar energy. These were all necessary, he said, 'to protect our jobs, our environment, our standard of living, and our future' It was simple: 'We must not be selfish or timid if we hope to have a decent world for our children and grandchildren.'[116]

Not enough people agreed with him. Part of the problem lay in geopolitical turbulence in the late 1970s which saw OPEC hiking oil prices once again by 50 per cent in 1979, threatening to tip global economies – including in the US – into recession. Matters were made worse by Carter's tin-eared response. 'In a nation that was proud of hard work, strong families, close-knit communities and our faith in God, too many of us now tend to worship self-indulgence and consumption,' he said in July 1979, chastising voters who were due to vote in presidential elections just a few months later. There was 'a growing disrespect for government and for churches and for schools', he added. 'It is the truth and it is a warning.' Personal greed was the problem. 'Human identity is no longer defined by what one does, but by what one owns.'[117] These admonitions went down badly, with Carter's popularity plunging to unprecedented levels. This was capitalised on by Ronald Reagan, whose defence of the 'common sense and common decency of ordinary men and women, working out their lives in their own way', fell on fertile ground and swept him into the White House.[118]

Looking back now, it is hard to see how the opportunities to engage with energy shifts were missed. In the summer of 1979, for example, Carter had announced a billion-dollar funding package for solar and other renewables while showing off new solar panels that had been fitted to the White House roof. 'In the year 2000, this solar water heater behind me, which is being dedicated today, will still be here supplying cheap, efficient energy,' he said. 'A generation from now, this solar heater can either be a curiosity, a museum piece, an example of a road not taken, or it can be just a small part of one of the greatest and most exciting adventures ever undertaken by the American people.'[119]

Rather than a world of clean, renewable energy, the pathway of the following decades was one that led to an extraordinary rise in energy consumption, fossil-fuel burning, carbon emissions and pollutants – propelled above all by rising and intensive international trade and globalisation. Ironically, one of the catalysts for this lay in the compromising of the US presidential system; and another in weather-modification programmes that had been developed during the Vietnam War.

A treasure trove of thousands of documents relating to US military activities in South-East Asia had been handed over to the press by Daniel Ellsberg, a disillusioned researcher who had been involved in compiling a history of the US role in Indo-China from the end of the Second World War to the late 1960s. Known as the Pentagon Papers, these began to appear in print in a series of scoops and revelations in the *New York Times* and *Washington Post* that had readers riveted – and outraged by the number and range of covert operations that had been authorised, and by the disclosure that the administration of Lyndon Johnson had systematically lied 'not only to the public but also to Congress'.[120]

Stories began to appear in the summer of 1971 and continued unabated after the Supreme Court ruled that efforts by the US government to block publication violated the First Amendment of the Constitution and could not be justified in law.[121] One of the most striking revelations was about long-term efforts to manipulate the weather in South-East Asia during the Vietnam War. A report that had appeared even before the Pentagon Papers were released had alleged that 'Air Force rainmakers' had 'succeeded in turning the weather against the North Vietnamese' in a hush-hush project codenamed Intermediary Compatriot that had begun in 1967 and had increased rainfall over jungle road networks during the wet seasons.[122]

This report had caused alarm in some quarters in Washington, not least within the Senate Subcommittee on Oceans and International Environment, which demanded a response from the Secretary of Defense, who first declined to reply on the grounds that it would threaten national security and then testified in the Senate that no weather modification had been attempted, stating that 'we have never engaged in that type of activity over North Vietnam'.[123]

It was highly embarrassing, therefore, when a detailed account of precisely that activity appeared in the summer of 1972 in a lengthy exposé in which the journalist Seymour Hersh set out how, where and why a concerted programme had been developed. As one official with detailed knowledge of the operation explained, 'We were trying to arrange the weather pattern to suit our convenience.'[124] This was a reference to the Project POPEYE, the new codename chosen for INTERMEDIARY COMPATRIOT after the latter's moniker had been revealed despite being classified.[125]

As it later emerged, the objective of POPEYE was 'to produce sufficient rainfall' along communication lines in North Vietnam and southern Laos by seeding clouds with iodine to 'interdict or at least interfere with truck traffic' and, in so doing, to interrupt supplies. Initial experiments were conducted in Laos without the knowledge of Lao authorities and in conditions of utmost secrecy, 'unknown to other than a severely limited number of U.S. officials'. During this test phase, 'more than 50 cloud seeding experiments were conducted'. The results were described by the Department of Defense as 'outstandingly successful', with more than 80 per cent of the clouds that were seeded producing rain soon afterwards, with sufficient downpours 'to have contributed substantially to rendering vehicular routes ... inoperable'. One US Special Forces camp in Vietnam was 'inundated with nine inches of rain in four hours', seemingly taking servicemen by surprise.[126]

The results were very promising, early reports claimed, and provided strong evidence that the techniques could have a significant impact on enemy movements, supplies and communications; this could be enhanced by bombing bridges and river crossings to create bottlenecks. There were potential side-effects, however. For one thing, 'the proposed program would drastically change the weather patterns over the next few months', according to the request presented to the Secretary of State, Dean Rusk, 'to inaugurate operations at once'. For another, as the memorandum conceded, 'There would be some hypothetical effect of an adverse nature in Thailand where ... normal rainfall might be somewhat diminished,' although this was not worth worrying about; changes to precipitation levels and timings were nonetheless likely to have 'appreciable consequences outside the target areas'. There were risks to confidentiality, such as 'planned weather experiments in India', the possibility of a US seeding aircraft being downed or of leaks in the press. This was why the proposal to move forward was dependent on the authorisation of the President.[127]

As it later emerged in Senate hearings, POPEYE was intended 'to deny the enemy the use of roads by: softening road services, causing landslides along roadways, washing out river crossings [and] maintaining saturated soil conditions beyond the normal time span'.[128] Over the five years between its inception and its termination two days after Hersh's revelations, more than 2,500 sorties were flown by US aircraft with target priorities set by intelligence-gathering officers based at Tan Son Nhut in South Vietnam, at a cost of around $3.6 million per year.[129] Aircraft whose roles 'were not dedicated exclusively to the cloudseeding missions' over Laos and northern Cambodia were operated out of Thailand, whose government was neither asked for approval nor informed of the purpose and nature of these sorties.[130]

These disclosures were the tip of the iceberg. In Senate hearings in April 1972, it emerged that the US had been involved in weather interventions in the Philippines three years earlier and had been in discussion with that country's government about 'potential hurricane modification'. US Navy personnel had selected, seeded and 'kept alive' rainclouds off Okinawa in 1971 to help with drought alleviation. Discussions had been held 'informally' with Canada about cloud-seeding over the Great Lakes and with the British about 'hurricane work' in the Bahamas.[131]

In addition to this there was Project GROMET, a top-secret plan to try to mitigate drought in India in 1966–7 in which President Johnson had taken a personal, even obsessive interest, examining weekly rainfall maps so closely that he knew 'exactly where the rain fell and where it failed to fall'. On this occasion, the US consulted with the Indian Prime Minister, Indira Gandhi, whose government gave approval, even though it took steps to ensure that American involvement and the initiatives were not made public. As it turned out, the interventions were not successful and, with the arrival of abundant monsoon rains in the summer of 1967, not needed.[132]

The nature and extent of these weather modifications came as a surprise. 'The seeding techniques' that had been developed by the US military 'are now in use in almost every country in the world', one senior official told shocked senators in a prepared statement. Collaborations had been undertaken with dozens of institutions in the US as well as with 'numerous private corporations', while there had also been 'contact with individuals interested in' climate changes in 'India, Philippines, Taiwan, Chile, Israel, Rhodesia, Mexico, Portugal, France, Italy, Argentina and Australia'.[133]

This was 'extremely distressing', protested Claiborne Pell, chair of the hearings in the Senate in April 1972, some of which were conducted behind closed doors. 'If we do not restrict military use of current environmental modification techniques,' he said, 'we risk the danger of the development of vastly more dangerous techniques whose consequences may be unknown and may cause irreparable damage to our global environment.'[134] The committee recommended that the

US government 'seek the agreement of other governments' in pursuit of an international treaty banning the 'use of any environmental or geophysical modification as a weapon of war'.[135]

That overture was of particular significance for the Soviet Union and for Washington's relations with Moscow. In the late 1960s the USSR had undertaken a huge programme to build up its nuclear arsenal, while investing in an anti-ballistic missile (ABM) defence system that threatened to offer a decisive advantage in models of how a military confrontation might play out. This eventually led to the Strategic Arms Limitation Talks (SALT) which began in 1969 and concluded in the signing of an initial agreement in Moscow three years later.[136]

Co-operation such as this was part of an encouraging thaw in relations that opened the door for other agreements. Leading policymakers and scientists had built up a rapport at events like the regular Dartmouth conferences held in New Hampshire which provided the opportunity for US–Soviet discussions.[137] These contacts were now parlayed into bilateral talks about anthropogenic climate interventions that started in the summer of 1974 and led to the outlines of an agreement a year later which was then submitted to the United Nations as the Environmental Modification Convention and adopted on 10 December 1976.[138]

The convention banned 'environmental modification techniques', namely 'techniques for changing ... the dynamics, composition or structure of the Earth, including its biota, lithosphere, hydrosphere and atmosphere, or of outer space' for military purposes. Examples were offered of phenomena that could be caused by such techniques: 'earthquakes, tsunamis; an upset in the ecological balance of a region; changes in weather patterns (clouds, precipitation, cyclones of various types and tornadic storms); changes in climate patterns; changes in ocean currents; changes in the state of the ozone layer; and changes in the state of the ionosphere'.[139]

In a further demonstration of co-operation and openness, a delegation of scientists had spent three weeks in the Soviet Union in May–June 1976 to discuss weather-modification research and cloud physics and to learn how cloud seeding was being used over 5 million hectares in the Caucasus, Moldavia, Central Asia and Soviet bloc countries such as Bulgaria and Hungary; Soviet administrators and scientists claimed that these were already 'very successful ... and that the operations will continue to enlarge in the years to come'. The talks were constructive and amicable with comparisons of hailstorms in the Caucasus and Colorado and discussions of the merits of using rockets to inject ice nuclei into supercooled clouds, and of the optimal techniques for fog dissipation. Access was granted to multiple research institutes, including in Moscow, Leningrad, Tbilisi, Kiev and Nalchik. What struck the visitors most, however, was the age of those leading the research programmes: they were all noticeably younger than their predecessors, which could be taken as a sign of the growing interest in and commitment to the field by the Soviet authorities.[140]

Personal relationships and scholarly interactions such as these between scientists were to prove important during the years that followed. Of central importance was not the deliberate use of weather-modification programmes, alterations in global temperatures or even the effect of emissions on the atmosphere, but rather modelling the possible impact of war between the US and the Soviet Union. As one study published in 1975 put it, even during a period of détente – a warming of ties between the two superpowers – it was important to underline what the consequences of nuclear warfare might be, so that such knowledge might act as a deterrent to the use of these powerful weapons.[141] The impact of a nuclear exchange would include extended periods, perhaps lasting months, when fires would 'strongly restrict the penetration of sunlight to the earth's surface'. In such a scenario, 'it is likely that agricultural production in the Northern Hemisphere would be almost totally eliminated, so that no food would be available for the survivors of the initial effects of the war'.[142] The effect of such studies was profound. As one scholar has written, 'In the United States, the sharpest dispute about climate change in the 1980s was not about carbon dioxide but rather about the possibility of "Nuclear Winter".'[143]

Anxieties were soon intensified by a series of publications that also modelled the impact of large-scale nuclear confrontation between the US and the Soviet Union. Scholars working in both countries presented hypotheses a few months apart in 1983 that grabbed public attention, and stoked fears of what was at stake in the rivalry between two starkly different socio-economic and political systems. A paper published by a group of scholars headed by Richard Turco in the journal *Science*, using 'new data and improved models' which also drew on studies of dust storms on Mars, looked at the effect of 'sooty smoke' that would follow from wildfires and extensive urban fires. While 'most of the world's population could probably survive the initial nuclear exchange', wrote the authors, a major military engagement would result in 'significant surface darkening over many weeks, subfreezing land temperatures persisting for up to several months … and dramatic changes in local weather and precipitation rates'. The result, they went on, would be a harsh 'nuclear winter'. Even more worrying was the observation that 'relatively large climatic effects could result even from relatively small nuclear exchanges … if urban areas were heavily targeted' because of 'massive smoke emissions'.[144]

A paper published in the Soviet Union not long afterwards reached similar conclusions about the consequences of nuclear war: black smoke would envelop the whole earth, blocking out the sun's light and leading to dramatic drops in temperature so that all fresh water would freeze, all harvests would fail and ecologies would be destabilised. There would be mass extinctions of animals, plants and microorganisms; all terrestrial biota would be affected.[145] Intelligence analysts in the US were dubious about this. Soviet research on 'nuclear winter' was 'derived almost entirely from US ideas, data, and models'. Moreover, scientists in the USSR 'consistently reported more severe climatic changes than are usually found in similar research in the West'.[146]

US intelligence and scientific assessments were that the Soviets' calculations were not just derivative but crude, as well as being based on results that took forty hours on their clunky BESM-6 computer but only eight minutes on the latest Cray-1 machine used in the United States. As such, it was hard to avoid the conclusion that Moscow was developing a well-directed propaganda campaign to encourage disarmament movements, push for reductions in military spending and sow discord abroad – all in the USSR's interests.[147]

Nonetheless, it was worth considering how to address a scenario that Turco and his colleagues suggested could lead to reductions in surface temperatures to –17 °C within thirty days of a nuclear attack – with a base-case scenario of subfreezing conditions lasting for about three months and only returning to normal levels after about a year.[148] Although more accurate modelling in the US downgraded some of these assumptions and suggested a 'nuclear autumn' rather than a nuclear winter, the doomsday scenario caught the public imagination as 'an end-of-the-world story that fit the times' – sitting alongside disasters in the 1980s such as drought and famine in Ethiopia, the global AIDS epidemic, the catastrophe at the Chernobyl nuclear facility and more which framed humans as their own worst enemy.[149]

Certainly, there were well-placed figures within the Soviet Union who took the threat of nuclear disaster seriously. The physicist Sergei Kapitsa, whose father Pyotr won the Nobel prize for physics, took part in a round-table discussion in the Senate during a visit in December 1983 at the invitation of Senators Edward Kennedy and Mark Hatfield to consider the effects of nuclear war. A well-known TV broadcaster who hosted a weekly science programme in the USSR, Kapitsa reminded attendees of the explosion of Mount Tambora in 1815 – and of the poem 'Darkness' by Lord Byron that was written the following year, which he said was well known in the Soviet Union because it had been translated into Russian by the writer Ivan Turgenev. As the poem puts it:

> I had a dream, which was not all a dream.
> The bright sun was extinguish'd, and the stars
> Did wander darkling in the eternal space,
> Rayless, and pathless, and the ice earth
> Swung blind and blackening in the moonless air.[150]

'Nuclear weapons have ceased to be an instrument of war,' Kapitsa said in a speech at the United Nations a month later, in which he again referenced the Tambora eruption and Byron's poem which he called 'by far the best description of a nuclear winter to be found in literature'. He also cited Mary Shelley's *Frankenstein*, which showed, he said, the dangers of the interface between humans and technologies that were not fully understood.[151]

There were many factors that moved US–Soviet relations towards discussions and actions around disarmament. The characters of both Ronald Reagan and Mikhail Gorbachev – the latter blowing fresh air into the sclerotic Soviet system after he became leader – were important, as was the personal relationship between the two men. Advances in the development of other military offensive and defensive tools such as the Strategic Defense Initiative which were very prominent in the media and beyond were significant, even if some scholars argue that the role these played was both more complicated and more limited than is often thought.[152] The Soviet invasion of Afghanistan, the travails of the US in the Middle East, Iran and elsewhere and the domestic pressures they produced were also ingredients that went into the pot of geopolitical change, disarmament talks and superpower co-operation in the 1980s.

Concerns about pollution, stronger environmental laws, better waste disposal and natural resource depletion were not only prominent issues across the United States and many parts of the west, but consistently ranked as the most important ones locally – which fed into political agendas, campaigning and the strengthening of green movements.[153]

This was mirrored in the Soviet Union, where there had been growing calls to focus on ecological protection rather than economic growth, to improve air quality, to stop harmful mining, to clear up industrial waste, to stop and reverse desertification and to protect the Aral Sea and Lake Baikal (the two most high-profile cases of anthropogenic devastation).[154] These calls became easier – and louder – following Gorbachev's reforms based on the principles of *glasnost* ('openness') and *perestroika* ('rebuilding'), which made it easier for citizens in the Soviet Union to demand action from the government. Local, regional, national and global environmental issues ranked highly as concerns of the Soviet population and were evidenced by mass demonstrations.[155]

There were parallels in other parts of the world, with the Chipko movement in the Himalayan foothills, which drew international attention to the relentless loss of forest cover, serving as an inspiration to others and as a warning that degradation of the environment was happening everywhere on the planet – and needed to be slowed down if not stopped.[156] The 1980s marked the peak of global deforestation, with tens of millions of hectares of rainforest cut down during that decade, largely in the form of clearances in the Amazon.[157] Protest groups and movements, from Brazil to Mexico, from West Africa to the Middle East, from Canada to Central America, mushroomed at this time demanding changes to the way consumption in rich countries was driving exploitation of nature and natural resources and causing substantial damage too.[158]

Environmental protection provided an obvious, mutual and indeed easy overlap in US–Soviet relations, and offered benefits in the presentation of both superpowers as sympathetic and benign global leaders. At their first meeting, in Geneva in November 1985, Reagan and Gorbachev discussed and agreed a raft of issues, from reaffirming the complete prohibition of chemical weapons to undertaking to meet regularly in the future. They also gave commitments 'to contribute to the preservation of the environment – a global task – through joint research and practical measures'.[159]

The opportunity to do so in practice came quickly. On top of fears of the potential impact of nuclear war, of rising pollution and of ecological degradation came yet another worrying concern about climatic change that might have devastating consequences. Scientists working on the ozone layer, which sits in the earth's stratosphere and absorbs almost all of the ultraviolet light emitted by the sun, had been concerned about its depletion since the 1970s, although suggestions that the damage was being done by the release of CFCs, often used in aerosols and as solvents, refrigerants and fire-fighting agents, were savaged by those working in the commercial sector.[160] By the mid-1980s, further work had emphatically established the link, and the worry about the rate of change and its effects had made it into mainstream media.[161]

International efforts to reduce and phase out chlorofluorocarbons led to the Vienna and then the Montreal Protocols being signed in March 1985 and September 1987 respectively, with Kofi Annan, later UN Secretary General, referring to the latter as 'perhaps the single most successful international agreement' overseen by the United Nations. These efforts will have resulted in the likely recovery of the ozone layer to 1980 levels in the second half of the twenty-first century.[162] The subject was important in all US–Soviet discussions during this period, such as when Reagan and Gorbachev met in Washington just a few months after Montreal, when both leaders agreed to sponsor 'joint studies in global climate and environmental change through cooperation in areas of mutual concern, such as protection and conservation of stratospheric ozone'.[163]

While such expressions of co-operation are often more meaningful as positive gestures and statements of intentions not backed up by results, looking back now the 1980s in many ways were a golden age, a time of genuine resolve to try to counter major problems through working together and even a time of optimism that it might be possible to find global solutions to global problems. The agreement about ozone in 1987 was one example; but other major initiatives were also adopted, such as the Convention on Biological Diversity after the United Nations agreed in 1988 to hold a major conference on environment and development. The UN General Assembly declared that it was 'deeply concerned by the continuing deterioration of the state of the environment and the serious degradation of the global life-support systems, as well as by trends that, if allowed to continue, could disrupt the global ecological balance, jeopardize the life-sustaining qualities of the Earth and lead to an ecological catastrophe'. It recognised that 'decisive, urgent and global action is vital to protecting the ecological balance of the Earth'. The only question was what steps should be taken, where and by whom.[164]

As some realised, the problem was not just ozone. In June 1988, James Hansen, chief scientist of NASA's Goddard Institute for Space Studies, gave testimony to the US Senate, stating that 'The greenhouse effect has been detected and it is changing our climate now.' Addressing the media straight afterwards, he put it bluntly: 'It's time to stop waffling.' Action was needed, not words.[165]

Words, though, were what followed. Climate change had been important enough an issue for Vice-President George H. W. Bush to talk about it on the campaign trail during his successful run for the presidency in 1988. 'I'm an environmentalist – always have been,' he said. 'Those who think we are powerless to do anything about the "greenhouse effect" are forgetting about the "White House effect",' he said, promising to bring the Soviet Union, China and others to the table to discuss global warming.[166]

It sounded encouraging: the United Nations Conference on Environment and Development – known as the Earth Summit – held in Rio de Janeiro in June 1992, set out to agree ways of dealing with 'dangerous human interference with the climate system', including through regulating greenhouse-gas emissions, while making allowances for 'those countries, especially developing countries, whose economies are particularly dependent on fossil fuel production, use and exportation'. Although the nations gathered in Rio noted that 'there are many uncertainties in predictions of climate change', they acknowledged that 'the global nature of climate change calls for the widest possible cooperation by all countries and their participation in an effective and appropriate international response'.[167] This was a real breakthrough, said Bush. It was time, he said, to take 'concrete action to protect the planet'. After all, 'our children' would 'judge us by the actions we take from this day forward. Let us not disappoint them.'[168]

Like the conference on 'Changing the Atmosphere: Implications for Global Security' in 1988 that led to non-binding commitments by the participating governments to seek a 20 per cent reduction in CO_2 emissions by 2005, the UN Framework Convention on Climate Change promised much but delivered little. Although seen as a landmark agreement at the time, the reality was that the commitments eventually signed by 192 countries to stabilise 'greenhouse gas concentrations in the atmosphere at a level that would prevent dangerous anthropogenic interference with the climate system' produced few tangible results.[169]

This was a missed trick, not least because of the extraordinary opportunity provided by the end of the Cold War. The early 1990s, said President George H. W. Bush at Rio, was an 'unprecedented era of peace, freedom, and stability [which] makes concerted action on the environment possible as never before'. There was no progress and there were few breakthroughs. Further rounds of negotiation, such as a new protocol agreed in 1992 and signed at Kyoto in 1997 which set out the principle that countries had 'differentiated responsibilities' – namely, that richer countries which had industrialised first should bear a heavier burden – had flawed outcomes too.[170] Not only that, but after the conference the United States, the world's biggest consumer of fossil fuels, refused to ratify the Kyoto agreement without amendments, with the Senate rejecting it by a vote of 95–0.[171]

Part of the reason was that, while Bush had seemed keen on climate in public, he was fundamentally 'uninterested' in it personally, as a journalist reported in 1992. 'He never sat for a full-dress scientific briefing on it or exercised control over administration policy, even after infighting among administration officials became public, or leaders of other industrialised nations pledged action,' delegating all decisions to an official whose 'personal belief' was that 'global warming projections were alarmist'.[172]

It was not just President Bush who realised that saying the right things about climate helped win votes while doing something about it lost them. Bill Clinton faced the same equation when he became president. Climate change was 'a real problem', he admitted in calls to British Prime Minister Tony Blair in recently declassified telephone records. There was no question, he said, that 'something has to be done', from using cleaner energy supplies to working with the car industry to improve efficiency and thereby reduce emissions. But political divisions in Congress meant that although there was 'broad acknowledgement' that climate change was a problem, the chances of taking meaningful steps to address it were slim indeed.[173]

It might be naïve to think that any initiatives might ever have reached a successful conclusion, although recent research suggests that the Montreal agreement on ozone will be responsible for global temperatures being at least 1 °C cooler – and 3–4 °C lower in the Arctic – by the middle of the twenty-first century than they would otherwise have been.[174] As it happened, however, the prospect of their doing so melted away. In the summer of 1989, a chain reaction began that eventually led to the fall of the Berlin Wall, to the reunification of Germany, to the replacement of communist regimes across Europe by democratically elected governments, to the splintering of the Soviet Union as its former constituent republics declared independence one by one and ultimately to the USSR ceasing to exist in December 1991.

In retrospect, this was a golden opportunity for bridge-building, rapprochement and mutual agreements; in the event, history took a different course. For Russian nationalists, the fall of the Soviet Union became a subject of shame and distress – the 'greatest catastrophe' of the twentieth century, according to President Vladimir Putin speaking in 2005.[175] It had major significance for the rest of the world, albeit for different reasons, because the collapse of the USSR gave rise to two seismic shifts. The first was that the vast natural resources that had been under the economic and political control of Moscow for decades suddenly became available for intensive exploitation on global markets. The second was that, in the months and years that followed, the shock of what had happened in 1989 in eastern Europe and the USSR had a profound effect on thinking in Beijing, which had almost succumbed to similar demands for reforms and freedoms when mass protests led by students reached Tiananmen Square in the heart of the Chinese capital that summer. The outcome was a fundamental reorientation in Chinese domestic, foreign and economic policies.

It is hard to overstate the importance of these two developments; taken together, they helped mould a new world that arguably rivalled the European arrival in the Americas in the 1490s. The decades that followed 1989 have proved to be by far the most intensive period of commercial exchange and integration – or globalisation – in history, in terms of scale and intensity as well as of speed. Needless to say, the impacts on local, regional and global climate have been epic – and unprecedented in human and natural history. The consequences will shape the world for generations to come.

24

On the Edge of Ecological Limits
(c.1990–today)

Широкий простор для мечты и для жизни
Грядущие нам открывают года.
(Wide spaces for dreams and for life
Are being opened up by the coming years.)

Russian National Anthem

The last decades of the twentieth century saw massive and relentless growth in demand for natural resources and raw materials. The search for supplies that were plentiful, cheap and easy to exploit was a global one. Bauxite mines were developed as reserves ran out in the US and France, with Jamaica and Guinea both becoming leading suppliers for industrial production in the developed world. To this could be added iron from West Africa, phosphates from Morocco and Senegal, copper and gold from Papua New Guinea and nickel from New Caledonia in the South Pacific, where exports rose at least a hundredfold between 1950 and 1976.[1]

The intensification of global trade has not only continued, it has accelerated rapidly, spurred on by the development of transportation networks that have enabled goods to be moved around the world more quickly and more cheaply than at any time in the past. New technologies have brought people closer together through the sharing of knowledge, harmonisation of information (and prices) and enhanced ability to do business. These trends were evident well

before the digital revolution which enabled orders to be placed in real
time: the opening up of Soviet airspace for civil aviation in 1985, which
was partly the result of a softening of enmities between the USSR and
the west and partly a reflection of Moscow's need to find sources of
foreign currency during a period of deteriorating economic fortunes,
did not just dramatically reduce typical journey times between Europe
and Asia, but helped pave the way for rapid economic co-operation
and investment by bringing businesses and business leaders into more
regular contact.[2]

The end of the Cold War was the catalyst for further deepening
of commercial ties and economic growth. The collapse of the Soviet
Union provided opportunities for its rich resources to be sold on global
markets, free of political doctrine and with no strings attached, while
investment in the USSR by western businesses improved efficiency,
boosted production, funded new operations, fields, mines and pipelines
and helped put pressure on global commodity prices. The collapse had
consequences in the post-Soviet world, with a large fall in industrial
emissions as inefficient plants were shut down or went out of business.
Standards of living fell, as did purchasing power which in turn severely
reduced consumption of livestock products, resulting in a halving of
cattle and pig numbers in the former Soviet Union between 1992 and
2011. The reconfiguration of food systems, including the abandonment
of cropland, was dropped and the restructuring of agricultural trade
resulted in a massive reduction in global greenhouse emissions.[3]
Uncertainties that impacted fossil-fuel production in the twenty years
after the 1991 dissolution of the USSR may also have been responsible
for a very significant fall in methane emissions that correlates closely
with the economic and political turmoil in the years that followed.[4]
The fall of the Soviet Union was bad for political idealists, but it seems
to have been good for global climate – at least in the short term.

Seismic shifts were taking place in other parts of the world too.
Although the impetus for China's integration into global markets
can be traced back to President Nixon's visit to Beijing in the early
1970s and then the award of Most Favoured Nation status under the
Carter administration, it was the decision of the leadership in Beijing
in the early 1990s to open up to foreign investment that proved
transformational: suddenly, a massive pool of cheap labour that could
make goods at a low price became available, alongside the tantalising

prospects offered by a market of more than a billion people that appeared certain to grow quickly in the years that followed.

The pace of that growth has proved mind-boggling. The global economy quadrupled in size in the three decades after 1990, an astonishing rate spread over such a short period and despite regular punctures caused by the Iraq War of 1990, the 'Asian crisis' of 1998, the attacks of 9/11 and their aftermath, the financial crisis of 2008 and the coronavirus pandemic (some estimates suggest that the latter cost the US economy alone some $16 trillion).[5] The expansion of international and global trade agreements, along with investment in logistical capabilities – such as enormous cargo ships, the largest of which now carry over 20,000 containers, and port facilities able to handle traffic – spurred the closer integration of markets, stimulated the creation of intercontinental supply chains and boosted global economic growth.

Membership of the World Trade Organisation (WTO) brought down tax barriers, putting downward pressure on prices that triggered a series of booms – at least for those well placed to benefit from them. For example, average taxes on imports in Colombia fell by more than three-quarters after it was admitted to the WTO in 1995, and those in India fell from over 80 per cent to an average of 30 per cent. While the US economy has more than tripled in size in the three decades after 1990, that of China grew by forty-five times in the same period, with nominal GDP rising from around $310 billion in 1990 to over $14 trillion in 2020. To give just one indication of the scale of change, there were an estimated 20,000 private vehicles on the road in China in 1985; today, there are more than 240 million.[6]

A series of political, economic, social, technological and digital revolutions since the early 1990s have transformed the world we live in. But perhaps the most important of all the changes has been the transformation of the natural world, the remodelling of ecological systems and the impacts on current and future local, regional and global climates. For example, intensive farming in the midwest of the United States has lost almost 60 billion metric tons of topsoil since 1860 – a process that degrades soils by removing organic matter and nutrients, reduces crop yields and increases agricultural costs.[7] An estimated 36 billion metric tons of soil are eroded globally every year, with losses predicted to rise quickly in sub-Saharan Africa, South America and South-East Asia.[8] Heavy use of pesticides in soils, along with contamination by heavy metals and plastics, has not only affected land quality and yields, but when ingested also damages the human cardiovascular system.[9]

Unsustainable exploitation of resources has obvious consequences once those resources become exhausted. For example, the High Plains Aquifer in the western portion of the Great Plains region of North America was formed around 65 million years ago from sediment eroded from the Rocky Mountains and carried by streams flowing towards the Mississippi River. The Aquifer contains enough water to cover the state of Colorado to a depth of almost fourteen metres. Today, it is recharged only by rainfall that seeps into the soil. Its deposits are being used faster than they are being replaced, largely to provide irrigation for what is often referred to as the 'breadbasket of the world'. This is not news: even in 1978, the Lieutenant Governor of Kansas, Shelby Smith, warned that Kansas had 'major water problems, and a crisis is on the horizon'. Although the rate of water decline slowed down, some sections of the Aquifer are now completely exhausted. As one recent report observed, the High Plains Aquifer was formed over millions of years; it is 'being depleted in the span of one human lifetime'.[10]

If this is damaging, so too is deforestation. Tens of millions of hectares of forest have been lost, with tropical forests the primary source of new agricultural land in the last years of the twentieth century.[11] Indeed, as much as 99 per cent of deforestation in the tropics is driven by agricultural expansion.[12] In South-East Asia, much land has been cleared for the cultivation of palm oil, the most common vegetable oil in the world and an ingredient found in more than half of all packaged products sold in the United States, including lipstick, soap and ice cream.[13] Ironically, in a classic case of one step forward and two steps back, forest clearance for plantations was also encouraged by the identification of palm oil by the European Union as a suitable biofuel to help reduce reliance on fossil fuels.[14]

It is a similar story in the Amazon, where massive land clearances have taken place in response to global demand for consumables. Some 63 per cent of the tree-cover loss across all forest types in the Amazon was the result of opening up land for beef and dairy production, with land conversion for cattle pasture in Brazil particularly acute.[15] Although deforestation slowed in the early years of the twenty-first century, rates hit a fifteen-year high in 2021, driven by fires lit to clear land illegally even before enforcement of environmental laws was scaled back under President Jair Bolsonaro.[16] In the countries of the Congo basin – Cameroon, Central African Republic, Democratic Republic of the Congo, Equatorial Guinea, Gabon and Republic of Congo – home to the second-largest area of humid tropical forest in the world, an estimated 16 million acres were lost in the years 2000–14 as a result of clearances for agriculture and logging; these losses would seem likely to accelerate given that the region is forecast to see a fivefold population increase by 2100.[17]

Major land clearances that are connected to beef, meat and dairy have also driven forest loss. Since 2000, for example, the amount of soy plantation in Brazil has doubled to 34 million hectares, resulting not only in loss of tropical rainforest but also in the negative impact of the Cerrado, a global biodiversity hotspot which distributes fresh water to the largest basins in South America.[18] Around three-quarters of global soy production is used for animal feed, particularly for chicken and pig consumption, helping meet demand in a world that has become richer, more interconnected and more populous in recent decades, with the number of people sharing the planet and its resources rising from 3 billion in 1960 to almost 8 billion in 2019.[19] It is not hard to understand the link between environmental degradation and the collapse of ecosystems, the loss of biodiversity and the obvious long-term consequences of decisions made when we choose what to eat. These pressures are likely to become more intense, with demand for meat, milk and eggs projected to grow strongly in coming decades.[20]

The impact of forest loss goes far beyond the dinner table in parts of the world that import palm oil, soy or beef planted, grown or reared in newly cleared lands. For one thing, around 1.6 billion people depend on forests for their livelihoods, above all in developing countries.[21] For another, the loss of forest cover has dramatic impacts on plant and animal habitats and on biodiversity: according to the United Nations Environment Programme, forests are home to around 80 per cent of all amphibian, 75 per cent of bird and 68 per cent of mammal species.[22] As such, dramatic remodelling of landscapes has deep and serious consequences for all forms of life.

Deforestation has a significant influence on climate patterns. For example, in the decade after 2000, palm cultivation was thought to be responsible for between 2 and 9 per cent of total worldwide emissions from tropical land use.[23] Average losses of more than 3 million hectares of forest per year in South-East Asia alone in the period 2001–19 have meant that more than 420 million metric tons of carbon are released into the atmosphere every year – a state of affairs made worse by the fact that forests are being cut down at increasingly high altitudes and on steeper slopes, where carbon densities are higher than in the lowlands.[24]

The ecosystem productivity of the Amazon rainforest has been declining since the 1980s, with increased tree mortality, fire feedbacks that amplify drought and reduced evapotranspiration resulting in shifts in moisture patterns, precipitation levels and forest viability.[25] These changes have been so pronounced that the Amazon has gone from being an absorber into a source of carbon emissions – with release much greater in the eastern regions where anthropogenic activities are concentrated.[26] Understanding the impact of other significant factors, such as methane, nitrous oxide, biogenic volatile organic compounds and aerosols is also important, as these have often been overlooked in assessments of current and projected changes across the diverse Amazon region.[27]

Although there are wide discrepancies between models of future changes, a series of recent studies have suggested that the major role that the Amazon system plays in global climate is at increasing risk of sudden collapse, with serious consequences for us all.[28] In addition to deliberate forest clearances, forest fires have already become increasingly frequent, fierce and extensive. In 2021, for example, the equivalent of around ten football pitches was lost to forest fires every minute – with over 5 million hectares burned in Russia alone (in addition to a million hectares that were cleared deliberately).[29] One recent report suggests that extreme fires will increase sharply in coming decades, with higher burning incidence in the Arctic accelerating the thawing of permafrost peatlands that are then susceptible to fire and to the release of irrecoverable terrestrial carbon, which is then released into the atmosphere, exacerbating global warming.[30]

At the COP26/UN Climate Change Conference in Glasgow in November 2021, a total of 141 countries vowed to 'halt and reverse forest loss and land degradation by 2030'.[31] On the evidence of recent years, decades and centuries, that would seem like an ambitious target. Just as repurposing of land and forests has dramatically changed the world we live in, so too has globalised trade altered ecosystems and plant and animal habitats – including our own. To be sure, and as we have seen, the opening up of trade and transport networks in the past has introduced new species and led to major ecological transformations.

However, the hyperconnected globalised trade and transport networks of the modern world that have brought producers and consumers closer together and made it easier for humans to move around the world have also led to the expansion, relocation and invasion of alien species into new locations that have resulted in the extinction of native species, disrupted food chains and produced significant changes in the natural environment on an unprecedented scale and at an astonishing speed.[32]

These networks also carry significant implications for human health, welfare and finances, given many have a major impact on activities related to economic productivity, or have financial consequences: for example, an infestation of brown tree snakes in Guam has caused thousands of hours of power outages as the creatures disturb or become trapped in electrical power lines and service stations, causing as many as 200 outages in a single year – at a cost of millions of dollars in damage, repair costs and lost productivity.[33] The spread of the common coqui frog in Hawai'i, meanwhile, has been linked to a decline in real estate values in places where infestation levels are high because its loud mating song is such a disturbance to local (human) residents.[34]

The emerald ash borer, native to East Asia, was first identified killing trees in Michigan and Ontario in 2002; two years later, 15 million trees were dead or dying. The entire stock of 8 billion ash trees in the US are vulnerable to this pest – which if lost would cost forestry more than $280 billion, not to mention the costs of removal of dead urban trees for an additional $20–$60 billion.[35] Then there are the Asian and citrus long-horned beetles, originally from South-East Asia, which some estimate might together kill 30 per cent of trees in US cities (more than a billion, in other words), valued at $669 billion, and which are already killing ash

trees in Russia and spreading westwards towards Europe.[36] An outbreak
in Kent in the UK, most likely from infested imported wood packaging,
required a six-year programme of trapping and surveillance by the
Animal and Plant Health Agency and the Forestry Commission, which
eventually led to a triumphant and relieved statement by the Minister
for Biosecurity.[37] Or there is the fall armyworm moth which is highly
damaging to many crops and to maize, sorghum and rice in particular,
and which despite being native to the Americas was first reported in
West Africa in 2016 and a year later was the fastest-spreading major pest
in the world.[38]

Taken as a whole, the consequences for world agriculture if the
spread of invasive pests and pathogens is not stopped runs to hundreds
of billions of dollars per year – with China, the United States, India
and Brazil facing the highest potential costs from around 1,300 species
that are particularly dangerous. Relative to GDP, however, the majority
of the twenty countries most at risk are located in sub-Saharan Africa.[39]
Pests do not just destroy ecosystems: animals such as termites can also
destroy built infrastructure, and are likely to do so more often as colony
sizes and geographic distribution rise sharply in the coming decades;
this has climate implications too since these animals are a substantial
source of methane emissions.[40]

On top of this comes crop disease from parasitic plants, viruses, fungi
and bacteria, which are thought to be the cause of the loss of around 10
per cent of global food production each year. These are typically dealt
with through the application of chemicals or antibiotics, though these
are losing their efficiency thanks to the natural development of bacterial
resistance.[41] This is in addition to measurable declines in nutritional
values in multiple crops in recent decades, presumably as a result of use
of fertilisers, pesticides and the more aggressive farming methods that
place a premium on crop size and yield.[42]

However, the most important environmental transformations have
resulted from the ways in which human populations have become
concentrated in recent decades. Cities and urban areas are the
source of most major climate-change impacts: although living close
together encourages high velocities of exchange, cities are centres of
consumption rather than production. Urban populations require food,
water and energy which need to be brought over distance – sometimes
considerable ones – and rarely only from the hinterland or immediate

vicinity. Factories and manufacturing are located in or near cities in order to access labour forces or transportation, power and digital networks, but the materials that supply them are usually brought from other locations – again often ones that are distant.

Transportation is responsible for around a quarter of global energy-related CO_2 emissions, and even more in many developed countries, accounting for about 29 per cent of emissions in the US as a whole and around 41 per cent of those in California. In some cases, more emissions are produced by inefficient production than by the transportation of goods over long distances; indeed, as well as the socio-economic benefits of trade, there are sometimes cases where transportation actually decreases emissions even when goods are not produced in a 'cleaner' way in different locations.[43]

Nevertheless, urbanisation is a driver of multiple environmental problems and not only because of the needs of high-density populations. Despite cities being synonymous with 'civilisation' in the literal meaning of the world (Latin *civitas* – city), the reality is that cities occupy only a tiny fraction of the world's land surface – perhaps as little as 3 per cent.[44] Despite this, rising urbanisation has meant that more than half the global population now live in urban areas, with this number projected to rise to 70 per cent by 2050. Given current demographic trends, this means that an additional 2.5 billion people will be living in cities within the next thirty years – which has obvious implications for demands on resources of all kinds, for infrastructure natural and otherwise and for emissions of carbon, greenhouse gases and heat.[45]

This comes on top of the rate of urbanisation over the last thirty years which has been jaw-dropping in scale. Construction in urban areas in China quadrupled between 1990 and 2010, with time-series satellite imagery showing that built-up land areas in parts of Chengdu grew by 300 per cent in the six years after 1996.[46] China's urban population rose from around 18 per cent to almost 60 per cent in forty years, with hundreds of millions of people moving to cities in what is the largest and fastest rate of urbanisation in history.[47] China is estimated to have used more concrete in 2011–13 than the United States did in the whole of the twentieth century.[48]

Urbanisation in other parts of the world – including Nigeria, India, Brazil and Indonesia – has been epic too, with cities such as Mexico

City, Lagos, Manila, Mumbai, Jakarta, Dhaka and Cairo now each having populations of at least 20 million and in some cases considerably more. Before the coronavirus took hold, all ten of the fastest-growing cities in the world by GDP in the next fifteen years were forecast to be in India; it is still too early to say whether these projections will hold.[49] What is clearer is that the distribution of urbanisation levels and growth rates across continents and regions is uneven, with more than 80 per cent of people living in urban areas in North America, Latin America and the Caribbean, compared to just 40 per cent in sub-Saharan Africa.[50]

Cities give rise to profound socio-economic inequalities, with more than two-thirds of the world's urban population experiencing widening income inequality since 1980 – which means that almost 3 billion people live in cities where realities and prospects are worse than they were a generation ago.[51] Although slum conditions have improved over the last fifteen years, an estimated 1 billion people – or one in four of all the world's urban population – live in housing that is non-durable or overcrowded, or lack access to safe water and sanitation or security against eviction.[52]

As well as being centres of consumption, cities and their populations generate vast amounts of waste, putting pressure on infrastructure as well as on the natural environment as a result of heat stress, water scarcity, energy demands and difficulties of sanitation treatment.[53] Sometimes, the challenges are not only overwhelming, but a source of health hazards. In Lagos, for example, less than half the population in high-density areas of the city have access to flush toilets, resorting to open pit latrines – with limited handwashing facilities available.[54] Cities in high-income countries are able to pay to remove waste out of sight and out of mind, as is the case with New York City, where more than 3 million metric tons of rubbish every year are moved out of the city to incineration and recycling plants and to landfill at a cost of more than $450 million.[55] These options are not available, however, in most of the world's largest cities, where disposal is often either rudimentary, hazardous or non-existent.

While the impacts of oil and gas on climate receive a great deal of attention, those of the waste sector are often ignored – despite the fact that the latter account for almost 20 per cent of total anthropogenic emissions, primarily as a result of anaerobic decay of organic materials in landfill.[56] Landfill waste is expected to grow at more than double the rate of population growth in the years leading up to 2050 – which is likely to lead to a significant rise in methane concentrations that in turn have a major impact on global warming. Recent research using satellite data on landfill sites in Delhi, Mumbai, Lahore and Buenos Aires reveals that city-level emissions are considerably higher than usually reported, which underlines the importance of the development and implementation of policies to mitigate the effect that they have on atmospheric conditions.[57]

This research also confirmed the importance of identifying 'superemitter' sites that produce a disproportionately large volume of emissions, where the rewards for reduction can be high. There are obvious lessons, therefore, for by-products created by cities around the world, not least since cities generate more than 70 per cent of greenhouse-gas emissions, which are closely linked to global warming. More than half of these are produced by just twenty-five 'megacities', all of which are located in China, apart from Moscow and Tokyo.[58]

Urbanisation, combined with rapid industrialisation in low-income countries and the intensification of international trade, has led to falls in air quality, which, as we have already seen, leads to a range of detrimental health outcomes. Particulate pollution in India, Pakistan, Bangladesh and Nepal has risen by 47 per cent since the start of the century, with some parts of India recording $PM_{2.5}$ concentration levels of 107 µg/m³ – more than twenty-one times World Health Organisation guidelines. Poor air quality means that life expectancy in several provinces east of Kinshasa in the Democratic Republic of Congo or in Mixco in Guatemala is more than three and a half years less than it should be, while across South-East Asia, where 99.9 per cent of the population live in areas that exceed the WHO guidelines, average life expectancy is reduced by one and a half years for every man, woman and child – a total loss of almost 960 million person-years. Taking a global average, the impact of particulate pollution on life expectancy is three times greater than that of alcohol, six times that of HIV/AIDS and eighty-nine times that of conflict and terrorism.[59]

Potential gain in life expectancy from permanently reducing PM$_{2.5}$ from 2020
concentrations to the WHO Guideline in the 10 most populated countries

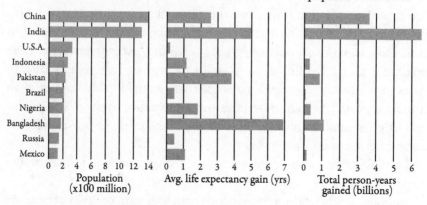

Life expectancy impact of PM$_{2.5}$ and unassociated causes/risks of death, Global

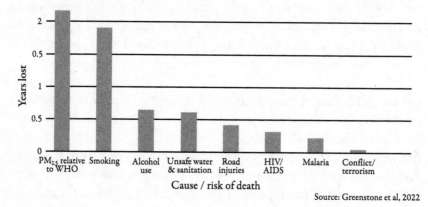

Source: Greenstone et al, 2022

That air quality is significantly better in high-income countries is partly a
testimony to legislation, but it owes much to the fact that manufacturing –
and its dirty by-products – have been outsourced to other parts of the world
that as a result bear the environmental, health and human consequences.
Despite this, the per capita greenhouse-gas emissions of cities in Europe,
the United States and Australia are significantly higher, because of higher
consumption patterns, lifestyle choices and energy availability.[60] Those
who are wealthy, in other words, do more direct and indirect damage to
the environment and to climate than those who are not.

Underpinning this has been the development and adoption of one
technology in particular: air conditioning. According to one prominent

economist, air conditioning ranks alongside the civil rights movement in the reshaping of US demographics and politics in the latter part of the twentieth century.[61] The ability to control temperature and to maintain it at a constant level transformed not only how people lived but where, with offices, factories, homes and leisure facilities springing up in locations that had previously been too hot to be practical as major urban settlements. Such has been the impact of air conditioning that studies suggest it has been responsible for up to 80 per cent of the decline in heat-related mortality in the United States alone.[62] The ability to cool classrooms and workplaces in warm regions is vital for productivity and cognition, with studies showing that student performances in standardised maths tests fall as temperatures rise from the low 20s °C, while thermal stress also affects simple cognitive and manual tasks such as handling a steering wheel.[63]

Air conditioning, and the associated energy requirements, has allowed humans to create artificial climate environments that are independent of their natural surroundings. Put differently, this has meant that urban metropolises could be built in locations that would not otherwise be sustainable – and whose futures are dependent on air-conditioning units and on power being supplied to them. As one leading commentator has pointed out, there are already more than 350 major cities today where the average maximum summer temperature is over 35 °C; given current warming patterns, more than 600 more cities might well fall into the same category by 2050 – resulting in 1.6 billion people living in challenging conditions that far exceed the optimal temperature for the human body that medical literature consistently places in the range of 15–20 °C.[64]

Air conditioning requires vast amounts of energy: in Saudi Arabia, for example, 700,000 barrels of oil per day are burned, primarily to keep premises cooler than they would otherwise be. Around 70 per cent of all energy consumption in Saudi Arabia is spent on air conditioning. Although there has been some progress in clean energy production, as recently as 2017 all of this came from fossil fuel energy sources.[65] As things stand today, air conditioners and fans account for 10 per cent of global electricity consumption – with demand likely to triple or even quadruple by the end of this decade; some estimate that there will be more than 9 billion cooling appliances being used by 2050.[66] In this context, rapid population growth in drought-prone places – such as the south and mountainous west of the United States – that adds to existing pressures on water, as well as on infrastructure, transportation and energy, is not just hard to understand but difficult to believe.[67]

This falls into a wider pattern of behaviour relating to the choices that are made on the basis of apparent certainties and assumptions that tomorrow's world will be as benign as today's. For example, across the United States, almost 4 million houses have been built in flood zones which are exposed to tropical cyclones, floods and other hazards; in Florida alone, one in six single-family houses have been built in floodplains. Taken as a whole, this means that US housing stock located on floodplains is overvalued by almost $44 billion, even before implications of rising sea levels, more frequent and more extreme storms and other hazards are considered.[68]

If decisions to move towards rather than away from risk seem perplexing, so too do countless other choices we make. Nearly three-quarters of all the energy produced by humans is squandered as waste heat; although this does not directly warm the atmosphere and has a negligible impact on climate change, the wastage is testimony to the profligate way we treat the world around us.[69] In the UK, for example, almost 10 million metric tons of food, worth almost £20 billion, are thrown away each year. Getting this food from the field, pasture or coop to the rubbish bin required energy to sow, tend, reap, harvest, feed, pasture, slaughter and transport not only to the point of sale and then to the larder or fridge, but also to compost heap or rubbish bin. As such, it represents 36 million metric tons of greenhouse-gas emissions that were generated because of inefficiency.[70]

When widened out with data from elsewhere, the UN Environment Programme and the climate action group WRAP suggest that 931 million metric tons of food were wasted globally in 2019 – the equivalent of 23 million fully loaded forty-ton trucks, which if lined up bumper to bumper would circle the earth seven times. Almost two-thirds of this food is wasted at household level, with a global average of 74 kilos per capita of food thrown away or not eaten each year – with little difference between lower-middle and high-income countries.[71] An estimated 8–10 per cent of global greenhouse-gas emissions are associated with food that is not consumed.[72] Our inability to budget and plan meals efficiently comes at a cost to plants and animals, to soil, to the atmosphere – and to each other.

So, for example, the fashion industry as a whole is estimated to contribute around 10 per cent of global greenhouse-gas emissions – more than the aviation and shipping industries combined.[73] Some believe that clothes production may account for a quarter of the world's carbon budget by 2050.[74] A significant part of the energy and resources that go into clothes production is squandered: in the UK alone, it is estimated that consumers have unworn clothes in their cupboards that are worth almost $50 billion.[75] This is the tip of the iceberg: around the world, the equivalent of a rubbish truck of textiles is burned or deposited in landfill every second of every minute of every hour of every day of the year because of clothes and fabrics that are no longer wanted or cannot find a buyer.[76] At least 39,000 metric tons of this end up being dumped in the Atacama desert, one of the world's driest regions, where it can take hundreds of years to biodegrade, if it biodegrades at all.[77]

Moreover, clothes production is extremely water intensive: making a single cotton shirt requires 2,700 litres of fresh water – the equivalent of a person's drinking needs for two and a half years – and 7,500 litres to make a single pair of jeans, drinking water for someone for seven years.[78] This clearly has implications in the context of a world where by 2010 a quarter of the world's population was already affected by water scarcity, with 17 per cent of the European Union's territory and more than 10 per cent of its population already also exposed.[79] These pressures are rising quickly, with demand for water projected to exceed supply by 40 per cent by the end of this decade – before factoring in what climate change will do to availability.[80] It does not help that poor infrastructure means that in some places water is wasted – with more than 3 billion litres estimated lost to leaks every day in England and Wales in the twelve months to March 2021, and an estimated 20–40 per cent of Europe's available water likewise lost unnecessarily.[81]

A long list of other examples could be given to illustrate the consequences of lifestyle choices. A study of drinking-water habits in Barcelona showed that the environmental impact of bottled water was up to 3,500 times greater than tap water, and that switching either to tap or filtered water had major benefits because of reduced use of resources and raw materials.[82] Around 500 billion plastic bottles are produced every year – the equivalent of more than sixty for every man, woman and child in the world. According to one study a few years ago, fewer than half of the 20,000 bottles or so that are sold every second around the world were collected for recycling, with just 7 per cent of those turned into new bottles.[83] One senior executive at Coca-Cola, whose factories are thought to produce more than 100 billion throwaway bottles a year, commented that the company would not abandon single-use plastic bottles because they were popular with customers, noting that 'Business won't be in business if we don't accommodate consumers.'[84]

Consumer demands may sate short-term desires – at costs that can be overlooked or obscured. The popularity of ship cruises, for example, has helped drive the construction of more and larger vessels that take customers on voyages of a lifetime. According to recent analysis, in 2017 emissions of noxious oxides around the coasts of Europe by the world's largest cruise operator, Carnival Corporation, were nearly ten times greater than those of all 260 million cars in Europe put together – with obvious consequences for marine and coastal life of all kinds, as well as enhanced risks to human health, especially in the main cruise ports such as Venice, Palma and Barcelona.[85] Then there is the boom in space tourism which threatens to deplete the ozone layer and to produce soot emissions that are 500 times greater than those of the aviation industry.[86]

The digital age that has powered a revolution in communication, information and connectivity has also served to boost consumption by putting pressure on prices: internet retailers do not need to have expensive shops in town centres, to employ staff to restock shelves that look attractive to customers or to be based physically or even legally in countries where orders are placed. Given also the ease with which one can order from mobile devices, it is perhaps not surprising that some of the largest companies in the world – from Apple to Alphabet, from Alibaba to Amazon, from Meta to Verizon – are directly and indirectly linked to boosting patterns of exchange. To give one example, in 2019, Alibaba's annual Singles' Day sale on 11 November saw sales of around $13 billion in a little over ninety seconds; the day as a whole resulted in the dispatch of 1.29 billion parcels.[87] In 2020, nearly 4 billion packages were shipped as a result of Singles' Day sales.[88] In the course of the previous year, China had consumed more than 9 million tons of plastic packaging – equivalent to the weight of 130 million adults, and carbon emissions that would take 700 million trees to neutralise.[89]

Inefficiency and wastage plague other sectors too. Many medicines that are produced become useless or have to be destroyed because they are susceptible to physical or chemical deterioration – a feature that makes them difficult to formulate, store and transport; as a result, almost half of all biologics, including vaccines, are ruined before they can be used.[90] As well as the cost of ingredients, manufacture and transportation, the difficulties of cold-chain and logistics systems means that countries that have poor transport and energy infrastructure, have challenging climates or geographies or both, or have large rural or semi-urban populations experience considerably more problems in delivering medicines such as vaccines. This leads to a range of outcomes that impact incomes, social development and political liberalisation.[91]

Or there is the military, whose energy expenditure is vast – even in times of peace. A new F-35A fighter plane consumes almost six litres per kilometre on a typical training sortie, and generates almost twenty-eight metric tons of carbon dioxide emissions if it uses its full capacity without refuelling. Financial and environmental costs rise sharply during periods of conflict: since 2001, when the United States has been heavily committed to operations in Syria, Iraq and Afghanistan, as well as in other arenas such as the South China Sea, the Department of Defense has been responsible for around 80 per cent of the entire energy consumption of the US government. Indeed, the Department is the world's largest institutional user of petroleum and the single largest institutional producer of greenhouse gases in the world, with energy primarily expended on fuel for aircraft, but also for other vehicles and vessels, as well as on heating, lighting and powering military facilities – which include 560,000 facilities with over 275,000 buildings at 800 bases, located on 11 million hectares of land in the US and elsewhere around the world.[92]

While the scale of the US military dwarfs those of other countries, many of whose energy expenditures are difficult to calculate accurately, it is nevertheless clear that defence comes with a very high price tag. Ironically, the link between climate change and war was at the forefront of the minds of participants at the Toronto Conference on Changing Atmosphere in 1988, where the delegates declared in an agreed statement that human activities amounted to an 'unintended, uncontrolled, globally pervasive experiment whose ultimate consequences could be second only to a global nuclear war'.[93]

It has taken a long time for this message to sink in. A 2019 study by the British Academy of Film and Television Arts (BAFTA) and the environmental organisation Albert, supported by Deloitte, looked at almost 130,000 TV programmes shown by four major broadcasters in the UK between September 2017 and September 2018. 'Climate change' and 'global warming' were mentioned 3,125 times and 799 times respectively. To put that in perspective, 'gravy' was mentioned 3,942 times, 'cheese' almost 33,000 times and 'dog' 105,245 times.[94] In a follow-up report published in the autumn of 2021, climate change was mentioned marginally more often than 'goldfish', slightly less often than 'Shakespeare' and just over twice as often as the word 'motherfucker'.[95] This would suggest that awareness of and concern about the world of tomorrow – and today – are not as high as they could and perhaps should be.

According to the World Meteorological Association, the decade 2010–20 was the warmest since modern record-keeping began in the 1880s.[96] This included a drought event in Europe that was the most intense in centuries and reduced crop yields in countries and regions most affected by as much as 40 per cent.[97] In the south-western United States, meanwhile, the years since 2000 have been the driest since at least AD 800 thanks to what some scientists refer to as a megadrought, with temperatures above average and precipitation well below, leading to the decline of two of North America's largest reservoirs, Lake Mead and Lake Powell, both of which reached their lowest levels on record.[98] The year 2019 was the warmest on record in the Arctic, resulting in the loss of almost 660 billion metric tons of ice, more than double the average of the previous two decades.[99] Glacial retreat in the Peruvian Andes that is almost certainly linked to human activities has raised the threat to towns and communities as a result of the increased probabilities of major flooding.[100] Severe storms and extreme rainfall that occurred when three tropical cyclones and two tropical storms hit Madagascar, Malawi and Mozambique in the space of just six weeks in early 2022 are also part of shifting weather conditions that can be associated with anthropogenic climate change.[101]

What makes these so significant is that they are part of a globally coherent pattern of warming. As we have seen, while there have been many epochs in the past that have experienced particularly cold or warm periods – such as the Little Ice Age, the Medieval Climate Anomaly or the Roman Warm Period – these were not universal, but rather were particularly pronounced in one or more regions, and even on one or more continents. In contrast, the last 150 years show near coherence all around the world. For 98 per cent of the planet, the twentieth century was the warmest period of the last two millennia. This is not only unprecedented; it is also not a coincidence.[102]

At the heart of the acceleration of warming have been resource depletion, manufacturing, urbanisation and population growth, all of which have relied heavily on vast consumptions of energy and of power produced by carbon in particular. Even though recent years have seen large investments in renewable energy sources (such as hydro, geothermal and above all wind and solar power), around 80 per cent of the energy used around the world comes from the burning of fossil fuels.[103] The burning of fossil fuels releases carbon dioxide into the air which intensifies the greenhouse effect by trapping heat in the atmosphere, causing the earth's average air temperatures to increase.

Rising population levels, greater urbanisation and new production and transport technologies as well as more frequent commercial interaction have all played a part in driving demand for more energy in recent decades: as the influential writer David Wallace-Wells put it, around 85 per cent of the burning of carbon-based fuel has been carried out since the end of the Second World War – while more than half has been since the first episode of *Seinfeld* was broadcast. The result is that there is more carbon in the atmosphere than there has been for millions of years.[104]

How human activities interact with, affect or mitigate natural changes in the climate is a matter of considerable discussion – primarily because of the technical difficulties of weighing up complex evidence and data. For example, the rapid warming of global surface temperatures was reported to have slowed down in the early part of the twenty-first century, with unusually low growth trends in surface and tropospheric temperatures, as well as those of upper ocean heat content and sea level – in what has become known as the 'global warming slowdown' (or 'hiatus'). It is not only the reasons for this that are contested; so too is the premise, with disagreement about whether selection or measurement bias (or both) influenced conclusions, or even if changes in instrumentation might have distorted data.[105]

Despite the spillover from such discussions, which can convince those who want to be convinced that there is no anthropogenic impact on climate, a survey of almost 90,000 peer-reviewed climate-related papers published since 2012 suggests that the consensus about human-caused climate change among scientists working in this field exceeds 99 per cent. This stands in sharp contrast to members of the 117th Congress of the United States, of whom more than a quarter both in the House of Representatives and in the Senate – including more than half the

Republican members of the former and 60 per cent of the latter – have made statements either doubting or refusing to accept the scientific evidence for anthropogenic climate change.[106]

In fact, the chances that some of the most significant changes – such as the earth's growing energy imbalance – are occurring naturally rather than as a result of human activities is extremely slim, to be measured as below 1 per cent probability.[107] Study after study reveals rapid transformations that when taken together paint not just a compelling but an ominous picture. For example, two major glaciers in the Antarctic are losing ice at the fastest rate for 5,500 years.[108] Climate modelling drawing on a period from around 15 million years ago when atmospheric carbon dioxide levels and global temperatures were similar to those expected at the end of this century suggests that a chain reaction may result in Antarctica being unable to maintain a large ice sheet.[109] This means that there is an elevated risk of the collapse of the West Antarctic Ice Sheet, which would bring with it projected global sea-level rises of three to four metres – if not more – as a result of the mass melting of ice.[110]

Although high geothermal heat flow beneath the Thwaites glacier is a contributor to ice loss, unusual mean temperature anomalies, such as in the spring of 2020 when the highest temperature over the Antarctic was recorded – some +4.5 °C above average – clearly add to the risks of sudden collapse.[111] So great have been the impacts of glacial ice melt since the 1990s that the earth's axis has shifted as a result of the redistribution of the planet's water; moreover, as well as showing that this is the result of human factors that cause global warming and the heating of polar regions, new research suggests that the movement of the pole has also been affected by groundwater depletion of terrestrial water storage, for example in northern India – where 351 billion cubic metres of water was drawn in 2010.[112]

Not all the problems are caused by humans. Tundra fires in Arctic Alaska are accelerating permafrost degradation – which in turn unlocks a vast store of frozen plant and animal matter that magnifies other warming factors through the release of carbon stocks into the atmosphere.[113] Massive fires in Siberia in the summer of 2021 dwarfed all the others on the planet combined and formed what NASA called a 'vast, thick, and acrid blanket' that covered most of Russia, releasing the estimated equivalent of 505 million metric tons of carbon.[114] This was broadly equal to fires that took hold in the Amazon five years earlier, when the 2015–16 El Niño resulted in extreme drought that paved the way for megafires.[115]

Such events intertwine with the wide range of other drivers of warming processes, which in turn create feedback loops that make bad situations even worse: as the atmosphere warms up, more water evaporates from oceans, rivers and lakes, rising up into the atmosphere where water vapour traps more heat, amplifying the initial warming. In this way, it is possible to reach spirals from which there is little or no chance of escape. If carbon dioxide concentrations reach high enough levels, stratocumulus clouds that are prevalent in the subtropics and also cover 20 per cent of the low-latitude oceans may become unstable and vanish, triggering further global warming.[116] There is much discussion about and research into similar 'tipping points' by which precarious situations are made worse and worse in a series of climate domino effects.[117]

Many examples could be given – including collapse of ice caps, the loss of almost all mountain glaciers, permafrost melt and carbon release, disruption to ocean circulation and massive northern hemisphere tree loss due to fire.[118] Warming conditions in Greenland have already resulted in 3.5 trillion tons of ice melt from the surface of the island into the ocean in the last decade alone.[119] The point of no return has already been passed for future ice loss – meaning that change is 'baked in', regardless of any steps taken to address emissions. The prognosis is ominous, according to one recent study, which projects a best-case scenario of global sea-level rises of almost eleven inches (twenty-eight centimetres) by the end of the century, and almost treble that in the worst-case. This has dire consequences for the hundreds of millions who currently live in coastal communities less than one metre above sea level.[120]

The oceans and other bodies of water offer another example. Heatwaves affecting the oceans have become more frequent, last longer and are more intense, resulting in damage to and loss of coral reefs, kelp forests and seagrass meadows.[121] Ocean warming has had a significant and detrimental effect on fish and invertebrate populations.[122] These losses come on top of those caused by industrialised fishing, which has reduced the biomass of large predatory fish to about 10 per cent of pre-industrial levels – with serious consequences for ecosystems.[123] Warming oceans seem likely to change these further, both because warming water temperatures increases metabolic rates which leads to predators eating more, and because of the likelihood of large species moving away from their historical ranges and latitudes.[124]

Oxygen levels in the world's oceans and freshwater lakes are declining rapidly, falling by 5.5 per cent at the surface and almost 20 per cent in deep waters of the latter since 1980. This is important since the concentration of dissolved oxygen in aquatic systems plays an important role in supporting biodiversity as well as in regulating greenhouse-gas emissions and the quality of drinking water.[125] Similarly, since oceans trap and absorb 90 per cent of greenhouse gases, the deoxygenation that has followed as a result of global warming not only has obvious implications for ecosystems, but will continue for centuries even if all carbon dioxide emissions were to stop immediately.[126] Indeed, some studies suggest that changes to the oceans are such that they have already passed the point of no return.[127] Climate change is not something to worry about in the future, in other words, nor even in the present: rather, it has already created problems that cannot be avoided.

Some of these challenges are now obvious. For example, field experiments suggest that warmer night-time temperatures reduce rice grain yield and quality.[128] Even very modest changes to two major field crops in North America, corn and soy, show that yields fall by half a per cent for each day when the temperature is a single degree warmer than the optimal 29 °C. This means that as the world gets hotter, growing food becomes more difficult, and much more expensive. Some studies suggest that there will be huge financial and production losses by 2050 based on most current climate models.[129]

Photosynthetic algae, which flourish in oceans, lakes and rivers when exposed to more light and to hotter weather conditions, can form blooms that are not only toxic but affect other organisms in the food web.[130] These blooms, which can cover thousands of square kilometres, can also be triggered and fuelled by the iron aerosols carried in smoke and ash – as demonstrated by the widespread bloom formation in the Southern Ocean following the Australian wildfires of 2019–20.[131]

The warming world is forcing the redistribution of life on earth in other ways too, and at unprecedented speed. For example, migratory birds play a fundamental role in long-distance seed dispersal, with seeds carried towards either warmer or cooler latitudes depending on plant fruiting periods and on northward or southward avian migration. One recent study showed that 86 per cent of plant species in a sample set were dispersed by birds migrating south, and only 35 per cent carried north. Another found that earlier onset of spring increases the risk of massive bumblebee declines because of shrinking food availability – which has obvious implications for the 'vital pollination services' bees (and other species) provide, not least for vegetables, fruits and many kinds of crops.[132] With climatic boundaries already shifting towards cooler latitudes, plant as well as animal habitats and ecosystems and food-chain networks are already being reconfigured.[133]

Of course, these include the spread of infectious diseases. Mosquitoes are active in tropical regions all year round, but in other locations go into diapause or seasonal dormancy in response to cold exposure. As autumns and winters become warmer, mosquitoes may increase their annual activity time as well as their distribution northwards, an important development given that many species carry viruses that induce human and/or wildlife diseases, including St Louis and Eastern equine encephalitis, dengue and West Nile virus.[134] Some estimate that in the worst case scenario about 8.5 billion people – or about 90 per cent of the world's projected population – will be at risk of malaria and dengue fever by 2078 as a result of global warming and the expansion of disease belts northwards.[135]

As the coronavirus pandemic that took hold in 2020 reminds us, diseases that jump species can be devastating. Specialists in this field already knew long before the pandemic that about 60 per cent of human infections are estimated to have an animal origin, while around 75 per cent of all new and emerging infectious diseases are ones that jump from animals to humans. The trade and transport networks that form the backbone of modern globalisation serve not only to move goods and people around the world faster than ever before, but also allow diseases to spread more quickly than ever before – as the recent coronavirus pandemic showed all too clearly.[136]

While important lessons will have been learned about future prevention, in a warming world it is also crucial to understand that not only do disease environments alter as temperatures and climates change; so too do diseases themselves. Almost 60 per cent of known infectious diseases have at some point been aggravated by climate hazards, such as drought, wildfire, extreme rainfall, floods and rising sea levels, which enhance specific aspects of pathogens, such as improving climate suitability for reproduction, accelerating the life cycle, increasing seasons or length of potential exposure, improving the chances of pathogen vector interactions and increasing virulence. In addition, climatic hazards can be responsible for diminishing human capacity to cope with pathogens, for example by compromising immunity to disease through heat stress, reducing access to medical care as a result of extreme weather events or through malnutrition, whether through lack of access to food or because of reduced concentration of nutrients in crops as a result of exposure to elevated carbon dioxide levels.[137]

Taking all these factors together, then, the world of today and tomorrow looks terrifying. Multiple models predict rises in sea levels as a result of extreme ice-melting events in Greenland as well as from melting Antarctic ice sheets.[138] Others predict that the threat of intense tropical cyclones will have doubled by 2050, extending into areas where exposure is currently low and where populations densities are high, putting millions of people at risk – not least in low-income countries such as Cambodia, Laos, Mozambique and many Pacific island nations.[139] The climate in Alaska may be so profoundly affected later this century that not only will the number of thunderstorms treble, but extreme conditions will lead to flash flooding, landslides and wildfires caused by lightning.[140] Slow-moving, intense storms that produce high levels of rainfall may be fourteen times more frequent across Europe by the end of the century.[141]

Then there are assessments which suggest that by 2053 a quarter of the land area of the United States, stretching from the northern border of Texas and Louisiana through Iowa, Indiana and Illinois – currently home to more than 100 million Americans – will be exposed to summer temperatures above 125 °F (just over 50 °C) if current demographic distribution patterns are maintained.[142] Increasing evidence suggests that summers in northern hemisphere mid-latitudes have already lengthened, whereas spring, autumn and winter have shortened – a pattern that will continue without climate-change mitigation to the point where six-month-long summers may become commonplace.[143]

The likelihood of multiple heatwaves covering regions the size of Iran or Mongolia (the eighteenth- and nineteenth-largest countries in the world respectively) occurring simultaneously across the northern hemisphere have become six times greater since the 1980s, with concurrent events growing larger and more intense during the warm season of May–September.[144] If the current trajectory of emissions and warming is maintained, the average temperature in Milan may reach 50 °C by 2100, according to Luca Mercalli, President of the Italian Meteorological Society – and 8 °C higher across Italy compared to today, as the Mediterranean as a whole bears the brunt of extreme weather.[145]

Such conditions present extraordinary challenges for human life in the absence of artificial cooling that is not affordable or available to all. Hot weather is closely correlated with rises in suicide rates, deterioration in mental health and sharp declines in cognitive abilities, with verbal reasoning, spatial awareness and attention span in particular affected by the heat.[146] When combined with humidity, things become precarious. Even modest global warming will put hundreds of millions of people at risk in densely populated regions of South-West Asia, in South Asia in the Indus and Ganges river valleys and in eastern China as a result of the 'wet bulb' effect which makes the fittest humans unable to tolerate more than a few hours of exposure even under shaded and ventilated conditions.[147]

The composition and nature of oceans and seas that are already being transformed are likely to be further affected in the future, according to scientists. Freshwater is accumulating in the Arctic Ocean, for example, with the Beaufort Sea, the largest oceanic freshwater reservoir in the northern hemisphere, increasing its freshwater content by over 40 per cent since the start of this century. The timing and scale of release of this water into the North Atlantic, and the magnitude of salinity anomalies, have major consequences for the strength of the Atlantic Meridional Overturning Circulation (AMOC) which itself has a significant influence on the climates of the northern hemisphere.[148] This is particularly important given the slowdown of the AMOC – also known as the Gulf Stream System – in recent decades to the point where it is now weaker than at any time in the last millennium, a downturn that has itself been linked with anthropogenic global warming.[149]

Or there is the Caspian Sea, the world's biggest inland body of water, which some projections suggest will experience a decline in water level of at least nine metres by the end of the twenty-first century – resulting in 'a loss of 25 per cent of the Caspian's total area and the uncovering of some 93,000 km² of dry land, an area roughly the size of Portugal'. Ironically, the melting of glaciers across Central Asia, a region with severe water stress, has provided a short-term boost to water availability that is expected to peak during this decade. There are already signs of things to come, with a drought in Kazakhstan in 2012 resulting in the loss of 1 million hectares of crops and hot conditions across a broad swathe of this region in 2021 leading to mass livestock deaths.[150]

Then there are the ice fields in the Himalaya mountain range which are projected to decline by one-third if carbon emissions fall rapidly and by as much as two-thirds if they don't. The glaciers store water for some 240 million hill and mountain people who live across the eight countries of the Hindu Kush Himalayan region, and for 1.65 billion who live in the ten river basins downstream and benefit directly and indirectly from the glaciers' resources and whose livelihoods, rich and diverse cultures, languages, religions, traditional knowledge systems – and lives – are at risk.[151]

Some models provide disturbing forecasts of what is in store for humanity in the decades ahead. Even if strong climate-mitigation measures are put in place, according to one recent study, around 1.5 billion people will be living in a climate that has historically been considered beyond the temperature niche sustainable for human life – a figure which in worst-case scenarios for population growth and climate change rises to 3.5 billion people, or around 30 per cent of the projected global population.[152] Perhaps not surprisingly, thoughts have turned to what the responses might be, in particular to the prospect of massive migrations – and of violent conflict.[153]

Guessing which regions and areas might be worst – or first – affected is difficult, as is assessing the ways that problems might brew, escalate or proliferate. Twelve of the seventeen countries most at risk from water stress are located in the Middle East and North Africa, where mortality risk because of excessive heat stress is forecast to intensify by a factor of 2–7 by the last decades of this century if global warming is limited, and by significantly more if it is not.[154] Then there is South-East Asia, where by 2050 daily high tides will flood areas that are currently home to almost 50 million people – to say nothing of the 77 per cent of the population of the region as a whole who live along the coast or in low-lying river deltas.[155] Or perhaps one should focus on the countries of the western Sahel, ranked among the most vulnerable in the world, where a rise in temperatures over the last five decades, combined with less frequent rainfall and an increase in the frequency and severity of extreme weather events, exacerbates high levels of extreme poverty, rapid population growth and fragile governance – a blend that helps explain a power vacuum that is being filled by organised-crime networks, terrorist organisations and a range of violent non-state actors, including affiliates of both ISIS and Al-Qaida.[156]

The concerns are such that many governments have started to engage with planning how to deal with the effects of climate change. For example, in a major review published in 2010, the US Department of Defense noted that climate change would 'shape the operating environment, roles, and missions that we undertake', and warned that 'assessments conducted by the intelligence community indicate that climate change could have significant geopolitical impact around the world, contributing to poverty, environmental degradation, and the further weakening of fragile governments. Climate change will contribute to food and water scarcity, will increase the spread of disease, and may spur or exacerbate mass migration.'[157] Significant resources have gone into forming committees, writing reports and preparing plans to deal with the fact that, as President Obama put it in 2016, 'Climate change poses a significant and growing threat to national security, both at home and abroad.'[158]

Not least among the problems is the cost of maintaining US military facilities in the face of climate change. As early as 1990, some were asking what the implications were for the armed forces, for their operations, facilities and systems. By 2018, the Pentagon was reporting that around half its installations had experienced climate-change-related effects, while a year later it was admitting to recurrent flooding, drought, wildfires and desertification across dozens of its facilities.[159]

Other countries face major challenges too. More than 80 per cent of Russia's gas and 15 per cent of its oil production, as well as non-ferrous and rare earth metals, are located in the Arctic regions, as are almost $250 billion of fixed assets. Changes in climate are projected to affect 20 per cent of buildings and other infrastructure, requiring a total spend of almost $85 billion to mitigate them as the permafrost that covers much of the region melts.[160] This is partly because of the reduced load-bearing capacity of the standard reinforced concrete piles used in the second half of the twentieth century, which are being undermined by permafrost degradation and changes to the strength properties of the soil.[161] One survey suggested that an increase in temperature of 1.5 °C above present-day averages could potentially result in the deformation of almost the whole of the city of Yakutsk, as well as affecting roads and railways in the surrounding area.[162] Studies on other parts of the world have likewise flagged the risks posed to the structural integrity of buildings by changing temperatures.[163]

As a recent major report by the Ministry of Natural Resources and the Environment in Russia observed, in addition to Moscow and other major cities being at serious risk because of rising temperatures, water contamination and pollution, melting permafrost in the country's northern regions threatens to release 'dangerous chemical, biological and radioactive substances into the human habitat'.[164] The growing number of cases of anthrax and of tick-borne encephalitis and borreliosis (Lyme disease) in Siberia provides yet another reason to be fearful about what the future has in store.[165]

Those fears play a real and increasing role in our daily lives – and particularly those of the next generation. A major survey of 10,000 children and young people (aged 16–25) across ten countries found that more than half felt sad, anxious, angry, powerless, helpless and guilty about climate change, although relative levels of emotion about each varies between individuals. Almost 50 per cent said that their feelings about climate change negatively affected their day-to-day lives and functioning, while 75 per cent admitted that they found thinking about the future frightening. Investigators found that while anxiety among young people is a complex topic, the climate crisis produced psychological stress that resulted in responses that included confusion and a sense of betrayal and abandonment because of adult inaction in relation to climate change.[166]

These concerns do not always extend to taking positive steps: music festivals in the UK, whose audiences are generally young people, generate 23,500 tonnes of waste a year, including plastic bottles, abandoned tents and food waste. Despite all ticket holders at Glastonbury in 2022 signing a 'Green Pledge' and many cheering Greta Thunberg as she warned that the world was 'approaching the precipice' and 'total natural catastrophe', festivalgoers left around 2,000 tonnes of waste behind – or 10 kilos per person.[167]

Nevertheless, some polls and surveys have even suggested that young adults are having fewer children because of concerns about climate change.[168] These seem to include the Duke and Duchess of Sussex, whose comments in an interview in *Vogue* that 'this place is borrowed', that 'we should be able to leave something better behind for the next generation' and that 'we've got to try to pay a little of it back. And get together to try and heal some of the harm, and at least slow down climate change' were widely interpreted as a conscious decision not to have more than two children for the earth's long-term benefit – not least by a British charity that focuses on population control, which hailed them as 'a role model for other families' and announced that they would be given a special award.[169]

With 40 per cent of the world's land degraded and half its population suffering from the impacts, it is hard not to be gloomy about the future.[170] The chances of limiting global warming to 1.5 °C – the goal set in the Paris climate agreements of 2015 – are slim, if they have not been missed already.[171] One recent study put the chances of limiting global average temperature change to the Paris targets at 0.1 per cent based on current projections.[172] In the words of António Guterres, Secretary General of the United Nations, in the summer of 2022, the Paris target is 'on life support' and in recent months 'its pulse has weakened further'.[173] Around 30 per cent of the world's population is already currently exposed to climatic conditions that put lives at risk for more than twenty days a year. By 2100, this will rise to almost half if drastic reductions in greenhouse-gas emissions take place, and to almost three-quarters if emissions continue to rise.[174]

With scientists testing hypotheses about the temperature-change thresholds for mass extinctions, declaring that even modest climate modifications to habitats have dramatic effects on plant and animal species, assessing whether the sixth mass extinction is not just possible but already under way or warning that more thought should be given to humanity's 'endgame', it is not entirely surprising that many feel the time has come for more dramatic action to force politicians and decision-makers to address these problems head-on, rather than just paying lip service.[175]

Nor is it difficult to be sympathetic. During the summer of 2022, when drought conditions, water shortages and heat stress developed into what the European Commission Joint Research Centre called the worst for at least 500 years, temperatures across Europe reached record levels, hitting 40 °C in the UK, France and the Iberian peninsula, and the month of July was the driest on record.[176] With the source of the River Thames drying up, the depth of the Rhine falling below thirty centimetres and forcing barge traffic to be hauled by road, and nuclear reactors in France shutting down or running on limited capacity because of the lack of water required to cool them, the responses from European leaders included the Prime Minister of Spain advising business people not to wear ties to meetings; the Swiss Prime Minister suggesting couples take showers together to conserve water; lights being turned off at major tourist attractions in Greece at 3 a.m.; and no hot water being allowed in showers in public buildings, swimming pools, sports halls and gyms in some cities in Germany. Although in some countries more stringent measures were later proposed (in any event largely in response to surging energy prices), initial reactions hardly provided assurance that effective measures were being put in place or that political leaders understood the scale of the current challenges – to say nothing of those of the future.[177]

The anticipated economic shocks of the future connected to crop failures, food and water shortages, rising prices, mass migration, rising levels of violence and increased prospects of warfare, and the apocalyptic ways in which they are described, are enough to turn blood to ice. Without further action to reduce greenhouse gases, said the IMF in October 2020, the planet would reach temperatures 'not seen in millions of years, potentially with catastrophic consequences'. These include disruption to economic activity and lower productivity, as well as more severe natural disasters, increased prevalence of infectious disease, deterioration in human health outcomes and possible loss of life.[178] 'Who cares if Miami is six metres underwater in 100 years?' said one senior investment banker in the summer of 2022. The 'average loan length' at his bank was seven years, he said, adding that 'What happens to the planet in year seven is actually irrelevant to our loan book.'[179] Some deliver the news with more sangfroid. 'Climate change is likely to affect monetary policy one way or the other,' said one senior member of the European Central Bank.[180] Putting it simply, said Larry Fink of BlackRock, 'climate risk is investment risk'.[181]

Just how great the financial risks are, on top of threats to the existence of humans, other animals and plants, is hard to estimate given the variables and uncertainties. However, some scholars have suggested that the average income in many countries will decline by 75 per cent by 2100 if climate change continues on current trajectories. The countries that will be worst hit are those that are already poorest, with more basic medical care, less well-developed infrastructure and institutions that are not as robust as in wealthier parts of the world.[182]

Ironically, many of the regions most exposed are ones that have suffered from high levels of pollution thanks to manufacturing that has moved away from high-income states, attracted by cheap labour forces and low environmental controls, in a form of neo-colonialism that has seen the needs of the wealthy provided for at the expense of the poor. It has been estimated, for example, that nearly 2.5 trillion metric tons of materials were extracted and used globally between 1970 and 2017; while high-income countries (as classified by World Bank criteria) consumed 75 per cent of this, lower-middle-income and low-income countries collectively consumed less than 1 per cent.[183]

Such are consumption inequalities that it has been estimated that New York City uses more energy than all of sub-Saharan Africa put together.[184] Despite being home to one-fifth of the world's population, Africa accounts for less than 3 per cent of energy-related carbon dioxide emissions – with the lowest per capita emissions of any region. In a telling sign of how resources continue to be exploited in the modern world, Africa contains 60 per cent of the best solar resources globally, but only 1 per cent of installed solar power, despite this already being the cheapest source of energy across the continent.[185]

This forms part of a wider story of climate, resource, geographic and socio-economic inequality: for example, while many in high-income countries worry about the future financial consequences of climate change, the reality is that lower-income countries are already suffering the pain of reduced circumstances both because of shifting patterns and because of the lack of climate finance. According to African Development Bank assessments, Africa as a whole has been losing 5–15 per cent of its GDP per capita growth because of climate change and related impacts. Despite being home to almost a fifth of the world's population (as well as the location of substantial natural resources), Africa has accounted for just 3 per cent of historic worldwide emissions of carbon dioxide.[186] Globally, the bottom half of the world's population by income produces the same amount of emissions as the richest 1 per cent.[187] Wealthy Americans have substantially bigger carbon footprints than those with fewer means, primarily because they own larger homes which require more heat, light and energy; in very affluent suburbs, emissions can be fifteen times higher than those in nearby neighbourhoods.[188] There are also significant implications about race and ethnicity, with research showing that people of colour are disproportionately exposed to lower air quality and higher health hazards, at least in the United States, because they typically live closer to emission sources.[189]

It seems particularly unfair, therefore, that damage from natural disasters, extreme weather and rising temperatures is greater in developing countries than in those that are wealthy and therefore more able to deal with individual and multiple setbacks.[190] That they can do so owes much to their own ecological and climatological good fortune in being located at latitudes that are particularly well suited for productivity: it has been shown that 13 °C is the optimal temperature for human productivity – which declines at higher temperatures. It is even more galling, then, that while incomes in poor countries are likely to fall sharply as a result of climate change, those in rich ones may actually rise. Apart from everything else, global warming increases inequality.[191] Indeed, it even increases inequality within low-income states, where the poor are most badly affected by climate change.[192]

Even if the Paris Agreement goals that were adopted in 2015 are met, especially the aim to limit global warming to 2 °C, exposure to dangerous heat levels will increase by 50–100 per cent across much of the tropics, and by a factor of 3–10 in many regions in mid-latitudes, which will experience deadly heatwaves that have thus far been rare on an annual basis.[193] As it happens, global warming of 1 °C, a threshold we have already passed, puts us at risk of triggering a series of catastrophic chain reactions in the natural world.[194] As if that was not enough, climate change seems likely to reorder global geopolitics. While many will suffer in the coming decades, new opportunities will also open up as habitats are transformed. Of twenty-two countries that stand to benefit, more than half are located in the former USSR and in central and eastern Europe including the Baltic States, Ukraine, Armenia, Belarus and Russia.[195]

These benefits will not flow quickly or easily, as a draft report by the Russian Ministry of Natural Resources and the Environment made clear: Russians should prepare for epidemics, crop failures, famine, pest infestations, forest fires and exposure to chemical, biological and radioactive substances – as well as heavy costs for fixing and repairing damage done to cities and to infrastructure in the largest country on earth.[196]

Some, however, have focused on the upside rather than the downside. Vladimir Putin, for example, once quipped that climate change would be a good thing, enabling Russians to spend less on fur coats, while being able to benefit from bigger and better harvests.[197] As it happens, investment in technologies, efficiencies and soil science meant that Russia's agricultural exports rose sixteenfold in the period 2000–18. Wheat exports doubled in the five years after 2015, making Russia the largest wheat exporter in the world and responsible for a quarter of the global market.[198]

While it is important not to oversimplify, it would not be unreasonable to argue that Russia's past, present and future ecological bonanzas played a role in strategic calculations when the decision was taken to invade Ukraine in February 2022 and to weaponise the dividends of its natural environment in order to put pressure on Ukraine, on Europe and beyond. These bonanzas were the oil, gas and natural resources formed by ancient climatic change, as well as the crops and other foods that are vitally important for calorie intake around the world today, all enhanced by the confidence that Russia's environmental hand of cards will improve while those of others become more difficult. As one former senior intelligence official put it, a year before the assault on Ukraine, 'Global ecological disruption is arguably the 21st Century's most underappreciated security threat.'[199] If, how, when and where resources are withheld, fought over or interrupted by states or for that matter by non-state actors will serve to exacerbate other stresses and challenges – not least those presented by climatic change. They deserve considerable attention.

Conclusion

The summer of 2022 might have been enough to convince even the most sceptical of people that something strange is afoot within global weather systems. Record heatwaves in Europe, the worst drought in many decades in Africa, nearly eight times the average rainfall in Pakistan that displaced tens of millions of people, flash floods in Death Valley in the USA that saw three-quarters of a year's average rainfall in three hours, the highest-ever recorded rate of rainfall in South Korea with almost 15 centimetres falling per hour, the wettest year in Australia's modern history, winter temperatures in the southern hemisphere in Paraguay that soared above 40 °C and almost as high in South Africa, and a long and severe drought in China that followed the hottest summer on record, which was called the most severe heatwave ever recorded anywhere and was unparalleled in world climatic history, brought the issue of climate change on to the front pages of newspapers around the world.[1]

Some, however, are unwilling to be swayed. 'Current evidence', stated the UK's Brexit negotiator Lord Frost, 'does not support the assertion that we are in a climate "emergency".' It was preposterous, he went on, that 'medieval technology like wind power' was being recommended for use in the modern world. 'We have all got used to being hectored by the government and by a huge body of intellectual and NGO [non-governmental organisation] opinion to make sacrifices to save the planet. We are told to stop travelling, live local, eat less, stop eating meat, turn our lights out, and generally to stop being a burden.' This was all nonsense, in other words.[2]

Similar views are not hard to find in the corridors of power, especially in democratic countries where freedoms of opinion, of expression and of the press are not only highly prized but of signal importance. 'For the past eighteen years', Senator Ted Cruz stated while chairing his first hearing on climate change, 'there has been no significant warming whatsoever.' Current models used to understand global climate trends 'are profoundly wrong' and, he went on, 'inconsistent with the evidence and the data'.[3] Climate change is 'a hoax', said former US President Donald Trump, who had previously said it was not a hoax. People now 'just talk about climate change. The climate's always been changing.' He insisted that there was nothing to worry about, and added, echoing Cruz, that scientists who said otherwise were not to be trusted.[4]

Two British prime ministers who took office in 2022 agreed that clean energy was not a priority. We should make 'sure our fields are used for food production and not solar panels', intoned Rishi Sunak. 'I'm somebody who wants to see farmers producing food, not filling in forms,' said Liz Truss, 'not filling fields with paraphernalia like solar farms. What we want is crops.'[5] It was not clear why it had to be one or the other or why it could not be both; and in any event, even if the UK government were to quintuple all the solar capacity it has committed to establishing by 2035, the amount of land covered by solar farms would be around 0.5 per cent of the land currently used for farming – or just over half the amount of land currently covered by golf courses.[6] Still, the message was clear: renewable energy, climate change and concerns about the future need not be taken seriously.

Some of these beliefs are fuelled by an energetic industry that has grown up to misinform, to magnify and distort reports about climate and climate science for a range of reasons that includes mischief, domestic political agendas, foreign interventions and genuine misgivings. Analysis of millions of social media accounts shows that in the course of a normal day around a quarter of original postings about climate change on Twitter are made by automated bots operated by a range of 'bad actors' seeking to sow discord through social media. Bots are disproportionately likely to promote 'denial research', with as many as 38 per cent of tweets about 'fake science' being generated automatically – and then often retweeted by human users, as well as by other bots to reach bigger and bigger audiences.[7]

Some climate sceptics point out – rightly – that forecasts that look into the future can be highly speculative, and they also seek to dampen alarm by noting, again quite correctly, that economic growth, new technologies and adaptation may alleviate the problems that lie ahead and, in some cases, may even solve them.[8] That too, however, requires faith and confidence; moreover, what history in general and this book in particular show is that there have been a great many times in the past when societies, peoples and cultures have proved unable to adapt. Indeed, in some respects, the human story of progress is about batons being repeatedly dropped and picked up by others.

The question, then, is not so much whether to adapt, but how, where and when to do so. And in that sense it is certainly true that there is plenty of good news, much to celebrate and reasons to be optimistic. For example, it would not take much for progressive decision-making, supported, if necessary, by appropriate investment and regulation, to reduce some of the astonishing wastage of resources that occurs every year – of food, of water, of energy and more. This does not need to come at a cost. For example, total water consumption in Australia declined by 40 per cent between 2001 and 2009 while GDP grew by almost a third during the same period.[9] A National Air Quality Plan introduced in China in 2013, and the announcement of a 'war against pollution' the following summer helped spark a decline in particulate pollution in China of almost 40 per cent in the years up to 2020. Beijing's air pollution alone fell by 55 per cent, adding around four and a half years to the average inhabitant's life expectancy as a result.[10]

A switch to low-energy LEDs (light-emitting diodes) has already generated major savings in the European Union, where policy measures have been introduced to ban inefficient lighting systems; some estimates suggest that similar steps in low-income countries could save $40 billion of electricity and prevent 320 million metric tons of carbon pollution annually.[11] In fact, and despite often acute pessimism, many countries have made considerable progress. Members of the European Union, for example, collectively reduced greenhouse-gas emissions by almost 25 per cent between 1990 and 2019, and seem likely to reduce these by another 15 per cent in the next decade.[12]

Or there is the United States, where progress over the last twenty years combined with current federal and state policies means that the world's largest economy is on track to reduce emissions so substantially that, despite population growth, industrial productivity and higher levels of international and global trade, greenhouse-gas emissions in 2030 are likely to be at least 24 per cent and perhaps as much as 35 per cent below what they were in 2005. Although this falls short of the pledges and ambitions set out in the Paris Agreement of 2015, it is nevertheless a sign of considerable progress.[13] The Inflation Reduction Act of 2022 which included by far the biggest climate and energy investments in American history was hailed by some commentators as likely to play a crucial role in speeding up a 'green transition' on which much depends in the future.[14]

Steps like these can be built on quickly, with solar and wind capable of meeting as much as 90 per cent of energy demands in many industrialised countries, even without new storage capacity being built.[15] Indeed, in the spring of 2021, California was producing enough clean power from renewable energy sources to supply nearly 95 per cent of its in-state needs and, if the Diablo Canyon nuclear plant is included, produced more than 100 per cent of its electricity needs for the first time.[16] In 2019 the UK energy grid went for more than two weeks without burning coal to make power for the first time in almost 150 years – a sequence apparently broken by millions tuning in on their televisions and turning on their kettles to watch the finale of the TV show *Love Island*.[17]

A survey of 29,000 fossil-fuel power stations in more than 200 countries showed that a small group of 'superemitter' plants – just 5 per cent of the total – produced almost 75 per cent of global emissions produced by electricity generation, a figure that would fall dramatically if they were made more efficient and/or switched from coal or oil to natural gas, while carbon-capture technologies that are already available could cut emissions fully in half.[18]

There are other good reasons to remain upbeat. Some scientists argue that many climate models are overly pessimistic and over-rely on worst-case scenarios, especially when it comes to key green technology costs. In fact, one recent study that draws on empirically grounded research shows that a rapid green energy transition would not only produce a 'greener, healthier and safer global energy system with reduced air pollution, more stable pricing and reduced climate damage', but that it would 'likely result in overall net savings of many trillions of dollars'. A very good outcome, in other words.[19]

Then there are the new ideas that are emerging about how to reduce damage to the environment. For example, recent research shows that raising or lowering the altitudes of less than 2 per cent of aircraft flights could reduce the climate impact inflicted by vapour trails made up of hot exhaust gases as they meet cold, low-pressure air: a small proportion of flights are responsible for 80 per cent of radiative forcing in the atmosphere by altering the balance between radiation emitted by the sun and heat emitted by the earth, resulting in changes in the climate. The costs of doing so are minimal, with researchers calculating that the diversion would add less than 0.1 per cent to fuel consumption.[20]

Similarly, studies on mitigating the climate impacts and carbon emissions of ships – which carry more than nine-tenths of global trade – suggest that reducing speeds by 10 per cent would result in a 13 per cent reduction in emissions as a result of reduced demands on engine power, which in turn would mean that voyages might require as much as 40 per cent less energy. Reductions in speed would reduce underwater noise for the benefit of aquatic life and greatly lessen the chances of whale strikes, thereby improving marine biodiversity.[21]

These are just a handful of examples where good research and clear thinking can help find low-hanging fruit that can make significant and immediate differences. Many, many others can also be cited: for instance, indiscriminate voiding of excreta by cattle contributes to greenhouse-gas emission as well as resulting in soil and water contamination, with emissions higher when animals are given more space to graze in. It has been shown, however, that cattle can be trained to control their micturition reflex and use a latrine for urination, with significant environmental and climatic benefits.[22]

Scientists have developed a plant-based emulsifier rich in protein and antioxidants that could replace eggs and dairy in foods such as mayonnaise, soups and sauces and would help reduce the impact of poultry.[23] Substituting 20 per cent of meat from cattle with imitation meats made from plants or with cultured meats made from animal cells or from fermentation-derived microbial protein could cut both annual global deforestation and related carbon dioxide emissions by half.[24]

New methods have been proposed and successfully tested which break down substances known as 'forever chemicals' that are used as water-, oil- and stain-resistant barriers for fabrics, make-up and food containers such as frying pans and which both avoid the very harsh techniques currently used to break them down and do so without creating damaging by-products – thereby helping the natural environment.[25] Enriching waters of lakes with particular bacteria can break down plastic pollution, helping to remove it from ecosystems.[26] Bioengineered acceleration of the photosynthesis process of soybean plants increases the efficiency of the chemical reaction, resulting in higher yields with no fall in quality, with obvious implications not only for soybean but for other crops too.[27]

Such then are the wonders of science and of scientific research. Sometimes, it is true, the process is one of two steps forward and one step back. Switching away from fossil fuels, for example, can increase pressures on different resources and materials that can be easily obscured by the elation of switching to notionally carbon-neutral technologies. Thus, while the declared intention to move to renewable energy such as wind power is laudable, it is easy to forget – or never to realise – that to produce a quarter of global energy in this way would require at least 450 million metric tons of steel, which in turn need the fossil-fuel equivalent of more than 600 million metric tons of coal.[28] Perhaps one answer lies in new materials: researchers at Michigan State University have made composite resin for turbine blades that can be broken down and recycled into gummy bears.[29]

Likewise, the switch to electric vehicles (EVs) in many countries, states and cities can make it easy to forget both that these need to charge and recharge, bringing about a higher demand for electricity, and that EVs produce high levels of pollution. As it is, synthetic tyres are a leading source of microplastics which are found in high concentrations not only next to roads but in rivers, in seas and even in the Arctic, with almost 7 million tons of particles from tyres and non-exhaust sources such as vehicle brakes emitted each year.[30] In fact, car tyres produce far more particle pollution than exhausts of modern cars – perhaps as much as 1,800 times more.[31] This is significant, since models with large batteries, capable of travelling up to 500 kilometres between charges, are much heavier than petrol and diesel models, and emit up to 8 per cent more fine particles as a result.[32]

Certainly, thinking about climate change, about the overexploitation of natural resources and about a world transforming in front of our eyes can even extend into mundane and seemingly petty issues. Will we still be playing or watching sports that take place outdoors and depend on good weather, such as cricket – the second most followed sport in the world, but which will be the hardest hit of all the major pitch sports by changing climates?[33] Will winter sports in Europe all but disappear, given the projected collapse of Alpine snow?[34] Should anyone buy a seaside property given that sea-level rises of between twenty-seven and seventy-eight centimetres are apparently all but guaranteed because of the impact of 110 trillion metric tons of melt from Greenland's ice sheet?[35]

Predicted CO_2 atmospheric concentration and
temperature increase to 2100 AD

Source: O'Neill et al, 2016

Will the famous nightclubs of Ibiza be a thing of the past, as temperatures in Spain soar to levels that make other destinations not only more sensible and enjoyable for dancing and partying, but also survivable?[36] Will the blue seas of the Caribbean still be picture-postcard beautiful if toxic algal blooms that have already spread in a chain from Africa right across the Atlantic Ocean continue to expand?[37] Will what we know of as the 'Mediterranean diet' disappear thanks to the soaring cost of olive oil as crops fail in the absence of rain and in scorching heat – and if so, will this affect life expectancy of populations whose longevity has long been attributed to how they live and what they eat?[38]

Will tourists still visit the Taj Mahal, the Great Wall of China or Petra if doing so comes at a risk to life? What will happen to the meaning of Islam if the hajj – the pilgrimage to Mecca, which involves walking around the sacred Ka'ba and is one of the pillars of the Muslim faith – becomes not only impractical but dangerous because of excessively high temperatures? What will become of the Kumbh Mela in India, where the sacred Ganges itself may reduce to a trickle, or indeed dry up altogether?[39]

How should one factor rapidly falling birth rates in most of the developed world into decisions that may substantially change as global populations levels fall – or at least do so in the more temperate regions of the earth? Will collapsing demographics lead to much lower demand on resources, less pressure on the environment and lower energy requirements, and if so, how and when will they do so?

It may well be that such questions become irrelevant, for it may be that what causes the greatest threat to our existence is not climate change and the horrors that lie ahead later in this century. Climate projections rely on assumptions that progress up to 2100 and beyond will be linear – that is to say, they look ahead at where current factors and trends are likely to lead us. However, there are plenty of things that could happen that render all projections about climate change redundant at a stroke. One is the prospect of major warfare.

There has been much discussion in recent years about the link between a warming world and violence, and in particular about 'water wars' in the future, as states fight over resources that become limited by global warming, overconsumption or both. What few had thought hard about was the prospect of the use of nuclear weapons, at least until President Putin put Russia's nuclear arsenal on alert early on during the invasion of

Ukraine. In the last thirty years, the use of nuclear weapons has seemed inconceivable, largely because of the dreadful consequences it would bring, but also because many had assumed that the end of the Cold War marked a decisive termination to the threat of 'mutually assured destruction' that had characterised strategic thinking in both Washington and Moscow for much of the period after the Second World War.

And yet, as António Guterres, Secretary General of the United Nations, said in a speech in August 2022, the threat of nuclear confrontation has become as high as it was during the Cold War. Indeed, as he underlined, 'humanity is just one misunderstanding, one miscalculation away from nuclear annihilation'.[40] New analysis using the latest climate modelling shows that the smoke from the use of significant numbers of warheads would destroy much of the ozone layer over the course of fifteen years.[41] Even a relatively small-scale nuclear scenario, such as a localised war between India and Pakistan, would have an effect on crops – reducing global calorie consumption by 7 per cent, and particularly affecting agriculture in mid- and high-latitude countries such as the United States and Russia.[42]

Then there are other events that could have severe effects on human life on earth. These include solar winds that have major impact on earth's magnetosphere, solar storms that could destroy power-grid transformers, costing the US economy more than $1 trillion and taking years to recover from. So high are risk levels that the US has formally passed laws to 'predict and detect' space weather events in the form of 'solar flares, solar energetic particles and geomagnetic disturbances'. President Obama's Executive Order of 2016 on 'Coordinating Efforts to Prepare the Nation for Space Weather Events' requires collaboration between the secretaries of Defense, Homeland Security, Interior, Commerce and Energy, along with the Administrator of NASA, to 'minimize the extent of economic loss and human hardship'.[43]

Solar storms represent major challenges. In July 2012, a coronal mass ejection was measured at an initial speed of around 2,500 kilometres per second. Had this taken place a week earlier, it would have passed much closer to earth, with grave consequences for satellites, aircraft, power systems and possibly human society as a whole – because of our reliance on modern technologies, which are extremely vulnerable to any form of electromagnetic disruption.[44]

Then there are risks of lunar flooding which NASA expects to occur in the mid-2030s when the lunar cycle amplifies rising sea levels to create increasing high-tide floods that surge in low-lying regions, and earthquakes which can cause major disruption; or there are tsunami events in the world's great oceans or even in the Mediterranean where UNESCO has said the chance of a tsunami striking at-risk coastal areas by 2030 is nothing less than 100 per cent.[45]

The threat of impactors is also a very real one. For example, the asteroid first named 1989FC and then rechristened 4581 Asclepius passed 680,000 kilometres from the earth in March 1989; had it approached six hours earlier, it would have struck the planet, with devastating consequences (the angle and the location of impact would both have helped determine how bad things would have been).[46] Another asteroid, 99942 Apophis, has been identified as one of the most hazardous threats to the earth after its discovery in 2004. Recently, NASA concluded that it certainly would not strike our planet – at least not for another hundred years.[47] Nevertheless, the development and deployment by NASA of the Double Asteroid Redirection Test (DART), a planetary defence system, at the end of 2021 provides an insight into the level of threats posed by extra-terrestrial objects.[48]

But by far the biggest risk to global climate comes from volcanoes. Although considerable thought and attention have gone and are going into planning for a warming world, almost no investment of time, planning or funding has been spent on the potential implications of major volcanic eruptions. This is despite the fact that new evidence shows that eruptions with a volcanic explosivity index (VEI) of magnitude 7 occur around once every 625 years – and those with VEI of magnitude 8 around every 14,300 years – much more frequently than previously thought.[49]

With the last VEI 7 taking place in 1815 at Mount Tambora, the clock is ticking, and indeed may be ticking faster than in the past: again, recent investigations have shown that volcanism and volcanic activity are closely connected to the melting of ice sheets and to sea-level rise, suggesting a causal link between pressure on the earth's crust and mantle.[50] If this is indeed correct, we may be facing an accelerated timetable for the next mega-eruption that would inject vast amounts of ash and gases into the atmosphere and make discussions about climate change redundant, possibly at the cost of millions or even hundreds of millions of lives as temperatures suddenly cool, harvests fail and plants and animals die. Some estimate the chance of a major eruption before

2100 as one in six – in other words hundreds of times more likely than asteroid and comet impact combined.[51]

In fact, so poor is general knowledge about the impact of volcanoes that while the news of a big underwater eruption of Hunga Tonga-Hunga Ha'apai in Tonga on 15 January 2022 was reported around the world, less well covered was the scientific analysis of what had happened. The eruption had more explosive force than a hundred simultaneous detonations of the atomic bomb dropped on Hiroshima in 1945.[52] A plume of ash and dust was forced into the atmosphere, reaching higher than any other on record, with volcanic heat and superheated moisture from the ocean acting like 'hyper-fuel for a mega-thunderstorm', according to NASA scientists – which generated almost 600,000 lightning strikes in three days.[53] Not only was it the most powerful eruption ever observed in the modern era; unlike most terrestrial volcanoes that create cooling events, in the case of Hunga Tonga-Hunga Ha'apai so much water was vaporised and sent into the atmosphere that it might lead to surface warming and intensify existing climate trends.[54]

The question is not if a major volcanic eruption will take place, but when. In March 2021, the US Geological Survey warned that Mauna Loa, the world's largest active volcano, 'continues to awaken from its slumber'. While noting that an eruption 'is not imminent, now is the time to revisit personal eruption plans'. Ominously, its update quoted Benjamin Franklin and was headed 'Volcano Watch – "By failing to prepare, you are preparing to fail."'[55] Moreover, although massive eruptions by single volcanoes feature vividly in public imagination, it may be that clusters of even relatively minor explosions from smaller active volcanoes are of greater concern and potentially pose a more serious threat to global temperatures, supply chains and travel and communication networks.[56]

Other phenomena may also affect the current projections, directions and realities of climatic change. More than fifty countries have ongoing climate-modification programmes, for example to disperse fog, enhance rain and snowfall and suppress hail.[57] 'Cloud seeding is one of the most promising solutions in Saudi Arabia,' commented Ayman Ghulam, Chief Executive Officer of the National Centre of Meteorology in Riyadh, speaking of a country which has an average of ten centimetres of rain per year and is likely to be one of the worst affected by climate change.[58] Russia used cloud-seeding technology to ensure that skies were clear above Moscow for May Day celebrations in May 2016, as did China before the Beijing Olympics to make conditions ideal for athletes and viewers and to show the country in the best possible light.[59]

China has been particularly active in climate and weather modification, developing schemes that energise rain particles by firing 'low-frequency, high-intensity acoustic waves [to] excite the cloud body, vibrating it', a process that can be controlled remotely and at low cost.[60] With artificial rainfall responsible for an increase in precipitation of more than 230 billion cubic metres between 2012 and 2017, Chinese state authorities have announced major subsequent upgrades to and expansions of such activities, including a bold announcement in 2020 of a series of 'breakthroughs in fundamental research and R&D in key technologies that would create rain and snow across five and a half million square kilometres – while also suppressing hail across 580,000 square kilometres.' Taken together, these would help with 'ecological protection and restoration', as well as being fundamental to major emergency responses to forest fires and mitigating the effects of high temperatures and droughts.[61]

An important element in this is the framing of nature as something to be 'tamed' – in other words, as something that can be controlled at will, rather than setting environmental boundaries or requiring adaptation. Describing clouds as 'airborne water reservoirs', for example, sets up the idea that the weather, climate and nature as a whole can and should be used to meet human requirements; this sits within wider suggestions that ecologies are being 'modernised', and that ingenuity, technology and sheer determination can mitigate and even remove large-scale environmental challenges.[62]

Choice of language has proved important with scientists elsewhere, with a major study produced by the US National Academy of Sciences in 2015 choosing to use the term 'climate intervention' rather than 'geoengineering' or 'climate engineering' or modification, not to assert the ability to control the weather but rather because the term better captures the imprecision and uncertainties of the outcomes when changes are made to atmospheric conditions.[63]

There is considerable activity in the field of weather modification in the United States, with over 850 non-federal programmes initiated (and in many cases completed) since 2000, as recorded by the office of the US Secretary of Commerce via the National Oceanic Atmospheric Administration, as required by law.[64] Although some scholars note that reports often seem to overstate the amount of precipitation that has been generated, and that such activities create little environmental downside (including contamination by the silver iodide used in cloud seeding), studies suggest that costs are relatively low compared to the value of water. In other words, overall benefits seem to outweigh the negatives.[65]

Many cloud-seeding operations are classified, and it is unclear what their aims, scope and methods are, or what their implications will be for other countries and regions whose weather systems – and precipitation levels in particular – may be affected by non-natural changes.[66] It goes without saying that programmes to 'modify weather patterns to divert massive amounts of water through the sky from a place where it's abundant to another that's dry' can have major implications for ecosystems in both locations, and indeed in others as well.[67]

One recent report noted that, leaving aside the fact that 'current observing systems are insufficient to quantify the effects' of modification programmes, there were also considerable 'political, social, legal, economic, and ethical dimensions' that were unclear. Above all, concluded a high-level group of researchers, 'there is significant potential for unanticipated, unmanageable, and regrettable consequences' that might result from human interference with natural weather systems.[68] In other words, attempts to influence, modify or control the climate might end up making things worse.

It is not clear quite what the status is of the latest and edgiest research conducted behind closed doors. Even thirty years ago there were proposals to look at ways of countering the effects of global warming. In a wide-ranging policy document produced in 1992 by scientists in the US collaborating across multiple disciplines, many different mitigation proposals were set out – from the introduction of energy-efficient lighting to fuel-efficiency programmes for cars and heavy trucks, from the use of clean and renewable energy to eliminating paddy-rice production, reducing the use of nitrogenous fertiliser, bringing down the numbers of ruminant animals and fixing leaks in natural-gas pipelines.

A host of other more creative suggestions were also put forward. These include the creation of 'space mirrors' in earth's orbit to reflect sunlight; using guns to fire particles to form and maintain a dust cloud in the stratosphere in order to increase sunlight reflection and therefore keep the earth cool; the placing of 'billions of aluminized, hydrogen-filled balloons in the stratosphere' to do something similar by providing a reflective screen that stops sunlight from reaching earth; burning sulphur 'in ships or power plants' to stimulate low marine clouds in order to reflect sunlight; and dumping iron in the oceans to stimulate the generation and growth of CO_2-absorbing phytoplankton. 'We need to know more about them because measures of this kind may be crucial if greenhouse warming occurs,' said the US report. Moreover, it went on, while 'all have large unknowns concerning possible environmental side-effects', it was worth noting that 'some of these options are relatively inexpensive to implement'.[69]

By 2003, few of the new ideas seemed to show promise, with the National Research Council guardedly declaring that 'there still is no convincing scientific proof of the efficacy of intentional weather modification efforts', and that while there were some hopeful signs, the evidence had yet to be robustly tested. Nevertheless, recent years had yielded 'many promising developments and advances'.[70] The Council therefore recommended that a co-ordinated national programme be developed to look into weather modification, and also at the inadvertent consequences that might flow from the use of new techniques and tools.[71]

Just over a decade later, sufficient progress had been made to allow a technical evaluation of the impacts of climate modification with a major survey offering glimpses into ideas about the removal of carbon dioxide from the atmosphere (known as carbon capture), and also into the latest thinking about ways to cool the planet by increasing the reflection of solar radiation. Attempts to achieve the latter 'should not be deployed at this time', according to this report, published in 2015. Other steps to address global warming should be looked at urgently, advised the researchers, who recommended that further work be undertaken on how to slow down or reverse global warming. It was clear, however, that dramatic reductions in carbon dioxide emissions remained the best route.[72]

As a follow-up study published in 2021 put it, even if solar geoengineering were possible it neither addresses the cause of climate change and global warming – namely, the increasing concentrations of greenhouse gases in the atmosphere – nor solves related problems, such as ocean acidification.[73] Although this is true, more research into large-scale modification is ongoing, including into its possible effects for all life on earth.[74] It would be reasonable to suppose that the volume of work will rise sharply in coming years, leading to more and perhaps better suggestions for what might be done to prevent worst-case scenarios emerging. It would be reasonable too to conclude that in the years and decades ahead some states, and even non-state actors, may choose to test and even implement strategies that offer (or seem to offer) solutions that have detrimental consequences for others on the planet.

The problems and challenges that we face today are of course in many ways completely different to those faced by our earliest human ancestors. One thing, however, remains the same: the fact that the natural environment around us, and the climate that supports it, frames our existence. Nonetheless, we have come to believe that technology can overcome our limitations, that our reshaping and remodelling of nature can alleviate, circumvent or overcome all obstacles and barriers that determine not only where we live but how we live.

That feeling of confidence is not without cost. According to the UN, as much as 40 per cent of the world's land is now degraded. At current rates, an area the size of South America will be degraded by 2050.[75] 'Overshoot Day' – a notional benchmark that flags the date each year on which consumption of resources surpasses the earth's capacity to regenerate, as a way to focus minds on sustainability – keeps falling earlier and earlier in the year, moving from around October in the 1990s to late July in 2022.[76]

It is not impossible, of course, that humanity proves able to adapt, perhaps by changing the way we live and the choices we make; perhaps thanks to new technologies and ideas; perhaps through higher levels of collaboration, brought on either by enlightened governance or by crisis and necessity. It is worth bearing in mind, however, that much of human history has been about the failure to understand or adapt to changing circumstances in the physical and natural world around us, and about the consequences that ensue.

In that sense, then, environmental factors – including the climate – are not actors in the story of our species, which sometimes make interventions that bring down empires, lead to societal collapse or catch people by surprise. Rather, they provide the very stage on which our existence plays out, shaping everything we do, who we are, where and how we live. And like most stages, it can be all too easy to think only about what happens on them – what the protagonists do and say – without thinking about the fabric of the set itself. Actors come and go; but if the theatre closes or collapses, that marks the end for us all.

Like all animals and plants, humans flourish in habitats that suit them; as for all animals and plants, life is difficult, if not impossible, in those habitats that do not. We have been uniquely resourceful in finding ways to modulate those habitats and to refashion nature to suit our own needs – whether through building cities, through creating artificial water systems that enable the growing of crops where they otherwise could not grow, or through creating artificial ecosystems by means of innovation, trial and error and by developing technologies that allow us to manipulate hostile conditions. Ours is a story of resourcefulness, resilience and adaptation.

Those qualities, however, can lead to a false sense of security, to the illusion that challenging times will one day revert to normal – which a historian can help remind us is nothing more than a mirage of wishful thinking. At the heart of the problems of today and tomorrow lies the fact that we are living on the edge of our means and indeed beyond them, reliant on everything to go right and with little margin of error for things to go wrong. This has created vulnerabilities, fragility and risk in the past, as this book has tried to show, and does so in the present as well.

Our species has been a part of this planet's history for a tiny sliver of its existence. While we conceptualise the natural past in terms of mass extinctions and think of these as almighty horrors, the reality is that nature does not care who wins or loses and does not choose one form of biota over another: the issue is always about adaptation and survival. It is worth remembering that we owe our presence on earth to previous, dramatic climate change and to the fluke circumstances that have made this planet well suited to our own existence.

Climate has shaped the earth since the beginning of time, with shifts in long-term patterns dictating the location of resources and materials that humans have found useful – such as coal, oil and gas – but which have no relevance to other plants and animals, except as a result of the ways in which we use them. It goes without saying, or at least it should do, that the world will keep spinning on its axis and rotating around the sun, however many of us – or however few of us – are around to witness and enjoy it. One thing is certainly clear: if we and future generations cannot prevent or adapt to global warming, then we go the way of a vast number of other species from the past. Our loss will be the gain of other animals and plants.

In *Paradise Lost*, John Milton wrote that expulsion from the Garden of Eden brought consequences. The short-sightedness, greed and disobedience of Adam and Eve, humanity's common ancestors, so angered God that he 'called forth by name His mightie Angels' and ordered them to shift the earth's axis, to change climate so that the earth was affected 'with cold and heat Scarce tollerable', and to make thunder 'rowle with terror'. Paradise had been climatically perfect, never affected by 'pinching cold and scorching heate'. Humans have long been the architects of their own climatic downfall, in other words.[77]

Returning to paradise is something that Milton thought about too, as is clear from his *Paradise Regained*. He was following a long tradition. Writing in the second century, St Irenaeus imagined the time of Christ's second coming and the joys which that would bring and the abundance that would greet the faithful. The Lord would surely declare, he wrote, that 'a grain of wheat would produce ten thousand ears, and that every ear should have ten thousand grains, and every grain would yield ten pounds of clear, pure, fine flour. Vines would have ten thousand branches, each with ten thousand twigs, each with ten thousand shoots, each of which would have ten thousand clusters, each with ten thousand grapes' – each of which would be magnificently juicy. 'All other fruit-bearing trees and seeds and grass would produce in similar proportions.' No one, that is to say, would go hungry.[78]

Similar promises of plenty can be found in almost every major world literature, not least in the Upaniṣads, which declare that 'the earth is like honey for all the creatures and all the creatures are like honey for this earth', or the Qur'ān, which notes that in paradise there would be 'soft couches' set in 'shady groves', filled with couples in a state of bliss enjoying fruits and everything else they might care for.[79] In Thomas More's fictional *Utopia*, food 'groweth everywhere without labour', while for some indigenous populations in North America idealised locations of paradise were not in new or post-life locations, but rather situated in the same place but in earlier times – when food and resources were more abundant.[80]

* * *

In the real world, however, we are taking extraordinary risks with our futures. Humans take advantage of nature's resources for food, water and shelter, says a major report on the economics of biodiversity that was published in the spring of 2021. And yet we also use nature as a sink for our waste products, such as carbon dioxide, plastics and pollution. This is the result not just of short-sightedness, says Sir Partha Dasgupta, the author of the report, but of stupidity: 'Governments almost everywhere exacerbate the problem by paying people more to exploit Nature than to protect it, and to prioritise unsustainable economic activities.'[81]

In fact, as the report goes on to say, we are living well beyond our means. At present, 'we would require 1.6 Earths to maintain the world's current living standards', an assessment that reveals critical shortcomings in how little thought and how little action has gone into tackling problems that reveal 'deep-rooted, widespread institutional failure'.[82] In the final analysis, however, as the UK government's Office for Budget Responsibility recently put it, it is easy to answer the question of how the problem of climate change is solved: it will be nature, rather than human action, that ultimately brings net emissions towards zero.[83] It will do so through catastrophic depopulation, whether through hunger, disease or conflict. With fewer of us around to burn fuel, cut down forests and tear minerals from the earth's crust, the human footprint may become drastically reduced – and we will move closer to the sustainable, lush paradise of our fantasised past. Perhaps we will find our way back there through peaceful means; a historian would not bet on it.

Acknowledgements

This has been a wonderful project to work on. Like many people, I have been thinking about climate change for a long time and trying to understand how global warming will affect the coming years as well as future generations. Being able to think about, read about and study the history of climate change, the role that extreme weather and long stable periods have played in the past, has been not only a great pleasure but a very challenging and rewarding journey too.

The process of writing this book has forced me to engage with new types of source material, above all from the sciences, and to learn how to interpret them; it has pushed me to look at the histories of peoples and places in regions that I have not worked on in great detail before, broadening my horizons considerably and making me a better historian in the process. I am acutely aware of the privilege that I have as a historian based at one of the world's great universities to enjoy the freedom to develop my ideas, to interact with colleagues and to be ambitious about which subjects I engage with.

This book has been brewing for several decades. In the first instance, however, I owe a debt of thanks to Jim Cuno, President of the Getty Center in Los Angeles, who invited me to the United States in 2017 as President's Council Fellow to benefit from the Center's extraordinary resources, to work on climate history and to give a preliminary lecture that offered me a chance to order my thought and crystallise my ideas. All ideas need a catalyst and a context, and I am hugely grateful to the Getty Center and its trustees for providing a glorious and generous setting which proved fundamental in helping take a project from the drawing board into something more ambitious than I had ever imagined.

I am grateful to my colleagues and students at Oxford for providing a relentlessly stimulating environment and to the staff at the Bodleian Library for their world-class collection and near endless patience in helping me track down material that was often obscure and difficult to find. I must thank the Provost and Fellows of Worcester College, Oxford, my friends and colleagues who work on Late Antique and Byzantine Studies, on Global and Imperial History and related fields for their help and support over many years. A similar level of support from the A. G. Leventis Foundation and the Stavros Niarchos Foundation has been crucial. I am very grateful to the trustees for their continued encouragement and backing.

I am also extremely grateful to the Provost and Fellows of King's College, Cambridge and in particular to Professor Robin Osborne, Lorraine Headon, Professor David Good and Dr Katie Campbell, as well as Alison Traub and Professor Stephen Toope, who have been instrumental in supporting the studies of the peoples, cultures and countries of the Silk Roads and opening new doors that have helped with my research in general as well as with this book in particular.

As ever, my agents Catherine Clarke and Zoe Pagnamenta have been my guardian angels, as has the team at Knight Ayton, always encouraging me and making me think I am more interesting and able than I really am. My editors, Alexis Kirschbaum and Michael Fishwick, provided a perfect balance of warm cheer and cold steel that has worked wonders in the past. I am grateful to both and to my copyeditor, Peter James, whose keen eye for detail I learned to admire long ago.

Lauren Whybrow, Genista Tate-Alexander, Hayley Camis and Jonny Coward have been a pleasure to work with in making sure this book reads and looks as good as it does. I'm very grateful to Mike Athanson for his wonderful maps. Jo Carlill was fantastic at tracking down images for the plates. And a special thanks, as with my last books, to Emma Ewbank for a magnificent cover. They say one should not judge a book by its cover; with *The Earth Transformed*, I am happy if readers do.

I've had a happy time over the years at Bloomsbury, as well as at Knopf, Spectrum, Rowohlt Berlin and my other publishers around the world, and thank them all for their continued confidence and backing. I am grateful too to Professor Chris Stringer for his generosity in providing invaluable advice and comments about human evolution and to Dr Vicki Smith for her guidance on the date of the Ilopango eruption of AD 431. Thanks too to Michael Drolet, my colleague at Worcester College, for pointing me towards Michel Chevalier.

I could not have managed to write this book without our children's support, encouragement and occasional teasing about how long it was taking to write and about why the same topics seemed to come up again and again at meal times. Thank you Katarina, Flora, Francis and Luke: you are all a lot older than when I started this book. I will miss you asking how much more I still have left to write; and what date I've got up to in my most recent chapter. I suspect I'll break my promise that I won't ever write another book; but you have all supported me more than you can imagine.

I lost my father in 2018. Not a day has gone by when I have not missed him. I used to speak to him every day about history, politics and his real passion, geology. I'm so sorry that I have not been able to do so while writing this book. I also missed my mother and my siblings terribly during the lockdowns of the global pandemic of 2020–2 when I could not see them in person for many months at a time.

But this book is dedicated to Jessica, the love of my life. For more than thirty years she has seen me turn half-baked ideas into something more promising, always telling me I could do it, suggesting ways I could do it better and, above all, being a pillar of support, happiness and laughter day in and day out, come rain or shine. She has inspired me every day for many, many years, always saying the right thing at the right time; she is the rock on which this book and my life are built. I could never have done this without her support and encouragement. She has always looked to the future with optimism and courage and reminded me to do the same. I will clasp her hand, and those of our children, ever more tightly – inspired by her urging that we must always be hopeful, even as we all head into an uncertain future. For that, and for so much more, this book is dedicated to her.

Notes

INTRODUCTION

1 J. Milton, *Paradise Lost*, S. Orgel and J. Goldberg (Oxford, 2004), pp. 4–6.

2 United Nations Secretary General, 'Secretary General's remarks to Climate Summit Preparatory Meeting', 30 June 2019.

3 The White House, 'Remarks by the President in State of Union Address', 20 January 2015.

4 Address of His Holiness Pope Francis, 'The Energy Transition and Care of our Common Home', 14 June 2019.

5 'Statement by HE Xi Jinping, President of the People's Republic of China, at the General Debate of the 75th Session of the United Nations General Assembly', Ministry of Foreign Affairs of the People's Republic of China, 22 September 2020. https://www.fmprc.gov.cn/eng/wjdt_665385/zyjh_665391/202009/t20200922_678904.html.

6 'Transcript: Greta Thunberg's Speech at the UN Climate Action Summit', NPR, 23 September 2019.

7 F. Braudel, *La Méditerannée: L'espace et l'histoire*, vol. 1 (Paris, 1949), p. 27.

8 A. Watson, *Agricultural Innovation in the Early Islamic World: The Diffusion of Crops and Farming Techniques, 700–1100* (Cambridge, 2008).

9 D. Zhang et al., 'Climatic Change, Wars and Dynastic Cycles in China over the Last Millennium', *Climate Change* 76 (2006), 459–77; D. Zhang et al., 'Climate Change and War Frequency in Eastern China over the Last Millennium', *Human Ecology* 35 (2007), 403–14.

10 Kālidāsa, *Meghadūtam: With the Kātāyani Sanskrit Commentary and English translation*, tr. C. Maharaj (Varanasi, 1973); I. Rajamani, 'Monsoon Feelings', in I. Rajamani, M. Pernau and K. Butler Schofield

(eds.), *Monsoon Feelings: A History of Emotions in the Rain* (New Delhi, 2018), pp. 11–43.

11   For forced labour, see for example Uzbek Forum for Human Rights, *Tashkent's Reforms Have Not Yet Reached Us: Unfinished Work in the Fight against Forced Labour in Uzbekistan's 2019 Cotton Harvest* (Berlin, 2019).

12   Central Asia Bureau for Analytical Reporting, 'Насколько опасен воздух в Центральной Азии', 15 January 2020, https://cabar.asia/ru/naskolko-opasen-vozduh-v-tsentralnoj-azii-obyasnyaem-na-grafikah.

13   J. Iriarte, 'Geometry by Design: Contribution of Lidar to the Understanding of Settlement Patterns of the Mound Villages in SW Amazonia', *Journal of Computer Applications in Archaeology* 3.1 (2020), 151–69.

14   O. Lokwalo Thabeng, E. Adam and S. Merlo, 'Spectral Discrimination of Archaeological Sites Previously Occupied by Farming Communities Using In Situ Hyperspectral Data', *Journal of Spectroscopy* (2019), 1–21.

15   N. Athfield et al., 'Influence of marine sources on 14C Ages: isotopic data from Watom Island, Papua New Guinea inhumations and pig teeth in light of new dietary standards', *Journal of the Royal Society of New Zealand* 38.1 (2008), 1–23.

16   O. Amichay et al., 'A bazaar assemblage: reconstructing consumption, production and trade from mineralised seeds in Abbasid Jerusalem', *Antiquity* 93 (2019), 199–217.

17   J. Warner et al., 'Investigating the influence of temperature and seawater $\delta^{18}O$ on *Donax obesulus* (Reeve, 1854) shell $\delta^{18}O$', *Chemical Geology* 588 (2022), 1–14.

18   Y. Aono and K. Kazui, 'Phenological Data Series of Cherry Tree Flowering in Japan and its Application to Reconstruction of Springtime Temperatures since the ninth century', *International Journal of Climatology* 28 (2008), 905–14.

19   A. Tarand and O. Nordli, 'The Tallinn Temperature Series Reconstructed Back Half a Millennium by Use of Proxy Data', *Climatic Change* 48 (2001), 189–99.

20   G. Hole et al., 'A Driftwood-Based Record of Arctic Sea Ice during the Last 500 Years from Northern Svalbard Reveals Sea Ice Dynamics in the Arctic Ocean and Arctic Peripheral Seas', *Journal of Geophysical Research: Oceans* 126 (2021), 1–20.

21   C. Prud'homme et al., 'Central Asian Modulation of Northern Hemisphere moisture transfer over the Late Cenozoic', *Communications Earth and Environment* 2 (2021), 1–8.

22 S. Liu et al., 'Sedimentary ancient DNA reveals a threat of warming-induced alpine habitat loss to Tibetan plateau plant diversity', *Nature Communications* 12 (2021), 1–9.

23 M. McCormick et al., 'Climate Change during and after the Roman Empire: Reconstructing the Past from Scientific and Historical Evidence', *Journal of Interdisciplinary History* 43.2 (2012), 169–220.

24 R. Anderson, N. Johnson and M. Koyama, 'Jewish Persecutions and Weather Shocks: 1100–1800', *Economic Journal* 127 (2015), 924–58.

25 M. Waldinger, 'The Economic Effects of Long-Term Climate Change: Evidence from the Little Ice Age', Centre for Climate Change Economics and Policy, Working Paper 239 (2015); J. Martínez-González et al., 'Assessing climate impacts on English economic growth (1645–1740): an econometric approach', *Climatic Change* 160 (2020), 233–49.

26 N. Stehr and H. van Storch, 'Von der Macht des Klimas: Ist der Klimadeterminismus nur noch Ideengeschichte oder relevanter Faktor gegenwärtiger Klimapolitik?', *Gaia* 9 (2000), 187–95.

27 D. Arnold and R. Guha (eds.), *Nature, Culture, Imperialism: Essays on the Environmental History of South Asia* (Delhi, 1995), pp. 1–20.

28 Good examples include J. Diamond, *Collapse: How Societies Choose to Fail or Succeed* (London, 2005); D. Webster, *The Fall of the Ancient Maya: Solving the Mystery of the Maya Collapse* (London, 2002); K. Harper, *The Fate of Rome: Climate, Disease and the End of an Empire* (London, 2017).

29 J. Tainter, 'Collapse, sustainability, and the environment: how authors choose to fail or succeed', *Reviews in Anthropology* 37 (2008), 342–71.

30 D. Degroot et al., 'Towards a rigorous understanding of societal responses to climate change', *Nature* 591 (2021), 539–50.

31 Herodotus, *The Histories*, ed. R. Waterfield and C. Dewald (Oxford, 1998), 7.34–5, p. 420.

32 J. Woodruff et al., 'Depositional evidence for the Kamikaze typhoon changes in typhoon climatology', *Geology* 43 (2015), 91–4; R. Sasaki, *The Origin of the Lost Fleet of the Mongol Empire* (College Station, TX, 2008).

33 For Napoleon's attack, see A. Roberts, *Napoleon the Great* (London, 2014); D. Lieven, *Russia against Napoleon: The Battle for Europe, 1807–1814* (London, 2016); for Hitler, A. Beevor, *Stalingrad* (London, 2007) and D. Stahel, *Kiev, 1941: Hitler's Battle for Supremacy in the East* (London, 2012).

34 D. Chakrabarty, 'The Climate of History: Four Theses', *Critical Inquiry* 35 (2009), 197–222.

35 For a good introduction, see R. Hide, D. Boggs and J. Dickey, 'Angular momentum fluctuations within the earth's liquid core and torsional oscillations of the core–mantle system', *Geophysical Journal International* 143 (2000), 777–86; R. Gross, I. Fukumori and D. Menemenlis, 'Atmospheric and oceanic excitation of decadal-scale Earth orientation variations', *Journal of Geophysical Research* 110 (2005), 1–15; J.-E. Lee et al., 'Hemispheric sea ice distribution sets the glacial tempo', *Geophysical Research Letters* 44.2 (2017), 1008–14; E. Zorita, S. Wagner and F. Schenk, 'The Global Climate System', in S. White, C. Pfister and F. Mauelshagen (eds), *The Palgrave Handbook of Climate History* (London, 2018), pp. 21–6.

36 M. McPhaden, S. Zebiak and M. Glantz, 'ENSO as an Integrating Concept in Earth Science', *Science* 314 (2006), 1740–5.

37 A. Turner and H. Annamalai, 'Climate change and the South Asian summer monsoon', *Nature Climate Change* 2 (2012), 587–95; P. Borah et al., 'Indian monsoon derailed by a North Atlantic wavetrain', *Science* 370 (2020), 1335–8.

38 J. Luterbacher et al., 'Circulation Dynamics and its Influence on European and Mediterranean January–April Climate over the Past Half Millennium: Results and Insights from Instrumental Data, Documentary Evidence and Coupled Climate Models', *Climatic Change* 101 (2010), 201–34.

39 J. Hansen et al., 'Ice melt, sea level rise and superstorms: evidence from paleoclimate data, climate modelling and modern observations that 2°C global warming could be dangerous', *Atmospheric Chemistry and Physics* 16 (2016), 3761–812.

40 S. McIntosh, 'Deciphering solar magnetic activity: on the relationship between the sunspot cycle and the evolution of small magnetic features', *Astrophysical Journal* 792 (2014), 1–19.

41 I. Usoskin, 'A history of solar activity over millennia', *Living Reviews in Solar Physics* 14.3 (2017), 1–97.

42 I. Usoskin et al., 'The Maunder minimum (1645–1715) was indeed a grand minimum: a reassessment of multiple datasets', *Astronomy and Astrophysics* 581 (2015), 1–19.

43 J. Proctor et al., 'Estimating global agricultural effects of geoengineering using volcanic eruptions', *Nature* 560 (2018), 480–3.

44 C. Oppenheimer, *Eruptions that Shook the World* (Cambridge, 2011), pp. 64–6; G. Stenchikov et al., 'Volcanic signals in oceans', *Journal of Geophysical Research* 114 (2009), 1–13.

45 S. Wilson et al., 'Kīlauea lava fuels phytoplankton bloom in the North Pacific Ocean', *Science* 365 (2019), 1040–4.

46 D. Arnold, 'The Indian Ocean as a Disease Zone, 1500–1950', *South Asia: Journal of South Asian Studies* 14.2 (1991), 1–21.

47 E. Predybaylo et al., 'El Niño/Southern Oscillation response to low-latitude volcanic eruptions depends on ocean pre-conditions and eruption timing', *Communications Earth & Environment* 1 (2020), 1–13.

48 M. Toohey et al., 'Disproportionately strong climate forcing from extratropical explosive volcanic eruptions', *Nature Geoscience* 12 (2019), 100–7.

49 M. Burton, G. Sawyer and D. Granieri, 'Deep Carbon Emissions from Volcanoes', *Reviews in Mineralogy & Geochemistry* 75 (2013), 323–54.

50 L. Bollinger et al., 'Seasonal modulation of seismicity in the Himalaya of Nepal', *Geophysical Research Letters* 34 (2007), 1–5; D. Panda et al., 'Seasonal modulation of deep slow-slip and earthquakes on the Main Himalayan Thrust', *Nature Communications* 9 (2018), 1–8.

51 C. Liu, A. Linde and S. Sacks, 'Slow earthquakes triggered by typhoons', *Nature* 459 (2009), 833–6.

52 M. Harvey et al., 'The evolution of a tropical biodiversity hotspot', *Science* 370 (2020), 1343–8.

53 McCormick et al., 'Climate Change during and after the Roman Empire', esp. 174–91; E. Roy Ladurie, *Times of Feast, Times of Famine: A History of Climate since the Year 1000* (1971); H. Lamb, *Climate: Past, Present and Future* (London, 1972); G. Parker, *Global Crisis: War, Climate Change and Catastrophe in the Seventeenth Century* (London, 2013).

54 R. Neukom, 'No evidence for globally coherent warm and cold periods over the preindustrial Common Era', *Nature* 571 (2019), 550–4.

55 E. Xoplaki et al., 'Modelling Climate and Societal Resilience in the Eastern Mediterranean in the Last Millennium', *Human Ecology* 46 (2018), 363–79.

56 B. Stenni et al., 'Antarctic climate variability on regional and continental scales over the last 2000 years', *Climate of the Past* 13 (2017), 1609–34.

57 IMF, *World Economic Outlook: Seeking Sustainable Growth – Short-term Recovery, Long-term Challenges* (Washington, DC, 2017), pp. 117–83.

58 D. Thornalley, et al., 'Anomalously weak Labrador sea convection and Atlantic overturning during the past 150 years', *Nature* 556 (2018), 227–30; for the AMOC, see M. Buckley and J. Marshall, 'Observations, inferences, and mechanisms of the Atlantic Meridional Overturning Circulation: A Review', *Reviews of Geophysics* 54 (2016), 5–63.

59 N. Boers, 'Observation-based early-warning signals for a collapse of the Atlantic Meridional Overturning Circulation', *Nature Climate Change* 11 (2021), 680–8; L. Jackson et al., 'Global and European climate impacts

of a slowdown of the AMOC in a high resolution GCM', *Climate Dynamics* 45 (2015), 3299–316.

60 T. Lenton et al., 'Climate', *Nature* 575 (2019), 592–5.

61 P. Crutzen, 'Geology of mankind', *Nature* 415 (2002), 23.

62 Subcommission on Quaternary Stratigraphy, Working Group on the 'Anthropocene', Results of Binding Vote by AWG, 21 May 2019.

63 J. Brigham-Grette, 'Pliocene Warmth, Polar Amplification, and Stepped Pleistocene Cooling Recorded in NE Arctic Russia', *Science* 340 (2013), 1421–7; J. Blundon and S. Arndt (eds), *State of the Climate in 2019*, Special Supplement, *Bulletin of the American Meteorological Society* 101 (2020); O. Dumitru, 'Constraints on global mean sea level during Pliocene warmth', *Nature* 574 (2019), 233–6.

64 US National Oceanic and Atmospheric Administration, 'Carbon dioxide peak for 2022 more than 50% higher than pre-industrial levels', 3 June 2022, https://phys.org/news/2022-06-carbon-dioxide-peak-higher-pre-industrial.html.

65 A. Braakmann-Folgmann et al., 'Observing the disintegration of the A68A iceberg from space', *Remote Sensing of Environment* 270 (2022), 1–9.

66 S. Kulp and B. Strauss, 'New elevation data triple estimates of global vulnerability to sea-level rise and coastal flooding', *Nature Communications* 10 (2019), 1–12.

67 UK Office for Science, *Future of the Sea: Current and Future Impacts of Sea Level Rise on the UK* (London, 2017).

68 J. Hinkel et al., 'Coast Flood Damage and Adaptation Costs under 21st Century Sea-Level Rise', *Proceedings of the National Academy of Sciences of the USA* 111 (2014), 3292–7.

69 IMF, *World Economic Outlook: A Long and Difficult Ascent* (Washington, DC, 2020), p. 86.

70 UNICEF, *The Climate Crisis is a Child Rights Crisis* (New York, 2021).

71 D. Welsbey et al., 'Unextractable fossil fuels in a 1.5 °C world', *Nature* 597 (2021), 230–4.

72 B. Lomborg, 'Welfare in the 21st century: increasing development, reducing inequality, the impact of climate change and the cost of climate policies', *Technological Forecasting & Social Change* 156 (2020), 1–35.

73 US Department of Transportation, National Highway Traffic Safety Administration, 'The Safer Affordable Fuel-Efficient (SAFE) Vehicles Rule for Model Year 2021–2026 Passenger Cars and Light Trucks', Draft Environmental Impact Statement, 5–30.

74 For example, 'Trump administration sees a 7-degree rise in global temperatures by 2100', *Washington Post*, 28 September 2018.

75 World Health Organisation, *Ambient Air Pollution: A Global Assessment of Exposure and Burden of Disease* (Geneva, 2016), p. 33.

76 C. Wiedinmyer, R. Yokelson and B. Gullett, 'Global emissions of trace gases, particulate matter and hazardous air pollutants from open burning of domestic waste', *Environmental Science & Technology* 18 (2014), 9523–30.

77 P. Landrigan, 'The Lancet Commission on pollution and health', *Lancet* 391 (2018), 462–512.

78 India State-Level Disease Burden Initiative Air Pollution Collaborators, 'Health and economic impact of air pollution in the states of India: The Global Burden of Disease Study 2019', *Lancet Planetary Health* (2020), 1–14.

79 I. Flores, 'Afghanistan's Air is Deadlier Than its War', *Foreign Policy*, 26 May 2019.

80 S. Khomenko, 'Premature mortality due to air pollution in European cities: a health impact assessment', *Lancet Planetary Health* (2021), pp. 1–14.

81 K. Vohra et al., 'Global mortality from outdoor particulate pollution generated by fossil fuel combustion: results from GEOM-Chem', *Environmental Research* 195 (2021), 1–8.

82 D. De Silva et al., 'Firm behaviour and pollution in small geographies', *European Economic Review* 136 (2021), 1–33.

83 M. Shehab and F. Pope, 'Effects of short-term exposure to particulate matter air pollution on cognitive performance', *Scientific Reports* 9 (2019), 1–10.

84 A. Reuben et al., 'Association of Air Pollution Exposure in Childhood and Adolescence with Psychopathology at the Transition to Adulthood', *JAMA Network Open* 4 (2021), 1–14; P. Mok et al., 'Exposure to ambient air pollution during childhood and subsequent risk of self-harm: a national cohort study', *Preventive Medicine* 152 (2021), 1–7.

85 M. Prunicki et al., 'Air pollution exposure is linked with methylation of immunoregulatory genes, altered immune cell profile and increased blood pressure in children', *Scientific Reports* 11 (2021), 1–12.

86 World Bank, *The Global Health Cost of PM2.5 Air Pollution: A Case for Action Beyond 2021* (Washington, D.C., 2022).

87 IPBES (Intergovernmental Platform for Biodiversity and Ecosystem Services), *The Global Assessment Report on Biodiversity and Ecosystem Services* (Paris, 2019).

88 N. Welden, 'The environmental impacts of plastic pollution', in T. Letcher (ed.), *Plastic Waste and Recycling: Environmental Impact, Societal Issues, Prevention, and Solutions* (London, 2020), pp. 195–222.

89 National Federation of Women's Institutes, *In a Spin: How our Laundry is Contributing to Plastic Pollution* (2018).

90 P. Ross et al., 'Pervasive distribution of polyester fibres in the Arctic Ocean is driven by Atlantic inputs', *Nature Communications* 12 (2021), 1–9.

91 K. Cox et al., 'Human Consumption of Microplastics', *Environment, Science and Technology* 53 (2019), 7068–74; A. Ragusa et al., 'Plasticenta: First evidence of microplastics in human placenta', *Environment International* 146 (2021), 1–8; J. Zhang et al., 'Occurrence of Polyethylene Terephthalate and Polycarbonate Microplastics in Infant and Adult Feces', *Environment, Science and Technology* (2021), 989–994; H. Leslie et al., 'Discovery and quantification of plastic particle pollution in human blood', *Environment International* 163 (2022), 1–8.

92 E. Nic Lughadha, 'Extinction risk and threats to plants and fungi', *Plants People Planet* 2 (2020), 389–408.

93 F. Sánchez-Bayo and K. Wyckhuys, 'Worldwide decline of the entomofauna: a review of its drivers', *Biological Conservation* 232 (2019), 8–27.

94 IPBES, *Global Assessment Report*, p. 11.

95 S. Ryding et al., 'Shape-shifting: changing animal morphologies as a response to climatic warming', *Trends in Ecology & Evolution* (2021), 1–13.

96 B. Ahmed et al., 'Maternal heat stress reduces body and organ growth in calves: relationship to immune status', *JDS Communications* 2 (2021), 295–9.

97 E. Stokstad, 'New global study reveals the "staggering" loss of forests caused by industrial agriculture', *Science* (2018); G. Pecl et al., 'Biodiversity redistribution under climate change: impacts on ecosystems and human well-being', *Science* 355 (2017).

98 K. Chandra et al., *Assemblages of Lepidoptera in Indian Himalaya through Long Term Monitoring Plots* (Kolkata, 2019).

99 S. Chaikin, S. Dubiner and J. Belmaker, 'Cold-water species deepen to escape warm water temperatures', *Global Ecology & Biogeography* (2021).

100 R. Almond, M. Grooten and T. Peterson (eds), *Living Planet Report 2020: Bending the Curve of Biodiversity Loss*, World Wildlife Fund (Gland, 2020).

101 K. Rosenberg, 'Decline of the North American avifauna', *Science* 366 (2019), 120–4; IPBES, *Global Assessment Report*, p. 24.

102 E. Schulz et al., 'Climate-driven, but dynamic and complex? A reconciliation of competing hypotheses for species' distributions', *Ecology Letters* 25 (2022), 1–14.

103 V. Danneyrolles et al., 'Scale-dependent changes in tree diversity over more than a century in eastern Canada: Landscape diversification and regional homogenization', *Journal of Ecology* 109 (2021), 273–83.

104 M. Dornelas et al., 'A balance of winners and losers in the Anthropocene', *Ecology Letters* 22 (2019), 847–54.

105 B. Leung et al., 'Clustered versus catastrophic global vertebrate declines', *Nature* 588 (2020), 267–71.

106 G. Ceballos, P. Ehrlich and R. Dirzo, 'Biological annihilation via the ongoing sixth mass extinction signaled by vertebrate population losses and declines', *Proceedings of the National Academy of Sciences of the USA* 114 (2017), 6089–96.

107 K. Martin et al., 'The biogeographic differentiation of algal microbiomes in the upper ocean from pole to pole', *Nature Communications* 12 (2021), 1–14.

108 See for example J. Memmott, N. Waser and M. Price, 'Tolerance of pollination networks to species extinctions', *Philosophical Transactions of the Royal Society of London B* 271 (2004), 2605–11; J. Herrera, J. Moody and C. Nunn, 'Predictions of primate–parasite coextinction', *Philosophical Transactions of the Royal Society of London B* 376 (2021), 1–8.

109 R. Cowie, P. Bouchet and B. Fontaine, 'The Sixth Mass Extinction: fact, fiction or speculation?', *Biological Reviews* (2022), 1–24.

110 C. Bradshaw et al., 'Underestimating the Challenges of Avoiding a Ghastly Future', *Frontiers in Conservation Science* 1 (2021), 1–10.

111 L. Borck, 'Constructing the Future History: Prefiguration as Historical Epistemology and the Chronopolitics of Archaeology', *Journal of Contemporary Archaeology* 5.2 (2018), 229–38.

112 L. Zhao, 'Global multi-model projections of local urban climates', *Nature Climate Change* (2021), 1–19.

113 N. Grimm, 'Global change and the ecology of cities', *Science* 319 (2008), 756–60.

114 L. Sun et al., 'Dramatic uneven urbanisation of large cities throughout the world in recent decades', *Nature Communications* 11 (2020), 1–9.

115 H. Ritchie and M. Roser, 'Urbanization', Our World in Data, https://ourworldindata.org/urbanization#how-many-people-will-live-in-urba n-areas-in-the-future.

116 R. Geyer, J. Jambeck and K. Law, 'Production, use, and fate of all plastics ever made', *Science Advances* 3 (2017), 1–5.

117 E. Elhacham et al., 'Global human-made mass exceeds all living
     biomass', *Nature* 588 (2020), 442–4.

118 Aristotle, *Politics*, tr. H. Rackham (Cambridge, MA, 1932), I.2,
     pp. 10–12.

119 Ibn Faḍlān, *Book of Ahmad ibn Faḍlān*, in *Ibn Fadlan and the Land of
     Darkness: Arab Travellers in the Far North*, tr. P. Lunde and C. Stone
     (London, 2011), p. 12.

120 R. Keller and M. Turek, *American Indians and National Parks* (Tucson,
     AZ, 1998), pp. 132–84.

121 S. Garnett et al., 'A spatial overview of the global importance of
     Indigenous lands for conservation', *Nature Sustainability* 1 (2018),
     369–74; W. Walker, 'The role of forest conversion, degradation, and
     disturbance in the carbon dynamics of amazon Indigenous territories
     and protected areas', *Proceedings of the National Academy of Sciences of
     the USA* 117 (2020), 3015–25.

122 J. Cruikshank, *Do Glaciers Listen? Local Knowledge, Colonial Encounter
     and Social Imagination* (Vancouver, 2005); N. Zappia, *Traders and
     Raiders: The Indigenous World of the Colorado Basin, 1540–1859* (Chapel
     Hill, NC, 2014), esp. pp. 28–42.

123 M. Burke, S. Hsiang and E. Miguel, 'Global non-linear effect of
     temperature on economic production', *Nature* 527 (2015), 235–9.

I THE WORLD FROM THE DAWN OF TIME (C.4.5BN–C.7M BC)

1 N. Sleep et al., 'Initiation of Clement Surface Conditions on the Earliest
   Earth', *Proceedings of the National Academy of Sciences of the USA* 98
   (2001), 3666–72.

2 C. Stringer, 'The origin and evolution of *Homo sapiens*', *Philosophical
   Transactions of the Royal Society B*, 371 (2016), 1–12.

3 E. Crist, 'On the Poverty of Our Nomenclature', *Environmental
   Humanities* 3 (2013), 129–47; D. Kidner, 'Why 'anthropocentrism' is not
   anthropocentric', *Dialectical Anthropology* 38 (2014), 465–80.

4 P. Sossi, 'Redox state of Earth's magma ocean and its Venus-like early
   atmosphere', *Science Advances* 6 (2020), 1–8; A. Morbidelli, 'Building
   Terrestrial Planets', *Annual Review of Earth and Planetary Sciences* 40
   (2012), 251–75.

5 B. Hess, S. Piazolo and J. Harvey, 'Lightning strikes as a major
   facilitator of prebiotic phosphorus reduction on early Earth', *Nature
   Communications* 12 (2021), 1–8.

6 R. Riding, P. Fralick and L. Liang, 'Identification of an Archean
   marine oxygen oasis', *Precambrian Research* 251 (2014), 232–7. Also see

Y. Bromberg, 'Quantifying structural relationships of metal-binding sites suggests origins of biological electron transfer', *Science Advances* 8 (2022), 1–13.

7 B. Schirrmeister, M. Gugger and P. Donoghue, 'Cyanobacteria and the Great Oxidation Event: evidence from genes and fossils', *Paleontology* 58 (2015), 769–85; J. Klatt et al., 'Possible link between Earth's rotation rate and oxygenation', *Nature Geoscience* 14 (2021), 564–70; J. Meixnerová et al., 'Mercury abundance and isotopic composition indicate subaerial volcanism prior to the end-Achean "whiff" of oxygen', *PNAS (Proceedings of the National Academy of Sciences)* 118 (2021), 1–6.

8 A. Walsh, T. Ball and D. Schultz, 'Extreme sensitivity in Snowball Earth formation to mountains on PaleoProterozoic supercontinents', *Scientific Reports* 9 (2019), 1–7.

9 R. Mitchell et al., 'Orbital forcing of ice sheets during Snowball Earth', *Nature Communications* 12 (2021), 1–9.

10 T. Johnson et al., 'Giant impacts and the origin and evolution of continents', *Nature* 608 (2022), 330–5.

11 R. Kopp et al., 'The Paleoproterozoic Snowball Earth: A climate disaster triggered by the evolution of oxygenic photosynthesis', *Proceedings of the National Academy of Sciences of the USA* 102 (2005), 11,131–6.

12 S. Poulton et al., 'A 200-million-year delay in permanent atmospheric oxygenation', *Nature* 592 (2021), 232–6.

13 P. Hoffman et al., 'Snowball Earth climate dynamic sand Cryogenian geology-geobiology', *Science Advances*, 3 (2017), 1–43.

14 C. Simpson, 'Adaptation to a viscous Snowball Earth Ocean as a path to complex multicellularity', *American Naturalist* (2021), 590–609.

15 Z. Zhu, 'The temporal distribution of Earth's supermountains and their potential link to the rise of atmospheric oxygen and biological evolution', *Earth and Planetary Science Letters* 580 (2022), 1–10.

16 S. Evans, M. Droser and D. Erwin, 'Developmental processes in Ediacara macrofossils', *Proceedings of the Royal Society B* 288 (2021), 1–10.

17 J.-B. Hou, N. Hughes and M. Hopkins, 'The trilobite upper limb branch is a well-developed gill', *Science Advances* 7 (2021), 1–8.

18 P. Sheehan, 'The Late Ordovician Mass Extinction', *Annual Review of Earth and Planetary Sciences* 29 (2001), 331–64.

19 D. Bond and S. Grasby, 'Late Ordovician Mass Extinction caused by volcanism, warming, and anoxia, not cooling and glaciation', *Geology* 48 (2020), 777–81; though also see A. Pohl, 'Vertical decoupling in Late Ordovician anoxia due to reorganization of ocean circulation', *Nature Geoscience* (2021), 1–13.

20 H. Yndestad, 'The influence of the lunar nodal cycle on Arctic climate', *Journal of Marine Science* 63 (2006), 401–20. For the moon's formation, S. Desch and K. Robinson, 'A unified model for hydrogen in the Earth and Moon: No one expects the Theia contribution', *Geochemistry* 79 (2019), 1–16.

21 H. Byrne et al., 'Tides: a key environmental driver of osteichthyan evolution and the fish-tetrapod transition?', *Proceedings of the Royal Society A* 476 (2020), 1–19.

22 N. Kronfeld-Schor, 'Chronobiology by moonlight', *Proceedings of the Royal Society B* 280 (2013), 1–11; A. Takemura, M. Rahman and Y. Park, 'External and internal controls of lunar-related reproductive rhythms in fishes', *Journal of Fish Biology* 76 (2010), 7–26.

23 P. Kaniewska et al., 'Signaling cascades and the importance of moonlight in coral broadcast mass spawning', *ELife* 4 (2015), 1–14.

24 A. Sinclair, 'Lunar cycle and timing of mating season in Serengeti wildebeest', *Nature* 267 (1977), 832–3; T. Yonezawa et al., 'Lunar cycle influences spontaneous delivery in cows', *PLOS ONE* 11 (2016), 1–8.

25 L. Nash, 'Moonlight and behaviour in nocturnal and cathemeral primates, especially Lepilemur leucopus: illuminating possible anti-predator efforts', in S. Gursky and K. Nekaris (eds), *Developments in Primatology: Progress and Prospects* (Berlin, 2007), pp. 173–87.

26 B. Phalan, 'Foraging behaviour of four albatross species by night and day', *Marine Ecology Progress Series* 340 (2007), 271–86.

27 G. Norevik, 'The lunar cycle drives migration of a nocturnal bird', *PLOS Biology* 17 (2019), 1–13; P. Pinet et al., 'Celestial Moderation of Tropical Seabird Behaviour', *PLOS One* 6 (2011) 1–7; K. Last et al., 'Moonlight Drives Ocean-Scale Mass Vertical Migration of Zooplankton during the Arctic Winter', *Current Biology* 26 (2016), 244–51.

28 L. Casiraghi et al., 'Moonstruck sleep: synchronization of human sleep with the moon cycle under field conditions', *Science Advances* 7 (2021), 1–8.

29 C. Helfrich-Förster et al., 'Women temporarily synchronize their menstrual cycles with the luminance and gravimetric cycles of the Moon', *Science Advances* 7 (2021), 1–13.

30 R. Foster and T. Rosenberg, 'Human responses to the geophysical daily, annual and lunar cycles', *Current Biology* 18 (2008), 784–94.

31 T. Wehr, 'Bipolar mood cycles and lunar tidal cycles', *Molecular Psychiatry* 23 (2018), 923–31.

32 X. Zhang, 'Lunar tide in the thermosphere and weakening of the northern polar vortex', *Geophysical Research Letters* 41 (2014), 8201–7.

33 S. Burgess, 'High-precision geochronology confirms voluminous magmatism before, during and after Earth's most severe extinction', *Science Advances* 1 (2015), 1–14.

34 S. Burgess, 'Initial pulse of Siberian Traps Sills as the trigger of the end-Permian mass extinction', *Nature Communications* 8 (2017), 1–6.

35 M. Benton, 'Hyperthermal-driven mass extinctions: killing models during the Permian-Triassic mass extinction', *Philosophical Transactions of the Royal Society A* 376 (2018), 1–19; J. Fan et al., 'A high-resolution summary of Cambrian to Early Triassic marine invertebrate biodiversity', *Science* 367 (2020), 272–7; P. Viglietti et al., 'Evidence from South Africa for a protracted end-Permian extinction on land', *PNAS* 118 (2021), 1–8; C. Mays and S. McLoughlin, 'End-Permian Burnout: The role of Permian-Triassic wildfires in extinction, carbon cycling and environmental change in Eastern Gondwana', *Palaios* 37 (2022), 292–317.

36 J. Davies et al., 'End-Triassic mass extinction started by intrusive CAMP activity', *Nature Communications* 8 (2017), 1–8.

37 C. Fox et al., 'Molecular and isotopic evidence reveals the end-Triassic carbon isotope excursion is not from massive exogenous light carbon', *PNAS* 117 (2020), 30,171–8; also see K. Kaiho et al., 'Volcanic temperature changes modulated volatile release and climate fluctuations at the end-Triassic mass extinction', *Earth and Planetary Science Letters* 579 (2022), 1–12; C. Fox et al., 'Flame out! End-Triassic mass extinction polycyclic aromatic hydrocarbons reflect more than just fire', *Earth and Planetary Science Letters* 584 (2022), 1–13.

38 See for example P. Olsen, 'Arctic ice and the ecological rise of the dinosaurs', *Science Advances* 8 (2022), 1–9.

39 S. Singh et al., 'Niche partitioning shaped herbivore macroevolution through the early Mesozoic', *Nature Communications* 12 (2021), 1–13.

40 L. Alvarez et al., 'Extraterrestrial Cause for the Cretaceous-Tertiary Extinction', *Science* 208 (1980), 1095–1108.

41 A. Garde et al., 'Searching for giant, ancient impact structures on Earth: The Mesoarchean Maniitsoq structure, West Greenland', *Earth and Planetary Science Letters* 337–8 (2012), 192–210.

42 N. Artemieva and J. Morgan, 'Quantifying the Release of Climate-Active Gases by Large Meteorite Impacts with a Case Study of Chicxulub', *Geophysical Research Letters* 44 (2017), 180–8; M. Range et al., 'The Chicxulub Impact Produced a Powerful Global Tsunami', *AGU Advances* 3 (2022), 1–21.

43 A. Siraj and A. Loeb, 'Breakup of a long-period comet as the origin of the dinosaur extinction', *Scientific Advances* 11 (2021), 1–5.

44 D. Degroot, ' "A Catastrophe Happening in Front of our Very Eyes": The Environmental History of a Comet Crash on Jupiter', *Environmental History* 22 (2017), 32–3.

45 A. Siraj and A. Loeb, 'Breakup of long-period comet as the origin of the dinosaur extinction', *Scientific Reports* 11 (2021), 1–5.

46 M. During et al., 'The Mesozoic terminated in boreal spring', *Nature* 603 (2022), 91–4.

47 R. DePalma et al., 'Seasonal calibration of the end-cretaceous Chicxulub impact event', *Scientific Reports* 11 (2021), 1–8.

48 B. Schoene, 'U-Pb constraints on pulsed eruption of the Deccan Traps across the end-Cretaceous mass extinction', *Science* 363 (2019), 862–6; C. Sprain et al., 'The eruptive tempo of Deccan volcanism in relation to the Cretaceous–Paleogene boundary', *Science* 363 (2019), 866–70.

49 K. Kaiho and N. Oshima, 'Site of asteroid impact changed the history of life on Earth: the low probability of mass extinction', *Scientific Reports* 7 (2017), 1–12.

50 M. Carvalho et al., 'Extinction at the end-Cretaceous and the origin of modern Neotropical rainforests', *Science* 372 (2021), 63–8.

51 M. Gutjahr et al., 'Very large release of mostly volcanic carbon during the Paleocene–Eocene Thermal Maximum', *Nature* 548 (2017), 573–7.

52 M. Huber, 'A hotter greenhouse?', *Science* 321 (2008), 353–4.

53 A. Winguth et al., 'Climate Response at the Paleocene–Eocene Thermal Maximum to Greenhouse Gas Forcing – A Model Study with CCSM3', *Journal of Climate* 23 (2010), 2564–84.

54 F. McInerney and S. Wing, 'The Paleocene–Eocene Thermal Maximum: A Perturbation of Carbon Cycle, Climate, and Biosphere with Implications for the Future', *Annual Review of Earth and Planetary Sciences* 39 (2011), 489–516; S. Kender, 'Paleocene/Eocene carbon feedbacks triggered by volcanic activity', *Nature Communications* 12 (2021), 1–10.

55 Ibid.; also C. Jaramillo et al., 'Effects of rapid global warming at the Paleocene–Eocene boundary on Neotropical vegetation', *Science* 330 (2010), 957–61.

56 V. Lauretano et al., 'Eocene to Oligocene terrestrial Southern Hemisphere cooling caused by declining pCO2', *Nature Geoscience* (2021), 1–16.

57 B. Mason, D. Pyle and C. Oppenheimer, 'The size and frequency of the largest explosive eruptions on Earth', *Bulletin of Volcanology* 66 (2004), 735–48; O. Bachmann, M. Dungan and P. Lipman, 'The Fish Canyon Magma Body, San Juan Volcanic Field, Colorado: Rejuvenation and Eruption of an Upper-Crustal Batholith', *Journal of Petrology* 43 (2002), 1469–1503.

58 K. Sieh et al., 'Australasian impact crater buried under the Bolaven volcanic field, Southern Laos', *Proceedings of the National Academy of Sciences of the USA* 117 (2020), 1346–53.

59 E. de la Vega et al., 'Atmospheric CO_2 during the Mid-Piacenzian
Warm Period and the M2 glaciation', *Scientific Reports* 10 (2020), 1–8;
R. Zhang, D. Jiang and Z. Zhang, 'Causes of mid-Pliocene strengthened
summer and weakened winter monsoons over East Asia', *Advances in
Atmospheric Sciences* 12 (2015), 1016–26.

60 N. Zhang et al., 'The dominant driving force for supercontinent
breakup: Plume push or subduction retreat?', *Geoscience Frontiers*
9 (2018), 997–1007; L. Chen et al., 'Subduction tectonics vs. Plume
tectonics – Discussion on driving forces for plate motion', *Science China
Earth Sciences* 63 (2020), 315–28.

61 D. van Hinsbergen, 'A record of plume-induced plate rotation triggering
subduction initiation', *Nature Geoscience* 14 (2021), 626–30.

62 N. Mortimer et al., 'Zealandia: Earth's Hidden Continent', *GSA Today*
27 (2017), 27–35.

63 D. van Hinsbergen et al., 'Orogenic architecture of the Mediterranean
region and kinematic reconstruction of its tectonic evolution since the
Triassic', *Gondwana Research* 81 (2020), 79–229.

64 J. Flynn et al., 'A new fossil mammal assemblage from the southern
Chilean Andes: implications for geology, geochronology, and tectonics',
Journal of South American Earth Sciences 15 (2002), 285–302; R. Enay
and E. Cariou, 'Ammonite faunas and palaeobiogeography of the
Himalayan belt during the Jurassic: Initiation of a Late Jurassic austral
ammonite fauna', *Palaeogeography, Palaeoclimatology, Palaeoecology* 134
(1997), 1–38.

65 M. Pontoppidan et al., 'Large-scale regional model biases in the
extratropical North Atlantic storm track and impacts on downstream
precipitation', *Quarterly Journal of the Royal Meteorological Society* 145
(2019), 2718–32.

66 G. Jung, M. Prange and M. Schulz, 'Influence of topography on tropical
African vegetation coverage', *Climate Dynamics* 46 (2016), 2535–49.

67 W. Sun, 'Northern Hemisphere Land Monsoon Precipitation Increased
by the Green Sahara During Middle Holocene', *Geophysical Research
Letters* 46 (2019), 9870–9.

68 A. Crosby, *The Columbian Exchange: Biological and Cultural
Consequences of 1492* (Westport, CT, 1972); J. Diamond, *Guns,
Germs and Steel: A Short History of Everybody for the Last 13,000 Years*
(London, 1997).

69 Z. Zhang et al., 'Aridification of the Sahara desert caused by the Tethys
Sea shrinkage during the Late Miocene', *Nature* 513 (2014), 401–4; A.
Holbourn et al., 'Changes in Pacific Ocean circulation following the

Miocene onset of permanent Antarctic ice cover', *Earth and Planetary Science Letters* 365 (2013), 38–50.

70 D. Garcia-Castellanos et al., 'Catastrophic flood of the Mediterranean after the Messinian salinity crisis', *Nature* 462 (2009), 778–81; P.-L. Blanc, 'The opening of the Plio-Quaternary Gibraltar Strait: assessing the size of a cataclysm', *Geodinamica Acta* 15 (2002), 303–17.

71 P. Mann, L. Gahagan and M. Gordon, 'Tectonic setting of the world's giant oil and gas fields', in M. Halbouty (ed.), *Giant Oil and Gas Fields of the Decade, 1990–1999* (Tulsa, OH, 2003), 15–105.

72 G. Feulner, 'Formation of most of our coal brought Earth close to global glaciation', *Proceedings of the National Academy of Sciences of the USA* 114 (2017), 11,333–7.

73 K. Pomeranz, *The Great Divergence: China, Europe and the Making of the Modern World Economy* (Princeton, 2000).

74 D. Zizzamia, 'Restoring the Paleo-West: Fossils, Coal, and Climate in Late Nineteenth Century America', *Environmental History* 24 (2019), esp. 133–9.

75 B. Rogers, 'The Processes and People in the Rise of Instant Cities and their Evolution into the Twenty-First Century', in G. Bakken (ed.), *The World of the American West* (London, 2010), pp. 267–307.

76 F. Egli, N. Schmid and T. Schmidt, 'Electoral response to the decline of coal mining in the United States', *SSRN* (2020), 1–54.

77 S. Dutch, 'Geology and Election 2000: Overview', https://stevedutch. net/research/elec2000/geolelec2000.htm.

78 For a good overview, see D. Yergin, *The Prize: The Epic Quest for Oil, Money & Power* (New York, 2008).

79 E. Rogan, *The Fall of the Ottomans: The Great War in the Middle East, 1914–1920* (London, 2015); D. Fieldhouse, *Western Imperialism in the Middle East, 1914–1958* (London, 2006).

80 P. Frankopan, *The Silk Roads: A New History of the World* (London, 2015), pp. 357–76; D. Yergin, *The Prize: The Epic Quest for Oil, Money and Power* (New York, 1991), pp. 333–49; J. Tolland, *The Rising Sun: The Decline and Fall of the Japanese Empire, 1936–1945* (London, 1970).

81 M. Willbold, T. Elliott and S. Moorbath, 'The tungsten isotopic composition of the Earth's mantle before the terminal bombardment', *Nature* 477 (2011), 195–8.

82 D. Siegel, J. Barnes and B. Metzger, 'Collapsars as a major source of r-process elements', *Nature* 569 (2019), 241–4.

83 C. McLeod and B. Shaulis, 'Rare Earth Elements in Planetary Crusts: Insights from the Chemically Evolved Igneous Suites on Earth and the Moon', *Minerals* 8 (2018), 1–24.

84 C. McLeod and M. Krekeler, 'Source of Extraterrestrial Rare Earth Elements: To the Moon and Beyond', *Resources* 6 (2017), 1–28.

85 See for example S. Sagan, *The Limits of Safety: Organisations, Accidents, and Nuclear Weapons* (Princeton, 1993).

86 S. Baum, R. de Neufville and A. Barrett, 'A Model for the Probability of Nuclear War', Global Catastrophic Risk Institute Working Paper 18-1 (2018), 1–39.

87 'Hawaii missile alert: False alarm sparks panic in US State', BBC News, 14 January 2018.

88 See below Chapter 23.

2 ON THE ORIGINS OF OUR SPECIES (C.7M–C.12,000 BC)

1 K. Langergraber et al., 'Generation times in wild chimpanzees and gorillas suggest earlier divergence times in great ape and human evolution', *Proceedings of the National Academy of Sciences of the USA* 109 (2012), 15,716–21; O. Venn et al., 'Strong male bias drives germline mutation in chimpanzees', *Science* 344 (2014), 1272–5.

2 A. Beaufet et al., 'Preliminary paleohistological observations of the StW 573 ("Little Foot") skull', *eLife* (2021), 1–8. Also D. Granger et al., 'Cosmogenic nuclide dating of *Australopithacus* at Sterkfontein, South Africa', *PNAS* 119 (2022), 1–7.

3 K. Carlson et al., 'The pectoral girdle of StW 573 ("Little Foot") and its implications for shoulder evolution in the Hominina', *Journal of Human Evolution* (2021), 1–28.

4 B. Villmoare et al., 'Early *Homo* at 2.8 Ma from Ledi-Geraru, Afar, Ethiopia', *Science* 347 (2015), 1352–5; although now see H. Püschel et al., 'Divergence-time estimates for hominins provide insight into encephalization and body mass trends in human evolution', *Nature Ecology & Evolution* (2021), 1–24.

5 W. Kimbel and B. Villmoare, 'From *Australopithacus* to *Homo*: the transition that wasn't', *Proceedings of the Royal Society B* 371 (2016), 1–10.

6 B. Wood and J. Baker, 'Evolution in the genus *Homo*', *Annual Review of Ecology, Evolution and Systematics* 42 (2011), 47–69.

7 F. Spoor et al., 'Reconstructed *Homo habilis* type OH 7 suggests deep-rooted species diversity in early *Homo*', *Nature* 519 (2015), 83–6.

8 See for example H. Dunsworth, 'Origin of the Genus *Homo*', *Evolution: Education and Outreach* 3 (2010), 353–66.

9 E. Vrba, 'Role of Environmental Stimuli in Hominid Origins', in W. Henke and I. Tattersall, *Handbook of Paleoanthropology* (New York, 2007), pp. 1837–86.

10 S. Antón, R. Potts and L. Aiello, 'Evolution of early *Homo*: An integrated biological perspective', *Science* 345 (2014), 1–15.

11 A. Harries et al., 'Contemporaneity of *Australopithacus*, *Paranthropus* and early *Homo erectus* in South Africa', *Science* 368 (2020), 1–21.

12 R. Potts et al., 'Environmental and behavioural evidence pertaining to the evolution of early *Homo*', *Current Anthropology* 53 (2012), 299–312.

13 J. Ferrero et al., 'Earliest archaeological evidence of persistent hominin carnivory', *PLOS ONE* 8 (2013), 1–10; D. Braun et al., 'Early hominin diet included diverse terrestrial and aquatic animals 1.95 Ma in East Turkana, Kenya', *Proceedings of the National Academy of Sciences of the USA* 107 (2010), 10,002–7.

14 C. Broadhurst, S. Cunnane and M. Crawford, 'Rift Valley lake fish and shellfish provided brain-specific nutrition for early Homo', *British Journal of Nutrition* 79 (1998), 3–21.

15 M. Raghanti et al., 'A neurochemical hypothesis for the origin of hominids', *Proceedings of the National Academy of Sciences of the USA* 115 (2018), 1108–16.

16 J. Gowlett, C. Gamble and R. Dunbar, 'Human evolution and the archaeology of the social brain', *Current Anthropology* 53 (2012), 693–722; A. Pinson et al., 'Human TKL1 implies greater neurogenesis in frontal neocortex of modern humans than Neanderthals', *Science* 377 (2022), 1–12. For an overview, see A. Barona, 'The archaeology of the social brain revisited: rethinking mind and material culture from a material engagement perspective', *Adaptive Behaviour* (2020), 1–16.

17 H. Christian, 'Global frequency and distribution of lightning as observed from space by the Optical Transient Detector', *Journal of Geophysical Research* 108 (2003), 4–15.

18 J. Gowlett, 'The discovery of fire by humans: a long and convoluted process', *Philosophical Transactions of the Royal Society B* 371 (2016), 1–12.

19 H. de Lumley, 'Il y a 400 000 ans: la domestication du feu, un formidable moteur d'hominisation', *Comptes Rendus Palevol* 5 (2006), 149–54.

20 M. Maslin and B. Christiansen, 'Tectonics, orbital forcing, global climate change and human evolution in Africa: introduction to the African paleoclimate', *Journal of Human Evolution* 53 (2007), 443–64; S. Kaboth-Bahr et al., 'Paleo-ENSO influence on African environments and early modern humans', *PNAS* 118 (2021), 1–6.

21 A. Bergström et al., 'Origins of modern human ancestry', *Nature* 590 (2021), 229–37.

22 A. Barash et al., 'The earliest Pleistocene record of a large-bodied hominin from the Levant supports two out-of-Africa dispersal events', *Scientific Reports* 12 (2022), 1–9.

23 R. Ferring et al., 'Earliest human occupations at Dmanisi (Georgian Caucasus) dated to 1.85–1.78Ma', *Proceedings of the National Academy of Sciences of the USA* 108 (2011), 10,432–6.

24 R. Zhu et al., 'Early evidence of the genus *Homo* in East Asia', *Journal of Human Evolution* 55 (2008), 1075–85; R. Larick, 'Early Pleistocene ⁴⁰Ar/³⁹Ar ages for Bapang Formation hominins, Central Jawa, Indonesia', *PNAS* 98 (2001), 4866–71; M. Arzarello, 'Evidence of earliest human occurrence in Europe: The site of Pirro Nord (Southern Italy)', *Naturwissenschaften* 94 (2007), 107–12.

25 Stringer, 'Origin and evolution of *Homo sapiens*', 1–12; A. Gómez-Robles, 'Dental evolutionary rates and its implications for the Neanderthal–modern human divergence', *Science Advances* 5 (2019), 1–9; X. Ni et al., 'Massive cranium from Harbin in northeastern China establishes a new Middle Pleistocene human lineage', *The Innovation* (2021), 1–7.

26 D. Reich et al., 'Genetic history of an archaic hominin group from Denisova Cave in Siberia', *Nature* 468 (2010), 1053–60; J. Krause et al., 'The complete mitochondrial DNA genome of an unknown hominin from southern Siberia', *Nature* 464 (2010) 894–7; I. Hershkovitz et al., 'A Middle Pleistocene *Homo* from Nesher Ramla, Israel', *Science* 372 (2021), 1424–8; J. Teixeira et al., 'Widespread Denisovian ancestry in Island Southeast Asia but no evidence of substantial super-archaic hominin admixture', *Nature Ecology & Evolution* (2021), 1–11.

27 P. Qin and M. Stoneking, 'Denisovan Ancestry in East Eurasian and Native American Populations', *Molecular Biology & Evolution* 32 (2015), 2665–74; B. Bonifante, 'A GWAS in Latin Americans identifies novel face shape loci, implicating VPS13B and a Denisovan introgressed region in facial variation', *Science Advances* 7 (2021), 1–18; P. Zhang et al., 'Denisovans and Homo sapiens on the Tibetan Plateau: dispersals and adaptations', *Trends in Ecology & Evolution* (2021); D. Zhang, 'Denisovan DNA in Late Pleistocene sediments from Baishiya Karst Cave on the Tibetan Plateau', *Science* 370 (2020), 584–7.

28 M. Hajdinjak et al., 'Initial Upper Paleolithic humans in Europe had recent Neanderthal ancestry', *Nature* 592 (2021), 253–7.

29 H. Zeberg and S. Pääbo, 'A genomic region associated with protection against severe COVID-19 is inherited from Neanderthals', *PNAS* 118 (2021), 1–5.

30 A. Steegmann, F. Cerny and T. Holliday, 'Neanderthal cold adaptation: Physiological and energetic factors', *American Journal of Human Biology* 14 (2002), 566–83.

31 J.-J. Hublin and W. Roebroeks, 'Ebb and flow or regional extinctions? On the character of Neanderthal occupation of northern environments', *Comptes Rendus Palevol* 8 (2009), 503–9.

32 B. Vernot et al., 'Unearthing Neanderthal population history using nuclear and mitochondrial DNA from cave sediments', *Science* 372 (2021), 1–8.

33 J.-J. Hublin, 'The origin of Neanderthals', *PNAS* 106 (2009), 16,022–7.

34 A. Bartsiokas and J.-L. Arsuaga, 'Hibernation in hominins from Atapuerca, Spain half a million years ago', *L'Anthropologie* 124 (2020), 1–34.

35 M. Conde-Valverde et al., 'Neanderthals and *Homo sapiens* had similar auditory and speech capacities', *Nature Ecology & Evolution* (2021), 1–19.

36 D. Garrigan and M. Hammer, 'Reconstructing human origins in the genomic era', *Nature Reviews Genetics* 7 (2006), 669–80.

37 M. Lahr, 'Genetic and Fossil Evidence for Modern Human Origins', in P. Mitchell and P. Lane (eds), *The Oxford Handbook of African Archaeology* (Oxford, 2013), pp. 327–9.

38 C. Stringer, *The Origins of our Species* (London, 2012), pp. 173–4, 229–33; A. Timmermann et al., 'Climate effects on archaic human habitats and species successions', *Nature* 604 (2022), 495–501.

39 Y. Kedar, G. Kedar and R. Barkai, 'The influence of smoke density on hearth location and activity areas at Lower Paleolithic Lazaret Cave, France', *Scientific Reports* 12 (2022), 1–14.

40 E. Böhm et al., 'Strong and deep Atlantic meridional overturning circulation during the last glacial cycle', *Nature* 517 (2015), 73–6; D. Sigman, M. Hain and G. Haug, 'The polar ocean and glacial cycles in atmospheric CO_2 concentration', *Nature* 466 (2010), 47–55.

41 J. Bassis, S. Petersen and L. MacCathles, 'Heinrich events triggered by ocean forcing and modulated by isostatic adjustment', *Nature* 542 (2017), 332–4.

42 S. Crump et al., 'Ancient plant DNA reveals High Arctic greening during the Last Interglacial', *PNAS* 118 (2021).

43 C. Turney et al., 'Early Last Interglacial ocean warming drove substantial ice mass loss from Antarctica', *PNAS* 117 (2020), 3996–4006.

44 R. Jones et al., 'Delayed maximum northern European summer temperatures during the Last Interglacial as a result of Greenland Ice Sheet melt', *Geology* 45 (2017), 23–6.

45 J. Smith, 'Palaeoenvironments of eastern North Africa and the Levant in the Late Pleistocene', in E. Garcea (ed.), *South-Eastern Mediterranean Peoples between 130,000 and 10,000 Years Ago* (Oxford, 2010), 6–17.

46 H. Groucutt et al., 'Multiple hominin dispersals into Southwest Asia over the past 400,000 years', *Nature* (2021), 1–15.

47 S. Tucci and J. Akey, 'A map of wanderlust', *Nature* 538 (2016), 179–80.

48 R. Dennell and W. Roebroeks, 'An Asian perspective on early human dispersal from Africa', *Nature* 438 (2005), 1099–104; J. Saarinen, 'Pliocene to Middle Pleistocene climate history in the Guadix-Baza Basin, and the environmental conditions of early *Homo* dispersal in Europe', *Quaternary Science Reviews* 268 (2021), 1–18; M. Belmaker, 'On the road to China: the environmental landscape of the Early Pleistocene in Western Eurasia and its implication for the dispersal of *Homo*', in C. Norton and D. R. Braun (eds), *Asian Palaeoanthropology: From Africa to China and Beyond*, Vertebrate Palaeobiology and Palaeoanthropology (London, 2011), pp. 31–40.

49 J. Shea, 'Transitions or turnovers? Climatically-forced extinctions of *Homo sapiens* and Neanderthals in the East Mediterranean Levant', *Quaternary Science Reviews* 27 (2008), 2253–70.

50 W. Rose and C. Chesner, 'Dispersal of ash in the great Toba eruption, 75 ka', *Geology* 15 (1987), 913–17.

51 M. Williams et al., 'Environmental impact of the 73 Ka Toba super-eruption in South Asia', *Palaeogeography, Palaeoclimatology, Palaeoecology* 284 (2009), 295–314; K. Acharyya and P. Basu, 'Toba Ash on Indian Subcontinent and its Implications for Correlation of Late Pleistocene Alluvium' *Quaternary Research* 40 (1993), 10–19.

52 A. Robock et al., 'Did the Toba Volcanic Eruption of ~74k BP produce Widespread Glaciation?', *Journal of Geophysical Research* 114 (2009), 1–9; S. Ambrose, 'Late Pleistocene Human Population Bottlenecks, Volcanic Winter, and Differentiation of Modern Humans', *Journal of Human Evolution* 34 (1998), 623–51.

53 B. Black et al., 'Global climate disruption and regional climate shelters after the Toba supereruption', *PNAS* 118 (2021), 1–8.

54 C. Lane, B. Chorn and T. Johnson, 'Ash from the Toba supereruption in Lake Malawi shows no volcanic winter in East Africa at 75ka', *PNAS* 110 (2013), 8025–9.

55 M. Haslam et al., 'The 74 ka Toba super-eruption and southern Indian hominins: archaeology, lithic technology and environments at Jwalapuram Locality 3', *Journal of Archaeological Science* 37 (2010), 3370–84.

56 C. Timmreck et al., 'Aerosol size confines climate response to volcanic super-eruptions', *Geophysical Research Letters* 37 (2010), 1–5.

57 C. Fernandes et al., 'Absence of Post-Miocene Red Sea Land
    Bridges: Biogeographic Implications', *Journal of Biogeography* 33 (2006),
    961–6; C. Clarkson et al., 'Human Occupation of Northern Australia by
    65,000 Years Ago', *Nature* 547 (2017), 306–10.

58 J. Thompson et al., 'Early human impacts and ecosystem reorganization
    in southern-central Africa', *Science Advances* 7 (2021), 1–13.

59 M. Ziegler et al., 'Development of Middle Stone Age innovation linked
    to rapid climate change', *Nature Communications* 4 (2013), 1–9.

60 F. d'Errico et al., '*Nassarius kraussianus* shell beads from Blombos
    Cave: evidence for symbolic behaviour in the Middle Stone Age', *Journal
    of Human Evolution* 48 (2005), 3–24.

61 R. Vogelsang, 'New Excavations of Middle Stone Age Deposits at Apollo
    11 Rockshelter, Namibia: Stratigraphy, Archaeology, Chronology and
    Past Environments', *Journal of African Archaeology* 8 (2010), 185–218;
    C. Henshilwood et al., 'An abstract drawing from the 73,000-year-old
    levels at Blombos Cave, South Africa', *Nature* 562 (2018), 115–8.

62 Y. Sahle and A. Brooks, 'Assessment of complex projectiles in the early
    Late Pleistocene at Aduma, Ethiopia', *PLOS ONE* 14 (2019), 1–18.

63 J. Shea, 'Neanderthals and early *Homo sapiens* in the Levant', in Garcea,
    South-Eastern Mediterranean Peoples, pp. 126–43.

64 J. Miller and Y. Wang, 'Ostrich eggshell beads reveal 50,000-year-old
    social network in Africa', *Nature* 601 (2022), 234–9.

65 D. Leder, 'A 51,000-year-old engraved bone reveals Neanderthals'
    capacity for symbolic behaviour', *Nature Ecology & Evolution* 5 (2021),
    1273–82; A. Pitarch Martí, 'The symbolic role of the underground world
    among Middle Paleolithic Neanderthals', *PNAS* 118 (2021), 1–6.

66 C. Henshilwood, F. d'Errico and I. Watts, 'Engraved ochres from the
    Middle Stone Age level at Blombos Cave, South Africa', *Journal of
    Human Evolution* 57 (2009), 27–47.

67 A. Brumm, 'Oldest cave art found in Sulawesi', *Science Advances*
    7 (2021), 1–13.

68 M. Aubert, 'Earliest hunting scene in prehistoric art', *Nature* 576
    (2019), 442–5; for the date of settlement in Europe, H. Fewlass, 'A 14C
    chronology for the Middle to Upper Palaeolithic transition at Bacho
    Kiro Cave, Bulgaria', *Nature Ecology & Evolution* 4 (2020), 794–801.

69 J. Kind et al., 'The smile of the Lion Man: Recent excavations in Stadel
    Cave (Baden-Württemberg, south-western Germany) and the restoration
    of the famous Upper Palaeolithic figurine', *Quartär* 61 (2014), 129–45.

70 A. Cooper et al., 'A global environmental crisis 42,000 years ago', *Science*
    371 (2021), 811–18; however, also see A. Picin et al., 'Comment on "A
    global environmental crisis 42,000 years ago"', *Science* 374 (2021), 1–4.

71 Y. Kedar, G. Kedar and R. Barkai, 'Hypoxia in Paleolithic decorated
    caves: the use of artificial light in deep caves reduces oxygen
    concentration and induces altered states of consciousness', *Time
    and Mind: The Journal of Archaeology, Consciousness and Culture* 14
    (2021), 1–36.

72 D. Wolf et al., 'Climate deteriorations and Neanderthal demise in
    interior Iberia', *Scientific Reports* 8 (2018), 1–10; S. Pederzani, 'Subarctic
    climate for the earliest *Homo sapiens* in Europe', *Science Advances*
    7 (2021), 1–10.

73 A. Timmermann, 'Quantifying the potential causes of Neanderthal
    extinction: Abrupt climate change versus competition and
    interbreeding', *Quaternary Science Reviews* 238 (2020), 1–14;
    L. Slimak et al., 'Modern human incursion into Neanderthal territories
    54,000 years ago at Mandrin, France', *Science Advances* 8 (2022), 1–16.

74 P. Mellars and J. French, 'Tenfold Population Increase in Western
    Europe at the Neanderthal-to-Modern Human Transition', *Science* 333
    (2011), 623–7.

75 Y. Kaifu, 'Archaic hominin populations in Asia before the arrival of
    modern humans: their phylogeny and implications for "Southern
    Denisovans"', *Current Anthropology* 58 (2017), 418–33.

76 P. Roberts et al., 'Isotopic evidence for initial coastal colonization and
    subsequent diversification in the human occupation of Wallacea', *Nature
    Communications* 11 (2020), 1–11; also N. Bourgon, 'Trophic ecology
    of a Late Pleistocene early modern human from tropical Southeast
    Asia inferred from zinc isotopes', *Journal of Human Evolution* 161
    (2021), 1–10.

77 J. Choin et al., 'Genomic insights into population history and biological
    adaptation in Oceania', *Nature* 592 (2021), 583–9; A. Fairbairn,
    'Pleistocene Occupation of New Guinea's Highland and Subalpine
    Environments', *World Archaeology* 38 (2017), 371–86.

78 Clarkson et al., 'Human Occupation of Northern Australia by 65,000
    Years Ago', 306–10; S. Florin et al., 'Pandanus nutshell generates a
    palaeoprecipitation record for human occupation at Madjedbebe,
    northern Australia,' *Nature Ecology & Evolution* 5 (2021), 295–303.

79 P. Clark et al., 'The Last Glacial Maximum', *Science* 325 (2009),
    710–14.

80 F. Sirocko, *Wetter, Klima, Menschheitsentwicklung: von der Eiszeit bis ins
    21. Jahrhundert* (Darmstadt, 2010), esp. 71–82.

81 S. Blockley et al., 'Dating human occupation and adaptation in the
    southern European last glacial refuge: The chronostratigraphy of Grotta
    del Romito (Italy)', *Quaternary Science Reviews* 184 (2018), 5–25.

82 G.-C. Weniger, 'Late Glacial rapid climate change and human response in the Westernmost Mediterranean (Iberia and Morocco)', *PLOS ONE* 14 (2019), 1–33.

83 C. Spötl et al., 'Increased autumn and winter precipitation during the Last Glacial Maximum in the European Alps', *Nature Communications* 12 (2021), 1–9.

84 R. Cheddadi et al., 'Putative glacial refugia of *Cedrus atlantica* deduced from Quaternary pollen records and modern genetic diversity', *Journal of Biogeography* 36 (2009), 1361–71; K. Cuffey et al., 'Large arctic temperature change at the Wisconsin-Holocene transition', *Science* 270 (1995), 455–8.

85 A. Seltzer et al., 'Widespread six degrees Celsius cooling on land during the Last Glacial Maximum', *Nature* 593 (2021), 228–32.

86 P. Wang and X. Sun, 'Last glacial maximum in China: comparison between land and sea', *Catena* 23 (1994), 341–53.

87 P. Zhao et al., 'Modeling the East Asian climate during the last glacial maximum', *Science in China Series D: Earth Sciences* 46 (2003), 1060–8.

88 A. Bouzouggar, R. Barton and S. Blockley, 'Re-evaluating the age of the Iberomaurusian in Morocco', *African Archaeological Review* 25 (2008), 3–19.

89 Mercader et al., 'Phytoliths from archaeological sites in the tropical forest of Ituri, Democratic Republic of Congo', *Quaternary Research* 54 (2000), 102–12.

90 J. Louys and P. Roberts, 'Environmental drivers of megafauna and hominin extinction in South East Asia', *Nature* 586 (2020), 402–24.

91 A. Barnosky, 'Assessing the Causes of Late Pleistocene Extinctions on the Continents', *Science* 306 (2004), 70–5; C. Bradshaw et al., 'Relative demographic susceptibility does not explain the extinction chronology of Sahul's megafauna', *ELife* (2021), 1–43; J. Cantalapiedra et al., 'The rise and fall of proboscidean ecological diversity', *Nature Ecology & Evolution* 5 (2021), 1266–72.

92 G. Prescott et al., 'Quantitative global analysis of the role of climate and people in explaining late Quaternary megafaunal extinctions', *PNAS* 109 (2012), 4257–31.

93 C. Sandom et al., 'Global late Quaternary megafauna extinctions linked to humans, not climate change', *Proceedings of the Royal Society B* 281 (2014), 1–9.

94 N. Munro et al., 'The emergence of animal management in the southern Levant', *Scientific Reports* 8 (2018), 1–11; J. Dembitzer et al., 'Levantine overkill: 1.5 million years of hunting down the body size distribution', *Quaternary Science Reviews* 276 (2022), 1–10.

95 D. Lorenzen et al., 'Species-specific responses of Late Quaternary megafauna to climate and humans', *Nature* 479 (2011), 359–64.

96 C. Doughty et al., 'Megafauna extinction, tree species range reduction, and carbon storage in Amazonian forests', *Ecography* 39 (2016), 194–203.

97 S. Rule et al., 'The Aftermath of Megafaunal Extinction: Ecosystem Transformation in Pleistocene Australia', *Science* 335 (2012), 1483–6.

98 T. Pico et al., 'Glacial isostatic adjustment directed incision of the Channeled Scabland by Ice Age megafloods', *PNAS* 119 (2022), 1–6.

99 P. Clark, 'Global climate evolution during the last deglaciation', *PNAS* 109 (2012), 1134–42.

100 J. Benjamin, 'Aboriginal artefacts on the continental shelf reveal ancient drowned cultural landscapes in northwest Australia', *PLOS ONE* 15 (2020), 1–31.

101 A. Williams, 'Sea-level change and demography during the last glacial termination and early Holocene across the Australian continent', *Quaternary Science Reviews* 182 (2018), 144–54.

102 G. Goebel, M. Waters and D. O'Rourke, 'The Late Pleistocene Dispersal of Modern Humans in the Americas', *Science* 319 (2008), 1497–1502; J. Dobson, G. Spada and G. Galassi, 'The Bering Transitory Archipelago: stepping stones for the first Americans', *Comptes Rendus Géoscience* 353 (2021), 55–65.

103 V. Moreno-Mayar, 'Early human dispersals within the Americas', *Science* 362 (2018), 1–11.

104 See for example W. Stinnesbeck, 'The earliest settlers of Mesoamerica date back to the late Pleistocene', *PLOS ONE* 12 (2017), 1–20; M. Araújo Castro e Silva et al., 'Deep genetic affinity between coastal Pacific and Amazonian natives evidenced by Australasian ancestry', *PNAS* 118 (2021), 1–3.

105 C. Ardelean et al., 'Evidence of human occupation in Mexico around the Last Glacial Maximum', *Nature* 584 (2020), 87–92; A. Somerville, I. Casar and J. Arroyo-Cabrales, 'New AMS Radiocarbon Ages from the Preceramic Levels of Coxcatlan Cave, Puebla, Mexico: A Pleistocene Occupation of the Tehuacan Valley', *Latin American Antiquity* (2021), 1–15; T. Rowe et al., 'Human Occupation of the North American Colorado Plateau ~37,000 Years Ago', *Frontiers in Ecology and Evolution* 10 (2022), 1–22.

106 J. Colella, 'Whole-genome resequencing reveals persistence of forest-associated mammals in Late Pleistocene refugia along North America's North Pacific coast', *Journal of Biogeography* 48 (2021), 1153–9.

107 For example, L. Bourgeon, A. Burke and T. Higham, 'Earliest human presence in North America dates to the Last Glacial Maximum: new radiocarbon dates from Bluefish Caves, Canada', *PLOS ONE* 12 (2017), 1–15; T. Dillehay et al., 'New Archaeological Evidence for an Early Human Presence at Monte Verde, Chile', *PLOS ONE* 10 (2015), 1–27; A. Campelo dos Santos et al, 'Genomic evidence for ancient human migration routes along South America's Atlantic coast', *Proceedings of the Royal Society B, Biological Sciences* 289 (2022), 1-11.

108 D. Shakun and A. Carlson, 'A global perspective on Last Glacial Maximum to Holocene Climate Change', *Quaternary Science Reviews* 29 (2010), 1801–16.

109 J. Annan and J. Hargreaves, 'A new global reconstruction of temperature changes at the Last Glacial Maximum', *Climate of the Past* 9 (2013), 367–76; P. Holden et al., 'A probabilistic calibration of climate sensitivity and terrestrial carbon change in GENIE-1', *Climate Dynamics* 35 (2009), 1–22; J. Shakun et al., 'Global warming preceded by increasing carbon dioxide concentrations during the last deglaciation', *Nature* 484 (2012), 49–54.

110 N. Thiagarajan et al., 'Abrupt pre-Bølling-Allerød warming and circulation changes in the deep ocean', *Nature* 511 (2014), 75.

111 W. van Ziest, U. Baruch and S. Bottema, 'Holocene palaeoecology of the Hula area, northern Israel', in E. Kaptijn and L. Petit (eds), *A Timeless Vale: Archaeological and Related Essays on the Jordan Valley in Honour of Gerrit van der Kooij on the Occasion of his Sixty-Fifth Birthday* (Leiden, 2009), pp. 29–64.

112 N. Goring Morris and A. Belfer-Cohen, 'Neolithization processes in the Levant: the outer envelope', *Current Anthropology* 52 (2011), 195–208.

113 C. Nolan et al., 'Past and future global transformation of terrestrial ecosystems under climate change', *Science* 361 (2018), 920–3.

114 C. Parmesan and M. Hanley, 'Plants and climate change: complexities and surprises', *Annals of Botany* 116 (2015), 849–64.

115 Ibid.; V. Masson-Delmotte et al., 'Information from paleoclimatic archives', in T. Stocker et al. (eds), *Climate Change 2013 – The Physical Science Basis* (Cambridge, 2013), pp. 383–464.

116 A. Carlson et al., 'Geochemical proxies of North American freshwater routing during the Younger Dryas cold event', *PNAS* 104 (2007), 6556–61.

117 G. Brakenridge, 'Core-collapse supernovae and the Younger Dryas/terminal Rancholabrean extinctions', *Icarus* 215 (2011), 101–6.

118 F. Thackeray, L. Scott and P. Pieterse, 'The Younger Dryas interval at Wonderkrater (South Africa) in the context of a platinum anomaly', *Palaeontologia Africana* 54 (2019), 30–5; M. Sweatman, 'The Younger Dryas impact hypothesis: Review of the impact evidence', *Earth-Science Reviews* 218 (2021), 1–22.

119 N. Sun, 'Volcanic origin for Younger Dryas geochemical anomalies ca. 12,000 cal BP', *Science Advances* 6 (2020), 1–9. For the date, see F. Reinig et al., 'Precise date for the Laacher See eruption synchronises the Younger Dryas', *Nature* 595 (2021), 66–9.

120 P. Schultz et al., 'Widespread glasses generated by cometary fireballs during the late Pleistocene in the Atacama Desert, Chile', *Geology* (2021), 1–5.

121 J. Steffensen et al., 'High-Resolution Greenland Ice Core Data Show Abrupt Climate Change Happens in Few Years', *Science* 321 (2008), 680–4; J. Severinghaus et al., 'Timing of abrupt climate change at the end of the Younger Dryas interval from thermally fractionated gases in polar ice', *Nature* 391 (1998), 141–6.

122 J. Fan et al., 'The manifestation of the Younger Dryas event in the East Asian summer monsoon margin: New evidence from carbonate geochemistry of the Dali Lake sediments in northern China', *The Holocene* 28 (2018), 1082–92.

123 R. Newnham and D. Lowe, 'Fine-resolution pollen record of late-glacial climate reversal from New Zealand', *Geology* 28 (2000), 759–62.

124 J. Bakke et al., 'Rapid oceanic and atmospheric changes during the Younger Dryas cold period', *Nature Geoscience* 2 (2009), 202–5.

125 Z. An, S. Colman and W. Zhou, 'Interplay between the Westerlies and Asian monsoon recorded in Lake Qinghai sediments since 32 ka', *Scientific Reports* 2 (2012), 1–7; X. Huang, P. Meyers and J. Yu, 'Moisture conditions during the Younger Dryas and the early Holocene in the middle reaches of the Yangtze River, central China', *The Holocene* 22 (2012), 1473–9.

126 M. Talbot et al., 'An abrupt change in the African monsoon at the end of the Younger Dryas', *Geochemistry, Geophysics, Geosystems* 8 (2007), 1–16.

127 R. Guthrie, 'New carbon dates link climate change with human colonization and Pleistocene extinctions', *Nature* 441 (2006), 207–9; also see J. Louys, 'No evidence for widespread island extinctions after Pleistocene hominin arrival', *PNAS* 118 (2021), 1–8.

128 M. Theuerkauf and H. Joosten, 'Younger Dryas cold stage vegetation patterns of central Europe – climate, soil and relief controls', *Boreas* 41 (2012), 391–407; F. Mayle and L. Cwynar, 'Impact of the Younger Dryas Cooling Event upon Lowland Vegetation of Maritime Canada', *Ecological Monographs* 65 (1995), 129–54; S. Mulitza et al., 'Sahel megadroughts triggered by glacial slowdowns of Atlantic meridional overturning', *Paleoceanography* 23 (2008), 1–11.

129 D. Anderson et al., 'Multiple lines of evidence for possible Human population decline/settlement reorganization during the early Younger Dryas', *Quaternary International* 242 (2011), 570–83.

130 J. Fernández-López de Pablo et al., 'Palaeodemographic modelling supports a population bottleneck during the Pleistocene–Holocene transition in Iberia', *Nature Communications* 10 (2019), 1–13.

131 Y. Nakazawa et al., 'Human responses to the Younger Dryas in Japan', *Quaternary International* 242 (2011), 416–33.

132 A. Belfer-Cohen and O. Bar-Yosef, 'Early sedentism in the Near East: a bumpy ride to village life', in I. Kujit (ed.), *Life in Neolothic Farming Communities: Social Organization, Identity and Differentiation* (New York, 2000), pp. 19–37.

133 H. Cheng et al., 'Timing and structure of the Younger Dyas event and its underlying climate dynamics', *PNAS* 117 (2020), 23,408–17; J. Partin et al., 'Gradual onset and recovery of the Younger Dryas abrupt climate event in the tropics', *Nature Communications* 6 (2015), 1–9.

134 Steffensen, 'High-Resolution Greenland Ice Core Data', 680–4.

135 M. Walker, 'Formal definition and dating of the GSSP (Global Stratotype Section and Point) for the base of the Holocene using the Greenland NGRIP ice core, and selected auxiliary records', *Journal of Quaternary Science* 24 (2009), 3–17.

136 S. Raghuraman, D. Paynter and V. Ramaswamy, 'Anthropogenic forcing and response yield observed positive trend in Earth's energy imbalance', *Nature Communications* 12 (2021), 1–10.

137 N. Barton and A. Bouzouggar, 'Hunter-gatherers of the Maghreb 25,000–6,000 Years Ago', in Mitchell and Lane (eds), *Oxford Handbook of African Archaeology*, pp. 432–4.

138 E. Rybin et al., 'Стратиграфия и культурная последовательность стоянки Толбор-21 (Северная Монголия): итоги работ 2014-2016

гг. И дальнейшие перспективы исследований' in Теория и практика археологических исследований 20 (2017), 15,868; Е. Novgorodova, Древняя Монголия: Некоторые проблемы хронологии и этнокультурной истории (Moscow, 1989).

139 L. Humphrey and E. Bocaege, 'Tooth Evulsion in the Maghreb: Chronological and Geographical Patterns', *African Archaeological Review* 25 (2008), 109–23. The earliest human burial, from a cave site in Kenya, has been dated to c.78,000 years ago: M. Martinón-Torres, 'Earliest known human burial in Africa', *Nature* 593 (2021), 95–100; N. Rahmani, *Le Capsien typique et le Capsien supérieur. Evolution ou contemporanité, les données technologiques* (Oxford, 2003), esp. 291–301.

140 G. Barker, *The Agricultural Revolution in Prehistory: Why Did Foragers Become Farmers?* (Oxford, 2006); L. Janz, D. Odsuren and D. Bukhchulun, 'Transitions in Palaeoecology and Technology: Hunter-Gatherers and Early Herders in the Gobi Desert', *Journal of World Prehistory* 30 (2017), 1–80.

141 A. Rosen and I. Rivera-Collazo, 'Climate change, adaptive cycles and the persistence of foraging economies during the late Pleistocene/ Holocene transition in the Levant', *PNAS* 109 (2012), 3640–5; G. Wilcox, 'The distribution, natural habitats and availability of wild cereals in relation to their domestication in the Near East: Multiple events, multiple centres', *Vegetation History and Archaeobotany* 14 (2005), 534–41.

142 O. Bar-Yosef and A. Belfer-Cohen, 'Facing environmental crisis. Societal and cultural changes at the transition from the Younger Dryas to the Holocene in the Levant', in R. Cappers and S. Bottema, *Transition from Foraging to Farming in South-West Asia: Dawn of Farming in the Near East* (Berlin, 2002), pp. 55–66.

143 E. Ellis et al., 'People have shaped most of territorial nature for at least 12,000 years', *PNAS* 118 (2021), 1–8.

144 M. Maslin, C. Stickley and V. Ettwein, 'Holocene Climate Variability', in K. Cochran et al. (eds), *Encyclopaedia of Ocean Sciences*, 6 vols (Amsterdam, 2019), 4, pp. 513–19.

3 HUMAN INTERACTIONS WITH ECOLOGIES
(C.12,000–C.3500 BC)

1 P. Richerson, R. Boyd and R. Bettinger, 'Was Agriculture Impossible during the Pleistocene but Mandatory during the Holocene? A Climate Change Hypothesis', *American Antiquity* 66 (2001), 387–411.

2 Z. Qiu et al., 'Late Pleistocene–Holocene vegetation history and anthropogenic activities deduced from pollen spectra and archaeological data at Guxu Lake, eastern China', *Scientific Reports* 10 (2020), 1–14.

3 E. Duarte et al., 'A progressively wetter early through middle Holocene climate in the eastern lowlands of Guatemala', *Earth and Planetary Science Letters* 561 (2021), 1–13.

4 H.-S. Park et al., 'Mid-Holocene Northern Hemisphere warming driven by Arctic amplification', *Science Advances* 5 (2019), 1–11.

5 S. Höpker et al., 'Pronounced northwest African monsoon discharge during the mid- to late-Holocene', *Frontiers in Earth Science* 7 (2019), 1–17.

6 R. Cheddadi et al., 'Early Holocene greening of the Sahara required Mediterranean winter rainfall', *PNAS* 118 (2021), 1–7.

7 N. Roberts et al. 'Human responses and non-responses to climatic variations during the last Glacial–Interglacial transition in the eastern Mediterranean', *Quaternary Science Reviews* 184 (2018), 47–67.

8 A. Moore and G. Hillman, 'The Pleistocene to Holocene transition and human economy in southwest Asia: the impact of the younger Dryas', *American Antiquity* 57 (1992), 482–94.

9 Rosen and Rivera-Collazo, 'Climate change', 3640–5.

10 K. Wright, 'The Social Origins of Cooking and Dining in Early Villages of Western Asia', *Proceedings of the Prehistoric Society* 66 (2000), 89–121.

11 T. Molleson, 'The Eloquent Bones of Any Hureyra', *Scientific American* 271 (1994), 60–5; T. Molleson, 'Bones of Work at the Origins of Labour', in S. Hamilton, R. Whitehouse and K. Wright (eds), *Archaeology and Women: Ancient and Modern Issues* (Walnut Creek, CA, 2007), pp. 185–98.

12 Y. Garfinkel, *Dancing at the Dawn of Agriculture* (Austin, TX, 2003).

13 A. Macintosh, R. Pinhasi and J. Stock, 'Prehistoric women's manual labor exceeded that of athletes through the first 5500 years of farming in Central Europe', *Science Advances* 3 (2017), 1–12; A. Macintosh, R. Pinhasi and J. Stock, 'Lower limb skeletal biomechanics track long-term decline in mobility across ~6150 years of agriculture in Central Europe', *Journal of Archaeological Science* 52 (2014), 376–90.

14 J. Peterson, *Sexual Revolutions: Gender and Labor at the Dawn of Agriculture* (Walnut Creek, CA, 2002); D. Bolger, 'The Dynamics of Gender in Early Agricultural Societies of the Near East', *Signs* 35 (2010), 503–31.

15 O. Bar Yosef, 'Facing climatic hazards: Palaeolithic foragers and Neolithic farmers', *Quaternary International* 428 (2017), 64–72; K. Daly et al., 'Herded and hunted goat genomes from the dawn of domestication in the Zagros Mountains', *PNAS* 118 (2021), 1–9.

16 S. Colledge, J. Conolly and S. Shennan, 'Archaeobotanical Evidence for the Spread of Farming in the Eastern Mediterranean', *Current Anthropology* 45 (2004), 35–58.

17 M. Zeder, 'The Origins of Agriculture in the Near East', *Current Anthropology* 52 (2011), 221–35.

18 B. Smith, 'Low-Level Food Production', *Journal of Archaeological Research* 9 (2001), 1–43.

19 C. Doughty, 'The development of agriculture in the Americas: an ecological perspective', *Ecosphere* 1 (2010), 1–11.

20 Z. Zhao, 'New Archaeobotanic Data for the Study of the Origins of Agriculture in China', *Current Anthropology* 52 (2011), 295–306.

21 D. Cohen, 'The Beginnings of Agriculture in China: A Multiregional View', 52 (2011), 273–94.

22 F. Bocquentin, 'Pour une approche anthropologique de la transition Épipaléolithique-Néolithique au Proche-Orient', *Bulletin du Centre de Recherche Français à Jérusalem* 17 (2006), 41–51; E. Schotsmans et al., 'New insights on commemoration of the dead through mortuary and architectural use of pigments at Neolithic Çatalhöyük, Turkey', *Scientific Reports* 12 (2022), 1–19.

23 N. Ezgi Altınışık et al, 'A genomic snapshot of demographic and cultural dynamism in Upper Mesopotamia during the Neolithic Transition', *Science Advances* 8 (2022), 1–17.

24 H. Hauptmann and K. Schmidt, 'Anatolien vor 12.000 Jahren. Die Skulpturen des Frühneolithikums', in Badisches Landesmuseum Karlsruhe, *Vor 12.000 Jahren in Anatolien. Die ältesten Monumente der Menschheit* (Karlsruhe, 2007), pp. 67–73.

25 O. Dietrich et al., 'The role of cult and feasting in the emergence of Neolithic communities: new evidence from Göbekli Tepe', *Antiquity* 86 (2012), 674–95; H. Hauptmann, 'Eine frühneolitische Kultfigur aus Urfa', in U. Esin et al. (eds), *Köyden Kente: yakındoğu'da ilk yerleşimler* (Istanbul, 2003), pp. 623–6; M. Zhilin et al., 'Early art in the Urals: new research on the wooden sculpture from Shigir', *Antiquity* 92 (2018), 334–50.

26 H. Thomas et al., 'The mustatils: cult and monumentality in Neolithic north-western Arabia', *Antiquity* (2021), 1–22.

27 R. Rainio et al., 'Prehistoric Pendants as Instigators of Sound and Body Movements: A Traceological Case Study from Northeast Europe, c.8200 cal BP', *Cambridge Archaeological Journal* 31 (2021), 639–60.

28 S. Bowles and J.-K. Choi, 'Coevolution of farming and private property during the early Holocene', *PNAS* 110 (2013), 8830–5.

29 A. Goring-Morris and A. Belfer-Cohen, 'A Roof over One's Head: Developments in Near Eastern Residential Architecture across the Epipalaeolithic–Neolithic Transition', in J. Boquet-Appel and O. Bar-Yosef (eds), *The Neolithic Demographic Transition and its Consequences* (Dordrecht, 2008), pp. 239–86.

30 W. Van Neer et al., 'Aquatic fauna from the Takarkori rock shelter reveals the Holocene central Saharan climate and palaeohydrography', *PLOS ONE* 15 (2020), 1–34.

31 B. Barich, 'Hunter-Gatherer-Fishers of the Sahara and the Sahel 12,000–4,000 Years Ago', in Mitchell and Lane (eds), *Oxford Handbook of African Archaeology*, p. 449.

32 R. Bettinger et al., 'The transition to agriculture in northwestern China', *Developments in Quaternary Sciences* 9 (2007), 83–101.

33 K. Prufer et al., 'Linking late Paleoindian stone tool technologies and populations in North, Central and South America', *PLOS ONE* 14 (2019), 1–20.

34 A. Luquin et al., 'Ancient lipids document continuity in the use of early hunter-gatherer pottery through 9,000 years of Japanese prehistory', *PNAS* 113 (2016), 3991–6.

35 J. Roset, 'Céramique et néolithisation en Afrique saharienne', in J. Guilaine (ed.), *Premiers Paysans du monde: naissances des agricultures* (Paris, 2000), 263–90; E. Huysecom et al., 'Ounjougou (Mali): a history of Holocene settlement at the southern edge of the Sahara', *Antiquity* 78 (2004), 579–93.

36 Barich, 'Hunter-Gatherer-Fishers', pp. 447–8.

37 R. Haaland and G. Haaland, 'Early Farming Societies along the Nile', in Mitchell and Lane (eds), *Oxford Handbook of African Archaeology*, pp. 541–53.

38 O. Nieuwenhuyse, M. Daskiewicz and G. Schneider, 'Investigating Late Neolithic ceramics in the northern Levant: the view from Shir', *Levant* (2018), 1–20.

39 S. Shoda, 'Pottery use by early Holocene hunter-gatherers of the Korean peninsula closely linked with exploitation of marine resources', *Quaternary Science Reviews* 170 (2017), 164–73.

40 E. Weiss, M. Kislev and A. Hartmann, 'Autonomous cultivation before domestication', *Science* 312 (2006), 1608–10.

41 F. Borrell, A. Junno and J. Barceló, 'Synchronous Environmental and Cultural Change in the Emergence of Agricultural Economies 10,000 Years Ago in the Levant', *PLOS ONE* 10 (2015), 1–19.

42 Bar-Yosef, 'Facing climatic hazards', 69; P. Roscoe, 'Settlement fortification in village and "tribal" society: evidence from contact-era New Guinea', *Journal of Anthropological Archaeology* 27 (2008), 507–19.

43 A. Palmisano et al., 'Holocene regional population dynamics and climatic trends in the Near East: A first comparison using archaeo-demographic proxies', *Quaternary Science Reviews* 252 (2021), 1–27.

44 J. Bocquet-Appel and S. Naji, 'Testing the hypothesis of a worldwide Neolithic demographic transition: Corroboration from American cemeteries', *Current Anthropology* 47 (2006), 341–64.

45 E. Page et al., 'Reproductive trade-offs in extant hunter-gatherers suggest adaptive mechanism for the Neolithic expansion', *PNAS* 113 (2016), 4694–9.

46 Ibid.; E. Herrera, 'Metabolic adaptations in pregnancy and their implications for the availability of substrates to the fetus', *European Journal of Clinical Nutrition* 54 (2000), 47–51.

47 R. McLaughlin et al., 'Late Glacial and Early Holocene human demographic responses to climatic and environmental change in Atlantic Iberia', *Proceedings of the Royal Society B* 376 (2020), 1–8.

48 J. Brooke, *Climate Change and the Course of Global History: A Rough Journey* (New York, 2014), p. 220.

49 J. Lukacs, 'Sex differences in dental caries rates with the origin of agriculture in South Asia', *Current Anthropology* 37 (1996), 147–53; J. Lukacs and L. Largaespada, 'Explaining sex differences in dental caries rates: Saliva, hormones and "life history" etiologies', *American Journal of Human Biology* 18 (2006), 540–55.

50 Brooke, *Climate Change*, p. 202.

51 A. Masclans et al., 'A sexual division of labour at the start of agriculture? A multi-proxy comparison through grave good stone tool technological and use-wear analysis', *PLOS One* 16 (2021), 1–41.

52 Brooke, *Climate Change*, pp. 192–3.

53 Bolger, 'Dynamics of Gender', 503–31; Masclans, 'Sexual division of labour', 1–41.

54 Page, 'Reproductive trade-offs', 4694.

55 G. Fournié, D. Pfeiffer and R. Bendrey, 'Early animal farming and zoonotic disease dynamics: modelling brucellosis transmission in Neolithic goat populations', *Royal Society Open Science* 4 (2017), 1–11; E. Dounias and A. Froment, 'When forest-based hunter-gatherers become sedentary: consequences for diet and health', *Unasylva* 224 (2006), 26–33.

56 J. Webb, 'Climate, Ecology and Infectious Human Disease', in White, Pfister and Mauelshagen (eds), *Palgrave Handbook of Climate History*, p. 358.

57 W. McNeill, *Plagues and Peoples* (Oxford, 1977).

58 R. McCaa, 'Spanish and Nahuatl Views on Smallpox and Demographic Catastrophe in Mexico', *Journal of Interdisciplinary History* 25 (1995), 397–431. In general, see A. Crosby, *The Columbian Exchange: Biological and Cultural Consequences of 1492* (Westport, CT, 2003).

59 J. Webb, *Humanity's Burden: A Global History of Malaria* (Cambridge, 2009), pp. 18–41; G. Berniell-Lee et al., 'Genetic and demographic implications of the Bantu expansion: insights from human paternal lineages', *Molecular Biology and Evolution* 26 (2009), 1581–9.

60 See below Chapter 16.

61 W. Taylor et al., 'Evidence for early dispersal of domestic sheep into Central Asia', *Nature Human Behaviour* (2021), 1–11.

62 D. Zohary, M. Hopf and E. Weiss, *Domestication of Plants in the Old World: The Origin and Spread of Domesticated Plants in Southwest Asia, Europe, and the Mediterranean Basin* (Oxford, 2012), pp. 4–5.

63 M. Furholt, 'Mobility and Social Change: Understanding the European Neolithic Period after the Archaeogenetic Revolution', *Journal of Archaeological Research* 29 (2021), 1–55.

64 I. Lazaridis, 'The evolutionary history of human populations in Europe', *Current Opinion in Genetics & Development* 53 (2018), 21–7.

65 D. Ju and I. Mathieson, 'The evolution of skin pigmentation-associated variation in West Eurasia', *PNAS* 118 (2021), 1–8.

66 N. Crawford et al., 'Loci associated with skin pigmentation identified in African populations', *Science* 358 (2017), 1–17.

67 P. Richerson and R. Boyd, 'Institutional evolution in the Holocene: the rise of complex societies', in W. Runciman (ed.), *The Origins of Social Institutions* (Oxford), pp. 218–19. Also see J. Diamond, *Guns, Germs, and Steel* (New York, 1997), pp. 137–8; J. Astini et al., 'Predicting Outcrossing in Maize Hybrid Seed Production', *Agronomy Journal* 101 (2009), 373–80.

68 S. di Lernia, 'The Emergence and Spread of Herding in Northern Africa', in Mitchell and Lane (eds), *Oxford Handbook of African Archaeology*, pp. 527–40.

69 D. Fuller, 'Finding Plant Domestication in the Indian Subcontinent', *Current Anthropology* 52 (2011), 347–62; B. Smith, 'The Cultural Context of Plant Domestication in Eastern North America', *Current Anthropology* 52 (2011), 471–84.

70 D. Cohen, 'The Beginnings of Agriculture in China', *Current Anthropology* 52 (2011), 273–93.

71 R. Bettinger, L. Barton and C. Morgan, 'The Origins of Food Production in North China: A Different Kind of Agricultural Revolution', *Evolutionary Anthropology* 19 (2010), 9–21.

72 G. Schmidt and A. LeGrande, 'The Goldilocks abrupt climate change event', *Quaternary Science Reviews* 24 (2005), 1109–10.

73 L. Matero et al., 'The 8.2 ka cooling event caused by Laurentide ice saddle collapse', *Earth and Planetary Science Letters* 473 (2017), 205–14.

74 A. Zerboni and K. Nicoli, 'Enhanced zoogeomorphological processes in arid North Africa on the human-impacted landscape of the Anthropocene', *Geomorphology* 331 (2019), 22–35.

75 Y. Dixit et al., 'Abrupt weakening of the Indian summer monsoon at 8.2 kyr BP', *Earth and Planetary Science Letters* 391 (2014), 16–23.

76 J. Park et al., 'The 8.2 ka cooling event in coastal East Asia: High-resolution pollen evidence from southwestern Korea', *Scientific Reports* 8 (2018), 1–9.

77 di Lernia, 'The Emergence and Spread of Herding in Northern Africa', pp. 534–5.

78 Crawford et al., 'Loci associated with skin pigmentation', 1–17.

79 P. Flohr et al., 'Evidence of resilience to past climate change in Southwest Asia: Early farming communities and the 9.2 and 8.2 ka events', *Quaternary Science Reviews* 136 (2016), 23–39.

80 M. Brami and V. Heyd, 'The origins of Europe's first farmers: the role of Hacılar and Western Anatolia, fifty years on', *Praehistorische Zeitschrift* 86 (2011), 165–206.

81 P. Biehl, 'Rapid change versus long-term social change during the Neolithic–Chalcolithic transition in Central Anatolia', *Interdisciplinaria Archaeologica. Natural Sciences in Archaeology* 3 (2012), 75–83.

82 M. Roffet-Salque et al., 'Evidence for the impact of the 8.2-kyBP climate event on Near Eastern early farmers', *PNAS* 115 (2018), 8705–9. For the adoption of plant fibres for textiles at this site, dating to c.6700 BC, see A. Rast-Eicher, S. Karg and L. Bender Jørgensen, 'The use of local fibres for textiles at Neolithic Çatalhöyük', *Antiquity* 95 (2021), 1129–44.

83 I. Hodder (ed.), *Religion in the Emergence of Civilization: Çatalhöyük as a Case Study* (Cambridge, 2010).

84 J. van der Plicht et al., 'Tell Sabi Abyad, Syria: Radiocarbon chronology, cultural change and the 8.2 ka event', *Radiocarbon* 53 (2011), 229–43.

85 O. Nieuwenhuyse et al., 'The 8.2 Event in Upper Mesopotamia',
    in P. Biehl and O. Nieuwenhuyse, *Climate and Cultural Change in
    Prehistoric Europe and the Near East* (Albany, NY, 2016), pp. 67–93.

86 P. Bryn et al., 'Explaining the Storegga Slide', *Marine and Petroleum
    Geology* 22 (2005), 11–19.

87 V. Gaffney et al., 'Multi-proxy evidence for the impact of the Storegga
    Slide Tsunami on the early Holocene landscapes of the southern North
    Sea', *Geosciences* 10 (2020), 1–19.

88 J. Walker et al., 'A great wave: the Storegga tsunami and the end of
    Doggerland?', *Antiquity* 94 (2020), 1409–25.

89 See for example I. Morris, *Geography is Destiny: Britain and the World, a
    10,000 year history* (London, 2022).

90 K. Rydgren and S. Bondevik, 'Moss growth patterns and timing of
    human exposure to a Mesolithic tsunamic in the North Atlantic',
    Geology 43 (2015), 111–14.

91 R. Sutter, 'The Pre-Columbian Peopling and Population Dispersals
    of South America', *Journal of Archaeological Research* 29 (2021),
    93–151.

92 A. Ortíz, 'Dental morphological variation among six pre-Hispanic
    South American populations', *Dental Anthropology* 26 (2013),
    20–32.

93 S. Perez et al., 'Peopling time, spatial occupation and demography
    of Late Pleistocene–Holocene human population from Patagonia',
    Quaternary International 425 (2016), 214–23.

94 Brooke, *Climate Change*, p. 175.

95 Ibid., p. 180; I. Usokin et al., 'Solar activity during the Holocene: the
    Hallstatt cycle and its consequence for grand minima and maxima',
    Astronomy & Astrophysics 587 (2016), 1–10.

96 Sutter, 'Pre-Columbian Peopling and Population Dispersals of South
    America', 100; M. Carré, 'Holocene history of ENSO variance and
    asymmetry in the eastern tropical Pacific', *Science* 345 (2014), 1045–8.

97 K. Cobb et al., 'Highly variable El Niño–Southern Oscillation
    throughout the Holocene', *Science* 339 (2013), 67–70.

98 Brooke, *Climate Change*, p. 177.

99 P. DeMenocal et al., 'Coherent and low latitude climate variability
    during the Holocene warm period', *Science* 288 (2000), 2198–202.

100 K. Manning and A. Timpson, 'The demographic response to
    Holocene climate change in the Sahara', *Quaternary Science Reviews*
    101 (2014), 28–35.

101 S. Kröpelin, 'The geomorphological and paleoclimatic framework of
    prehistoric occupation in the Wadi Bakht area', in J. Lindstädter (ed.),

Wadi Bakht. Landschaftsarchäologie einer Siedlungskammer im Gilf Kebir, Africa Praehistorica 18 (2005), 51–65.

102 S. di Lernia, 'Dry Climatic Events and Cultural Trajectories: Adjusting Middle Holocene Pastoral Economy of the Libyan Sahara', in F. Hassan (ed.), *Droughts, Food and Culture* (London, 2002), pp. 225–50.

103 S. Höpker et al., 'Pronounced Northwest African Monsoon Discharge during the Mid- to Late Holocene', *Frontiers in Earth Science* 7 (2019), 1–17.

104 E. D'Atanasio et al., 'The peopling of the last Green Sahara revealed by high-coverage resequencing of trans-Saharan patrilineages', *Genome Biology* 19 (2018), 2–15.

105 A. Holl, 'Dark Side Archaeology: Climate Change and Mid-Holocene Saharan Pastoral Adaptation', *African Archaeological Review* 37 (2020), 491–5.

106 B. Shuman et al., 'Predictable hydrological and ecological responses to Holocene North Atlantic variability', *PNAS* 116 (2019), 5985–90.

107 Y. Zhao, A. Yu and C. Zhao, 'Holocene climatic shifts in north eastern North America', *The Holocene* 20 (2010), 877–86.

108 N. Roberts et al., 'The mid-Holocene climatic transition in the Mediterranean: Causes and consequences', *The Holocene* 21 (2011), 3–13.

109 Y. Huo, W. Peltier and D. Chandan, 'Mid-Holocene in South and Southeast Asia: dynamically downscaled simulations and the influence of the Green Sahara', *Climate of the Past* (2021).

110 Z. Feng, C. An and H. Wang, 'Holocene climatic and environmental changes in the arid and semi-arid areas of China: a review', *The Holocene* 16 (2006), 119–30.

111 N. Leonard et al., 'Holocene sea level instability in the southern Great Barrier Reef, Australia: high-precision U-Th dating of fossil microatolls', *Coral Reefs* 35 (2016), 625–39.

4 THE FIRST CITIES AND TRADE NETWORKS
(C.3500–C.2500 BC)

1 L. Stephens et al., 'Archaeological assessment reveals Earth's early transformation through land use', *Science* 365 (2019), 897–902.

2 X. Li et al., 'Increases of population and expansion of rice agriculture in Asia, and anthropogenic methane emissions since 5000 BP', *Quaternary International* 202 (2008), 41–50; A. Timpson et al., 'Reconstructing regional population fluctuations in the European Neolithic using radiocarbon dates: a new case-study using an improved method', *Journal of Archaeological Science* 52 (2014), 549–57.

3 W. Ruddiman, 'Geographic Evidence of the Early Anthropogenic Hypothesis', *Anthropocene* 20 (2017), 4–14.

4 W. Ruddiman, 'The anthropogenic greenhouse era began thousands of years ago', *Climatic Change* 61 (2003), 261–93; P. Verburg et al., 'Upscaling Regional Emissions of Greenhouse Gases from Rice Cultivation: Methods and Sources of Uncertainty', *Plant Ecology* 182 (2006), 89–106.

5 W. Ruddiman et al., 'The early anthropogenic hypothesis: A review', *Quaternary Science Reviews* 240 (2020), 1–14.

6 S. Varvus et al., 'Glacial Inception in Marine Isotope Stage 19: An Orbital Analog for a Natural Holocene Climate', *Scientific Reports* 8 (2018), 1–12.

7 J. Chapman, B. Gaydarska and M. Nebbia, 'The Origins of Trypillia Megasites', *Frontiers in Digital Humanities* 6 (2019), 1–20.

8 A. Shukurov et al., 'Productivity of Premodern Agriculture in the Cucuteni-Trypillia Area', *Human Biology* 87 (2015), 235–82.

9 R. Hofmann et al., 'Governing Tripolye: Integrative architecture in Tripolye settlements', *PLOS One* 14 (2019), 1–54.

10 T. Harper, 'Demography and climate in Late Eneolithic Ukraine, Moldova and Romania: Multiproxy evidence and pollen-based regional corroboration', *Journal of Archaeological Science: Reports* 23 (2019), 973–82.

11 R. Ohlrau et al., 'Living on the Edge? Carrying Capacities of Trypillian Settlements in the Buh–Dnipro Interfluve', in J. Müller, K. Rassamann and M. Videlko (eds), *Trypillia Mega-Sites and European Prehistory, 4100–3400 BCE* (London, 2016), pp. 207–20.

12 J. Chapman and B. Gaydarska, 'Low-density urbanism: the case of the Trypillia group of Ukraine', in M. Fernández-Götz and D. Krausse (eds), *Eurasia at the Dawn of History: Urbanisation and Social Change* (Cambridge, 2016), pp. 81–105.

13 N. Rascovan et al., 'Emergence and Spread of Basal Lineages of *Yersinia pestis* during the Neolithic Decline', *Cell* 176 (2019), 295–305.

14 S. Kadrow, 'The concept of the "stage of reduction and concentration of settlements" in Neolithic studies: demography, settlements and social conflict', *Documenta Praehistorica* 47 (2020), 232–44.

15 R. Wuthnow, *Meaning and Moral Order: Explorations in Cultural Analysis* (Berkeley, 1987); S. Kadrow, 'The concept of the "stage of reduction and concentration of settlements" in Neolithic studies: demography, settlements and social conflict', *Documenta Praehistorica* 47 (2020), 232–44.

16 P. Bellwood, 'Holocene Population History in the Pacific Region as a
    Model for Worldwide Food Producer Dispersals', *Current Anthropology*
    52 (2011), 363–78.

17 M. Sanger and Q.-M. Ogden, 'Determining the use of Late Archaic
    shell rings using lithic data: "ceremonial villages" and the importance of
    stone', *Southeastern Archaeology* 37 (2018), 232–52.

18 M. Sanger, 'Coils, slabs, and molds: examining community affiliation
    between Late Archaic shell ring communities using radiographic
    imagery of pottery', *Southeastern Archaeology* 36 (2017), 95–109.

19 D. Davis et al., 'Deep learning reveals extent of Archaic Native
    American shell-ring building practices', *Journal of Archaeological Science*
    132 (2021), 1–12.

20 M. Robbeets et al., 'Triangulation supports agricultural spread of the
    Transeurasian languages', *Nature* (2021), 1–20.

21 W. Haak et al., 'Massive migration from the steppe was a source for
    Indo-European languages in Europe', *Nature* 522 (2015), pp. 207–11.

22 F. Clemente, 'The genomic history of the Aegean palatial civilizations',
    Cell 184 (2021), 2565–86.

23 A. Juras et al., 'Mitochondrial genomes reveal an east to west cline of
    steppe ancestry in Corded Ware populations', *Scientific Reports*
    8 (2018), 1–10.

24 I. Olalde et al., 'The Beaker phenomenon and the genomic
    transformation of northwest Europe', *Nature* 555 (2018), 190–6.

25 L. Papac et al., 'Dynamic changes in genomic and social structures
    in third millennium BCE central Europe', *Science Advances* 7
    (2021), 1–18.

26 M. Pearson, 'Beaker people in Britain: migration, mobility and diet',
    Antiquity 90 (2016), 620–37.

27 N. Yoffee, *Myths of the Archaic State: Evolution of the Earliest Cities, States
    and Civilisations* (Cambridge, 2004), pp. 62, 214.

28 See here R. Carneiro, 'A theory of the origin of the state', *Science* 169
    (1970), 733–8; Brooke, *Climate Change*, p. 189.

29 P. Akkermans and G. Schwartz, *The Archaeology of Syria: From
    Complex Hunter-Gatherers to Early Urban Societies (ca. 16,000–300
    BC)* (Cambridge, 2003), pp. 154ff.; D. Wengrow, 'The evolution of
    simplicity: aesthetic labour and social change in the Neolithic Near East
    World', *Archaeology* 33 (2001), 168–88; J. Clarke et al., 'Climatic changes
    and social transformations in the Near East and North Africa during the

"long" 4th millennium BC; A comparative study of environmental and archaeological evidence', *Quaternary Science Reviews* 136 (2016), 104.

30 A. Sherratt, 'Cash Crops before Cash: Organic Consumables and Trade', in C. Godsen and J. Hather (eds), *The Prehistory of Food: Appetites for Change* (London, 1999), pp. 13–34.

31 A. Stevenson, 'The Egyptian Predynastic and State Formation', *Journal of Archaeological Research* 24 (2016), 421–68; V. Tripathi, 'Metals and Metallurgy in the Harappan Civilization', *Indian Journal of History of Science* 53 (2018), 279–95; L. Liu and X. Chen, *The Archaeology of China: From the Late Paleolithic to the Early Bronze Age* (Cambridge, 2012), pp. 169–252.

32 T. Kohler, 'Greater post-Neolithic wealth disparities in Eurasia than in North America and Mesoamerica', *Nature* 551 (2017), 619–22.

33 M. Borgerhoff Mulder et al., 'Pastoralism and Wealth Inequality', *Current Anthropology* 51 (2010), 35–48.

34 E. Smith et al., 'Wealth Transmission and Inequality among Hunter-Gatherers', *Current Anthropology* 51 (2010), 19–34.

35 For the terminology, see M. Frangipane, 'Different Trajectories in State Formation in Greater Mesopotamia: A View from Arslantepe (Turkey)', *Journal of Archaeological Research* 26 (2018), 3–63.

36 J. Ur, 'Cycles of Civilization in Northern Mesopotamia', *Journal of Archaeological Research* 18 (2010), 390–3.

37 J. Ur, P. Karsgaard and J. Oates, 'Urban development in the ancient Near East', *Science* 317 (2007), 1188.

38 M. Liverani, *Uruk: The First City* (London, 2006).

39 D. Katz, 'Ups and Downs in the Career of Enmerkar, King of Uruk', in O. Drewnowska and M. Sandowicz (eds), *Fortune and Misfortune in the Ancient Near East* (Winona Lake, IN, 2017), pp. 201–10.

40 M. Civil, 'Remarks on D-GI₄ (AKA "Archaic World List C" or "Tribute")', *Journal of Cuneiform Studies* 65 (2013), 20.

41 M. Hudson, 'The dynamics of privatization, from the Bronze Age to the present', in M. Hudson and B. Levine (eds), *Privatization in the Ancient Near East and Classical World* (Cambridge, MA, 1996), p. 37.

42 G. Algaze, *Ancient Mesopotamia at the Dawn of Civilisation* (Chicago, 2008), p. 190.

43 C. Lamberg-Karlovsky, *Beyond the Tigris and Euphrates: Bronze Age Civilisations* (Beer-Sheva, 1996), pp. 82ff.

44 J. Manning, 'Egypt', in P. Bang and W. Scheidel (eds), *The Oxford Handbook of the State in the Ancient Near East and Mediterranean* (Oxford, 2013), pp. 61–93.

45 B. Pennington, 'Landscape change in the Nile Delta during the fourth millennium BC: A new perspective on the Egyptian Predynastic and Protodynastic periods', *World Archaeology* 52 (2021), 550–65.

46 C. Renfrew and L. Bin, 'The emergence of complex society in China: the case of Liangzhu', *Antiquity* 92 (2018), 975–90.

47 E. Childs-Johnson, 'The Art of Working Jade and the Rise of Civilization in China', in E. Childs-Johnson and F. Gu, *The Jade Age and Early Chinese Jade in American Museums* (Beijing, 2009), pp. 291–393.

48 P. Michalowski, 'The Domestication of Stranger Kings: Making History by List in Ancient Mesopotamia', in J. Bains, H. van der Blom, T. Rood and W. Chen (eds), *Historical Consciousness and the Use of the Past in the Ancient World* (Sheffield, 2019), pp. 15–38; Y. Pines, 'Chu identity as seen from its manuscripts: a re-evaluation', *Journal of Chinese History* 2 (2017), 1–26.

49 See for example D. Johnson, 'God's punishment and public goods: a test of the supernatural punishment hypothesis in 186 world cultures', *Human Nature* 16 (2005), 410–46; also see D. Johnson, 'The wrath of academics: criticisms, applications, and extensions of the supernatural punishment hypothesis', *Religion, Brain Behaviour* 8 (2018), 320–50.

50 H. Whitehouse et al., 'Complex societies precede moralizing gods throughout world history', *Nature* 568 (2019), 226–9.

51 D. Keightley, 'What did make the Chinese "Chinese"?', in D. Keightley, *These Bones Shall Rise Again: Selected Writings on Early* China (Albany, NY, 2014), pp. 78–9.

52 J. Watts et al., 'Ritual human sacrifice promoted and sustained the evolution of stratified societies', *Nature* 532 (2016), 228–31.

53 D. Katz, 'Sumerian funerary rituals in context', in N. Laneri, *Performing Death: Social Analyses of Funerary Traditions in the Ancient Near East and Mediterranean* (Chicago, 2007), pp. 169–70.

54 S. Pollock, 'Feast, funerals and fast food in early Mesopotamian states', in T. Bray (ed.), *The Archaeology and Politics of Food and Feasting in Early States and Empires* (New York, 2003), pp. 17–38.

55 A. Baadsgaard et al., 'Human sacrifice and intentional corpse preservation in the Royal Cemetery of Ur', *Antiquity* 85 (2011), 27–42.

56 J. Ur, 'Households and the Emergence of Cities in Ancient Mesopotamia', *Cambridge Archaeological Journal* 26 (2014), 249–68; H. Nissen, *The Early History of the Ancient Near East* (Chicago, 1988), pp. 83–5.

57 R. Redding, 'A Tale of Two Sites: Old Kingdom Subsistence Economy and the Infrastructure of Pyramid Construction', in B. De Cupere, V. Linseele and S. Hamilton-Dyer (eds), *Archaeology of the Near East X: Proceedings of*

the 10th Meeting of the ICAZ Working Group 'Archaeozoology of Southwest Asia and Adjacent Areas' (Leuven, 2014), pp. 307–22.

58 I. Diakonoff, 'The Rise of the Despotic State in Ancient Mesopotamia', in I. Diakonoff (ed.), *Ancient Mesopotamia: Socio-economic History* (Moscow, 1969), pp. 173–203.

59 R. Ellison, 'Diet in Mesopotamia: The Evidence of the Barley Ration Texts (c.3000–1400 BC), *Iraq* 43 (1981), 35–45.

60 B. Foster, 'The Sargonic Period: Two Historiographical Problems', in G. Barjamovic et al., *Akkade is King: A Collection of Papers by Friends and Colleagues Presented to Aage Westenholz* (Leiden, 2011), pp. 127–38.

61 S. Pollock, *Ancient Mesopotamia: The Eden That Never Was* (Cambridge, 1999), pp. 123–48.

62 Ibid., pp. 72–3.

63 O. Bar-Yosef, 'The Walls of Jericho: An Alternative Interpretation', *Current Anthropology* 27 (1986), 157–62.

64 P. Butterlin and S. Rey, 'Mari and the Development of Complex Defensive Systems in Mesopotamia at the Dawn of History', in R. Frederiksen, S. Müth, P. Schneider and M. Schnelle (eds), *Focus on Fortifications: New Research on Fortifications in the Ancient Mediterranean and the Near East* (Oxford, 2016), pp. 23–33.

65 A. Gnirs, 'Ancient Egypt', in K. Raaflaub and N. Rosenstein (eds), *War and Society in the Ancient and Medieval Worlds: Asia, the Mediterranean, Europe and Mesoamerica* (Washington, DC, 1999), pp. 71–104.

66 J. Scott, *Against the Grain: A Deep History of the Earliest States* (New Haven, 2017), pp. 152–3; also here see N. Clark, '(Un)Earthing Civilization: Holocene Climate Crisis, City-State Origins and the Birth of Writing', *Humanities* 9 (2020), 1–16.

67 Algaze, 'Demographic Trends in Early Mesopotamian Urbanism', in C. Maner, M. Horowitz and A. Gilbert (eds), *Overturning Certainties in Near Eastern Archaeology* (Leiden, 2017), pp. 25–33.

68 Ibid., p. 29.

69 Algaze, *Ancient Mesopotamia*, p. 81.

70 P. Beaujard, *The Worlds of the Indian Ocean: A Global History* (Cambridge, 2019), p. 56.

71 J.-J. Glassner, *La Mésopotamie* (Paris, 2002), pp. 13–19, 220–4.

72 D. Schmandt-Besserat, 'The Token System of the Ancient Near East: Its Role in Counting, Writing, the Economy and Cognition', in I. Morley and C. Renfrew (eds), *The Archaeology of Measurement Comprehending Heaven, Earth and Time in Ancient Societies* (Cambridge, 2010), pp. 27–34.

73 D. Schmandt-Besserat, 'The Evolution of Writing', in J. Wright (ed.), *International Encyclopedia of Social and Behavioural Sciences* (Amsterdam, 2014).

74 See however B. Ansumali Mukhopadhyay, 'Interrogating Indus inscriptions to unravel their mechanisms of meaning conveyance', *Humanities and Social Sciences Communications* 5 (2019), 1–37.

75 M. Oelschlaeger, *The Idea of Wilderness: From Prehistory to the Age of Ecology* (New Haven, 1991), pp. 36–8.

76 See for example A. Gardiner, *Ancient Egyptian Onomastica* (London, 1947); R. Hallock and B. Landsberger, *Materialien zum Sumerischen Lexikon* (Rome, 1956).

77 Keightley, 'What did make the Chinese "Chinese"?', p. 80.

78 J. Clayton, A. De Trafford and M. Borda, 'A Hieroglyphic Inscription Found at Jebel Uweinat Mentioning Yam and Tekhebet', *Sahara* 19 (2008), 129–35; Bruce Beyer-Williams, 'Kush in the Wider World during the Kerma Period', in G. Emberling and B. Beyer-Williams (eds), *The Oxford Handbook of Ancient Nubia* (Oxford, 2021), pp. 179–200.

79 G. Possehl, 'Sociocultural complexity without the state: the Indus civilization', in G. Feinman and J. Marcus (eds), *Archaic States* (Santa Fe, NM, 1998), pp. 261–91.

80 A. Green, 'Killing the Priest-King: Addressing Egalitarianism in the Indus Civilization', *Journal of Archaeological Research* (2020), 1–50.

81 T. Pozorski and S. Pozorski, 'Early complex society on the north and central Peruvian coast: new archaeological discoveries and new insights', *Journal of Archaeological Research* 26 (2018), 353–86.

82 M. Moseley, 'Maritime foundations and multilinear evolution: Retrospect and prospect', *Andean Past* 3 (1992), 5–42; J. Haas and W. Creamer, 'Crucible of Andean Civilization: The Peruvian Coast from 3000 to 1800 BC', *Current Anthropology* 47 (2006), 745–75.

83 R. Shady, 'Los orígenes de la civilización y la formación del estado en el Perú: Las evidencias arqueológicas de Caral-Supe', in R. Shady and C. Leyva (eds), *La ciudad sagrado de Caral-Supe: Los orígenes de la civilización andina y la formación del estado prístino en el antiguo Perú* (Lima, 2003), pp. 93–100.

84 Pozorski and Pozorski, 'Early complex society', 375–6.

85 J. Haas et al., 'Evidence for maize (Zea mays) in the Late Archaic (3000–1800 BC) in the Norte Chico region of Peru', *PNAS* 110 (2013), 4945–9.

86 Brooke, *Climate Change*, pp. 183–4; J. Richardson and D. Sandweiss, 'Climate Change, El Niño and the Rise of Complex Society on the

Peruvian Coast during the Middle Holocene', in D. Sandweiss and J. Quilter (eds), *El Niño, Catastrophism and Culture Changes in Ancient America* (Washington, DC, 2008), pp. 59–75.

87 J. Haas and W. Creamer, 'Why do people build monuments? Late Archaic platform mounds in the Norte Chico', in R. Burger and R. Rosenweig (eds), *Early New World Monumentality* (Gainesville, FL, 2012), pp. 289–312; R. Burger and L. Salazar, 'Monumental public complexes and agricultural expansion on Peru's Central Coasts during the second millennium bc', in Burger and Rosenweig (eds), *Early New World Monumentality*, pp. 399–430.

88 Haas and Creamer, 'Crucible of Andean Civilization', 751.

89 R. Burger and K. Makowski, *Arqueología del Período Formativo en la Cuenca Baja de Lurín* (Lima, 2009).

90 B. Gaydarska, M. Nebbia and J. Chapman, 'Trypillia Megasites in Context: Independent Urban Development in Chalcolithic Eastern Europe', *Cambridge Archaeological Journal* 30 (2020), 97–121.

91 J. Cooper, *Reconstructing History from Ancient Inscriptions: The Lagash–Umma Border Conflict* (Malibu, CA, 1983).

92 E. Stone, 'Mesopotamian cities and countryside', in D. Snell (ed.), *A Companion to the Ancient Near East* (Malden, MA, 2005), pp. 141–54.

93 S. Dalley, 'The Natural World in Ancient Mesopotamian Literature', in Parham and Westling (eds), *A Global History of Literature and the Environment*, pp. 25–6.

94 See for example A. Frank and W. Thompson, 'Afro-Eurasian Bronze Age Economic Expansion and Contraction Revisited', *Journal of World History* 16 (2005), 115–72.

95 A. Podany, *Brotherhood of Kings: How International Relations Shaped the Ancient Near East* (Oxford, 2010).

96 M. Van De Mieroop, 'In Search of Prestige: Foreign contacts and the rise of an elite in early Dynastic Babylonia', in E. Ehrenberg (ed.), *Leaving No Stones Unturned: Essays on the Ancient Near East and Egypt in Honour of Donald P. Hansen* (Winona Lake, IN, 2002), pp. 125–38; L. Liu, ' "The products of minds as well as of hands": production of prestige goods in the Neolithic and Early States Periods of China', *Asian Perspectives* 42 (2003), 1–40.

97 W. Thompson, 'Climate, Water and Political-Economic Crises in Ancient Mesopotamia and Egypt', in A. Hornborg and C. Crumley (eds), *The World System and the Earth System: Global Socioenvironmental Change and Sustainability since the Neolithic* (Walnut Creek, CA, 2006), pp. 165–6.

98 G. Algaze, 'Initial Social Complexity in Southwestern Asia: The
    Mesopotamian Advantage', *Current Anthropology* 42 (2001), 199–233.

99 Lander, The King's Harvest, pp. 25–8.

100 *Enki and the World Order*, in J. Black et al. (eds), *The Literature of
    Ancient Sumer* (Oxford, 2004), pp. 220–1.

101 *Myths from Mesopotamia: Creation, the Flood, Gilgamesh, and Others*, tr.
    S. Dalley (Oxford, 2008), p. 257.

102 Ibid., pp. 269–70.

103 *The Rig Veda*, tr. W. Doniger (London, 2005), 1:32, pp. 150–6.

104 Genesis, 1:1–2, 1:26.

105 Genesis, 1:28–30, 3:17–19.

5 ON THE RISKS OF LIVING BEYOND ONE'S MEANS
(C.2500–C.2200 BC)

1   M. Van De Mieroop, *The Ancient Mesopotamian City* (Oxford, 1997),
    pp. 68–78.

2   D. Frayne, *The Royal Inscriptions of Mesopotamia: Sargonic and Gutian
    Periods*, 2 vols (Toronto, 1993), 2, p. 23.

3   A. Grayson, *Assyrian and Babylonian Chronicles* (Locust Valley, NY,
    1966), pp. 1–3.

4   J.-J. Glassner, *Mesopotamian Chronicles* (Atlanta, 2004), p. x.

5   A. Westenholz, 'The Old Akkadian Period: History and Culture', in
    P. Attinger and M. Wäfler (eds), *Mesopotamien: Akkade-Zeit und Ur
    III-Zeit* (Göttingen, 1999), 17.

6   J. Westenholz, *Legends of the Kings of Akkade: The Texts* (Winona Lake,
    IN, 1997), pp. 221–62, 300–31.

7   Black et al. (eds), *Literature of Ancient Sumer*, 'The Cursing of Agade',
    p. 122.

8   H. Arz, J. Kaiser and D. Fleitmann, 'Paleoceanographic and
    paleoclimatic changes around 2200 BC in sediment cores from the
    northern Red Sea', in H. Meller et al. (eds), *2200 BC: A Climatic
    Breakdown as a Cause for the Collapse of the Old World?* (Halle,
    2015), 53–61.

9   T. Watanabe, 'Oman corals suggest that a stronger winter shamal season
    caused the Akkadian Empire (Mesopotamia) collapse', *Geology* 47
    (2019), 1141–5.

10 S. Carolin et al., 'Precise timing of abrupt increase in dust activity
    in the Middle East coincident with 4.2ka social change', *PNAS* 116
    (2019), 67–72.

11 H. Weiss et al., 'The Genesis and Collapse of Third Millennium North Mesopotamian Civilization', *Science* 261 (1993), 995–1004.

12 H. Weiss, '4.2 ka BP Megadrought and the Akkadian Collapse', in H. Weiss (ed.), *Megadrought and Collapse: From Early Agriculture to Angkor* (Oxford, 2017), pp. 93–160.

13 Brooke, *Climate Change*, pp. 292–3. For the Gutians, 'Cursing of Agade', in Black et al. (eds), *Literature of Ancient Sumer*, p. 119.

14 See C. Kuzucuoğlu and C. Marro, *Sociétés humaines et changement climatique à la fin du troisième millénaire: une crise a-t-elle eu lieu en Haute Mésopotamie?: Actes du Colloque de Lyon, 5–8 Décembre 2005* (Lyon, 2007).

15 M. Walker et al., 'Formal subdivision of the Holocene Series/Epoch: A Discussion Paper by a Working Group of INTIMATE (Integration of ice-core marine and terrestrial records) and the Subcommission on Quaternary Stratigraphy (International Commission on Stratigraphy', *Journal of Quaternary Science* 27 (2012), 649–59.

16 The International Commission on Stratigraphy, 'Collapse of civilizations worldwide defines youngest unit of the Geologic Time Scale', www.stratigraphy.org. Also see M. Walker et al., 'Formal Subdivision of the Holocene Series/Epoch: A Summary', *Journal of the Geological Society of India* 93 (2019), 135–41.

17 For a good overview, see M. Ran and L. Chen, 'The 4.2 ka BP climatic event and its cultural responses', *Quaternary International* 521 (2019), 158–67.

18 R. Bradley and J. Bakke, 'Is there evidence for a 4.2 ka BP event in the northern Atlantic region?', *Climate of the Past* 15 (2019), 1665–76.

19 M. Griffiths et al., 'End of Green Sahara amplified mid- to late Holocene megadroughts in mainland Southeast Asia', *Nature Communications* 11 (2020), 1–12.

20 See here the collection of papers in D. Rousseau et al., 'Special issue – the 4.2ka climatic event', *Climate of the Past* 15 (2019).

21 G. Middleton, 'Bang or whimper', *Science* 361 (2018), 1204–5.

22 V. Villalba-Mouco, 'Genomic transformation and social organization during the Copper Age–Bronze Age transition in southern Iberia', *Science Advances* 7 (2021), 1–18.

23 Y. Dixit, D. Hodell and C. Petrie, 'Abrupt weakening of the summer monsoon in northwest India', *Geology* 42 (2014), 339–42; S. Prasad and Y. Enzel, 'Holocene paleoclimates of India', *Quaternary Research* 66 (2006), 442–53.

24 A. Singh et al., 'Counter-intuitive influence of Himalayan river morphodynamics on Indus Civilisation urban settlements', *Nature Communications* 8 (2017), 1–14.

25 L. Giosan et al., 'Fluvial landscapes of the Harappan civilization', *PNAS* 109 (2012), 1688–94.

26 L. Giosan et al., 'Neoglacial climate anomalies and the Harappan metamorphosis', *Climate of the Past* 14 (2018), 1669–86.

27 G. Robbins Schug et al., 'Infection, Disease and Biosocial Processes at the End of the Indus Civilization', *PLOS One* 8 (2013), 1–20.

28 N. Scroxton et al., 'Circum-Indian ocean hydroclimate at the mid- to late Holocene transition: The Double Drought hypothesis and consequences for the Harappan', *Climate of the Past* (2020).

29 T. Sengupta et al., 'Did the Harappan settlement of Dholavira (India) collapse during the onset of the Meghalayan stage drought?', *Journal of Quaternary Science* 35 (2020), 382–95.

30 Robbins Schug et al., 'Infection, Disease and Biosocial Processes', 9–11.

31 H. Zhang et al., 'Hydroclimatic variations in southeastern China during the 4.2 ka event reflected by stalagmite records', *Climate of the Past* 14 (2018), 1805–17.

32 H. Zhang, 'Collapse of the Liangzhu and other Neolithic cultures in the lower Yangtze region in response to climate change', *Science Advances* 7 (2021), 1–9.

33 Q. Qu et al., 'Outburst flood at 1920 BCE supports historicity of China's Great Flood and the Xia dynasty', *Science* 353 (2016), 579–82.

34 L. Tan et al., 'Great flood in the middle-lower Yellow River reaches at 4000 a BP inferred from accurately-dated stalagmite records', *Science Bulletin* 63 (2018), 206–8; Q. Sun et al., 'Climate as a factor for Neolithic cultural collapses approximately 4000 years BP in China', *Earth Science Reviews* 197 (2019), 1–21.

35 H. Xu, 'Introduction to the Xia period: Definitions, Themes and Debate', in E. Childs-Johnson, *The Oxford Handbook of Early China* (Oxford, 2020), pp. 161–75.

36 Ran and Chen, 'Climatic event', 164–5.

37 Carolin et al., 'Precise timing of abrupt increase in dust activity', 67–72.

38 See for example C. Renfrew, 'Systems Collapse as Social Transformation: Catastrophe and Anastrophe in Early State Societies', in C. Renfrew and K. Cooke (eds), *Transformations: Mathematical Approaches to Culture Change* (New York, 1979), pp. 481–506.

39 A. Otto, 'Archaeological perspectives of the localization of Naram-Sin's Armanum', *Journal of Cuneiform Studies* 58 (2006), 1–25.

40 P. Michalowski, 'The Kingdom of Akkad in Contact with the World', in K. Radner, N. Moeller and D. Potts (eds), *The Oxford History of the Ancient Near East*, vol. 1: *From the Beginnings of the Old Kingdom Egypt and the Dynasty of Akkad* (Oxford, 2020), p. 725.

41 G. Barjamovic, 'Mesopotamian Empires', in Bang and Scheidel (eds), *Oxford Handbook of the State in the Ancient Near East*, pp. 129–32.

42 G. Leick, *Mesopotamia: The Invention of the City* (London, 2003), pp. 106–7.

43 I. Schrakamp, 'The Kingdom of Akkad: A View from Within', in Radner, Moeller and Potts (eds), *Oxford History of the Ancient Near East*, pp. 627–32.

44 L. Oppenheim, *Ancient Near Eastern Texts Relating to the Old Testament* (Princeton, 1969), p. 266.

45 S. Dalley, 'The Epic of Atrahasis', in *Myths from Mesopotamia: Creation, the Flood, Gilgamesh, and Others*, tr. S. Dalley (London, 2009), pp. 7–31.

46 Genesis, 6–8.

47 Genesis, 18.6–19.27.

48 T. Buch et al., 'A Tunguska sized airburst destroyed Tall el-Hammam a Middle Bronze Age city in the Jordan Valley near the Dead Sea', *Scientific Reports* 11 (2021), 1–63.

49 K. El-Rouayheb, *Before Homosexuality in the Arab-Islamic World, 1500–1800* (Chicago, 2005), pp. 124–5.

50 'Cursing of Agade', in Black et al. (eds), *Literature of Ancient Sumer*, p. 119.

51 J. Assmann, *The Mind of Egypt: History and Meaning in the Time of the Pharaohs* (Cambridge, MA, 2002), pp. 109–13.

52 R. Faulkner, 'The Admonitions of an Egyptian Sage', *Journal of Egyptian Archaeology* 51 (1965), 56.

53 F. Welc and L. Marks, 'Climate change and the end of the Old Kingdom in Egypt around 4200 BP: New geoarchaeological evidence', *Quaternary international* 324 (2014), 124–33.

54 Ibid.; K. Butzer, 'Collapse, environment, and society', *PNAS* 109 (2012), 3632–9.

55 Faulkner, 'Admonitions', 53–62.

56 N. Strudwick, 'The Old Kingdom and First Intermediate Period', in I. Shaw and E. Bloxam, *Oxford Handbook of Egyptology* (Oxford, 2020), pp. 619–37.

57 X. Zhou et al., '5,200 year old cereal grains from the eastern Altai mountains redate the trans-Eurasian crop exchange', *Nature Plants* 6 (2020), 78–87; N. Boivin and D. Fuller, 'Shell middens, ships and

seeds: exploring coastal subsistence, maritime trade and the dispersal of domesticates in and around the ancient Arabian Peninsula', *Journal of World Prehistory* 22 (2009), 113–80; R. Gutaker et al., 'Genomic history and ecology of the geographic spread of rice', *Nature Plants* 6 (2020), 492–502.

58 D. Fuller and N. Boivin, 'Crops, cattle and commensals across the Indian Ocean: current and potential archaeobiological evidence', *Études Océan Indien* 42 (2009), 13–46; A. Scott et al., 'Exotic foods reveal contact between South Asia and the Near East during the second millennium BCE', *PNAS* 118 (2021), 1–10.

59 I. Pugach et al., 'Genome-wide data substantiates Holocene gene flow from India to Australia', *PNAS* 110 (2013), 1803–8.

60 H. Vandkilde, 'Bronzization: The Bronze Age as Pre-Modern Globalization', *Praehistorische Zeitschrift* 9 (2016), 103–23.

61 R. Wright, *The Ancient Indus: Urbanism, Economy, and Society* (Cambridge, 2010), pp. 221ff.

62 A. Sen, *Poverty and Famines: An Essay on Entitlement and Deprivation* (Oxford, 1981).

6 THE FIRST AGE OF CONNECTIVITY (C.2200–C.800 BC)

1 C. Woods, 'Sons of the Sun: The Mythological Foundations of the First Dynasty of Uruk', *Journal of Ancient Near Eastern Religions* 12 (2012), 78–96.

2 *Law Collections from Mesopotamia and Asia Minor*, ed. M. Roth (Atlanta, GA, 1997), p. 20.

3 P. Barmash, *The Laws of Hammurabi: At the Confluence of Royal and Scribal Traditions* (Oxford, 2020), p. 138.

4 T. Bryce, 'Anatolian States', in Bang and Scheidel (eds), *Oxford Handbook of the State in the Ancient Near East*, pp. 161–79.

5 C. Broodbank, *An Island Archaeology of the Early Cyclades* (Cambridge, 2000), pp. 341–9.

6 J. Bennet, 'Bronze Age Greece', in Bang and Scheidel (eds), *Oxford Handbook of the State in the Ancient Near East*, pp. 235–58.

7 Ibid.; C. Renfrew, *The Emergence of Civilisation: The Cyclades and the Aegean in the Third Millennium* BC (London, 1972), pp. 225ff.; E. Tsafou and J. García-Granero, 'Beyond staple crops: exploring the use of "invisible" plant ingredients in Minoan cuisine through starch grain analysis on ceramic vessels', *Archaeological and Anthropological Sciences* 13 (2021), 1–16.

8 Beyer-Williams, 'Kush in the Wider World', pp. 183–5.

9   A. Grayson, *Assyrian Royal Inscriptions* (Wiesbaden, 1972), p. 8.

10 A. Holl, 'The dawn of African pastoralisms: An introductory note',
    Journal of Anthropological Archaeology 17 (1998), 81–96.

11 E. Coulibaly, *Savoir et savoir-faire des anciens métallurgistes d'Afrique*
    (Paris, 2006).

12 M. Brass, 'The Emergence of Mobile Pastoral Elites during the Middle
    to Late Holocene in the Sahara', *Journal of African Archaeology* 17
    (2019), 53–75.

13 V. Shinde, 'An Ancient Harappan Genome Lacks Ancestry from Steppe
    Pastoralists or Iranian Farmers', *Cell* 179 (2019), 729–35.

14 M. Silva et al., 'A genetic chronology for the Indian Subcontinent
    points to heavily sex-biased dispersals', *BMC Ecology and Evolution* 17
    (2017), 1–18.

15 See for example Haak et al., 'Massive migration from the steppe was a
    source for Indo-European languages in Europe', 207–11; W. Chang et al.,
    'Ancestry-constrained phylogenetic analysis supports the Indo-European
    steppe hypothesis', *Language* 91 (2015), 194–244.

16 V. Naraimhan et al., 'The formation of human populations in South and
    Central Asia', *Science* 365 (2019), 1–16.

17 P. de Barros Damgaard, 'The first horse herders and the impact of
    early Bronze Age steppe expansions into Asia', *Science* 360 (2018), 1–9.
    Above all see C. Renfrew, *Archaeology and Language: The Puzzle of Indo-
    European Origins* (Cambridge, 1987).

18 For genetic markers of multiple waves of migration, see G. Poznik
    et al., 'Punctuated bursts in human male demography inferred
    from 1,244 worldwide Y-chromosome sequences', *Nature Genetics* 58
    (2016), 593–9.

19 A. Basu et al., 'Ethnic India: a genomic view, with special reference to
    peopling and structure', *Genome Research* 13 (2003), 2277–90.

20 See A. Parpola, *The Roots of Hinduism: The Early Aryans and the Indus
    Civilization* (New York, 2015).

21 K. Plofker, 'Humans, Demons, Gods and their Worlds: The Sacred and
    Scientific Cosmologies of India', in K. Raaflaub and R. Talbert (eds),
    *Geography and Ethnography: Perceptions of the World in Pre-Modern
    Studies* (Chichester 2010), p. 33; *Atharva Veda*, 12.1.

22 F. Baird and R. Helmbeck, *Philosophic Classics: Asian Philosophy*
    (London, 2005), p. 3.

23 J. Bourriau, 'The Second Intermediate Period (c.1650–1550 BC)', in
    I. Shaw (ed.), *The Oxford History of Ancient Egypt* (Oxford, 2000),
    p. 188ff.

24 C. Stantis et al., 'Who were the Hyksos? Challenging traditional narratives using strontium isotope (^{87}Sr/^{86}Sr) analysis of human remains from ancient Egypt', *Plos ONE* 15 (2020), 1–14; A.-L. Mourad, *Rise of the Hyskos: Egypt and the Levant from the Middle Kingdom to the Early Second Intermediate Period* (Oxford, 2015).

25 M. Bietak, 'From where came the Hyskos and where did they go?', in M. Marée (ed.), *The Second Intermediate Period (Thirteenth–Seventeenth Dynasties)* (Leuven, 2010), pp. 139–81.

26 T. Schneider, *Ausländer in Ägypten während des Mittleren Reiches und der Hyskoszeit* (Wiesbaden, 1998).

27 O. Goldwasser, 'How the Alphabet was Born from Hieroglyphs', *Biblical Archaeology Review* 36 (2010), 40–53; F. Höflmayer, 'Early alphabetic writing in the ancient Near East: the "missing link" from Tel Lachish', *Antiquity* (2021), 1–15.

28 I. Shaw, 'Egyptians, Hyskos and military technology: Causes, effects or catalysts?', in A. Shortland (ed.), *The Social Context of Technological Change: Egypt and the Near East, 1650–1550 BC* (Oxford, 2001), pp. 59–71.

29 G. Beckman and H. Hoffner, *Hittite Diplomatic Texts* (Atlanta, GA, 1996), pp. 42–3; P. Thieme, 'The "Aryan" Gods of the Mitanni Texts', *Journal of the American Oriental Society* 80 (1960), 301–17.

30 D. Anthony, *The Horse, the Wheel and Language: How Bronze-Age Riders from the Eurasian Steppes Shaped the Modern World* (Princeton, 2007), pp. 48–50.

31 de Barros Damgaard, 'First horse herders', 1–9.

32 W. Kirfel, *Das Purāṇa vom Weltgebäude (Bhuvanavinyāsa)* (Bonn, 1954); D. Pingree, 'The Mesopotamian Origin of Early Indian Mathematical Astronomy', *Journal for the History of Astronomy* 4 (1973), 1–12.

33 R. Burger, *Chavín and the Origins of Andean Civilization* (London, 1992); R. Burger and L. Salazar, 'The Manchay culture and the coastal inspiration for highland Chavín civilization', in W. Conklin and J. Quilter, *Chavín Art, Architecture and Culture* (Los Angeles, CA, 2008), pp. 85–105.

34 P. Heggarty and D. Beresford-Jones, 'Agriculture and language dispersals: limitations, refinements, and an Andean exception', *Current Anthropology* 51 (2010), 163–91.

35 S. Pozorski and T. Pozorski, 'Early Cultural Complexity on the Coast of Peru', in H. Silverman and W. Isbell (eds), *The Handbook of South American Archaeology* (New York, 2008), pp. 607–31.

36 J. Li et al., 'When peripheries were centres: a preliminary study of the
    Shimao-centred polity in the loess highland, China', *Antiquity* 92 (2018),
    1008–22.

37 R. Bagley, 'Shang Archaeology', in M. Loewe and E. Shaughnessy (eds),
    *The Cambridge History of Ancient China from the Origins of Civilization
    to 221 BC* (Cambridge, 1999), pp. 124–231.

38 R. Thorp, *Early Bronze Age China: Shang Civilization*
    (Philadelphia, 2006).

39 D. Keightley, 'The Shang: China's first historical dynasty', in Loewe and
    Shaughnessy (eds), *Cambridge History of Ancient China*, pp. 232–91.

40 M. Vlok et al., 'Forager and farmer evolutionary adaptations to malaria
    evidenced by 7000 years of thalassemia in Southeast Asia', *Scientific
    Reports* 11 (2021), 1–15; K. Klieman, *The Pygmies Were our Compass: Bantu
    and Batwa in the History of Western Central Africa: Early Times to
    c.1900 CE* (Portsmouth, NH, 2003).

41 J. Webb, 'Malaria and the Peopling of Early Tropical Africa', *Journal of
    World History* 16 (2005), 269–91.

42 D. Dominey-Howes, 'A re-analysis of the Late Bronze Age eruption
    and tsunami of Santorini, Greece, and the implications for the
    volcano–tsunami hazard', *Journal of Volcanology and Geothermal
    Research* 130 (2004), 107–32; P. Nomikou et al., 'Post-eruptive flooding
    of Santorini caldera and implications for tsunami generation', *Nature
    Communications* 7 (2016), 1–10; NASA Jet Propulsion Laboratory,
    'Thera/Santorini Eruption', 1 December 1989.

43 I. Babkin and I. Babkina, 'The Origin of the Variola Virus', *Viruses*
    7 (2015), 1100–12.

44 D. Henderson, 'The eradication of smallpox – An overview of the past,
    present and future', *Vaccine* 29 (2011), 7–9; D. Henderson, *Smallpox: The
    Death of a Disease* (Amherst, NY, 2009).

45 E. Strouhal, 'Traces of a smallpox epidemic in the family of Ramesses V
    of the Egyptian 20th dynasty', *Anthropologie* 34 (1996), 315–19.

46 R. Grove, *Green Imperialism: Colonial Expansion, Tropical Island Edens
    and the Origins of Environmentalism* (Cambridge, 1995), p. 19.

47 S. Liu et al., 'Did China import metals from Africa in the Bronze
    Age?', *Archaeometry* 60 (2018), 105–17; W.-d. Sun et al., 'Origin of the
    mysterious Yin-Shang bronzes in China indicated by lead isotopes',
    Scientific Reports 6 (2016), 1–9.

48 I. Singer, 'A political history of Ugarit', in W. Watson and N. Wyatt
    (eds), *Handbook of Ugaritic Studies* (Leiden, 1999), pp. 603–733.

49 E. Cline, *1177 BC: The Year Civilisation Collapsed* (Princeton, 2014),
    pp. 154–5.

50 Ibid., p. 155.

51 Y. Cohen and I. Singer, 'Late synchronism between Ugarit and Emar', in
    Y. Amit et al. (eds), *Essays on Ancient Israel in its Near Eastern Context: A
    Tribute to Nadav Na'aman* (Winona Lake, IN, 2006), pp. 123–39.

52 W. Edgerton, 'The Strikes in Ramses III's Twenty-Ninth Year', *Journal of
    Near Eastern Studies* 10 (1951), 137–45.

53 J. Janssen, *Commodity Prices from the Ramessid Period: An Economic
    Study of the Village of the Necropolis Workmen at Thebes* (Leiden, 1975).

54 M. Feldman et al., 'Ancient DNA sheds light on the genetic origins of
    early Iron Age Philistines', *Science Advances* 5 (2019), 1–10.

55 Singer, 'Political history of Ugarit', p. 3.

56 J. Pritchard, *Ancient Near Eastern Texts Related to the Old Testament*
    (Princeton, 1969).

57 J. Wilson, 'The War against the Peoples of the Sea', in J. Pritchard (ed.),
    Ancient Near Eastern Texts Relating to the Old Testament (Princeton,
    1969), pp. 262–3.

58 B. Drake, 'The influence of climatic change on the Late Bronze Age
    Collapse and the Greek Dark Ages', *Journal of Archaeological Science*
    39 (2012), 1862–70; D. Kaniewski and E. Van Campo, '3.2 ka BP
    Megadrought and the Late Bronze Age Collapse', in Weiss (ed.),
    Megadrought and Collapse, pp. 161–82.

59 P. LaMoreaux, 'Worldwide environmental impacts from the eruption of
    Thera', *Environmental Geology* 26 (1995), 172–81.

60 R. Drews, *The End of the Bronze Age* (Princeton, 1993).

61 T. Bryce, 'The secession of Tarhuntassa', in D. Groddek and M. Zorman
    (eds), *Tabularia Hethaeorum. Hethitologische Beiträge Silvin Košak zum
    65. Geburtstag* (Wiesbaden, 2007), pp. 119–29.

62 E. Cline, *1177 BC*, pp. 167–80; A. Nur, 'Poseidon's Horses: Plate
    Tectonics and Earthquake Storms in the Late Bronze Age Aegean
    and Eastern Mediterranean', *Journal of Archaeological Science* 27
    (2000), 43–63.

63 On fragility, see: N. Taleb, *Antifragile: Things that Gain from Disorder*
    (New York, 2012).

64 H. Lee, V. Padmanabhan and S. Whang, 'The Bullwhip Effect in Supply
    Chains', *MIT Sloan Management Review* 38 (1997), 93–102.

65 T. Kidder et al., 'Multi-method geoarchaeological analyses demonstrates
    exceptionally rapid construction of Ridge West 3 at Poverty Point',
    Southeastern Archaeology 40 (2021), 1–16.

66　T. Kidder, E. Henry and L. Arco, 'Rapid climate change-induced collapse of hunter-gatherer societies in the lower Mississippi River valley between ca.3300 and 2780 cal yr BP', *Science China Earth Sciences* 61 (2018), 178–89.

67　T. Kidder, 'Climate change and the archaic to woodland transition (3000–2500 BP) in the Mississippi River Basin', *American Antiquity* 71 (2006), 195–231.

68　N. Heller et al., *The Tchefuncte Culture and the Early Woodland Period in Coastal Louisiana: New Analysis of the Tchefuncte, Bayou Jasmine, Lafayette Mounds, and Little Woods Sites* (New Orleans, 2013).

69　Li Feng, *Early China: A Social and Cultural History* (Cambridge, 2014), pp. 66–111.

70　S. Bedford and M. Spriggs (eds), *Debating Lapita: Distribution, Chronology, Society and Subsistence* (Canberra, 2019); M. Spriggs and D. Reich, 'An ancient DNA Pacific journey: a case study of collaboration between archaeologists and geneticists', *World Archaeology* 51 (2019), 620–39.

71　E. Cochrane, 'The Evolution of Migration: the Case of Lapita in the Southwest Pacific', *Journal of Archaeological Method and Theory* 25 (2018), 520–58.

7　REGARDING NATURE AND THE DIVINE (C.1700–C.300 BC)

1　Dalley, 'Epic of Atrahasis', pp. 9–14.

2　Ibid., pp. 13–14.

3　Ibid., pp. 9–35.

4　I. Finkel, *The Ark before Noah: Decoding the Story of the Flood* (London, 2014).

5　Dalley, 'Epic of Atrahasis', pp. 31–5.

6　N. Yoffee, 'The collapse of ancient Mesopotamian states and civilization', in N. Yoffee and G. Cowgill (eds), *The Collapse of Ancient States and Civilizations* (Tucson, AZ, 1988), pp. 44–68.

7　J. Tainter, 'Problem Solving: Complexity, History, Sustainability', *Population and Environment: A Journal of Interdisciplinary Studies* 22 (2000), 3–41.

8　Grove, *Green Imperialism*, pp. 18–19.

9　Oelschlaeger, *Idea of Wilderness*, p. 39.

10　Dalley, 'The Natural World in Ancient Mesopotamian Literature', pp. 21–36.

11　Brooke, *Climate Change*, pp. 211–12.

12　F. Rochberg, 'Observing and Describing the World through Divination and Astronomy', in Radner and Robson (eds), *Oxford Handbook of Cuneiform Culture*, p. 631.

13 J. Ur, 'Physical and Cultural Landscapes of Assyria', in E. Frahm (ed.), *A Companion to Assyria* (Hoboken, NJ, 2017), pp. 21–2.

14 J. Ur, 'Sennacherib's Northern Assyrian Canals: New Insights from Satellite Imagery and Aerial Photography', *Iraq* 67 (2005), 317–45.

15 P. Ergenzinger et al., 'The Reconstruction of Environment, Irrigation and Development of Settlement in the Habur in North-east Syria', in J. Bintliff, D. Davidson and E. Grant (eds), *Conceptual Issues in Environmental Archaeology* (Edinburgh, 1988), pp. 108–28.

16 E. Frahm, *Einleitung in die Sanherrib-Inschriften* (Vienna, 1997), pp. 151–4.

17 K. Kessler, 'Royal Roads and Other Questions of the Neo-Assyrian Communication System', in S. Parpola and R. Whiting (eds), *Assyria 1995: Proceedings of the 10th Anniversary Symposium of the Neo-Assyrian Text Corpus Project* (Helsinki, 1997), pp. 129–36; Ur, 'Physical and Cultural Landscapes of Assyria', p. 26.

18 B. Oded, *Mass Deportations and Deportees in the Neo-Assyrian Empire* (Wiesbaden, 1979), p. 20.

19 S. Parpola (ed.), *State Archives of Assyria*, 19 vols (Helsinki, 1987–), 19, p. 17.

20 K. Radner, 'How Did the Neo-Assyrian King Perceive his Land and its Resource?', in R. Jas (ed.), *Rainfall and Agriculture in Northern Mesopotamia* (Leiden, 2000), pp. 233–46.

21 2 Kings 18; K. Radner, 'Economy, Society and Daily Life in the Neo-Assyrian Period', in Frahm (ed.), *Companion to Assyria*, p. 211.

22 British Museum, *I Am Ashurbanipal: King of the World, King of Assyria* (London, 2018).

23 D. Wiseman, 'A new stela of Aššur-nasir-pal II', *Iraq* 14 (1952), 31–2; J. Bottéro, *Everyday Life in Ancient Mesopotamia* (Edinburgh, 2001), p. 70.

24 A. Thomason, 'Representations of the North Syrian Landscape in Neo-Assyrian Art', *Bulletin of the American Schools of Oriental Research* 323 (2001), 63–96.

25 S. Dalley, *The Mystery of the Hanging Garden of Babylon: An Elusive World Wonder Traced* (Oxford, 2013).

26 T. Wilkinson, *Archaeological Landscapes of the Near East* (Tucson, AZ, 2003), pp. 11–14. For the decline of Assyrian, J. Postgate, 'The Four "Neo-Assyrian" Tablets from Šēḫ Ḥamad', *State Archives of Assyria Bulletin* 7 (1993), 109–24.

27 A. Sachs and H. Hunger, *Astronomical Diaries and Related Texts from Babylon*, 3 vols (Vienna, 1988–96).

28 U. Koch, 'Sheep and Sky: Systems of Divinatory Interpretation', *Oxford Handbook of Cuneiform Culture*, pp. 448–69.

29 E. Leichty, *The Omen Series Šumma Izbu* (Locust Valley, NY, 1970), p. 202.

30 H. Hunger, *Astrological Reports to Assyrian Kings* (Helsinki, 1992), p. 82; Koch, 'Sheep and Sky', p. 449.

31 D. Keightley, 'Oracle-Bone Inscriptions of the Late Shang Dynasty', in W. T. de Bary (ed.), *Sources of East Asian Tradition*, 2 vols (New York, 2008), 1, p. 15.

32 K. Chang, 'Shang Shamans', in W. Peterson (ed.), *The Power of Culture: Studies in Chinese Cultural History* (Hong Kong, 1994), pp. 10–36.

33 F. Li, *Early China: A Social and Cultural History* (Cambridge, 2013), p. 102.

34 Keightley, 'The Shang: China's First Historical Dynasty', pp. 266–8, 281.

35 Ibid., p. 252.

36 Keightley, 'Oracle-Bone Inscriptions', p. 19.

37 Ibid., p. 21.

38 Karlgren, *Book of Odes*, pp. 276, 244.

39 E. Shaughnessy, 'Western Zhou History', in Loewe and Shaughnessy (eds), *Cambridge History of Ancient China*, pp. 310–11.

40 B. Watson, D. Nivison and I. Bloom, 'Classical Sources of Chinese Tradition', in de Bary (ed.), *Sources of East Asian Tradition*, 1, p. 28.

41 B. Karlgren, *The Book of Odes* (Stockholm 1950), 258, p. 225.

42 C. Hsu and K. Linduff, *Western Chou Civilisation* (New Haven, 1988), pp. 283–4.

43 Sima Qian, *Records of the Grand Historian*, tr. B. Watson (New York, 1969), pp. 230–1.

44 F. Rochberg, 'Observing and Describing the World through Divination and Astronomy', in K. Radner and E. Robson (eds), *The Oxford Handbook of Cuneiform Culture* (Oxford, 2011), pp. 629–30.

45 W. van Soldt, *Solar Omens of Enuma Anu Enlil: Tablets 23(24)–29(30)* (Leiden, 1995), p. 29.

46 S. Parpola, *Letters from Assyrian and Babylonian Scholars* (Helsinki, 1993), Text 100: r-1-11, p. 77.

47 J. Hughes, 'Sustainable agriculture in Ancient Egypt', *Agricultural History* 66 (1992), 17.

48 Hesiod, *Works and Days*, tr. A. Stallings (London, 2018), pp. 12–14.

49 L. von Falkenhausen, 'The economic role of cities in Eastern Zhou China', *Archaeological Research in Asia* 14 (2018), 161–9.

50 I. Bloom, 'Confucius and the *Analects*', in de Bary (ed.), *Sources of East Asian Tradition*, 1, pp. 29–32, Confucius, *Analects*, 16.8.

51 Laozi, 'The Ways', in F. Müller (ed.), *The Sacred Books of the East*, 50 vols (Oxford, 1879–1910), 39, p. 50.

52 W. de Bary and I. Bloom (eds), *Sources of Chinese Tradition*, 2 vols (1999), 1, p. 93.

53 Mozi, 'Honoring the Worthy', in de Bary (ed.), *Sources of East Asian Tradition*, 1, p. 43.

54 M. Sivaramakrishnan, 'Ecopoetics and the Literature of Ancient India', in Parham and Westling (eds), *Global History of Literature and the Environment*, p. 68.

55 *Rig Veda*, 6.70. For the composition and date of the *Rig Veda*, M. Witzel, 'Early Indian history: linguistic and textual parameters', in G. Erdosy (ed.), *The Indo-Aryans of Ancient South Asia: Language, Material Culture and Ethnicity* (Berlin, 1995), pp. 307–52.

56 *Atharva Veda*, 6.56, 4.17.

57 Ibid., 8.7.10, 8.2.25, 19.38.1.

58 *Yajur Veda*, 38.22, 6.33.

59 Ibid., 13.37, 13.49.

60 *Rig Veda*, 5.80.

61 P. Olivelle, *The Early Upanisads: Annotated Text and Translation* (New York, 1998). For the date of the Upaniṣads, also see S. Radhakrishnan, *The Principal Upaniṣads* (New Delhi, 2004), pp. 20–2.

62 *The Thirteen Principal Upanishads*, tr. F. Müller, rev. S. Navlakha (London, 2000), *Maitrāyaṇīya Upaniṣad*, p. 132.

63 N. Sundaraswaran, 'Environmental and ecological awareness in the Rāmāyana and the Mahābhārata', *Samskrtakairali* 5 (2016–17), 105–16.

64 *Rig Veda*, 10.121.

65 G. Flood, *An Introduction to Hinduism* (Cambridge, 1996), p. 6.

66 H. Bechert, *Die Datierung des historischen Buddha*, 3 vols (Göttingen, 1991–7).

67 R. Gethin, *The Foundations of Buddhism* (Oxford, 1998), pp. 59–84.

68 See for example J. Bronkhorst, *Greater Magadha: Studies in the Culture of Early India* (Leiden, 2007).

69 *The Sutta-Nipāta*, tr. H. Saddhatissa (London, 1994), verse 136.

70 *Dīgha Nikāya, Dialogues of the Buddha*, ed. and tr. T. Rhys Davids and E. Carpenter, 3 vols (1890–1911), 3, pp. 180ff.

71 *Khuddakapāṭha*, 8.

72 L. Schmithausen, 'The early Buddhist tradition and ecological ethics', *Journal of Buddhist Ethics* 4 (1997), 11.

73 A. Embree, S. Hay and W. de Bary (eds), *Sources of Indian Tradition*, 2 vols (New York, 1988), 1, p. 46.

74 *Ācārāṅga Sūtra*, 1.1

75 *Sūtrakṛtāriga*, 1.1–9.

76 *Uttarādhyayana Sūtra*, 19.67–74.

77 *Sūtrakṛtāriga*, 1.1–9.

78 R. Thapar, *Early India: From the Origins to* AD *1300* (Berkeley, 2005), pp. 29–32, 280–2.

79 M. Fisher, *An Environmental History of India: From the Earliest Times to the Twenty-First Century* (Cambridge, 2018), pp. 60–1.

80 P. Sahni, *Environmental Ethics in Buddhism: A Virtues Approach* (London, 2008), pp. 47–8.

81 *Dīgha Nikāya*, I, p. 141.

82 M. McLish, *The History of the Arthaśāstra: Sovereignty and Sacred Law in Ancient India* (Cambridge, 2019), pp. 1–2.

83 *Kaustiaki Upaniṣad*, 1.2.

84 M. Burley, *Rebirth and the Stream of Life* (London, 2016), p. 102.

85 Herodotus, *The Histories*, 2.123, p. 144.

86 Iamblichus, *De vita Pythagorica*, ed. G. Clark (Liverpool, 1989), 14, p. 25.

87 Empedokles, *The Extant Fragments*, ed. M. Wright (London, 1995), 108, p. 139.

88 R. Stoneman, *The Greek Experience of India: From Alexander to the Indo-Greeks* (Princeton, 2019), pp. 341–2.

89 J. Hughes, *Environmental Problems of the Greeks and Romans* (Baltimore, 2014), p. 53.

90 Plato, *Timaeus*, ed. R. Bury (Cambridge, MA, 2014), p. 54.

91 Hughes, *Environmental Problems*, p. 53.

92 Aristotle, *Politics*, ed. H. Rackham (Cambridge, MA, 2014), 1.3, pp. 20–2, 36.

93 Plato, *Phaedrus*, ed. C. Emlyn-Jones (Cambridge, MA, 2022), p. 356.

94 Xenophon, *Oikonomikos*, ed. E. Marchant (Cambridge, MA, 1997), 5, p. 428.

95 J. Hughes, 'Theophrastus as Ecologist', in W. Fortenbaugh and R. Sharples (eds), *Theophrastean Studies: Fifteen Papers on Natural Science, Physics and Metaphysics, Ethics, Religion and Rhetoric* (New Brunswick, NJ, 1998), pp. 67–75.

96 R. Foltz and M. Saadi-nejad, 'Is Zoroastrianism an Ecological Religion?', *Journal for the Study of Religion, Nature and Culture* I (2007), 414–17.

97 Herodotus, *Histories*, 1.138, p. 63; Strabo, *Geography*, 15.3, ed. and tr. H. Jones, 2 vols (Cambridge, MA, 1930), 2, p. 178; Foltz and Saadi-nejad, 'Is Zoroastrianism an Ecological Religion?', 421.

98 M. Boyce, *Zoroastrianism: Its Antiquity and Constant Vigour* (Costa Mesa, CA, 1992).

99 J. Duchesne-Guillemin, *The Hymn of Zarathustra, Being a Translation of the Gathas Together with Introduction and Commentary*, tr. M. Henning (Boston, MA, 1992), *Yasna*, 50, pp. 29–33.

100 L. Mills. 'The Pahlavi Texts of the Yasna Haptanghaiti (Y. XXXV–XLI (XLII)) for the First Time Critically Translated', *The Journal of the Royal Asiatic Society of Great Britain and Ireland* (1905), 37.1, 64.

101 Duchesne-Guillemin, *Yasna*, 29, pp. 56–61.

102 Mills, 'The Pahlavi Texts of the Yasna Haptanghaiti', 66–9.

103 A. Amanat, 'Environment and Culture: An Introduction', *Iranian Studies* 49 (2016), 928–9.

104 E. Frahm, 'Assyria in the Hebrew Bible', in E. Frahm (ed.), *A Companion to Assyria* (Hoboken, NJ, 2017), pp. 556–69.

105 Isaiah, 45.1. Note, however, the problems of chronology in associating the Jews' release from captivity with the reign of Cyrus: D. Edelman, *The Origins of the Second Temple: Persian Imperial Policy and the Rebuilding of Jerusalem* (London, 2005).

106 E. Davis, *Scripture, Culture, and Agriculture: An Agrarian Reading of the Bible* (Cambridge, 2009).

107 S. Kramer, 'The "Babel of Tongues": A Sumerian Version', *Journal of the American Oriental Society* 88 (1968), 108–11.

108 E. Bloom and K. Weingart, 'The Joseph Story: Diaspora Novella or North-Israelite Narrative?', *Zeitschrift für die alttestamentliche Wissenschaft* 129 (2017), 501–21; I. Knohl, 'Joseph and the Famine: The Story's Origins in Egyptian History', https://www.thetorah.com/article/joseph-and-the-famine-the-storys-origins-in-egyptian-history.

109 T. Levy, T. Schneider and W. Propp (eds), *Israel's Exodus in Transdisciplinary Perspective: Text, Archaeology, Culture and Geoscience* (New York, 2015).

110 S. Garfinkle, 'Ancient Near Eastern City States', in Bang and Scheidel (eds), *Oxford Handbook of the State in the Ancient Near East*, p. 107.

111 Oelschlaeger, *Idea of Wilderness*, p. 47.

112 D. Green, 'The Garden of Eden in the Hebrew Bible', in Parham and Westling (eds), *Global History of Literature and the Environment*, pp. 55–6.

113 Psalm 23:1–4.

114 Oelschlaeger, *Idea of Wilderness*, p. 49.

115 Deuteronomy, 11:8–12; Green, 'The Garden of Eden in the Hebrew Bible', pp. 56–7.

116 Exodus, 24:15–16; Genesis, 9:13.

117 Ezekiel, 29:9–15; T. Tvedt, *The Nile: History's Greatest River* (London, 2021), pp. 22–3.

118 K. Jaspers, *Vom Ursprung und Ziel der Geschichte* (Munich, 1949)

119 B. Wittrock, 'Social Theory and Global History: The Three Cultural Crystallizations', *Thesis Eleven* 65 (2001), 27–50; B. Schwartz, 'The age of transcendence', *Daedalus* 104 (1975), 1–7.

120 S. Eisenstadt, 'Introduction: The Axial Age Breakthroughs – Their Characteristics and Origins', in S. Eisenstadt (ed.), *The Origins and Diversity of Axial Age Civilizations* (Albany, NY, 1986), pp. 29–39.

121 I. Provan, *Convenient Myths: The Axial Age, Dark Green Religion and the World That Never Was* (Waco, TX, 2013); D. Mullins et al., 'Systematic Assessment of "Axial Age" Proposals Using Global Comparative Historical Evidence', *American Sociological Review* 83 (2018), 596–626.

122 R. Bellah, *Religion in Human Evolution: From the Paleolithic to the Axial Age* (Cambridge, MA, 2011).

123 N. Baumard et al., 'Increased Affluence Explains the Emergence of Ascetic Wisdoms and Moralizing Religions', *Current Biology* 25 (2015), 10–15.

124 B. Karlgren, 'The Book of Documents', *Museum of Far Eastern Antiquities, Bulletin* 20 (Stockholm, 1950), p. 33.

125 Ibid., pp. 29, 85–91; J. Li, *Chinese Civilisation in the Making, 1766–221 BC* (New York, 1996), pp. 85–91.

126 M. Elvin, 'Who Was Responsible for the Weather? Moral Meteorology in Late Imperial China', *Osiris* (1998), 213–27.

127 J. Snyder-Reinke, *Dry Spells: State Rainmaking and Local Government in Late Imperial China* (Cambridge, MA, 2009), pp. 24ff.; A. Birrell, *Chinese Mythology: An Introduction* (Baltimore, 1993), pp. 72–83.

128 M. Bloch, 'The Ritual of the Royal Bath in Madagascar: The Dissolution of Death, Birth and Fertility into Authority', in D. Cannadine and S. Price (eds), *Rituals of Royalty: Power and Ceremonial in Traditional Societies* (Cambridge, 1987), pp. 271–97; C. Geertz, *Negara: The Theatre State in Nineteenth-Century Bali* (Princeton, 1980).

129 T. Lewis, 'A History of Buddhist Ritual', in J. Powers (ed.), *The Buddhist World* (London, 2015), p. 325.

130 B. Ruppert, 'Buddhist Rainmaking in Early Japan: The Dragon King and the Ritual Careers of Esoteric Monks', *History of Religions* 42 (2002), 143–74.

131 V. Scarborough, 'Ecology and Ritual: Water Management and the Maya', *Latin American Antiquity* 9 (1998), 135–59.

132 R. Ishihara, 'Rising clouds, blowing winds: Late Classic Maya rain rituals in the Main Chasm, Aguateca, Guatemala', *World Archaeology* 40 (2008), 178–9.

133 P. Schaafsma and K. Taube, 'Bringing the rain: an ideology of rain making in the Pueblo Southwest and Mesoamerica', in J. Quilter (ed.), *A Pre-Columbian World: Searching for a Unitary Vision of Ancient America* (Washington, DC, 2006), pp. 235–6, 256–9.

134 M. Cosmopoulos, *Bronze Age Eleusis and the Origins of the Eleusinian Mysteries* (Cambridge, 2015), pp. 1–2.

135 Elvin, 'Three thousand years of unsustainable growth', 17.

136 S. Dhammika, *Nature and the Environment in Early Buddhism* (Singapore, 2015), p. 14.

137 P. Brancaccio, 'Aṅgulimāla or the Taming of the Forest', *East and West* 49, 105–18.

138 Dhammika, *Nature and the Environment*, p. 30.

139 R. Thapar, *Ashoka and the Decline of the Mauryas* (New Delhi, 1997), p. 250.

140 S. Dhammika, *Middle Land, Middle Way: A Pilgrim's Guide to the Buddha's India* (Kandy, 2008), p. 150.

141 P. Olivelle, *King, Governance and Law in Ancient India: Kauṭilya's Arthaśāstra* (Oxford, 2013), pp. 25–31; M. McLish, *The History of the Arthaśāstra: Sovereignty and Sacred Law in Ancient India* (Cambridge, 2019), pp. 140–54.

142 Olivelle, *Kauṭilya's Arthaśāstra*, pp. 99–100.

143 Ibid., pp. 152–4.

144 Fisher, *Environmental History of India*, pp. 68–9.

145 X. Yang, 'Idealizing Wilderness in Medieval Chinese Poetry', in K. Gaul and J. Hiltz (eds), *Landscapes and Communities on the Pacific Rim: Cultural Perspectives from Asia to the Pacific Northwest* (Armonk, NY, 2000), pp. 94–5.

146 Xenophon, *Oikonomikos*, 5, p. 429.

8 THE STEPPE FRONTIER AND FORMATION OF EMPIRES
(C.1700–C.300 BC)

1 P. Crossley Kyle, *Hammer and Anvil: Nomad Rulers at the Forge of the Modern World* (Lanham, MD, 2019), pp. 10–11.

2 P. Mitchell, *Horse Nations: The Worldwide Impact of the Horse on Indigenous Societies* (Oxford, 2015).

3 R. Sommer et al., 'Holocene survival of the wild horse in Europe: A matter of open landscape?', *Journal of Quaternary Science* 26 (2011), 805–12; E. Sandoval-Castellanos et al., 'Coat colour adaptation of post-glacial horses to increasing forest vegetation', *Nature Ecology & Evolution* 1 (2017), 1816–19.

4 M. Leonardi et al., 'Late Quaternary horses in Eurasia in the face of climate and vegetation change', *Science Advances* 4 (2018), 1–11.

5 A. Outram et al., 'The Earliest Horse Harnessing and Milking', *Science* 323 (2009), 1332–5.

6 I. Lazaridis et al., 'Genomic insights into the origin of farming in the ancient Near East', *Nature* 536 (2016), 419–24; W. Taylor et al., 'Early Pastoral Economies and Herding Transitions in Eastern Eurasia', *Scientific Reports* 10 (2020), 3; P. Librado et al., 'The origins and spread of domestic horses from the Western Eurasian steppes', *Nature* (2021), 1–19.

7 V. Gening, G. Zdanovich and V. Gening, Синташта: археологические памятники арийских племен Урало-Казахстанских степей (Chelyabinsk, 1992); K. Jones-Bley, 'The Sintashta "chariots"', in J. Davis-Kimball et al. (eds), *Kurgans, Ritual Sites, and Settlements: Eurasian Bronze and Iron Age* (Oxford, 2000), pp. 135–40; D. Owen, 'The first equestrian: An Ur III glyptic scene', *Acta Sumerologica* 13 (1991), 259–73; D. Owen, 'A Thirteen Month Summary Account from Ur', *Bulletin of the School of Oriental and African Studies* 43 (1980), 57–70.

8 J. Oates, 'A note on the early evidence for horse and the riding of equids in Western Asia', in M. Levine, C. Renfrew and K. Boyle (eds), *Prehistoric Steppe Adaptation and the Horse* (Cambridge, 2003), pp. 115–38.

9 S. Smith, 'State and empire in the Middle and New Kingdoms', in J. Lustig (ed.), *Anthropology and Egyptology: A Developing Dialogue* (Sheffield, 1997), pp. 66–89.

10 B. Sandor, 'Tutankhamun's chariots: secret treasures of engineering mechanics', *Fatigue & Fracture of Engineering Materials & Structures* 27 (2004), 637–46.

11 A. Waley, *The Book of Songs* (London, 1954), pp. 128–9.

12 E. Shaughnessy, 'Historical Perspectives on the Introduction of the Chariot into China', *Harvard Journal of Asiatic Studies* 48 (1988), 189–237.

13 E. Shaughnessy, *Sources of the Western Zhou* (Berkeley, 1991), p. 81.

14 A. Pleterski, *Mitska stvarnost koroških knežjih kamnov* (Ljubljana, 1997), pp. 39–40.

15 R. Sharma, 'The Aryan Problem and the Horse', *Social Scientist* 21 (1993), 3–16.

16 W. Taylor, 'A Bayesian chronology for early domestic horse use in the
    Eastern Steppe', *Journal of Archaeological Science* 81 (2017), 49–58.

17 R. Zaroff, 'Aśvamedha – A Vedic horse sacrifice', *Studia Mythologica
    Slavica* 8 (2005), 74–8.

18 Taylor et al., 'Early pastoral economies', 12; W. Taylor and
    T. Tuvshinjragal, 'Horseback riding, asymmetry, and changes to the
    equine skull: evidence for mounted riding in Mongolia's late Bronze
    Age', in L. Bartosiewicz and E. Gal (eds), *Care or Neglect? Evidence of
    Animal Disease in Archaeology* (Oxford, 2018), pp. 134–54; U. Dietz,
    'Horseback riding: Man's access to speed', in Levine, Renfrew and Boyle
    (eds), *Prehistoric Steppe Adaptation and the Horse*, pp. 189–99.

19 Taylor et al., 'Early pastoral economies', 11.

20 K. Privat, 'Preliminary report of palaeodietary analysis of human
    and faunal remains from Bolshekaragansky kurgan 25', in
    D. Zdanovich (ed.), Аркаим Некрополь *(по материалам кургана 25
    больщекараганского могильника)* (Chelyabinsk, 2002), pp. 166–71.

21 Taylor et al., 'Early pastoral economies', 13; X. Belauzaran et al.,
    'Horse-meat for human consumption – Current research and future
    opportunities', *Meat Science* 108 (2015), 74–81.

22 E. Kuz'mina, Откуда пришли индоарии? (Moscow, 1994).

23 A. Ventresca Miller et al., 'Pasture usage by ancient pastoralists in the
    northern Kazakh steppe informed by carbon and nitrogen isoscapes of
    contemporary floral biomes', *Archaeological and Anthropological Sciences*
    11 (2019), 2151–66.

24 W. Honeychurch, *Inner Asia and the Spatial Politics of Empire: Archaeology,
    Mobility, and Culture Contact* (New York, 2015), pp. 109–220.

25 M. Frachetti, 'Seeds for the soul: ideology and diffusion of domesticated
    grains across inner Asia', in V. Mair and J. Hickman, *Reconfiguring
    the Silk Road: New Research on East–West Exchange in Antiquity*
    (Philadelphia, 2014), pp. 41–53.

26 R. Spengler et al., 'Early agriculture and crop transmission among
    Bronze Age mobile pastoralists of Central Eurasia', *Proceedings of the
    Royal Society B* 281 (2014), 1–7.

27 N. Miller, R. Spengler and M. Frachetti, 'Millet cultivation across
    Eurasia: Origins, spread, and the influence of seasonal climate', *The
    Holocene* 26 (2016), 1566–75.

28 M. Frachetti, 'Multiregional emergence of mobile pastoralism
    and nonuniform institutional complexity across Eurasia', *Current
    Anthropology* 53 (2012), 2–38. For the concept of the 'Trans-Eurasian
    Exchange' network, see A. Sherratt, 'The Trans-Eurasian Exchange: The

Prehistory of Chinese Relations with the West', in V. Mair (ed.), *Contact and Exchange in the Ancient World* (Honolulu, 2006), pp. 30–61; T. Larsen Høisæter, 'Polities and nomads: the emergence of the Silk Road exchange in the Tarim Basin region during late prehistory (2000–400 BCE)', *Bulletin of the School of Oriental and African Studies* 80 (2017), 339–63.

29 G. Gnecchi-Ruscone et al., 'Ancient genomic time transect from the Central Asian steppe unravels the history of the Scythians', *Science Advances* 7 (2021), 1–14.

30 J. Davis-Kimball, V. Bashilov and L. Yablonsky (eds), *Nomads of the Eurasian Steppe in the Early Iron Age* (Berkeley, 1995).

31 R. Spengler et al., 'Linking agriculture and exchange to social developments of the Central Asian Iron Age', *Journal of Anthropological Archaeology* 48 (2017), 295–308.

32 B. Hanks, 'Archaeology of the Eurasian Steppes and Mongolia', *Annual Review of Anthropology* 39 (2010), 469–86.

33 S. Wilkin et al., 'Economic Diversification Supported the Growth of Mongolia's Nomadic Empires', *Scientific Reports* 10 (2020), 1–12.

34 W. Honeychurch and C. Makarewicz, 'The Archaeology of Pastoral Nomadism', *Annual Review of Anthropology* 45 (2016), 341–59.

35 Herodotus, *The Histories*, 4.127, p. 277.

36 See for example U. Brosseder and B. Miller, *Xiongnu Archaeology: Multidisciplinary Perspectives of the First Steppe Empire in Inner Asia* (Bonn, 2011); N. Kradin and A. Ivliev, Истопия Киданьской Империи ляо (Moscow, 2014); H.-G. Hüttel and U. Erdenebat, *Karabalgasun and Karakorum: Two Late Nomadic Urban Settlements in the Orkhon Valley* (Ulaanbaatar, 2010).

37 L. Yuqi et al., 'Early irrigation and agropastoralism at Mohuchahangoukou (MGK), Xinjiang, China', *Archaeological Research in Asia* 12 (2017), 23–32.

38 Huan Kuan, *Yan Tie Lun*, cited by Y. Yu, *Trade and Expansion in Han China: A Study in the Structure of Sino-Barbarian Economic Relations* (Berkeley, 1967), p. 40; Ammianus Marcellinus, *Rerum Gestarum Libri*, 31.2, 3, p. 382. Priscus, *Testimonia*, fragment 49, in R. Blockley (ed. and tr.), *The Fragmentary Classicising Historians of the Later Roman Empire: Eunapius, Olympiodorus, Priscus, and Malchus*, 2 vols (Liverpool, 1981–3), 2, p. 356; A. Porter, *Mobile Pastoralism and the Formation of Near Eastern Civilisations* (Cambridge, 2011), p. 293.

39 B. Yang, 'The Rise and Fall of Cowrie Shells: The Asian Story', *Journal of World History* 22 (2011), 1–25.

40 Herodotus, *The Histories*, 4.46, p. 250.

41 D. Sneath, *The Headless State: Aristocratic Orders, Kinship Society, and Misrepresentations of Nomadic Inner Asia* (New York, 2007).

42 N. Anfinset, *Metal, Nomads and Culture Contacts, the Middle East and North Africa* (London, 2010), pp. 81–7.

43 W. Honeychurch, 'From Steppe Roads to Silk Roads, in R. Amitai and M. Biran (eds), *Nomads as Agents of Cultural Change: The Mongols and their Eurasian Predecessors* (Honolulu, 2014), pp. 50–89. Also see T. Barfield, *The Perilous Frontier: Nomadic Empires and China* (Oxford, 1992).

44 A. Khazanov, *Nomads and the Outside World* (Cambridge, 1984), pp. 75, 104.

45 Y. Liu, 'Exotica as prestige technology: the production of luxury gold in Western Han society', *Antiquity* 91 2017), 1588–1602; P. Andreeva, 'Re-making Animal Bodies in the Arts of Early China and North Asia: Perspectives from the Steppe', *Early China* (2022), 1–53.

46 J. Rawson, 'Carnelian beads, animal figures and exotic vessels: traces of contact between the Chinese states and Inner Asia, *ca.* 1000–650 BC', in M. Wagner and W. Wang (eds), *Bridging Eurasia* (Berlin, 2010), pp. 1–41.

47 T. Hayashi, 'Griffin motif: from the West to East Asia via the Altai', *Parthica* 14 (2012), 49–64.

48 K. Peng and Y. Zhu, 'New Research on the Origins of Cowries used in Ancient China', *Sino-Platonic Papers* 68 (1995), 1–18; M. Dwivedi, 'Tools of Economic Connectivity in Early Historic South Asia', in S. von Reden (ed.), *Handbook of Ancient Afro-Eurasian Economies*, 2 vols (Berlin, 2022), 2, esp. pp. 501–3.

49 Y. Li, 'On the Function of Cowries in Shang and Western Zhou China', *Journal of East Asian Archaeology* 5 (2003), 1–26.

50 D. Potts, *Nomadism in Iran: From Antiquity to the Modern Era* (Oxford, 2014), pp. 419–30.

51 For example B. Arbuckle and E. Hammer, 'The Rise of Pastoralism in the Ancient Near East', *Journal of Archaeological Research* 27 (2019), 391–449.

52 N. Di Cosmo, 'The "Birth" of the Silk Road between Ecological Frontiers and Military Innovation', in J. Lerner and Y. Shi (eds), *Silk Roads: From Local Realities to Global Narratives* (Oxford, 2020), pp. 11–20.

53 Ibid.

54 Ibid. Also see here N. Di Cosmo, *Ancient China and its Enemies: The Rise of Nomadic Power in East Asian History* (Cambridge, 2002), esp. pp. 138–55.

55 Thucydides, *History of the Peloponnesian War*, ed. C. Smith, 4 vols (Cambridge, 2014), 1, I.7–8, pp. 12–14.

56 J. Bennett, 'Retrodicting the Rise, Spread, and Fall of Large-Scale States in the Old World', *SocArViv Preprint* (2021), 1–27.

57 P. Turchin et al., 'War, space, and the evolution of Old World complex societies', *PNAS* 110 (2013), 16,384–9.

58 P. Turchin, 'A theory for the formation of large empires', *Journal of Global History* 4 (2009), 191–217.

59 N. Kradin, 'Nomadism, Evolution and World-Systems: Pastoral Societies in Theories of Historical Development', *Journal of World-Systems Research* 8 (2002), 368–88; W. Scheidel, 'The Xiongnu and the comparative study of empire', in B. Miller and U. Brosseder (eds), *Xiongnu Archaeology: Multidisciplinary Perspectives of the First Steppe Empire in Inner Asia* (Bonn, 2011), pp. 111–20; M. Byington, *The Ancient State of Puyö in Northeast Asia: Archaeology and Historical Memory* (Cambridge, MA, 2020).

60 Turchin et al., 'War, space, and the evolution of Old World complex societies', 16,384–9.

61 W. Scheidel, 'The scale of empire: territory, population, distribution', in P. Bang, C. Bayly and W. Scheidel (eds), *The Oxford World History of Empire*, 2 vols (Oxford, 2021), 2, pp. 91–110.

62 T. Currie et al., 'Duration of agriculture and distance from the steppe predict the evolution of large-scale human societies in Afro-Eurasia', *Humanities and Social Sciences Communications* 7 (2020), 1–8.

63 W. Scheidel, *Escape from Rome: The Failure of Empire and the Road to Prosperity* (Princeton, 2019), pp. 281–90. T. Currie et al., 'Duration of agriculture and distance from the steppe predict the evolution of large-scale human societies in Afro-Eurasia', *Humanities and Social Sciences Communications* 7 (2020), 1–8.

64 C. Yu Ko, M. Koyama and T.-H. Sng, 'Unified China and Divided Europe', *International Economic Review* 59 (2018), 285–327.

65 H. Root, 'Network assemblage of regime stability and resilience: Comparing Europe and China', *Journal of Institutional Economics* 13 (2017), 523–48.

66 J. Fernández-Villaverde et al., 'The Fractured Land Hypothesis', National Bureau of Economic Research (NBER) Working Paper 27774 (2020), 1–64; for Lü Buwei, see J. Sellmann, *Master Lü's Spring and Autumn Annals (Lüshi Chunqiu)* (Albany, NY, 2002), p. 107.

67 D. Gifford-Gonzalez, 'Animal disease challenges to the emergence of pastoralism in sub-Saharan Africa', *African Archaeological Review* 17 (2000), 95–139.

68 A. Wink, *The Making of the Indo-Islamic World, c.700–1800* CE (Cambridge, 2020), esp. pp. 8–123.

69 A. Wink, 'From the Mediterranean to the Indian Ocean: Medieval History in Geographic Perspective', *Comparative Studies in Society and History* 44 (2002), 419.

70 Wink, *Making of the Indo-Islamic World*, pp. 18–20.

71 A. Wink, *Al-Hind: The Making of the Indo-Islamic World*, vol. 1: *Medieval India and the Expansion of Islam, 7th–11th Centuries* (Leiden 1990), pp. 181–5.

72 Wink, 'From the Mediterranean to the Indian Ocean', 420.

73 M. Mate, *A History of Water Management and Hydraulic Technology in India (1500 BC to 1800 AD)* (Delhi, 1998), p. 77.

74 *Ibn Baṭṭūṭa, al-Riḥla*, tr. H. Gibb, *The Travels of Ibn Battuta*, 4 vols (Cambridge, 1994), 3, p. 204; Wink, *Making of the Indo-Islamic World*, p. 21.

75 *Bābur-Nāma*, tr. W. Thackston, *Memoirs of Babur, Prince and Emperor* (London, 2006), p. 334; Wink, *Making of the Indo-Islamic World*, p. 20.

76 J. Das Gupta, *Bengal in the Sixteenth Century* AD (Calcutta, 1914), p. 104.

77 Kauṭilya, *Arthaśāstra* 2.30; P. Olivelle, 'Long-distance trade in ancient India: Evidence from Kauṭilya, *Arthaśāstra*', *Indian Economic and Social History Review* 57 (2020), 41–2.

78 J. Gommans, *Mughal Warfare: Indian Frontiers and High Roads to Empire, 1500–1700* (London, 2002); J. Gommans, 'Warhorse and Post-Nomadic Empire in Asia, c. 1000–1800', *Journal of Global History* 2 (2007), 1–21.

79 Frankopan, *Silk Roads*, pp. 234ff.

80 M. Green, 'Climate and Disease in Medieval Eurasia', in *Oxford Research Encyclopedia of Asian History* (Oxford, 2018).

81 M. Green and L. Jones, 'The Evolution and Spread of Major Human Diseases in the Indian Ocean World', in G. Campbell and E. Knoll (eds), *Disease Dispersion and Impact in the Indian Ocean World* (London, 2020), pp. 25–57.

82 C. Stanish, 'The origin of state societies in South America', *Annual Review of Anthropology* 30 (2001), 41–64.

83 C. Spencer and E. Redmond, 'Primary State Formation in Mesoamerica', *Annual Review of Anthropology* 33 (2004), 173–99.

84 R. González Lauck, 'Observaciones en torno a los contextos de la escultura olmeca en La Venta, Tabasco', in M. Uriarte and K. Staines Cicero (eds), *Acercarse y mirar: Homenaje a Beatriz de la Fuente* (Mexico City, 2004), pp. 75–106.

85 K. Taube, 'Los significados del jade', *Arqueología Mexicana* 133 (2015), 49–55.

86 A. Christenson, *Popul Vuh: Literal Poetic Version Translation and Transcription* (Winchester, 2003), pp. 77–8.

87 Ibid., pp. 197–8.

88 S. McIntosh, 'Pathways to Complexity: An African Perspective', in S. McIntosh (ed.), *Beyond Chiefdoms: Pathways to Complexity in Africa* (Cambridge, 1999), pp. 1–30.

89 S. Falconer and C. Redman, *Polities and Power: Archaeological Perspectives on the Landscapes of Early States* (Tucson, AZ, 2009); R. McIntosh, *Ancient Middle Niger: Urbanism and the Self-Organising Landscape* (Cambridge, 2005); G. Connah, *African Civilisations: An Archaeological Perspective* (Cambridge, 2001).

90 McIntosh, 'Pathways to Complexity: An African Perspective', pp. 1–30.

91 K. Macdonald et al., 'New Light on the Tichitt Tradition: A Preliminary Report on Survey and Excavation at Dhar Nema', in P. Mitchell, A. Haour and J. Hobart (eds), *Researching Africa's Past: New Contributions from British Archaeologists* (Oxford, 2003), pp. 73–80.

92 A. LaViolette and J. Fleischer, 'The Archaeology of Sub-Saharan Urbanism: Cities and their Countrysides', in A. Brower Stahl (ed.), *African Archaeology: A Critical Introduction* (Oxford, 2004), pp. 327–52.

93 A. Holl, 'Coping with uncertainty: Neolithic life in the Dhar Tichitt-Walata, Mauritania (ca. 4000–2300 BP)', *Comptes Rendus Geoscience* 341 (2009), 703–12.

94 M. Ould Kattar, 'Les sites Gangara, la fin de la culture Tichitt et l'origine de Ghana', *Journal des Africanistes* 65 (1995), 31–41; K. Macdonald, 'Before the Empire of Ghana: pastoralism and the origins of cultural complexity in the Sahel', in G. Connah (ed.), *Transformations in Africa: Essays on Africa's Later Past* (Leicester, 1998), pp. 71–103.

95 J. Monroe, '"Elephants for Want of Towns": Archaeological Perspectives on West Africa Cities and Their Hinterlands', *Journal of Archaeological Research* 26 (2018), 387–446.

96 Herodotus, *The Histories*, 4.45, pp. 249–50.

97 Cicero, *De legibus*, ed. C. Keyes (Cambridge, 2000), I, p. 300; Plutarch, 'De Herodoti malignitate', *Moralia*, tr. L. Pearson (Loeb, 1965).

98   D. Leslie and K. Gardiner, *The Roman Empire in Chinese Sources* (Rome, 1996), p. 98.

99   Zaroff, 'Aśvamedha', 79–81.

100 Polybius, *Histories*, ed. C. Habicht, 8 vols (Cambridge, MA, 2014), 1.3, pp. 8–9.

101 R. Sugirtharajah, *The Bible and Asia: From the Pre-Christian Era to the Postcolonial Age* (Cambridge, MA, 2013), pp. 28–30.

102 N. Cohn, *Cosmos, Chaos, and the World to Come: The Ancient Roots of Apocalyptic Faith* (New Haven, 1993), pp. 77–104, 26–8.

103 A. Mariaselvam, *The Song of Songs and Ancient Tamil Love Poems: Poetry and Symbolism* (Rome, 1988).

104 Sugirtharajah, *The Bible and Asia*, pp. 44–5.

105 Pausanias, 4.32.4.

106 Empedokles, *Extant Fragments*, 117, 127.

107 D. Pingree, 'On the Greek Origin of the Indian Planetary Model Employing a Double Epicycle', *Journal for the History of Astronomy* 2 (1971), 80–5; D. Pingree, 'The Recovery of Early Greek Astronomy from India', *Journal for the History of Astronomy* 7 (1976), 109–23.

108 A. Deter-Wolf et al., 'The world's oldest tattoos', *Journal of Archaeological Science: Reports* 5 (2016), 19–24.

109 C. Reed, 'Tattoo in Early China', *Journal of the American Oriental Society* 120 (2000), 360–76.

110 J. Huehnergard and H. Liebowitz, 'The Biblical Prohibition against Tattooing', *Vetus Testamentum* 63 (2013), 59–77.

111 C. Jones, 'Stigma: Tattooing and Branding in Graeco-Roman Antiquity', *Journal of Roman Studies* 77 (1987), 139–55.

112 P. Temin, 'Price Behavior in Ancient Babylon', *Explorations in Economic History* 39 (2002), 46–60.

113 G. Campbell, *Africa and the Indian Ocean World from Early Times to Circa 1900* (Cambridge, 2019), pp. 43–4.

114 J. Dorey et al., 'Holocene population expansion of a tropical bee coincides with real human colonisation of Fiji rather than climate change', *Molecular Ecology* 30 (2021), 1–38.

115 M. Hurles et al., 'The Dual Origin of the Malagasy in Island Southeast Asia and East Africa: Evidence from Maternal and Paternal Lineages', *American Journal of Human Genetics* 76 (2005), 894–901.

116 A. Adelaar, 'Asian roots of the Malagasy: a linguistic perspective', *Bijdragen tot de Taal-, Land- en Volkenkunde*, 151 (1995), 325–56.

117 R. Grollemund et al., 'Bantu expansion shows that habitat alters the route and pace of human dispersals', *PNAS* 112 (2015), 13,296–301.

118 P. Pollegoni et al., 'Ancient Humans Influenced the Current Spatial Genetic Structure of Common Walnut Populations in Asia', *Plos One* 10 (2015), 1–16.

119 R. Spengler et al., 'Exaptation Traits for Megafaunal Mutualisms as a Factor in Plant Domestication', *Frontiers in Plant Science* 12 (2021), 1–14.

120 R. Power et al., 'Asian Crop Dispersal in Africa and Late Holocene Human Adaptation to Tropical Environments', *Journal of World Prehistory* 32 (2019), 353–92.

121 F. Aubalie, 'Pathways of diffusion of some plants and animals between Asia and the Mediterranean region', *Revue d'Ethnoécologie* 1 (2012), 1–26.

9 THE ROMAN WARM PERIOD (C.300 BC–AD C.500)

1 D. Crawford, 'Ptolemy, Ptah and Apis in Hellenistic Memphis', in D. Crawford, J. Quaegebeur and W. Clarysse (eds), *Studies on Ptolemaic Memphis* (Leuven, 1980), pp. 1–41.

2 J. Manning et al., 'Volcanic suppression of Nile summer flooding triggers revolt and constrains interstate conflict in ancient Egypt', *Nature Communications* 8 (2017), 1–9. For land sales, J. Manning, *Land and Power in Ptolemaic Egypt: The Structure of Land Tenure* (Cambridge, 2003).

3 G. Hölbl, *Geschichte des Ptolemäerreiches: Politik, Ideologie und religiöse Kultur von Alexander dem Grossen bis zur römischen Eroberung* (Darmstadt, 1994).

4 P. Hordern and N. Purcell, *The Corrupting Sea: A Mediterranean History* (Oxford, 2000).

5 J. McConnell et al., 'Extreme climate after massive eruption of Alaska's Okmok volcano in 43 BCE and effects on the late Roman Republic and Ptolemaic Kingdom', *PNAS* 117 (2020), 15,443–9.

6 Manning, 'Volcanic suppression of Nile summer flooding', 1–9.

7 McConnell, 'Extreme climate after massive eruption', 15,443–9.

8 S. Strunz and O. Braeckel, 'Did volcano eruptions alter the trajectories of the Roman Republic and the Ptolemaic Kingdom? Moving beyond black-box determinism', *PNAS* 117 (2020), 32,207–8.

9 S. Ager, 'Familiarity Breeds: Incest and the Ptolemaic Dynasty', *Journal of Hellenic Studies* 125 (2005), 1–34; S. Ager, 'The Power of Excess: Royal Incest and the Ptolemaic Dynasty', *Anthropologica* 48 (2006), 165–86.

10 M. Chauveau, *L'Egypte au temps de Cléopatre* (Paris, 1997).

11 L. Borges, *Le Conflit propagandiste entre Octavien et Marc Antoine: De l'usage politique de la uituperatio entre 44 et 30 a. C. n* (Brussels, 2016).

12 Scheidel, *Escape from Rome*, pp. 51–88.

13 Hordern and Purcell, *The Corrupting Sea*.

14 N. Roberts et al., 'The Mid-Holocene climatic transition in the Mediterranean: Causes and consequences', *The Holocene* 21 (2011), 3–13.

15 V. Nieto-Moreno et al., 'Tracking climate variability in the western Mediterranean during the Late Holocene: a multiproxy approach', *Climate of the Past* 7 (2011), 1395–1411; also G. Margaritelli et al., 'Persistent warm Mediterranean surface waters during the Roman period', *Scientific Reports* 10 (2020), 1–10.

16 K. Harper and M. McCormick, 'Reconstructing the Roman climate', in W. Scheidel (ed.), *The Science of Roman History* (Princeton, 2018), pp. 11–52.

17 McCormick et al., 'Climate Change during and after the Roman Empire', 169–220; also see P. Erdkamp, *The Grain Market in the Roman Empire* (Cambridge, 2005).

18 Fan Ye, in J. Holmgren, *Chinese Colonisation of Northern Vietnam: Administrative Geography and Political Development in the Tongking Delta, First to Sixth Centuries* AD (Canberra, 1980), pp. 5–6.

19 K. DuVal, 'The Mississippian Peoples' Worldview', in Raaflaub and Talbert, *Geography and Ethnography*, p. 90; B. Smith, 'The Archaeology of the Southeastern United States: from Dalton to De Soto, 10,500–500 BP', *Advances in World Archaeology* 5 (1986), 21, 35–51.

20 G. Cowgill, *Ancient Teotihuacan: Early Urbanism in Central Mexico* (New York, 2015).

21 D. Carballo, 'The Social Organisation of Craft Production and Interregional Exchange at Teotihuacan', in K. Hirth and J. Pilsbury (eds), *Merchants, Markets, and Exchange in the Pre-Columbian World* (Washington, DC, 2013), pp. 113–40.

22 L. Manzanilla, 'Cooperation and tensions in multiethnic corporate societies using Teotihuacan, Central Mexico, as a case study', *PNAS* 112 (2015), 9210–15; B. Álvarez-Sandoval et al., 'Genetic evidence supports the multiethnic character of Teopancazco, a neighborhood center of Teotihuacan, Mexico (AD 200–600)', *Plos ONE* 10 (2015), 1–19; N. Sugiyama et al., '¿Artistas mayas en Teotihuacan?', *Arqueología Mexicana* 142 (2016), 8.

23 J. Urrutia-Fucugauchi, 'Archaeomagnetic dating of the eruption of Xitle volcano, Basin of Mexico: Implications for the Mesoamerican centers of Cuicuilco and Teotihuacan', *Arqueología Iberoamericana* 30 (2016), 23–9.

24 S. Sugiyama, 'Worldview Materialized in Teotihuacan, Mexico', *Latin American Antiquity* 4 (1993), 103–29.

25 Scheidel, *Escape from Rome*, pp. 308–11.

26 Cicero, *De officiis*, 1.72–3.

27 Cicero, *Letter to his Brother Quintus*, 1.1.44.

28 Cicero, *De officiis*, 1.42.

29 M. van der Veen and J. Morales, 'Food Globalisation and the Red Sea: New Evidence from the Ancient Ports at Quseir al-Qadim, Egypt', in D. Agius et al (eds), *Human Interaction with the Environment in the Red Sea* (Leiden, 2017), pp. 254–89.

30 Pliny, *Historia naturalis*, ed. D. Eichholz, 10 vols (Cambridge, MA, 1962), 36.109–10.

31 J. Watts Belser, 'Opulence and Oblivion: Talmudic Feasting, Famine and the Social Politics of Disaster', *AJS Review* 38 (2014), 89–107.

32 E. Zanda, *Fighting Hydra-Like Luxury: Sumptuary Regulation in the Roman Republic* (London, 2011).

33 K. Lapatin, 'Luxus', in C. Mattusch (ed.), *Pompeii and the Roman Villa: Art and Culture around the Bay of Naples* (London, 2008), pp. 31–52.

34 Juvenal, Satire, 10, in S. Morton Brand (ed.), *Juvenal and Persius* (Cambridge, MA, 2004), p. 372.

35 Seneca, *Epistles* 104, p. 192.

36 J. McConnell et al., 'Lead pollution recorded in Greenland ice indicates European emissions tracked plagues, wars, and imperial expansion during antiquity', *PNAS* 115 (2018), 5726–31.

37 Hughes, *Environmental Problems*, p. 68.

38 Pliny, *Historia naturalis*, 33.1, 9, pp. 2–4.

39 Plato, *Critias*, III, in R. Bury (ed. and tr.), *Plato: Timaeus. Critias. Cleitophon. Menexenus. Epistles* (Cambridge, MA, 1929); A. Taylor, *Plato: Timaeus and Critias* (Cambridge, MA, 1929), p. 272.

40 Theophrastus, *De causis plantarum*, 5.14, ed. and tr. B. Einarson and G. Link, 3 vols (Cambridge, 2014), 3, pp. 130–2.

41 Lucretius, *De natura rerum*, V.821–30.

42 Columella, *De re rustica*, Preface, ed. and tr. H. Boyd, *On Agriculture* (Cambridge, MA, 1941), pp. 2–4; in general here, see Hughes, *Environmental Problems*, passim.

43 J. Mehta, *Vedic Wisdom: Selected Verses from the Vedas for Material Gain and Spiritual Happiness* (New Delhi, 2012), p. 140.

44 *Kohelet Rabbah*, 7:13.

45 Kaibara Ekken, *Ekken zenshū*, in W. T. de Bary, C. Gluck and
     A. Tiedemann (eds), *Sources of Japanese Tradition, 1600–1868* (New York,
     2006), pp. 99–100.

46 W. Harris, 'Defining and detecting Mediterranean deforestation,
     800 BCE to 700 CE', in W. Harris (ed.), *The Ancient Mediterranean
     Environment between Science and History* (Leiden, 2013), pp. 173–94.

47 N. Morley, 'Population size and social structure', in P. Erdkamp
     (ed.), *The Cambridge Companion to Ancient Rome* (Cambridge, 2013),
     pp. 29–44.

48 For some comments on demographic trends in Rome, P. Turchin and
     W. Scheidel, 'Coin hoards speak of population declines in Ancient
     Rome', *PNAS* 106 (2009), 17,276–9.

49 B. Graham and R. Van Dam, 'Modelling the Supply of Wood Fuel in
     Ancient Rome', in A. Izdebski and M. Mulryan (eds), *Environment and
     Society in the Long Late Antiquity* (Leiden, 2018), pp. 148–9.

50 P. Squatriti, 'Rye's Rise and Rome's Fall: Agriculture and Climate in
     Europe during Late Antiquity', in ibid., p. 160.

51 P. Halstead, 'The Contribution of Zooarchaeology', in P. Erdkamp and
     C. Holleran (eds), *The Routledge Handbook to Diet and Nutrition in the
     Roman World* (London, 2019), pp. 64–76.

52 P. Malanima, 'Energy Consumption in the Roman World', in Harris
     (ed.), *Ancient Mediterranean Environment*, pp. 13–36. Also V. Smil,
     America is Not a New Rome (Cambridge, MA, 2010), pp. 107–13;
     D. Van Limbergen, 'What Romans ate and how much they ate of it: Old
     and new research on eating habits and dietary proportions in classical
     antiquity', *Revue Belge de Philologie et d'Histoire* 96 (2018), 1049–92.

53 W. Harris, 'Bois et déboisement dans la Méditerranée antique', *Annales
     Histoire Science Sociale* 66 (2011), 105–40.

54 R. Gowland and P. Garnsey, 'Skeletal evidence for health, nutritional
     status and malaria in Rome and the Empire', in H. Eckardt (ed.), *Roman
     Diasporas: Archaeological Approaches to Mobility and Diversity in the Roman
     Empire* (Portsmouth, 2010), 1; Hughes, *Environmental Problems*, p. 104.

55 Van Limbergen, 'What Romans ate'.

56 Cassius Dio, 58.15.

57 C. Lo Giudice, 'L'impiego degli animali negli spettacoli romani: *venatio
     e damnatio ad bestias*', *Italies* 12 (2008), 361–95.

58 A. Rodrigues et al., 'Forgotten Mediterranean calving grounds of grey
     and North Atlantic right whales: evidence from Roman archaeological
     records', *Proceedings of the Royal Society B* 285 (2018), 1–9.

59 Malanima, 'Energy Consumption', pp. 22–4.

60 A. Marzano, *Harvesting the Sea: The Exploitation of Marine Resources in the Roman Mediterranean* (Oxford, 2013).

61 S. Corcoran, *The Empire of the Tetrarchs: Imperial Pronouncements and Government* AD *284–324* (Oxford, 2000), pp. 205–33.

62 P. Erdkamp, 'War, Food, Climate Change, and the Decline of the Roman Empire', *Journal of Late Antiquity* 12 (2019), 437–8.

63 D. Edwards, 'Meroe and the Sudanic Kingdoms', *Journal of African History* 39 (1998), 175–93. Also J. Goody, *Technology, Tradition and the State in Africa* (London, 1971).

64 Elvin, 'Three thousand years of unsustainable growth', 24.

65 Ibid., 25.

66 M. Hinkel, 'The water reservoirs in Ancient Sudan', in C. Bonnet (ed.), *Etudes Nubiennes*, 2 vols (Geneva, 1994), 2, pp. 171–6.

67 J. Crow, J. Bardill and R. Bayliss, *The Water Supply of Byzantine Constantinople* (London, 2008).

68 G. Sürmelihindi et al., 'Carbonates from the ancient world's longest aqueduct: A testament of Byzantine water management', *Geoarchaeology* 36 (2021), 643–59.

69 Scarborough, 'Ecology and Ritual', 139–45.

70 P. Harrison, 'Aspects of Water Management in the Southern Lowlands', in V. Scarborough and B. Isaac (eds), *Aspects of Water Management in the Prehispanic New World* (Greenwich, CT, 1993), pp. 71–120.

71 W. Jongman, J. Jacobs and G. Klein Goldewijk, 'Health and wealth in the Roman Empire', *Economics and Human Biology* 34 (2019), 138–50.

72 W. Scheidel, 'Germs for Rome', in C. Edwards and G. Woolf (eds), *Rome the Cosmopolis* (Cambridge, 2003), pp. 158–76.

73 McConnell et al., 'Lead pollution', 5728; K. Harper, 'Pandemics and passages to late antiquity; rethinking the plague of c.249–270 described by Cyprian', *Journal of Roman Archaeology* 28 (2015), 223–60; S. Huebner, 'The "Plague of Cyprian": A revised view of the origin and spread of a 3rd CE pandemic', *Journal of Roman Archaeology* 34 (2021), 151–74.

74 W. Scheidel, 'A model of demographic and economic change in Roman Egypt after the Antonine plague', *Journal of Roman Archaeology* 15 (2002), 97–114; though also W. Scheidel, 'Roman wellbeing and the economic consequences of the Antonine Plague', in E. Lo Cascio (ed.), *L'impatto della 'peste Antonina'* (Bari, 2012), 265–95.

75 B. Rossignol and S. Durost, 'Volcanisme global et variations climatiques de courte durée dans l'histoire romaine (Ier s. av. J.-C.–IVème s. ap.

J.-C.): leçons d'une archive glaciaire (GISP2)', *Jahrbuch des römisch-germanischen Zentralmuseums Mainz* 54 (2007), 395–438.

76 McCormick et al., 'Climate Change during and after the Roman Empire', 169–220.

77 K. Harper, *The Fate of Rome: Climate, Disease and the End of an Empire* (Princeton, 2017), p. 15.

78 A. Izdebski, 'Realising Consilience: How Better Communication between Archaeologists, Historians and Natural Scientists Can Transform the Study of Past Climate Change in the Mediterranean', *Quaternary Science Reviews* 136 (2016), 5–22.

79 Erdkamp, 'War, Food, Climate Change', 448.

80 F. Virkus, *Politische Strukturen im Guptareich (300–500 n. Chr)* (Wiesbaden, 2014); D. Bhandarkar, B. Chhabra and G. Gai, *Inscriptions of the Gupta Kings* (New Delhi, 1981).

81 In general, see N. Taleb, *Fooled by Randomness: The Hidden Role of Chance in Life and the Markets* (London, 2007).

82 S. Esmonde Cleary, *The Roman West, AD 200–500: An Archaeological Study* (Cambridge, 2013), pp. 455–82.

83 P. Sarris, 'Integration and Disintegration in the Late Roman Economy: The Role of Markets, Emperors, and Aristocrats', in L. Lavan (ed.), *Local Economies? Production and Exchange of Inland Regions in Late Antiquity* (Leiden, 2015), pp. 167–88.

84 P. Erdkamp, 'War, Food, Climate Change, and the Decline of the Roman Empire', *Journal of Late Antiquity* 12 (2019), 422–65.

85 P. Sheppard et al., 'Annual precipitation since 515 BC reconstructed from living and fossil juniper growth of Northeast Qinghai Province, China', *Climate Dynamics* 23 (2004), 869–81.

86 P. Heather, *The Fall of the Roman Empire: A New History of Rome and the Barbarians* (Oxford, 2006).

87 M. McCormick, 'Radiocarbon dating the end of urban services in a late Roman town', *PNAS* 116 (2019), 8096–8; A. Boozer, 'The Urbanisation of Egypt's Western Desert under Roman Rule', in M. Sterry and D. Mattingly (eds), *Urbanisation and State Formation in the Ancient Sahara and Beyond* (Cambridge, 2020), pp. 147–86; C. Wickham, *The Inheritance of Rome: A History of Europe from 400 to 1000* (London, 2009).

88 D. Edwards, 'The Archaeology of Sudan and Nubia', *Annual Review of Anthropology* 36 (2007), 211–28.

89 S. Hakenbeck et al., 'Practising pastoralism in an agricultural
    environment: An isotopic analysis of the Hunnic incursions on
    Pannonian populations', *PLOS One* 12 (2017), 1–25.

90 Science Daily, 'Tiller the Hun? Farmers in Roman Empire converted to
    Hun lifestyle – and vice versa', 22 March 2017.

91 Z. Safrai, 'Where are the fifth-century coins?', *Israel Numismatic Journal*
    18 (2011–14), 198–208.

92 D. Fuks et al., 'Dust clouds, climate change and coins: consiliences
    of paleoclimate and economy in the Late Antique southern Levant',
    Levant 49 (2017), 205–33.

93 U. Schamiloglu, 'Climate Change in Central Eurasia and the Golden
    Horde', *Golden Horde Review* 1 (2016), 9–10.

94 E. Cook, 'Megadroughts, ENSO, and the invasion of Late-Roman
    Europe by the Huns and Avars', in Harris (ed.), *Ancient Mediterranean
    Environment*, pp. 89–102.

95 Y. Zou, 'The decline of Pingcheng: climate change impact, vulnerability
    and adaptation in the Northern Wei dynasty, China', *Journal of
    Historical Geography* 58 (2017), 12–22.

96 D. Wang, *Longmen's Stone Buddhas and Cultural Heritage: When
    Antiquity Met Modernity in China* (Lanham, MD, 2020).

97 Sen, *Poverty and Famines*. Also see P. Slavin, 'Climate and famines: a
    historical reassessment', *WIREs Climate Change* 7 (2016), 433–47.

10 THE CRISIS OF LATE ANTIQUITY (AD C.500–C.600)

 1 Z. Zhang et al., 'Periodic temperature-associated drought/flood
    drives locust plagues in China', *Proceedings of the Royal Society B* 276
    (2009), 823–31; L. Stige et al., 'Thousand-year-long Chinese time series
    reveals climatic forcing of decadal locust dynamics', *PNAS* 103 (2007),
    16,188–93.

 2 Z. Zhang and D. Li, 'A possible relationship between outbreaks of the
    oriental migratory locust (*Locusta migratoria manilensis* Meyen) in China
    and the El Niño episodes', *Ecological Research* 14 (1999), 267–70.

 3 R. Brázdil et al., 'Past locust outbreaks in the Czech lands: do they
    indicate particular climate patterns?', *Theoretical and Applied Climatology*
    116 (2014), 343–57.

 4 *Chandogyya Upaniṣad.*

 5 Pliny, *Historia naturalis*, 1.11; D. Camuffo and S. Enzi, 'Locust Invasions
    and Climatic Factors from the Middle Ages to 1800', *Theoretical and
    Applied Climatology* 43 (1991), 48–9.

6 N. Stenseth, 'Plague dynamics are driven by climate variation', *PNAS* 103 (2006), 13,110–15.

7 R. Sallares, A. Bouwman and C. Anderung, 'The Spread of Malaria to Southern Europe in Antiquity: New Approaches to Old Problems', *Medical History* 48 (2004), 311–28.

8 A. Richter, *Letters and Epistolary Culture in Early Medieval China* (Seattle, 2013), pp. 90–1.

9 A. Chepstow-Lusty et al., 'Tracing 4,000 Years of Environmental History in the Cuzco Area, Peru, from the Pollen Record', *Mountain Research and Development* 18 (1998), 159–72.

10 S. Ingram, 'Settlement ecology in the precontact North American Southwest', in L. Kellett and E. Jones (eds), *Settlement Ecology of the Ancient Americas* (Abingdon, 2017), pp. 85–110.

11 A. Osborn, 'Snowblind in the Desert Southwest: Moisture Islands, Ungulate Ecology, and Alternative Prehistoric Overwintering Strategies', *Journal of Anthropological Research* 49 (1993), 135–64.

12 B. Onac et al., 'Late Holocene droughts and cave ice harvesting by Ancestral Puebloans', *Scientific Reports* 10 (2020).

13 F. Ljungqvist, 'A new reconstruction of temperature variability in the extra-tropical northern hemisphere during the last two millennia', *Geografiska Annaler: Series A. Physical Geography* 92 (2010), 339–51.

14 M. Lachniet, 'ad 550–600 Collapse at Teotihuacan', in Weiss (ed.), *Megadrought*, pp. 195–6.

15 G. Cowgill, 'An update on Teotihuacan', *Antiquity* 82 (2008), 962–75; P. Plunket and G. Urunuela, 'Social and cultural consequences of a late Holocene eruption of Popocatepetl in central Mexico', *Quaternary International* 151 (2006), 19–28.

16 U. Büntgen, '2500 Years of European Climate Variability and Human Susceptibility', *Science* 331 (2011), 578–82.

17 Y. Garcin et al., 'Early anthropogenic impact on Western Central African rainforests 2,600 y ago', *PNAS* 115 (2018), 3261–6; P. Giresse et al., 'Understanding the 2500 yr BP rainforest crisis in West and Central Africa in the framework of the Late Holocene: Pluridisciplinary analysis and multi-archive reconstruction', *Global Planetary Change* 192 (2020), 1–19.

18 D. Seidensticker et al., 'Population collapse in Congo rainforest from 400 CE urges reassessment of the Bantu Expansion', *Science Advances* 7 (2021), 1–13.

19 Squatriti, 'Rye's Rise and Rome's Fall', 160–9.

20 W. Swartz, 'There's No Place Like Home: Xie Lingyun's representation of his estate in "Rhapsody on Dwelling in the Mountains"', *Early Medieval China* 21 (2015), 21–37.

21 I. Nilsson, 'Nature Controlled by Artistry: The Poetics of the Literary Garden in Byzantium', in H. Bodin and R. Hedlund (eds), *Byzantine Gardens and Beyond* (Uppsala, 2013), pp. 15–29; D. Fairchild Ruggles, *Islamic Gardens and Landscapes* (Philadelphia, 2008); HRH The Prince of Wales, *Highgrove: An English Country Garden* (London, 2015); C. Clunas, *Fruitful Sites: Garden Culture in Ming Dynasty China* (London, 2004), pp. 16–18.

22 J. Elverskog, *The Buddha's Footprint: An Environmental History of Asia* (Philadelphia, 2020), pp. 51–3; L. Lancaster, 'Buddhism and Ecology: Collective Cultural Perceptions', in M. Tucker and D. Williams (eds), *The Interconnection of Dharma and Deeds* (Cambridge, MA, 1997), pp. 3–20.

23 F. Orsini, 'Clouds, Cuckoos and an Empty Bed: Emotions in Hindi–Urdu *Barahmasas*', in I. Rajamani, M. Pernau and K. Butler Schofield (eds), *Monsoon Feelings: A History of Emotions in the Rain* (New Delhi, 2018), pp. 97–136.

24 R. Inden, 'Imperial Purānas. Kashmire as Vaisnava Center of the World', in R. Inden, J. Walters and D. Ali (eds), *Querying the Medieval: Texts and the History of Practices in South Asia* (Oxford, 2000), pp. 29–98.

25 G. Hidas, *Mahāpratisarā-Mahāvidyārājñī: The Great Amulet, Great Queen of Spells. Introduction, Critical Editions and Annotated Translation* (New Delhi, 2012), p. 224.

26 G. Hidas, *A Buddhist ritual manual on agriculture: Vajratuṇḍasamayakalparāja – critical edition* (Berlin, 2019), pp. 11–22.

27 B. Goodman-Tchernov and J. Austin, 'Deterioration of Israel's Caesarea Maritima's ancient harbor linked to repeated tsunami events identified in geophysical mapping of offshore stratigraphy', *Journal of Archaeological Science: Reports* 3 (2015), 444–54; H. Dey, B. Goodman-Tchernov and J. Sharvit, 'Archaeological evidence for the tsunami of January 18, A.D. 749: a chapter in the history of Early Islamic Qâysariyah (Caesarea Maritima)', *Journal of Roman Archaeology* 27 (2014), 357–73.

28 Ammianus Marcellinus, *Res gestae*, 26.10.

29 R. Ott et al., 'Reassessing Eastern Mediterranean Tectonics and Earthquake Hazard from the 365 CE Earthquake', *AGU Advances* 2 (2021), 1–18. For a discussion of the trustworthiness of Ammianus'

account, G. Kelly, 'Ammianus and the Great Tsunami', *Journal of Roman Studies* 94 (2004), 141–67.

30 D. Pedrazzi et al., 'The Ilopango Tierra Blanca Joven (TBJ) eruption, El Salvador: Volcano-stratigraphy and physical characterization of the major Holocene event of Central America', *Journal of Volcanology and Geothermal Research* 377 (2019), 81–102.

31 V. Smith et al., 'The magnitude and impact of the 431 CE Tierra Blanca Joven eruption of Ilopango, El Salvador', *PNAS* 117 (2020), 26,061–8.

32 E. Cook, 'Megadroughts, ENSO and the invasion of Late-Roman Europe by the Huns and Avars', in W. Harris (ed.), *The Ancient Mediterranean Environment between Science and History* (Leiden, 2013), pp. 88–102.

33 McCormick, 'Climate Change during and after the Roman Empire', 191–5.

34 Polybius, *The Rise of the Roman Empire*, tr. I. Scott-Kilvert (London, 1979), VI, p. 304.

35 U. Büntgen et al., 'Cooling and societal change during the Late Antique Little Ice Age from 536 to around 660 AD', *Nature Geoscience* 9 (2016), 231–6.

36 M. Sigl et al., 'Timing and climate forcing of volcanic eruptions for the past 2500 years', *Nature* 523 (2015), 543–9.

37 S. Helama et al., 'Volcanic dust veils from sixth century tree-ring isotopes linked to reduced irradiance, primary production and human health', *Scientific Reports* 8 (2018), 1–11.

38 Smith, 'Magnitude and impact', 26061–8; R. Dull et al., 'Radiocarbon and geologic evidence reveal Ilopango volcano as source of the colossal "mystery" eruption of 539/40 CE', *Quaternary Science Reviews* 222 (2019), 1–17.

39 Mani, Tzachor and Cole, 'Global catastrophic risk', 1–5.

40 S. Fecht, 'Undersea volcanism may help explain medieval year of darkness', https://phys.org/news/2019-12-undersea-volcanism-medieval-year-darkness.html

41 Harper, *Fate of Rome*, p. 219. See Sigl et al., 'Timing and climate forcing of volcanic eruptions', 543–9.

42 P. Sarris, 'Climate and Disease', in E. Hermans (ed.), *A Companion to the Global Early Middle Ages* (Amsterdam, 2020), pp. 511–38.

43 M. Toohey et al., 'Climatic and societal impacts of a volcanic double event at the dawn of the Middle Ages', *Climatic Change* 136 (2016), 401–12.

44 Helama, 'Volcanic dust veils', 2.

45 T. Newfield, 'The Climate Downturn of 536–50', in S. White,
    C. Pfister and F. Mauelshagen (eds), *Palgrave Handbook of Climate
    History* (London, 2018), p. 470.

46 Sarris, 'Climate and Disease', p. 515.

47 R. Sinesky et al., 'Volcanic climate forcing, extreme cold and the
    Neolithic Transition in the northern US Southwest', *Antiquity 96*
    (2021), 123–41.

48 M. Axboe, 'Guld og gulder', in T. Capelle and C. Fischer (eds),
    Ragnarok. Odins verden (Silkeborg, 2005), pp. 41–56.

49 B. Gräslund and N. Price, 'Twilight of the gods? The "dust veil
    event" of AD 536 in critical perspective', *Antiquity* 86 (2012), 431–2;
    Price and Gräslund, 'Excavating the Fimbulwinter? Archaeology,
    geomythology and the climate event(s) of AD 536', in F. Riede (ed.), *Past
    Vulnerability: Volcanic Eruptions and Human Vulnerability in Traditional
    Societies Past and Present* (Aarhus, 2015), pp. 109–32.

50 B. Gräslund, 'Fimbulvintern. Ragnarök och klimatkrisen år 536–7', *Saga
    och Sed* (2007), 93–123.

51 See Büntgen, 'Cooling and societal change', 231–6; Harper, *Fate of Rome*,
    pp. 42–5.

52 F. Virkus, *Politische Strukturen im Guptareich (300–550 n. Chr)*
    (Wiesbaden, 2004), pp. 166–9; J. Mitchiner (ed.), *The Yuga Purana*
    (Kolkata, 2002), p. 103; Stoneman, *Greek Experience*, pp. 163–4.

53 J. Elverskog, *Buddhism and Islam on the Silk Road* (Philadelphia, 2010),
    pp. 29–30; R. Singh, 'The decline of Pātaliputra with special reference
    to geographical factors', *Proceedings of the Indian History Congress* 36
    (1975), 51–62.

54 Sarris, 'Climate and Disease', p. 528.

55 F. Chayette, 'The Disappearance of the Ancient Landscape and the
    Climate Anomaly of the Early Middle Ages: A Question to be Pursued?',
    Early Medieval Europe 16 (2008), 127–65; P. Sarris, *Empires of Faith: The
    Fall of Rome to the Rise of Islam* (Oxford, 2011), pp. 75–6.

56 M. Ramakrishna Bhat, *Varāhamihira's Brhat Samhita* (Delhi, 1997),
    pp. 328–9.

57 P. Dundas, 'Floods, Taxes, and a Stone Cow: A Jain Apocalyptic
    Account of the Gupta Period', *South Asian Studies* 30 (2014), 230–44.

58 Wu Jing, *The Essentials of Governance*, ed. and tr. H. De Weerdt, G.
    Dudbridge and G. van Beijeren (Cambridge, 2020), VIII, pp. 279–80.

59 A. Wright, *The Sui Dynasty: The Unification of China, AD 581–617*
    (New York, 1978), pp. 48–149; Lewis, *The Earliest Chinese Empires*
    (Cambridge, 2009).

60 L. Beramendi-Orosco et al., 'High-resolution chronology for the Mesoamerican urban center of Teotihuacan derived from Bayesian statistics of radiocarbon and archaeological data', *Quaternary Research* 71 (2009), 99–107; D. Carballo and M. Robb, 'Lighting the World: Teotihuacan and Urbanism in Central Mexico', in M. Robb et al., *Teotihuacan: City of Water, City of Fire* (San Francisco, 2017), pp. 12–19.

61 D. Pedrazzi et al., 'The Ilopango Tierra Blanca Joven (TBJ) eruption, El Salvador: Volcano-stratigraphy and physical characterization of the major Holocene event of Central America', *Journal of Volcanology and Geothermal Research* 377 (2019), 81–102.

62 J. Blomster (ed.), *After Monte Albán: Transformation and Negotiation in Oaxaca, Mexico* (Boulder, CO, 2008); C. Elson, 'Su transición de un centro secundario a un cacicazgo en la época Clásica tardía', in N. Robles Garcia and A. Riviera Guzmán (eds), *Memoria de la quinta mesa redonda de Monte Albán* (Mexico City, 2011), pp. 345–74.

63 M. Etayo-Cadavid et al., 'Marine radiocarbon reservoir age variation in *Donax obesulus* shells from northern Peru: Late Holocene evidence for extended El Niño', *Geology* 41 (2013), 599–602; B. Culleton, 'Intrashell Radiocarbon Variability in Marine Mollusks', *Radiocarbon* 48 (2006), 387–400.

64 I. Shimada, *Pampa Grande and the Mochica Culture* (Austin, TX, 1994), pp. 122–31.

65 Galvez and Briceño, 'The Moche in the Chicama Valley', in J. Pillsbury (ed.), *Moche Art and Archaeology in Ancient Peru* (Washington, DC, 2001), pp. 141–57.

66 T. Dillehay, 'Town and Country in Late Moche Times: A View from Two Northern Valleys', in Pillsbury, *Moche Art*, pp. 259–83; L. Castillo, 'Los últimos Mochicas en Jequetepeque', in S. Uzeda and E. Mujica (eds), *Moche: Hacia el final del milenio*, 2 vols (Lima, 2003), 2, pp. 65–123.

67 See here E. Benson, *The Worlds of the Moche on the North Coast of Peru* (Austin, TX, 2012), pp. 131–2.

68 I. Shimada et al., 'Cultural impacts of severe droughts in the prehistoric Andes: application of a 1,500-year ice core precipitation record', *World Archaeology* 22 (1991), 247–71.

69 C. Robin, 'Joseph, dernier roi de Himyar (de 522 à 525, ou une des années suivantes)', *Jerusalem Studies in Arabic and Islam* 34 (2008), 1–124.

70 D. Phillipson, *Foundations of an African Civilisation: Aksum and the Northern Horn, 1000 BC–AD 1300* (Addis Ababa, 2012), p. 209.

71 J. Phillips and J. Ford, 'The Aksumite quarries at Gobedra Hill and Adi Tsehafi', in D. Phillipson (ed.), *Archaeology at Aksum, Ethiopia, 1993–7* (London, 2000), pp. 229–46.

72 K. Butzer, 'Rise and Fall of Axum, Ethiopia: a geo-archaeological interpretation', *American Antiquity* 46 (1981), 471–95; F. Sulas et al., 'State formation and water resources management in the Horn of Africa', *World Archaeology* 41 (2009), 2–15.

73 Phillipson, *Foundations of an African Civilisation*, p. 210; D. Phillipson, 'Aksum, the entrepot, and highland Ethiopia, 3rd–12th centuries', in M. Mango, *Byzantine Trade, 4th–12th Centuries: The Archaeology of Local, Regional, and International Exchange* (Farnham, 2009), 353–68.

74 Phillipson, *Foundations of an African Civilisation*, pp. 210–12.

75 Helama, 'Volcanic dust veils', 7–9.

76 Procopius, *Hyper ton polemon*, 2.22–3, in *History of the Wars, Secret History, Buildings*, ed. and tr. H. Dewing, 7 vols (Cambridge, MA), I, pp. 450–72.

77 M. Meier, 'The "Justinianic Plague": The Economic Consequences of the Pandemic in the Eastern Roman Empire and its Cultural and Religious Effects', *Early Medieval Europe* 24 (2016), 267–92; K. Harper, 'Invisible Environmental History: Infectious Disease in Late Antiquity', *Late Antique Archaeology* 12 (2018), 116–31.

78 J. Haldon et al., 'Lessons from the past, policies for the future: resilience and sustainability in past crises', *Environment Systems and Decisions* 40 (2020), 287–97.

79 L. Mordechai et al., 'The Justinianic Plague: An inconsequential pandemic?', *PNAS* 116 (2019), 25,546–54; L. Mordechai and M. Eisenberg, 'Rejecting Catastrophe: The Case of the Justinianic Plague', *Past & Present* 244 (2019), 3–50.

80 C. Wickham, *Framing the Middle Ages: Europe and the Mediterranean, 400–800* (Oxford, 2004), p. 548; also here see J. Banaji, *Agrarian Change in Late Antiquity: Gold, Labour and Aristocratic Dominance* (Oxford, 2001).

81 M. Keller et al., 'Ancient Yersinia pestis genomes from across Western Europe reveal early diversification during the First Pandemic (541–750)', *PNAS* 116 (2019), 12,363–72.

82 P. Sarris, 'Viewpoint New Approaches to the "Plague of Justinian"', *Past & Present* 254 (2022), 333.

83 G. Bar-Oz et al., 'Ancient trash mounds unravel urban collapse a century before the end of Byzantine hegemony in the southern Levant', *PNAS* 116 (2019), 8239–48.

84 D. Fuks, 'The rise and fall of viticulture in the Late Antique Negev Highlands reconstructed from archaeobotanical and ceramic data', *PNAS* 117 (2020), 1–12; Sarris, 'Climate and Disease', p. 517.

85 R. Hoyland, *Arabia and the Arabs from the Bronze Age to the Coming of Islam* (London, 2001), pp. 231–2; H. Munt, 'Arabic and Persian Sources for Pre-Islamic Arabia', in G. Fisher (ed.), *Arabs and Empires before Islam* (Oxford, 2015), pp. 441–2.

86 Y. Park et al., 'Ancient familial Mediterranean fever mutations in human pyrin and resistance to *Yersinia pestis*', *Nature Immunology* 21 (2020), 857–67.

87 H. Yu et al., 'Palaeogenomic analysis of black rat (*Rattus rattus*) reveals multiple European introductions associated with human economic history', *Nature Communications* 13 (2022), 1–12.

88 Seidensticker, 'Population collapse in Congo rainforest', 7.

89 For the Central Asian origin, see P. de Barros Damgaard et al., '137 ancient human genomes from across the Eurasian steppes', *Nature* 557 (2018), 369–74; M. Keller, C. Paulus and E. Xoplaki, 'Die Justinianische Pest. Grenzen und Chancen naturwissenschaftlicher Ansätze für ein integratives Geschichtsverständnis', *Evangelische Theologie* 81 (2021), 385–400.

90 M. Green, 'Taking "pandemic" seriously: Making the Black Death global', in M. Green (ed.), *Pandemic Disease in the Medieval World: Rethinking the Black Death* (Kalamazoo, MI, 2014), pp. 1–61.

91 Sarris, 'Climate and Disease', pp. 511–37; H.-S. Lee, 'Environment, Epidemic and Power in Early Medieval China – Epidemics and Authority's Strategies in the Northern and Southern Dynasties of Early Medieval China', *Saengtae Whankyeong kwa Yeoksa (Korean Journal of Ecological and Environmental History)* 1 (2015), 82–128.

92 H. Pottier, 'L'empereur Justinien survivant à la peste bubonique (542)', *Travaux et Mémoires* 16 (2010), 685–91; Harper, *Fate of Rome*, p. 260.

93 M. Morony, 'Economic Boundaries? Late Antiquity and Early Islam', *Journal of the Economic and Social History of the Orient* 47 (2004), pp. 166–94.

94 D. Stathakopoulos, 'Population, Demography, and Disease', in E. Jeffreys, J. Haldon and R. Cormack (eds), *The Oxford Handbook of Byzantine Studies* (Oxford, 2008), pp. 310–12.

95 J. Koder, 'Climatic Change in Fifth and Sixth Centuries?', in P. Allen and E. Jeffreys, *The Sixth Century: End or Beginning* (Sydney, 1996).

96 J. Haldon et al., 'Lessons from the past, policies for the future: resilience and sustainability in past crises', *Environment Systems and Decisions* 40 (2020), 287–97.

97 F. Iversen, 'Big bang, lordship or inheritance? Changes in the settlement structure on the threshold of the Merovingian Period,

South-Eastern Norway', in J. Klapste (ed.), *Hierarchies in Rural Settlements* (Turnhout, 2013), pp. 341–58; A. Tvauri, 'The impact of the climate catastrophe of 536–7 AD in Estonia and neighbouring areas', *Estonian Journal of Archaeology* 18 (2014), 30–56.

98 F. Iversen, 'Estate Division: Social Cohesion in the Aftermath of AD 536-7', in F. Iversen and H. Petersson (eds), *The Agrarian Life of the North, 2000 BC–AD 1000* (Oslo, 2017), 41–75. See also D. Löwenborg, 'An Iron Age shock doctrine – did the AD 536–7 event trigger large-scale social changes in the Mälaren valley area?', *Journal of Archaeology and Ancient History* 4 (2012), 3–29; Büntgen et al., 'Cooling and societal change', 23–6.

99 A. Cameron, 'The Theotokos in Sixth-Century Constantinople: A City Finds its Symbol', *Journal of Theological Studies* 29 (1978), 79–108; K. Sessa, 'The New Environmental Fall of Rome: A Methodological Consideration', *Journal of Late Antiquity* 12 (2019), 240–3.

100 M. Meier, *Das andere Zeitalter Justinians. Kontingenzerfahrung und Kontingenzbewältigung im 6. Jahrhundert n. Chr* (Göttingen, 2004), pp. 307ff.; M. Meier, 'The "Justinianic Plague"', 284–7.

101 Gregory of Tours, *Historia Francorum*, ed. B. Krusch and W. Levison (Bonn, 1884), 6.46, pp. 286–7; C. Rohr, 'Signa apparuerunt, quae aut regis obitum adnunciare solent aut regiones excidium: Naturerscheinungen und ihre "Funktion" in der *Historia Francorum* Gregors von Tours', in D. Groh, M. Kempe and F. Mauelshagen (eds), *Naturkatastrophen: Beiträge zu ihrer Deutung, Wahrnemung und Darstellung in Text und Bild von der Antike bis ins 20. Jahrhundert* (Tübingen, 2003), pp. 65–78.

102 A. Stoclet, '*Consilia humana, ops divina, superstitio*: Seeking Succor and Solace in Times of Plague, with Particular Reference to Gaul in the Early Middle Ages', in L. Little (ed.), *Plague and the End of Antiquity: The Pandemic of 541–750* (Cambridge, 2012), pp. 137–8; Sarris, 'Climate and Disease', p. 529.

103 K. Høilund Nielsen, 'Endzeiterwartung – expecting the End of the World', *Neue Studien zur Sachsenforschung* 5 (2015), pp. 23–50.

104 K. Sessa, 'The New Environmental Fall of Rome: A Methodological Consideration', *Journal of Late Antiquity* 12 (2019), 240–3.

105 P. Jenkins, *Climate, Catastrophe, and Faith: How Changes in Climate Drive Religious Upheaval* (New York, 2021).

106 D. Keys, *Catastrophe: An Investigation into the Origins of the Modern World* (London, 1999), pp. 182–3; Sarris, 'Climate and Disease', pp. 531–2.

107 *The History of Theophylact Simocatta: An English Translation with Introduction and Notes*, ed. and tr. M. Whitby and M. Whitby (Oxford, 1986), 5.10, p. 147.

108 G. Zanchetta et al., 'Beyond one-way determinism: San Frediano's miracle and climate change in Central and Northern Italy in late antiquity', *Climatic Change* 165 (2021), 1–21.

11 THE GOLDEN AGE OF EMPIRE (c.600–c.900)

1 J. Howard-Johnston, *The Last Great War of Antiquity* (Oxford, 2021).

2 J. Howard-Johnston, *Witnesses to a World Crisis: Historians and Histories of the Middle East in the Seventh Century* (Oxford, 2010), esp. pp. 436–44.

3 *Chronicon Pascale 284–628 ad*, tr. M. Whitby and M. Whitby (Liverpool, 1989), pp. 169–70.

4 *The Armenian History attributed to Sebeos*, ed. and tr. R. Thomson and J. Howard-Johnston, 2 vols (Liverpool, 2001), 131, p. 90.

5 C. Zuckerman, 'La Petite Augusta et le Turc: Epiphania-Eudocie sur les monnaies d'Héraclius', *Revue Numismatique* 150 (1995), 113–26.

6 M. Whittow, 'Byzantium's Eurasian Policy in the Age of the Türk Empire', in N. Di Cosmo and M. Maas (eds), *Empires and Exchanges in Eurasian Late Antiquity: Rome, China, Iran and the Steppe, ca. 250–750* (Cambridge, 2018), pp. 271–86; J. Howard-Johnston, 'The Sasanians' Strategic Dilemma', in H. Börm and J. Wiesehöfer (eds), *Commutatio et contentio: Studies in the Late Roman, Sasanian and Early Islamic Near East in Memory of Zeev Rubin* (Düsseldorf, 2010), pp. 37–70.

7 D. Sinor, 'The establishment and dissolution of the Türk empire', in D. Sinor (ed.), *The Cambridge History of Early Inner Asia* (Cambridge, 1990), pp. 285–316.

8 R. Payne, 'The Silk Road and the Iranian Political Economy in Late Antiquity', *Bulletin of the School of Oriental and African Studies* 81 (2018), 227–50; E. de la Vaissière, *Sogdian Traders* (Leiden, 2005), pp. 209–10.

9 K. Rezakhani, *ReOrienting the Sasanians: East Iran in Late Antiquity* (Edinburgh, 2017); Sarris, *Empire of Faith*, pp. 157–8.

10 W. Pohl, *Die Awaren: Ein Steppenvolk im Mitteleuropa, 567–822 n. Chr* (Munich, 1988); W. Pohl (ed.), *Von den Hunnen zu den Türken – Reiterkrieger in Europa und Zentralasien* (Berlin, 2021).

11 D. Graff, 'Strategy and contingency: The Tang defeat of the Eastern Turks', in N. Di Cosmo (ed.), *Warfare in Inner Asian History* (Leiden, 2002), pp. 33–71.

12 J. Skaff, *Sui-Tang China and its Turko-Mongol Neighbours: Culture, Power, and Connections, 580–800* (Oxford, 2012), p. 58.

13 Ibid., pp. 149–50.

14 E. de la Vaissière, 'Oncles et frères: Les qaghans Ashinas et le vocabulaire turc de la parenté', *Turcica* 42 (2010), 267–78.

15 N. Di Cosmo, C. Oppenheimer and U. Büntgen, 'Interplay of environmental and socio-political factors in the downfall of the Eastern Türk Empire in 630 CE', *Climatic Change* 145 (2017), 390.

16 Ibid., 383–95; J. Fei, J. Zhou and Y. Hou, 'Circa AD 626 volcanic eruption, climatic cooling, and the collapse of the Eastern Turkic Empire', *Climatic Change* 81 (2007), 469–75.

17 Q. Ma, 'How Do Multiscale Interactions Affect Extreme Precipitation in Eastern Central Asia?', *Journal of Climate* 34 (2021), 7475–91.

18 S. Begzsuren et al., 'Livestock responses to droughts and severe winter weather in the Gobi Three Beauty National Park, Mongolia', *Journal of Arid Environment* 59 (2004), 785–96; Di Cosmo, Oppenheimer and Büntgen, 'Interplay of environmental and socio-political factors', 388.

19 C. Field et al., *Managing the Risks of Extreme Events and Disasters to Advance Climate Change Adaptation: Special Report of the Intergovernmental Panel on Climate Change* (Cambridge, 2012), p. 501; Di Cosmo, Oppenheimer and Büntgen, 'Interplay of environmental and socio-political factors', 392.

20 J.-Q. Fang and G. Liu, 'Relationship between climatic change and the nomadic southward migrations in eastern Asia during historical times', *Climatic Change* 22 (2012), 151–68.

21 Di Cosmo, Oppenheimer and Büntgen, 'Interplay of environmental and socio-political factors', 391–2; Z. Batjargal, *Lessons learnt from the Dzud, 1999–2000*, UNDP (United Nations Development Programme) (Ulanbaatar, 2000), p. 41.

22 Field et al., *Managing the Risks*, p. 502; J. Janzen, 'Mobile Livestock-Keeping in Mongolia: Present Problems, Spatial Organisation, Interactions between Mobile and Sedentary Population Groups and Perspectives for Pastoral Development', *Geography Research Forum* 25 (2005), pp. 62–82.

23 Di Cosmo, Oppenheimer and Büntgen, 'Interplay of environmental and socio-political factors', 390.

24 Ibid., 392. Also see Sneath, *Headless State*; N. Di Cosmo, *Ancient China and its Enemies: The Rise of Nomadic Power in East Asian History* (Cambridge, 2004).

25 J. Skaff, 'Ideological interweaving in Eastern Eurasia: Simultaneous Kingship and Dynastic Competition, 580–755', in Di Cosmo and Maas, *Empires and Exchanges*, pp. 386–7.

26 E. de la Vaissière, 'Ziebel Qaghan Identified', in C. Zuckerman, *Constructing the Seventh Century: Travaux et Mémoires* 17 (2014), 741–8.

27 de la Vaissière, 'Oncles et frères', 267–78; Skaff, *Sui-Tang China*, pp. 150–3.

28 See Skaff, *Sui-Tang China*, pp. 105ff.

29 D. McMullen, 'Bureaucrats and Cosmology: The Ritual Code of T'ang China', in Cannadine and Price (eds), *Rituals of Royalty*, pp. 181–236.

30 G. Heck, '"Arabia without Spices": An Alternative Hypothesis: The Issue of "Makkan Trade and the Rise of Islam"', *Journal of the American Oriental Society* 123 (2003), 547–76; J. Schiettecatte and C. Robin, *L'Arabie à la veille de l'Islam: un bilan clinique* (Paris, 2009).

31 U. von Freeden, 'Das Ende engzelligen Cloisonnés und die Eroberung Südarabiens durch die Sasaniden', *Germania* 78 (2000), 97–122.

32 Howard-Johnston, *Witnesses to a World Crisis*, pp. 436–60.

33 L. Little, 'Life and Afterlife of the First Plague Pandemic', in Little (ed.), *Plague and the End of Antiquity*, p. 8.

34 Qur'ān, 81–2.

35 T. Macintosh-Smith, *Arabs: A Three Thousand Year History of Peoples, Tribes and Empires* (New Haven, 2019), pp. 340–6.

36 R. Nicholson, *A Literary History of the Arabs* (London, 1998), p. 418.

37 A. Watson, 'The Arab agricultural revolution and its diffusion, 700–1100', *Journal of Economic History* 34 (1974), 8–35; A. Watson, *Agricultural Innovation in the Early Islamic World: The Diffusion of Crops and Farming Techniques* (Cambridge, 1983).

38 M. Decker, 'Plants and progress: rethinking the Islamic agricultural revolution', *Journal of World History* 20 (2009), 187–206.

39 D. Fuks, O. Amichay and E. Weiss, 'Innovation or preservation? Abbasid aubergines, archaeobotany, and the Islamic Green Revolution', *Archaeological and Anthropological Sciences* 12 (2020), 1–16.

40 John of Nikiu, *Khronike*, tr. R. Charles, *The Chronicle of John of Nikiu* (London, 1916), 120.17–28, pp. 193–4.

41 A. Gariboldi, 'Social conditions in Egypt under the Sasanian Occupation (619-29 AD)', *La Parola del Passato: Rivista di Studi Antichi* 64 (2009), 321–50.

42 S. Stark, *Die Alttürkenzeit in Mittel- und Zentralasien: Archäologische und historische Studien* (Wiesbaden, 2008), pp. 287–314.

43 R. Bulliet, *Conversion to Islam in the Medieval History: An Essay in Quantitative History* (Cambridge, 1979); R. Bulliet, *Islam: The View from the Edge* (New York, 1994).

44 A. Izdebski, *A Rural Economy in Transition: Asia Minor from Late Antiquity into the Early Middle Ages* (Warsaw, 2013).

45 M. Jacq-Hergoualc'h, *Malay Peninsula: Crossroads of the Maritime Silk Road* (Leiden, 2002), pp. 233–57.

46 O. Wolters, *Fall of Srivijaya* (Ithaca, NY, 1970), pp. 9–10.

47 H. Kulke, ' "Kadātuan Śrīvijaya" – Empire or Kraton of Śrīvijaya? A Reassessment of the Epigraphical Evidence', *Bulletin de l'École Française d'Extrême-Orient* 80 (1993), 159–81.

48 V. Lieberman, *Strange Parallels: Southeast Asia in Global Context, c.800–1830*, 2 vols (Cambridge, 2010–12). Also T. Maxwell, *The Gods of Asia: Image, Text and Meaning* (Delhi, 2007).

49 R. Hoyland, 'Physiognomy in Islam', *Jerusalem Studies in Arabic and Islam* 30 (2005), 361–402; A. Al-Azmeh, 'Barbarians in Arab Eyes', *Past & Present* 134 (1992), 3–18; Hippocrates, 'Airs, Waters, Places', in W. Jones (ed. and tr.), *Ancient Medicine*, 11 vols (Cambridge, 1923–2018), I, 12–23, pp. 104–32; Aristotle, *Politics*, 7.6, pp. 564–6.

50 D. Phillipson, 'Trans-Saharan gold trade and Byzantine coinage', *Antiquaries Journal* 97 (2017), 145–69.

51 Ibn Qutayba, *Kitāb al-maʿarif*, in J. Hopkins and N. Levtzion (eds), *Corpus of Early Arabic Sources for West African History* (Cambridge, 1981), p. 15.

52 al-Yaʾqūbī, *Tārīkh*, in Hopkins and Levtzion (eds), *Corpus of Early Arabic Sources*, p. 21.

53 P. Hardy, 'Medieval Muslim Philosophers on Race', in J. Ward and T. Scott (eds), *Philosophers on Race: Critical Essays* (Oxford, 2002), pp. 38–62; R. Schine, 'Conceiving the Pre-Modern Black-Arab hero: On the Gendered Production of Racial Difference in *Sīrat al-amīrah dhāt al-himmah*', *Journal of Arabic Literature* 48 (2017), 298–326.

54 D. Goldenberg, *The Curse of Ham: Race and Slavery in Early Judaism, Christianity, and Islam* (Princeton, 2003).

55 al-Yaʾqūbī, *Tārīkh*, p. 21.

56 Al-Masʿūdī, *Akhbār al-zamān*, in Hopkins and Levtzion (eds), *Corpus of Early Arabic Sources*, p. 35.

57 Al-Iṣṭakhrī, *Kitāb Masālik al-mamālik*, in ibid., p. 40.

58 Al-Bīrunī, *Kitāb al-jamāhir fī maʿrifat al- jamāhir*, in ibid., p. 59.

59 T. Insoll, 'Syncretism, Time and Identity: Islamic Archaeology in West Africa', in D. Whitcomb (ed.), *Changing Social Identity with the Spread of Islam: Archaeological Perspectives* (Chicago, 2004), pp. 89–101.

60 S. Takezawa and M. Cissé, 'Discovery of the Earliest Royal Palace in Gao and its Implications for the History of West Africa', *Cahiers d'Études Africaines* 208 (2012), 813–44.

61 M. Gomez, *African Dominion: A New History of Empire in Early and Medieval Africa* (Princeton, 2018), pp. 21–3; D. Lange, 'From Mande to Songhay: Towards a Political and Ethnic History of Medieval Gao', *Journal of African History* 35 (1994), 275–301.

62 L. Sebastian, *The Chaco Anasazi: Sociopolitical Evolution in the Prehistoric Southwest* (Cambridge, 1992).

63 T. Pauketat, *Ancient Cahokia and the Mississippians* (Cambridge, 2004).

64 L. Carlson and W. Keegan, 'Resource depletion in the prehistoric northern West Indies', in S. Fitzpatrick (ed.), *Voyages of Discovery: The Archaeology of Islands* (Westport, CT, 2004), pp. 85–107.

65 L. Newsom and E. Wing, *On Land and Sea: Native American Uses of Biological Resources in the West Indies* (Tuscaloosa, AL, 2004).

66 B. Marshak, 'The archaeology of Sogdiana', *Silk Road* 1 (2003), 2–8; G. Azarpay, *Sogdian Painting: The Pictorial Epic in Oriental Art* (Berkeley, 1981).

67 P. Owczarek et al., 'Relationships between loess and the Silk Road reflected by environmental change and its implications for human societies in the area of ancient Panjikent, central Asia', *Quaternary Research* (2017), 1–11.

68 J. Safran, *The Second Umayyad Caliphate: The Articulation of Caliphal Legitimacy in al-Andalus* (Cambridge, MA, 2000).

69 F. Grenet and E. de la Vaissière, 'The Last Days of Panjikent', *Silk Road Art and Archaeology* 8 (2002), 155–96.

70 D. Gutas, *Greek Thought, Arabic Culture: The Graeco-Arabic Translation Movement in Baghdad and Early ʿAbbāsid Society (2nd–4th/8th–10th Centuries* (London, 1998), pp. 34–52.

71 K. van Bladel, 'Eighth-Century Indian Astronomy in the Two Cities of Peace', in B. Sadeghi et al. (eds), *Islamic Cultures, Islamic Contexts: Essays in Honor of Professor Patricia Crone* (Leiden, 2015), pp. 257–94; R. Yoeli-Tlaim, *Reorienting Histories of Medicine: Encounters along the Silk Roads* (London, 2021).

72 J. Montgomery, *Al-Jāḥiẓ: In Praise of Books* (Edinburgh, 2013), p. 5.

73 R. Irwin (ed.), *Night and Horses and the Desert: An Anthology of Classical Arabic Literature* (London, 2002), p. 68.

74 Qu'ian, 13.23.

75 Ibid., 9:2.

76 Ibid., 36:56–7.

77 Exodus, 16:1–36; Numbers, 11.1–9.

78 Luke, 5:1–11.

79 Matthew, 14:13–21; Mark, 6:31–44; Luke, 9:12–17; John, 6:1–14.

80 S. Tlili, 'I Invoke God, Therefore I Am: Creation's Spirituality and
    its Ecologic Impact in Islamic Texts', in Parham and Westling, *Global
    History of Literature and the Environment*, pp. 107–22.

81 I. Özdemir, 'Toward an Understanding of Environmental Ethics from a
    Qur'anic Perspective', in R. Foltz, M. Denny and A. Baharuddin (eds),
    Islam and Ecology: A Bestowed Trust (Cambridge, 2003), pp. 3–38.

82 al-Buḥturī, *Dīwān al-Buḥturī* (Cairo, 1963), 3, 1631.

83 G. Dawkes et al., 'The form and abandonment of the city of Kuik-Marden,
    Otrar oasis, Kazakhstan in the Early Islamic period', *Journal of Islamic
    Archaeology* 6 (2020), 137–52; R. Sala and J. Deom, 'Medieval tortkuls of
    northern Tienshan and mid-low Syrdarya in the role of Eurasian steppes
    nomads in the development of world military art', in N. Masanov (ed.),
    *Proceedings of the International Conference Almaty 22–23 April 2010, in
    Commemoration of N. E. Masanov* (Almaty, 2010), pp. 263–86.

84 B. Yang et al., 'Late Holocene climatic and environmental changes in
    arid central Asia', *Quaternary International* 194 (2009), 68–78; Q. Pei
    and D. Zhang, 'Long-term relationship between climate change and
    nomadic migration in historical China', *Ecology and Society* 19 (2014),
    1–17; S. Fowell et al., 'Mid- to late Holocene paleoclimate evolution of
    the Lake Telmen Basin, north central Mongolia, based on palynological
    data', *Quaternary Research* 59 (2003), 353–63.

85 D. Waines, '"Luxury goods" in medieval Islamic societies', *World
    Archaeology* 34 (2003), 571–80.

86 *Annals of the Caliphs' Kitchens: Ibn Sayyār Al-Warrāq's Tenth-Century
    Baghdadi Cookbook*, tr. N. Nasrallah (Leiden, 2007), p. 29.

87 Ibid., pp. 35–6.

88 Ibid., p. 30.

89 Al-Thaʿālibī, *Yatīmat al-Dahr*, quoted in *Annals of the Caliphs'
    Kitchens*, p. 42.

90 M. Campopiano, 'Fiscalité et structures économiques et sociales en Irak
    de la conquête arabe à la crise du califat abbasside (VIIe–Xe siècles)',
    in S. Gilotte and E. Voguet (eds), *Terroirs d'Al-Andalus et du Maghreb
    VIIIe–XVe siècle: Peuplements, ressources et sainteté* (Paris, 2015), pp. 51–77.

91 C. Robinson, *Empire and Elites after the Muslim Conquest: The
    Transformation of Northern Mesopotamia* (Cambridge, 2000), pp. 44–50.

92 B. van Bavel, 'New Perspectives on Factor Markets and Ancient Middle Eastern Economies: A Survey', *Journal of the Economic and Social History of the Orient* 57 (2014), 145–72.

93 W. Toonen et al., 'A hydromorphic reevaluation of the forgotten river civilisations of Central Asia', *PNAS* 117 (2020), 32,982–8.

94 R. Adams, *Land behind Baghdad: A History of Settlement on the Diyala Plain* (Chicago, 1965), pp. 97–106.

95 M. Campopiano, 'Cooperation and Private Enterprise in Water Management in Iraq: Continuity and Change between the Sasanian and Early Islamic Periods (Sixth to Tenth Centuries)', *Environment and History* 23 (2017), 385–407.

96 A. Popovic, *La Révolte des esclaves in Iraq au IIIe/IXe siècle* (Paris, 1976), pp. 13–25.

97 B. van Bavel, M. Campopiano and J. Dijkman, 'Factor Markets in Early Islamic Iraq, c. 600–1100 AD', *Journal of the Economic and Social History of the Orient*, 57 (2014), 262–89.

98 Campopiano, 'Fiscalité', pp. 51–77.

99 H. Kennedy, *When Baghdad Ruled the World: The Rise and Fall of Islam's Greatest Dynasty* (Cambridge, MA, 2005), p. 194.

100 M. Canard, 'Baghdad au IVe siècle de l'Hégire (Xe siècle de l'ère chrétienne)', *Arabica* 9 (1962), 267–87.

101 Bar Hebraeus, *Ktābā d-maktbānūt zabnē*, ed. and tr. E. Budge, *The Chronography of Gregory Abul Faraj*, 2 vols (Oxford, 1932), 1, p. 164.

102 Matthew of Edessa, *Armenia and the Crusades, Tenth to Twelfth Centuries: The Chronicle of Matthew of Edessa* (Lanham, MD, 1993), 1, p. 19.

103 R. Ellenblum, *The Collapse of the Eastern Mediterranean: Climate Change and the Decline of the East, 950–1072* (Cambridge, 2012); X. Wang, 'Climate, Desertification, and the Rise and Collapse of China's Historical Dynasties', *Human Ecology* 38 (2010), 157–72.

104 S. Raphael, *Climate and Political Climate: Environmental Disasters in the Medieval Levant* (Leiden, 2013), p. 60.

105 A. Allouche, *Mamluk Economics: A Study and Translation of al-Maqrīzī's Ighāthah* (Salt Lake City, UT, 1994), p. 33; Raphael, *Climate and Political Climate*, p. 60.

106 Campopiano, 'Cooperation and Private Enterprise', 385–407.

107 Brooke, *Climate Change*, p. 358. For climate as a factor in the fall of the Tang dynasty, see for example P. Zheng et al., 'A Test of Climate, Sun, and Culture Relationships from an 1810-Year Chinese Cave Record', *Science* 322 (2008), 940–2; H. Cheng, R. Edwards and

G. Haug, 'Simulated warm periods of climate over China during the last two millennia: The Sui-Tang warm period versus the Song-Yuan warm period', *JGR Atmospheres* 120 (2015), 2229–41.

108 D. Johnson, 'The Last Years of a Great Clan: The Li Family of Chao Chün in Late T'ang and Early Sung', *Harvard Journal of Asiatic Studies* 37 (1977), 5–102.

109 N. Tackett, *The Destruction of the Medieval Chinese Aristocracy* (Cambridge, MA, 2014).

110 Ibid., p. 188.

111 P. Lorge, 'Introduction', in P. Lorge (ed.), *Five Dynasties and Ten Kingdoms* (Hong Kong, 2011), pp. 5–7.

112 R. Gill et al., 'Drought and the Maya collapse', *Ancient Mesoamerica* 18 (2007), 283–302.

113 S. Thompson et al. (eds), *Mayan Folktales; Cuentos folklóricos mayas* (Westport, CT, 2007), p. xxv.

114 G. Jones, *The Conquest of the Last Maya Kingdom* (Stanford, 1998).

115 P. Harrison, *The Lords of Tikal: Rulers of an Ancient Maya City* (London, 1999).

116 M. Golitko et al., 'Complexities of collapse: the evidence of Maya obsidian as revealed by social network graphical analysis', *Antiquity* 86 (2012), 507–23.

117 M. Golitko and G. Feinman, 'Procurement and Distribution of Pre-Hispanic Mesoamerican Obsidian 900 BC–AD 1250: A Social Network Analysis', *Journal of Archaeological Method and Theory* 22 (2015), 206–47.

118 S. Houston et al., 'A Teotihuacan complex at the Classic Maya city of Tikal, Guatemala', *Antiquity* 95 (2021), 1–9.

119 A. Thompson, G. Feinman and K. Prufer, 'Assessing Classic Maya multi-scalar household inequality in southern Belize', *Plos ONE* 16 (2021), 1–30.

120 K. Tsukamoto and O. Olguín, 'Ajpach' Waal: The hieroglyphic stairway at the Guzmán Group of El Palmar, Campeche, Mexico', in C. Golden, S. Houston and J. Skidmore (eds), *Maya Archaeology* (San Francisco, 2015), 30–55; O. Esparza Olguín and K. Tsukamoto, 'Espacios de la escenografía ritual', in A. Velasco Martínez and M. Vega (eds), *Los Mayas: Voces de piedra* (Mexico City, 2011), pp. 393–9.

121 V. Tiesler, A. Cucina, and M. Ramírez-Salomon, 'Permanent Dental Modifications among the Ancient Maya: Procedures, Health Risks, and Social Identities', in S. Burnett and J. Irish (eds), *A World View of Bioculturally Modified Teeth* (Gainesville, FL, 2017), pp. 270–84.

122 J. Cerezo-Román and K. Tsukamoto, 'The Life Course of a Standard-Bearer: A Nonroyal Elite Burial at the Maya Archaeological Site of El Palmar, Mexico', *Latin American Antiquity* 32 (2021), 274–91.

123 M. Canuto et al., 'Ancient lowland Maya complexity as revealed by airborne laser scanning of northern Guatemala', *Science* 361 (2018), 1–17.

124 P. Harrison, *The Lords of Tikal: Rulers of an Ancient Maya City* (London, 1999).

125 S. Martin, 'In the line of the founder: a view of dynastic politics at Tikal', in J. Sabloff (ed.), *Tikal: Dynasties, Foreigners, and Affairs of State – Advancing Maya Archaeology* (Santa Fe, NM, 2003), pp. 3–45.

126 T. Culbert et al., 'The population of Tikal, Guatemala', in T. Culbert and D. Rice (eds), *Precolumbian Population History in the Maya Lowlands* (Albuquerque, 1990), pp. 103–21.

127 V. Scarborough et al., 'Water and sustainable land use at the ancient tropical city of Tikal, Guatemala', *PNAS* 109 (2012), 12,408–13.

128 D. Lentz et al., 'Imperial resource management at the ancient Maya city of Tikal: A resilience model of sustainability and collapse', *Journal of Anthropological Archaeology* 52 (2018), 113–22.

129 D. Lentz et al., 'Forests, Fields, and the Edge of Sustainability at the Ancient Maya City of Tikal', *PNAS* 111 (2014), 18,513–18.

130 D. Kennett et al., 'Development and Disintegration of Maya Political Systems in Response to Climate Change', *Science* 338 (2012), 788–91; D. Hodell, J. Curtis and M. Brenner, 'Terminal Classic drought in the northern Maya lowlands inferred from multiple sediment cores in Lake Chichancanab (Mexico)', *Quaternary Science Reviews* 24 (2005), 1413–27; G. Haug et al., 'Climate and the collapse of Maya civilisation', *Science* 299 (2003), 1731–5.

131 N. Evans et al., 'Quantification of drought during the collapse of the classic Maya civilisation', *Science* 361 (2018), 498–501; N. Hodell et al., 'Solar forcing of drought frequency in the Maya lowlands', *Science* 292 (2001), 1367–70.

132 B. Turner and J. Sabloff, 'Classic Period collapse of the Central Maya Lowlands: Insights about human-environment relationships for sustainability', *PNAS* 109 (2012), 13,908–14.

133 B. Cook et al., 'Pre-Columbian deforestation as an amplifier of drought in Mesoamerica', *Geophysical Research Letters* 39 (2012), 1–5.

134 S. Fedick and L. Santiago, 'Large variation in availability of Maya food plant sources during ancient droughts', *PNAS* 119 (2021), 1–7.

135 D. Lobell et al., 'Nonlinear heat effects on African maize as evidenced
     by historical yield trials', *Nature Climate Change* 1 (2011), 42–5.

136 W. Carleton, D. Campbell and M. Collard, 'Increasing temperatures
     exacerbated Classic Maya conflict over the long term', *Quaternary
     Science Reviews* 163 (2017), 209–18.

137 C. Ting, 'Continuity and change in fine-ware production in the eastern
     Maya lowlands during the Classic to Postclassic transition (AD 800–1250)',
     Archaeological and Anthropological Sciences 10 (2018), 1913–31.

138 H. Haines, P. Willink and D. Maxwell, 'Stingray Spine Use and Maya
     Bloodletting Rituals: A Cautionary Tale', *Latin American Antiquity* 19
     (2008), 83–98.

139 G. Feinman and L. Nicholas, 'After Monte Albán in the Central
     Valleys of Oaxaca: A Reassessment', in R. Faulseit (ed.), *Beyond
     Collapse: Archaeological Perspectives on Resilience, Revitalization, and the
     Transformation in Complex Societies* (Carbondale, IL, 2016), pp. 43–69.

140 D. Lentz et al., 'Molecular genetic and geochemical assays reveal severe
     contamination of drinking water resevoirs at the ancient Maya city of
     Tikal', *Scientific Reports* 10 (2020), 1–9.

141 R. Markens, M. Winter and C. Martínez López, 'Ethnohistory,
     Oral History, and Archaeology at Macuilxochitl: Perspectives on the
     Postclassic Period (800–1521 CE) in the Valley of Oaxaca', in J. Blomster
     (ed.), *After Monte Albán: Transformation and Negotiation in Oaxaca,
     Mexico* (Boulder, CO, 2008), pp. 193–215.

142 J. Aimers, 'What Maya collapse? Terminal Classic variation in the
     Maya lowlands', *Journal of Archaeological Research* 15 (2007), 329–77;
     A. Demarest, P. Rice and D. Rice, 'The Terminal Classic in the Maya
     Lowlands: Assessing collapses, terminations, and transformations', in
     A. Demarest, P. Rice and D. Rice (eds), *The Terminal Classic in the
     Maya Lowlands: Collapse, Transition, and Transformation* (Boulder, CO,
     2004), pp. 545–72.

143 L. Lucero, J. Gunn and V. Scarborough, 'Climate change and Classic
     Maya water management', *Water* 3 (2011), 479–94.

144 Lentz et al., 'Imperial resource management', 119.

12 THE MEDIEVAL WARM PERIOD (C.900–C.1250)

1 H. Lamb, 'The Early Medieval Warm Epoch and its Sequel',
    Palaeogeography, Palaeoclimatology, Palaeoecology 1 (1965), 13–37.

2 T. Wozniak, *Naturereignisse im frühen Mittelalter. Das Zeugnis der
   Geschichtsschreibung vom 6. bis 11. Jahrhundert* (Berlin, 2020); H.
   Wanner, *Klima und Mensch. Eine 12000-jährige Geschichte* (Bern, 2016).

3 R. Hoffmann, *An Environmental History of Medieval Europe* (Cambridge, 2014).

4 C. Rohr, C. Camenisch and K. Pribyl, 'European Middle Ages', in White, Pfister and Mauelshagen (eds), *Palgrave Handbook of Climate History*, pp. 247–63.

5 J. Preiser-Kapeller, 'A collapse of the Eastern Mediterranean?', *Jahrbuch der Österreichischen Byzantinistik* 65 (2015), 201.

6 A. Izdebski, T. Słocyński and G. Koloch, 'Exploring Byzantine and Ottoman economic history with the use of palynological data: a quantitative approach', *Jahrbuch der Österreichischen Byzantinistik* 65 (2015), 67–110.

7 Cook, 'Megadroughts, ENSO, and the Invasion of Late-Roman Europe', 89–92.

8 D. Zhang, 'Evidence for the existence of the Medieval Warm Period in China', *Climatic Change* 24 (1994), 289–97.

9 Y. Zhang et al., ' "Medieval Warm Period": on the northern slope of central Tianshan Mountains, Xinjiang, NW China', *Geophysical Research Letters*, 36 (2009), 1–5.

10 F. Wang et al., 'Northern Hemisphere millennial temperature variability', *Nature Communications* 13 (2022), 1–10.

11 V. Trouet et al., 'Persistent positive North Atlantic oscillation mode dominated the medieval climate anomaly', *Science* 324 (2009), 78–80; J. Chen et al., 'Hydroclimatic changes in China and surroundings during the Medieval Climate Anomaly and Little Ice Age: spatial patterns and possible mechanisms', *Quaternary Science Reviews* 107 (2015), 98–111.

12 M. Mann et al., 'Global Signatures and Dynamical Origins of the Little Ice Age and Medieval Climate Anomaly', *Science* 326 (2009), 1256–60; H. Diaz et al., 'Spatial and Temporal Characteristics of Climate in Medieval Times Revisited', *Bulletin of the American Meteorological Society* 92 (2011), 1487–1500.

13 J. Vaquero and R. Trigo, 'A Note on Solar Cycle Length during the Medieval Climate Anomaly', *Solar Physics* 279 (2012), 289–94.

14 N. Graham et al., 'Support for global climate reorganization during the "Medieval Climate Anomaly" ', *Climate Dynamics* 37 (2011), 2017–45.

15 F. Malamud-Roam et al., 'Holocene paleoclimate records from a large California estuarine system and its watershed region: linking watershed climate and bay conditions', *Quaternary Science Reviews* 25 (2006), 1570–98; I. Shimada, 'The Late Prehistoric Coastal States', in L. Minelli (ed.), *The Inca World: The Development of Pre-Columbian Peru, AD 1000–1534* (Norman, OK, 2000), pp. 49–110.

16 S. Stine, 'Extreme and persistent drought in California and Patagonia
    during mediaeval time', *Nature* 369 (1994), 546–9; D. Meko et al.,
    'Medieval drought in the upper Colorado River Basin', *Geophysical
    Research Letters* 34 (2007), 1–5; H. Mackay, 'Spatial variation of
    hydroclimate in north-eastern North America during the last
    millennium', *Quaternary Science Reviews* 256 (2021), 1019.

17 S. Lüning et al., 'Hydroclimate in Africa during the Medieval Climate
    Anomaly', *Palaeography, Palaeoclimatology, Palaeoecology* 495 (2018),
    309–22.

18 S. Lüning, M. Gałka and F. Vahrenholt, 'Warming and Cooling: The
    Medieval Climate Anomaly in Africa and Arabia', *Paleoceanography*
    32 (2017), 1219–35.

19 C. Wickham, *Framing the Early Middle Ages: Europe and the
    Mediterranean, 400–800* (Oxford, 2005).

20 Wickham, *Inheritance of Rome*, pp. 229–230.

21 In general see M. McCormick, *Origins of the European
    Economy: Communications and Commerce*, AD *300–900* (Cambridge,
    2001); J. Zanden, *The Long Road to the Industrial Revolution: The
    European Economy in a Global Perspective, 1000–1800* (Leiden, 2009),
    pp. 1–91.

22 S. Reynolds, *Kingdoms and Communities in Western Europe, 900–1300*
    (Oxford, 1984); S. Reynolds, *Fiefs and Vassals: The Medieval Evidence
    Reinterpreted* (Oxford, 1994); C. West, *Reframing the Feudal
    Revolution: Political and Social Transformation between Marne and
    Moselle, c.800–c.1100* (Cambridge, 2013).

23 West, *Reframing the Feudal Revolution*.

24 P. Brown, *Through the Eye of a Needle: Wealth, the Fall of Rome, and the
    Making of Christianity in the West, 350–550* AD (Princeton, 2012).

25 A. Verhulst and G. Bublot, *L'Agriculture en Belgique – hier et aujourd'hui*
    (Brussels, 1980), p. 7.

26 G. Duby, *L'Économie rurale et la vie des campagnes dans l'Occident
    médiéval* (Paris, 1962), pp. 160–4; L. White, *Medieval Technology and
    Social Change* (Oxford, 1962).

27 S. Mileson and S. Brookes, *Peasant Perceptions of Landscape: Ewelme
    Hundred, South Oxfordshire, 500–1650* (Oxford, 2022).

28 T. Andersen, P. Jensen and C. Skovsgaard, 'The Heavy Plough and
    the Agricultural Revolution in Medieval Europe', EHES (European
    Historical Economics Society) Working Paper 70 (2014), 1–44;
    D. Acemoglu, S. Johnson and J. Robinson, 'Institutions as a Fundamental

Cause of Economic Growth', in P. Aghion and S. Durlauf (eds), *Handbook of Economic Growth* (Amsterdam, 2005), pp. 386–472.

29 M. Rubin, *Cities of Strangers: Making Lives in Medieval Europe* (Cambridge, 2020).

30 T. Whited et al., *Northern Europe: An Environmental History* (Denver, CO, 2005), pp. 45–72; A. Demandt, *Der Baum: Eine Kulturgeschichte* (Cologne, 2014).

31 C. Goodson, *Cultivating the City in Early Medieval Italy* (Cambridge, 2021), p. 1.

32 Schamiloglu, 'Climate Change in Central Eurasia', 12–14.

33 Ibn Faḍlān, *Book of Ahmad ibn Faḍlān*, in *Ibn Fadlan and the Land of Darkness: Arab Travellers in the Far North*, tr. P. Lunde and C. Stone (London, 2011), p. 12.

34 P. Raposeiro, 'Climate change facilitated the early colonization of the Azores Archipelago during medieval times', *PNAS* 118 (2021), 1–7.

35 A. Dugmore, C. Keller and T. McGovern, 'Norse Greenland settlement: reflections on climate change, trade, and the contrasting fates of human settlements in the North Atlantic islands', *Arctic Anthropology* 44 (2007), 12–36; B. Campbell, *The Great Transition: Climate, Disease and Society in the Late-Medieval World* (Cambridge, 2016), pp. 34–8.

36 K. Smith, G. Ólafsson and A. Pálsdóttir, 'Ritual responses to catastrophic volcanism in Viking Age Iceland: Reconsidering Surtshellir Cave through Bayesian analyses of AMS dates, tephrochronology, and texts', *Journal of Archaeological Science* 126 (2021), 1–19.

37 T. Thordarson et al., 'New estimates of sulfur degassing and atmospheric mass-loading by the 934a.d. Eldgjá eruption, Iceland', *Journal of Volcanology and Geothermal Research* 108 (2001), 33–54.

38 S. Goodacre, 'Genetic evidence for a family-based Scandinavian settlement of Shetland and Orkney during the Viking periods', *Heredity* 95 (2005) 129–35; S. Ebenesersdóttir, 'Ancient genomes from Iceland reveal the making of a human population', *Science* 360 (2018), 1028–32.

39 B. Wallace, 'The Norse in Newfoundland: L'Anse aux Meadows and Vinland', *Newfoundland and Labrador Studies* 19 (2003), 5–43.

40 M. Kuitens, 'Evidence for European presence in the Americas in AD 1021', *Nature* (2021), 1–13; P. Ledger, L. Girdland-Flink and V. Forbes, 'New horizons and L'Anse aux Meadows', *PNAS* 116 (2019), 15,341–3.

41 B. Einarsson, *Skálar og rúst á Stöð í Stöðvarfirði Stöð* (Reykjavik, 2019).

42 Dugmore, Keller and McGovern, 'Norse Greenland settlement', 14–16.

43 S. Sindbæk, 'Urbanism and exchange in the North Atlantic/Baltic, 600–1000 CE', in T. Hodos (ed.), *The Routledge Handbook of Archaeology*

and Globalisation (London, 2017), pp. 553–65; Frankopan, *Silk Roads*, pp. 102–35.

44 R. Kovalev, 'The Infrastructure of the Northern Part of the "Fur Road" between the Middle Volga and the East during the Middle Ages', *Archivum Eurasiæ Medii Aevi* 10 (2001), 25–64.

45 S. Socin et al., 'High-resolution dietary reconstruction of victims of the 79 CE Vesuvius eruption at Herculaneum by compound-specific isotope analysis', *Science Advances* 7 (2021), 1–9; L. Maravall Buckwalter and J. Baten, 'Valkyries: Was gender equality high in the Scandinavian periphery since Viking times? Evidence from enamel hypoplasia and height ratios', *Economics and Human Biology* 34 (2019), 181–93.

46 M. Church et al., 'Timing and mechanisms of deforestation of the settlement period in Eyjafjallsveit, southern Iceland', *Radiocarbon* 49 (2007), 659–72.

47 X. Keighley et al., 'Disappearance of Icelandic Walruses Coincided with Norse Settlement', *Molecular Biology and Evolution* 36 (2019), 2656–67.

48 T. McGovern et al., 'Landscapes of settlement in northern Iceland: Historical ecology of human impact and climate fluctuation on the millennial scale', *American Anthropologist* 109 (2007), 27–51.

49 J. Arnebord, N. Lynnerup and J. Heinemeier, 'Human Diet and Subsistence Patterns in Norse Greenland AD c.980–AD c.1450: Archaeological Interpretations', *Journal of the North Atlantic* 3 (2012), 119–33.

50 P. Kirch, 'The impact of the prehistoric Polynesians on the Hawaiian ecosytem', *Pacific Science* 36 (1982), 1–14.

51 P. Kirch, 'Hawaii as a Model System for Human Ecodynamics', *American Anthropologist* 109 (2007), 8–26.

52 I. Goodwin, S. Browning and A. Anderson, 'Climate windows for Polynesian voyaging to New Zealand and Easter Island', *PNAS* 111 (2014), 14,716–21; P. Kirch, *On the Road of the Winds: An Archaeological History of the Pacific Islands before European Contact* (Berkeley, 2017), pp. 191–208.

53 J. Flexner, *Oceania, 800–1800 CE* (Cambridge, 2021), pp. 11–12.

54 P. Kirch (ed.), *Tangatatau Rockshelter: The Evolution of an Eastern Polynesian Socio-Ecosystem* (Los Angeles, 2017).

55 D. Sear et al., 'Human settlement of East Polynesia earlier, incremental, and coincident with prolonged South Pacific drought', *PNAS* 117 (2021), 8813–19.

56 P. Nunn and J. Britton, 'Human–Environment Relationships in the Pacific Islands around AD 1300', *Environment and History* 7 (2001), 6–8.

57 P. Nunn, 'Nature–society interactions in the Pacific Islands', *Human Geography* 85 (2003), 219–29.

58 S. Lüning et al., 'The Medieval Climate Anomaly in Oceania', *Environmental Reviews* 28 (2020), 45–54.

59 M. Smith et al., 'Human-environment interactions in Australian drylands: exploratory time-series analysis of archaeological records', *The Holocene* 18 (2008), 389–401; I. Davidson, 'Prehistoric Australian demography', in B. Meehan and N White (eds), *Hunter-Gatherer Demography: Past and Present* (Sydney, 1990), pp. 41–58.

60 P. Veth, 'Cycles of aridity and human mobility: risk minimisation among Late Pleistocene foragers of the Western Desert, Australia', in P. Veth, M. Smith and P. Hiscock (eds), *Desert Peoples: Archaeological Perspectives* (Melbourne, 2005), pp. 100–15.

61 A. Williams et al., 'Hunter-gatherer response to late Holocene climatic variability in northern and central Australia', *Journal of Quaternary Science* 25 (2010), 831–8.

62 M. Smith and J. Ross, 'What happened at 1500–1000 cal. BP in central Australia? Timing, impact and archaeological signatures', *The Holocene* 18 (2008), 379–88.

63 T. Cohen et al., 'A pluvial episode identified in arid Australia during the Medieval Climatic Anomaly', *Quaternary Science Reviews* 56 (2012), 167–71.

64 S. Scudder, 'Evidence of Sea Level Rise at the Early Ostionan Coralie Site (GT-3), c. AD 700, Grand Turk, Turks and Caicos Islands', *Journal of Archaeological Science* 28 (2001), 1221–33.

65 I. Rouse, *The Tainos: Rise and Decline of the People Who Greeted Columbus* (New Haven, 1992); W. Keegan, 'Islands of Chaos', in A. Delpuech and C. Hofman (eds), *Late Ceramic Age Societies in the Eastern Caribbean* (Oxford, 1994), pp. 33–44; H. Petitjean Roget, 'Contribution à l'Étude du Troumassoïde et du Suazoïde (600–1200 AD). Une hypothèse sur les causes de la régression du Saladoïde aux Petites Antilles', in L. Alofs and R. Dijkhoff (eds), *Proceedings of the 19th International Congress for Caribbean Archaeology* (Aruba, 2001), pp. 227–38.

66 S. Fitzpatrick and W. Keegan, 'Human impacts and adaptations in the Caribbean Islands: An historical ecology approach', *Earth and Environmental Science Transactions of the Royal Society of Edinburgh* 98 (2007), 29–45.

67 R. Hall, 'Exploring the Mississippian Big Bang at Cahokia', in J. Quilter and M. Miller (eds), *A Pre-Columbian World* (Washington, DC, 2006),

pp. 187–229; C. Antoniuk, 'Cahokia's influence in the Yazoo Basin: a ceramic analysis of Early Mississippian features at the Carson site', *Southeastern Archaeology* 40 (2021), 116–34.

68 L. Benson, T. Pauketat and E. Cook, 'Cahokia's Boom and Bust in the Context of Climate Change', *American Antiquity* 74 (2009), 467–83.

69 D. McLean, 'Identification of the Changbaishan "Millennium" (B-Tm) eruption deposit in the Lake Suigetsu (SG06) sedimentary archive, Japan: Synchronisation of hemispheric-wide palaeoclimate archives', *Quaternary Science Reviews* 150 (2016), 301–7; S. Guillet et al., 'Climatic and societal impacts of a "forgotten" cluster of volcanic eruptions in 1108–1110 CE', *Scientific Reports* 10 (2020), 1–10.

70 L. Zhang, *The River, The Plain, and the State: An Environmental Drama in Northern Song China, 1048-1128* (Cambridge, 2016), pp. 1–19.

71 R. Bocinsky, 'Exploration and exploitation in the macrohistory of the pre-Hispanic Pueblo Southwest', *Science Advances* 2 (2016), 1–11; J. Coltrain and J. Janetski, 'The stable and radio-isotope chemistry of southeastern Utah Basketmaker II burials: Dietary analysis using the linear mixing model SISUS, age and sex patterning, geolocation and temporal patterning', *Journal of Archaeological Science* 40 (2013), 4711–30.

72 S. Ortman, 'Uniform Probability Density Analysis and Population History in the Northern Rio Grande', *Journal of Archaeological Method and Theory* 23 (2016), 95–126.

73 K. Roler Durand, 'Function of Chaco-Era Great Houses', *Kiva* 69 (2003), 141–69; T. Windes, 'This Old House: Construction and Abandonment at Pueblo Bonito', in J. Neitzel (ed.), *Pueblo Bonito: Center of the Chacoan World* (Washington, DC, 2003), pp. 14–32.

74 A. Watson et al., 'Early procurement of scarlet macaws and the emergence of social complexity in Chaco Canyon, NM', *PNAS* 112 (2015), 8238–43.

75 R. Reina and M. Kensinger, *The Gift of Birds: Featherworking of Native South American Peoples* (Philadelphia, 1991); H. King, *Peruvian Featherworks: Art of the Precolumbian Era* (New York, 2012); B. Malinowski, *Magic, Science and Religion and Other Essays* (Prospect Heights, IL, 1948).

76 A. Chepstow-Lusty et al., 'Tracing 4,000 Years of Environmental History in the Cuzco Area, Peru, from the Pollen Record', *Mountain Research and Development* 18 (1998), pp. 159–72; A. Chepstow-Lusty and M. Winfield, 'Inca agroforestry: Lessons from the past', *Ambio* 29 (2000), 322–8; K. Lane, 'Engineered highlands: the social organization of water in the ancient north-central Andes (AD 1000–1480)', *World Archaeology* 41 (2009), 169–90.

77 P. Williams and D. Nash, 'Imperial Interaction in the Andes: Huari and Tiwanaku at Cerro Baúl', in W. Isbell and H. Silverman (eds), *Andean Archaeology I* (New York, 2002), pp. 243–65.

78 P. Williams, 'Rethinking Disaster-Induced Collapse in the Demise of the Andean Highland States: Wari and Tiwanaku', *World Archaeology* 33 (2002), 361–74.

79 B. Owen, 'Distant Colonies and Explosive Collapse: The Two Stages of the Tiwanaku Diaspora in the Osmore Drainage', *Latin American Antiquity* 16 (2005), 45–80; also A. Kolata and C. Ortloff, 'Climate and collapse: agro-ecological perspectives on the drought of the Tiwanku State', *Journal of Archaeological Science* 20 (1993), 195–221.

80 G. Huckleberry, A. Caramanica and J. Quilter, 'Dating the Ascope Canal System: Competition for Water during the Late Intermediate Period in the Chicama Valley, North Coast of Peru', *Journal of Field Archaeology* 43 (2018), 17–30.

81 L. Thompson et al., 'Late Glacial Stage and Holocene tropical ice core records from Huascarán, Peru', *Science* 269 (1995), 46–50.

82 M. Bush, N. Mosblech and W. Church, 'Climate change and the agricultural history of a mid-elevation Andean montane forest', *The Holocene* 25 (2015), 1522–32.

83 W. Church and A. von Hagen, 'Chachapoyas: Cultural development at an Andean cloud forest crossroads', in Silverman and Isbell (eds), *Handbook of South American Archaeology*, pp. 903–26; W. Isbell, 'Wari and Tiwanaku: International identities in the central Andean Middle Horizon', in Silverman and Isbell (eds), *Handbook of South American Archaeology*, pp. 731–59; M. Biwer et al., 'Hallucinogens, alcohol and shifting leadership strategies in the ancient Peruvian Andes', *Antiquity* 96 (2022), 142–58.

84 V. Azevedo, 'Medieval climate Variability in the eastern Amazon-Cerrado regions and its archeological implications', *Scientific Reports* 9 (2019), 1–9.

85 A. Toso et al., 'Fishing intensification as response to Late Holocene socio-ecological instability in southeastern South America', *Scientific Reports* 11 (2021), 1–14. Also here see S. Lüning et al., 'The Medieval Climate Anomaly in South America', *Quaternary International* 508 (2019), 70–87.

86 T. Nagaoka et al., 'Paleodemography of a medieval population in Japan: analysis of human skeletal remains from the Yuigahama-minami site', *American Journal of Physical Anthropology* 131 (2006), 1–14.

87　I. Nagaoka et al., 'A health crisis during the Japanese Medieval Period: A new paleodemographic perspective', *International Journal of Paleopathology* 26 (2019), 145–56.

88　V. Lieberman and B. Buckley, 'The Impact of Climate Change on Southeast Asia, circa 950–1820: New Findings', *Modern Asian Studies* 46 (2012), 1063.

89　M. Green and L. Jones, 'The Evolution and Spread of Major Human Diseases in the Indian Ocean World', in G. Campbell and E.-M. Knoll (eds), *Disease Dispersion and Impact in the Indian Ocean World* (Basingstoke, 2020), p. 39; R. Nicholas, 'The Goddess Śītalā and Epidemic Smallpox in Bengal', *Journal of Asian Studies* 41 (1981), 21–44.

90　Green 'Climate and Disease in Medieval Eurasia'.

91　A. Winterbottom and F. Tesfaye (eds), *Histories of Medicine and Healing in the Indian Ocean World*, 2 vols (Basingstoke, 2016).

92　C. Ferrier, 'Sri Lanka and North India during the Gupta Period: Facts and fancy', *Indian Economic and Social History Review* 55 (2018), 249–81.

93　M. Smith, 'Indianization from the Indian Point of View: Trade and Cultural Contacts with Southeast Asia in the Early First Millennium C.E.', *Journal of the Economic and Social History of the Orient* 42 (1999), 1–26; C. Cereti, 'The Pahlavi Signatures on the Quilon Copper Plates (Tabula Quilonensis)', in W. Sundermann, A. Hintze and F. de Blois (eds), *Exegisti monumenta: Festschrift in Honour of Nicholas Sims-Williams* (Wiesbaden, 2009), pp. 31–50.

94　Campbell, *Great Transition*, pp. 41–3.

95　A. Pokharia et al., 'Variable monsoons and human adaptations: Archaeological and palaeoenvironmental records during the last 1400 years in north-western India', *The Holocene* 30 (2020), 1332–44; V. Lieberman, *Strange Parallels: Southeast Asia in Global Context, c. 800–1830*, 2 vols (Cambridge 2003–9), 2, pp. 77–92.

96　Lieberman and Buckley, 'Impact of Climate Change on Southeast Asia', 1065.

97　M. Bryson, *Goddess on the Frontier: Religion, Ethnicity, and Gender in Southwest China* (Stanford, 2016).

98　See R. von Glahn, *An Economic History of China: From Antiquity to the Nineteenth Century* (Cambridge, 2016); J. Chafee, 'Sung education: Schools, academies, and examinations', in J. Chafee and D. Twitchett (eds), *The Cambridge History of China: Sung China, 960–1279 AD*, 5.2 (Cambridge, 2015), pp. 286–320; W. Liu, 'The making of a fiscal state in Song China, 960–1279', *Economic History Review* 68 (2015), 48–78.

99 For Song dynasty, R. von Glahn, 'Revisiting the Song monetary revolution: A review essay', *International Journal of Asian Studies* 1 (2004), 159–78.

100 G. Wade, 'An Early Age of Commerce in Southeast Asia, 900–1300 CE', *Journal of Southeast Asian Studies* 40 (2009), 232.

101 M. Pirazzoli-t'Serstevens, 'Une Denrée recherchée: La céramique chinoise importée dans le golfe arabo-persique, IXe–XIVe siècles', *Mirabilia Asiatica* 2 (2005), 77.

102 Wink, *Al-Hind*, 1, p. 354.

103 Fisher, *Environmental History of India*, pp. 75–6.

104 A. Kanisetti, *Lords of the Deccan: Southern India from the Chalukyas to the Cholas* (Delhi, 2022), pp. 247ff.

105 S. Chemburkar, 'Dancing architecture at Angkor: "Halls with dancers" in Jayavarman VII's temples', *Journal of Southeast Asian Studies* 46 (2015), 518.

106 T. Sen, *Buddhism, Diplomacy and Trade: The Realignment of Sino-Indian Relations, 600–1400* (Honolulu, 2003), pp. 156–8.

107 J. Miksic and G. Yian Goh, *Ancient South East Asia* (Abingdon, 2017), pp. 397–8.

108 S. Balasubrahmanyam et al., *Later Chola Temples: Kulottunga I to Rajendra III (AD 1070–1280)* (Chennai, 1979); R. Champakalakshmi, 'The Study of Settlement Patterns in the Cola Periods: Some Perspectives', *Man and Environment* 14 (1989), 91–101.

109 R. Nagaswamy, 'Archaeological finds in South India: Esālam bronzes and copper-plates', *Bulletin de l'École Française d'Extrême Orient* 76 (1987), 1–68; D. Ali, 'Royal Eulogy as World History: Rethinking Copper-plate Inscriptions in Cōḷa India', in R. Inden, J. Walters and D. Ali (eds), *Querying the Medieval: Texts and the History of Practices in South Asia* (New York, 2000), pp. 165–229.

110 T. Srinavasan, *Irrigation and Water Supply: South India, 200 B.C.–1600 A.D.* (Chennai, 1991); K. Rajan, *Ancient Irrigation Technology: Sluice Technology in Tamil Nadu* (Thanjavur, 2008); Y. Subbarayalu, *South India under the Cholas* (New Delhi, 2012).

111 J. Shanmugasundaram et al., 'Societal response to monsoon variability in Medieval South India: Lessons from the past for adapting to climate change', *Anthropocene Review* 4 (2017), 110–35.

112 D. Evans et al., 'A comprehensive archaeological map of the world's largest preindustrial settlement complex at Angkor, Cambodia', *PNAS* 104 (2007), 14277–82; R. Fletcher et al., 'Redefining Angkor: Structure

and Environment in the Largest Low Density Urban Complex of the Pre-Industrial World', *Udaya* 4 (2003), 107–25.

113 P. Sharrock, 'Garuḍa, Vajrapāṇi and religious change in Jayavarman VII's Angkor', *Journal of Southeast Asian Studies* 40 (2009), 111–51.

114 C. Pottier, 'Some Evidence of an Inter-relationship between Hydraulic Features and Rice Field Patterns at Angkor during Ancient Times', *Journal of Sophia Asian Studies* 18 (2000), 99–119; R. Fletcher et al., 'The Development of the Water Management System of Angkor: A Provisional Model', *Indo-Pacific Prehistory Association Bulletin* 28 (2008), 57–66. Also see T. Maxwell and J. Poncar, *Of Gods, Kings and Men: The Reliefs of Angkor Wat* (London, 2007); Chemburkar, 'Dancing architecture at Angkor', 514–36.

115 D. Penny et al., 'Vegetation and Land-Use at Angkor, Cambodia', *Antiquity* 80 (2006), 599–614.

116 Liberman and Buckley, 'Impact of Climate Change on Southeast Asia', 1067; for Pagan, see M. Aung-Thwin, *Pagan: The Origins of Modern Burma* (Honolulu, 1985).

117 Wade, 'Early Age of Commerce', 245–6.

118 P.-Y. Manguin, 'Trading ships of the South China Sea: Shipbuilding techniques and their role in the development of Asian trade networks', *Journal of the Economic and Social History of the Orient* 36 (1993), 253–80; J. Stargardt, 'Indian Ocean Trade in the Ninth and Tenth Centuries: Demand, Distance, and Profit', *South Asian Studies* 30 (2014), 35–55.

119 *Chau Ju-kua: His work on the Chinese and Arab Trade in the Twelfth and Thirteenth Centuries, Entitled Chu-fan-chï*, tr. F. Hirth and W. Rockhill (St Petersburg, 1911), pp. 117–18.

120 Wade, 'Early Age of Commerce', 244, 261.

121 M. Hall, *Farmers, Kings and Traders: The People of Southern Africa, 200 AD–1860 AD* (Chicago, 1987), pp. 74–90.

122 T. Huffman, 'The Soapstone Birds from Great Zimbabwe', *African Arts* 18 (1985), 68–100.

123 S. Chirikure, *Great Zimbabwe: Reclaiming a 'Confiscated' Past* (Abingdon, 2021), pp. 26–50, 78–107.

124 B. Zhao, 'Global Trade and Swahili Cosmopolitan Material Culture: Chinese-Style Ceramic Shards from Sanje ya Kati and Songo Mnara (Kilwa, Tanzania)', *Journal of World History* 23 (2012), 49; S. Wynne-Jones, *A Material Culture: Consumption and Materiality on the Coast of Precolonial East Africa* (Oxford, 2016), pp. 12–16, 63.

125 J. Fleischer, 'Behind the Sultan of Kilwa's Rebellious Conduct: Local Perspectives on an International East African Town', in A. Reid and P. Lane (eds), *African Historical Archaeologies* (New York, 2004), pp. 91–124.

126 A. LaViolette and J. Fleischer, 'Developments in Rural Life on the Eastern African Coast, AD 700–1500', *Journal of Field Archaeology* 43 (2018), 380–98.

127 M. Wood et al., 'Zanzibar and Indian Ocean Trade in the First Millennium CE: The Glass Bead Evidence', *Archaeological and Anthropological Sciences* 9 (2017), 879–901; M. Horton, N. Boivin and A. Crowther, 'Eastern Africa and the Early Indian Ocean: Understanding Mobility in a Globalising World', *Journal of Egyptian History* 13 (2020), 393–4.

128 J. Fleischer, 'Rituals of Consumption and the Politics of Feasting on the Eastern African Coast, AD 700–1500', *Journal of World Prehistory* 23 (2010), 195–217.

129 A. Kotarba-Morley et al., 'Coastal landscape changes at Unguja Ukuu, Zanzibar: Contextualizing the archaeology of an early Islamic port of trade', *Journal of Island and Coastal Archaeology* (2022), 1–35.

130 H. De Weerdt, *Information, Territory, and Networks: The Crisis and Maintenance of Empire in Song China* (Cambridge, MA, 2016).

131 C.-C. Wu et al., 'Impact of the early-ripening Champa rice – Are they aus?', *Research Square* (2020), 1–12; E. Anderson, *The Food of China* (New Haven, 1988); P.-T. Ho, 'Early-ripening rice in Chinese history', *Economic History Review* 9 (1956), 200–18. For rice in general, see R. Spengler et al., 'A Journey to the West: The Ancient Dispersal of Rice Out of East Asia', *Rice* 14 (2021), 1–18.

132 E. Anderson, *Food and Environment in Early and Medieval China* (Philadelphia, 2014).

133 D. Herlihy, 'Outline of Population Developments in the Middle Ages', in B. Herrmann, R. Sprandel and U. Dirlmeier (eds), *Determinanten der Bevölkerungsentwicklung im Mittelalter* (Weinheim, 1987), pp. 1–23.

134 P. Schuster, 'Die Krise des Spätmittelalters. Zur Evidenz eines sozial- und wirtschaftgeschichtlichen Paradigmas in der Geschichtsschreibung des 20. Jahrhunderts', *Historische Zeitschrift* 269 (1999), 19–55.

135 For example, J. Xie, *Chinese Urbanism: Urban Form and Life in the Tang–Song Dynasties* (Singapore, 2020), p. 81.

136 Lieberman, *Strange Parallels*, 2, p. 91.

137 R. Fletcher, 'The water management network of Angkor, Cambodia', *Antiquity* 82 (2008), 658–70.

138 Ibn al-Qalānisī, *The Damascus Chronicle of the Crusades, Extracted and Translated from the Chronicle of Ibn al-Qalānisī*, tr. H. Gibb (London, 1932), p. 297.

139 Ibn al-Athīr, *The Annals of the Saljuq Turks: Selections from al-Kāmil fī l-taʾrīkh of ʿIzz al-Dīn Ibn al-Ibn al-Athīr*, tr. D. Richards (London, 2002), pp. 268–70; Raphael, *Climate and Political Climate*, pp. 41–2, 74.

140 ʿAbd al-Laṭīf al-Baghdādī, *A Physician on the Nile: A Description of Egypt and Journal of the Famine Years*, tr. T. Mackintosh-Smith (New York, 2021).

141 William of Tyre, *Chronicon*, ed. R. Huygens, *Willelmi Tyrensis Archiepiscopi*, 2 vols (Turnhout, 1986), 1, 8.4, pp. 388–9.

142 E. Xoplaki, 'The medieval climate anomaly and Byzantium: a review of the evidence on climatic fluctuations, economic performance and societal change', *Quaternary Science Reviews* 136 (2016), 229–52; J. Preiser-Kapeller and E. Mitsiou, 'The Little Ice Age and Byzantium within the Eastern Mediterranean, ca. 1200–1350: An Essay on Old Debates and New Scenarios', in M. Bauch and G. Schenk (eds), *The Crisis of the 14th Century: Teleconnections between Environmental and Societal Change?* (Berlin, 2019), pp. 203–4.

143 W. Klein, 'A Christian Heritage on the Northern Silk Road: Archaeological and Epigraphic Evidence of Christianity in Kyrgyzstan', *Journal of the Canadian Society for Syriac Studies* 1 (2001), 85–100; Toonen et al., 'Hydromorphic reevaluation', 32982–8; N. Pederson et al., 'Pluvials, droughts, the Mongol Empire, and modern Mongolia', *PNAS* 111 (2014), 4375–9.

144 N. Di Cosmo, 'State formation and periodization in Inner Asian History', *Journal of World History* 10 (1999), 1–40.

145 A. Lagamma, *Sahel: Art and Empires on the Shores of the Sahara* (New York, 2020), pp. 60, 87–91.

146 E. Bovill, *The Golden Trade of the Moors: West African Kingdoms in the Fourteenth Century* (Princeton, NJ, 1995); S. McIntosh, S. Gallaher and R. McIntosh, 'Tobacco Pipes from Excavations at the Museum Site, Jenne, Mali', *Journal of African Archaeology* 1 (2003), 171–99.

147 C. Rankin, C. Barrier and T. Horsley, 'Evaluating narratives of ecocide with the stratigraphic record at Cahokia Mounds State Historic Site, Illinois, USA', *Geoarchaeology* 36 (2021), 369–87; A. White et al., 'Fecal stanols show simultaneous flooding and seasonal precipitation change correlate with Cahokia's population decline', *PNAS* 116 (2019), 5461–6; DuVal, 'Mississippian Peoples' Worldview', 99–100.

148 B. Buckley et al., 'Climate as a contributing factor in the demise of Angkor, Cambodia', *PNAS* 107 (2010), 6748–52.

149 L. Briggs, *The Ancient Khmer Empire* (Banglamung, 1999), p. 258; P. Sharrock and C. Jacques, '"The Grief of Kings is the Suffering of their Subjects": A Cambodian King's Twelfth-Century Network of Hospitals', in C. Salguero (ed.), *In Buddhism and Medicine: An Anthology of Premodern Sources* (New York, 2017), pp. 226–32.

150 Sharrock, 'Garuḍa, Vajrapāṇi and religious change', 116; M. Masako, 'The Discovery of Buddhist Statues at Banteay Kdei Temple, an Angkor Monument in Cambodia: Report on the Discovery of 274 Discarded Buddhist Statues and Four-sided Buddhist Stone Pillar', *Journal of Sophia Asian Studies* 19 (2001), 267–84.

151 D. Penny et al., 'Geoarchaeological evidence from Angkor, Cambodia, reveals a gradual decline rather than a catastrophic 15th-century collapse', *PNAS* 116 (2019), 4871–6.

152 P. Strachan, *Imperial Pagan: Art and Architecture of Old Burma* (Honolulu, 1989); R. Wicks, *Money, Markets, and Trade in Early Southeast Asia: The Development of Indigenous Monetary Systems to* AD *1400* (Ithaca, NY, 1992).

153 Aung-Thwin, *Pagan: The Origins of Modern Burma*, pp. 167–81.

154 M. Htin Aung, *A History of Burma* (Cambridge, 1967).

155 Briggs, *Ancient Khmer Empire*, p. 258; also Sharrock, 'Garuḍa, Vajrapāṇi and religious change', 111–51.

156 Dugmore, Keller and McGovern, 'Norse Greenland Settlement', 14–15; A. Dugmore et al., 'Cultural adaptation, compounding vulnerabilities and conjunctures in Norse Greenland', *PNAS* 109 (2012), 3658–63; B. Zhao et al., 'Prolonged drying trend coincident with the demise of Norse settlements in southern Greenland', *Science Advances* 8 (2022), 1–8.

157 Dugmore, Keller and McGovern, 'Norse Greenland Settlement', 18; E. Roesdahl, 'Walrus Ivory: Demand, Supply, Workshops, and Vikings', in A. Mortensen and S. Arge (eds), *Viking and Norse in the North Atlantic: Select Papers from the Proceedings of the 14th Viking Congress* (Torshavn, 2005), pp. 182–91.

158 H. Gulløv, 'The nature of contact between native Greenlanders and Norse', *Journal of the North Atlantic* 1 (2008), 16–24; Dugmore et al., 'Cultural adaptation', 3662.

159 D. Sear et al., 'Human settlement of East Polynesia earlier, incremental, and coincident with prolonged South Pacific drought', *PNAS* 117 (2020), 8813–19; Flexner, *Oceania*, p. 13.

160 A. Ioannidis et al., 'Native American gene flow into Polynesia predating
    Easter Island settlement', *Nature* 583 (2020), 572–7.
161 Flexner, *Oceania*, p. 14.
162 Ibid., p. 15.
163 P. Nunn, 'The A.D. 1300 Event in the Pacific Basin', *Geographical
    Review* 97 (2007), 1–23.

13 DISEASE AND THE FORMATION OF A NEW WORLD
(C.1250–C.1450)

 1 M. Drompp, *Tang China and the Collapse of the Uighur Empire: A
   Documentary History* (Leiden, 2005); J. Shi, *The Economy of Western
   Xia: A Study of 11th to 13th Century Tangut Records* (Leiden, 2021).
 2 S. West, 'The Interpretation of a Dream: The Sources, Evaluation, and
   Influence of the "Dongjing Meng Hua Lu"', *T'oung Pao* 71 (1985), 67–8.
 3 P. Ebrey, *Emperor Huizong* (Cambridge, MA, 2014), pp. 480–1.
 4 Li Qingzhao, *Complete poems of Li Ch'ing-chao*, tr. K. Rexroth
   (New York, 1982), p. 57.
 5 M. Favreau, *The Horde: How the Mongols Changed the World*
   (Cambridge, MA, 2020); C. Atwood, 'Six Pre-Chinggisid Genealogies in
   the Mongol Empire', *Archivum Eurasiae Medii Aevi* 19 (2012), 5–58.
 6 Favreau, *The Horde*, pp. 33–6.
 7 N. Di Cosmo, 'State formation and periodization in Inner Asian
   history', *Journal of World History* 10 (1999), 1–40.
 8 T. Tekin, *A Grammar of Orkhon Turkic* (Bloomington, IN, 1968),
   pp. 229–95; for the *sülde*, S. Dmitriev, 'Istoriko-etnograficheskie aspekty
   politicheskoy kul'tury turko-mongolskikh kochevnikov' (unpublished
   PhD thesis, Museum of Anthropology and Ethnology, Russian Academy
   of Sciences, St Petersburg, 2000); Favreau, *The Horde*.
 9 D. Sneath (ed.), *Imperial Statecraft: Political Forms and Techniques of
   Governance in Inner Asia, Sixth–Twentieth Centuries*, (Bellingham, WA,
   2006), *passim*.
10 I. de Rachewiltz, 'The Title Činggis Qan/Qayan Reconsidered', in
   W. Heissig and K. Sagaster (eds), *Gedanke und Wirkung. Festschrift zum
   90. Geburtstag von Nikolaus Poppe* (Wiesbaden, 1989), pp. 281–98.
11 *The Secret History of the Mongols: A Mongolian Epic Chronicle of the
   Thirteenth Century*, tr. I. de Rachewiltz (Leiden, 2004), ch. 154. See
   D. Durand-Guédy and T. Masuya, 'Seasonal capitals with permanent
   buildings in the Mongol empire', in D. Durand-Guédy (ed.), *Turko-
   Mongol Rulers, Cities and City Life* (Leiden, 2013), pp. 224–35.

12 T. May, *The Mongol Art of War: Chinggis Khan and the Mongol Military System* (Barnsley, 2007).

13 N. Pederson et al., 'Pluvials, droughts, the Mongol Empire, and modern Mongolia', *PNAS* 111 (2014), 4375–9.

14 W. Atwell, 'Volcanism and Short-Term Climatic Change in East Asian and World History, c.1200–1699', *Journal of World History* 12 (2001), 42–3.

15 W. Farris, *Japan's Medieval Population: Famine, Fertility, and Warfare in a Transformative Age* (Cambridge, MA, 2006), 34ff.

16 Atwell, 'Volcanism and Short-Term Climatic Change in East Asian and World History', pp. 43–4.

17 A. Silverstein, *Postal Systems in the Pre-Modern Islamic World* (Cambridge, 2009), pp. 141–64.

18 See for example G. Lane, 'Persian Notables and the Families Who Underpinned the Ilkhanate', in R. Amitai and M. Biran (eds), *Nomads as Agents of Cultural Change: The Mongols and their Eurasian Predecessors* (Honolulu, 2015), pp. 182–213.

19 J. Pongratz et al., 'Coupled climate–carbon simulations indicate minor global effects of wars and epidemics on atmospheric CO_2 between AD 800 and 1850', *The Holocene* 21 (2011), 843–51; Jon Henley, 'Why Genghis Khan was good for the planet', *Guardian*, 26 January 2011.

20 Toonen et al., 'Hydromorphic reevaluation', 32,982–8; T. Khaidarov et al., 'Biological Aspects of the Historical Urban Development of Jochi Ulus (Latter Half of the 13th–14th Centuries)', *Social Sciences* 10 (2015), 1047–53; K. Campbell, 'The City of Otrar, Kazakhstan: Using Archaeology to Better Understand the Impact of the Mongol Conquest of Central Asia', in A. Otto, M. Herles and K. Kaniuth (eds), *Proceedings of the 11th International Congress on the Archaeology of the Ancient Near East* (Wiesbaded, 2020)

21 See above all T. Allsen, *Commodity and Exchange in the Mongol Empire: A Cultural History of Islamic Textiles* (Cambridge, 1997), and T. Allsen, *Culture and Conquest in Mongol Eurasia* (Cambridge, 2001).

22 G. Lane, *Genghis Khan and Mongol Rule* (Westport, CT, 2004), pp. 57–8.

23 T. Masuya, 'Seasonal capitals with permanent buildings in the Mongol Empire', in Durand-Guédy (ed.), *Turko-Mongol Rulers, Cities and City Life*, pp. 223–56; P. Jackson, *The Mongols and the Islamic World: From Conquest to Conversion* (New Haven, 2017), pp. 95ff.

24 J. Bemmann et al., 'Mapping Karakorum, the capital of the Mongol Empire', *Antiquity* (2021), 1–20; K. Sagaster, 'Die mongolische

Hauptstadt Karakorum', *Beiträge zur Allgemeinen und Vergleichenden Archäologie* 19 (1999), 113–28.

25 V. Egorov, Историческая география Золотая Орды в *XIII–XIV* вв (Moscow, 1985).

26 A. Goldschmidt, 'Epidemics and Medicine during the Northern Song Dynasty: Revival of Cold Damage Disorders (Shanghan)', *T'oung Pao* 93 (2007), 53–109.

27 Green, 'Climate and Disease in Medieval Eurasia'.

28 M. Green, 'The Four Black Deaths', *American Historical Review* 125 (2020), 1601–31.

29 N. Fancy and M. Green, 'Plague and the Fall of Baghdad (1258)', *Medical History* 65 (2021), 157–77.

30 M.-L. Derat, 'Du lexique aux talismans: occurrences de la peste dans la Corne de l'Afrique du XIIIe au XVe siècle', *Afriques* 9 (2018).

31 G. Chouin and C. Decorse, 'Prelude to the Atlantic Trade: New Perspectives', *Journal of African History* 51 (2010), 142–3; R. Hymes, 'A Hypothesis on the East Asian Beginnings of the Yersinia pestis Polytomy', in Green (ed.), *Pandemic Disease in the Medieval World*, pp. 285–308; G. Chouin, 'Reflections on Plague in African History (14th–19th c.)', *Afriques* 9 (2018), 1–47.

32 Green, 'Four Black Deaths', 1616–18; J. Masson Smith, 'Mongol Campaign Rations: Milk, Marmots, and Blood?', *Journal of Turkish Studies* 8 (1984), 223–8.

33 M. Green, 'How a microbe becomes a pandemic: a new story of the Black Death', *Lancet Microbe* 1 (2020), 311–12.

34 Y. Cui et al., 'Historical Variations in Mutation Rate in an Epidemic Pathogen, Yersinia pestis', *PNAS* 110 (2013), 577–82.

35 Fancy and Green, 'Plague and the Fall of Baghdad', 158.

36 Green, 'Four Black Deaths', 1629–30.

37 B. Alloway et al., 'Archaeological implications of a widespread 13th Century tephra marker across the central Indonesian Archipelago', *Quaternary Science Reviews* 155 (2017), 86–99.

38 R. Mutaqin, 'Landscape evolution on the eastern part of Lombok (Indonesia) related to the 1257 CE eruption of the Samalas Volcano', *Geomorphology* 327 (2019), 338–50.

39 J. Emile-Geay et al., 'Volcanoes and ENSO over the past millennium', *Journal of Climate* 21 (2008), 3134–48; M. Toohey et al., 'Disproportionately strong climate forcing from extratropical explosive volcanic eruptions', *Nature Geoscience* 12 (2019), 100–7; M. Baroni et al., 'Persistent Draining of the Stratospheric 10 Be Reservoir after

the Samalas Volcanic Eruption (1257 CE)', *Journal of Geophysical Research: Atmospheres* 29 (2016), 7082–97.

40 B. Campbell, 'Global Climates, the 1257 Mega-Eruption of Samalas volcano, Indonesia, and the English Food Crisis of 1258', *Transactions of the RHS* 27 (2017), 87–121.

41 Campbell, *Great Transition*, pp. 57–8.

42 M. Sigl et al., 'A new bipolar ice core record of volcanism from WAIS Divide and NEEM and implications for climate forcing of the last 2000 years', *Journal of Geophysical Research: Atmospheres* 118 (2013), 1151–69; Sigl et al., 'Timing and climate forcing of volcanic eruptions', 543–54.

43 S. Guillet et al., 'Climate response to the Samalas volcanic eruption in 1257 revealed by proxy records', *Nature Geoscience* 10 (2017), 123–8.

44 I. Usokin et al., 'Solar activity during the Holocene: the Hallstatt cycle and its consequence for grand minima and maxima', *Astronomy & Astrophysics* 587 (2016), 1–10.

45 M. Stuiver and P. Quay, 'Changes in atmospheric carbon-14 attributed to a variable Sun', *Science* 207 (1980), 11–19; W. Hong et al., 'Calibration curve from AD 1250 to AD 1650 by measurements of tree-rings grown on the Korean peninsula', *Nuclear Instruments and Methods in Physics Research Section B: Beam Interactions with Materials and Atoms* 294 (2013), 425–9; A. Fogtmann-Schulz et al., 'Changes in Solar Activity during the Wolf Minimum – New Insights from a High Resolution (^{14}C) Record Based on Danish Oak', *Radiocarbon* 63 (2021), 91–104.

46 G. Feulner and S. Rahmstorf, 'On the effect of a new grand minimum of solar activity on the future climate of the Earth', *Geophysical Research Letters* 37 (2010), 1–5.

47 Campbell, *Great Transition*, pp. 199–201.

48 E. Koster, 'Aeolian Environments', in E. Koster (ed.), *The Physical Geography of Western Europe* (Oxford, 2007), pp. 139–60.

49 M. De Keyzer, 'All we are is dust in the wind: The social causes of a "subculture of coping" in the late medieval coversand belt', *Journal for the History of Environment and Society* 1 (2016), 1–35; T. Soens, 'Threatened by the Sea, Condemned by Man? Flood Risk and Environmental Inequalities along the North Sea Coast, 1200–1800', in G. Massard-Guilbaud and R. Rodger (eds), *Environmental and Social Justice in the City: Historical Perspectives* (Cambridge, 2011), pp. 91–111.

50 Campbell, *Great Transition*, pp. 253–61.

51 R. Oram, ' "The Worst Disaster Suffered by the People of Scotland in Recorded History": Climate Change, Dearth and Pathogens in the Long Fourteenth Century', *Proceedings of the Society of Antiquaries of Scotland* 144 (2015), 223–44.

52 B. Campbell, 'The European Mortality Crises of 1346–52 and Advent of the Little Ice Age', in D. Collett and M. Schuh (eds), *Famines during the 'Little Ice Age' (1300–1800): Socionatural Entanglements in Premodern Societies* (Cham, 2018), pp. 23–4.

53 Campbell, *Great Transition*, p. 6.

54 Oram, 'Worst Disaster', 228.

55 J.-H. Lee, 'Climate Change in East Asia and Agricultural Production Activities in Koryŏ and Japan during the Twelfth–Thirteenth Centuries', *International Journal of Korean History* 12 (2008), 133–56; Q. Yan et al., 'Simulated warm periods of climate over China during the last two millennia: The Sui-Tang warm period versus the Song-Yuan warm period', *JGR Atmospheres* 120 (2015), 2229–41.

56 T. Li, 'The Mongol Yuan Dynasty and the Climate, 1260–1360', in Bauch and Schenk (eds), *Crisis of the 14th Century*, pp. 157–8, 163.

57 T. Brook, *Troubled Empire: China in the Yuan and Ming Dynasties* (Cambridge, MA, 2010), p. 72.

58 Li, 'The Mongol Yuan Dynasty', pp. 158, 165.

59 W. Tucker, 'Environmental Hazards, Natural Disasters, Economic Loss, and Mortality in Mamluk Syria', *Mamlūk Studies* 3 (1999), 114–15.

60 M. Bauch et al., 'A prequel to the Dantean Anomaly: the precipitation seesaw and droughts of 1302 to 1307 in Europe', *Climate of the Past* 16 (2020), 2343–58.

61 P. Slavin, 'The Great Bovine Pestilence and its economic and environmental consequences in England and Wales, 1318–50', *Economic History Review* 65 (2021), 1239.

62 D. Camuffo, et al., 'When the Lagoon was frozen over', *Journal of Mediterranean Geography* 21–2.

63 B. Campbell, 'Nature as historical protagonist', *Economic History Review* 63 (2010), 281–314.

64 P. Frankopan, *The First Crusade: The Call from the East* (London, 2012), pp. 119–20.

65 Anderson, Johnson and Koyama, 'Jewish Persecutions and Weather Shocks', 928–9.

66 S. van Hees, 'The Great Fire in Cairo in 1321', in G. Schenk (ed.), *Historical Disaster Experiences: Towards a Comparative and Transcultural History of Disasters across Asia and Europe* (Cham, 2017), 307–26.

67 R. Irwin, *The Middle East in the Middle Ages: The Early Mamluk Sultanate, 1250–1382* (Carbondale, IL, 1986), pp. 49–50.

68 E. Chaney, 'Revolt on the Nile: Economic Shocks, Religion, and Political Power', *Econometrica* 81 (2013), 2033–53.

69 A. Kors and E. Peters (eds), *Witchcraft in Europe, 400–1700: A Documentary History* (Philadelphia, 2001), p. 119.

70 P. Nanni, 'Facing the Crisis in Medieval Florence: Climate Variability, *Carestie*, and Forms of Adaptation in the First Half of the Fourteenth Century', in Bauch and Schenk (eds), *Crisis of the 14th Century*, pp. 169–89.

71 Ibn Faḍl Allāh al-ʿUmarī, *Masālik al-abṣār fī mamālik al-amṣār*, in Hopkins and Levtzion (eds), *Corpus of Early Arabic Sources*, pp. 268–71.

72 For Mansa Musa's journey, see H. Collet, 'Échos d'Arabie. Le Pèlerinage à La Mecque de Mansa Musa (724–725/1324–1325) d'après des Nouvelles Sources', *History in Africa* 46 (2019), 105–35.

73 Slavin, 'Great Bovine Pestilence', 1242–3.

74 Ibid., 1246–7.

75 W. Jordan, *The Great Famine: Northern Europe in the Early Fourteenth Century* (Princeton, 1996), pp. 182–8.

76 P. Slavin, 'Death by the Lake: Mortality Crisis in Early Fourteenth-Century Central Asia', *Journal of Interdisciplinary History* 50 (2019), 69–71.

77 D. Chwolson, *Syrisch-nestorianische Grabinschriften aus Semirjetschie, Neue Folge* (St Petersburg, 1897); Slavin, 'Death by the Lake', 62–4, 73; M. Spyrou et al., 'The source of the Black Death in fourteenth-century central Eurasia', *Nature* 606 (2022), 718–24.

78 M. Green, 'Taking "Pandemic" Seriously: Making the Black Death Global', in Green (ed.), *Pandemic Disease in the Medieval World*, pp. 32–4.

79 N. Varlık, 'New Science and Old Sources: Why the Ottoman Experience of Plague Matters', in Green (ed.), *Pandemic Disease in the Medieval World*, pp. 193–228.

80 Green, 'Four Black Deaths', 1622–3; Varlık, 'New Science and Old Sources', pp. 207–16.

81 V. Ciocîltan, *The Mongols and the Black Sea Trade in the Thirteenth and Fourteenth Centuries* (Leiden, 2012), pp. 150ff.

82 Campbell, 'European Mortality Crises', pp. 21–3.

83 Brázdil et al., 'Past locust outbreaks', 345.

84 Camuffo and Enzi, 'Locust Invasions', 56.

85 Campbell, 'European Mortality Crises', p. 19.

86 H. Helama et al., 'Something old, something new, something borrowed. New insights to human–environment interaction in medieval Novgorod inferred from tree rings', *Journal of Archaeological Science: Reports* 13 (2017), 341–50.

87 N. Di Cosmo, 'Black Sea Emporia and the Mongol Empire: A Reassessment of the Pax Mongolica', *Journal of the Economic and Social History of the Orient* 53 (2010), 83–108; M. Grinberg, 'Janibeg's Last Siege of Caffa (1346–1347) and the Black Death: The Evidence and Chronology Revisited', *Tiurkologicheskie Issledovaniya* 1/2 (2018), 19–32.

88 M. Bauch and A. Engel, 'Die 1340er Jahre als Schlüsseljahrzehnt der "Great Transition". Eine klimahistorische Perspektive auf den Vorabend des Schwarzen Todes', in A. Berner, S. Leenen and S. Maus (eds), *Pest! Sonderausstellung des LWL-Museums für Archäologie in Herne* (2019), pp. 76–82; J. Esper et al., 'Eastern Mediterranean summer temperatures since 730 CE from Mt. Smolikas tree-ring densities', *Climate Dynamics* 54 (2020), 1367–82.

89 H. Barker, 'Laying the Corpses to Rest: Grain, Embargoes, and *Yersinia pestis* in the Black Sea, 1346–48', *Speculum* 96 (2021), 91–126.

90 R. Hymes, 'Buboes in Thirteenth Century China: Evidence from Chinese Medical Writings', The Medieval Globe 8 (2022), 3–59

91 M. Green, 'Editor's Introduction', in Green (ed.), *Pandemic Disease in the Medieval World*, p. 9.

92 A. More, 'Next-generation ice core technology reveals true minimum natural levels of lead (Pb) in the atmosphere: Insights from the Black Death', *GeoHealth* 1 (2017), 211–19.

93 Petrarch, *Epistolae de rebus familiaribus et variae*, in R. Horrox (ed. and tr.), *The Black Death* (Manchester, 1994), p. 248.

94 G. Christakos et al., *Interdisciplinary Public Health Reasoning and Epidemic Modelling: The Case of Black Death* (Berlin, 2005), pp. 217–18.

95 P. Guzowski, C. Kuklo and R. Poniat, 'O metodach pomiaru natężenia epidemii i zaraz w preindustrialnej Europie w demografii historycznej', in K. Polek and Ł. Sroka (eds), *Epidemie w dziejach Europy: konsekwencje społeczne, gospodarcze i kulturowe* (Kraków, 2016), pp. 119–44; P. Guzowski et al., 'Did the Black Death reach the Kingdom of Poland in the middle of the 14th century?', *arXiv*: 2111/02714 – preprint (2021), 1–36; A. Izdebski et al., 'Palaeoecological data indicates land-use changes across Europe

linked to spatial heterogeneity in mortality during the Black Death pandemic', *Nature Ecology & Evolution* 6 (2022), 297–306.

96   O. Benedictow, *The Black Death: The Complete History, 1346–1354* (Woodbridge, 2004), p. 95.

97   A. Izdebski et al., 'Big Data Palaeoecology reveals significant variation in Black Death mortality in Europe', *Research Square* (2021).

98   S. Borsch, 'Plague Depopulation and Irrigation Decay in Medieval Egypt', in Green (ed.), *Pandemic Disease in the Medieval World*, pp. 125–6.

99   G. Geltner, 'The Path to Pistoia; Urban Hygiene before the Black Death', *Past & Present* 246 (2020), 3–33.

100 S. Borsch and T. Sabraa, 'Refugees of the Black Death: Quantifying Rural Migration for Plague and Other Environmental Disasters', *Annales de Démographie Historique* 134 (2017), 63–93; D. Davis, 'The Characteristics of Global Rat Populations', *American Journal of Public Health* 41 (1951), 158–63.

101 J. Loiseau, *Reconstruire la maison du sultan, 1350–1450: ruine et recomposition de l'ordre urbain au Caire*, 2 vols (Cairo, 2011).

102 Borsch and Sabraa, 'Refugees of the Black Death', 63–93; P. Galanaud et al., 'Mortality and demographic recovery in early post-black death epidemics: Role of recent emigrants in medieval Dijon', *PLOS One* 15 (2020), 1–20.

103 W. McNeill, *Plagues and Peoples* (New York, 1967), pp. 64–5.

104 K. Puchstein, 'Zur Vogelwelt der schleswig-holsteinischen Knicklandschaft mit einer ornitho-ökologischen Bewertung der Knickstrukturen', *Corax* 8 (1980), 62–106; R. Schreg, 'Plague and Desertion – A Consequence of Anthropogenic Landscape Change? Archaeological Studies in Southern Germany', in Bauch and Schenk (eds), *Crisis of the 14th Century*, p. 240.

105 Campbell, 'European Mortality Crises', pp. 26–9.

106 Frankopan, *Silk Roads*, pp. 191–2.

107 P. Slavin, 'Out of the West: Formation of Permanent Plague Reservoir in South-Central Germany (1349–1356) and its Implications', *Past & Present* 252 (2021), 3–51.

108 Henry Knighton, *Chronicon Henrici Knighton vel Cnitthon monachi Leycestrensis*, 2 vols (London, 1889–95), 2, pp. 62–3.

109 Oram, 'Worst Disaster', 230–1.

110 Al-Maqrīzī, *Kitāb al-Sulūk li-Maʿrifat Duwal al-Mulūk*, 12 vols (Cairo, 1956–73), 4, p. 227, cited in Borsch, 'Plague Depopulation', pp. 129–35.

111 Varlık, 'New Science and Old Sources', pp. 201–7.

112 Ibid., p. 196; N. Varlık, *Plague and Empire in the Early Modern Mediterranean World: The Ottoman Experience, 1347–1600* (Cambridge, 2015); R. Singer, 'The Black Death in the Maghreb: A Call to Action', *Journal of Medieval Worlds* 2 (2020), 115–23.

113 Chouin, 'Reflections on Plague in African History', 1–47; D. Gallagher and S. Dueppen, 'Recognizing plague epidemics in the archaeological record of West Africa', *Afriques* 9 (2018); Derat, 'Du lexique aux talismans: occurrences de la peste dans la Corne de l'Afrique du XIIIe au XIe siècle'; M. Green, 'Putting Africa on the Black Death Map: Narratives from Genetics and History', *Afriques* 9 (2018).

114 N. Langer, 'The Black Death in Russia: Its Effects upon Urban Labor', *Russian History* 2 (1975), 53–67.

115 G. Mellinger, 'The Silver Coins of the Golden Horde: 1310–1358', *Archivum Eurasiae Medii Aevi* 7 (1987–91), 153–211; U. Schamiloglu, 'The Impact of the Black Death on the Golden Horde: Politics, Economy, Society, Civilisation', *Golden Horde Review* 5 (2017), 331.

116 M. Balard, 'Black Sea Slavery in Genoese Notarial Sources, 13th–15th Centuries', in F. Roşu (ed.), *Slavery in the Black Sea Region, c.900–1900: Forms of Unfreedom at the Intersection between Christianity and Islam* (Leiden, 2022), pp. 19–40.

117 Schamiloglu, 'Impact of the Black Death', 328.

118 R. Frost, *The Oxford History of Poland–Lithuania*, vol. 1: *The Making of the Polish–Lithuanian Union, 1385–1569* (Oxford, 2015), pp. 18–35.

119 M. Biran, 'The Chaghadaids and Islam: The Conversion of Tarmashirin Khan (1331–34)', *Journal of the American Oriental Society* 122 (2002), 742–52; W. Klein, 'Nestorianische Inschriften in Kirgizistan: Ein Situationsbericht', *Symposium Syriacum* 7 (1998), 661–9.

120 Schamiloglu, 'Impact of the Black Death', 336–7.

121 Borsch, 'Plague Depopulation', pp. 135–48.

122 Mileson and Brookes, *Peasant Perceptions of Landscape*, pp. 245–50.

123 S. Borsch, *The Black Death in England and Egypt: A Comparative Study* (Cairo, 2005).

124 J. Klunk et al, 'Evolution of immune genes is associated with Black Death', Nature 611 (2022), 312–9.

125 S. DeWitte and M. Lewis, 'Medieval menarche: Changes in pubertal timing before and after the Black Death', *American Journal of Human Biology* 33 (2020), 1–15. Also see S. DeWitte, 'Stress, sex, and plague: Patterns of developmental stress and survival in pre- and post-Black Death London', *American Journal of Human Biology* 30 (2018), 1–15.

126 J. Teplitsky, 'Imagined Immunities: Medieval Myths and Modern Histories of Jews and the Black Death,' *AJS Review: The Journal of the Association for Jewish Studies* 46 (2022), 32–46.

127 J. Belich, *The World the Plague Made. The Black Death and the Rise of Europe* (Princeton, 2022), p. 50.

128 S. Einbinder, *After the Black Death: Plague and Commemoration among Iberian Jews* (Philadelphia, 2018).

129 Belich, *The World the Plague Made*, p. 99.

130 Ibid.

131 Ibid. pp. 131–5.

132 A. Carmichael, 'Plague Persistence in Western Europe: A Hypothesis', in Green (ed.), *Pandemic Disease in the Medieval World*, pp. 157–91.

133 G. Pinto, 'Dalla tarda antichità alla metà del XVI secolo', in L. Del Panta et al. (eds), *La popolazione italiana dal Medioevo ad oggi* (Rome, 1996), pp. 17–71, at pp. 59–61.

134 Belich, *The World the Plague Made*, pp. 88, 123.

135 Ibid, passim.

136 C. Camenisch, *Endlose Kälte. Witterungsverlauf und Getreidepreise in den Burgundischen Niederlanden im 15. Jahrhundert* (Basel, 2015).

137 C. Camenisch, 'The 1430s: a cold period of extraordinary internal climate variability during the early Spörer Minimum with social and economic impacts in north-western and central Europe', *Climate of the Past* 12 (2016), 2107–26.

138 M. Bauch, 'The Day the Sun Turned Blue: A Volcanic Eruption in the Early 1460s and its Possible Climatic Impact – a Natural Disaster Perceived Globally in the Late Middle Ages?', in G. Schenk (ed.), *Historical Disaster Experiences: A Comparative and Transcultural Survey across Asia and Europe* (Heidelberg, 2017), pp. 107–38.

139 C. Ballard, 'The Lizard in the Volcano: Narratives of the Kuwae Eruption', *Contemporary Pacific* 32 (2020), 98–123.

140 P. Brown, '*Ventus vehemens et terribilis per totam Angliam*: Responses and Reactions to a Short-term Crisis in the British Isles', in Bauch and Schenk (eds), *Crisis of the 14th Century*, pp. 24–42.

141 C. Pfister et al., 'The meteorological framework and the cultural memory of three severe winter-storms in early eighteenth-century Europe', *Climate Change* 101 (2010), 281.

142 J. Allen, 'Trees and their response to wind: mid Flandrian strong winds, Severn Estuary and inner Bristol Channel, southwest Britain', *Philosophical Transactions of the Royal Society B* 338 (1992), 335–64.

14 ON THE EXPANSION OF ECOLOGICAL HORIZONS
(C.1400–C.1500)

1 I. Wallerstein, *The Modern World System*, vol. 1: *Capitalist Agriculture and the Origins of the European World Economy in the Sixteenth Century* (New York, 1974).

2 B. Van Loo, *De Bourgondiërs* (Amsterdam, 2019).

3 R. Frost, *The Oxford History of Poland–Lithuania*, vol. 1: *The Making of the Polish Lithuanian Union, 1385–1569* (Oxford, 2015), pp. 28–30; D. Kołodziejczyk, *The Crimean Khanate and Poland–Lithuania* (Leiden, 2011), pp. 5–8.

4 Y. Ayalon, *Natural Disasters in the Ottoman Empire: Plague, Famine, and Other Misfortunes* (Cambridge, 2014), pp. 21–60; J. Emmert, 'The Battle of Kosovo: Early Reports of Victory and Defeat', in W. Vucinich and J. Emmert (eds), *Kosovo: Legacy of a Medieval Battle* (Minneapolis, MN, 1991), pp. 19–40; D. Djoedjević, 'The Tradition of Kosovo and the Formation of Modern Serbian Statehood in the Nineteenth Century', in Vucinich and Emmert (eds), *Kosovo*, pp. 309–30.

5 R. Bulliet, 'The Other Siege of Vienna and the Ottoman Threat: An Essay in Counter-Factual History', *ReOrient* 1 (2015), 11–22.

6 M. Iyigun, 'Luther and Suleyman', *Quarterly Journal of Economics* 123 (2008), 1465–94.

7 A. Mikhail, *God's Shadow: Sultan Selim, his Ottoman Empire, and the Making of the Modern World* (New York, 2020), pp. 248, 265.

8 V. Aksan and D. Goffman (eds), *The Early Modern Ottomans: Remapping the Empire* (New York, 2007).

9 P. Beaujard, 'Un seul système-monde avant le 16e siècle? L'océan Indien au cœur de l'intégration de l'hémisphère Afro-Eurasien', in P. Beaujard, L. Berger and P. Norel (eds), *Histoire globale, mondialisations et capitalisme* (Paris, 2009), pp. 82–148.

10 G. Casale, 'The Ottoman "Discovery" of the Indian Ocean in the Sixteenth Century', in J. Bentley, R. Bridenthal and K. Wagen (eds), *Seascapes: Maritime Histories, Littoral Cultures, and Transoceanic Exchanges* (Honolulu, 2007), pp. 87–104.

11 A. Peacock, 'Jeddah and the Red Sea Trade in the Sixteenth Century: Arabian Contexts and Imperial Policy', in D. Agius et al. (eds), *Human Interaction with the Environment in the Red Sea* (Leiden, 2017), pp. 290–322; S. Özbaran, 'A Turkish Report on the Red Sea and the Portuguese in the Indian Ocean (1525)', in S. Özbaran (ed.), *The Ottoman Response to European Expansion: Studies on Ottoman–Portuguese Relations*

in the Indian Ocean and Ottoman Administration in the Arab Lands during the Sixteenth Century (Istanbul, 1994), pp. 99–110.

12 V. Krebs, *Medieval Ethiopian Kingship, Craft, and Diplomacy with Latin Europe* (Cham, 2021).

13 Ibid., pp. 78–9.

14 Flexner, *Oceania*, p. 4.

15 T. Brook, 'Nine Sloughs: Profiling the Climate History of the Yuan and Ming Dynasties, 1260–1644', *Journal of Chinese History* 1 (2017), 54–5.

16 E. Farmer, *Zhu Yuanzhang and Early Ming Legislation: The Reordering of Chinese Society Following the Era of Mongol Rule* (Leiden, 1995); T. Brook, *The Troubled Empire: China in the Yuan and Ming Dynasties* (Cambridge, MA, 2010).

17 J. Yin, X. Fang and Y. Su, 'Correlation between climate and grain harvest fluctuations and the dynastic transitions and prosperity in China over the past two millennia', *The Holocene* 26 (2016), 1914–23; C. Gao et al., 'Volcanic climate impacts can act as ultimate and proximate causes of Chinese dynastic collapse', *Communications Earth and Environment* 2 (2021), 1–11.

18 Y. Tai et al., 'The impact of Ming and Qing dynasty maritime bans on trade ceramics recovered from coastal settlements in northern Sumatra, Indonesia', *Archaeological Research in Asia* 21 (2020), 1–18.

19 Ma Huan, *Ying-yai sheng-lan: The Overall Survey of the Ocean's Shores (1433), translated from the Chinese text edited by Feng Ch'eng-chün*, tr. J. Mills (Cambridge, 1970), pp. 127–8.

20 G. Wade, 'Early Muslim expansion in South-East Asia, eighth to fifteenth centuries', in D. Morgan and A. Reid (eds), *New Cambridge History of Islam*, vol. 3: *The Eastern Islamic World, Eleventh to Eighteenth Centuries* (Cambridge, 2011), pp. 388–9. For early gravestones, L. Kalus and C. Guillot, 'Réinterprétation des plus anciennes stèles funéraires islamiques nousantariennes: II. La stèle de Leran (Java) datée de 475/1082 et les stèles associées', *Archipel* 67 (2004), 17–36.

21 Wade, 'Early Muslim expansion', pp. 391–2.

22 T. Pigeaud, *Nagarakertagama: Java in the 14th Century: A Study in Cultural History: The Nāgara-Kĕrtāgama by Rakawi Prapañca of Majapahit, 1365 A.D.* (The Hague, 1960).

23 *Bābur-Nāma*, p. 186.

24 Ibid., p. 350.

25 Though see G. Miller et al., 'Abrupt onset of the Little Ice Age triggered by volcanism and sustained by sea-ice/ocean feedbacks', *Geophysical Research Letters* 39 (2012), 1–5.

26 Flexner, *Oceania*, pp. 28–30; G. Clark, D. Burley and T. Murray, 'Monumentality and the development of the Tongan maritime chiefdom', *Antiquity* 82 (2008), 994–1008.

27 G. Peterson, 'Indigenous island empires: Yap and Tonga considered', *Journal of Pacific History* 35 (2000), 5–27.

28 Flexner, *Oceania*, pp. 30–1.

29 C. Townsend, *Fifth Sun: A New History of the Aztecs* (New York, 2019).

30 R. Hassig, *Trade, Tribute, and Transportation: The Sixteenth-Century Political Economy of the Valley of Mexico* (Norman, OH, 1993), pp. 56–64.

31 C. Dodds Pennock, 'Aztecs Abroad? Uncovering the Early Indigenous Atlantic', *American Historical Review* 125 (2020), 787–814.

32 D. Kennett and N. Marwan, 'Climatic volatility, agricultural uncertainty, and the formation, consolidation and breakdown of preindustrial agrarian states', *Philosophical Transactions of the Royal Society A* 373 (2015), 1–17; C. Morehart, 'Mapping ancient chinampa landscapes in the Basin of Mexico: a remote sensing and GIS approach', *Journal of Archaeological Science* 39 (2012), 2541–51.

33 A. Palerm, *Obras Hidráulicas Prehispanicas en el Sistema Lacustre del Valle de México* (Mexico City, 1973), pp. 16–20.

34 R. Hassig, 'The Famine of One Rabbit: Ecological Causes and Social Consequences of a Pre-Columbian Calamity', *Journal of Anthropological Research* 37 (1981), 172–82; D. Stahle et al., 'Major Mesoamerican droughts of the past millennium', *Geophysical Research Letters* 38 (2011), 1–4.

35 A. Kovar, 'The Physical and Biological Environment of the Basin of Mexico', in W. Sanders et al. (eds), *The Teotihuacán Valley Project, Final Report, 1: The Natural Environment, Contemporary Occupation and 16th Century Population of the Valley* (University Park, PA, 1970), pp. 28–9; Gill et al., 'Drought and the Maya Collapse', p. 298.

36 N. Davies, *The Aztecs* (London, 1977), pp. 93–102.

37 Hassig, 'Famine of One Rabbit', 178–80.

38 M. Masson, 'Resiliency and cultural reconstitution of the postclassic Mayapan confederacy and its aftermath', in J. Hendon, L. Overholtzer and R. Joyce (eds), *Mesoamerican Archaeology: Theory and Practice* (Oxford, 2021), pp. 278–314; D. Kennett et al., 'Drought-Induced

Civil Conflict among the Ancient Maya', *Nature Communications* 13 (2022), 1–10.

39 R. Keating, 'Urban Settlement Systems and Rural Sustaining Communities: An Example from Chan Chan's Hinterland', *Journal of Field Archaeology* 2 (1975), 215–27; J. Moore and C. Mackey, 'The Chimú empire', in Silverman and Isbell (eds), *Handbook of South American Archaeology*, pp. 784–6.

40 Moore and Mackey, 'Chimú empire', pp. 789–96.

41 G. Prieto et al., 'A mass sacrifice of children and camelids at the Huanchaquito-Las Llamas site, Moche Valley, Peru', *Plos One* 14 (2019), 1–29.

42 D. Socha, J. Reinhard and R. Chávez Perea, 'Inca Human Sacrifices on Misti Volcano (Peru)', *Latin American Antiquity* 32 (2021), 138–53.

43 R. Covey, 'The Inca Empire', in Silverman and Isbell, *Handbook of South American Archaeology*, pp. 814–16; T. Bray, 'The role of chicha in Inca state expansion: a distributional study of Inca aríbalos', in J. Jennings and B. Bowser (eds), *Drink, Power and Society in the Andes* (Tallahassee, FL, 2009), pp. 108–32.

44 A. Chepstow-Lusty et al., 'Putting the rise of the Inca Empire within a climatic and land management context', *Climate of the Past* 5 (2009), 375–88.

45 M. Malpass and S. Alconini (eds), *Distant Provinces in the Inka Empire: Toward a Deeper Understanding of Inka Imperialism* (Iowa City, 2010), pp. 75–107, 151–71, 260–78.

46 J. Hyslop, *The Inca Road System* (New York, 1984).

47 Covey, 'Inca Empire', p. 816.

48 G. Atchison et al., 'Lost crops of the Incas: Origins of domestication of the Andean pulse crop tarwi, *Lupinus mutabilis*', *American Journal of Botany* 103 (2016), 1592–1606; A. Chepstow-Lusty et al., 'Evaluating social economic change in the Andes using oribatid mite abundances as indicators of domestic animal densities', *Journal of Archaeological Science* 34 (2007), 1178–86.

49 A. Franco, S. Galiani and P. Lavado, 'Long-term effects of the Inca Road', NBER Working Paper 28, 979 (2021), 1–30.

50 Z. Ahmad and L. Chicoine, 'Silk Roads to Riches: Persistence along an Ancient Trade Network', MPRA (Munich Personal RePEc Archive) Paper 105146, University Library of Munich (2021), 1–96.

51 W. Woods, 'Population nucleation, intensive agriculture, and environmental degradation: The Cahokia example', *Agriculture and Human Values* 21 (2004), 255–61; D. Pompeiani et al., 'The

environmental impact of a pre-Columbian city based on geochemical insights from lake sediment cores recovered near Cahokia', *Quaternary Research* 91 (2019), 714–28; S. Munoz et al., 'Cahokia's emergence and decline coincided with shifts of flood frequency on the Mississippi River', *PNAS* 112 (2015), 6319–24; White et al., 'Fecal stanols show simultaneous flooding and seasonal precipitation change correlate with Cahokia's population decline', 5461–6.

52 Rankin, Barrier and Horsley, 'Evaluating narratives of ecocide with the stratigraphic record at Cahokia Mounds State Historic Site, Illinois, USA', 369–87.

53 Benson, Pauketat and Cook, 'Cahokia's Boom and Bust', 467–83; T. Pauketat and T. Emerson (eds), *Cahokia: Domination and Ideology in the Mississippian World* (Lincoln, NB, 1997).

54 Toso, 'Fishing intensification', 1–14; R. Grove, 'El Niño Chronology and the Little Ice Age', in R. Grove and G. Adamson, *El Niño in World History* (London, 2018), pp. 53–5.

55 L. Golombek and D. Wilber, *The Timurid Architecture of Iran and Turan*, 2 vols (Princeton, 1988).

56 P. Jackson, *The Delhi Sultanate: A Political and Military History* (Cambridge, 1999); E. Barbier, *Emergence of the World Economy: How Economies have Developed through Natural Resource Exploitation* (Cambridge, 2011), pp. 189–90.

57 Chŏng Tojŏn, 'Sambong Chip', cited by P. Lee and W. de Bary (eds), *Sources of Korean Tradition*, vol. 1: *From Early Times through the Sixteenth Century* (New York, 1997), p. 328.

58 King Sejong, 'Sejong silip', in ibid., p. 331.

59 Ung Qua, 'Bing Ngo dai cao', cited by G. Dutton, J. Werner and J. Whitmore (eds), *Sources of Vietnamese Tradition* (New York, 2012), p. 93.

60 N. Huy and T. Tai, *The Lê Code*, 2 vols (Athens, OH, 1987), 1, pp. 98–9.

61 L. Jardine, *Worldly Goods* (London, 1996), pp. 14–17.

62 Barbier, *Emergence of the World Economy*, pp. 165–6.

63 E. Hobsbawm, *Industry and Empire: From 1750 to the Present Day* (London, 1968).

64 F. Trentmann, *Empire of Things: How We Became a World of Consumers, from the Fifteenth Century to the Twenty-First* (London, 2016), pp. 23–4.

65 C. Glacken, *Traces on the Rhodian Shore: Nature and Culture in Western Thought from Ancient Times to the End of the Eighteenth Century* (Berkeley, 1967), p. 338.

66 Grove, *Green Imperialism*, p. 26.

67 P. Warde, *The Invention of Sustainability: Nature and Destiny, c.1500–1870* (Cambridge, 2018), pp. 80, 67–8.

68 J. Richards, *The Unending Frontier: An Environmental History of the Early Modern World* (Berkeley, 2003), p. 25.

69 A. Murali, 'Whose Trees? Forest Practices and Local Communities in Andhra, 1600–1922', in Arnold and Guha (eds), *Nature, Culture, Imperialism*, pp. 88–9.

70 A. Massing, 'Mapping the Malagueta coast: A history of the Lower Guinea Coast, 1460–1510, through Portuguese maps and accounts', *History in Africa* 36 (2009), 331–65.

71 P. Freedman, *Out of the East: Spices and the Medieval Imagination* (New Haven, 2008), pp. 195–6.

72 W. Phillips and C. Rahn Phillips, *Worlds of Christopher Columbus* (Cambridge, 1992); F. Fernández-Armesto, *1492: The Year the World Began* (London, 2010).

73 *The Four Voyages of Columbus*, ed. and tr. J. Cohen (London, 1969), p. 117.

74 Ibid., pp. 79–80.

75 Ibid., p. 67.

76 L. Collingham, *The Hungry Empire: How Britain's Quest for Food Shaped the Modern World* (London, 2017), pp. 6–7.

77 J. de Acosta, *Natural and Moral History of the Indies*, ed. J. Mangan, tr. F. López-Morillas (Durham, NC, 2002), p. 89; S. White, 'Unpuzzling American Climate: New World Experience and the Foundations of a New Science', *Isis* 106 (2015), 550.

78 B. Cobo, *Obras del Padre Bernabé Cobo*, ed. F. Mateos (Madrid, 1964), p. 55; White, 'Unpuzzling American Climate', 550.

79 See S. White, *A Cold Welcome: The Little Ice Age and Europe's Encounter with North America* (Cambridge, MA, 2017), pp. 9–11.

80 R. Eden, *The Decades of the Newe Worlde of West India*, in E. Arber (ed.), *The First Three English Books on America* (Birmingham, 1885), pp. 87, 104; J. Chaplin, 'Natural Philosophy and an Early Racial Idiom in North America: Comparing English and Indian Bodies', *William and Mary Quarterly* 54 (1997), 239.

81 W. Bradford, *Of Plymouth Plantation, 1620–1647*, ed. S. Morison (New York, 1953), pp. 26, 14; Chaplin, 'Natural Philosophy and an Early Racial Idiom', 239.

82 Chaplin, 'Natural Philosophy and an Early Racial Idiom', 239.

83 Comte de Buffon, *Buffon's Natural History, Containing a Theory of the Earth, a General History of Man, of the Brute Creation and of Vegetables, Minerals &c. &c.*, 10 vols (London, 1910), 9, p. 86.

84 G. Raynal, *Histoire philosophique et politique des établissements et du commerce des Européens dans les deux Indes* (Paris, 1780).

85 Buffon, *Buffon's Natural History*, 4, pp. 332–3.

86 L. Dugatkin, 'Buffon, Jefferson and the theory of New World degeneracy', *Evolution: Education and Outreach* 12 (2019), 1–8.

87 Jefferson, *Notes on the State of Virginia*, p. 64.

88 D. Eltis, *The Rise of African Slavery in the Americas* (Cambridge, 2012), p. 143.

89 W. Beinart and L. Hughes, *Environment and Empire* (Oxford, 2020), p. 27.

90 A. Crosby, *Ecological Imperialism: The Biological Expansion of Europe, 900–1900* (Cambridge, 1986), pp. 138–9.

91 H. French, *Born in Blackness: Africa, Africans, and the Making of the Modern World, 1471 to the Second World War* (New York, 2021), pp. 113–15.

92 Beinart and Hughes, *Environment and Empire*, pp. 29–30; also see P. Lovejoy, *Transformations in Slavery: A History of Slavery in Africa* (Cambridge, 2012), pp. 37–44.

93 Eltis, *Rise of African Slavery*, p. 148.

15 THE FUSION OF THE OLD AND THE NEW WORLDS (c.1500–c.1700)

1 N. Nunn and N. Qian, 'The Columbian Exchange: A History of Disease, Food, and Ideas', *Journal of Economic Perspectives* 24 (2010), 163–88.

2 J. de Vries, 'The Industrial Revolution and the Industrious Revolution', *Journal of Economic History* 54.2 (1994), 249–70; J. van Zanden, 'The "Revolt of the Early Modernists" and the "First Modern Economy": An Assessment', *Economic History Review* 55 (2002), 619–41.

3 S. Beckert et al., 'Commodity frontiers and the transformation of the global countryside: a research agenda', *Journal of Global History* 16 (2021), 435–6.

4 D. Acemoglu, S. Johnson and J. Robinson, 'The Rise of Europe: Atlantic Trade, Institutional Change, and Economic Growth', *American Economic Review* 95 (2005), 546–79.

5 A. Smith, *An Inquiry into the Nature and Causes of the Wealth of Nations*, 2 vols (London, 1961), 2, p. 161.

6 Acemoglu, Johnson and Robinson, 'Rise of Europe', 546–79.

7 J. van Zanden, E. Buringh and M. Bosker, 'The rise and decline of European parliaments, 1188–1789', *Economic History Review* 65 (2011), 835–61.

8 C. Álvarez-Nogal and L. de la Escosura, 'The rise and fall of Spain (1270–1850)', *Economic History Review* 66 (2013), 1–37; A. Henriques and N. Palma, 'Comparative European Institutions and the Little Divergence, 1385–1800', Centre for Economic Policy Research, Discussion Paper 14124 (2019).

9 H. Thomas, *Conquest: Montezuma, Cortés and the Fall of Old Mexico* (London, 1993); K. MacQuarrie, *The Last Days of the Incas* (London, 2007).

10 A. Reséndez, *A Land So Strange: The Epic Journey of Cabeza de Vaca* (New York, 2007), p. 25.

11 R. Pike, 'Seville in the Sixteenth Century', *Hispanic American Historical Review* 41.1 (1961), 6, 12–13.

12 H. Cross, 'South American Bullion Production and Export, 1550–1750', in J. Richards (ed.), *Precious Metals in the Later Medieval and Early Modern Worlds* (Durham, NC, 1983), pp. 402–4.

13 O. Dunn and J. Kelley (ed. and tr.), *The Diario of Christopher Columbus's First Voyage to America, 1492–3* (Norman, OK, 1989), pp. 235–7.

14 A. Reséndez, *Other Slavery: The Uncovered Story of Indian Enslavement in America* (New York, 2016), p. 324.

15 N. Van Deusen, *Global Indios: The Indigenous Struggle for Justice in Sixteenth-Century Spain* (Durham, NC, 2015), p. 2.

16 Ibid., p. 3.

17 L. Paravisini-Gebert, 'Food, Biodiversity, Extinctions: Caribbean Fauna and the Struggle for Food Security during the Conquest of the New World', *Journal of West Indian Literature* 24 (2016), 12–13.

18 Ibid., 14.

19 E. Melville, *A Plague of Sheep: Environmental Consequences of the Conquest of Mexico* (Cambridge, 1994), pp. 99–102.

20 Crosby, *Ecological Imperialism*, p. 159.

21 Ibid., p. 1.

22 R. McCaa, 'Spanish and Nahuatl Views on Smallpox and Demographic Catastrophe in Mexico', *Journal of Interdisciplinary History* 25 (1995), 397–431; Crosby, *Columbian Exchange*; N. Cook, *Born to Die: Disease and New World Conquest, 1492–1650* (Cambridge, 1998), pp. 15–59.

23 C. Warriner et al., 'Disease, Demography, and Diet in Early Colonial New Spain: Investigation of a Sixteenth-Century Mixtec Cemetery at Teposcolula Yucundaa', *Latin American Quarterly* 23 (2012), 467–89.

24 Å. Vågene, '*Salmonella enterica* genomes from victims of a major sixteenth-century epidemic in Mexico', *Nature Ecology & Evolution* 2 (2018), 520–8.

25 R. Herzog, 'How Aztecs reacted to Colonial Epidemics', *JSTOR Daily*, 23 September 2020.

26 F. de Torquemada, *Monarquia Indiana*, 7 vols (Mexico City, 1969), 5, pp. 642–3.

27 R. Acuña-Soto, L. Calderon Romero and J. Maguire, 'Large epidemics of Hemorrhagic Fevers in Mexico, 1545–1815', *American Journal of Tropical Medicine and Hygiene* 62 (2000), 733–4.

28 F. Cervantes, *The Devil in the New World: The Impact of Diabolism in New Spain* (New Haven, 1997).

29 H. Cagle, *Assembling the Tropics: Science and Medicine in Portugal's Empire, 1450–1700* (Cambridge, 2018), pp. 169–207.

30 Bartolomé de las Casas, *Historia de las Indias*, in P. Sullivan (ed. and tr.), *Indian Freedom: The Cause of Bartolomé de las Casas, 1484–1566: A Reader* (Kansas City, 1995), 3.29, p. 146.

31 G. Icazbalceta, *Códice Mendieta: documentos franciscanos, siglos XVI y XVII*, 2 vols (Guadalajara, Mexico, 1971), 1, pp. 26–7, 245.

32 C. Riley, *The Kachina and the Cross: Indians and Spaniards in the Early Southwest* (Salt Lake City, UT, 1999), pp. 52, 60.

33 R. Flint (ed.), *Great Cruelties Have Been Reported: The 1544 Investigation of the Coronado Expedition* (Dallas, 2002), pp. 250–69.

34 S. White, 'Cold, Drought, and Disaster: The Little Ice Age and the Spanish Conquest of New Mexico', *New Mexico Historical Review* 89 (2014), 425–58.

35 S. White, 'North American Climate History (1500–1800)', in White, Pfister and Mauelshagen (eds), *Palgrave Handbook of Climate History*, p. 302.

36 White, *Cold Welcome*, pp. 57–8.

37 Ibid., pp. 75–7.

38 Ibid., p. 62.

39 A. Lawler, *The Secret Token: Myth, Obsession, and the Search for the Lost Colony of Roanoke* (New York, 2018).

40 White, *Cold Welcome*, p. 62.

41 W. Fitzgerald, 'Contact, Neural Iroquoian Transformation, and the Little Ice Age', in D. Brose et al. (eds), *Societies in Eclipse: Archaeology of the Eastern Woodland Indians, AD 1400–1700* (Washington, DC, 2012), pp. 37–48.

42 J. Rice, *Nature and History in the Potomac Country: From Hunter-Gatherers to the Age of Jefferson* (Baltimore, 2009), pp. 26–71.

43 S. Potter, *Commoners, Tribute, and Chiefs: The Development of Algonquian Culture in the Potomac Valley* (Charlottesville, VA, 1993),

pp. 165–8; S. White, ' "Shewing the difference betweene their conjuration, and our invocation on the name of God for rayne": Weather, Prayer, and Magic in Early American Encounters', *William and Mary Quarterly* 72 (2015), 45–6.

44 See here T. Wickham, 'Narrating Indigenous Histories of Climate Change in the Americas and Pacific', in White, Pfister and Mauelshagen (eds), *Palgrave Handbook of Climate History*, pp. 390–2.

45 See for example S. Sabol, *'The Touch of Civilization': Comparing American and Russian Internal Colonization* (Boulder, CO, 2017).

46 D. Jones, 'Virgin Soils Revisited', *William and Mary Quarterly* 60 (2003), 703–42.

47 Townsend, *Fifth Sun*, pp. 11–12.

48 M. Rubino, 'Low atmospheric CO_2 levels during the Little Ice Age due to cooling-induced terrestrial uptake', *Nature Geoscience* 9 (2016), 691–4; A. Koch et al., 'Earth system impacts of the European arrival and Great Dying in the Americas after 1492', *Quaternary Science Reviews* 207 (2019), 13–36.

49 Koch et al., 'Earth system impacts', 13.

50 D. Degroot, 'Climate change and society in the 15th to 18th centuries', *WIREs Climate Change* 9 (2018), 10.

51 P. Mancall and T. Weiss, 'Was Economic Growth Likely in Colonial British North America?', *Journal of Economic History* 59 (1999), 17–40.

52 Richards, *Unending Frontier*, p. 389.

53 Ibid., pp. 391–2; also here J. Hemming, *Red Gold: The Conquest of the Brazilian Indians* (London, 1978), pp. 143–5.

54 Hemming, *Red Gold*, p. 144.

55 P. Kelton, *Epidemics and Enslavement: Biological Catastrophe in the Native Southeast, 1492–1715* (Lincoln, NB, 2007), pp. 76–7.

56 P. Curtin, *The Rise and Fall of the Plantation Complex: Essays in Atlantic History* (Cambridge, 1990).

57 D. Eltis, *The Rise of African Slavery in the Americas* (Cambridge, 2000), pp. 57ff.

58 Quoted in A. Mikhail, *God's Shadow: Sultan Selim, his Ottoman Empire, and the Making of the Modern World* (New York, 2020), pp. 151–2.

59 J. Borja, 'Barbarización y redes de endoctrinamiento en los negros: cosmovisiones en Cartagena, siglos XVII y XVIII', in A. Ulloa (ed.), *Contribución africana a la cultura de las Americas* (Bogotá, 1992), p. 249.

60 S. Diouf, *Servants of Allah: African Muslims Enslaved in the Americas*, 15th anniversary edn (New York, 2013), pp. 35–6.

61 Ibid., pp. 210–11.

62 Ibid., pp. 36–7, 210–50.

63 J. Montaño, *The Roots of English Colonialism in Ireland* (Cambridge, 2011), p. 18; Collingham, *Hungry Empire*, pp. 17–22.

64 G. Milton, *Big Chief Elizabeth: The Adventures and Fate of the First English Colonists in America* (London, 2000).

65 N. Matar, *Turks, Moors, and Englishmen in the Age of Discovery* (New York, 1999), p. 92.

66 F. Thorpe, *The Federal and State Constitutions: Colonial Charters, and Other Organic Laws of the States, Territories, and Colonies, Now or Heretofore Forming the United States of America*, 7 vols (Washington, DC, 1909), 3, pp. 1828–9.

67 J. Marr and J. Cathey, 'New Hypothesis for Cause of Epidemic among Native Americans, New England, 1616–1619', *Emerging Infectious Diseases* 16 (2010), 281–6.

68 P. Mancall, *Nature and Culture in the Early Modern Atlantic* (Philadelphia, 2017).

69 A. Zilberstein, *A Temperate Empire: Making Climate Change in Early America* (Oxford, 2016), pp. 97–8.

70 Ibid., p. 98.

71 D. Galenson, *White Servitude in Colonial America: An Economic Analysis* (Cambridge, 1981); S. Salinger, 'Labor, Markets, and Opportunity: Indentured Servitude in Early America', *Labor History* 38 (1997), 311–38.

72 M. Parker, *The Sugar Barons: Family, Corruption and War in the West Indies* (London, 2011), pp. 2–3.

73 Collingham, *Hungry Empire*, pp. 49–50; for France, see P. Boucher, *France and the American Tropics to 1700: Tropics of Discontent?* (Baltimore, 2008).

74 F. Bacon, 'Of Plantations', in S. Reynolds, *The Essays or Counsels, Civil and Moral of Francis Bacon* (Oxford, 1890), p. 237; D. Souden, ' "Rogues, Whores and Vagabonds"? Indentured Servant Emigrants to North America and the Case of Mid-Seventeenth Century Bristol', *Social History* 3 (1978), 23–41.

75 R. Hakluyt, *The Original Writings and Correspondence of the Two Richard Hakluyts* (London, 1935), pp. 317–19.

76 H. Beckles, 'The Hub of Empire: The Caribbean and Britain in the Seventeenth Century', in N. Canny (ed.), *The Oxford History of the British Empire*, vol. 1: *The Origins of Empire: British Overseas Enterprise to the Close of the Seventeenth Century* (Oxford, 1998), pp. 222–3.

77 H. Beckles, *White Servitude and Black Slavery in Barbados, 1627–1715* (Knoxville, TN, 1989), pp. 59–79.

78 Beinart and Hughes, *Environment and Empire*, p. 34.

79 Beckles, 'Hub of Empire', p. 232.

80 J. Thornton, *Africa and Africans in the Making of the Atlantic World, 1400–1600* (Cambridge, 1992), pp. 134–6; J. García Zaldúa and D. Hosler, 'Copper Smelting at the Archaeological Site of El Manchón, Guerrero: From Indigenous Practice to Colonial-Scale Production', *Latin American Antiquity* 31 (2020), 558–75; L. Gragg, *Englishmen Transplanted: The English Colonization of Barbados, 1627–1660* (Oxford, 2003), pp. 19–20.

81 C. Gardina Pestana, *The English Conquest of Jamaica: Oliver Cromwell's Bid for Empire* (New York, 2017).

82 B. Higman, *Concise History of the Caribbean* (Cambridge, 2011), pp. 64–80.

83 C. Palmer, *Slaves of the White God: Blacks in Mexico, 1570–1650* (Cambridge, MA, 1976), pp. 69–70.

84 T. Burnard, ' "The Countrie Continues Sicklie": White Mortality in Jamaica, 1655–1780', *Social History of Medicine* 12 (1999), 45–72.

85 M. Espinosa, 'The Question of Racial Immunity to Yellow Fever in History and Historiography', *Social Science History* 38 (2014), 437–53.

86 J. Powell, A. Gloria-Soria and P. Kotsakiozi, 'Recent History of *Aedes aegypti*: Vector Genomics and Epidemiology Records', *BioScience* 68 (2018), 854–60; J.-P. Chippaux and A. Chippaux, 'Yellow fever in Africa and the Americas: a historical and epidemiological perspective', *Journal of Venomous Animals and Toxins Including Tropical Diseases* 24 (2018), 1–14.

87 J. McNeil, *Ecology and War in the Greater Caribbean, 1620–1914* (Cambridge, 2010).

88 Beinart and Hughes, *Environment and Empire*, p. 23.

89 K. Kiple and B. Higgins, 'Yellow Fever and the Africanization of the Caribbean', in J. Verano and D. Uberlaker (eds), *Disease and Demography in the Americas* (Washington, DC, 1992), pp. 237–48.

16 ON THE EXPLOITATION OF NATURE AND PEOPLE (C.1650–C.1750)

1 W. Pettigrew, *A Freedom's Debt: The Royal African Company and the Politics of the Atlantic Slave Trade* (Chapel Hill, NC, 2014).

2 D. Eltis and L. Jennings, 'Trade between Western Africa and the Atlantic World in the Pre-Colonial Era', *American Historical Review* 93 (1988), 944.

3 L. Gragg, ' "To procure Negroes". The English slave trade to Barbados, 1627–60', *Slavery and Abolition: A Journal of Slave and Post-Slave Studies* 16 (1995), 65.

4 M. Rediker, *The Slave Ship: A Human History* (New York, 2007), pp. 291–301.

5 R. Davis, *The Rise of the English Shipping Industry in the Seventeenth and Eighteenth Centuries* (Oxford, 2017), pp. 373–4.

6 Lovejoy, *Transformations in Slavery*, p. 48.

7 A. de Pleijt and J. van Zanden, 'Accounting for the "Little Divergence": What drove economic growth in pre-industrial Europe, 1300–1800?', *European Review of Economic History* 20 (2016), 387–409; Gragg, ' "To procure Negroes" ', 65.

8 Davis, *Rise of the English Shipping Industry*, p. 374.

9 H. Raynes, *A History of British Insurance* (London, 1948).

10 H. Beckles, 'The Economic Origins of Black Slavery in the British West Indies, 1640–1680: A Tentative Analysis of the Barbados Model', *Journal of Caribbean History* (1982), 16, 52–3; H. Beckles and A. Downes, 'The Economics of Transition to the Black Labor System in Barbados', *Journal of Interdisciplinary History* (1987) 37, 225–47.

11 A. Hochschild, *Bury the Chains: The British Struggle to Abolish Slavery* (London, 2005), pp. 65–8.

12 Ibid., p. 67.

13 J. Krikler, 'A Chain of Murder in the Slave Trade: A Wider Context of the Zong Massacre', *International Review of Social History* 57 (2012), 402–3.

14 H. Klein and S. Engerman, 'Long-Term Trends in African Mortality in the Transatlantic Slave Trade', *Slavery and Abolition* 18 (1997), 40–4.

15 Krikler, 'Chain of Murder in the Slave Trade', 410–11.

16 Hochschild, *Bury the Chains*, p. 22.

17 N. Rodger, 'War as Economic Activity in the "Long" Eighteenth Century', *International Journal of Maritime History* 22 (2010), 1–18.

18 F. Arteaga, D. Desierto and M. Koyama, 'Shipwrecked by Rents', Centre for Economic Policy Research, Discussion Paper DP15300 (2020), 1–50.

19 'The First Voyage of the right worshipfull and valiant knight, Sir John Hawkins', in C. Markham (ed.), *The Hawkins Voyages* (London, 1878), p. 5. Also H. Kelsey, *Sir John Hawkins: Queen Elizabeth's Slave Trader* (New Haven, 2003), pp. 52–69.

20 See for example G. Heng, *The Invention of Race in the European Middle Ages* (Cambridge, 2018).

21 Herodotus, *The Histories*, 9.122, pp. 589–90.

22 Montesquieu, *The Spirit of the Laws*, ed. A. Cohler, B. Miller and H. Stone (Cambridge, 1998), 14, pp. 231–2, 235.

23 Ibid., p. 233.

24 Ibid., pp. 243–4.

25 Ibid., 17, p. 278.

26 Krikler, 'Chain of Murder in the Slave Trade', 414.

27 W. Wiecek, *The Sources of Antislavery Constitutionalism in America, 1760–1848* (Ithaca, NY, 1977), p. 24.

28 J. Krikler, 'The *Zong* and the Lord Chief Justice', *History Workshop Journal* 64 (2007), 36.

29 See for example M. Dresser and A. Hann (eds), *Slavery and the British Country House* (London, 2013).

30 Krikler, 'The *Zong* and the Lord Chief Justice', 33; Anonymous, "Great Newes from the Barbadoes" (London: L. Curtis, 1676), 6–7, cited by N. Oatsvall and V. Scribner, '"The Devil Was in the Englishman that He Makes Everything Work": Implementing the Concept of "Work" to Reevaluate Sugar Production and Consumption in the Early Modern British Atlantic World' Agricultural History Society 92 (2018), 464.

31 French, *Born in Blackness*, pp. 124, 190.

32 Krikler, 'Chain of Murder in the Slave Trade', 409–10.

33 Gomes Eanes de Zurara, *Crónica de Guiné*, in C. Beazley and E. Prestage (eds and trs), *The Chronicle of the Discovery and Conquest of Guinea*, 2 vols (London, 1896–9), 1, pp. 81–2.

34 Vincent Brown, *Tacky's Revolt: The Story of an Atlantic Slave War* (Cambridge, MA, 2020), p. 57.

35 Ibid., p. 58.

36 Cited by B. Philips, *Loot: Britain and the Benin Bronzes* (London, 2021), p. 52.

37 Krikler, 'Chain of Murder in the Slave Trade', 407.

38 Ibid., 398, 401.

39 Collingham, *Hungry Empire*, p. 43.

40 K. Davies, *Royal African Company* (London, 1957), p. 299.

41 Beinart and Hughes, *Environment and Empire*, pp. 28–9.

42 E. Esposito, 'Side Effects of Immunity: The Rise of African Slavery in the US South', *Cahiers de Recherches Économiques du Département d'Économie* (2018), 3–5.

43 E. Yalcindag et al., 'Multiple independent introductions of *Plasmodium falciparum* in South America', *PNAS* 109 (2012), 511–16.

44 F. Cambournac, 'Contribution to the history of malaria epidemiology and control in Portugal and some other places', *Parassitologia* 36 (1994), 215–22.

45 J. Taylor et al., 'The evolutionary history of *Plasmodium vivax* as inferred from mitochondrial genomes: parasite genetic diversity in the Americas', *Molecular Biology Evolution* 30 (2013), 2050–64. See P. Rodrigues et al., 'Human migration and the spread of malaria parasites to the New World', *Scientific Reports* 8 (2018), 1–13.

46 Esposito, 'Side Effects of Immunity', 20, 9.

47 Grove, 'El Niño Chronology and the Little Ice Age', pp. 58–9; M. Amoroso et al., 'South American dendroecological field week 2016: exploring dendrochronological research in Northern Patagonia', *Tree Ring Research* 74 (2018), 120–31.

48 Esposito, 'Side Effects of Immunity', 7–10.

49 M. Humphreys, *Malaria: Poverty, Race and Public Health in the United States* (Baltimore, 2003).

50 D. Brown and C. Webb, *Race in the American South: From Slavery to Civil Rights* (Edinburgh, 2007), pp. 28–31.

51 Esposito, 'Side Effects of Immunity', 33.

52 Ibid., 18–19, 26; W. Wiecek, 'The statutory law of slavery and race in the thirteen mainland colonies of British America', *William and Mary Quarterly* (1977), 258–80.

53 J. Morgan, 'Partus sequitur ventrem: law, race, and reproduction in colonial slavery', *Small Axe* 22 (2018), 1–17.

54 M. Schwartz, *Birthing a Slave: Motherhood and Medicine in the Antebellum South* (Cambridge, MA, 2006).

55 R. Steckel, 'A Dreadful Childhood: The Excess Mortality of American Slaves', *Social Science History* 10 (1986), 427–65.

56 D. Owens, *Medical Bondage: Race, Gender, and the Origins of American Gynecology* (Athens, GA, 2017).

57 D. Cooper Owens and S. Fett, 'Black Maternal and Infant Health: Historical Legacies of Slavery', *American Journal of Public Health* 109 (2019), 1342–5.

58 R. Packard, *The Making of a Tropical Disease: A Short History of Malaria* (Baltimore, 2007), pp. 53–66; J. McNeill, 'Ecology, Epidemics and Empires: Environmental Change and the Geopolitics of Tropical America', *Environment and History* 5 (1999), 177–9.

59 A. Acharya, M. Blackwell and M. Sen, 'The Political Legacy of American Slavery', *Journal of Politics* 78 (2016), 621–41.

60 J. Hersh and H.-J. Voth, 'Sweet Diversity: Colonial Goods and the Welfare Gains from Trade after 1492', CEPR Working Paper (2011), 1.

61 M. Mulcahy, *Hubs of Empire: The Southeastern Lowcountry and British Caribbean* (Baltimore, 2014), pp. 103, 82, 46.

62 G. Beer, *British Colonial Policy, 1754–65* (New York, 1907), p. 144.

63 R. Naylor, *Canada in the European Age, 1453–1919* (Montreal, 2006), p. 121.

64 Brown, *Tacky's Revolt*, p. 76.

65 P. Gauci, *William Beckford: First Prime Minister of the London Empire* (New Haven, 2013), p. 96.

66 Ibid., p. 91.

67 de Vries, 'The Industrial Revolution and the Industrious Revolution', 249–70; van Zanden, 'The "Revolt of the Early Modernists" and the "First Modern Economy" ', 619–41; R. Allen, *The British Industrial Revolution in Global Perspective* (Cambridge, 2009), pp. 1–22.

68 J. de Vries, 'Between Purchasing Power and the World of Goods: Understanding the Household Economy in Early Modern Europe', in J. Brewer and R. Porter (eds), *Consumption and the World of Goods* (London, 1993), pp. 85–132; H.-J. Voth, 'Time and Work in Eighteenth-Century London', *Journal of Economic History* 58 (1998), 29–58.

69 Hersh and Voth, 'Sweet Diversity', 1–52.

70 F. Braudel, *Civilisation and Capitalism, 15th–18th Century*, 3 vols (London, 1981–5), p. 130; R. Sheridan, *Sugar and Slavery* (Barbados, 1974), p. 21.

71 Hersh and Voth, 'Sweet Diversity', 21–2; Nunn and Qian, 'The Columbian Exchange', 178.

72 Beinart and Hughes, *Environment and Empire*, p. 25.

73 J. Mokyr, 'Is There Still Life in the Pessimist Case? Consumption during the Industrial Revolution, 1790–1850', *Journal of Economic History* 48 (1988), 69–92.

74 G. Faulconbridge, 'British "Treasure Island" tax havens face a tempest', Reuters, 7 June 2021.

75 J. Lovell, *The Opium War: Drugs, Dreams and the Making of China* (London, 2011).

76 See for example Z. Wang, *Never Forget National Humiliation: Historical Memory in Chinese Politics and Foreign Relations* (New York, 2014).

77 Braudel, *Civilisation and Capitalism,* 1, p. 258.

78 B. Cowan, *The Social Life of Coffee: The Emergence of the British Coffeehouse* (New Haven, 2005); J. van Horn Melton, *The Rise of the Public in Enlightenment Europe* (Cambridge, 2001), and most famously,

J. Habermas, *Strukturwandel der Öffentlichkeit: Untersuchungen zu einer Kategorie der bürgerlichen Gesellschaft* (Berlin, 1962).

79 The classic study here is Hobsbawm, *Industry and Empire*; also see S. Mintz, *Sweetness and Power: The Place of Sugar in Modern History* (New York, 1986).

80 S. Beckert, *Empire of Cotton: A Global History* (New York, 2014), p. 63.

81 Ibid., pp. 98–106.

82 D. Watts, 'Cycles of Famine in Islands of Plenty', in B. Currey and G. Hugo (eds), *Famine as a Geographical Phenomenon* (Dordrecht, 1984), p. 62. Also see J. Weaver, *The Great Land Rush and the Making of the Modern World, 1650–1900* (Montreal, 2003).

83 D. Watts, *The West Indies: Patterns of Development, Culture and Environmental Change since 1492* (Cambridge, 1987), p. 397.

84 E. Williams, *Capitalism and Slavery* (London, 1944).

85 M. Mulcahy, *Hurricanes and Society in the British Greater Caribbean, 1624–1783* (Baltimore, 2006); S. Schwartz, *Sea of Storms: A History of Hurricanes in the Greater Caribbean to Katrina* (Princeton, 2015).

86 K. Douglass and J. Cooper, 'Archaeology, environmental justice, and climate change on islands in the Caribbean and southwestern Indian Ocean', *PNAS* 117 (2020), 8254–62.

87 Pomeranz, *The Great Divergence*, pp. 211ff.

88 A. Carlos and F. Lewis, *Commerce by a Frozen Sea: Native Americans and the European Fur Trade* (Philadelphia, 2010), pp. 22–6.

89 V. Hyden-Hanscho, 'Invisible Globalization: French Hats in Habsburg Vienna, 1650–1750', *Journal of European Economic History* 45 (2016), 11–54. Also M. Francis, 'The Strange Career of the Canadian Beaver: Anthropomorphic Discourses and Imperial History', *Journal of Historical Sociology* 17 (2004), 202–32.

90 See for example E. Dolin, *Leviathan: The History of Whaling in America* (New York, 2007).

91 J. Jackson, *The Thief at the End of the World: Rubber, Power and the Seeds of Empire* (New York, 2008).

92 See J. Miller, *Way of Death: Merchant Capitalism and the Angolan Slave Trade, 1730–1830* (London, 1988); Klein and Engerman, 'Long-Term Trends in African Mortality', 36–48.

93 S. Behrendt, 'Ecology, Seasonality, and the Transatlantic Slave Trade', in B. Bailyn and P. Denault (eds), *Soundings in Atlantic History: Latent Structures and Intellectual Currents, 1500–1830* (Cambridge, 2009), pp. 44–85.

94 J. Fenske and N. Kala, 'Climate and the slave trade', *Journal of Development Economics* 112 (2015), 19–32.

95 C. Wilder, *Ebony and Ivy: Race, Slavery, and the Troubled History of America's Universities* (New York, 2013).

96 J. Weil, 'More than 1700 congressmen once enslaved Black People', *Washington Post*, 10 January 2021.

97 Cited by Collingham, *Hungry Empire*, p. 49.

98 Cited by Brown, *Tacky's Revolt*, p. 27.

99 P. Earle, *The World of Defoe* (New York, 1977); V. Brown, *The Reaper's Garden: Death and Power in the World of Atlantic Slavery* (Cambridge, MA, 2008), p. 24.

100 J. Holcomb, *Moral Commerce: Quakers and the Transatlantic Boycott of the Slave Labor Economy* (Ithaca, NY, 2016), p. 33.

101 J. Holcomb, 'Rejecting the Gain of Oppression: Quaker Abstention and the Abolitionist Cause', in M. Jackson and S. Kozel (eds), *Quakers and their Allies in the Abolitionist Cause, 1754–1808* (Abingdon, 2015), pp. 99–110.

102 R. Hall and C. Jones, 'Why Do Some Countries Produce So Much More Output Per Worker Than Others?', *Quarterly Journal of Economics* 114 (1999), 83–116.

103 Belich, *The World the Plague Made*, p. 281.

104 P. Manning, *Slavery and African Life: Occidental, Oriental, and African Slave Trades* (Cambridge, 1990), pp. 60–85.

105 W. Whatley and R. Gillezeau, 'The fundamental impact of the slave trade on African economies', in P. Rhode, J. Rosenbloom and D. Weiman (eds), *Economic Evolution and Revolution in Historical Time* (Stanford, 2011), pp. 86–110.

106 G. Ugo Nwokeji, *The Slave Trade and Culture in the Bight of Biafra: An African Society in the Atlantic World* (Cambridge, 2010), pp. 117–43.

107 Lovejoy, *Transformations in Slavery*, pp. 106–11; W. Hawthorne, 'The Production of Slaves Where There Was No State: The Guinea-Bissau Region, 1450–1815', *Slavery & Abolition*, 20 (1999), 97–124.

108 Brown, *Tacky's Revolt*, pp. 22–5; W. Richards, 'The Import of Firearms into West Africa in the Eighteenth Century', *Journal of African History* 21 (1980), 43–59.

109 N. Nunn, 'The long-term effects of Africa's slave trades', *Quarterly Journal of Economics* 123 (2008), 139–76.

110 N. Nunn and L. Wantchekon, 'The slave trade and the origins of mistrust in Africa', *American Economic Review* 101 (2011), 3221–52.

111 W. Whatley and R. Gillezeau, 'The Impact of the Transatlantic Slave Trade on Ethnic Stratification in Africa', *American Economic Review* 101 (2010), 571–6.

112 H. Thomas, *The Story of the Atlantic Slave Trade: 1440–1870* (New York, 1997), pp. 810–11.

113 G. Ugo Nwokeji, 'African Conceptions of Gender and the Slave Traffic', *William and Mary Quarterly* 58 (2001), 47–68.

114 K. Wood, 'Gender and Slavery', in M. Smith and R. Paquette (eds), *The Oxford Handbook of Slavery in the Americas* (Oxford, 2010), pp. 515–17.

115 E. Baptist, ' "Cuffy," "Fancy Maids," and "One-Eyed Men": Rape, Commodification, and the Domestic Slave Trade in the United States', *American Historical Review* 106 (2001), 1619–50; M. Smith (ed.), *Sex without Consent: Rape and Sexual Coercion in America* (New York, 2001).

116 Wood, 'Gender and Slavery', pp. 518–19.

117 D. Berry, *'Swing the Sickle for the Harvest is Ripe': Gender and Slavery in Antebellum Georgia* (Chicago, 2007), pp. 35–51 and *passim*.

118 J. Dalton and T. Leung, 'Why Is Polygyny More Prevalent in Western Africa? An African Slave Trade Perspective', *Economic Development and Cultural Change* 62 (2014), 599–632.

119 J. Fenske, 'African polygamy: Past and present', *Journal of Development Economics* 117 (2015), 58–73.

120 M. Tertilt, 'Polygyny, Fertility, and Savings', *Journal of Political Economy* 113(6) (2005), 1341–71; M. Tertilt, 'Polygyny, Women's Rights, and Development', *Journal of the European Economic Association* 4 (2006), 523–30.

121 G. Bertocchi and A. Dimico, 'The long-term determinants of female HIV infection in Africa: The slave trade, polygyny, and sexual behavior', *Journal of Development Economics* 140 (2019), 90–105.

122 L. Walters, C. Chisadza and M. Chance, 'Slave Trade and Women Political Participation in Africa', University of Pretoria, Department of Economics Working Paper Series 56 (2021), 1–34.

123 J. Locke, *Second Treatise of Government and A Letter Concerning Toleration*, ed. M. Goldie (Oxford, 2016), 5, pp. 15–16.

124 S. Stoddard, *An Answer to Some Cases of Conscience Respecting the Country* (1722).

125 Y. Hale Hendlin, 'From Terra Nullius to Terra Communis', *Environmental Philosophy* 11 (2014), 141–74.

126 G. Chakraborty, 'Roots and Ramifications of a Colonial "Construct": The Wastelands in Assam', Occasional Paper 39, Institute of Development Studies Kolkata (2012), 9–10.

127 C. Saunt, *Unworthy Republic: The Dispossession of Native Americans and the Road to Indian Territory* (New York, 2020).

17 THE LITTLE ICE AGE (C.1550–C.1800)

1 F. Matthes, 'Report of Committee on Glaciers, April 1939', *Eos, Transactions, American Geophysical Union* 20 (1939), 520.

2 Some historians suggest dating the onset to c.1300, for example Miller et al., 'Abrupt onset of the Little Ice Age triggered by volcanism and sustained by sea-ice/ocean feedbacks', 1–5; B. Fagan, *The Little Ice Age: How Climate Made History, 1300–1850* (New York, 2000).

3 J. Grove, 'The Onset of the Little Ice Age', in P. Jones et al. (eds), *History and Climate: Memories of the Future?* (London, 2001), pp. 153–85; J. Grove, *Little Ice Ages: Ancient and Modern*, 2 vols (London, 2004).

4 C. Camenisch, 'The 1430s: A cold period of extraordinary internal climate variability during the early Spörer Minimum with social and economic impacts in north-western and central Europe', *Climate of the Past* 12 (2016), 2107–26; J. Luterbacher, 'The Late Maunder Minimum (1675–1715) – Climax of the "Little Ice Age" in Europe', in Jones et al. (eds), *History and Climate*, pp. 29–54; A. Schurer, S. Tett and G. Hegerl, 'Small influence of solar variability on climate over the past millennium', *Nature Geoscience* 7 (2014), 104–8.

5 J. de Vries, 'Histoire et climat et économie: Des faits nouveaux, une interpretation différente', *Annales Histoire, Sciences Sociales* 32 (1977), 198–226; Tarand and Nordli, 'The Tallinn Temperature Series', 360–407; L. Leijonhufvud et al., 'Five Centuries of Stockholm Winter/Spring Temperatures Reconstructed from Documentary Evidence and Instrumental Observations', *Climatic Change* 101 (2010), 109–41.

6 F. Charpentier Ljunqvist, *Global nedkylning: klimatet och människan under 10000 år* (Stockholm, 2009).

7 C. Pfister et al., 'Early Modern Europe', in White, Pfister and Mauelshagen (eds), *Palgrave Handbook of Climate History*, p. 268.

8 A. Gupta, D. Anderson and J. Overpeck, 'Abrupt changes in the Asian southwest monsoon during the Holocene and their links to the North Atlantic Ocean', *Nature* 421 (2003), 354–7; D. Oppo, Y. Rosenthal and B. Linsley, '2,000 year-long temperature and hydrology reconstructions from the Indo-Pacific warm pool', *Nature* 460 (2009), 1113–16; W. Liu et al., 'Wet climate during the "Little Ice

Age" in the arid Tarim Basin, northwestern China', *The Holocene* 21 (2010), 409–16.

9 J. Matthews and K. Briffa, 'The "little ice age": re-evaluation of an evolving concept', *Geografiska Annaler: Series A, Physical Geography* 87 (2005), 17–36.

10 G. Parker, 'Crisis and Catastrophe: Global Crisis of the 17th Century Reconsidered', *American Historical Review* 113 (2008), 1056.

11 G. Parker, *Global Crisis: War, Climate and Catastrophe in the Seventeenth Century* (London, 2013).

12 G. Eriksdotter, 'Did the Little Ice Age Affect Indoor Climate and Comfort?: Re-theorizing Climate History and Architecture from the Early Modern Period', *Journal for Early Modern Cultural Studies* 13 (2013), 24–42; R.-M. Söderström, *Bostadskultur, informationsflöden och hantverkare 1740–1820 med utgånspunkt i Bålby (Närke) och Skottberska gården (Blekinge)* (Lund, 2009).

13 H. Neuberger, 'Climate in Art', *Weather* 25 (1970), 46–66.

14 Fagan, *Little Ice Age*, p. 48; M. Kelly and Ó Gráda, 'The Waning of the Little Ice Age: Climate Change in Early Modern Europe', *Journal of Interdisciplinary History* 44 (2014), 313.

15 W. Behringer, '"Kleine Eiszeit" und Frühe Neuzeit', in W. Behringer, H. Lehmann and C. Pfister (eds), *Kulturelle Konsequenzen der 'Kleinen Eiszeit'* (Göttingen, 2005), pp. 415–508.

16 W. Behringer, 'Climatic Change and Witch-hunting: The Impact of the Little Ice Age on Mentalities', *Climatic Change* 43 (1999), 335–51.

17 B. Levack, 'Introduction', in *The Oxford Handbook of Witchcraft in Early Modern Europe and Colonial America* (Oxford, 2013), p. 5.

18 E. Midelfort, *Witch Hunting in Southwestern Germany, 1562–1684* (Stanford, 1972), 201–30.

19 Ibid., pp. 31–3.

20 A. Meier, 'Natural disasters? Droughts and epidemics in pre-colonial Sudanaic Africa', *Medieval History Journal* 10 (2006), 209–36; A. Ashforth, 'AIDS, Witchcraft, and the Problem of Power in Post-Apartheid South Africa', Institute for Advanced Study, Occasional Papers of the School of Social Science 10 (2001), 1–30.

21 W. Behringer, *Kulturgeschichte des Klimas: von der Eiszeit bis zur globalen Erwärmung* (Munich, 2007), pp. 115–18.

22 C. Pfister, 'Die Schwankungen des Unteren Grindelwaldgletschers im Vergleich mit historischen Witttrungsbeobachtungen und messungen', *Zeitschrift für Gletscherkunde* 11 (1976), 74–90; W. Behringer, 'Weather, Hunger and Fear: Origins of the European Witch-Hunts in Climate, Society and Mentality', *German History* 13 (1995), 1–27.

23 C. Pfister, 'Climatic Extremes, Recurrent Crises and Witch
    Hunts: Strategies of European Societies in Coping with Exogenous
    Shocks in the Late Sixteenth and Early Seventeenth Centuries', *Medieval
    History Journal* 10 (2006), 33–73.

24 P. Malanima, 'Energy crisis and growth, 1650–1850: the European deviation
    in a comparative perspective', *Journal of Global History* 1 (2006), 112.

25 H. Lee et al., 'Climate change and epidemics in Chinese history: A
    multi-scalar analysis', *Social Science & Medicine* 174 (2017), 53–63;
    A. McMichael, 'Insights from past millennia into climatic impacts on
    human health and survival', *PNAS* 109 (2012), 4730–7.

26 Fagan, *Little Ice Age*, pp. 92–3, xvi; D. Degroot, 'Climate Change and
    Conflict', in White, Pfister and Mauelshagen (eds), *Palgrave Handbook
    of Climate History*, pp. 377–8.

27 N. Brown, *History and Climate Change: A Eurocentric Perspective*
    (London, 2001), p. 296.

28 Contribution of Working Groups I, II and III to the Third Assessment
    Report of the Intergovernmental Panel on Climate Change, *Climate
    Change 2001: Synthesis Report* (Cambridge, 2001), p. 173.

29 Kelly and Ó Gráda, 'Waning of the Little Ice Age', 303–4.

30 G. Alfani, 'Climate, Population and Famine in Northern Italy: General
    Tendencies and Malthusian Crisis, ca. 1450–1800', *Annales de
    Démographie Historique* 120 (2010), 24–6.

31 A. Ogilvie and T. Jónsson (eds), *The Iceberg in the Mist: Northern
    Research in Pursuit of a 'Little Ice Age'* (Dordrecht, 2001).

32 J. Grove, 'The Initiation of the "Little Ice Age" in Regions round the
    North Atlantic', *Climatic Change* 48 (2001), 53–82.

33 A. Ogilvie and T. Jónsson, '"Little Ice Age" Research: A Perspective from
    Iceland', *Climate Change* 48 (2001); A. Ogilvie, 'Historical climatology,
    Climatic Change, and implications for climate science in the twenty-first
    century', *Climatic Change* 100 (2010), 33–47.

34 M. Owens et al., 'The Maunder minimum and the Little Ice Age: an
    update from recent reconstructions and climate simulations', *Journal of
    Space Weather and Space Climate* 7 (2013), 1–10.

35 G. Feulner and S. Rahmstorf, 'On the effect of a new grand minimum
    of solar activity on the future climate on Earth', *Geophysical Research
    Letters* 37 (2010), 1–5; Schurer, Tett and Hegerl, 'Small influence of solar
    variability', 104–8.

36 Kelly and Ó Gráda, 'Waning of the Little Ice Age', 313–15.

37 D. Degroot, *The Frigid Golden Age: Climate Change, the Little Ice Age,
    and the Dutch Republic, 1560–1720* (Cambridge, 2018), pp. 263–6.

38 *Malleus Maleficurum*, tr. M. Summers and P. Hughes (London, 1968), p. 6.

39 E. Ross, 'Syphilis, Misogyny and Witchcraft in 16th-Century Europe', *Current Anthropology* 36 (1995), 333–7, N. Ben-Yehuda, 'The European Witch Craze of the 14th to 17th Centuries: A Sociologist's Perspective', *American Journal of Sociology* 86 (1980), 1–31; Also E. Oster, 'Witchcraft, Weather and Economic Growth in Renaissance Europe', *Journal of Economic Perspectives* 18 (2004), 215–28.

40 H. Trevor-Roper, *The European Witch-Craze of the 16th and 17th Centuries* (London, 1969). For psychogenic illness, R. Bartholomew and R. Baloh, *Havana Syndrome: Mass Psychogenic Illness and the Real Story behind the Embassy Mystery and Hysteria* (Cham, 2020).

41 P. Brown, 'Gazing Anew at Poltava: Perspectives from the Military Revolution Controversy, Comparative History, and Decision-Making Doctrines', *Harvard Ukrainian Studies* 31 (2009), 107–33.

42 F. Lapointe and R. Bradley, 'Little Ice Age abruptly triggered by intrusion of Atlantic waters into the Nordic Seas', *Science Advances* 7 (2021), 1–12.

43 Miller et al., 'Abrupt onset of the Little Ice Age triggered by volcanism and sustained by sea-ice/ocean feedbacks', 1–5; A. Condron, A. Joyce and R. Bradley, 'Arctic sea ice export as a driver of deglacial climate', *Geology* 48 (2020), 395–9.

44 N. Gennaoili and H.-J. Voth, 'State Capacity and Military Conflict', *Review of Economic Studies* 82 (2015), 1413–14.

45 G. Parker, *The Military Revolution, Military Innovation and the Rise of the West, 1500–1800* (Cambridge, 1996), p. 1.

46 Gennaioli and Voth, 'State Capacity and Military Conflict', 1412–14; H. Bowen, 'The Bank of England during the Long Eighteenth Century: 1694–1820', in R. Roberts and D. Kynaston (eds), *The Bank of England* (Oxford, 1995), pp. 1–18; J. Kim, 'How Modern Banking Originated: The London Goldsmith-Bankers' Institutionalisation of Trust', *Business History*, 53 (2011), 939–59.

47 Gennaioli and Voth, 'State Capacity and Military Conflict', 1409–48.

48 J. Elliott, *Empires of the Atlantic World: Britain and Spain in America, 1492–1830* (New Haven, 2006), p. 112.

49 J. de Vries, 'The Economic Crisis of the Seventeenth Century after Fifty Years', *Journal of Interdisciplinary History* 40 (2009), 173.

50 Acemoglu, Johnson and Robinson, 'Rise of Europe', 546–79.

51 C. McIlwain (ed.), *The Political Works of James I* (Cambridge, 1918), pp. 343–4; Parker, *Global Crisis*, p. 60.

52 J. Evelyn, *Fumifugium; or the Inconvenience of the Aer and Smoak of London Dissipated* (London, 1661), p. 5.

53 Letter of James Howell to Captain Francis Bacon, 30 March 1630, cited by Parker, *Global Crisis*, p. 60.

54 N. Voigtländer and H.-J. Voth, 'The Three Horsemen of Riches: Plague, War, and Urbanization in Early Modern Europe', *Review of Economic Studies* 80 (2013), 774–811.

55 M. Waldinger, 'The Economic Effects of Long-Term Climate Change: Evidence from the Little Ice Age', Centre for Climate Change Economics and Policy Working Paper No. 239 (2015), 1–46.

56 S. White, *The Climate of Rebellion in the Early Modern Ottoman Empire* (New York, 2011), pp. 204–11; 'Winter is Coming: The Long-Run Effects of Climate Change on Conflict, 1400–1900', NBER Working Paper 23033 (2017).

57 Waldinger, 'Economic Effects of Long-Term Climate Change', 1–46.

58 O. Wetter et al., 'The year-long unprecedented European heat and drought of 1540 – a worst case', *Climatic Change* 125 (2014), 349–63.

59 Abū al-Faẓl, *Ain-I Akbari*, tr. H. Blochmann and H. Jarrett, 3 vols (Calcutta, 1873–94), 2, p. 57.

60 Degroot, *Frigid Golden Age*, pp. 33–4.

61 I. Pikirayi, 'Environmental data and historical process: historical climatic reconstruction & the Mutapa state, 1450–1862', in W. Beinart and J. McGregor (eds), *Social History and African Environments* (Oxford, 2003), pp. 60–71; M. Hannaford and D. Nash, 'Climate, history, society over the last millennium in southeast Africa', *WIREs Climate Change*, 7 (2016), 379.

62 Brook, 'Nine Sloughs', 40.

63 Degroot, *Frigid Golden Age*, p. 34.

64 S. White, J. Brooke and C. Pfister, 'Climate, Weather, Agriculture, and Food', in White, Pfister and Mauelshagen (eds), *Palgrave Handbook of Climate History*, p. 340; W. Behringer, 'Die Krise von 1570. Ein Beitrag zur Krisengeschichte der Neuzeit', in M. Jakubowski and H. Lehmann (eds), *Um Himmels Willen: Religion in Krisenzeiten* (Göttingen, 2003), pp. 58–136.

65 White, *Climate of Rebellion*, pp. 80–1.

66 Camuffo and Enzi, 'Locust invasions and climatic factors', 63–4.

67 R. Endres, 'Zur wirtschaftlichen und sozialen Lage Franken vor dem Dreissigjährigen Krieg', *Jahrbuch für fränkische Landesforschung* 28 (1968), 5–52.

68 Pfister, 'Climatic Extremes, Recurrent Crises and Witch Hunts', 51–2.

69 Degroot, *Frigid Golden Age*, pp. 161–73, R. Suykerbuyk, *The Matter of Piety: Zoutleeuw's Church of Saint Leonard and Religious Material Culture in the Low Countries (c. 1450–1620)* (Leiden, 2021), pp. 240–55.

70 Pfister, 'Climatic Extremes, Recurrent Crises and Witch Hunts', 33–73; A. Palmer et al., 'Antarctic volcanic flux ratios from Law Dome ice cores', *Annals of Glaciology* 35 (2002), 329–32.

71 S. de Silva and G. Zielinski, 'Global influence of the AD 1600 eruption of Huaynaputina, Peru', *Nature* 393 (1998), 455–8.

72 K. Briffa et al., 'Influence of volcanic eruptions on Northern Hemisphere summer temperature over the past 600 years', *Nature* 393 (1998), 450–5.

73 C. Pfister, 'Weeping in the Snow: The Second Period of Little Ice Age-type Impacts, 1570–1630', in Behringer, Lehmann and Pfister (eds), *Kulturelle Konsequenzen der 'Kleinen Eiszeit'*, pp. 31–74.

74 G. Alfani, 'The Famine of the 1590s in Northern Italy: An Analysis of the Greatest "System Shock" of Sixteenth Century', *Histoire et Mesure* 26 (2011), 17–49.

75 Alfani, 'Climate, Population and Famine in Northern Italy', 34–5.

76 Ibid., 36–40; G. Alfani, *Il Grand Tour dei Cavalieri dell'Apocalisse. L'Italia del 'lungo Cinquecento' (1494–1629)* (Venice, 2010).

77 M.-C. Engels, *Merchants, Interlopers, Seamen and Corsairs: The 'Flemish' Community in Livorno and Genoa (1615–1635)* (Hilversum, 1997), pp. 74–5.

78 G. Darby, *Spain in the Seventeenth Century* (London, 2013), p. 22; White, *Cold Welcome*, pp. 81–2.

79 M. Clarke and H. Rendell, 'The impact of North Atlantic storminess on western European coasts: a review', *Quaternary International* 195 (2009), 31–41.

80 S. Chairi, 'Climatic issues in early modern England: Shakespeare's views of the sky', *WIREs Climate Change* 10 (2019), 1–10; also see P. Armstrong, 'Preposterous nature in Shakespeare's tragedies', in M. Neill and D. Schalkwyk (eds), *The Oxford Handbook of Shakespearean Tragedy* (Oxford, 2016), pp. 104–19; M. Hulme, 'Climate', in B. Smith (ed.), *The Cambridge Guide to the Worlds of Shakespeare: Shakespeare's World, 1500–1660* (Cambridge, 2016), i, pp. 29–34.

81 Hannaford and Nash, 'Climate, history, society', 379–80; D. Beach, *The Shona and their Neighbours* (Oxford, 1994).

82 See for example Degroot, *Frigid Golden Age*, p. 36; Brooke, *Climate Change*, pp. 442–4.

83 J. Hunwick, *Timbuktu and the Songhay Empire: Al-Saʿdī's Taʾrīkh al-sūdān down to 1613 and Other Contemporary Documents* (Leiden, 1999); D. Yahya, *Morocco in the Sixteenth Century: Problems and Patterns in African Foreign Policy* (Atlantic Highlands, NJ, 1981).

84 J. Webb, *Desert Frontier: Ecological Change and Economic Change along the Western Sahel, 1600–1850* (Madison, WI, 1995), pp. 47–8; E. Saad, *Social History of Timbuktu: The Role of Muslim Scholars and Notables, 1400–1900* (Cambridge, 1993).

85 H. Dunstan, 'The Late Ming Epidemics: A Preliminary Survey', *Late Imperial China* 3 (1975), 1–59.

86 Brook, 'Nine Sloughs', 40; T. Brook, 'Comparative pandemics: The Tudor–Stuart and Wanli–Chongzhen years of pestilence, 1567–1666', *Journal of Global History* 15 (2020), 363–79.

87 A. Reid, 'The Seventeenth Century Crisis in Southeast Asia', *Modern Asian Studies* 24 (1990), 649.

88 Grove and Adamson, *El Niño in World History*, p. 61.

89 Nurul Haq, *Zubdat al tawārīkh*, in H. Elliot (ed. and tr.), *The History of India, as Told by its Own Historians: The Muhammadan Period*, 8 vols (Cambridge, 1873), 6, p. 193.

90 White, *Climate of Rebellion*, pp. 78–226.

91 C. Bussov, *Moskovskaia Kronika, 1584–1613* (Moscow, 1961), pp. 222–4. For death rates, see Parker, *Global Crisis*, p. 152.

92 White, *Climate of Rebellion*, pp. 146–7.

93 R. Smith and D. Christian, *Bread and Salt: A Social and Economic History of Food and Drink in Russia* (Cambridge, 1984), p. 110.

94 'A rhyme set upon the wall of the Dutch churchyard in London, 5 May 1593', in J. Strype (ed.), *Annals of the Reformation*, 4 vols (Oxford, 1824), 4, pp. 234–5.

95 C. Dunning, *Russia's First Civil War: The Time of Troubles and the Founding of the Romanov Dynasty* (University Park, PA, 2001), pp. 13ff.; P. Bushkovitch, 'Taxation, Tax Farming, and Merchants in Sixteenth-Century Russia', *Slavic Review* 37 (1978), 381–98.

96 G. Hamburg, *Russia's Path toward Enlightenment: Faith, Politics, and Reason, 1500–1801* (New Haven, 2016), p. 255.

97 See for example G. Agoston, 'Habsburgs and Ottomans: Defense, Military Change, and Shifts in Power', *Turkish Studies Association Bulletin* 22 (1998), 126–41.

98 Lieberman, *Strange Parallels*, 1, pp. 154–63.

99 Cited in Reid, 'Seventeenth Century Crisis in Southeast Asia', 652.

100 J. Healey, *The First Century of Welfare: Poverty and Poor Relief in Lancashire, 1620–1730* (Cambridge, 2014), pp. 4–26.

101 White, ' "Shewing the difference betweene their conjuration, and our invocation', 51–2.

102 K. Kupperman, 'Environmental Stress and Rainmaking: Cosmic Struggles in Early Colonial Times', *ReVista: Harvard Review of Latin America* 6 (2007).

103 Grove and Adamson, *El Niño in World History*, pp. 63–4.

104 A. Reid, *A History of Southeast Asia: Critical Crossroads* (Chichester, 2015), pp. 114–15.

105 M. Ricklefs, 'Islamising Java: The Long Shadow of Sultan Agung', *Archipel* 56 (1998), 469–82.

106 See N. van Doom-Harder and K. de Jong, 'The Pilgrimage to Tembayat: Tradition and Revival in Indonesian Islam', *Muslim World* 91 (2001), 325–54; C. Guillot, 'The Tembayat hill: clergy and royal power in Central Java from the 15th to the 17th century', in H. Chambert-Noir and A. Reid (eds), *The Potent Dead: Ancestors, Saints and Heroes in Contemporary Indonesia* (Honolulu, 2002), pp. 141–59.

107 T. Wickman, 'Narrating Indigenous Histories of Climate Change in the Americas and Pacific', in White, Pfister and Mauelshagen (eds), *Palgrave Handbook of Climate History*, p. 397; B. Skopyk, 'Rivers of God, Rivers of Empire: Climate Extremes, Environmental Transformation and Agroecology in Colonial Mexico', *Environment and History* 23 (2017), 497–500.

108 Pfister, 'Climatic Extremes, Recurrent Crises and Witch Hunts', 46–63.

109 Parker, 'Crisis and Catastrophe', 1062. Above all see Parker, *Global Crisis, passim.*

110 G. Alfani, 'Plague in seventeenth century Europe and the decline of Italy: an epidemiological hypothesis', *European Review of Economic History* 17 (2013), 408–30.

111 de Vries, 'Economic Crisis of the Seventeenth Century', pp. 159–60.

112 G. Alfani, V. Gierok and F. Schaff, 'Economic Inequality in Preindustrial Germany, ca. 1300–1850', *Journal of Economic History* 82 (2022), 1–39. Also see B. Milanovic, *Global Inequality: A New Approach for the Age of Globalization* (Cambridge, MA, 2016), pp. 46–53.

113 G. Alfani and M. Percoco, 'Plague and long-term development: the lasting effects of the 1629–30 epidemic on the Italian cities', *Economic History Review* 72 (2019), 1175–1201.

114 P. Malanima, 'When did England overtake Italy? Medieval and early modern divergence in prices and wages', *European Review of Economic*

History 17 (2013), 45–70; M. Rota and J. Weisdorf, 'Italy and the little divergence in wages and prices: evidence from stable employment in rural areas', *Economic History Review* 74 (2021), 449–70.

115 de Vries, 'Economic Crisis of the Seventeenth Century', 167–74.

116 Parker, *Global Crisis*, p. xxiii.

117 Ibid., pp. xxxv–xxxvi, 14.

118 K. Chen et al., 'One Drought and One Volcanic Eruption Influenced the History of China: The Late Ming Dynasty Mega-drought', *Geophysical Research Letters* 47 (2020), 1–9.

119 See for example J. Zheng et al., 'How climate change impacted the collapse of the Ming dynasty', *Climatic Change* 127 (2013), 169–82; Chen et al., 'One Drought and One Volcanic Eruption', 1–9; J. Zhao et al., 'Role of the Summer Monsoon Variability in the Collapse of the Ming Dynasty: Evidences from Speleothem Records', *Geophysical Research Letters* 48 (2021), 1–12.

120 M. Marmé, 'Survival through transformation: how China's Suzhou-centred world economy weathered the general crisis of the seventeenth century', *Social History* 32 (2007), 144–65.

121 Brook, 'Nine Sloughs', 28–9.

122 A. Chan, *The Glory and Fall of the Ming Dynasty* (Norman, OK, 1982), pp. 235–6. For Ye Shaoyuan, see G. Fong, 'Reclaiming Subjectivity in a Time of Loss: Ye Shaoyuan (1589–1648) and Autobiographical Writing in the Ming-Qing Transition', *Ming Studies* 1 (2009), 21–41.

123 T. Brook, *The Troubled Empire: China in the Yuan and Ming Dynasties* (Cambridge, MA, 2010), pp. 253–5; J. Parsons, *Peasant Rebellions of the Late Ming Dynasty* (Tucson, AZ, 1970), pp. 17–21.

124 K. Swope, *The Military Collapse of China's Ming Dynasty, 1618–44* (Abingdon, 2014), p. 190.

125 Swope, *Military Collapse*, pp. 200–1.

126 T. Brook, *Confusions of Pleasure, Commerce and Culture in Ming China* (Berkeley, 1998), pp. 153ff.; M. Marmé, 'Locating Linkages or Painting Bull's-Eyes around Bullet Holes? An East Asian Perspective on the Seventeenth-Century Crisis', *American Historical Review* 113 (2008), 1062.

127 F. Wakeman, *The Great Enterprise: The Manchu Reconstruction of Imperial Order in Seventeenth Century China* (Berkeley, 1985), p. 16.

128 F. Wakeman, 'China and the Seventeenth Century Crisis', *Late Imperial China* 7 (1986), 10–14.

129 J. Warren, 'Weather, History and Empire: The Typhoon Factor and the Manila Galleon Trade, 1565–1815', in G. Wade and L. Tana (eds),

Anthony Reid and the Study of the Southeast Asian Past (Singapore, 2012), pp. 183–220.

130 Arteaga, Desierto and Koyama, 'Shipwrecked by Rents', CEPR Discussion Paper 15300 (2020), 1–50.

131 Parker, 'Crisis and Catastrophe', 1059.

132 Marmé, 'Survival through transformation', 1086; de Vries, 'The Crisis of the Seventeenth Century: The Little Ice Age and the Mystery of the Great Divergence', *Journal of Interdisciplinary History* 44 (2014), 374–5.

133 J. Schlesinger, *A World Trimmed with Fur: Wild Things, Pristine Places and the Natural Fringes of Qing* (Stanford, 2017), pp. 18ff.

18 CONCERNING GREAT AND LITTLE DIVERGENCES (C.1600–C.1800)

1 Degroot, *Frigid Golden Age*, pp. 82–91.

2 Jardine, *Worldly Goods*, p. 107.

3 M. Loderichs, *Rumphius' wonderwereld – Zeventiende-eeuwse natuurbeschrijvingen uit Ambon* (Zutphen, 2004).

4 J. Pimentel, 'The Iberian Vision: Science and Empire in the Framework of a Universal Monarchy, 1500–1800', *Osiris* 15 (2000), 21.

5 Ibid., 26.

6 Ibid., 23.

7 Jardine, *Worldly Goods*, pp. 172–3.

8 M. Pearson, 'First Contacts between Indian and European Medical Systems: Goa in the Sixteenth Century', in D. Arnold (ed.), *Warm Climates and Western Medicine: The Emergence of Tropical Medicine, 1500–1900* (Leiden, 1996), pp. 25–6.

9 J. Osterhammel, *Unfabling the East: The Enlightenment's Encounter with Asia* (Princeton, 2018), p. 58.

10 Ibid., pp. 99, 147.

11 Ibid., pp. 139, 311–13.

12 Varlık, 'New Science and Old Sources', pp. 205–6; H.-U. Lammel, 'Western European Perception and Representation of Plagues in Eastern Europe, the Ottoman Empire and the Near East, 1650–1800', in S. Cavaciocchi (ed.), *Economic and Biological Interactions in Pre-Industrial Europe from the Thirteenth to Eighteenth Centuries* (Florence, 2010), pp. 399–421.

13 P. Mansel, *Constantinople: City of the World's Desire, 1453–1924* (London, 1995), p. 382.

14 T. Babington Macaulay, 'Minute on Education', in M. Young (ed.), *Prose and Poetry* (Cambridge, MA, 1967), pp. 71–4.

15 S. Miller, *An Environmental History of Latin America* (Cambridge, 2015), pp. 84–5.

16 Ibid., pp. 95–7.

17 Ibid., pp. 93–4.

18 Ibid., pp. 57–8.

19 Zappia, *Traders and Raiders*, pp. 1–20; N. Blackhawk, *Violence over the Land: Indians and Empires in the Early American West* (Cambridge, MA, 2006).

20 Abū al-Faẓl, *Ain-I Akbari*, 3, pp. 394–6.

21 See for example K. Morrison and M. Lycett, 'Forest products in a wider world: Early historic connections across southern India', in S. A. Abraham et al. (eds), *Connections and Complexity: New Approaches to the Archaeology of South Asia* (Walnut Creek, CA, 2013), pp. 127–42.

22 Richards, *Unending Frontier*, pp. 17–38.

23 Y.-C. Wang, *Land Taxation in Imperial China, 1750–1911* (Cambridge, MA, 1973), pp. 6–7; Schlesinger, *World Trimmed with Fur*, p. 51.

24 K. Pomeranz, 'The Transformation of China's Environment, 1500–2000', in E. Burke and K. Pomeranz (eds), *The Environment and World History* (Berkeley, 2009), pp. 118–64; R. Mostern, 'Sediment and State in Imperial China: The Yellow River Watershed as an Earth System and a World System', *Nature and Culture* 11 (2016), 125.

25 P. Crossley, H. Siu and D. Sutton, *Empire at the Margins: Culture, Ethnicity, and Frontier in Early Modern China* (Berkeley, 2006).

26 P. Perdue, *China Marches West: The Qing Conquest of Central Eurasia* (Cambridge, MA, 2005), pp. 256–99.

27 Schlesinger, *World Trimmed with Fur*, pp. 1–15.

28 Miller, *Environmental History of Latin America*, p. 87.

29 Warde, *Invention of Sustainability*, p. 102.

30 Ibid., p. 160.

31 D. Blackbourn, *The Conquest of Nature: Water, Landscape, and the Making of Modern Germany* (London, 2006), pp. 31–2.

32 Warde, *Invention of Sustainability*, p. 172.

33 A. Rahman, M.-A. Clarke and S. Byrne, 'The Art of Breaking People Down: The British Colonial Model in Ireland and Canada', *Peace Research* 49 (2017), 15–38.

34 Pomeranz, *The Great Divergence*; K. Pomeranz, 'Re-Thinking the Late Imperial Chinese Economy: Development, Disaggregation and Decline, circa 1730–1930', *Itinerario* 22 (2000), 29–74.

35 J. Elliott, *The Old World and the New, 1492–1650* (Cambridge, 1970); K. Kupperman (ed.), *America in European Consciousness, 1493–1750* (Chapel Hill, NC, 1995).

36 Miller, *Environmental History of Latin America*, pp. 61–3.

37 Ibid., p. 64.

38 Collingham, *Hungry Empire*, pp. 67–71; J. Cock, 'Cassava: A Basic Energy Source in the Tropics', *Science* 218 (1982), 755–62.

39 Collingham, *Hungry Empire*, p. 67.

40 N. Nunn and N. Qian, 'The Potato's Contribution to Population and Urbanization: Evidence from a Historical Experiment', *Quarterly Journal of Economics* 126 (2011), 599–600.

41 Ibid., 601–2.

42 Ibid., 602–4.

43 Smith and Christian, *Bread and Salt*, pp. 199–200.

44 Smith, *Wealth of Nations*, pp. 67–8.

45 Nunn and Qian, 'Potato's Contribution to Population and Urbanization', 640–3.

46 M. Iyigun, N. Nunn and N. Qian, 'The Long-Run Effects of Agricultural Productivity on Conflict', NBER Working Paper 24066 (2017), 1–50; D. Acemoglu et al., 'A Dynamic Theory of Resource Wars', *Quarterly Journal of Economics* 127 (2012), 283–331.

47 R. Jia, 'Weather Shocks, Sweet Potatoes and Peasant Revolts in Historical China', *Economic Journal* 124 (2013), 92–118.

48 Nunn and Qian, 'Potato's Contribution to Population and Urbanization', 593–650.

49 E. Jones, *The European Miracle: Environments, Economies and Geopolitics in the History of Europe and Asia* (Cambridge, 2003).

50 S. Becker and L. Woessman, 'Was Weber Wrong? A Human Capital Theory of Protestant Economic History', *Quarterly Journal of Economics* 124 (2009), 531–96.

51 P. Lindert, 'Voice and Growth: Was Churchill Right?', *Journal of Economic History* 63 (2003), 315–50. For the effect of the Reformation outside Europe more generally see the important new work of U. Rublack (ed.), *Protestant Empires: Globalizing the Reformation* (Cambridge, 2020).

52 D. Cantoni and N. Yuchtman, 'Medieval Universities, Legal Institutions and the Commercial Revolution', NBER Working Paper 17979 (2012), 1–59.

53 Smith, *Wealth of Nations*, p. 198.

54 J. Shih, *Chinese Rural Society in Transition: A Case Study of the Lake Tai Area, 1368–1800* (Berkeley, 1992), pp. 29–30.

55 J. van Zanden, 'The road to the Industrial Revolution: hypotheses and conjectures about the medieval origins of the "European Miracle"', *Journal of Global History* 3 (2008), 342–3.

56 J. Mokyr, *The Enlightened Economy: Britain and the Industrial Revolution, 1700–1850* (London, 2011).

57 M. Wiesner-Hanks, *Women and Gender in Early Modern Europe* (Cambridge, 2019), pp. 159–228.

58 M. Kelly, J. Mokyr and C. Ó Gráda, 'The Mechanics of the Industrial Revolution', *Journal of Political Economy* 131 (2023).

59 J. Mokyr, A. Sarid, and K. van der Beek, 'The Wheels of Change: Technology Adoption, Millwrights and the Persistence in Britain's Industrialisation', *The Economic Journal* 132 (2022), 1894–1926.

60 M. Kelly, J. Mokyr and C. Ó Gráda, 'Precocious Albion: A New Interpretation of the British Industrial Revolution', *Annual Review of Economics* 6 (2014), 363–89.

61 Allen, *British Industrial Revolution in Global Perspective*, pp. 25–56.

62 van Zanden, 'Road to the Industrial Revolution', 348.

63 Wakeman, 'China and the Seventeenth Century Crisis', 1–26, 237ff.

64 Pfister et al., 'The meteorological framework and the cultural memory of three severe winter-storms in early eighteenth-century Europe', 281–310.

65 S. Johnson, 'El Niño, Environmental Crisis, and the Emergence of Alternative Markets in the Hispanic Caribbean, 1760s–70s', *William and Mary Quarterly* 62 (2005), 365–410.

66 Ibid., 398–400.

67 'Declaration and Resolves of the First Continental Congress', in J. Vile (ed.), *The Declaration of Independence: America's First Founding Document in US History and Culture* (Santa Barbara, CA, 2019), p. 14.

68 'John Adams to Isaac Smith Sr, 1 June 1776', in W. Morgan (ed.), *Naval Documents of the American Revolution*, 12 vols (Washington, DC, 1964–2013), 5, p. 338.

69 S. Johnson, *Climate and Catastrophe in Cuba and the Atlantic World in the Age of Revolution* (Chapel Hill, NC, 2011), pp. 132–6.

70 D. Clingingsmith and J. Williamson, 'Mughal Decline, Climate Change and Britain's Industrial Ascent: An Integrated Perspective on India's 18th and 19th Century Deindustrialization', NBER Working Paper 11730 (2005), 1–38.

71 J. Nehru, *The Discovery of India* (London, 1947); T. Roy, 'Economic History and Modern India: Redefining the Link', *Journal of Economic Perspectives* 16 (2002), 109–30.

72 W. Hunter, *The Annals of Rural Bengal* (London, 1868), p. 21.

73 Ibid., p. 22.

74 W. Dalrymple, *The Anarchy* (London, 2019), pp. 215–17.

75 Ibid., pp. 220–3.

76 Ibid., pp. 229–32; Frankopan, *Silk Roads*, pp. 275–9.

77 T. Thordarson and S. Self, 'Atmospheric and environmental effects of the 1783–1784 Laki eruption: A review and reassessment', *Journal of Geophysical Research* 108 (2003), 1–29.

78 L. Oman et al., 'Modeling the distribution of the volcanic aerosol cloud from the 1783–1784 Laki eruption', *Journal of Geophysical Research* 111 (2006), 1–15.

79 B. Franklin, 'Meteorological imaginations and conjectures', *Memoirs and Proceedings of the Manchester Literary and Philosophical Society* 122 (1784), 237–40.

80 Thordarson and Self, 'Atmospheric and environmental effects of the 1783–1784 Laki eruption', 21–2.

81 B. Zambri et al., 'Modeling the 1783–4 Laki Eruption in Iceland: 2. Climate Impacts', *JGR Atmospheres* 124 (2019), 6770–90.

82 Thordarson and Self, 'Atmospheric and environmental effects of the 1783–1784 Laki eruption', 15–16. For Mexico, see G. Endfield, *Climate and Society in Colonial Mexico: A Study in Vulnerability* (Malden, MA, 2008), pp. 177–81.

83 R. D'Arrigo et al., 'The anomalous winter of 1783–1784: Was the Laki eruption or an analog of the 2009–2010 winter to blame?', *Geophysical Research Letters* 38 (2011).

84 F. Pausata et al., 'Impacts of a high-latitude volcanic eruption on ENSO and AMOC', *PNAS* 112 (2015), 13,784–8; F. Pausata et al., 'ENSO response to high-latitude volcanic eruptions in the Northern Hemisphere: the role of the initial conditions', *Geophysical Research Letters* 43 (2016) 8694–702.

85 Grove and Adamson, *El Niño in World History*, pp. 82–3; also E. Cook et al., 'Asian Monsoon Failure and Megadrought during the Last Millennium', *Science* 328 (2010), 486–9.

86 V. Damodaran, 'The East India Company, Famines and Ecological Conditions in Eighteenth Century Bengal', in A. Winterbottom, A. Lester and V. Damodaran (eds), *The East India Company and the Natural World* (London, 2015), pp. 80–101.

87 A. Janetta, 'Famine Mortality in Nineteenth-Century Japan: The Evidence from a Temple Death Register', *Population Studies* 46 (1992), 428–9.

88 C. Totman, *Early Modern Japan* (Berkeley, 1993), p. 239.

89 M. Ravina, *Land and Lordship in Early Modern Japan* (Stanford, 1999), pp. 128–41.

90 D. Crecelius, *The Roots of Modern Egypt: A Study of the Regimes of 'Ali Bey al-Kabir and Muhammad Bey Abu al-Dhahab, 1760–1775* (Minneapolis, MN, 1981), pp. 79–91, 159–68.

91 A. Mikhail, 'Ottoman Iceland: A Climate History', *Environmental History* 20 (2015), 262–84; Endfield, *Climate and Society in Colonial Mexico.*

92 Mikhail, 'Ottoman Iceland', 275–6.

93 V. Damodaran et al., 'The 1780s: Global Climate Anomalies, Floods, Droughts and Famines', in White, Pfister and Mauelshagen (eds), *Palgrave Handbook of Climate History*, pp. 539–40.

94 R. Grove, 'The Great El Niño of 1789–93 and its Global Consequences: Reconstructing an Extreme Climate Event in World Environmental History', *Medieval History Journal* 10 (2007), 75–98; for mega-Niños, see B. Meggers, 'Archaeological Evidence for the Impact of Mega-Niño Events in Amazonia during the Past Two Millennia', *Climatic Change* 28 (1994), 321–38.

95 J. Gergis and A. Fowler, 'A History of El Niño–Southern Oscillation (ENSO) Events since A.D. 1525: Implications for Future Climate Change', *Climatic Change* 92 (2009), 343–87.

96 Damodaran et al., 'The 1780s', pp. 531–2.

97 Ibid., pp. 532–4.

98 Grove, 'The Great El Niño of 1789–93', 84–5.

99 Ibid., 80–3.

100 Ibid., pp. 86–7; J. Guy, 'Ecological factors in the rise of Shaka and the Zulu kingdom', in S. Marks and A. Atmore (eds), *Economy and Society in Pre-Industrial South Africa* (London, 1980), pp. 102–19.

101 Wickman, 'Narrating Indigenous Histories', pp. 393-4.

102 L. Abad and N. Maurer, 'Do Pandemics Shape Elections? Retrospective Voting in the 1918 Spanish Flu Pandemic in the United States', CEPR Discussion Paper 15678 (2021), 13–14.

103 Grove, 'The Great El Niño of 1789–93', 91–3.

104 A. Taylor, 'The Hungry Year: 1789 on the Northern Border of Revolutionary America', in A. Johns, *Dreadful Visitations: Confronting Natural Catastrophe in the Age of Enlightenment* (London, 2013), pp. 156–7.

105 Ibid., p. 158.

106 S. Schama, *Citizens: A Chronicle of the French Revolution* (London, 1989), pp. 330–2.

107 S. Hazareesingh, *Black Spartacus: The Epic Life of Toussaint Louverture*
(New York, 2020); J. Marr and J. Cathey, 'The 1802 Saint-Domingue
Yellow Fever Epidemic and the Louisiana Purchase', *Journal of
Public Health Management and Practice* 19 (2013), 77–82; Miller,
Environmental History of Latin America, pp. 112–13.

108 J. Belich, *Replenishing the Earth: The Settler Revolution and the Rise of
the Angloworld* (Oxford, 2001), p. 55.

19 INDUSTRY, EXTRACTION AND THE NATURAL WORLD
(c.1800–c.1870)

1 D. Hume, 'On the Populousness of Ancient Nations', in *Essays: Moral;
Political; and Literary*, ed. E. Miller (Indianapolis, 1987), pp. 448–9.

2 Ibid., pp. 450–1; J. Fleming, *Historical Perspectives on Climate Change*
(Oxford, 1998), p. 17.

3 B. Franklin, 'From Benjamin Franklin to Ezra Stiles, 29 May 1763', in
L. Labaree (ed.), *The Papers of Benjamin Franklin*, 43 vols (New Haven,
1959), 10, pp. 264–7.

4 S. Johnson, 'Review of Du Halde's *Description of China*', *Gentleman's
Magazine* 12 (1742), 320; Fleming, *Historical Perspectives*, p. 24.

5 See P. Moore, *Endeavour: The Ship and the Attitude that Changed the
World* (London, 2018).

6 S. Bedini, *Jefferson: Statesman of Science* (New York, 1990), pp. 73–5.

7 J. Boyd (ed.), *The Papers of Thomas Jefferson*, 27 vols (Princeton,
1955–97), 7, p. 31.

8 Bedini, *Jefferson*, p. 75.

9 T. Jefferson, *Notes on the State of Virginia* (Boston, MA, 1832),
pp. 80–90; Fleming, *Historical Perspectives*, p. 26.

10 H. Williamson, 'An Attempt to Account for the Change of Climate
Which Has Been Observed in the Middle Colonies in North America',
Transactions of the American Philosophical Society 1 (1769–71), 272–80.

11 S. Williams, *The Natural and Civil History of Vermont* (Walpole, NH,
1794), p. 57; Fleming, *Historical Perspectives*, p. 25.

12 N. Webster, 'On the Supposed Change in the Temperature of Winter',
in N. Webster, *A Collection of Papers on Political, Literary and Moral
Subjects* (New York, 1843), pp. 144–8.

13 G.-L. de Buffon, 'Époques de la nature', in G.-L. de Buffon, *Histoire
Naturelle*, 5 vols (Paris, 1778), 1, pp. 240–1. Also see Fleming, *Historical
Perspectives*, pp. 21–32.

14 T. Malthus, *An Essay on the Principle of Population, as it Affects the Future Improvement of Society. With Remarks on the Speculations of Mr Godwin, M. Condorcet and Other Writers* (London, 1798), p. 44.

15 J. Leslie, *An Experimental Inquiry into the Nature and Propagation of Heat* (London, 1804), pp. 180–2. Also see A. Zilberstein, 'Inured to Empire: Wild Rice and Climate Change', *William and Mary Quarterly* 72.1 (2015), 149–50.

16 Zilberstein, 'Inured to Empire: Wild Rice and Climate Change', 127–58.

17 D. Blackbourn, *The Conquest of Nature: Water, Landscape and the Making of Modern Germany* (London, 2006), p. 21.

18 J. Frey, 'The Indian Saltpeter Trade, the Military Revolution, and the Rise of Britain as a Global Superpower', *The Historian* 71 (2009), 507–54.

19 J. Houdaille, 'Pertes de l'armée de terre sous l'empire', *Population* 27 (1972), 27–50; also J. Houdaille, 'Le problème des pertes de la guerre', *Revue d'Histoire Moderne et Contemporaine* 17 (1970), 411–23.

20 C. Hall, 'The Royal Navy and the Peninsular War', *Mariner's Mirror* 4 (1993), 413–18.

21 J-J. Voth, B. Caprettini and A. Trew, 'Fighting for Growth: Labor Scarcity and Technological Progress During the British Industrial Revolution', University of Glasgow, Adam Smith Business School, Working Paper Series 2022–15 (2022), 1–62.

22 R. Poole, *Peterloo: The English Uprising* (Oxford, 2019).

23 Ibid., p. 18.

24 P. Pickering and A. Tyrrell, *The People's Bread: A History of the Anti–Corn Law League* (London, 2000).

25 C. Oppenheimer, *Eruptions that Shook the World* (Cambridge, 2011), pp. 299–311.

26 S. Cao, Y. Li and B. Yang, 'Mt. Tambora, Climatic Changes, and China's Decline in the Nineteenth Century', *Journal of World History* 23 (2012), 587–607; E. Skeen, ' "The Year without a Summer": A Historical View', *Journal of the Early Republic* 1 (1981), 51–67; W. Soon and S. Yaskell, 'Year without a Summer', *Mercury* 32 (2003), 13–22.

27 Oppenheimer, *Eruptions*, pp. 313–17.

28 G. D'Arcy Wood, *Tambora: The Eruption that Changed the World* (Princeton, 2014), pp. 45–71.

29 Ibid., pp. 220–4.

30 Ibid., pp. 78–84.

31 Ibid., pp. 72–96.

32 M. Harrison, 'A Dreadful Scourge. Cholera in early nineteenth-century India', *Modern Asian Studies* 54 (2020), 502–53.

33 R. Peckham, *Epidemics in Modern Asia* (Cambridge, 2016), pp. 53–63.

34 C. Emsley, *British Society and the French Wars, 1793–1815* (Basingstoke, 1979), p. 173.

35 J. Tosh, 'Jeremiah Goldswain's Farewell: Family and Fortune in Early Nineteenth-Century English Emigration', *History Workshop Journal* 77 (2014), 26–44; Belich, *Replenishing the Earth*, p. 147.

36 Belich, *Replenishing the Earth*, p. 146.

37 E. Richards, *Britannia's Children: Emigration from England, Scotland, Wales and Ireland since 1600* (London, 2004), pp. 110–11.

38 Ibid., p. 118.

39 A. Crosby, *The Columbian Exchange: Biological and Cultural Consequences of 1492* (Westport, CT, 2003), pp. 214–15.

40 L. Putterman and D. Weil, 'Post-1500 Population Flows and the Long-Run Determinants of Economic Growth and Inequality', *Quarterly Journal of Economics* 125 (2010), 1627–82.

41 T. Hatton and J. Williamson, 'Migrations from Europe in the Late Nineteenth Century?', *Population and Development Review* 20 (1994), 533–59; for a case study, see R. Abramitzky, L. Platt Boustan and K. Eriksson, 'Europe's Tired, Poor, Huddled Masses: Self-Selection and Economic Outcomes in the Age of Mass Migration', NBER Working Paper 15684 (2010), 1–50.

42 T. Dublin (ed.), *Immigrant Voices: New Lives in America, 1773–1986* (Urbana, IL, 1993), pp. 78–80.

43 C. Saunt, *Unworthy Republic: The Dispossession of Native Americans and the Road to Indian Territory* (New York, 2020), pp. xii, 5.

44 D. Martinez, *Documents of American Indian Removal* (Santa Barbara, CA, 2019), pp. 41–4.

45 Saunt, *Unworthy Republic*, pp. 14, 17.

46 Collingham, *Hungry Empire*, pp. 190–2.

47 Belich, *Replenishing the Earth*, pp. 82–3.

48 Ibid., p. 1.

49 D. Moon, *The Plough that Broke the Steppes: Agriculture and Environment on Russia's Grasslands, 1700–1914* (Oxford, 2013), pp. 16–17.

50 D. Moon, 'Peasant Migration, the Abolition of Serfdom and the Internal Passport System in the Russian Empire, c.1800–1914', in D. Eltis (ed.), *Free and Coerced Migration: Global Perspectives* (Stanford, 2002), pp. 324–57.

51 J. Reardon-Anderson, *Reluctant Pioneers: China's Expansion Northward, 1644–1937* (Stanford, 2005), pp. 18–84.

52 J. Goldstone, 'Efflorescences and Economic Growth in World History: Rethinking the "Rise of the West" and the Industrial Revolution', *Journal of World History* 13 (2002), p. 364.

53 G. Clark and D. Jacks, 'Coal and the Industrial Revolution', *European Review of Economic History* 11 (2007), 39–72; also see P. Vries, 'Are Coal and Colonies Really Crucial? Kenneth Pomeranz and the Great Divergence', *Journal of World History* 12 (2001), 407–46.

54 D. Theodoridis, P. Warde and A. Kander, 'Trade and overcoming land constraints in British industrialization: an empirical assessment', *Journal of Global History* 13 (2018), 328–51.

55 Pomeranz, *Great Divergence,* pp. 60–4; Goldstone, 'Efflorescences and Economic Growth', 348–53; Belich, *Replenishing the Earth*, pp. 11–13; P. Malanima, 'Energy crisis and growth, 1650–1850: the European deviation in a comparative perspective', *Journal of Global History* 1 (2006), 101–21; P. O'Brien, 'Colonies in a Globalizing Economy, 1815–1948', in B. Gillis and W. Thompson (eds), *Globalization and Global History* (London, 2006), pp. 248–91; P. O'Brien, 'European Economic Development: The Contribution of the Periphery', *Economic History Review* 35 (1982), 1–18.

56 Belich, *The World the Plague Made*, pp. 420, 428.

57 R. Allen, *The British Industrial Revolution in Global Perspective* (Cambridge, 2014), pp. 80–4; Belich, *The World the Plague Made*, p. 421.

58 O'Brien, 'Colonies in a Globalizing Economy', pp. 248–91; O'Brien, 'European Economic Development', 1–18.

59 S. Heblich, S. Redding and H.-J. Voth, 'Slavery and the British Industrial Revolution', NBER Working Paper 30451 (2022), 1–45.

60 O'Brien, 'Colonies in a Globalizing Economy', pp. 90–8.

61 Pomeranz, 'Re-Thinking the Late Imperial Chinese Economy', 29–74.

62 Belich, *The World the Plague Made*, pp. 289-300.

63 J. Brolin and A. Kander, 'Environmental factors in trade during the great transformation: advancing the geographical coverage before 1950', *Journal of Global History* 15 (2020), 250–1.

64 A. Stone, *Petrified Intelligence: Nature in Hegel's Philosophy* (Albany, NY, 2005).

65 H. Spalding, 'The Black Sea and the Caspian', *Van Nostrand's Eclectic Engineering Magazine* 15 (1876), 122–7; Moon, *Plough that Broke the Steppes*, p. 212.

66 'The Black Sea and the Caspian', *The Age*, 7 August 1876.

67 D. Davis, *The Arid Lands: History, Power, Knowledge* (Cambridge, MA, 2015), p. 83.

68 M. Williams, *Deforesting the Earth: From Prehistory to Global Crisis* (Chicago, 2003), p. 430. Also see Wulf, *The Invention of Nature: Alexander von Humboldt's New World* (New York, 2016), pp. 59–60.

69 H. Haygarth, *Recollections of Bush Life in Australia during a Residence of Eight Years in the Interior* (London, 1864), p. 121.

70 Moon, *Plough that Broke the Steppes*, pp. 173–4.

71 Ibid., p. 120.

72 D. Moon, 'Agriculture and the Environment on the Steppes in the Nineteenth Century', in N. Bregfoyle, A. Schrader and W. Sunderland (eds), *Peopling the Russian Periphery: Borderland Colonisation in Eurasian History* (Abingdon, 2007), pp. 88–90.

73 Moon, *Plough that Broke the Steppes*, pp. 118–38.

74 M. Drolet, 'Nature, Science and the Environment in Nineteenth-Century French Political Economy: The Case of Michel Chevalier (1805–79)', *Modern Intellectual History* 15 (2018), 711–45.

75 M. Gadgill and R. Guha, *This Fissured Land: An Ecological History of India* (New Delhi, 1992), pp. 116ff.

76 Williams, *Deforesting the Earth*, p. 335.

77 Ibid., pp. 354ff.; Gadgil and Guha, *This Fissured Land, passim.*

78 J. Weaver, *The Great Land Rush and the Making of the Modern World* (Montreal, 2003), pp. 88–90.

79 K. Sivaramakrishnan, 'Colonialism and Forestry in India: Imagining the Past in Present Politics', *Comparative Studies in Society and History* 37 (1995), 16–17.

80 C. Ross, *Ecology and Power in the Age of Empire: Europe and the Transformation of the Tropical World* (Oxford, 2017), p. 246.

81 Ibid., p. 297.

82 P. Hirst and G. Thompson, *Globalization in Question: The International Economy and the Possibilities of Governance* (Cambridge, MA, 1996), p. 31.

83 Survival International, 'How will we survive? The destruction of the Congo Basin tribes in the name of conservation' (London, 2017); Middle East Monitor, 'Tanzania: Maasai people face violence, eviction amid protests over UAE-owned game reserve', 22 June 2022.

84 H. Wauchope et al., 'Protected areas have a mixed impact on waterbirds, but management helps', *Nature* 605 (2022), 1–19.

85 Beinart and Hughes, *Environment and Empire*, pp. 66–7.

86 P. Lane, 'Material Desires, Ecological Anxieties and East African Elephant Ivory from a Long-Term Perspective', *Environmental History* 24 (2019), 688–94.

87 Beinart and Hughes, *Environment and Empire*, pp. 67–8.

88 Ibid., pp. 64–9.

89 Belich, *Replenishing the Earth*, pp. 3–4.

90 Ibid., p. 49.

91 Ibid., p. 79.

92 P. Sharp and J. Weisdorf, 'Globalization revisited: Market integration and the wheat trade between North America and Britain from the eighteenth century', *Explorations in Economic History* 50 (2013), 90.

93 O. Barak, *Powering Empire: How Coal Made the Middle East and Sparked Global Carbonization* (Oakland, CA, 2020).

94 O. Figes, *The Europeans: Three Lives and the Making of a Cosmopolitan Culture* (London, 2019), pp. 43–52, 220–6.

95 H.-J. Teuteberg, *Die Rolle des Fleischextrakts für die Ernährungswissenschaften und den Aufstieg der Suppenindustrie* (Stuttgart, 1990); R. Woods, *The Herds Shot around the World: Native Breeds and the British Empire, 1800–1900* (Chapel Hill, NC, 2017).

96 I. Gazeley and A. Newell, 'The First World War and working-class food consumption in Britain', *European Review of Economic History* 17 (2013), pp. 71–94.

97 T. Dubois, 'Many roads from pasture to plate: a commodity chain approach to China's beef trade, 1732–1931', *Journal of Global History* 14 (2019), 22–43.

98 Collingham, *Hungry Empire*, pp. 232, 223.

99 Ibid., p. 223.

100 Ibid., p. 218.

101 Belich, *Replenishing the Earth*, pp. 3, 113.

102 Clingingsmith and Williamson, 'Mughal Decline, Climate Change and Britain's Industrial Ascent', 1–38.

103 S. Ambirajan, 'Malthusian population theory and Indian famine policy in the nineteenth century', *Population Studies* 30 (1976), 8.

104 Collingham, *Hungry Empire*, p. 221; S. Sweeney, 'Indian Railways and Famine, 1875–1914: Magic Wheels and Empty Stomachs', *Essays in Economic and Business History* 26 (2008), p. 146; W. Cosgrove et al., 'Colonialism international trade and the nation state', in L. Newman (ed.), *Hunger in History* (Oxford, 1990), p. 234; D. Hall-Matthews, *Peasants, Famine and the State in Colonial Western India* (London, 2005), p. 8.

105 S. Beckert, 'Emancipation and Empire: Reconstructing the worldwide web of cotton production in the age of the American Civil War', *American Historical Review* 109 (2004), 1409–10.

106 Ibid., 1410; Ross, *Ecology and Power*, p. 29.

107 H. Berger and M. Spoerer, 'Economic Crises and the European Revolutions of 1848', *Journal of Economic History* 61 (2001), 293–326; P. Wilson (ed.), *1848: The Age of Revolutions* (Burlington, VT, 2006).

108 E. Baptist, *The Half Has Never Been Told: Slavery and the Making of American Capitalism* (New York, 2016), pp. 111–44; French, *Born in Blackness*, p. 8.

109 Beckert, 'Emancipation and Empire', 1424–5.

110 Ibid., 1430.

111 Ibid., 1414.

112 K. Cuno, 'African Slaves in 19th-Century Rural Egypt', *International Journal of Middle East Studies* 41 (2009), 186–8.

113 L. Satya, *Cotton and Famine in Berar, 1850–1900* (New Delhi, 1997), 273–300.

114 Ross, *Ecology and Power*, pp. 37–9; R. Owen, *Cotton and the Egyptian Economy, 1820–1914: A Study in Trade and Development* (Oxford, 1969), pp. 98–102.

20 THE AGE OF TURBULENCE (C.1870–C.1920)

1 Ross, *Ecology and Power*, pp. 35–6.

2 Ibid., p. 99.

3 Ibid.; also H. Hobhouse, *Seeds of Wealth: Four Plants that Made Men Rich* (Oxford, 2003), pp. 125–37.

4 Ross, *Ecology and Power*, pp. 106–7, 100–1, 111–12; J. Drabble, *Rubber in Malaya, 1876–1922: The Genesis of the Industry* (Oxford, 1973), pp. 216–19.

5 L. Córdoba, 'White Blood, Black Gold: The Commodification of Wild Rubber in the Bolivian Amazon, 1870–1920', *Environmental History* 24 (2019), 695–702.

6 C. Elkins, *Legacy of Violence: A History of the British Empire* (London, 2022), p. 73.

7 J. Mackenzie, 'The Popular Culture of Empire in Britain', in J. Brown and W. Louis (eds) *The Twentieth Century* (Oxford, 1999), pp. 212-31.

8 P. Curtin, *Disease and Empire: The Health of European Troops in the Conquest of Africa* (Cambridge, 1998), pp. 23–8; R. Deb Roy, *Malarial Subjects: Empire, Medicine and Nonhumans in British India, 1820–1909* (Cambridge, 2017); S. Gänger, 'Cinchona Harvest, Deforestation,

and "Extinction" in the Viceroyalty of New Granada, 1752–1811',
Environmental History 24 (2019), 673–9.

9 Ross, *Ecology and Power*, pp. 138–9.

10 Ibid., pp. 139–55; S. Naylor, 'Spacing the Can: Empire, Modernity, and
the Globalisation of Food', *Environment and Planning A* 32 (2000),
1625–39.

11 Richards, *Unending Frontier*, pp. 576–622.

12 E. Dolin, *Leviathan: The History of Whaling in America*
(New York, 2007).

13 G. Cushman, *Guano and the Opening of the Pacific World: A Global
Ecological History* (Cambridge, 2013), p. 41–3.

14 Ibid., p. 8.

15 Ibid., pp. 23–7.

16 Ibid., pp. 45–8.

17 J. Conrad, *Lord Jim* (London, 1993), ch. 14, pp. 102–8; Cushman,
Guano, p. 98.

18 R. Dillon, 'A Tall Tale about the Guano Trade', *Western Folklore* 11
(1952), 125.

19 Cushman, *Guano*, pp. 60–74.

20 P. Guardino, The Dead March. A History of the Mexican American War
(Cambridge, MA: 2017).

21 Ibid., pp. 54–6, 81–2; C. Duffy Barnett, 'The Edges of Empire and the
Limits of Sovereignty: American Guano Islands', *American Quarterly* 57
(2005).

22 C. Simmonds, *The Battle of Midway* (Oxford, 2011).

23 Cushman, *Guano*, p. 73. For Peru's sovereign debt, see C. Vizcarra,
'Guano, Credible Commitments, and Sovereign Debt Repayment
in Nineteenth-Century Peru', *Journal of Economic History* 69 (2009),
358–87.

24 J. Baron, 'Sailors' scurvy before and after James Lind – a reassessment',
Nutrition Reviews 67 (2009), 315–32.

25 A. Dimico, A. Isopi and O. Olsson, 'Origins of the Sicilian Mafia: The
Market for Lemons', *Journal of Economic History* 77 (2017), 1083–1115.

26 C. Ó Gráda, 'The Next World and the New World: Relief, Migration, and
the Great Irish Famine', *Journal of Economic History* 79 (2019), 319–55.

27 L. Goméz-Alpizar, I. Carbone and J. Ristaino, 'An Andean origin
for *Phytophthora infestans* inferred from nuclear and mitochondrial
DNA sequences', *PNAS* 104 (2007), 3306–11; M. Martin et al.,
'Genomic characterization of a South American *Phytophthora* hybrid
mandates reassessment of the geographic origins of *Phytophthora*

infestans', *Molecular Biology and Evolution* 33 (2016), 478–91; P. Bourke, 'Emergence of potato blight, 1843–46', *Nature* 203 (1964), 805–8.

28 C. O Gráda, *Black '47 and Beyond: The Great Irish Famine in History, Economy, and Memory* (Princeton, 1999).

29 C. Kinealy, *The Great Irish Famine: Impact, Ideology and Rebellion* (Basingstoke, 2001).

30 H. Hirota, *Expelling the Poor: Atlantic Seaboard States and the Nineteenth-Century Origins of American Immigration Policy* (New York, 2017).

31 W. Collins and A. Zimran, 'The Economic Assimilation of Famine Irish Migrants to the United States', NBER Working Paper 25287 (2017), 1–48.

32 See for example https://www.irishcentral.com/news/census-shows-alm ost-seven-times-more-irish-americans-than-population-of-ireland-218344 001-237779801.

33 A. McKeown, 'Global migration, 1846–1940', *Journal of World History* 15 (2004), 157–8.

34 Ibid., 159.

35 J. Osterhammel, *Die Verwandlung der Welt: Eine Geschichte des 19. Jahrhunderts* (Munich, 2009), p. 159.

36 H. Maude, *Slavers in Paradise: The Peruvian Slave Trade in Polynesia, 1862–1864* (Stanford, 1981).

37 D. Northrop, *Indentured Labor in the Age of Imperialism, 1834–1922* (Cambridge, 1995); E. Collingham, *Imperial Bodies: The Physical Experience of the Raj, c.1800–1947* (Cambridge, 2001), pp. 140–1.

38 Driscoll, *The Whites Are the Enemies of Heaven*, p. 35.

39 Collingham, *Imperial Bodies*, pp. 140–1.

40 J. Loadman, *Tears of the Tree* (Oxford, 2005), pp. 140–1.

41 K. Marx, *Capital*, tr. B. Fowkes, 3 vols (London, 1976–81), I, p. 895.

42 Ross, *Ecology and Power*, p. 112.

43 Belich, *Replenishing the Earth*, pp. 507–8.

44 A. Knighton, 'River adjustment to changes in sediment load: The effects of tin mining on the Ringarooma River, Tasmania, 1875–1984', *Earth Surface Processes and Landforms* 14 (1989), 333–59; Ross, *Ecology and Power*, pp. 149–50.

45 U. Förstner and G. Wittman, *Metal Pollution in the Aquatic Environment* (Berlin, 1979).

46 R. Eisler and S. Wiemeyer, 'Cyanide Hazards to Plants and Animals from Gold Mining and Related Water Issues', *Reviews of Environmental Contamination and Toxicology* 183 (2004), 21–54.

47 Ibid., 22; S. Fields, 'Tarnishing the Earth: Gold Mining's Dirty Secret', *Environmental Health Perspectives* 109 (2001), 474–81.

48 Beinart and Hughes, *Environment and Empire*, pp. 133–6.

49 *Report of the Indian Famine Commission* (London, 1880), p. 9.

50 S. Amrith, *Unruly Waters: How Mountain Rivers and Monsoons Have Shaped South Asia's History* (London, 2018), pp. 40–1.

51 Ibid., p. 18.

52 Ibid., p. 45; R. Guha (ed.), *Subaltern Studies III: Writings on South Asian History and Society* (Delhi, 1984), p. 69.

53 Beinart and Hughes, *Environment and Empire*, pp. 134–5.

54 Amrith, *Unruly Waters*, pp. 52–5.

55 Ross, *Ecology and Power*, pp. 312–14; for the Dutch East Indies, see W. Ravesteijn, *De zegenrijke heeren der wateren: irrigatie en staat op Java, 1832–1942* (Delft, 1997), pp. 111–206.

56 Guha (ed.), *Subaltern Studies III*, p. 69; Beinart and Hughes, *Environment and Empire*, p. 137.

57 Dalhousie to the Court of Directors, 20 April 1853, in S. Settar (ed.), *Railway Construction in India: Select Documents*, 2 vols (Delhi, 1999), 2, pp. 23–57; Amrith, *Unruly Waters*, p. 52.

58 See for example R. White, *Railroaded: The Transcontinentals and the Making of Modern America* (New York, 2011); C. Leonard et al., 'Land prices and railroad building in European Russia, 1860s to the early 1900s', *Russian Journal of Economics* 7 (2021), 93–104; J. Atack et al., 'Did railroads induce or follow economic growth? Urbanization and population growth in the American Midwest, 1850–1860', *Social Science History* 34 (2010), 171–97; B. Elleman and S. Kotkin, *Manchurian Railways and the Opening of China: An International History* (Armonk, NY, 2010).

59 Ross, *Ecology and Power*, p. 314; P. Schmitthenner, 'Colonial Hydraulic Projects in South India: Environmental and Cultural Legacy', in D. Kumar, V. Damodaran and R. D'Souza (eds), *The British Empire and the Natural World: Environmental Encounters in South Asia* (Oxford, 2011), pp. 181–201.

60 Amrith, *Unruly Waters*, pp. 121–4.

61 D. Gilmartin, *Blood and Water: The Indus River Basin in Modern History* (Berkeley, 2015), pp. 27–68.

62 *Indian Industrial Commission, 1916–1918* (Calcutta, 1918), p. 4; Amrith, *Unruly Waters*, p. 137.

63 C. Brantley, 'Kikuyu-Maasai Nutrition and Colonial Science: The Orr and Gilks Study in Late 1920s Kenya Revisited', *International Journal of African History* 30 (1997), 84.

64 J. Voelcker, *Report on the Improvement of Indian Agriculture* (London, 1893), p. vi; also Ross, *Ecology and Power*, p. 331.

65 Ross, *Ecology and Power*, p. 314.

66 A. Church, *East Africa, a New Dominion: A Crucial Experiment in Tropical Development and its Significance to the British Empire* (London, 1927), p. 114.

67 Ross, *Ecology and Power*, p. 314; D. Biggs, *Quagmire: Nation-Building and Nature in the Mekong Delta* (Seattle, 2010), pp. 91–195.

68 E. Whitcombe, 'The Environmental Costs of Irrigation in British India: Waterlogging, Salinity and Malaria', in Arnold and Guha (eds), *Nature, Culture, Imperialism*, pp. 254–6.

69 Beinart and Hughes, *Environment and Empire*, p. 135.

70 Whitcombe, 'Environmental Costs', p. 255.

71 Ross, *Ecology and Power*, p. 317.

72 Ibid., pp. 71–95.

73 S. Barraclough and K. Ghimire, *Forests and Livelihoods: The Social Dynamics of Deforestation in Developing Countries* (Basingstoke, 1995); M. Tiffen, M. Mortimore and F. Gichuki, *More People, Less Erosion: Environmental Recovery in Kenya* (Chichester, 1994).

74 L. Hughes, *Moving the Maasai: A Colonial Misadventure* (Basingstoke, 2006).

75 H. Tilley, *Africa as Living Laboratory: Empire, Development, and the Problem of Scientific Knowledge, 1870–1950* (Chicago, 2011), pp. 124–6.

76 W. Churchill, 'My African Journey', National Liberal Club Speech, 18 January 1908, in W. Churchill (ed.), *Winston S. Churchill: Never Give In!* (London, 1990), p. 22.

77 R. Roosevelt, *African Game Trails: An Account of the African Wanderings of an American Hunter-Naturalist* (New York, 1910), p. 148.

78 R. Horsman, *Race and Manifest Destiny: The Origins of Racial Anglo-Saxonism* (Cambridge, MA, 1981), p. 227 and *passim*; also Belich, *Replenishing the Earth*, pp. 5–6.

79 Horsman, *Race and Manifest Destiny*, p. 292.

80 Ibid., p. 227.

81 F. Harcourt, 'Disraeli's imperialism, 1866–1868: a question of timing', *Historical Journal* 23 (1980), 96.

82 See here P. Satia, 'Developing Iraq: Britain, India and the Redemption of Empire and Technology in the First World War', *Past & Present* 197 (2007), 228–30.

83 A. Anand, *Patient Assassin* (London, 2019), p. 218.

84 T. Coogan, *The Famine Plot: England's Role in Ireland's Greatest Tragedy* (New York, 2012), p. 229.

85 G. Morton-Jack, *The Indian Empire at War: From Jihad to Victory. The Untold Story of the Indian Army in the First World War* (London, 2018), p. 58.

86 E. Frankema, J. Williamson and P. Woltjer, 'An economic rationale for the West African scramble? The commercial transition and the commodity price boom of 1835–1885', *Journal of Economic History* (2018), 231–67.

87 G. Vanthemsche, *Belgium and the Congo, 1885–1980* (Cambridge, 2012).

88 B. Phillips, *Loot: Britain and the Benin Bronzes* (London, 2021), p. 56.

89 Ibid., p. 67.

90 D. Hicks, *The Brutish Museums: The Benin Bronzes, Colonial Violence and Cultural Restitution* (London, 2020).

91 R. Home, *City of Blood Revisited: A New Look at the Benin Expedition of 1897* (London, 1982), p. 111–3.

92 Tonkin Free School, 'A Civilisation of New Learning', in Dutton, Werner and Whitmore (eds), *Sources of Vietnamese Tradition*, pp. 369–71.

93 C. Stolte and H. Fischer-Tiné, 'Imagining Asia in India: Nationalism and Internationalism (ca. 1905–1940)', *Comparative Studies in Society and History* 54 (2012), 65–92.

94 A. Khalid, 'The Fascination of Revolution: Central Asian Intellectuals, 1917–27', *Slavic Eurasian Studies* 14 (2007), 137–52.

95 M. Gandhi, *Hind Swaraj and Other Writings*, ed. A. Patel (Cambridge, 1997), p. 131.

96 M. Gandhi, *The Collected Works of Mahatma Gandhi* (Delhi, 1999), p. 409.

97 G. Morton-Jack, *The Indian Empire at War: From Jihad to Victory, the Untold Story of the Indian Army in the First World War* (London, 2018), p. 58.

98 D. Weiner, *Models of Nature: Ecology, Conservation and Cultural Revolution in Soviet Russia* (Bloomington, IN, 1988), p. 37.

99 T. Weber, *Gandhi as Disciple and Mentor* (Cambridge, 2004), pp. 69–83.

100 M. Gandhi, *Hind Swaraj – or Indian Home Rule* (Madras, 1921).

101 G. Marsh, *Man and Nature; or, Physical Geography as Modified by Human Action* (New York, 1867), p. 41.

102 L. White, 'The Historical Roots of our Ecologic Crisis', *Science* 155 (1967), 1203–7.

103 M. Rangarajan, 'Imperial Agendas and India's Forests: The Early History of Indian Forests, 1800–1878', *Indian Economic and Social History Review* 21 (1994), 147–67.

104 P. McElwee, *Forests Are Gold: Trees, People and Environmental Rule in Vietnam* (Seattle, 2017); also see R. Guha, *The Unquiet Woods: Ecological Change and Peasant Resistance in the Himalaya* (Berkeley, 2000).

105 Saunt, *Unworthy Republic*, pp. 136–40.

106 Ibid., p. 145.

107 M. Hightower, *1889: The Boomer Movement, the Land Run and Early Oklahoma City* (Norman, OK, 2018).

108 A. Chekhov, *The Cherry Orchard,* tr. P. Carson (London, 2002).

109 Marx, *Kapital*, 3, p. 909.

110 Saunt, *Unworthy Republic*, pp. 189–97.

111 C. Totman, *The Green Archipelago: Forestry in Preindustrial Japan* (Berkeley, 1989); D. Fedman, *Seeds of Control: Japan's Empire of Forestry in Colonial Korea* (Seattle, 2020).

112 Beinart and Hughes, *Environment and Empire*, pp. 149–50; J. Darwin, *Unlocking the World: Port Cities and Globalization in the Age of Steam, 1830–1930* (London, 2020).

113 J. Daughton, *In the Forest of No Joy: The Congo-Océan Railroad and the Tragedy of French Colonialism* (New York, 2021).

114 For the Carrington Event, see D. Boteler, 'The super storms of August/September 1859 and their effects on the telegraph system', *Advances in Space Research* 38 (2006), 159–72; R. Braithwaite, *Armageddon and Paranoia: The Nuclear Confrontation since 1945* (Oxford, 2018), p. 276.

115 W. Herschel, 'Observations tending to investigate the nature of the sun, in order to find the causes or symptoms of its variable emission of light and heat, with remarks on the use that may possibly be drawn from solar observations', *Royal Society, Philosophical Transactions* 91 (1801), 265–320.

116 W. Jevons, *Investigations in Currency and Finance* (London, 1884), pp. 195–243.

117 J. Fourier, 'General Remarks on the Temperature of the Terrestrial Globe and the Planetary Spaces', *American Journal of Science and Arts* 32 (1837), 1–20.

118 E. Foote, 'On the Heat of the Sun's Rays', *American Journal of Science and Arts*, Second series, 22 (1856), 377–83.

119 S. Arrhenius, 'On the Influence of Carbonic Acid in the Air upon the Temperature on the Ground', *Philosophical Magazine and Journal of Science*, 5th series, 41 (1896), 8–11. See Fleming, *Historical Perspectives*, pp. 55–82.

120 D. Hart and D. Victor, 'Scientific Elites and the Making of US Policy for Climate Change Research, 1957–74', *Social Studies of Science* 23 (1993), 658.

121 S. Asselin, 'A Climate of Competition: Climate Change as Political Economy in Speculative Fiction, 1889–1915', *Science Fiction Studies* 45 (2018), 440–53.

122 J. Fleming, *Fixing the Sky: The Checkered History of Weather and Climate Control* (New York, 2010), pp. 31–3.

123 Asselin, 'A Climate of Competition', 443–4.

124 J. Church and N. White, 'A 20th century acceleration in global sea-level rise', *Geophysical Research Letters* 33 (2006), 1–4; J. Church and N. White, 'Sea-level rise from the late 19th to the early 21st century', *Surveys in Geophysics* 32 (2011), 585–602.

125 J. Walker et al., 'Timing of emergence of modern rates of sea-level rise by 1863', *Nature Communications* 13 (2022), 1–8.

126 B. Beach and W. Hanlon, 'Coal Smoke and Mortality in an Early Industrial Economy', *The Economic Journal* 128 (2018), 2652–75.

127 R. Bailey, T. Hatton and K. Inwood, 'Atmospheric Pollution, Health, and Height in Late Nineteenth Century Britain', *Journal of Economic History* 78 (2018), 1210–47.

128 A. Kidd, *Manchester: A History* (Keele, 2007), pp. 21–7.

129 F. Engels, *The Condition of the Working Class in England in 1844* (London, 1892), pp. 163, 150–2, 209–10.

130 H. Platt, *Shock Cities: The Environmental Transformation and Reform of Manchester and Chicago* (Chicago, 2005), p. 24.

131 J. McHugh and P. Mackowiak, 'Death in the White House: President William Henry Harrison's Atypical Pneumonia', *Clinical Infectious Diseases* 59 (2014), 990–5.

132 W. Wordsworth, *The Excursion* (London, 1853), pp. 300–2; J. Moore and J. Strachan, *Key Concepts in Romantic Literature* (London, 2010), pp. 46–51.

133 S. Kheraj, 'The Great Epizootic of 1872–73: Networks of Animal Disease in North American Urban Environments', *Environmental History* 23 (2018), 495–521.

134 H. Kjekshus, *Ecology Control and Economic Development in East African History: The Case of Tanganyika, 1850 to 1950* (London, 1977), pp. 127–43; P. Phoofolo, 'Epidemics and Revolutions: The Rinderpest Epidemic in

Late Nineteenth-Century Southern Africa', *Past & Present* 138 (1992), 112–43.

135 R. Shlomowitz, 'Morality and the Pacific labour trade', *Journal of Pacific History* 22 (1987), 34–5.

136 D. Northrop, *Indentured Labor in the Age of Imperialism, 1843–1922* (Cambridge, 1995); R. Shlomowitz, 'Epidemiology and the Pacific labor trade', *Journal of Interdisciplinary History* 19 (1989), 585–610.

137 E. Ladurie, 'A concept: the unification of the globe by disease', *The Mind and Method of the Historian* (Brighton, 1981), pp. 28–91.

138 Arnold, 'Indian Ocean as a disease zone', 10–11; D. Arnold, *Colonizing the Body: State Medicine and Epidemic Disease in Nineteenth-Century India* (Berkeley, 1993).

139 Arnold, 'Indian Ocean as a disease zone', 8–9.

140 A. Khalid, 'Of cholera, colonialism and pilgrimage sites: Rethinking popular responses to state sanitation, c.1867–1900', in B. Pati and M. Harrison (eds), *Society, Medicine and Politics in Colonial India* (London, 2018), p. 74.

141 Ibid., p. 75.

142 'Mauni Amavasya: Five crore pilgrims take holy dip at Kumbh till 5 pm', *Times of India*, 4 February 2019.

143 Khalid, 'Of cholera', pp. 78–87.

144 D. Lippi and E. Gotuzzo, 'The greatest steps towards the discovery of Vibrio cholerae', *Clinical Microbiology and Infection* 60 (2014), 191–5.

145 M. Harrison, 'A Question of Locality: The Identity of Cholera in British India, 1860–1890', in D. Arnold (ed.), *Warm Climates and Western Medicine: The Emergence of Tropical Medicine, 1500–1900* (Amsterdam, 1996), pp. 133–59.

146 A. Valeron et al., 'Transmissibility and geographic spread of the 1889 influenza pandemic', *PNAS* 107 (2010), 8778–81.

147 P. Berche, 'The enigma of the 1889 Russian flu pandemic: A coronavirus?', *La Press Médicale* 51 (2022), 1–9.

148 M. Brazleton, 'Viral Reflections: Placing China in Global Health Histories', *Journal of Asian Studies* 79 (2020), 581–3.

149 N. Johnson and J. Mueller, 'Updating the accounts: global mortality of the 1918–1920 "Spanish" influenza pandemic', *Bulletin of the History of Medicine* 76 (2002), 105–15; P. Spreeuwenberg, 'Reassessing the Global Mortality of the 1918 Influenza Pandemic', *American Journal of Epidemiology* 187 (2018), 2561–7.

150 M. Nickol and J. Kindrachuk, 'A year of terror and a century of reflection: perspectives on the great influenza pandemic of 1918–1919', *BMC Infectious Diseases* 19 (2019), 1–10.

151 D. Morens, J. Taubenberger and A. Fauci, 'Predominant Role of Bacterial Pneumonia as a Cause of Death in Pandemic Influenza: Implications for Pandemic Influenza Preparedness', *Journal of Infectious Diseases* 198 (2008), 962–70.

152 K. Clay, J. Lewis and E. Severini, 'Pollution, Infectious Disease, and Mortality: Evidence from the 1918 Spanish Influenza Pandemic', *Journal of Economic History* 78 (2018), 1179–209.

153 A. Guimbeau, N. Menon and A. Musacchio, 'The Brazilian Bombshell? The Long-Term Impact of the 1918 Influenza Pandemic the South American Way', NBER Working Paper 26929 (2020), 1–46.

154 D. Almond, 'Is the 1918 Influenza Pandemic Over? Long-Term Effects of *In Utero* Influenza Exposure in the Post-1940 U.S. Population', *Journal of Political Economy* 114 (2006), 672–712.

155 K. Blickle, 'Pandemics Change Cities: Municipal Spending and Voter Extremism in Germany, 1918–1933', Federal Reserve Bank of New York, *Staff Reports* 921 (2020), 1–34.

156 M. Humphries, 'Paths of infection: The First World War and the origins of the 1918 influenza pandemic', *War in History* 21 (2014), 55–81; A. Erkoreka, 'Origins of the Spanish influenza pandemic (1918–1920) and its relation to the First World War', *Journal of Molecular and Genetic Medicine: An International Journal of Biomedical Research* 3 (2009), 190–4.

157 A. More et al., 'The Impact of a Six-Year Climate Anomaly on the "Spanish Flu" Pandemic and WWI', *GeoHealth* 4 (2020), 1–8; also J. Brown et al., 'Avian influenza virus in water: Infectivity is dependent on pH, salinity and temperature', *Veterinary Microbiology* 136 (2009), 20–6.

158 N. Lambert, *The War Lords and the Gallipoli Disaster: How Globalised Trade Led Britain to its Worst Defeat of the First World War* (London, 2021), pp. 42–61, 202–9.

159 P. Castañeda Dower and A. Markevich, 'Labor Misallocation and Mass Mobilization: Russian Agriculture during the Great War', *Review of Economics and Statistics* 100 (2018), 245–59.

160 T. Uyama, 'Why in Central Asia, why in 1916? The revolt as an interface of the Russian colonial crisis and the World War', in A. Morrison, C. Drieu and A. Chokobaeva, *The Central Asian Revolt of*

1916: A Collapsing Empire in the Age of War and Revolution (Manchester, 2020), pp. 27–45.

161 See for example M. Saul and P. Royer, *West African Challenge to Empire, Culture and History in the Volta-Bani Anticolonial war* (Athens, OH, 2001); S. Chowdhury, *The First World War, Anticolonialism and Imperial Authority in British India, 1914–1924* (London, 2019); K. Jeffery, *1916: A Global History* (London, 2016), pp. 362–4.

162 Jeffery, *1916*, pp. 190–210.

163 D. Brantz, 'Environments of Death: Trench Warfare on the Western Front, 1914–1918', in C. Closmann (ed.), *War and the Environment: Military Destruction in the Modern Age* (College Station, TX, 2009), p. 82.

164 M. Van Meirvenne et al., 'Could shelling in the First World War have increased copper concentrations in the soil around Ypres?', *European Journal of Soil Science* 59 (2008), 372–9.

165 T. Harper, *Underground Asia: Global Revolutionaries and the Assault on Empire* (London, 2020), esp. 207ff.

166 A. Link, 'That Cobb Interview', *Journal of American History* 72 (1985), 9; R. Lansing, *War Memoirs of Robert Lansing, Secretary of State* (Indianapolis, IN, 1935), pp. 212–13.

167 P. Mishra, *Bland Fanatics: Liberals, Race and Empire* (London, 2020), p. 45; for Weber's legacy, see Allen, *British Industrial Revolution*, pp. 7–8.

168 D. Low, *Eclipse of Empire* (Cambridge, 1991), p. 11.

21 FASHIONING NEW UTOPIAS (C.1920–C.1950)

1 O. Melsted and I. Pallua, 'The Historical Transition from Coal to Hydrocarbons: Previous Explanations and the Need for an Integrative Perspective', *Canadian Journal of History* 53 (2018), 398–401.

2 D. Yergin, *The Prize: The Epic Quest for Oil, Money and Power* (New York, 1991), pp. 62–149.

3 US Department of Commerce, *Historical Statistics of the United States, Colonial Times to 1957*, cited by G. Babcock, *History of the United States Rubber Company* (Bloomington, IN, 1966), p. 419.

4 R. Casey, *The Model T: A Centennial History* (Baltimore, 2008).

5 Hankey to Balfour, 1 August 1918, Public Records Office, Foreign Office file 800/204.

6 BP, *Statistical Review of World Energy* (2021), p. 16.

7 'Summary of Report on Near Eastern Oil', 3 February 1943, in Yergin, *The Prize*, p. 375.

8 J. Marshall, *To Have and to Have Not: Materials and the Origins of the Pacific War* (Berkeley, 2021); A. Toprani, *Oil and the Great Powers: Britain and Germany, 1914–1945* (Oxford, 2019), esp. pp. 231–52.

9 US Energy Information Administration, 'Petroleum & Other Liquids', https://www.eia.gov/dnav/pet/hist/LeafHandler.ashx?n=pet&s=wttntus2&f=4.

10 A. Bhandari, *India and the Changing Geopolitics of Oil* (Abingdon, 2022); A. Kutuleva, *China's Energy Security and Relations with Petrostates: Oil as an Idea* (Abingdon, 2022).

11 M. Willibold, T. Elliott and S. Moorbath, 'The tungsten isotopic composition of the Earth's mantle before the terminal bombardment', *Nature* 477 (2011), 195–8.

12 Oxfam America, 'Valentine's Gold Jewelry Sales Generate 34 Million Tons of Mine Waste', 11 February 2005; also A. Septoff, 'How the 20 tons of mine waste per gold ring was calculated', *Earthworks*, 21 May 2004.

13 WALHI: Indonesian Forum for Environment, *The Environmental Impacts of Freeport–Rio Tinto's Copper and Gold Mining Operation in Papua* (Jakarta, 2006); M. Alonzo, J. Van Den Hoek and N. Ahmed, 'Capturing coupled riparian and coastal disturbance from industrial mining using cloud-resilient satellite time series analysis', *Scientific Reports* 6 (2016), 1–12.

14 M. Rohrbough, *Days of Gold: The California Gold Rush and the American Nation* (Berkeley, 1997); E. West, *The Contested Plains: Indians, Goldseekers, and the Rush to Colorado* (Lawrence, KS, 1998); W. Frost, 'The Environmental Impacts of the Victorian Gold Rushes: Miners' Accounts during the First Five Years', *Australian Economic History Review* 53 (2013), 72–90; M. Singer et al., 'Enduring legacy of a toxic fan via episodic redistribution of California gold mining debris', *PNAS* 110 (2013), 18,336–41; M. Bassin, 'Expansion and Colonialism on the Eastern Frontier: Views of Siberia and the Far East in Pre-Petrine Russia', *Journal of Historical Geography* 14 (1988), 3–21.

15 Goldman Sachs, 'Copper is the New Oil' (2021); Hinrich Foundation, 'Lithium is "new oil" as vehicle market expands', 21 September 2021; The Diplomat, 'Rare Earths: Fighting for the Fuel of the Future', 8 January 2022; WSP, 'Net Zero: Is Hydrogen the New Oil?', 10 November 2021; Forbes, 'Data is the New Oil', 15 November 2019.

16 American Resources Policy Network, cited by 'Wire Harness Recycling', *Assembly*, 1 July 2014; International Energy Agency, 'The Role of Critical Minerals in Clean Energy Transitions' (2021).

17 Ross, *Ecology and Power*, pp. 177–82.

18 Ibid., pp. 191–3.

19 Ibid., p. 197.

20 A. Williams, *White Malice: The CIA and the Neocolonisation of Africa* (London, 2021), pp. 31–2, 127.

21 Natural Resource Governance Institute, 'The Resource Curse: The Political and Economic Challenges of Natural Resource Wealth', *NRGI Reader* (March 2015), 1–6; M. Humphreys, J. Sachs and J. Stiglitz (eds), *Escaping the Resource Curse* (New York, 2007).

22 Williams, *White Malice*, p. 127.

23 K. Ochieng' Opalo, *Legislative Development in Africa: Politics and Postcolonial Legacies* (Cambridge, 2019).

24 M. Wilfahrt, *Precolonial Legacies in Postcolonial Politics: Representation and Redistribution in Decentralised West Africa* (Cambridge, 2021).

25 D. Euraque, *Reinterpreting the Banana Republic: Region and State in Honduras, 1870–1972* (Chapel Hill, NC, 1996), p. 44; L. Langley and T. Schoonover, *The Banana Men: American Mercenaries and Entrepreneurs in Central America* (Lexington, KY, 1995).

26 R. Immerman, *The CIA in Guatemala: The Foreign Policy of Intervention* (Austin, TX, 1982), pp. 73–5.

27 Ibid., pp. 75–80.

28 N. Cullather, *Secret History: The CIA's Classified Account of its Operations in Guatemala, 1952–1954* (Stanford, 1999).

29 G. Hegerl et al., 'The early 20th century warming: Anomalies, causes, and consequences', *WIREs Climate Change* 9 (2018), 1–20.

30 Ibid., 5–6; also see L. Svendsen et al., 'Pacific contribution to decadal surface temperature trends in the Arctic during the twentieth century', *Climate Dynamics* 57 (2021), 3223–43.

31 J. McConnell et al., '20th-Century Industrial Black Carbon Emissions Altered Arctic Climate Forcing', *Science* 317 (2007), 1381–4; S. Brönnimann, 'Early twentieth century warming', *Nature Geoscience* 2 (2009), 735–6.

32 S. Kim, 'Expansion of Markets and the Geographic Distribution of Economic Activities: The Trends in U.S. Regional Manufacturing Structure, 1860–1987', *Quarterly Journal of Economics* 110 (1995), pp. 881–908.

33 S. Kim, 'Urban Development in the United States, 1690–1990', *Southern Economic Journal* 66 (2000), 855–80.

34 S. Broadberry and D. Irwin, 'Labor productivity in the United States and the United Kingdom during the nineteenth century', *Explorations in Economic History* 43 (2006), 257–79.

35 J. Atack, R. Margo and P. Rhode, 'Industrialization and Urbanization in Nineteenth Century America', NBER Working Paper 25897 (2021), 1–52.

36 J. Tang, 'Railroad Expansion and Industrialization: Evidence from Meiji Japan', *Journal of Economic History* 74 (2014), 863–86; T. Berger, 'Railroads and Rural Industrialization: Evidence from a Historical Project Experiment', *Exploration in Economic History* 74 (2019), 1–18; C. Lundh, L. Schön and L. Svensson, 'Regional wages in industry and labour market integration in Sweden, 1861–1913', *Scandinavian Economic History Review* 53 (2005), 71–84; E. Hornung, 'Railroads and Growth in Prussia', *Journal of the European Economic Association* 13 (2015), 699–736.

37 V. Lenin, *Development of Capitalism in Russia* (Moscow, 1956).

38 V. Lenin, *Полное собрание сочинений*, 55 vols (Moscow, 1975–9), 36, pp. 15–16.

39 P. Josephson, 'War on Nature as Part of the Cold War: The Strategic and Ideological Roots of Environmental Degradation in the Soviet Union', in J. McNeill and C. Unger (eds), *Environmental Histories of the Cold War* (Cambridge, 2010), pp. 21–32.

40 D. Weiner, *A Little Corner of Freedom: Russian Nature Protection from Stalin to Gorbachev* (Berkeley, 1999), p. 133.

41 L. Trotsky, *Literature and Revolution*, ed. W. Keach (Chicago, 2005), p. 204.

42 P. Josephson, ' "Projects of the Century" in Soviet History: Large-Scale Technologies from Lenin to Gorbachev', *Technology and Culture* 36 (1995), 531–2.

43 Josephson, 'War on Nature', pp. 31–2; D. Nordlander, 'Origins of a Gulag Capital: Magadan and Stalinist Control in the Early 1930s', *Slavic Review* 57 (1998), 791–812.

44 S. Fitzpatrick, *Stalin's Peasants: Resistance and Survival in the Russian Village after Collectivization* (Oxford, 1994); R. Davies and S. Wheatcroft, *The Years of Hunger: Soviet Agriculture, 1931–1933* (New York, 2004); A. Applebaum, *Red Famine: Stalin's War on Ukraine* (London, 2017), p. 285.

45 For bias in the 1930s, see A. Markevich, N. Naumenko and N. Qian, 'The Political-Economic Causes of the Soviet Great Famine, 1932–3', NBER Working Papers 29089 (2021), 1–34.

46 R. Davies, *The Socialist Offensive: The Collectivization of Soviet Agriculture, 1929–1930* (Cambridge, MA, 1980), p. 395.

47 N. Dronin and E. Bellinger, *Climate Dependence and Food Problems in Russia, 1900–1990: The Interaction of Climate and Agricultural Policy and their Effect on Food Problems* (Budapest, 2005), p. 116; Josephson, ' "Projects of the Century" ', 533.

48 B. Bonhomme, *Forests, Peasants, and Revolutionaries: Forest Conservation and Organization in Soviet Russia, 1917–1929* (Boulder, CO, 2005), p. 128.

49 S. Brain, 'Stalin's Environmentalism', *Russian Review* 69 (2010), 98.

50 S. Brain, 'The Environmental History of the Soviet Union', in J. McNeill and E. Mauldin (eds), *A Companion to Global Environmental History* (Chichester, 2012), p. 232.

51 Brain, 'Stalin's Environmentalism', 104.

52 Ibid., 102.

53 Weiner, *Models of Nature*, pp. 252–6.

54 Brain, 'Stalin's Environmentalism', 110–11.

55 Josephson, ' "Projects of the Century" ', 536.

56 I. Gazeley and A. Newell, 'The First World War and working-class food consumption in Britain', *European Review of Economic History* 17 (2013), 71–94.

57 E. Frankema, 'Raising revenue in the British empire, 1870–1940: how "extractive" were colonial taxes?', *Journal of Global History* 5 (2010), 447–77.

58 K. O'Connor, 'The King's Christmas pudding: globalization, recipes, and the commodities of empire', *Journal of Global History* 4 (2009), 127–55.

59 Ross, *Ecology and Power*, p. 339; M. van Beusekom, *Negotiating Development: African Farmers and Colonial Experts at the Office du Niger* (Portsmouth, NH, 2002), pp. 40–2.

60 Ross, *Ecology and Power*, p. 339.

61 J. van Zanden, 'The first green revolution: the growth of production and productivity in European agriculture, 1870–1914', *Economic History Review* 44 (1991), 21–39.

62 N. Cullather, *The Hungry World: America's Cold War Battle against Poverty in Asia* (Cambridge, MA, 2022), pp. 46–7.

63 B. Cook, R. Miller and R. Seager, 'Amplification of the North American "Dust Bowl" drought through human-induced land degradation', *PNAS* 106 (2009), 4997–5001.

64 W. Lockeretz, 'The Lessons of the Dust Bowl: Several Decades before the Current Concern with Environmental Problems, Dust Storms Ravaged the Great Plains, and the Threat of More Dust Storms Still Hangs over Us', *American Scientist* 66 (1978), 560.

65 T. Egan, *The Worst Hard Time: The Untold Story of Those Who Survived the Great American Dust Bowl* (Boston, MA, 2006), p. 8; D. Worster, *Dust Bowl: The Southern Plains in the 1930s* (New York, 1979).

66 R. Hornbeck, 'The Enduring Impact of the American Dust Bowl: Short- and Long-Run Adjustments to Environmental Catastrophe', *American Economic Review* 102 (2012), 1477–1507.

67 K. Sylvester and E. Rupley, 'Revising the Dust Bowl: High above the Kansas Grasslands', *Environmental History* 17 (2012), 603–33.

68 T. Cowan et al., 'Factors Contributing to Record-Breaking Heat Waves over the Great Plains during the 1930s Dust Bowl', *Journal of Climate* 30 (2017), 2437–61.

69 M. Sanderson and R. Scott Frey, 'From desert to breadbasket … to desert again? A metabolic rift in the High Plains Aquifer', *Journal of Political Ecology* 21 (2014), 516–32.

70 H. Holleman, *Dust Bowls of Empire: Imperialism, Environmental Politics, and the Injustice of 'Green' Capitalism* (New Haven, 2018), pp. 47–50.

71 G. Jacks and R. Whyte, *The Rape of the Earth: A World Survey of Soil Erosion* (London, 1939), pp. 18–26.

72 L. Stamp, 'Land Utilization and Soil Erosion in Nigeria', *Geographical Review* 28 (1938), 32–45.

73 G. Callendar, 'The artificial production of carbon dioxide and its influence on temperature', *Quarterly Journal of the Royal Meteorological Society* 64 (1938), 223–37.

74 E. Hawkins and P. Jones, 'On increasing global temperatures: 75 years after Callendar', *Quarterly Journal of the Royal Meteorological Society* 139 (2013), 1961–3.

75 S. Tett et al., 'Causes of twentieth-century temperature change near the Earth's surface', *Nature* 399 (1999), 569–72; T. Delworth and T. Knutson, 'Simulation of early 20th century global warming', *Science* 287 (2000), 2246–50.

76 O. Johannsson, 'Die Temperaturverhältnisse Spitzbergens (Svalbard)', *Annalen der Hydrographie und Maritimen Meteorologie* 64 (1936), 81–96; R. Scherhag, 'Die Erwärmung des Polargebiets', *Annalen der Hydrographie* 67 (1939), 57–67.

77 J. Williams, *Turning to Nature in Germany: Hiking, Nudism and Conservation* (Stanford, 2007), p. 220.

78 Ibid.

79 F. Uekoetter, *The Green and the Brown: A History of Conservation in Nazi Germany* (Cambridge, 2009), p. 22; Williams, *Turning to Nature*, p. 219.

80 Williams, *Turning to Nature*, p. 231.

81 Uekoetter, *The Green and the Brown*, p. 31.

82 Williams, *Turning to Nature*, p. 252.

83 Uekoetter, *The Green and the Brown*, p. 1.

84 Ibid., p. 41.

85 Williams, *Turning to Nature*, p. 252.

86 Uekoetter, *The Green and the Brown*, p. 41.

87 T. Zeller, 'Molding the Landscape of Nazi Environmentalism: Alwin Seifert and the Third Reich', in Brüggemeier, Cioc and Zeller (eds), *How Green Were the Nazis?*, pp. 147–70.

88 Uekoetter, *The Green and the Brown*, pp. 32, 171–6; T. Zeller, '"The Landscape's Crown": Landscape, Perceptions, and Modernizing Effects of the German Autobahn System, 1934 to 1941', in D. Nye (ed.), *Technologies of Landscape: From Reaping to Recycling* (Amherst, MA, 1999), p. 230.

89 Uekoetter, *The Green and the Brown*, p. 32.

90 A. Whiston Spirn, *The Language of Landscape* (New Haven, 1998), p. 246.

91 Ibid.

92 J. Wolschke-Bulmahn, 'Violence as the Basis of National Socialist Landscape Planning in the "Annexed Eastern Areas"', in Brüggemeier, Cioc and Zeller (eds), *How Green Were the Nazis?*, p. 244.

93 Ibid., p. 247.

94 Ibid., pp. 246–7.

95 J. Wolschke-Bulmahn and G. Gröning, 'The National Socialist Garden and Landscape Ideal: *Bodenständigkeit* (Rootedness in the Soil)', in R. Eltin (ed.), *Art, Culture, and Media under the Third Reich* (Chicago, 2002), pp. 80–3; Wolschke-Bulmahn, 'Violence as the Basis', p. 245.

96 Reichsführer SS, *Der Untermensch* (Berlin, 1942), p. 2; Wolschke-Bulmahn, 'Violence as the Basis', p. 245.

97 H. Klemann and S. Kudryashov, *Occupied Economies: An Economic History of Nazi-Occupied Europe, 1939–1945* (London, 2012); also see A. Tooze, *The Wages of Destruction: The Making and Breaking of the Nazi Economy* (London, 2006).

98 Frankopan, *Silk Roads*, pp. 386–7. Above all see here A. Beevor, *Stalingrad* (London, 1998); D. Stahel; *Kiev 1941: Hitler's Battle for Supremacy in the East* (Cambridge, 2012).

99 J. Nehru, *The Discovery of India* (New Delhi, 1981), p. 16.

100 A. Sen, *Poverty and Famines: An Essay on Entitlement and Deprivation* (Oxford, 1983), esp. pp. 52–85.

101 A. Sen, 'Famines as failures of exchange entitlements', *Economic and Political Weekly* 11 (1976), 1273–80; M. Tauger, 'The British famine crises of World War II', *British Scholar* 1 (2009), 421–40; V. Mishra et al., 'Drought and Famine in India, 1870–2016', *Geophysical Research Letters* 46 (2019), 2075–83.

102 A. Speer, *Inside the Third Reich*, trans. R. and C. Winston (London, 1970), p. 317.

103 E. Konopinski, C. Marvin and E. Teller, *Ignition of the Atmosphere with Nuclear Bombs: Report LA-602* (Los Alamos, NM, 1946), p. 1.

104 Y. Fujii, 'The role of atmospheric nuclear explosions on the stagnation of global warming in the mid 20th century', *Journal of Atmospheric and Solar-Terrestrial Physics* 73 (2011), 643–52.

105 T. Higuchi, 'Atmospheric Nuclear Weapons Testing and the Debate on Risk Knowledge in Cold War America, 1945–1963', in McNeill and Unger (eds), *Environmental Histories*, pp. 305–6.

106 A. Robock and B. Zambri, 'Did Smoke from City Fires in World War II Cause Global Cooling?', *Journal of Geophysical Research: Atmospheres* 123 (2018), 314–25.

107 Ibid.

108 Hegerl, 'The early 20th century warming', 15; J. Hansen et al., 'Global Surface Temperature Change', *Reviews of Geophysics* 48 (2010), 1–29.

109 M. Dörries, 'The Politics of Atmospheric Sciences: "Nuclear Winter" and Global Climate Change', *Osiris* 26 (2011), 202.

110 H. Arakawa, 'Possible Atmospheric Disturbances and Damages to the Rice-Crops in Northern Japan That May Be Caused by Experimentation with Nuclear Weapons', *Geophysical Magazine* (1954), 125–34; H. Arakawa and K. Tsutsumi, 'A decrease in the normal incidence radiation values for 1953 and 1954 and its possible cause', *Geophysical Magazine* 27 (1956), 205–8; Dörries, 'Politics of Atmospheric Sciences', 202.

111 US Atomic Energy Commission, US Air Force and Rand Corporation, 'Worldwide Effects of Atomic Weapons: Project Sunshine', R-251-AEC, 6 August 1953.

112 P. Edwards, 'Entangled histories: Climate science and nuclear weapons research', *Bulletin of the Atomic Scientists* 68 (2012), 29–30.

113 Committee on Meteorological Aspects of the Effects of Atomic Radiation, 'Meteorological Aspects of Atomic Radiation', *Science* 124 (1956), 105–12.

114 Ibid., 112.

115 For the wider discussion about climate change discussions in the 20th century, see M. Dörries, 'In the public eye: Volcanology and climate change studies in the 20th century', *Historical Studies in the Physical and Biological Sciences* 37 (2006), 87–125.

116 Fujii, 'Role of atmospheric nuclear explosions', 643–52.

22 RESHAPING THE GLOBAL ENVIRONMENT (THE MID-TWENTIETH CENTURY)

1 J. Fleming, 'The pathological history of weather and climate modification: Three cycles of promise and hype', *Historical Studies in the Physical and Biological Sciences* 37 (2006), 6–7. The epigraph at the start of the chapter is from J. Shapiro, *Mao's War against Nature: Politics and the Environment in Revolutionary China* (Cambridge, 2001), p. 87.

2 Fleming, 'Pathological history', 6-7; J. Townsend, *Making Rain in America: A History* (Lubbock, TX, 1975), pp. 11–13.

3 Fleming, *Fixing the Sky*, pp. 77–108.

4 P. Heazell, *Most Secret: The Hidden History of Orford Ness* (Stroud, 2013), p. 40.

5 K. Harper, *Weather by the Numbers: The Genesis of Modern Meteorology* (Oxford, 2008), pp. 49–52.

6 K. Harper, *Make It Rain: State Control of the Atmosphere in Twentieth-Century America* (Chicago, 2017), pp. 37–48.

7 Harper, *Weather by the Numbers*, p. 58.

8 Harper, *Make It Rain*, pp. 46–8.

9 V. Schaefer, 'A Method for Making Snowflake Replicas', *Science* 93 (1941), 239–40; B. Vonnegut, 'The nucleation of ice formation by silver iodide', *Journal of Applied Physics* 18 (1947), 593–5; I. Langmuir, 'The production of a chain reaction in cumulus clouds at temperatures above freezing', *Journal of Meteorology* 5 (1948), 175–81.

10 K. Harper, 'Climate control: United States weather modification in the cold war and beyond', *Endeavour* 32 (2008), 20; Harper, *Make It Rain*, p. 49.

11 Harper, 'Climate control', 20.

12 K. Harper and R. Doel, 'Environmental Diplomacy in the Cold War: Weather Control, the United States and India, 1966–1967', in McNeill and Unger (eds), *Environmental Histories*, p. 119; Harper, *Make It Rain*, pp. 56–62.

13 *Time*, 28 August 1950.

14 *Charleston Daily Mail*, 11 December 1950.

15 'Weather Made to Order?', *Collier's*, 28 May 1954.

16 *Brainerd Daily Dispatch*, 6 July 1954.

17 H. Orville, 'The Impact of Weather Control on the Cold War', in 85th Congress, *Weather Modification Research: Hearings of the House Committee on Interstate and Foreign Commerce* (Washington, DC, 1958), 53.

18 Harper, *Make It Rain*, pp. 62–80.

19 'US is Urged to Seek Methods to Control the World's Weather', *New York Times*, 1 January 1958.

20 H. Houghton, 'Present position and future possibilities of weather control', in *Final Report of the United States Advisory Committee on Weather Control*, cited in Fleming, 'Pathological history', 17–18.

21 M. Brzezinski, *Red Moon Rising: Sputnik and the Hidden Rivalries that Ignited the Space Age* (New York, 2007).

22 H. Plaster, 'Snooping on Space Pictures', *Studies in Intelligence* 8 (1964), 31–9.

23 S. Wheatcroft, 'The Soviet Famine of 1946–1947, the Weather and Human Agency in Historical Perspective', *Europe-Asia Studies* 64 (2012), 987–1005.

24 M. Ellman, 'The 1947 Soviet famine and the entitlement approach to famines', *Cambridge Journal of Economics* 24 (2000), 603–30. Also see see V. Zima, Голод в СССР *1946–1947* годов: происхождение и последствия (Moscow, 1996) and N. Ganson, *The Soviet Famine of 1946–1947 in Global and Historical Perspective* (New York, 2009).

25 S. Brain, 'The Great Stalin Plan for the Transformation of Nature', *Environmental History* 15 (2010), 680.

26 J. Stalin, 'Dialectical and Historical Materialism', in J. Stalin, *Problems of Leninism* (Moscow, 1947), p. 482.

27 S. Brain, *Song of the Forest: Russian Forestry and Stalinist Environmentalism, 1905–1953* (Pittsburgh, 2011), pp. 148–67.

28 Ibid., p. 149.

29 D. Shaw, 'Mastering Nature through Science: Soviet Geographers and the Great Stalin Plan for the Transformation of Nature, 1948–53', *Slavonic and East European Review* 93 (2015), 120–46.

30 Brain, 'The Great Stalin Plan', 691–2.

31 N. Khrushchev, 'Special Report to the 20th Congress of the Communist Party of the Soviet Union', Closed session, 24–25 February 1956, in B. Nicolaevsky (ed.), *The Crimes of the Stalin Era* (New York, 1956), p. 58.

32 Ibid., pp. 13, 59.

33 L. Battan, 'A Survey of Recent Cloud Physics Research in the Soviet Union', *Bulletin of the American Meteorological Society* 44 (1963), 755–71.

34 Orville, 'Impact of Weather Control on the Cold War', pp. 51–3.

35 Ibid.

36 Draft of Presidential Speech before the General Assembly of the United Nations, 28 November 1953, Eisenhower Presidential Library.

37 G. DeGroot, *Dark Side of the Moon: The Magnificent Madness of the American Lunar Quest* (London, 2007), p. 10.

38 E. Teller, 'We're Going to Work Miracles', *Popular Mechanics* 113 (1960), 97–101; 'How to Be an Optimist in a Nuclear Age', in E. Teller with A. Brown, *The Legacy of Hiroshima* (New York, 1962), pp. 84–9; D. O'Neill, *The Firecracker Boys: H-Bombs, Inupiat Eskimos, and the Roots of the Environmental Movement* (New York, 1994), p. 28.

39 Fleming, 'Pathological history', 18–19. See A. Markin, *Soviet Electric Power: Developments and Prospects* (Moscow, 1956), pp. 133–5; N. Rusin and L. Flit, *Man versus Climate* (Moscow, 1960).

40 I. Adabashev, *Global Engineering* (Moscow, 1966), pp. 191–7, 162.

41 Ibid., pp. 198–201; 237.

42 J. Hamblin, 'A Global Contamination Zone: Early Cold War Planning for Environmental Warfare', in McNeill and Unger (eds), *Environmental Histories*, pp. 103, 95.

43 US Senate Committee on Interior and Insular Affairs, *Hearings before the Subcommittee on Water and Power Resources* (Washington, DC, 1966), pp. 26–7.

44 L. Reiffel et al, 'A Study of Lunar Research Flights', Air Force Special Weapons Center (Defense Documentation Center for Air Research and Development Command, Kirtland Air Force Base, NM, 1959), pp. 2–3.

45 J. von Neumann, 'Can We Survive Technology?' in A. Taub (ed.), *John von Neumann: The Collected Works*, 6 vols (Oxford, 1961–3), 6, pp. 504–19.

46 S. Weart, *The Rise of Nuclear War* (Cambridge, MA, 1988), p. 121.

47 Teller, 'How to Be an Optimist in a Nuclear Age', pp. 86–7.

48 J. Fay, *Inhospitable World: Cinema in the Time of the Anthropocene* (Oxford, 2018), pp. 80–3.

49 Harper and Doel, 'Environmental Diplomacy', pp. 120–1; 'High Altitude Tests Stirs Wide Protests', *New York Times*, 1 June 1962; J. Manaffey, *Atomic Awakening: A New Look at the History and Future of Nuclear Power* (New York, 2009), pp. 233–4.

50 V. Tolstikov and V. Kuznetsov, 'Кыштымская радиационная авария 1957 года: первые оценки радиационной обстановки после взрыва', *Глобальная ядерная безопасность* 201 (2018), 107–15.

51 G. Seaborg, *Kennedy, Khrushchev and the Test Ban* (Berkeley, 1981);
    R. Strode, 'Soviet Policy toward a Nuclear Test Ban: 1958–1963', in
    M. Mandelbaum (ed.), *The Other Side of the Table: The Soviet Approach
    to Arms Control* (New York, 1990), pp. 5–40.

52 R. Huschke, 'A Brief History of Weather Modification since 1946',
    Bulletin of the American Meteorological Society 44 (1963), 428.

53 J. Perkins, *Geopolitics and the Green Revolution: Wheat, Genes and the
    Cold War* (Oxford, 1997), p. 163.

54 Ibid., pp. 166–8.

55 Amrith, *Unruly Waters*, pp. 194–6; Central Water and Power
    Commission, *Major Water and Power Projects of India* (Delhi, 1957), p. 3.

56 S. Singh, *Taming the Waters: The Political Economy of Large Dams in
    India* (Oxford, 1997), pp. 133–203; Amrith, *Unruly Waters*, pp. 212–13.

57 Cullather, *Hungry World*, p. 109.

58 A. Rotter, *Comrades at Odds: The United States and India, 1947–64*
    (Ithaca, NY, 2000), pp. 15–16.

59 Cullather, *Hungry World*, p. 112.

60 Amrith, *Unruly Waters*, pp. 184–6; 'Certified Statement in Accordance
    with Article 2 of the Regulations Adopted by the General Assembly on
    14 December 1946 (Resolution No. 97 (I)) to Give Effect to Article 102
    of the Charter of the United Nations'.

61 D. Lilienthal, 'Another Korea in the Making?', *Collier's*, 4 August
    1951, 18–20.

62 Cullather, *Hungry World*, pp. 112–13.

63 V. Nemenchok, ' "That So Fair a Thing Should Be So Frail": The Ford
    Foundation and the Failure of Rural Development in Iran, 1953–1964',
    Middle East Journal 63 (2009), 263–71.

64 Ibid., 263–84.

65 See Frankopan, *Silk Roads*, pp. 433ff.

66 Cullather, *Hungry World*, pp. 125–7.

67 S. Hamilton, *Supermarket USA: Food and Power in the Cold War Farms
    Race* (New Haven, 2018), pp. 70–96.

68 Harry S. Truman, 'Inaugural address', *New York Times*, 21 January 1949.

69 Ibid.

70 National Security Council, 'NSC-48, the Position of the United
    States with Respect to Asia', 30 December 1949, in T. Etzold and
    J. Gaddis (eds), *Containment: Documents on American Policy and Strategy*
    (New York, 1978), p. 265.

71 H. Byroade, 'The World's Colonies and Ex-Colonies: A Challenge to
    America', quoted in Cullather, *Hungry World*, p. 74.

72 Cullather, *Hungry World*, p. 79.

73 Perkins, *Geopolitics and the Green Revolution*, pp. 146–54, 119–20.

74 Hamilton, *Supermarket USA*, pp. 18–19; P. Conkin, *A Revolution Down on the Farm: The Transformation of American Agriculture since 1929* (Lexington, KY, 2008).

75 A. Hale-Dorrell, *Corn Crusade: Khrushchev's Farming Revolution in the Post-Stalin Soviet Union* (New York, 2019), p. 36.

76 H. Luce, 'The American Century', *Life*, 17 February 1941; Hamilton, *Supermarket USA*, p. 44.

77 Truman, 'Inaugural address'.

78 Cullather, *Hungry World*, p. 142.

79 M. Swaminathan, *Wheat Revolution: A Dialogue* (London, 1993), p. vii; Perkins, *Geopolitics and the Green Revolution*, p. 168.

80 Cullather, *Hungry World*, pp. 144–5.

81 Ibid., pp. 142–6.

82 Perkins, *Geopolitics and the Green Revolution*, p. 175.

83 Ibid.; F. Frankel, *India's Political Economy, 1947–1977* (Princeton, 1978), pp. 157–9.

84 J. Nehru, *Toward Freedom: The Autobiography of J. Nehru* (New York, 1941), pp. 230–1; S. Gopal, *Jawaharlal Nehru: A Biography*, 3 vols (London, 1975–84), 2, p. 202; D. Engerman, 'The Romance of Economic Development and New Histories of the Cold War', *Diplomatic History* 28 (2004), 30–1.

85 Hale-Dorrell, *Corn Crusade*, pp. 48–52.

86 See for example N. Telepneva, *Cold War Liberation: The Soviet Union and the Collapse of the Portuguese Empire in Africa, 1961–75* (Chapel Hill, NC, 2022); R. Yordanov, *The Soviet Union and the Horn of Africa during the Cold War* (Lanham, MD, 2016); C. Katsakoris, 'The Soviet-South Encounter: Tensions in the Friendship with Afro-Asian Partners, 1945–65' in P. Babiracki and K. Zimmer (eds) *Cold War Crossings: International Travel and Exchange across the Soviet Bloc* (College Station, TX, 2014), pp. 134–65.

87 Ibid., pp. 48–54; for Aswan, A. Shokr, 'Hydropolitics, Economy, and the Aswan High Dam in Mid-Century Egypt', *Arab Studies Journal* 17 (2009), 9–31; S. Borzutzky and D. Berger, 'Dammed If You Do, Dammed If You Don't: The Eisenhower Administration and the Aswan Dam', *Middle East Journal* 64 (2010), 84–102.

88 Hale-Dorrell, *Corn Crusade*, p. 20.

89 Ibid., p. 22.

90 Dronin and Bellinger, *Climate Dependence and Food Problems in Russia*, pp. 171–217.

91 A. Hale-Dorrell, 'The Soviet Union, the United States, and Industrial Agriculture', *Journal of World History* 26 (2015), 305–6.

92 Hale-Dorrell, *Corn Crusade*, pp. 21, 61–2.

93 Ibid., pp. 1–2.

94 W. Taubman, *Khrushchev: The Man and his Era* (New York, 2003), p. 609; Hale-Dorrell, 'The Soviet Union, the United States, and Industrial Agriculture', 318.

95 N. Khrushchev, *Строительство коммунизма в СССР и сельское хозяйство*, 8 vols (Moscow, 1962–8), 1, p. 342; Hale-Dorrell, *Corn Crusade*, p. 29.

96 Khrushchev, *Строительство коммунизма*, 2, p. 69; Hale-Dorrell, *Corn Crusade*, p. 95.

97 Hale-Dorrell, 'The Soviet Union, the United States, and Industrial Agriculture', 309–11.

98 Dronin and Bellinger, *Climate Dependence and Food Problems in Russia*, pp. 1–2; P. Gatrell, *The Tsarist Economy, 1850–1917* (London, 1986); N. Field, 'Environmental quality and land productivity: a comparison of the agricultural land base of the USSR and North America', *Canadian Geographer* 12 (1968), 1–14.

99 Hamilton, *Supermarket USA*, p. 126.

100 Hale-Dorrell, *Corn Crusade*, p. 33.

101 Hamilton, *Supermarket USA*, p. 119.

102 N. Khrushchev, *Memoirs of Nikita Khrushchev*, ed. S. Khrushchev, tr. G. Shriver, 3 vols (University Park, PA, 2004–7), 2, p. 558; also Hale-Dorrell, *Corn Crusade*, p. 30.

103 Engerman, 'Romance of Economic Development', 23–54.

104 S. Reid, 'The Khrushchev Kitchen: Domesticating the Scientific-Technological Revolution', *Journal of Contemporary History* 40 (2005), 289–316.

105 D. Stone, *Goodbye to All That: A History of Europe since 1945* (Oxford, 2014), pp. 153–5; S. Reid, '"Our Kitchen Is Just as Good": Soviet Responses to the American Kitchen', in R. Oldenziel and K. Zachmann (eds), *Cold War Kitchen: Americanization, Technology and European Users* (Cambridge, MA, 2009), pp. 83–112.

106 See M. Feshbach and A. Friendly, *Ecocide in the USSR: Health and Nature under Siege* (New York, 1992), pp. 73–90.

107 B. Gaybullaev, S.-C. Chen and D. Gaybullaev, 'Changes in water volume of the Aral Sea after 1960', *Applied Water Science* 2 (2012), 285–6; B. Alikhanov, 'Environmental challenges of the Aral Sea and the Aral Sea area', International Meeting Report, Tashkent (2010), 5–27;

D. Alieva, 'Reproductive health in females of the Aral Sea area',
Gynecology and Obstetrics 70 (2008), 64.

108 P. Micklin, 'The Aral Sea Disaster', *Annual Review of Earth Planetary Sciences* 35 (2007), 47–72.

109 Hamilton, *Supermarket USA*, p. 181.

110 Hale-Dorrell, *Corn Crusade*, pp. 1–9.

111 F. Dikötter, *The Tragedy of Liberation: A History of the Chinese Revolution, 1945–1957* (London, 2013), pp. 9–11.

112 Ibid., pp. 15–16.

113 Ibid., pp. 17–19.

114 Ibid., pp. 22–6.

115 Sun Yat-sen, *The True Solution for the Chinese Question* (New York, 1904), pp. 9–10.

116 Sun Yat-sen, *Vital Problems of China* (Taipei, 1953), p. 80.

117 Sun Yat-sen, 'The Three Principles of the People', in D. Chetham, *Before the Deluge* (London, 2002), p. 117.

118 Dikötter, *Tragedy of Liberation*, p. 137.

119 Ibid.

120 Ibid., pp. 126–7.

121 Ibid., pp. 49, 87–94.

122 Ibid., pp. 275–7.

123 Ibid., pp. 281–91.

124 L. Cabral, P. Pandey and X. Xu, 'Epic narratives of the Green Revolution in Brazil, China, and India', *Agriculture and Human Values* 39 (2022), 253.

125 Shapiro, *Mao's War against Nature*, pp. 67–8.

126 Ibid., p. 68.

127 Ibid., pp. 67–8.

128 Ibid., pp. 71–2.

129 Ibid., pp. 67–74.

130 Ibid., pp. 23, 49, 63.

131 Ibid., pp. 106–7.

132 M. Muscolino, ' "Water Has Aroused the Girls' Hearts": Gendering Water and Soil Conservation in 1950s China', *Past & Present* 255 (2022), 351–87.

133 Ibid., 29–31.

134 Dikötter, *Tragedy of Liberation*, p. 144.

135 Shapiro, *Mao's War against Nature*, pp. 86–7.

136 B. Maohong, 'The Evolution of Environmental Problems and Environmental Policy in China: The Interaction of Internal and

External Forces', in McNeill and Unger (eds), *Environmental Histories of the Cold War*, p. 328.

137 Shapiro, *Mao's War against Nature*, pp. 75, 89.

138 J. Becker, *Hungry Ghosts: Mao's Secret Famine* (New York, 1996), p. 2.

139 Ball, *Water Kingdom*, p. 225.

140 M. Wolfe, ' "A Revolution is a Force More Powerful Than Nature": Extreme Weather and the Cuban Revolution, 1959–64', *Environmental History* 24 (2020), 469–91.

141 E. Borgwardt, *A New Deal for the World: America's Vision for Human Rights* (Cambridge, MA, 2005), p. 157.

23 THE SHARPENING OF ANXIETIES (C.1960–C.1990)

1 M. Mead, *Culture and Commitment: A Study of the Generation Gap* (Garden City, NY, 1970), pp. 58–9.

2 J. Kerouac, *The Dharma Bums* (New York, 1958), pp. 97–8; A. Rome, ' "Give Earth a Chance": The Environmental Movement and the Sixties', *Journal of American History* 90 (2003), 543.

3 J. Willis, *Daily Life in the 1960s Counterculture* (Santa Barbara, CA, 2019), p. 147.

4 V. Intondi, *African Americans Against the Bomb: Nuclear Weapons, Colonialism and the Black Freedom Movement* (Stanford, CA, 2015).

5 Rome, ' "Give Earth a Chance" ', 536.

6 Higuchi, 'Atmospheric Nuclear Weapons Testing', p. 318.

7 C. Merchant, 'Women of the Progressive Conservation Movement: 1900–1916', *Environmental Review* 8 (1984), 57–85.

8 V. Norwood, *Made from This Earth: American Women and Nature* (Chapel Hill, NC, 1993), pp. 143–4; M. Ryan, *Womanhood in America, from Colonial Times to the Present* (New York, 1975), p. 281.

9 Rome, ' "Give Earth a Chance" ', 538.

10 R. Carson, *Silent Spring* (Cambridge, MA, 1962), pp. 1–2. Also see Norwood, *Made from This Earth*, pp. 143–71.

11 Carson, *Silent Spring*, p. 16.

12 Norwood, *Made from This Earth*, p. 154; Rome, ' "Give Earth a Chance" ', 532.

13 Rome, ' "Give Earth a Chance" ', 528.

14 J. Galbraith, *The Affluent Society* (Cambridge, MA, 1958), p. 253.

15 E. Dale, 'Big Debate: Public vs Private Spending', *New York Times*, 13 March 1960.

16 S. Udall, *The Quiet Crisis* (New York, 1963), pp. viii, 189; Rome, ' "Give Earth a Chance" ', 532.

17 President Johnson, 'Remarks at the University of Michigan, May 22, 1964', in *Public Papers of the Presidents of the United States: Lyndon B. Johnson, 1963–1964*, 2 vols (Washington, DC, 1965), 1, pp. 704–5.

18 US Army, 'Proceedings of the First Defoliation Conference, 29–30 July 1963' (Fort Detrick, MD, 1964).

19 E. Pfeiffer and G. Orians, 'Military Use of Herbicides in Vietnam', in J. Neilands et al. (eds), *Harvest of Death: Chemical Warfare in Vietnam and Cambodia* (New York, 1972), pp. 120–1.

20 J. Cookson and J. Nottingham, *A Survey of Chemical and Biological Warfare* (New York, 1969), pp. 42–3.

21 A. Galston et al., 'Scientists' Petition to President Johnson against Herbicidal Warfare', *Bioscience* 17 (1967), 10.

22 D. Zierler, *The Invention of Ecocide: Agent Orange, Vietnam and the Scientists who Changed the Way We Think about the Environment* (Athens, GA, 2011), pp. 14–32.

23 R. Olney, *Black Tide: The Santa Barbara Oil Spill and its Consequences* (New York, 1972).

24 G. Hill, 'One Year Later, Impact of Great Oil Slick is Still Felt', *New York Times*, 25 January 1970.

25 D. Stradling and R. Stradling, 'Perceptions of the Burning River: Deindustrialization and Cleveland's Cuyahoga River', *Environmental History* 13 (2008), 517–18.

26 Ibid., 523.

27 President Nixon, 'Remarks Following Inspection of Oil Damage at Santa Barbara Beach', 21 March 1969, in *Public Papers of the Presidents of the United States: Richard Nixon, 1969* (Washington, DC, 1971), p. 234.

28 G. Hill, ' "Environmental Crisis" May Eclipse Vietnam as College Issue', *New York Times*, 30 November 1969.

29 A. Rome, *The Genius of Earth Day: How a 1970 Teach-in Unexpectedly Made the First Green Generation* (New York, 2013).

30 A. Cooke, 'The US gets down to Earth', *Guardian*, 24 April 1970.

31 Hill, 'One Year Later'.

32 M. Melosi, 'Lyndon Johnson and Environmental Policy', in R. Divine (ed.), *The Johnson Years*, vol. 2: *Vietnam, the Environment, and Science* (Lawrence, KA, 1987), pp. 113–49; M. Melosi, *Effluent America: Cities, Industry, Energy and the Environment* (Pittsburgh, 2001), pp. 254–5.

33 J. Brooks Flippen, *Nixon and the Environment* (Albuquerque, NM, 2000); Rome, ' "Give Earth a Chance" ', 551–2; T. Spezio, *Slick Policy: Environmental and Science Policy in the Aftermath of the Santa Barbara Oil Spill* (Pittsburgh, 2018).

34 Dörries, 'Politics of Atmospheric Sciences', 206.

35 B. Ward and R. Dubos, *Only One Earth: The Care and Maintenance of a Small Planet* (New York, 1972), pp. 192, 195.

36 Ibid., p. xiii.

37 I. Gandhi, 'Man and Environment', in I. Gandhi, *Safeguarding Environment* (New Delhi, 1984), pp. 13–22.

38 John F. Kennedy Presidential Library, 'Inaugural Address', 20 January 1961.

39 President Truman, 'America's Solemn Obligation in World Famine Crisis', *Department of State Bulletin* 14 (1946), 716.

40 Central Intelligence Agency, *Potential Implications of Trends in World Population, Food Production, and Climate* (Washington, DC, 1974), p. 6.

41 E. East, *Mankind at the Crossroads* (New York, 1923); W. Thompson, *Population and Peace in the Pacific* (Chicago, 1946); W. Vogt, *Road to Survival* (New York, 1948); F. Osborn, *Our Plundered Planet* (Boston, MA, 1948); also see Perkins, *Geopolitics and the Green Revolution*, pp. 135–7.

42 Perkins, *Geopolitics and the Green Revolution*, p. 137.

43 Ibid., p. 138; Cullather, *Hungry World*, p. 154.

44 President Truman, 'America's Solemn Obligation', p. 716.

45 P. Ehrlich, *The Population Bomb* (New York, 1968), p. xii; also see P. Ehrlich and A. Ehrlich, 'The Population Bomb Revisited', *Electronic Journal of Sustainable Development* 1 (2009), 63–71.

46 S. Sontag, 'What's Happening to America', *Partisan Review* 34 (1967), 57–8; Rome, ' "Give Earth a Chance" ', 547.

47 N. Cullather, 'Parable of Seeds: The Green Revolution in the Modernizing Imagination', in M. Frey, R. Pruessen and T. Yong (eds), *The Transformation of Southeast Asia: International Perspectives on Decolonization* (Armonk, NY, 2003), p. 265.

48 G. Stone, 'Commentary: New histories of the Indian green revolution', *Geographical Journal* 185 (2019), 243–50.

49 G. Conway, *The Doubly Green Revolution: Food for All in the 21st Century* (London, 1997), pp. 44–65.

50 Nobel Prize, 'Peace Prize 1970: Norman Borlaug', https://www.nobelprize.org/prizes/peace/1970/borlaug/biographical/.

51 Cullather, *Hungry World*, p. 239.

52 P. Pingali, 'Green revolution: impacts, limits, and the path ahead', *PNAS* 109 (2012), 12,302–8.

53 G. Federico, *Feeding the World: An Economic History of Agriculture, 1800–2000* (Princeton, 2000), pp. 69–82; Ross, *Ecology and Power*, p. 384.

54 Cullather, *Hungry World*, pp. 236–7.

55 Ibid., pp. 241–2; Ross, *Ecology and Power*, p. 386.

56 United Nations Research Institute for Social Development, *The Social and Economic Implications of Large-Scale Introduction of New Varieties of Food Grain: Summary of Conclusion of a Global Research Project* (Geneva, 1974), p. 24.

57 Z. Poppel, 'Quick rice: International development and the Green Revolution in Sierra Leone, 1960–1976', in C. Helstosky (ed.), *The Routledge History of Food* (Abingdon, 2015), pp. 332–51.

58 N. Cullather, 'Miracles of Modernization: The Green Revolution and the Apotheosis of Technology', *Diplomatic History* 28 (2004), 243–5.

59 V. Shiva, *The Violence of the Green Revolution: Third World Agriculture, Ecology and Politics* (London, 1991), p. 47; Ross, *Ecology and Power*, p. 386.

60 P. Srivastava, M. Balhara and B. Giri, 'Soil health in India: past history and future perspective', in B. Giri and A. Varma (eds), *Soil Health* (New Delhi, 2020), pp. 1–19.

61 N. Sharma and R. Singhvi, 'Effects of chemical fertilizers and pesticides on human health and environment: a review', *International Journal of Agriculture, Environment and Biotechnology* 10 (2017), 675–80.

62 Ross, *Ecology and Power*, pp. 386–9.

63 D. Nierenberg, 'Agriculture: growing food and solutions', in E. Assadourian et al. (eds), *State of the World 2013: Is Sustainability Still Possible?* (Washington, DC, 2013), pp. 190–200.

64 K. Davis et al., 'Alternative cereals can improve water use and nutrient supply in India', *Science Advances* 4 (2018), 1–11.

65 I. Singh, 'Changes of agriculture production in India after Green Revolution', *Journal of the Gujarat Research Society* 21 (2019), 2290–4.

66 S. Prasad, 'Innovating at the margins: the system of rice intensification in India and transformative social innovation', *Ecology & Society* 21 (2016), 1–9.

67 E. Brainerd and N. Menon, 'Seasonal effects of water quality: the hidden costs of the Green Revolution to infant and child health in India', *Journal of Development Economics* 107 (2014), 49–64. For a full overview, see D. John and G. Babu, 'Lessons from the Aftermaths of Green Revolution on Food System and Health', *Frontiers in Sustainable Food Systems* 22 (2021), 1–6.

68 Ross, *Ecology and Power*, p. 393; M. Fearnside, 'Transmigration in Indonesia: Lessons from its Environmental and Social Impacts', *Environmental Management* 21 (1997), 553–70.

69 Cullather, *Hungry World*, pp. 242, 247–8.

70 Ehrlich, *Population Bomb*, p. xii.

71 A. Coale and E. Hoover, *Population Growth and Economic Development in Low-Income Countries* (Princeton, 1958), 128–31.

72 Cullather, *Hungry World*, p. 245.

73 M. Connelly, *Fatal Misconception: The Struggle to Control World Population* (Cambridge, MA, 2008).

74 Shapiro, *Mao's War against Nature*, pp. 29–32.

75 Ibid., p. 34.

76 Ibid., pp. 35–6; J. Banister, *China's Changing Population* (Stanford, 1987), pp. 148–50.

77 J. Zhang, 'The Evolution of China's One-Child Policy and its Effects on Family Outcomes', *Journal of Economic Perspectives* 31 (2017), 141–59.

78 S. Greenhalgh, 'Science, Modernity, and the Making of China's One-Child Policy', *Population and Development Review* 29 (2003), 163–96; S. Greenhalgh, *Just One Child: Science and Policy in Deng's China* (Berkeley, 2008).

79 R. Revelle and H. Suess, 'Carbon dioxide exchange between the atmosphere and ocean and the question of an increase of atmospheric CO_2 during the past decades', *Tellus* 9 (1957), 18–27.

80 R. Doel, 'Constituting the Postwar Earth Sciences: The Military's Influence on the Environmental Sciences in the USA after 1945', *Social Studies of Science* 33 (2003), 636–8.

81 Edwards, 'Entangled histories', 32; C. Keeling, 'Atmospheric carbon dioxide variations at the Mauna Loa Observatory', *Tellus* 28 (1976), 538–51.

82 J. Gertner, *The Ice at the End of the World: An Epic Journey into Greenland's Buried Past and our Perilous Future* (London, 2019); E. Weiss, 'Cold War under the ice: The Army's bid for a long-range nuclear role, 1959–1963', *Journal of Cold War Studies* 3 (2011), 31–58. For Operation HIGHJUMP, see US Navy Antarctic Development Program, *Report of Operation Highjump 1947* (2020).

83 E. Fedorov, 'Воздействие человека на метеорологические процессы', *Вопросы философии* 4 (1958), 137–44.

84 M. Budyko, *Современные проблемы метеорологии приземного слоя воздуха* (Leningrad, 1956); M. Budyko, 'Метеорологические исследования в СССР', *Метеорология и Гидрология* 11 (1957), 7–16; M. Budyko, *Тепловой баланс земной поверхности* (Leningrad, 1958). See J. Oldfield, 'Mikhail Budyko's (1920–2001)

contributions to Global Climate Science: from heat balances to climate change and global ecology', *WIREs Climate Change* 7 (2016), 682–92.

85 M. Budyko, 'The effect of solar radiation variations on the climate of the Earth', *Tellus* 21 (1969), 611–19; also see H. Kashiwase et al., 'Evidence for ice-ocean albedo feedback in the Arctic Ocean shifting to a seasonal ice zone', *Scientific Reports* 7 (2017), 1–10.

86 M. Budyko, *Изменения климата* (Leningrad, 1969), pp. 35–6. Also M. Budyko, *Влияние человека на климат* (Leningrad, 1972).

87 E. Robinson and R. Robbins, 'Sources, adundance, and fate of gaseous atmospheric pollutants. Final report and supplement', Stanford Research Institute (1968).

88 H. Landsberg, 'Man-Made Climatic Changes', *Science* 170 (1970), 1265–74.

89 G. Kukla and R. Matthews, 'When will the present inter-glacial end?', *Science* 178 (1972), 190–1; N.-A. Mörner, 'When will the present interglacial end?', *Quaternary Research* 2 (1972), 341–9.

90 P. Crutzen, 'The Influence of Nitrogen Oxides on the Atmospheric Ozone Content', *Quarterly Journal of the Royal Meteorological Society* 96 (1970), 320–5; H. Johnston, 'Reduction of Stratospheric Ozone by Nitrogen Oxide Catalysts from Supersonic Transport Exhaust', *Science* 173 (1971), 517–22; M. Molina and F. Rowland, 'Stratospheric Sink for Chlorofluoromethanes: Chlorine Atom-Catalyzed Destruction of Ozone', *Nature* 249 (1974), 810–12.

91 See for example B. Friedan, 'The Coming Ice Age', *Harper's Magazine*, September 1958; Fleming, *Historical Perspectives on Climate Change*, pp. 131–2.

92 For an extensive list of reports from the 1970s, see R. Cordato, 'Climate experts believe the next ice age is on its way … within a lifetime', John Locke Foundation, 6 March 2013, https://www.johnlo cke.org/climate-experts-believe-the-next-ice-age-is-on-its-way-within-a-lifetime/.

93 *Time*, 'Another Ice Age?', 24 June 1974.

94 *Newsweek*, 'The Cooling World', 28 April 1975.

95 National Science Board, *Science and the Challenges Ahead* (Washington, DC, 1974), pp. 25–6.

96 National Academy of Sciences, *Understanding Climatic Change: A Program for Action* (Washington, DC, 1975), p. 1.

97 Kissinger, 'Address to the Sixth Special Session of the United Nations General Assembly', 580–1.

98 R. Reeves and D. Gemmill (eds), *Climate Prediction Center: Reflections on 25 Years of Analysis, Diagnosis and Prediction* (Washington, DC, 2004), pp. 2–3.

99 Central Intelligence Agency, *A Study of Climatological Research as it Pertains to Intelligence Problems* (Washington, DC, 1974), p. 1.

100 Ibid., pp. 2–3.

101 Ibid., p. 7.

102 Central Intelligence Agency, *Potential Implications of Trends in World Population, Food Production, and Climate*, pp. 2, 30–1.

103 Ibid., pp. 4, 31–4.

104 Ibid., pp. 36–7.

105 Ibid., pp. 41–2.

106 G. Will, 'A Change in the Weather', *Washington Post*, 24 January 1975.

107 Cullather, *Hungry World*, p. 256.

108 'The Energy Emergency', in *Presidential Documents. Richard Nixon, 1973* (Washington, DC, 1973), p. 1312.

109 Ibid., p. 1313.

110 Ibid., pp. 1313–14.

111 Ibid., p. 1317.

112 C. Clotfelter and J. Hahn, 'Assessing the National 55mph Speed Limit', *Policy Sciences* 9 (1978), 281–94.

113 President Carter, 'Address to the Nation on the National Energy Plan', 8 November 1977, *Public Papers of the Presidents of the United States: Jimmy Carter*, 9 vols (Washington, DC, 1977–82), 1, pp. 1981–2.

114 Ibid., pp. 1982–7.

115 President Nixon, 'Special Message to the Congress on Energy Resources', 4 June 1971, in *Public Papers of the Presidents of the United States: Richard Nixon, 1971* (Washington, DC, 1971), pp. 703–14; Nixon, 'Special Message to the Congress on Energy Policy', 18 April 1973, *Public Papers of the Presidents of the United States: Richard Nixon, 1973* (Washington, DC, 1975), pp. 302–9.

116 President Carter, 'The Energy Problem', 18 April 1977, *Public Papers of the Presidents of the United States: Jimmy Carter*, 1, pp. 656–62.

117 President Carter, 'Energy and National Goals', 13 July April 1979, *Public Papers of the Presidents of the United States: Jimmy Carter*, 5, pp. 1235–41.

118 Ronald Reagan, 'Fourth Annual Conservative Political Action Conference', in *The Last Best Hope: The Greatest Speeches of Ronald Reagan* (West Palm Beach, 2016), p. 50.

119 President Carter, 'Solar Energy, 20 June 1979', *Public Papers of the Presidents of the United States: Jimmy Carter*, 5, pp. 1095–6.

120 R. Apple, '25 Years Later: Lessons from the Pentagon Papers', *New York Times*, 23 June 1996; *The Pentagon Papers: The Secret History of the Vietnam War* (New York, 2017).

121 *New York Times Co.* vs *United States*, 403 US 713 (1971); A. Aviki et al., 'The Pentagon Papers Framework, Fifty Years Later', in L. Bollinger and G. Stone (eds), *National Security, Leaks & Freedom of the Press: The Pentagon Papers Fifty Years On* (Oxford, New York, 2021), pp. 1–2.

122 J. Anderson, 'Air Force Has its Rainmakers', *Washington Post*, 18 March 1971.

123 United States Senate, *Hearings before the Subcommittee on Oceans and International Environment: Weather Modification*, 29 January 1974, referring to testimony delivered on 18 April 1972, pp. 109–10.

124 S. Hersh, 'Rainmaking is Used as Weapon by US', *New York Times*, 3 July 1972.

125 United States Senate, 'Weather Modification', p. 112.

126 'Memorandum from the Deputy Under Secretary of State for Political Affairs (Kohler) to Secretary of State Rusk', 13 January 1967, in *Foreign Relations of the United States, 1964–1968*, vol. 28: *Laos* (Washington, DC, 1998), Document 274.

127 Ibid.

128 United States Senate, 'Weather Modification', p. 89.

129 Ibid., pp. 93, 102.

130 Ibid., pp. 93, 115–16.

131 Ibid., pp. 36, 77.

132 R. Doel and K. Harper, 'Prometheus Unleashed: Science as a Diplomatic Weapon in the Lyndon B. Johnson Administration', *Osiris* 21 (2006), 66–85.

133 United States Senate, 'Weather Modification', p. 47.

134 Ibid., pp. 1–2.

135 93rd Congress, S. Res. 71, 'Resolution', 11 July 1973, pp. 3–8.

136 L. Freedman, 'SALT 50 Years On: Strategic Theory and Arms Control', *Global Politics and Strategy* 62 (2022).

137 Y. Richmond, *Cultural Exchange and the Cold War: Raising the Iron Curtain* (University Park, PA, 2003); J. Voorhees, *Dialogue Sustained: The Multilevel Peace Process and the Dartmouth Conference* (Washington, DC, 2002).

138 I. Detter, *The Law of War* (Cambridge, 2003), pp. 295–8.

139 United Nations, 'Convention on the Prohibition of Military or Any Other Hostile Use of Environmental Modification Techniques', 10 December 1976.

140 L. Battan, 'Weather Modification in the Soviet Union – 1976', *Bulletin of the American Meteorological Society* 58 (1977), 4–19.

141 United States Arms Control, *Worldwide Effects of Nuclear War: Some Perspectives* (Washington, DC, 1975).

142 P. Crutzen and J. Birks, 'The atmosphere after a nuclear war: Twilight at noon', *Ambio* 11 (1982), 114–25.

143 M. Hamblin, *Arming Mother Nature: The Birth of Catastrophic Environmentalism* (Oxford, 2013), p. 137.

144 R. Turco et al., 'Nuclear Winter: Global Consequences of Multiple Nuclear Explosions', *Science* 222 (1983), 1283–92.

145 V. Aleksandrov and G. Stenchikov, 'О моделировании климатических последствий ядерной войны', Журнал вычислительной математики и математической физики 24 (1984), 87–90.

146 Director of Central Intelligence, Interagency Intelligence Assessment, 'The Soviet Approach to Nuclear Winter', 10 December 1984 (1984), pp. 1–2.

147 Ibid., pp. 11–17.

148 Ibid., p. 6.

149 Hamblin, *Arming Mother Nature*, pp. 237–41.

150 'Science and the Citizen', *Scientific American* 240 (1984), 74.

151 S. Kapitsa, 'A Soviet view of nuclear winter', *Bulletin of the Atomic Scientists* 41 (1985), 37–8.

152 See for example P. Podvig, 'Did Star Wars Help End the Cold War? Soviet Response to the SDI Program', *Science & Global Security* 25 (2017), 3–27.

153 D. Kuzmiak, 'The American Environmental Movement', *Geographical Journal* 157 (1991), 265–78.

154 I. Gerasimov, *Ресурсы биосферы на территории СССР: научные основы их рационального использования и охраны* (Moscow, 1971); P. Oldak, 'Равновесное природополь зование и экономический рос', *Экономика и организация промышленного производства* 8 (1984), 44–61.

155 K. Gestwa, "Ökologischer Notstand und sozialer Protest. Ein umwelthistorischer Blick auf die Reformunfähigkeit und den Zerfall der Sowjetunion', *Archiv für Sozialgeschichte* 43 (2003), 325–48; L. Coumel and M. Elie, 'A Belated and Tragic Ecological Revolution: Nature, Disasters, and Green Activists in the Soviet Union and the Post-Soviet States, 1960s–2010s', *Soviet and Post-Soviet Review* 40 (2013), 157–65.

156 Guha, *Unquiet Woods*, pp. 152–84.

157 H. Gibbs et al., 'Tropical forests were the primary sources of new agricultural land in the 1980s and 1990s', *PNAS* 107 (2010), 16,732–7; S. Schwartzman et al., 'Social Movements and Large-Scale Tropical Forest Protection on the Amazon Frontier: Conservation from Chaos', *Journal of Environment & Development* 19 (2010), 274–99.

158 T. Doyle and S. MacGregor (eds), *Environmental Movements around the World: Shades of Green in Politics and Culture*, 2 vols (Santa Barbara, CA, 2014).

159 'Joint Statement: President Ronald Reagan and General Secretary Mikhail Gorbachev, 21 November, 1985', *Department of State Bulletin* 86 (1986), 10.

160 See for example Molina and Rowland, 'Stratospheric Sink for Chlorofluoromethanes', 810–12; S. Roan, *Ozone Crisis: The 15-Year Evolution of a Sudden Global Emergency* (New York, 1989).

161 J. Farman, B. Gardiner and J. Shanklin, 'Large losses of total ozone in Antarctica reveal seasonal ClO_x/NO_x interaction', *Nature* 315 (1985), 207–10.

162 K. Annan, *We the Peoples: The Role of the United Nations in the 21st Century* (New York, 2005), p. 56; A. Douglass, P. Newman and S. Solomon, 'The Antarctic ozone hole: an update', *Physics Today* 67 (2014), 42–8.

163 House of Representatives, Committee on Foreign Affairs, 'The Reagan–Gorbachev Summit and its Implications for United States–Soviet Relations' (Washington, DC, 1987), p. 79.

164 United Nations, 'Resolution adopted by the General Assembly 43/196. UN Conference on Environment and Development', 20 December 1988; 'Resolution adopted by the General Assembly 44/228. UN Conference on Environment and Development', 22 December 1989.

165 US Senate, 'Testimony before US Senate Committee on Energy and Natural Resources: Greenhouse Effect and Climate Change', 100th Congress, 1st session, 23 June 1988; C. Moser and L. Dilling (eds), *Creating a Climate for Change: Communicating Climate Change and Facilitating Social Change* (Cambridge, 2007), p. 1.

166 S. Agrawala and S. Andresen, 'Indispensability and Indefensibility? The United States in the Climate Treaty Negotiations', *Global Governance* 5 (1999), 459.

167 United Nations, 'United Nations Framework Convention on Climate Change' (1992).

168 President George H. W. Bush, 'Address to the United Nations Conference on Environment and Development in Rio de Janeiro', 12

June 1992, *Public Papers of the Presidents of the United States. George Bush, 1992–93*, 2 vols (Washington, DC, 1993), 1, p. 925.

169 World Meteorological Organisation, *Proceedings: World Conference, Toronto 1988* (Geneva, 1989); United Nations, 'United Nations Framework Convention on Climate Change' (New York, 1992).

170 Kyoto Protocol to the United Nations Framework Convention on Climate Change, 9 May 1992.

171 US Congress, 'A resolution expressing the sense of the Senate regarding the conditions for the United States becoming a signatory to any international agreement on greenhouse gas emissions under the United Nations Framework Convention on Climate Change', 12 June 1997.

172 'Bush was aloof in warming debate', *Washington Post*, 31 October 1992.

173 Clinton Presidential Records: Mandatory Declassification Review, 'Telcon with Prime Minister Blair of the United Kingdom', 27 May 2000.

174 R. Goyal et al., 'Reduction in surface climate change achieved by the 1987 Montreal Protocol', *Environmental Research Letters* 14 (2019), 1–10.

175 BBC News, 'Putin deplores collapse of USSR', 25 April 2005.

24 ON THE EDGE OF ECOLOGICAL LIMITS (C.1990–TODAY)

1 Ross, *Ecology and Power*, pp. 398–9.

2 B. Söderlund, 'The Importance of Business Travel for Trade: Evidence from the Liberalisation of Soviet Airspace', Research Institute of Industrial Economics, Working Paper Series 1355 (2020), 1–51.

3 F. Schierhorn et al., 'Large greenhouse gas savings due to changes in the post-Soviet food systems', *Environmental Research Letters* 14 (2019), 1–12.

4 H. Schaefer, 'On the Causes and Consequences of Recent Trends in Atmospheric Methane', *Current Climate Change Reports* 5 (2019), 259–74.

5 D. Cutler and L. Summers, 'The COVID-19 pandemic and the $16 trillion virus', *Journal of the American Medical Association* 324 (2020), 1495–6.

6 F. Dikötter, *China after Mao: The Rise of a Superpower* (London, 2022), p. vii.

7 E. Thaler et al., 'Rates of Historical Anthropogenic Soil Erosion in the Midwestern United States', *Earth's Future* 10 (2022), 1–16.

8 P. Borelli et al., 'An assessment of the global impact of 21st century land use change on soil erosion', *Nature Communications* 8 (2017), 1–13.

9 T. Münzel et al., 'Soil and water pollution and human health: what should cardiologists worry about?', *Cardiovascular Research* (2022), 1–10.

10 Sanderson and Frey, 'From desert to breadbasket', 516–32.

11 K. Gibbs et al., 'Tropical forests were the primary sources of new agricultural land in the 1980s and 1990s', *PNAS* 107 (2010), 16,732–7.

12 F. Pendrill et al., 'Disentangling the numbers behind agriculture-driven tropical deforestation', *Science* 377 (2022), 1–12.

13 World Wildlife Fund, 'Palm Oil: Overview', https://www.worldwildlife.org/industries/palm-oil.

14 E. Fitzherbert, 'How will oil palm expansion affect biodiversity?', *Trends in Ecology & Evolution* 23 (2008), 538–45; Commission of the European Communities, *An EU Strategy for Biofuels* (Brussels, 2006).

15 E. Tyukavina et al., 'Types and rates of forest disturbance in Brazilian Legal Amazon, 2000–2013', *Science Advances* 3 (2017), 1–15; P. Vale et al., 'The Expansion of Intensive Beef Farming to the Brazilian Amazon', *Global Environmental Change* 57 (2019), 1–11.

16 W. Carvalho et al., 'Deforestation control in the Brazilian Amazon: a conservation struggle being lost as agreements and regulations are subverted and bypassed', *Perspectives in Ecology and Conservation* 17 (2019), 122–30; D. Roy, 'Deforestation of Brazil's Amazon Has Reached a Record High. What's Being Done?', Council for Foreign Relations, 17 March 2022.

17 A. Tyukavina et al., 'Congo Basin forest loss dominated by increasing smallholder clearing', *Science Advances* 4 (2018), 1–12.

18 E. zu Ermgassen et al., 'Using supply chain data to monitor zero deforestation commitments: an assessment of progress in the Brazilian soy sector', *Environmental Research Letters* 15 (2020), 1–12; P. Oliveira, 'Trends in water balance components across the Brazilian Cerrado', *Water Resources Research* 50 (2014), 7100–14; N. Myers et al., 'Biodiversity hotspots for conservation priorities', *Nature* 403 (2000), 853–8.

19 R. Estoque et al., 'Spatiotemporal pattern of global forest change over the past 60 years and the forest transition theory', *Environmental Research Letters* 17 (2022), 6.

20 United Nations Food and Agriculture Organisation, *The Future of Food and Agriculture – Trends and Challenges* (Rome, 2017); T. Garnett et al., *Grazed and Confused?*, Food Climate Research Network (Oxford, 2017).

21 P. Newton et al., 'The number and spatial distribution of forest-proximate people globally', *One Earth* 3 (2020), 363–70.

22 United Nations Environment Programme (UNEP), *The State of the World's Forests: Forests, Biodiversity and People* (Rome, 2020), p. vi.

23 K. Carlson and L. Curran, 'Refined carbon accounting for oil palm agriculture: Disentangling potential contributions of indirect emissions and smallholder farmers', *Carbon Management* 4 (2013), 347–9.

24 Y. Fung et al., 'Upward expansion and acceleration of forest clearance in the mountains of Southeast Asia', *Nature Sustainability* 4 (2021), 892–9.

25 C. Boulton, T. Lenton and N. Boers, 'Pronounced loss of Amazon rainforest resilience since the early 2000s', *Nature Climate Change* 12 (2022), 271–8.

26 V. Gatti et al., 'Amazonia as a carbon source linked to deforestation and climate change', *Nature* 595 (2021), 388–93.

27 K. Covey et al., 'Carbon and Beyond: The Biogeochemistry of Climate in a Rapidly Changing Amazon', *Frontiers in Forests and Global Change* 11 (2021), 1–20.

28 N. Boers et al., 'A deforestation-induced tipping point for the South American monsoon system', *Scientific Reports* 7 (2017), 1–9; T. Lovejoy and C. Nobre, 'Amazon tipping point', *Science Advances* 4 (2018), 1.

29 M. Weisse and E. Goldman, 'Forest Pulse: The Latest on the World's Forests', *Global Forest Review*, 28 April 2022.

30 United Nations, *Spreading Like Wildfire: The Rising Threat of Extraordinary Landscape Fires* (Nairobi, 2022); S. Hugelius et al., 'Large stocks of peatland carbon and nitrogen are vulnerable to permafrost thaw', *PNAS* 17 (2020), 20,438–46; A. Goldstein et al., 'Protecting irrecoverable carbon in Earth's ecosystems', *Nature Climate Change* 10 (2020), 287–95.

31 UN Climate Change Conference, https://ukcop26.org/glasgow-leaders-declaration-on-forests-and-land-use/.

32 C. Bellard, P. Casey and T. Blackburn, 'Alien species as a driver of recent extinctions', *Biology Letters* 12 (2016), 1–4; H. Seebens et al., 'Projecting the continental accumulation of alien species through to 2050', *Global Change Biology* 27 (2021), 970–82.

33 T. Fritts, 'Economic costs of electrical system instability and power outages caused by snakes on the Island of Guam', *International Biodeterioration & Biodegradation* 49 (2002), 93–100.

34 B. Kaiser and K. Burnett, 'Economic impacts of E. Coqui frogs in Hawaii', *Interdisciplinary Environmental Review* 8 (2006), 1–11. In general, I. Soto et al., 'Global economic costs of herpetofauna invasions', *Scientific Reports* 12 (2022), 1–12.

35 K. Willis (ed.), *State of the World's Plants 2017* (London, 2017), p. 71.

36 A. Antonelli et al., *State of the World's Plants and Fungi 2020* (London, 2020), p. 51.

37 N. Straw et al., 'History and development of an isolated
    outbreak of Asian longhorn beetle *Anoplophora glabripennis*
    (Coleoptera: Cerambycidae) in southern England', *Agricultural and
    Forest Entomology* 18 (2016), 280–93; UK Government, 'Asian Longhorn
    Beetle eradicated in the UK', Press Release, 17 May 2019.

38 Willis (ed.), *State of the World's Plants*, p. 71.

39 D. Paini et al., 'Global threat to agriculture from invasive species', *PNAS*
    113 (2016), 7575–9; D. Renault et al., 'The magnitude, diversity, and
    distribution of the economic costs of invasive terrestrial invertebrates
    worldwide', *Science of the Total Environment* 835 (2022), 1–11.

40 G. Buczkowski and C. Bertelsmeier, 'Invasive termites in a changing
    climate: A global perspective', *Ecology and Evolution* 7 (2017), 974–85.

41 C. Buttimer et al., 'Bacteriophages and Bacterial Plant Diseases',
    Frontiers in Microbiology 8 (2017), 1–15.

42 D. Davis, M. Epp and H. Riordan, 'Changes in USDA Food
    Composition Data for 43 Garden Crops, 1950 to 1999', *Journal of the
    American College of Nutrition* 23 (1999), 669–82; D. Wallace-Wells, *The
    Uninhabitable Earth: A Story of the Future* (London, 2019), p. 57.

43 R. Forslid, 'Trade, Transportation and the Environment', Centre for
    Economic Policy Research, Discussion Papers 14228 (2019), 1.

44 A. Schneider, M. Friedl and D. Potere, 'A new map of global urban
    extent from MODIS satellite data', *Environmental Research Letters* 4
    (2009), 1–11.

45 United Nations, *World Urbanization Prospects: The 2018 Revision*
    (New York, 2019), pp. xix, 1.

46 Organisation for Economic Co-Operation and Development
    (OECD), *Urbanisation and Green Growth in China*, OECD Regional
    Development Working Papers 2013/07 (Paris, 2013), pp. 30–1.

47 Y. Guo and W. Qiao, 'Rural Migration and Urbanisation in
    China: Historical Evolution and Coupling Pattern', *Sustainability* 12
    (2020), 1; K. Farrell and H. Westlund, 'China's rapid urban ascent: an
    examination into the components of urban growth', *Asian Geographer* 35
    (2018), 85–106.

48 V. Smil, *Making the Modern World: Materials and Dematerialization*
    (London, 2018), p. 285.

49 Oxford Economics, *Global Cities: The Future of the World's Leading
    Urban Economies to 2035* (Oxford, 2018).

50 United Nations, *World Urbanization Prospects*, pp. 77, xix.

51 United Nations, *World Cities Report 2020: The Value of Sustainable
    Urbanization* (Nairobi, 2020), p. xvii.

52 Ibid., p. 3; United Nations, *The Sustainable Development Goals Report 2022* (New York, 2022), p. 47.

53 L. Zhao et al., 'Global multi-model projections of local urban climates', *Nature Climate Change* 11 (2021), 152–7.

54 O. Daramola and P. Olawuni, 'Assessing the water supply and sanitation sector for post-2015 development agenda: a focus on Lagos Metropolis, Nigeria', *Environment, Development and Sustainability* 21 (2019), 1127–38.

55 City of New York, *Annual Comprehensive Financial Report of the Comptroller for the Fiscal Years Ended June 30, 2021 and 2020* (New York, 2022), p. 355.

56 M. Saunois et al., 'The global methane budget 2000–2017', *Earth System Science Data* 12 (2020), 1561–623.

57 J. Maasakkers et al., 'Using satellites to uncover large methane emissions from landfills', *Science Advances* 8 (2022), 1–8.

58 T. Wei, J. Wu and S. Chen, 'Keeping Track of Greenhouse Gas Emission Reduction Progress and Targets in 167 Cities Worldwide', *Frontiers in Sustainable Cities* 3 (2021), 1–13.

59 M. Greenstone, C. Hasenkopf and K. Lee, *Air Quality Life Index: Annual Update 2022* (Chicago, 2022), pp. 10, 12, 6–7.

60 Wei, Wu and Chen, 'Keeping Track of Greenhouse Gas', 1–13.

61 J. Scott, 'Think Tank: Studying How the Defeat of Heat Changed Nearly Everything', *New York Times*, 29 August 1998.

62 O. Deschênes and M. Greenstone, 'Climate Change, Mortality, and Adaptation: Evidence from Annual Fluctuations in Weather in the US', NBER Working Papers 13178 (2007), 1–42.

63 G. Heal and J. Park, 'Feeling the Heat: Temperature, Physiology and the Wealth of Nations', Harvard Environmental Economics Discussion Paper 14-51 (2014), 3.

64 Wallace-Wells, *Uninhabitable Earth*, pp. 46–7; Heal and Park, 'Feeling the Heat', 19.

65 Wallace-Wells, *Uninhabitable Earth*, p. 42; A. Demirbas et al., 'The Cost Analysis of Electric Power Generation in Saudi Arabia', *Energy Sources* 12 (2017), 591–6.

66 Wallace-Wells, *Uninhabitable Earth*, p. 42.

67 D. Newman, 'New Residents Flood into the Most Drought-Stricken Counties', Economic Innovation Group, 27 July 2021.

68 M. Hino and M. Burke, 'The effect of information about climate risk on property values', *PNAS* 118 (2021), 1–9.

69 N. Jones, 'Waste Heat: Innovators Turn to an Overlooked Renewable Resource', *Yale Environment 360*, 29 May 2018.

70 WRAP, *Food Surplus and Waste in the UK – Key Facts* (2021), p. 1.

71 UNEP, *Food Waste Index Report 2021* (Nairobi, 2021); Reuters, 'U.N. report says 17% of food wasted at consumer level', 4 March 2021.

72 C. Mbow et al., 'Food security', in IPCC, *Special Report: Climate Change and Land* (2019), pp. 475–80.

73 United Nations Climate Change, 'UN Helps Fashion Industry Shift to Low Carbon', 6 September 2018.

74 UNEP, 'Putting the Brakes on Fast Fashion', 12 November 2018.

75 J. Leonard, '30 Shocking Figures and Facts in Global Textile and Apparel Industry', Business 2 Community, 7 May 2015.

76 UNEP, 'Putting the Brakes on Fast Fashion'.

77 Agence France Presse, 'Chile's desert dumping ground for fast fashion leftovers', 12 November 2021.

78 L. van Woensel and S. Lipp, 'What if fashion were good for the planet?', European Parliamentary Research Service, Scientific Foresight Unit (2020); United Nations, 'UN launches drive to highlight environmental cost of staying fashionable', 25 March 2019.

79 European Commission, 'Water Scarcity and Drought', August 2010, p. 2.

80 UNEP, *Options for Decoupling Economic Growth from Water Use and Water Pollution: Report of the International Resource Panel Working Group on Sustainable Water Management* (2015), pp. 2, 17.

81 Discover Water, 'Leaking Pipes', https://discoverwater.co.uk/leak ing-pipes; European Commission, 'Water Scarcity and Drought', p. 2.

82 C. Villanueva et al., 'Health and environmental impacts of drinking water choices in Barcelona, Spain: A modelling study', *Science of the Total Environment* 795 (2021), 1–10.

83 S. Laville and M. Taylor, 'A million bottles a minute: world's plastic binge "as dangerous as climate change"', *Guardian*, 28 June 2017; Euromonitor International, *Global Trends in Food and Drinks Packaging* (2017).

84 C. Jones, 'Davos: Coke insists customers still want plastic bottles', *The Times*, 22 January 2020.

85 Transport & Environment, *One Corporation to Pollute Them All: Luxury Cruise Air Emissions in Europe* (Brussels, 2019); M. Sofiev et al., 'Cleaner fuels for ships provide public health benefits with climate tradeoffs', *Nature Communications* 9 (2018), 1–12.

86 R. Ryan et al., 'Impact of Rocket Launch and Space Debris Air Pollutant Emissions on Stratospheric Ozone and Global Climate', *Earth's Future* 10 (2022), 1–13.

87 WARC, 'Alibaba attempts to green Singles Day', 12 November 2019.

88 X. Yue and H. Minghe, 'Singles' Day: Record-breaking sales focus new attention on packaging waste as almost 4 billion parcels are shipped', *South China Morning Post*, 18 November 2019.

89 Greenpeace China, '环保团体:电商、快递行业包装绿色化进程亟待推进', 11 November 2019.

90 W. Wang, 'Protein aggregation and its inhibition in biopharmaceutics', *International Journal of Pharmaceutics* 289 (2005), 1–30.

91 R. Mihigo et al., 'Improving access to affordable vaccines for middle-income countries in the African Region', *Vaccine* 37 (2019), 2838–42; A. Ashok, M. Brinson and Y. LeTallec, 'Improving cold chain systems: Challenges and solutions', *Vaccine* 35 (2017), 2217–23.

92 N. Crawford, 'The Pentagon Fuel Use, Climate Change and the Costs of War', Watson Institute, Brown University (2019), pp. 13, 4, 6.

93 World Meteorological Organisation, *The Changing Atmosphere: Implications for Global Security* (Geneva, 1989), p. 292.

94 Albert, *Subtitles to Save the World: Understanding How the Broadcasting Community is Covering Climate* (2019).

95 Albert, *Subtitles to Save the World: An Analysis of How UK Broadcasters are Exposing Audiences to Climate Change through their Content* (2021).

96 Met Office, '2020 ends earth's warmest 10 years on record', 14 January 2021.

97 O. Rakovec et al., 'The 2018–2020 Multi-Year Drought Sets a New Benchmark in Europe', *Earth's Future* 10 (2022), 1–11.

98 A. Park Williams, B. Cook and J. Smerdon, 'Rapid intensification of the emerging southwestern North American megadrought in 2020–2021', *Nature Climate Change* 12 (2022), 232–4.

99 I. Velicogna et al., 'Continuity of Ice Sheet Mass Loss in Greenland and Antarctica from the GRACE and GRACE Follow-On Missions', *Geophysical Research Letters* 47 (2020), 1–8.

100 R. Stuart-Smith et al., 'Increased outburst flood hazard from Lake Palcacocha due to human-induced glacier retreat', *Nature Geoscience* 14 (2021), 85–90.

101 F. Otto et al., 'Climate change increased rainfall associated with tropical cyclones hitting highly vulnerable communities in Madagascar, Mozambique & Malawi', World Weather Attribution, 11 April 2022.

102 R. Neukom et al., 'No evidence for globally coherent warm and cold periods', *Nature* 571 (2019), 550–4.

103 T. Ahmad and D. Zhang, 'A critical review of comparative global historical energy consumption and future demand: The story told so far', *Energy Reports* 6 (2020), 1973–91.

104 Wallace-Wells, *Uninhabitable Earth*, p. 4.

105 M. Wei et al., 'Revisiting the Existence of the Global Warming Slowdown during the Early Twenty-First Century', *Journal of Climate* 35 (2022), 1853–71.

106 M. Lynas, B. Houlton and S. Perry, 'Greater than 99% consensus on human caused climate change in the peer-reviewed scientific literature', *Environmental Research Letters* 16 (2021), 1–8.

107 S. Raghuraman, D. Paynter and V. Ramaswamy, 'Anthropogenic forcing and response yield observed positive trend in Earth's energy imbalance', *Nature Communications* 12 (2021), 1–10.

108 S. Braddock et al., 'Relative sea-level data preclude major late Holocene ice-mass change in Pine Island Bay', *Nature Geoscience* 15 (2022), 568–72.

109 C. Bradshaw et al., 'Hydrological impact of Middle Miocene Antarctic ice-free areas coupled to deep ocean temperatures', *Nature Geoscience* 14 (2021), 429–36.

110 L. Pan, 'Rapid postglacial rebound amplifies global sea level rise following West Antarctic Ice Sheet collapse', *Science Advances* 7 (2021), 1–9.

111 R. Dziadek, F. Ferraccioli and K. Gohl, 'High geothermal heat flow beneath Thwaites Glacier in West Antarctica inferred from aeromagnetic data', *Communications Earth & Environment* 2 (2021), 1–6; S. González-Herrero et al., 'Climate warming amplified the 2020 record-breaking heatwave in the Antarctic Peninsula', *Communications Earth & Environment* 3 (2022), 1–9.

112 S. Deng et al., 'Polar Drift in the 1990s Explained by Terrestrial Water Storage Changes', *Geophysical Research Letters* 48 (2021), 1–10.

113 Y. Chen et al., 'Thermokarst acceleration in Arctic tundra driven by climate change and fire disturbance', *One Earth* 4 (2021), 1718–29.

114 NASA, Moderate Resolution Imaging Spectroradiometer, 'Smoke from Siberian Wildfires', 7 August 2021.

115 E. Berenguer et al., 'Tracking the impacts of El Niño drought and fire in human-modified Amazonian forests', *PNAS* 118 (2021), 1–8.

116 T. Schneider, C. Kaul and K. Pressel, 'Possible climate transitions from breakup of stratocumulus decks under greenhouse warming', *Nature Geoscience* 12 (2019), 163–7.

117 N. Wunderling, 'Interacting tipping elements increase risk of climate domino effects under global warming', *Earth System Dynamics* 12 (2021), 601–19.

118 D. Armstrong Mackay et al., 'Exceeding 1.5 °C global warming could trigger multiple climate tipping points', *Science* 377 (2020), 1–11.

119 T. Slater, 'Increased variability in Greenland Ice Sheet runoff from satellite observations', *Nature Communications* 12 (2021), 1–9.

120 J. Box, 'Greenland ice sheet climate disequilibrium and committed sea-level rise', *Nature Climate Change* (2022), 1–11.

121 D. Smale et al., 'Marine heatwaves threaten global biodiversity and the provision of ecosystem services', *Nature Climate Change* 9 (2019), 306–12; N. Gruber et al., 'Biogeochemical extremes and compound events in the ocean', *Nature* 600 (2021), 395–407.

122 C. Free et al., 'Impacts of historical warming on marine fisheries production', *Science* 363 (2019), 979–83.

123 R. Myers and B. Worm, 'Rapid worldwide depletion of predatory fish communities', *Nature* 423 (2003), 280–3; J. Jackson et al., 'Historical overfishing and the recent collapse of coastal ecosystems', *Science* 293 (2001), 629–38; S. Ohayon, I. Granot and J. Belmaker, 'A meta-analysis reveals edge effects within marine protected areas', *Nature Ecology & Evolution* 5 (2021), 1301–8.

124 G. Ashton, 'Predator control of marine communities increases with temperature across 115 degrees of latitude', *Science* 376 (2022), 1215–19; E. Tekwa, J. Watson and M. Pinsky, 'Body size and food–web interactions mediate species range shifts under warming', *Proceedings of the Royal Society B: Biological Sciences* 289 (2022).

125 S. Jane et al., 'Widespread deoxygenation of temperate lakes', *Nature* 594 (2021), 66–70.

126 A. Oschiles, 'A committed fourfold increase in ocean oxygen loss', *Nature Communications* 12 (2021), 1–8.

127 K. Tanaka and K. Van Houtan, 'The recent normalization of historical marine heat extremes', *Plos Climate* 1 (2022), 1–13.

128 W. Wang et al., 'Emergent constraint on crop yield response to warmer temperature from field experiments', *Nature Sustainability* 3 (2020), 908–16; J. Desai et al., 'Warm nights disrupt transcriptome rhythms in field-grown rice panicles', *PNAS* 118 (2021), 1–12.

129 M. Burke and K. Emerick, 'Adaptation to Climate Change: Evidence from US Agriculture', *American Economic Journal: Economic Policy* 8 (2016), 106–40.

130 K. Coyne et al., 'Interactive effects of light, CO_2 and temperature on growth and resource partitioning by the mixotrophic dinoflagellate, *Karlodinium veneficum*', *PLOS ONE* 16 (2021), 1–20.

131 W. Tang et al., 'Widespread phytoplankton blooms triggered by 2019–2020 Australian wildfires', *Nature* 597 (2021), 370–5.

132 O. Koppel and J. Kerr, 'Strong phenological shifts among bumblebee species in North America can help predict extinction risk', *Biological Conservation* 272 (2022), 1–7.

133 J. González-Varo et al., 'Limited potential for bird migration to disperse plants to cooler latitudes', *Nature* 595 (2021), 75–9.

134 B. Oliveira et al., 'Community-wide seasonal shifts in thermal tolerances of mosquitoes', *Ecology* 102 (2021), 1–12.

135 F. Colón-González et al., 'Projecting the risk of mosquito-borne diseases in a warmer and more populated world: a multi-model, multi-scenario intercomparison modelling study', *Lancet Planetary Health* 5 (2021), 1–11.

136 UNEP, *Preventing the Next Pandemic: Zoonotic Diseases and How to Break the Chain of Transmission* (Nairobi, 2020), pp. 11–12.

137 C. Mora et al., 'Over half of known human pathogenic diseases can be aggravated by climate change', *Nature Climate Change* (2022), 1–9.

138 T. Slater et al., 'Increased variability in Greenland Ice Sheet runoff from satellite observations', *Nature Communications* 12 (2021) 1–9; A. Grinset and J. Hesselbjerg Christiansen, 'The transient sensitivity of sea level rise', *Ocean Science* 17 (2021), 181–6; C. Chambers et al., 'Mass loss of the Antarctic ice sheet until the year 3000 under a sustained late-21st-century climate', *Journal of Glaciology* 68 (2022), 605–17.

139 N. Bloemendaal et al., 'A globally consistent local-scale assessment of future tropical cyclone risk', *Science Advances* 8 (2022), 1–13.

140 B. Poujol et al., 'Kilometer-scale modeling projects a tripling of Alaskan convective storms in future climate', *Climate Dynamics* 55 (2020), 3543–64; B. Poujol et al., 'Dynamic and thermodynamic impacts of climate change on organized convection in Alaska', *Climate Dynamics* 56 (2021), 2569–93.

141 A. Kahraman et al., 'Quasi-Stationary Intense Rainstorms Spread across Europe under Climate Change', *Geophysical Research Letters* 48 (2021), 1–11.

142 First Street Foundation, *The 6th National Risk Assessment: Hazardous Heat* (2022), pp. 4, 8–9.

143 J. Wang et al., 'Changing Lengths of the Four Seasons by Global Warming', *Geophysical Research Letters* 48 (2021), 1–9; N. Lorenzo, A. Díaz-Poso and D. Royé, 'Heatwave intensity on the Iberian Peninsula: Future climate projections', *Atmospheric Research* 258 (2021), 1–10.

144 C. Rogers et al., 'Sixfold Increase in Historical Northern Hemisphere Concurrent Large Heatwaves Driven by Warming and Changing Atmospheric Circulations', *Journal of Climate* 35 (2022), 1063–78.

145 I. Wilkinson and T. Kington, 'Mediterranean faces months of 50C heatwaves', *The Times*, 28 May 2021.

146 M. Burke et al., 'Higher temperatures increase suicide rates in the United States and Mexico', *Nature Climate Change* 8 (2018); L. Taylor et al., 'Environmental Conditions on Cognitive Function: A Focused Review', *Frontiers in Physiology* 6 (2016).

147 E.-S. Im, J. Pal and E. Eltahir, 'Deadly heat waves projected in the densely populated regions of South Asia', *Science Advances* 3.8 (2017), 1–7.

148 J. Zhang et al., 'Labrador Sea freshening linked to Beaufort Gyre freshwater release', *Nature Communications* 12 (2021), 1–8.

149 L. Caesar et al., 'Current Atlantic Meridional Overturning Circulation weakest in last millennium', *Nature Geoscience* 114 (2021), 118–20; D. Thornalley et al., 'Anomalously weak Labrador Sea convection and Atlantic overturning during the past 150 years', *Nature* 556 (2018), 227–30.

150 S. Peyrouse, *The Nexus of Environmental Issues, Poverty, and Political Authoritarianism in Central Asia*, Central Asia Program, IERES, George Washington University (2022), p. 10.

151 P. Wester et al., *The Hindu Kush Himalaya Assessment: Mountains, Climate Change, Sustainability and People* (Cham, 2019).

152 C. Xu et al., 'Future of the human climate niche', *PNAS* 117 (2020), 11,350–5.

153 R. Reuveny, 'Climate change-induced migration and violent conflict', *Political Geography* 26 (2007), 656–73; G. Abel, 'Climate, conflict and forced migration', *Global Environmental Change* 54 (2019), 239–49.

154 A. Ahmadalipour, H. Moradkhani and M. Kumar, 'Mortality risk from heat stress expected to hit poorest nations the hardest', *Climatic Change* 152 (2019), 569; T. Lok Wu, 'Countries with Water Scarcity Right Now', Earth.org, 13 August 2022.

155 D. Dennis, 'Southeast Asia's Coming Climate Crisis', Centre for Strategic and International Studies, 22 May 2020.

156 T. Middendorp and R. Bergema, 'The Warning Signs are Flashing Red: The Interplay between Climate Change and Violent Extremism in the Western Sahel', International Centre for Counter-Terrorism (2019), pp. 2–4.

157 Department of Defense, *Quadrennial Defense Review Report 2010* (Washington, DC, 2010), pp. 84–5.

158 President Obama, 'Presidential Memorandum – Climate Change and National Security', The White House, 21 September 2016.

159 Crawford, 'Pentagon Fuel Use, Climate Change and the Costs of War', pp. 27–8; Office of the Under Secretary of Defense for Acquisition and Sustainment, *Report on the Effects of a Changing Climate to the Department of Defense* (2019), p. 5.

160 D. Sterletskiy et al., 'Assessment of climate change impacts on buildings, structures and infrastructure in the Russian regions on permafrost', *Environmental Research Letters* 14 (2019), 1–15.

161 N. Nikiforova and A. Konnov, 'Influence of permafrost degradation on piles bearing capacity', *Journal of Physics: Conference Series* 1928 (2021), 1–6.

162 L. Khrustalev and I. Davidova, 'Forecast of climate warming and account of it at estimation of foundation reliability for buildings in permafrost zone', *Earth Cryosphere* 11 (2007), 68–75.

163 Y. Yau and S. Hasbi, 'A review of climate change impacts on commercial buildings and their technical services in the tropics', *Renewable and Sustainable Energy Reviews* 18 (2013), 430–41.

164 Ministry of Natural Resources and Ecology, Russian Federation, *О состоянии и об охране окружающей среды Российской Федерации в 2017 году* (Moscow, 2018).

165 E. Simonova et al., 'Сибирская язва на Ямале: Оценка эпизоотологических и эпидемиологических рисков', *Проблемы Особо Опасных Инфекций* 1 (2017), 89–93; V. Yastrebov, 'Эпидемиология трансмиссивных клещевых инфекций в России', *Здоровье Населения и Среда Обитания* 11 (2016), 8–12.

166 C. Hickman et al., 'Climate anxiety in children and young people and their beliefs about government responses to climate change: a global survey', *Lancet Planetary Health* 5 (2021), 863–73.

167 R. Gray, 'The people fighting the war on waste at music festivals', BBC News, 27 June 2019; E. Santillo, 'Glastonbury Festival clean-up: How much does it cost to clean Worthy Farm and what happens to the waste?', Somerset Live, 27 June 2022; J. Halliday, 'Greta Thunberg makes surprise appearance at Glastonbury festival', *Guardian*, 25 June 2022.

168 S. Helm, J. Kemper and S. White, 'No future, no kids – no kids, no future?', *Population and Environment* 43 (2021), 108–29; C. Cain Miller, 'Americans are Having Fewer Babies. They Told Us Why', *New York Times*, 5 July 2018.

169 'When the Duke of Sussex Interviewed Dr Jane Goodall about the Future of Sustainability', *Vogue*, September 2019; Population Matters, 'World Population Day 2021: Change Champions', 11 July 2021.

170 United Nations, *Global Land Outlook, Second Edition: Land Restoration for Recovery and Resilience* (Bonn, 2022).

171 L. Warszawski et al., 'All options, not silver bullets, needed to limit global warming to 1.5 °C: a scenario appraisal', *Environmental Research Letters* 16 (2021), 1–14.

172 L. Vargas Zeppetello, A. Raftery and D. Battisti, 'Probabilistic projections of increased heat stress driven by climate change', *Communications Earth & Environment* 3 (2022), 1–7.

173 United Nations, 'Secretary-General's video message to the Petersberg Dialogue', 18 July 2022.

174 C. Mora et al., 'Global risk of deadly heat', *Nature Climate Change* 7 (2017), 501–6.

175 H. Song et al., 'Thresholds of temperature change for mass extinctions', *Nature Communications* 12 (2021), 1–8; C. Román-Palacios and J. Wiens, 'Recent responses to climate change reveal the drivers of species extinction and survival', *PNAS* 117 (2020), 4211–17; R. Cowie, P. Bouchet and B. Fontaine, 'The Sixth Mass Extinction: fact, fiction or speculation?', *Biological Reviews* 97 (2022), 640–63; L. Kemp et al., 'Climate Endgame: Exploring catastrophic climate change scenarios', *PNAS* 119 (2022), 1–9.

176 European Commission, *Drought in Europe: August 2022* (Brussels, 2022); European Commission, 'Daily News', 23 August 2022.

177 Also note M. Cruz, 'El Gobierno limitará el aire acondicionado a 27 grados en transportes, centros de trabajo y comercios y la calefacción a 19', *El Mundo*, 29 July 2022.

178 International Monetary Fund, *World Economic Outlook: A Long and Difficult Ascent* (Washington, DC, 2020), p. 86.

179 D. Vetter, ' "Who Cares if Miami is 6 Meters Underwater in 100 Years?": HSBC Executive's Incendiary Climate Comments', *Forbes*, 20 May 2022.

180 D. Strauss, 'Central bankers plan for disruption caused by climate change', *Financial Times*, 22 January 2019.

181 L. Fink, 'A Fundamental Reshaping of Finance', Larry Fink's 2020 Letter to CEOs, Black Rock, 16 January 2020.

182 M. Burke, S. Hsiang and E. Miguel, 'Global non-linear effect of temperature on economic production', *Nature* 527 (2015), 235–9.

183 J. Hickel et al., 'National responsibility for ecological breakdown: a fair-shares assessment of resource use, 1970–2017', *Lancet Planetary Health* 6 (2022), 342–9.

184 M. Howerton, 'New York City Energy Use All over Map', *Wall Street Journal*, 1 February 2012.

185 International Energy Agency, *Africa Energy Outlook 2022* (2022), pp. 15–17.

186 African Development Bank, 'Africa Economic Outlook 2022', https://www.afdb.org/en/knowledge/publications/african-economic-outlook

187 S. Wagstyl, 'Climate change is becoming less a battle of nations than rich vs poor', *Financial Times*, 21 May 2021.

188 B. Goldstein, D. Gounaridis and J. Newell, 'The carbon footprint of household energy use in the United States', *PNAS* 117 (2020), 19,122–30.

189 C. Tessum et al., 'PM$_{2.5}$ polluters disproportionately and systemically affect people of color in the United States', *Science Advances* 7 (2021), 1–6.

190 T. Carleton et al., 'Valuing the Global Mortality Consequences of Climate Change Accounting for Adaptation Costs and Benefits', *Quarterly Journal of Economics* (2022).

191 Burke, Hsiang and Miguel, 'Global non-linear effect of temperature', 235; M. Dell, 'Temperature Shocks and Economic Growth: Evidence from the Last Half Century', *American Economic Journal: Macroeconomics* 4 (2012), 66–95.

192 S. Hallegatte and J. Rozenberg, 'Climate Change through a Poverty Lens', *Nature Climate Change* 4 (2017), 250–6.

193 Vargas Zeppetello, Raftery and Battisti, 'Probabilistic projections', 1–7.

194 Armstrong Mackay et al., 'Exceeding 1.5 °C global warming', 1–11.

195 'Mongolia, Russia to reap benefits from climate change', bne Intellinews, 3 October 2017, extrapolating from IMF, *World Economic Outlook: Seeking Sustainable Growth: Short-Term Recovery, Long-Term Challenges* (Washington, DC, 2017).

196 Ministry of Natural Resources and Environment of the Russian Federation, Об Охране Окружающей Среды (Moscow, 2018); 'Голод, мор, смерть и деформация железнодорожных путей', *Kommersant*, 6 September 2018.

197 F. Pearce, 'Global Warming will hurt Russia', *New Scientist*, 3 October 2003.

198 A. Lustgarten, 'The Big Thaw: How Russia Could Dominate a Warming World', *ProPublica*, 16 December 2020.

199 R. Schoonover, C. Cavallo and I. Caltabiano, *The Security Threat that Binds Us: The Unraveling of Ecological and Natural Security and What the United States Can Do about It*, ed. E. Femia and A. Rezzonico, The Converging Risks Lab, Council on Strategic Risks (Washington, DC, 2021), p. 6.

CONCLUSION

1 M. Le Page, 'Heatwave in China is the most severe ever recorded in the world', *New Scientist*, 23 August 2022.

2 Lord Frost, *Holy Illusions: Reality-Based Politics and Sustaining the Brexit Revolt*, Policy Exchange (2022), 14–15.

3 J. Mervis, 'From a bully pulpit, Ted Cruz offers his take on climate change', *Science Insider*, 9 December 2015.

4 M. Joyella, 'On Fox, Donald Trump Calls Climate Change A "Hoax": "In The 1920's They Were Talking About Global Freezing"', *Forbes*, 21 March 2022.

5 M. Lempriere, 'Energy sector hits back as Truss dubs solar farms "paraphernalia"', Solar Power Portal, 16 August 2022.

6 CarbonBrief, 'Factcheck: Is solar power a "threat" to UK farmland?', 25 August 2022.

7 T. Marlow, S. Miller and J. Timmons Roberts, 'Bots and online climate discourses: Twitter discourse on President Trump's announcement of U.S. withdrawal from the Paris Agreement', *Climate Policy* 21 (2021), 765–77.

8 See for example Lomborg, 'Welfare in the 21st century', 1–35.

9 M. Smith, *Water Efficiency Opportunities Best Practice Guides* (Canberra, 2011).

10 Greenstone, Hasenkopf and Lee, *Air Quality Life Index*, p. 18.

11 G. Zissis, P. Bertoldi and T. Serrenho, *Update on the status of LED-lighting world market since 2018*, European Commission, JRC Technical Report (Luxembourg, 2021), pp. 3–4, 35–6.

12 European Commission, 'Progress made in cutting emissions', https://ec.europa.eu/clima/policies/strategies/progress_en.

13 Rhodium Group, *Taking Stock 2022: US Greenhouse Gas Emissions Outlook in an Uncertain World*, 14 July 2022.

14 D. Wallace-Wells, 'The Green Transition is Happening Fast: The Climate Bill Will Only Speed It Up', *New York Times*, 17 August 2022.

15 D. Tong et al., 'Geophysical constraints on the reliability of solar and wind power worldwide', *Nature Communications* 12 (2021), 1–12.

16 S. Roth, 'California just hit 95% renewable energy. Will other states come along for the ride?', *Los Angeles Times*, 29 April 2021.

17 J. Bairstow, 'Coal-free period of generation comes to an end … and was it due to Love Island?', *Energy Live News*, 6 June 2019.

18 D. Grant, D. Zelinka and S. Mitova, 'Reducing CO_2 emissions by targeting the world's hyper-polluting power plants', *Environmental Research Letters* 16 (2021), 1–10.

19 R. Way et al., 'Empirically grounded technology forecasts and the energy transition', *Joule* 6 (2022), 1–26.

20 R. Teoh et al., 'Mitigating the Climate Forcing of Aircraft Contrails by Small-Scale Diversions and Technology Adoption', *Environmental Science & Technology* 54 (2020), 2941–50.

21 G.L. Reynolds, *The Multi-Issue Mitigation Potential of Reducing Ship Speeds* (London, 2019); R. Leaper, 'The Role of Slower Vessel Speeds in Reducing Greenhouse Gas Emissions, Underwater Noise and Collision Risk to Whales', *Frontiers in Marine Science* 6 (2019), 1–8.

22 N. Dirksen et al., 'Learned control of urinary reflexes in cattle to help reduce greenhouse gas emissions', *Current Biology* 31 (2021), 1033–4.

23 Y. Ling Chin, K. Fie Chai and W. Ning Chen, 'Upcycling of brewers' spent grains via solid-state fermentation for the production of protein hydrolysates with antioxidant and techno-functional properties', *Food Chemistry: X* 13 (2022), 1–10.

24 F. Humpenöder et al., 'Projected environmental benefits of replacing beef with microbial protein', *Nature* 605 (2022), 90–6.

25 B. Trang et al., 'Low-temperature mineralization of perfluorocarboxylic acids', *Science* 377 (2022), 839–45.

26 E. Sheridan et al., 'Plastic pollution fosters more microbial growth in lakes than natural organic matter', *Nature Communications* 13 (2022), 1–9.

27 A. de Souza et al., 'Soybean photosynthesis and crop yield are improved by accelerating recovery from photoprotection', *Science* 377 (2022), 851–4.

28 V. Smil, *Numbers Don't Lie* (London, 2020), pp. 147–8.

29 C. Henshaw, 'Wind turbine blades could be recycled into gummy bears, scientists say', *Guardian*, 23 August 2022.

30 F. Parker-Jurd et al., 'Investigating the sources and pathways of synthetic fibre and vehicle tyre wear contamination into the marine environment', Department for Environment, Food and Rural Affairs (2019), pp. 3–4; N. Evangeliou, 'Atmospheric transport is a major pathway of microplastics to remote regions', *Nature Communications* 11 (2020), 1–11.

31 Emissions Analytics, 'Tyres Not Tailpipe', 29 January 2020.

32 OECD, *Non-Exhaust Particulate Emissions from Road Transport: An Ignored Environmental Policy Challenge* (Paris, 2020).

33 BASIS, *Hit for Six: The Impact of Climate Change on Cricket* (2019).

34 M. Matiu and F. Hanzer, 'Bias adjustment and downscaling of snow cover fraction projections from regional climate models using remote sensing for the European Alps', *Hydrology and Earth Science Systems* 26 (2022), 3037–54.

35 Box, 'Greenland ice sheet climate disequilibrium', 1–11.

36 Lorenzo, Díaz-Poso and Royé, 'Heatwave intensity on the Iberian Peninsula', 1–10.

37 M. Wang et al., 'The great Atlantic *Sargassum* belt', *Science* 365 (2019), 83–7.

38 M. Lowen, 'Spain's olive oil producers devastated by worst ever drought', BBC News, 29 August 2022. For Mediterranean longevity, A. Campanella et al., 'The effect of the Mediterranean Diet on lifespan: a treatment-effect survival analysis of a population-based prospective cohort study in Southern Italy', *Journal of Epidemiology* 50 (2021), 245–55.

39 For Mecca, Wallace-Wells, *Uninhabitable Earth*, p. 42.

40 United Nations, 'Humanity's just one misunderstanding away from "nuclear annihilation" warns UN chief', 1 August 2022.

41 C. Bardeen et al., 'Extreme Ozone Loss Following Nuclear War Results in Enhanced Surface Ultraviolet Radiation', *JGR Atmospheres* 126 (2021), 1–22.

42 L. Xia et al., 'Global food insecurity and famine from reduced crop, marine fishery and livestock production due to climate disruption from nuclear war soot injection', *Nature Food* 3 (2022), 586–96; NASA, 'Study Projects a Surge in Coastal Flooding, Starting in 2030s', 7 July 2021; L. Li, 'A modest 0.5-m rise in sea level will double the tsunami hazard in Macau', *Science Advances* 4 (2018), 1–11; United Nations Educational, Scientific and Cultural Organisation (UNESCO), 'UNESCO will train 100% of at-risk coastal communities by 2030', 22 June 2022.

43 Presidential Executive Order, 'Executive Order – Coordinating Efforts to Prepare the Nation for Space Weather Events', The White House, President Obama, 13 October 2016.

44 D. Baker et al., 'A major solar eruptive event in July 2012: Defining extreme space weather scenarios', Space Weather 11 (2013), 585–91.

45 www.spaceweather.com; J. Liu, 'Solar flare effects in the Earth's magnetosphere', *Solar Physics* 17 (2021), 807–21; S. Shankland, 'We aren't ready for a solar storm smackdown', *CNET*, 14 June 2016.

46 J. Lin et al., 'Magnitude, frequency and climate forcing of global volcanism during the last glacial period as seen in Greenland and Antarctic ice cores (60–9 ka)', *Climate of the Past* 18 (2022), 485–506.

47 T. Gilchrist, 'Effects of an impact event: an analysis of asteroid 1989FC', https://www.brookes.ac.uk/geoverse/original-papers/effects-of-an-impact-event--an-analysis-of-asteroid-1989fc/

48 NASA, 'NASA Analysis: Earth is Safe from Asteroid Apophis for 100-Plus Years', 26 March 2021, https://www.nasa.gov/feature/jpl/nasa-analysis-earth-is-safe-from-asteroid-apophis-for-100-plus-years.

49 NASA, https://www.nasa.gov/planetarydefense/dart.

50 P. Huybers and C. Langmuir, 'Feedback between deglaciation, volcanism, and atmospheric CO_2', *Earth and Planetary Science Letters* 286 (2009), 470–91; S. Watt, D. Pyle and T. Mather, 'The volcanic response to deglaciation: Evidence from glaciated arcs and a re-assessment of global eruption records', *Earth-Science Reviews* 122 (2013), 77–102.

51 D. Trilling et al., 'The Size Distribution of Near-Earth Objects Larger Than 10m', *Astronomical Journal* 154 (2017), 1–10.

52 NASA Earth Observatory, 'Dramatic Changes at Hunga Tonga-Hunga Ha'apai', 10 April 2021.

53 NASA Earth Observatory, 'Tonga Volcano Plume Reached the Mesosphere', 15 January 2022; M. Sharma and S. Scarr, 'Tonga Eruption: The Perfect Storm', Reuters, 14 February 2022.

54 C. Wright et al., 'Surface-to-space atmospheric waves from Hunga Tonga-Hunga Ha'apai eruption', *Nature* (2022).

55 US Geological Survey, 'Volcano Watch', 11 March 2021, https://www.usgs.gov/center-news/volcano-watch-failing-prepare-you-are-preparing-fail-benjamin-franklin.

56 L. Mani, A. Tzachor and P. Cole, 'Global catastrophic risk from lower magnitude volcanic eruptions', Nature Communications 12 (2021), 1–5.

57 R. Munoz, 'Seeing Change in Weather Modification Globally', World Meteorological Organisation, *Bulletin* 66 (2017).

58 R. Radwan, 'How artificial rain can make a difference to Saudi Arabia and Gulf region's water situation', *Arab News*, 27 May 2022.

59 A. Griffin, 'Russia spends millions on "cloud seeding" technology to ensure it doesn't rain on May Day public holiday', *Independent*, 2 May 2016; M. Williams Pontin, 'Weather Engineering in China', *MIT Technology Review*, 25 March 2008.

60 J. Wei, 'Cloud and precipitation interference by strong low-frequency sound wave', *Science China Technological Sciences* 64 (2021), 261–72; also

G. Wang et al., 'Sky River: Discovery, concept, and implications for future research', *Science China Technological Sciences* 46 (2016), 649–56.

61 State Council of China, 'China allocates 8.8b yuan for weather modification program', 21 September 2017; State Council of China, 'China to forge ahead with weather modification service', 2 December 2020.

62 S.-S. Chien, D.-L. Hong and P.-H. Lin, 'Ideological and volume politics behind cloud water resource governance – Weather modification in China', *Geoforum* 85 (2017), 225–33.

63 National Research Council, *Climate Intervention: Reflecting Sunlight to Cool Earth* (Washington, DC, 2015), p. x.

64 NOAA, 'Weather Modification Project Reports', https://library.noaa.gov/Collections/Digital-Collections/Weather-Modification-Project-Reports.

65 J. Dineen, 'Can Cloud Seeding Help Quench the Thirst of the U.S. West?', *Yale Environment 360*, 3 March 2022; S. Tsiouris et al., 'Soil silver content of agricultural areas subjected to cloud seeding with Silver Iodide', *Fresenius Environmental Bulletin* 11 (2002), 697–702; B. Williams, 'An assessment of the environmental toxicity of silver iodide – with reference to a cloud seeding trial in the snowy mountains of Australia', *Journal of Weather Modification* 41 (2009), 75–96.

66 See for example the list of papers presented at the 22nd Conference on Planned and Inadvertent Weather Modification, University of Wyoming, 13–16 January 2020, https://ams.confex.com/ams/2020Annual/webprogram/22WXMOD.html.

67 J. Zheng, 'Project aims to divert water through the sky', *China Daily*, 14 September 2016.

68 National Research Council, *Climate Intervention*, p. 7.

69 Panel on Policy Implications of Greenhouse Warming Committee on Science, Engineering, and Public Policy, *Policy Implications of Greenhouse Warming: Mitigation, Adaptation, and the Science Base* (Washington, DC, 1992), pp. 48–65.

70 National Research Council, *Critical Issues in Weather Modification Research* (Washington, DC, 2003), pp. 3–5.

71 Ibid., p. 73.

72 National Research Council, *Climate Intervention*, pp. 1–12.

73 National Academies of Sciences, Engineering, and Medicine, *Reflecting Sunlight: Recommendations for Solar Geoengineering Research and Research Governance* (Washington, DC, 2021).

74 See for example M. Lawrence, 'Evaluating climate geoengineering proposals in the context of the Paris Agreement temperature goals',

Nature Communications 9 (2018), 1–19; D. MacMartin, K. Ricke and D. Keith, 'Solar geoengineering as part of an overall strategy for meeting the 1.5°C Paris target', *Philosophical Transactions of the Royal Society A: Mathematical, Physical and Engineering Sciences* 376 (2017), 1–19; Y. Fan et al., 'Solar geoengineering can alleviate climate change pressures on crop yields', *Nature Food* 2 (2021), 373–81.

75 United Nations, *Global Land Outlook*, p. 149.

76 https://www.overshootday.org/.

77 E. Crist, 'On the Poverty of our Nomenclature', *Environmental Humanities* 3 (2013), 129–47; D. Kidner, 'Why "anthropocentrism" is not anthropocentric', *Dialectical Anthropology* 38 (2014), 465–80.

78 Milton, *Paradise Lost*, 10, pp. 260–1.

79 *The Ante-Nicene Fathers: Translations of the Writings of the Fathers down to AD 325*, ed. and tr. A. Roberts and J. Donaldson, 10 vols (Edinburgh, 1869), 9, pp. 145–6.

80 Qur'ān, 36:56–7, tr. N. Dawood (London, 2014), p. 443; Brhad-aranyaka, *Upanishad* II, V Brahmana.1.

81 J. Mohawk, *Utopian Legacies: A History of Conquest and Oppression in the Western World* (Santa Fe, 2000), p. 17.

82 *The Economics of Biodiversity: The Dasgupta Review* (London, 2021), pp. 1–2.

83 Ibid., p. 2.

84 Office for Budget Responsibility, *Fiscal Risks Report* (London, 2021), p. 86.

Index

Image Credits

1. Lascaux. Photo: © 2023 RMN-Grand Palais/Dist. Photo SCALA, Florence
2. Mohenjo-daro, Indus valley. Photo: Saqib Qayyum, March 2014. CC BY-SA 3.0. https://commons.wikimedia.org/wiki/File·Mohenjodaro_-_view_of_the_stupa_mound.JPG
3. Female figurine from the Indus Valley Civilisation. Harappa. Photo: Ismoon, 2021. CC0 1.0.
   https://commons.wikimedia.org/wiki/File:Female_figurine_3._Mature_Harappan_period._Indus_civilization_(detail).jpg
4. Sargon of Akkad; materials: bone, gold and lapislazuli; Iraqi National Museum, Baghdad.
   Photo: INTERFOTO/Alamy
5. Chinese cracked turtle shell. 18.4cm (L)×11.1cm (W). Photo: Courtesy of the Institute of History and Philology, Academia Sinica
6. Detail of Painted Casket of Tutankhamun. Photo: Roger Wood/Corbis/VCG via Getty Images
7. Ramagrama stupa. Photo: Dharma from Sadao, Thailand, 2017. CC BY 2.0. https://commons.wikimedia.org/wiki/File:Ashoka%27s_visit_to_the_Ramagrama_stupa_Sanchi_Stupa_1_Southern_gateway.jpg
8. Gold adornment from Issyk Kul, modern Kazakhstan. Photo: Peter Frankopan
9. Mosaic from Saint-Romain-en-Gal.Photo: DEA/J. E. BULLOZ/De Agostini via Getty Images
10. Teotihuacan. Photo: Peter Frankopan
11. Vessel from the Moche Apogee Epoch, Larco Museum, Lima, Peru. Photo: Peter Frankopan
12. Detail from Codex Borbonicus Aztec manuscript, 1562-63. Palais Bourbon, Bibliothèque.
    Photo: De Agostini via Getty Images
13. Emperor Justinian, Byzantine mosaic in the presbytery of the Basilica of San Vitale.
    Photo: DEA/A. DAGLI ORTI/Getty Images
14. Temple V – Tikal. Photo: Peter Frankopan
15. Tlaloc from Templo Mayor. Photo: Peter Frankopan
16. Angkor Wat. Photo: Peter Frankopan

17. Monks Mound at Cahokia. Photo: Matthew Gush/Alamy

18. Segara Anak caldera. Photo: Robby Oktawianto/EyeEm/Getty Images

19. Pieter Bruegel The Elder, *Hunters in the Snow (Winter)*, 1565 (oil on wood). Kunsthistorisches Museum, Vienna, Austria. Photo: Fine Art Images/Heritage Images/Getty Images

20. Diagram of the 'Brookes' slave ship, 1801. British Library shelfmark 1881.d.8 (46) 522.f.23
    Photo: © British Library Board/Bridgeman Images

21. Sample of potato plant 1660, Sloan Herbarium. Photo: © The Trustees of the Natural History Museum, London.

22. Thomas Jefferson temperatures July 1776 from Massachusetts Historical Society: Coolidge Collection of Thomas Jefferson Manuscripts, 1705-1827, Reel 15, Memorandum book, 1776-1778, Image 128, Special Colls. Jefferson. Photo: Collection of the Massachusetts Historical Society

23. Cast brass plaques from Benin City. British Museum, London. Photo: Andreas Praefcke /Wikimedia.org. PD-self

24. Sailors visiting a guano bed off the coast of Peru 1880. Hand-coloured woodcut. Photo: North Wind Picture Archives/Alamy

25. Eunice Newton Foote, *Circumstances Affecting the Heat of the Sun's Rays* (1856), The American Journal of Science and Arts. Second Series, Vol XXII (1856). Photo: Archive.org/Missouri Botanical Garden/The Biodiversity Heritage Library.

26. Photogravure after a photograph of Svante August Arrhenius. Photo: SSPL/Getty Images

27. Construction of the White Sea-Baltic Canal (Belomorkanal).Photo: Laski Diffusion/Getty Images

28. Rubber-tapping. Photo: The Granger Collection/Alamy

29. Ding Hao, *Exterminate the four pests! Chu siha!* Feb 1958. Photo: Private Collection, International Institute of Social History, Amsterdam. pc-1958-025 (chineseposters .net, Private collection)

30. A mushroom cloud after the explosion of a French atomic bomb above the atoll of Mururoa, also known as Aopuni. 1971. Photo: Galerie Bilderwelt/Getty Images

31. Peace march on the United Nations Plaza. Photo: Bettmann/Getty Images

32. Collier's magazine cover, May 28, 1954. Weather made to order by HT Orville, Illustration by Frederick (Fritz) Siebel. Photo: Courtesy of Novak Archive/ Paleofuture

33. Film poster for the 1986 film *When the Wind Blows*. Photo: Courtesy of www .themoviedb.org

34. Propaganda poster from the Virgin Lands Campaign. Text reads, 'Let us gather vast crops from the virgin lands'. Photo: Universal History Archive/Universal Images Group via Getty Images

35. Earth Summit on 12 June 1992 at Rio de Janeiro. Photo: DANIEL GARCIA/AFP via Getty Images

36. Photo taken during an aerial survey mission by Greenpeace at East Kotawaringin district in Central Kalimantan province on Indonesia's Borneo Island on 24 February 2014 shows cleared trees in a forest located in the concession of Karya Makmur Abadi which is being developed for a palm oil plantation. Photo: BAY ISMOYO/AFP via Getty Images

37. A view of Kartavya Path engulfed in dense smog on 7 November 2022 in New Delhi, India.
Photo: Arvind Yadav/Hindustan Times via Getty Images
38. Aerial view of plumes of smoke rising out of the oil refinery towers in Houston, Texas, USA.
Photo: simonkr/Getty Images

Chart Credits

1. 'Synoptic frequency of extreme marine heat across ocean basins from 1900–2019', adapted from K. R. Tanaka and K. S. Van Houtan, 'The recent normalization of historical marine heat extremes', PLOS Clim 1(2) (2022)
2. 'Deep ocean temperature from the Palaeocene to the present', adapted from a graph by Hunter Allen and Michon Scott, using data from the NOAA National Climatic Data Center © NOAA Climate.gov
3. 'Global mean surface temperature change over the past 24 kyr' from M. B. Osman, J. E. Tierney, J. Zhu et al, 'Globally resolved surface temperatures since the Last Glacial Maximum', *Nature* 599, 239–244 (2021) © Springer Nature
4. 'Isotopic ratio profiles at three sites in Italy from 1000 BC–1000 AD', M. Bini, G. Zanchetta, E. Regattieri, I. Isola, R. N. Drysdale, F. Fabiani, S. Genovesi, J. C. Hellstrom, 'Hydrological changes during the Roman Climatic Optimum in northern Tuscany (Central Italy) as evidenced by speleothem records and archaeological data', *Journal of Quaternary Science* 35 (2020) © 2020 John Wiley & Sons, Ltd
5. 'Temperature profiles at six sites in the North Atlantic from 200 BC–2000 AD', used with permission from T. Cronin, K. Hayo, R. Thunell, G. S. Dwyer, C. Saenger, D. Willard, 'The Medieval Climate Anomaly and Little Ice Age in Chesapeake Bay and the North Atlantic Ocean', *Palaeogeography, Palaeoclimatology, Palaeoecology* 297 (2010)
6. 'Potential gain in life...' and 'Life expectancy impact...' adapted from B.C. O'Neill, C. Tebaldi, D. P. van Vuuren, V. Eyring, P. Friedlingstein, G. Hurtt, R. Knutti, E. Kriegler, J.-F. Lamarque, J. Lowe, G. A. Meehl, R. Moss, K. Riahi and B. M. Sanderson, 'The Scenario Model Intercomparison Project (ScenarioMIP) for CMIP6', *Geosci. Model Dev* 9 (2016)
7. 'Predicted CO_2 atmospheric concentration and temperature increase from 2100 AD', adapted from 'Air Quality Life Index Annual Update', June 2022, awli.epic .uchicago.edu